21st Century Canon of Lunar Eclipses

DELUXE
BLACK AND WHITE
EDITION

Fred Espenak

Edition 1.0
September 2020

21st Century Canon of Lunar Eclipses – Deluxe Black and White Edition

Astropixels Publishing
P.O. Box 16197
Portal, AZ 85632

www.astropixels.com/pubs

Printed in the United States of America

This book may be ordered at: *www.astropixels.com/pubs/21CCLE.html*

More about lunar eclipses of the 21st Century can be found at:

www.eclipsewise.com/lunar/LEcatalog/21CCLEcat.html

Astropixels Publication Number: AP021

ISBN 978-1-941983-20-1

First Edition (Version 1.0a)

Front Cover: A time sequence shows the partial phases and totality during the total lunar eclipse of 2000 July 16 from Maui, Hawaii (Copyright ©2000 by Fred Espenak).

Back Cover: Portrait of Fred Espenak (Copyright ©2016 by Fred Espenak).

Table of Contents

*Photo 1 – Five minutes before the start of totality (2018 Jan 31), the Moon is bathed in an orange-red light.
The narrow rim outside the umbra and still in sunlight appears brilliant white. ©2018 F. Espenak*

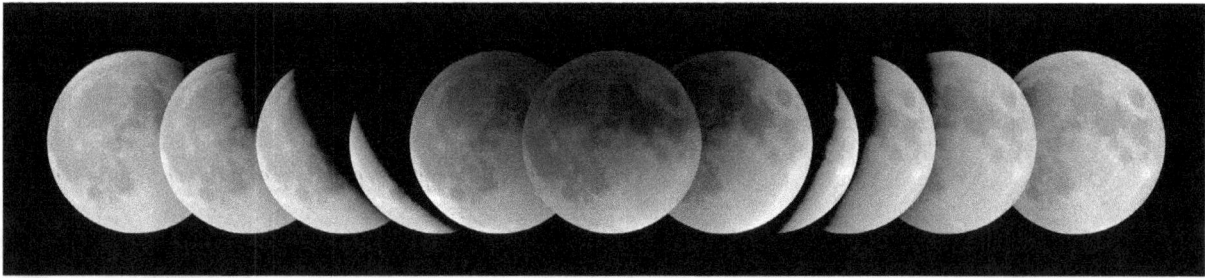

Photo 2 – Various phases of the total lunar eclipse of 2001 Jan 21. ©2001 F. Espenak

Section 1: Lunar Eclipse Fundamentals

1.1 Introduction

The Moon orbits Earth once every 29.5306 days with respect to the Sun. Over the course of its orbit, the Moon's changing position relative to the Sun results in its familiar phases: New Moon > New Crescent > First Quarter > Waxing Gibbous > Full Moon > Waning Gibbous > Last Quarter > Old Crescent > New Moon. The New Moon phase is not visible because the illuminated side of the Moon then points away from Earth. The other phases are easily seen as the Moon cycles through them month after month.

During Full Moon, the Moon appears opposite the Sun in the sky. It rises as the Sun sets and is visible throughout the night. The Full Moon sets in the morning just as the Sun rises. This unique geometry occurs when the Moon is 180° from the Sun as seen from Earth. It is the same direction that Earth casts its shadow into space.

The Moon's orbit is tilted about 5.1° to Earth's orbit around the Sun. As seen from Earth, the points where the two orbits appear to cross are called the nodes. When the Full Moon occurs near one of these nodes, Earth's shadow can fall on some portion of the Moon and a lunar eclipse takes place.

Earth's shadow is composed of two cone-shaped components, one nested inside the other. The outer or penumbral shadow is a zone where the Sun's rays are partially blocked, while the inner or umbral shadow is a region where direct rays from the Sun are completely blocked.

1.2 Classification of Lunar Eclipses

There are three basic types of lunar eclipses:

1. **Penumbral Lunar Eclipse** — The Moon passes through Earth's faint penumbral shadow. Penumbral eclipses are subtle and quite difficult to observe. On rare occasions, the Moon's entire disk can be enveloped within the penumbral shadow. Such events are called total penumbral eclipses.

2. **Partial Lunar Eclipse** — A portion of the Moon passes through Earth's dark umbral shadow. The remaining part of the Moon appears bright even though it lies deep within the penumbra. Partial eclipses are easy to see, even with the unaided eye.

3. **Total Lunar Eclipse** — The entire Moon passes through Earth's umbral shadow. Total eclipses are quite striking for the vibrant range of colors the Moon can take on during the total phase, referred to as totality.

Figure 1–1. Geometry of a Total Lunar Eclipse

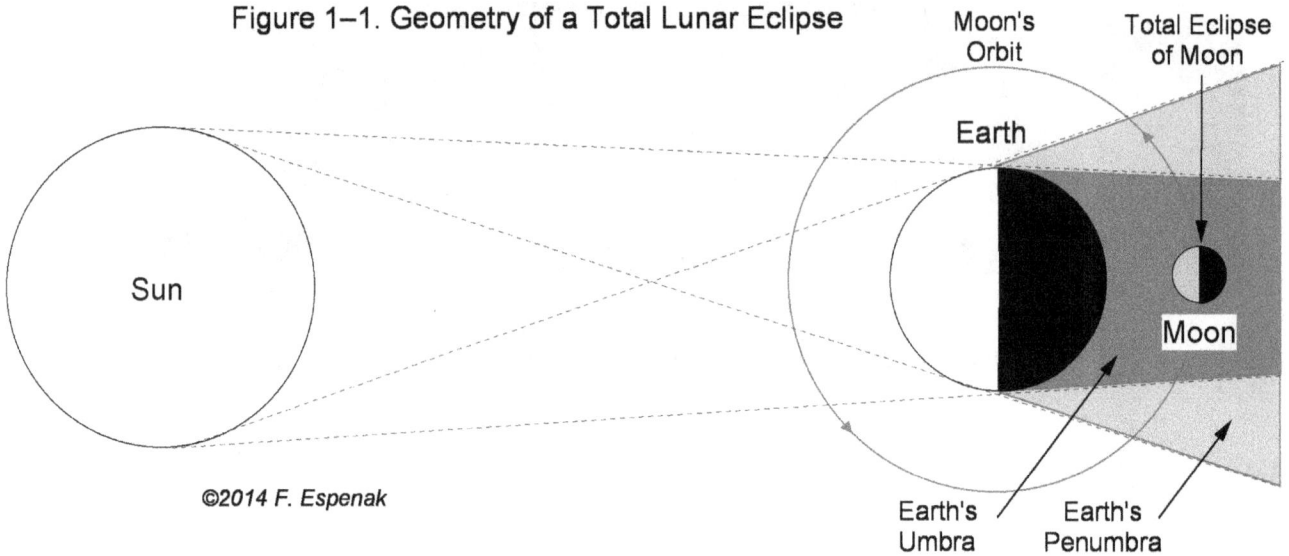

Figure 1–1 illustrates the geometry of a total lunar eclipse. A partial eclipse is visible if only part of the Moon enters Earth's umbral shadow. If the Moon passes through the penumbral shadow but misses the umbral shadow, then a penumbral eclipse occurs.

Figure 1–2. Types of Lunar Eclipses

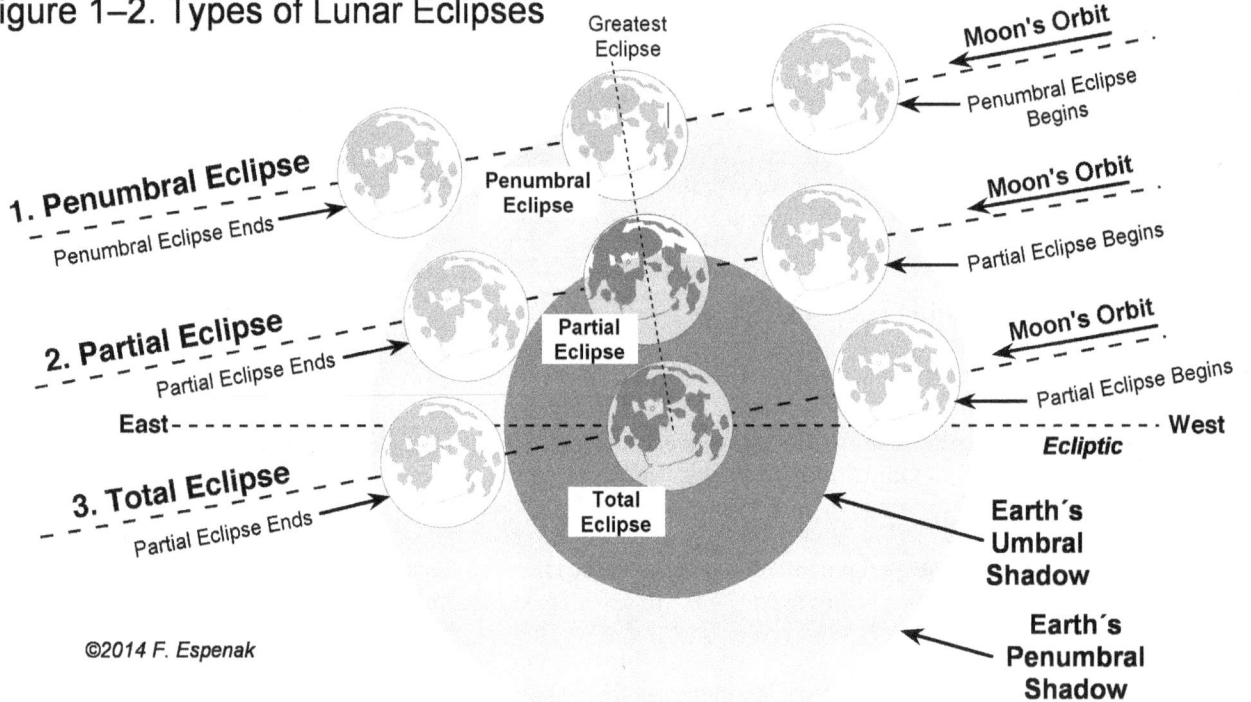

Figure 1–2 illustrates the three types of lunar eclipses as seen from Earth. 1) A penumbral eclipse occurs when the Moon passes through the penumbra but completely misses the umbra. 2) A partial eclipse occurs if some portion of the Moon enters the umbra. 3) A total lunar eclipse takes place when the entire disk of the Moon enters the umbral shadow.

Photo 3 – The visual appearance of penumbral, partial, and total lunar eclipses is illustrated. ©2020 F. Espenak

1.3 Visual Appearance of Penumbral Lunar Eclipses

The visual appearance of penumbral, partial and total lunar eclipses differs significantly. While penumbral eclipses are pale and difficult to see, partial eclipses are easy naked-eye events while total eclipses are colorful and dramatic.

Earth's penumbral shadow forms a diverging cone that expands into space in the direction away from the Sun. From within this zone, Earth blocks part but not the Sun's entire disk. Thus, some portion of the Sun's direct rays continues to reach the Moon during a penumbral eclipse.

The early and late stages of a penumbral eclipse are completely invisible to the eye. It is only when about 2/3 of the Moon's disk has entered the penumbral shadow that a skilled observer can detect a faint shading across the Moon.

Even when 90% of the Moon is immersed in the penumbra, about 10% of the Sun's rays still reach the deepest edge of the Moon. Under such conditions, the Moon remains relatively bright with only a subtle shadow gradient across its disk.

1.4 Visual Appearance of Partial Lunar Eclipses

Compared to penumbral eclipses, partial eclipses are easy to see with the naked eye. The lunar limb that extends into the umbral shadow appears very dark or black. This is due to a contrast effect since the remaining portion of the Moon in the penumbra may be 500 times brighter. Because the umbral shadow's diameter is about 2.7x times the Moon's diameter, it appears as though a semi-circular bite has been taken out of the Moon.

Aristotle (384–322 BCE) first proved that Earth was round using the curved umbral shadow seen at partial eclipses. In comparing observations of several eclipses, he noted that Earth's shadow was round no matter where the eclipse took place, whether the Moon was high in the sky or low near the horizon. Aristotle correctly reasoned that only a sphere casts a round shadow from every angle.

1.5 Visual Appearance of Total Lunar Eclipses

A total lunar eclipse is the most dramatic and visually compelling type of lunar eclipse. The Moon's appearance can vary enormously throughout the period of totality and from one eclipse to the next. The geometry of the Moon's path through the umbra plays a significant role in determining the appearance of totality. The effect that Earth's atmosphere has on a total eclipse is not as apparent. Although the physical mass of Earth blocks all direct sunlight from the umbra, the planet's atmosphere filters, attenuates and bends some of the Sun's rays into the shadow.

The molecules in Earth's atmosphere scatter short wavelength light (i.e., yellow, green, blue) more than long wavelength light (i.e., orange, red). The same process responsible for making sunsets red also gives total lunar eclipses their characteristic ruddy color. The exact appearance can vary considerably in both hue and brightness.

Because the lowest layers of the atmosphere are the thickest, they absorb more sunlight and refract it through larger angles. About 75% of the atmosphere's mass is concentrated in the bottom 10 kilometers (troposphere) as well as most of the water vapor, which can form massive clouds that block even more light. Just above the troposphere lies the stratosphere (10 to 50 kilometers), a rarified zone above most of the planet's weather systems. The stratosphere is subject to important photochemical reactions due to the high level of solar ultraviolet radiation that penetrates the region. The troposphere and stratosphere act together as a ring-shaped lens that refracts heavily reddened sunlight into Earth's umbral shadow. Since the higher stratospheric layers contain less gas, they refract sunlight through progressively smaller angles into the outer parts of the umbra. In contrast, denser tropospheric layers refract sunlight through larger angles to reach the inner parts of the umbra.

Photo 4 – The beginning, middle and end of totality during the total lunar eclipse of 2004 October 28. ©2004 F. Espenak

Because of this lensing effect, the amount of light refracted into the umbra tends to increase radially from center to edge. Inhomogeneities from varying amounts of cloud and dust at differing latitudes can cause significant variations in brightness throughout the umbra.

Besides water (cloud, mist, precipitation), Earth's atmosphere also contains aerosols or tiny particles of organic debris, meteoric dust, volcanic ash and photochemical droplets. This material attenuates sunlight before it is refracted into the umbra. For instance, major volcanic eruptions in 1963 (Agung) and 1982 (El Chichon) each dumped huge quantities of gas and ash into the stratosphere and were followed by several years of very dark eclipses (Keen, 1983).

The 1991 eruption of Pinatubo in the Philippines had a similar effect. While most of the solid ash fell to Earth several days after circulating through the troposphere, a sizable volume of sulphur dioxide (SO_2) reached the stratosphere where it combined with water vapor to produce sulphuric acid (H_2SO_4). This high-altitude volcanic haze layer severely attenuates sunlight that must travel several hundred kilometers horizontally through the layer before being refracted into the umbral shadow. Thus, total eclipses following large volcanic eruptions are particularly dark. The total lunar eclipse of 1992 Dec 09 (1½ years after Pinatubo) was so dark that it was difficult to see the Moon's dull gray disk with the naked eye.

All total eclipses begin with penumbral and partial phases. After the total phase, the eclipse ends with partial and penumbral phases. While *solar* eclipses require special filters for safe viewing, no such precautions are needed to watch *lunar* eclipses. The best views of a lunar eclipse are with binoculars and the naked eye.

1.6 Danjon Scale of Lunar Eclipse Brightness

The French astronomer A. Danjon proposed a five-point scale to evaluate the visual appearance and brightness of the Moon during a total lunar eclipse. The L values for various luminosities are defined in Table 1–1.

Table 1–1. Danjon Brightness Scale for Total Lunar Eclipses

Danjon Value	Visual Description
L=0	Very dark eclipse (Moon is almost invisible, especially at mid-totality)
L=1	Dark eclipse, grey or brownish in coloration (details are distinguishable only with difficulty)
L=2	Deep red or rust-colored eclipse (very dark central shadow, while outer umbra is relatively bright)
L=3	Brick-red eclipse (umbral shadow usually has a bright or yellow rim)
L=4	Very bright copper-red or orange eclipse (umbral shadow has a bluish, very bright rim)

The Danjon scale is a useful tool to characterize the range of colors and brightness the Moon takes on during a total lunar eclipse. Assigning an L value is best done with the naked eye or binoculars near the time of mid-totality. The Moon's appearance should also be evaluated just after the start and before the end of totality. The Moon is then near the edge of the shadow, providing an opportunity to assign an L value to the outer umbra. In making such determinations, the instrumentation and the time should be recorded.

Photo 5 – Various phases of the total lunar eclipse of 2014 Apr 15 are captured in this multiple exposure sequence.
©2014 F. Espenak

Section 2: Lunar Eclipse Predictions

2.1 Lunar Eclipse Contacts

During the course of a lunar eclipse, the instants when the Moon's disk becomes tangent to Earth's shadows are known as eclipse contacts. They mark the primary stages or phases of a lunar eclipse (see Figure 2–1).

Penumbral lunar eclipses have two major contacts although neither of these events is observable.

> **P1** — Instant of first exterior tangency of the Moon with the Penumbra
>   (Penumbral Eclipse Begins)
> **P4** — Instant of last exterior tangency of the Moon with the Penumbra
>   (Penumbral Eclipse Ends)

Partial lunar eclipses have four contacts. As the Moon's limb enters and exits the umbral shadow, contacts U1 and U4 mark the instants when the partial eclipse phase begins and ends, respectively.

> **P1** — Instant of first exterior tangency of the Moon with the Penumbra
>   (Penumbral Eclipse Begins)
> **U1** — Instant of first exterior tangency of the Moon with the Umbra
>   (Partial Umbral Eclipse Begins)
> **U4** — Instant of last exterior tangency of the Moon with the Umbra
>   (Partial Umbral Eclipse Ends)
> **P4** — Instant of last exterior tangency of the Moon with the Penumbra
>   (Penumbral Eclipse Ends)

Total lunar eclipses have six contacts. Contacts U2 and U3 mark the instants when the Moon's entire disk is first and last internally tangent to the umbra. These are the times when the total phase of the eclipse begins and ends, respectively.

> **P1** — Instant of first exterior tangency of the Moon with the Penumbra
>   (Penumbral Eclipse Begins)
> **U1** — Instant of first exterior tangency of the Moon with the Umbra
>   (Partial Umbral Eclipse Begins)
> **U2** — Instant of first interior tangency of the Moon with the Umbra
>   (Total Umbral Eclipse Begins)
> **U3** — Instant of last interior tangency of the Moon with the Umbra
>   (Total Umbral Eclipse Ends)
> **U4** — Instant of last exterior tangency of the Moon with the Umbra
>   (Partial Umbral Eclipse Ends)
> **P4** — Instant of last exterior tangency of the Moon with the Penumbra
>   (Penumbral Eclipse Ends)

The instant of greatest eclipse occurs when the Moon passes closest to the shadow axis. This corresponds to the maximum phase of the eclipse when the Moon is at its deepest position within either the penumbral or umbral shadow.

Figure 2–1. Lunar Eclipse Contacts

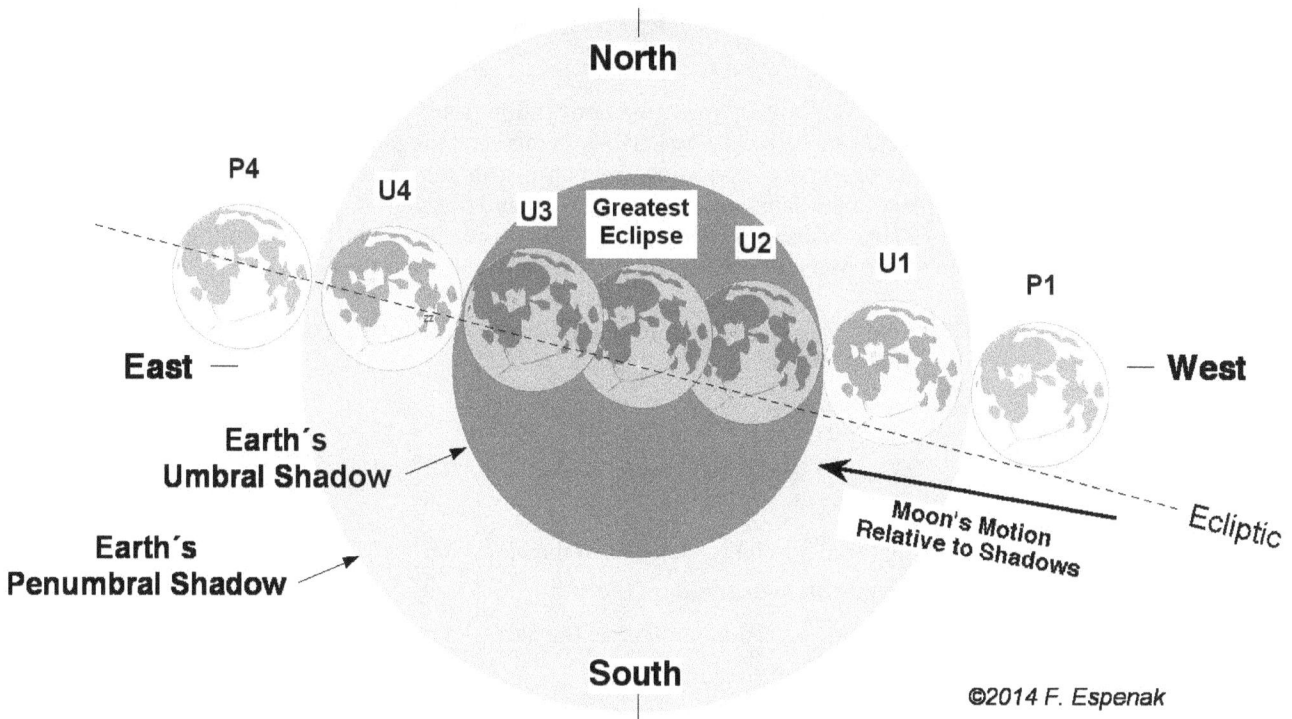

©2014 F. Espenak

Figure 2–1 illustrates the six contacts for a total lunar eclipse. These correspond to the instants when the Moon's disk is externally tangent to the penumbra (P1 and P4), or either externally or internally tangent to the umbra (U1, U2, U3, and U4). Partial eclipses do not have contacts U2 and U3, while penumbral eclipses only have contacts P1 and P4.

2.2 Enlargement of Earth's Shadows

In 1707, Philippe de La Hire made a curious observation about Earth's umbra. The predicted radius of the shadow needed to be enlarged by about 1/41 in order to fit timings made during a lunar eclipse. Additional observations over the next two centuries revealed that the shadow enlargement was somewhat variable from one eclipse to the next. According to William Chauvenet (1891):

"This fractional increase of the breath of the shadow was given by Lambert as 1/40, and by Mayer as 1/60. Beer and Maedler found 1/50 from a number of observations of eclipses of lunar spots in the very favorable eclipse of December 26, 1833."

Chauvenet adopted a value of 1/50, which has become the standard enlargement factor for lunar eclipse predictions published by many national institutes worldwide. The enlargement enters into the definitions of the penumbral and umbral shadow radii as follows.

penumbral radius: $\quad R_p = 1.02 \times (0.998340 \times \pi_m + S_s + \pi_s)$ (2–1)

umbral radius: $\quad R_u = 1.02 \times (0.998340 \times \pi_m - S_s + \pi_s)$ (2–2)

Where: $\quad \pi_m$ = Equatorial horizontal parallax of the Moon

$\quad S_s$ = Geocentric semi-diameter of the Sun

$\quad \pi_s$ = Equatorial horizontal parallax of the Sun

The factor 1.02 is the enlargement of the shadows by 1/50. Earth's true figure approximates that of an oblate ellipsoid with a flattening of ~1/300. Furthermore, the degree of axial tilt of the planet towards or away from the

Sun throughout the year means the shape of the penumbral and umbral shadows varies by a small amount. A mean radius of Earth at latitude 45° is used to approximate the departure from perfectly circular shadows. The *Astronomical Almanac* uses a factor of 0.998340 to scale the Moon's equatorial horizontal parallax to account for this (0.998340 ≈ 1.0 – 0.5 × 1/300).

Some authorities dispute Chauvenet's shadow enlargement convention. Danjon (1951) notes the only reasonable way of accounting for a layer of opaque air surrounding Earth is to increase the planet's radius by the altitude of the layer. This is accomplished by proportionally increasing the parallax of the Moon. The radii of the umbral and penumbral shadows are then subject to the same absolute correction and not the same relative correction employed in the traditional Chauvenet 1/50 convention. Danjon estimates the thickness of the occulting layer to be 75 kilometers and this results in an enlargement of Earth's radius and the Moon's parallax of about 1/85.

In 1951, the French almanac *Connaissance des Temps* adopted Danjon's method for the enlargement of Earth's shadows in their eclipse predictions as shown below.

Penumbral radius: $R_p = 1.01 \times \pi_m + S_s + \pi_s$ (2–3)

Umbral radius: $R_u = 1.01 \times \pi_m - S_s + \pi_s$ (2–4)

Where: π_m = Equatorial horizontal parallax of the Moon
 S_s = Geocentric semi-diameter of the Sun
 π_s = Equatorial horizontal parallax of the Sun

And $1.01 \approx 1 + 1/85 - 1/594$

The factor 1.01 combines the 1/85 shadow enlargement term with a 1/594 term to correct for Earth's oblateness at a latitude of 45°.

Danjon's method correctly models the geometric relationship between an enlargement of Earth's radius and the corresponding increase in the size of its shadows. Meeus and Mucke (1979) and Espenak & Meeus (2009a) both use Danjon's method. However, the resulting umbral and penumbral eclipse magnitudes are smaller by approximately 0.006 and 0.026, respectively, as compared to predictions using the traditional Chauvenet convention of 1/50.

For instance, the umbral magnitude of the partial lunar eclipse of 2008 Aug 16 was 0.813 according to the *Astronomical Almanac for 2008* (2008) using Chauvenet's method, but only 0.8076 according to Espenak & Meeus (2009a) using Danjon's method.

Chauvenet's method is still used by the *Astronomical Almanac* (published jointly by the USNO and HMNAO) to calculate lunar eclipse circumstances, while Danjon's method is used by Meeus and Mucke (1979), Espenak and Meeus (2009a) and *Connaissance des Temps* (published by the Bureau des Longitudes).

2.3 Earth's Elliptical Shadows

Both the Chaunenet and Danjon methods of accounting for the enlargement of Earth's two shadows assume circular shadows scaled at 45° latitude. However, Earth is flattened at the poles and bulges at the Equator, so an oblate spheroid more closely represents its shape. The projection of each of the planet's shadows is an ellipse rather than a circle. Furthermore, Earth's axial tilt towards or away from the Sun throughout the year means the elliptical shape of the penumbral and umbral shadows varies as well.

Herald and Sinnott performed an analysis of 22,539 observations made at 94 lunar eclipses from 1842 to 2011 (Herald and Sinnott, 2014). This is the largest collection of crater and contact timings ever compiled. The authors define the height of a 'notional eclipse-forming layer' in Earth's atmosphere (abbreviated as NEL) corresponding to the occulting layer height used by Danjon. Given the size and consistency of their dataset, they refine the NEL height to 87 kilometers (compared to Danjon's value of 75 kilometers).

Herald and Sinnott find that size and shape of the umbra are consistent with an oblate spheroid at the time of each eclipse, enlarged by the empirically determined NEL that uniformly surrounds Earth. They conclude that future lunar eclipse predictions should be based on a Danjon-like approach with full allowance for an oblate Earth, with the umbral radius r_u being computed using equation 2–5.

$$r_u = R_\oplus\, \pi m - Ss + \pi s \qquad\qquad (2\text{–}5)$$

Where:
$R_\oplus$ = Radius of Earth, where $R_\oplus = 1 + h - 0.003353 \sin^2 \psi \cos^2 (\delta s + f \sin \psi)$

$h = 0.0136$ is the height of the NEL in Earth radii ($h = 87 / 6378.137$)

ψ = Angular position angle (measured from the east–west direction, positive to the north) of the relevant contact point about the edge of the umbra

$f = Ss - R_\oplus\, \pi s$

δs = Declination of the Sun

πm = Equatorial horizontal parallax of the Moon

Ss = Geocentric semi-diameter of the Sun

πs = Equatorial horizontal parallax of the Sun

The calculation of r_u requires a single iteration between $R_\oplus$ and f to generate mutually consistent values for a given ψ. Similar adjustments can be made for the penumbral radius r_p although the resulting effects are not observable.

The Herald and Sinnott method of calculating Earth's shadow enlargement is the most rigorous and accurate procedure to date. It is superior to the methods of Chaunenet and Danjon because it uses a better determined value of the NEL and an elliptical cross section for Earth's shadow. The *21st Century Canon of Lunar Eclipses* uses the Herald and Sinnott method in the lunar eclipse predictions presented here.

2.4 Solar and Lunar Coordinates

The coordinates of the Sun and the Moon used in the eclipse predictions presented here have been calculated with the JPL DE430 (Jet Propulsion Laboratory Developmental Ephemeris 430). The DE430 is based upon the International Celestial Reference Frame (ICRF), the adopted reference frame of the IAU. The DE430 includes both nutation or libration and has an absolute accuracy of several kilometers for planetary positions. In most cases this corresponds to a small fraction of an arc-second.

The Moon's center of figure does not coincide with its center of mass. To compensate, an empirical correction is sometimes added to the Moon's center of mass position. Unfortunately, the large variation in lunar libration from one eclipse to the next minimizes the effectiveness of this empirical correction. Because of this, no correction has been made to the Moon's center of mass position in the *21st Century Canon*.

2.5 Measurement of Time

The most natural form of time measurement is the solar day (usually measured from solar noon to solar noon). The length of the solar day varies during the year because of the eccentricity of Earth's orbit around the Sun. Mean solar time resolves this problem by using an average to define the mean solar day.

In 1884, Greenwich Mean Time (GMT) — the mean solar time on the Greenwich Meridian (0° longitude) — was adopted as the standard reference time for clocks around the world. A fundamental basis of GMT is the assumption that Earth's rotation on its axis is constant. It wasn't until the mid-twentieth century that astronomers realized the rotation period is gradually increasing. Earth is slowing down because of tidal friction with the Moon.

For purposes of orbital calculations, time using Earth's rotation was abandoned for a more uniform time scale based on Earth's orbit about the Sun. In 1952, Ephemeris Time was introduced to address the problem. The ephemeris second was defined as a fraction of the tropical year[1] for 1900 Jan 01 as calculated from Newcomb's *Tables of the Sun* (1895). Ephemeris Time was used for Solar System ephemeris calculations until 1979.

Terrestrial Dynamical Time (TD) is the modern replacement for Ephemeris Time and is used in theories of planetary motion in the Solar System. TD is based on International Atomic Time (TAI), which is a high-precision standard using several hundred atomic clocks worldwide. To ensure continuity with Ephemeris Time, TD was defined to match ET for the date 1977 Jan 01. In 1991, the IAU refined the definition of TD to make it more precise. It was also renamed Terrestrial Time (TT) although the author prefers to use the older name Terrestrial Dynamical Time.

Civilian time used throughout the world is still based on mean solar time, although indirectly. While Greenwich Mean Time was determined though observations of the Sun, its modern day replacement, Universal Time (actually UT1) is based on Earth's rotation using observations of distant quasars. UT1 is a nonuniform time because Earth is gradually slowing down at an irregular rate. At present (2020), the accumulated error in the rotation of Earth in the course of one year is ~0.3 seconds.

Coordinated Universal Time (UTC) is derived from International Atomic Time (TAI). The length of the UTC second is defined in terms of an atomic transition of cesium and is accurate to approximately one nanosecond (billionth of a second) per day. UTC was defined to closely parallel UT1. However, the two time systems are intrinsically incompatible since UTC is uniform while UT1 is based on Earth's rotation, which is gradually slowing. In order to keep the two times within 0.9 seconds of each other, a leap second is added to UT1 as needed (currently once every few years).

Today, UTC is the time standard used to define time zones around the world. It is the time reference for GPS satellites and aviation, and is used to synchronize the clocks of computers across the Internet.

Photo 6 – Thirteen phases of the total lunar eclipse of 2000 January 21 appear in this composite image. ©2000 F. Espenak

---

[1] The tropical year is the length of time that the Sun takes to return to the same position in the cycle of seasons, as seen from Earth (e.g., the time from vernal equinox to vernal equinox).

2.6 ΔT (Delta T)

The orbital positions of the Sun and the Moon, required by eclipse predictions, are calculated using Terrestrial Dynamical Time (TD) because it is a uniform time scale. However, world time zones and daily life are based on Universal Time[2] (UT1). In order to convert predictions from TD to UT1, the difference between these two time scales must be known. The parameter ΔT (Delta T) is the arithmetic difference, in seconds, between the two as:

$$\Delta T = TD - UT1 \tag{2-1}$$

Past values of ΔT can be deduced from historical records. In spite of their relatively low precision, these data represent the only evidence for the value of ΔT prior to 1600. In the centuries following the introduction of the telescope (circa 1609), thousands of high quality observations have been made of lunar occultations of stars, affording valuable data with increased accuracy in the determination of ΔT.

In modern times, the determination of ΔT is made using atomic clocks and radio observations of quasars. From 1955 to 2010, the average 1-year change in ΔT ranges from 0.18 seconds to 1.06 seconds. Future changes in ΔT are unknown since theoretical models of the physical causes are imprecise. Extrapolations from the table weighted by the long period trend from tidal braking of the Moon offer estimates of +71 seconds in 2024, +85 seconds in 2050, and +128 seconds in 2100.

Polynomial expressions for ΔT based on this data can be found at: *eclipsewise.com/help/deltatpoly2014.html*

2.7 Date Format

There are a number of ways to write the calendar date through variations in the order of day, month, and year. The International Organization for Standardization's (ISO) 8601 advises a numeric date representation, which organizes the elements from the largest to the smallest. The exact format is YYYY–MM–DD where YYYY is the calendar year, MM is the month of the year between 01 (January) and 12 (December), and DD is the day of the month between 01 and 31. For example, the 27th day of April in the year 1943 would then be expressed as 1943–04–27. The ISO convention is adopted here, but the month number has been replaced with the three-letter English abbreviation of the month name for additional clarity. From the previous example, the date then is expressed as 1943 Apr 27.

---

[2] World time zones are actually based on Coordinated Universal Time (UTC). It is an atomic time synchronized and adjusted to stay within a second of astronomically determined Universal Time (UT1) through the addition of an occasional "leap second" to compensate for the gradual slowing of Earth's rotation.

Section 3: Lunar Eclipse Statistics

3.1 Statistical Distribution of Lunar Eclipse Types

There are three types of lunar eclipses:

1. **Penumbral** — The Moon passes partially or completely into Earth's penumbral shadow
2. **Partial** — The Moon passes through Earth's penumbral shadow and partially into the umbral shadow
3. **Total** — The Moon passes through Earth's penumbral shadow and completely into the umbral shadow

During the 100-year period from 2001 to 2100, Earth experiences 228 eclipses of the Moon. The statistical distribution of the three eclipse types over this interval is shown in Table 3–1.

Table 3–1: Distribution of Basic Lunar Eclipse Types During 21st Century

Eclipse Type	Abbreviation	Number	Percent
All Eclipses	—	228	100.0%
Penumbral	N	86	37.7%
Partial	P	57	25.0%
Total	T	85	37.3%

During most penumbral eclipses, only part of the Moon passes through Earth's penumbral shadow. Examples of such *partial* penumbral eclipses include: 2023 May 05, 2024 Mar 25, 2027 Feb 20 and 2027 Aug 17. It is also possible to have a penumbral eclipse in which the Moon passes completely within Earth's penumbral shadow while not entering the inner umbral shadow. Such *total* penumbral eclipses are quite rare compared the typical *partial* penumbral eclipses. During the 21st century, there are 81 *partial* penumbral eclipses, but only 5 *total* penumbral eclipses (5.8%): 2006 Mar 14, 2053 Aug 29, 2070 Apr 25, 2082 Aug 08, and 2099 Sep 29.

While there are no special classifications for partial eclipses, total lunar eclipses through Earth's umbral shadow can be categorized into two types:

1. **Central** — The Moon passes through the central axis of Earth's umbral shadow
2. **Non-Central** — The Moon misses the central axis of Earth's umbral shadow

Table 3–2: Statistics of Total Eclipses During 21st Century

Eclipse Type	Number	Percent
All Total Eclipses	85	100.0%
Central Total	24	28.2%
Non-Central Total	61	71.8%

Examples of central total eclipses include: 2022 May 16, 2022 Nov 08, 2029 Jun 26 and 2036 Aug 07. Several examples of non-central total eclipses are: 2025 Mar 14, 2025 Sep 07, 2026 Mar 03 and 2028 Dec 31. The track of central total eclipses through Earth's umbral shadow is longer than for non-central total eclipses, so the total phase durations of centrals are longer than non-centrals.

3.2 Lunar Eclipse Frequency and the Calendar Year

There are two to five lunar eclipses in every calendar year. Table 3–3 shows the distribution in the number of eclipses per year for the 100 years covered in the *21st Century Canon*.

Table 3–3: Number of Lunar Eclipses per Year

Number of Eclipses Per Year	Number of Years	Percent
2	78	78%
3	16	16%
4	6	6%
5	0	0%

When two eclipses occur in one calendar year, they can be in any combination of penumbral, partial, or total (N, P, or T, respectively).

The 16 years in which three lunar eclipses occur include: 2001, 2002, 2009, 2013, 2027, 2028, 2031, 2048, 2049, 2060, 2066, 2067, 2074, 2075, 2078, 2092, and 2094.

The 6 years in which four lunar eclipses occur include: 2009, 2020, 2038, 2056, 2085, and 2096.

During the 1000-year period 1501 to 2500, there are 7 years containing 5 lunar eclipses: 1676, 1694, 1749, 1879, 2132, 2262, and 2400.

3.3 Extremes in Duration and Eclipse Magnitude

The longest and shortest eclipses of the century appear in Table 3–4.

Table 3–4: Extremes in Duration

Extrema Type	Date (Dynamical Time)	Eclipse Duration	Eclipse Magnitude
Longest Total	2018 Jul 27	01h43m47s	1.6100
Shortest Total	2015 Apr 04	00h07m16s	1.0019
Longest Partial	2021 Nov 19	03h29m09s	0.9760
Shortest Partial	2082 Feb 13	00h27m20s	0.0153
Longest Penumbral	2099 Apr 05	04h49m05s	1.0360*
Shortest Penumbral	2027 Jul 18	00h17m39s	0.0032*

*penumbral eclipse magnitude

The largest and smallest eclipses (in umbral eclipse magnitude) appear in Table 3–5.

Table 3–5: Extremes in Eclipse Magnitude

Extrema Type	Date (Dynamical Time)	Eclipse Duration	Eclipse Magnitude
Largest Total	2029 Jun 26	01h42m40s	1.8452
Smallest Total	2015 Apr 04	00h07m16s	1.0019
Largest Partial	2086 Nov 20	03h08m52s	0.9877
Smallest Partial	2082 Feb 13	00h27m20s	0.0153
Largest Penumbral	2070 Apr 25	04h47m43s	1.0527*
Smallest Penumbral	2027 Jul 18	00h17m39s	0.0032*

*penumbral eclipse magnitude

3.4 Lunar Eclipse Tetrads

When four consecutive lunar eclipses are all total, the group is termed a tetrad. They occur because of the eccentricity of Earth's orbit in conjunction with the timing of eclipse seasons (Section 3.5). During the 21st Century the first eclipse of every tetrad occurs between January and June.

The umbral magnitudes of the total eclipses making up a tetrad are all relatively small. For the 300-year period 1901 to 2200, the largest umbral magnitude of a tetrad eclipse is 1.4251 on 1949 Apr 13. Table 3–6 lists the date of the first total eclipse in each of the 27 tetrads from 1501 to 2500.

Table 3–6. Date of First Total Eclipse in Lunar Eclipse Tetrads: 1501 to 2500

1504 Mar 01	1927 Jun 15	2043 Mar 25	2137 Mar 07
1515 Jan 30	1949 Apr 13	2050 May 06	2155 Mar 19
1522 Mar 12	1967 Apr 24	2061 Apr 04	2448 Jun 17
1533 Feb 09	1985 May 04	2072 Mar 04	2466 Jun 28
1562 Jan 20	2003 May 16	2090 Mar 15	2477 May 28
1580 Jan 31	2014 Apr 15	2101 Feb 14	2495 Jun 08
1909 Jun 04	2032 Apr 25	2119 Feb 25	

3.5 Eclipse Seasons

Because of its ~5.1° inclination, the Moon's orbit crosses the ecliptic at two points or nodes. If Full Moon takes place within approximately 17° of a node,[3] then a lunar eclipse will be visible from some region of Earth.

The Sun makes one complete circuit of the ecliptic in 365.24 days, so its average angular velocity is 0.99° per day. At this rate, it takes 34.5 days for the Sun and, at the opposite node, Earth's umbral and penumbral shadows to cross the 34° wide eclipse zone centered on each node. Because the Moon's orbit with respect to the Sun has a mean duration of 29.53 days, there will always be one and possibly two lunar eclipses during each 34.5-day interval when the Sun (and Earth's shadows) pass through the nodal eclipse zones. These time periods are called eclipse seasons.

The mid-point of each eclipse season is separated by 173.3 days because this is the mean time for the Sun to travel from one node to the next. The period is a little less than half a calendar year because the lunar nodes regress westward by 19.3° per year.

3.6 Quincena

The mean time interval between New Moon and Full Moon is 14.77 days. This is less than half the duration of an eclipse season. As a consequence, the same Sun–node alignment geometry responsible for producing a lunar eclipse always results in a complementary solar eclipse within a fortnight. The solar eclipse may either precede or succeed the lunar eclipse. In either case, the pair of eclipses is referred to here as a Quincena[4]. The QSE (Quincena Solar Eclipse parameter) identifies the type of the solar eclipse and whether it precedes or succeeds a particular lunar eclipse.

---

[3] The exact angular distance from the node depends of the distances of the Sun and Moon from Earth, which determine their angular diameters as well as the diameters of Earth's shadows.

[4] Quincena is a Spanish word meaning *a period of fifteen days*. This also happens to be the time, rounded to the nearest day, between New Moon and Full Moon, or Full Moon and New Moon. So *quincena* is a convenient and appropriate term for describing a pair of eclipses (one solar and one lunar) separated by this period.

Solar eclipses can be classified into four different types:

1. partial solar eclipse (p) — The Moon's penumbral shadow traverses Earth; the umbral/antumbral shadow misses Earth
2. annular solar eclipse (a) — The Moon's antumbral shadow traverses Earth; the Moon is too far to completely cover the Sun
3. total solar eclipse (t) — The Moon's umbral shadow traverses Earth; the Moon is close enough to completely cover the Sun
4. hybrid solar eclipse (h) — The Moon's umbral and antumbral shadows traverse different parts of Earth; the eclipse appears either total or annular along different sections of its path; hybrid eclipses are also known as annular-total eclipses

The QSE is a two-character string consisting of one or more of the above solar eclipse types. The first character in the QSE identifies the type of solar eclipse preceding a lunar eclipse. The second character identifies the type of solar eclipse succeeding a lunar eclipse. In most instances, one of the two characters is "–" indicating no solar eclipse occurs. For example, a QSE of "–p" means that no solar eclipse precedes the lunar eclipse, but the lunar eclipse is followed by a partial solar eclipse 15 days later.

On rare occasions, a double Quincena occurs in which a lunar eclipse is both preceded and succeeded by a solar eclipse. In every case in the *Canon*, the lunar eclipse is always total.

Photo 7 – A time lapse sequence captures the phases of a partial lunar eclipse on 2012 Jun 04 from Portal, AZ.
©2012 F. Espenak

Section 4: Explanation of Lunar Eclipse Catalog in Appendix A

4.1 Introduction

Earth experiences 228 eclipses of the Moon during the 21st Century. The catalog in *Appendix A* summarizes the principal characteristics of each eclipse and complements the figure in *Appendices B and C*.

Each line in the tables corresponds to a single lunar eclipse and provides parameters to characterize the eclipse. The calendar date and Dynamical Time of greatest eclipse (when the Moon passes closest to the axis of Earth's shadows) are given along with the adopted value of Delta T (ΔT). The lunation number and the Saros series are listed along with the eclipse type (N=Penumbral, P=Partial, or T=Total). Gamma is the distance from the Moon's center to the axis of Earth's shadow cones at greatest eclipse, while the penumbral and umbral eclipse magnitudes are defined as the fraction of the Moon's diameter immersed in each shadow at that instant. The duration of the penumbral, partial, and total eclipse phases is given in minutes. Finally, the geographic latitude and longitude are given for the location where the Moon lies in the zenith at greatest eclipse. A more detailed description of each field in the catalog appears in the following sections.

4.2 Cat Num (Catalog Number)

The catalog number is the sequential number assigned to each eclipse from 1 to 228.

4.3 Calendar Date

Gregorian calendar date at greatest eclipse is given in the ISO order of YEAR-MONTH-DAY.

4.4 TD of Greatest Eclipse (Terrestrial Dynamical Time of Greatest Eclipse)

The instant of greatest eclipse occurs when the distance between the axis of the Moon's shadow cone and the center of Earth reaches a minimum. This instant differs slightly from the instant of greatest magnitude due to ellipticity of Earth's shadows.

Greatest eclipse is given in Terrestrial Dynamical Time or TD (Sect. 2.5), which is a time system based on International Atomic Time. As such, TD is the modern equivalent to its predecessor Ephemeris Time and is used in theories of planetary motion in the Solar System. To determine the geographic visibility of an eclipse, TD is converted to Universal Time (UT1) using the parameter Delta T.

4.5 ΔT (Delta T)

ΔT (Delta T) is the arithmetic difference, in seconds, between Terrestrial Dynamical Time (TD) and Universal Time (UT1). For more information on ΔT, see Section 2.6.

4.6 Luna Num (Lunation Number)

The lunation number is the number of synodic months or lunations since New Moon on 2000 Jan 06. It can be converted to the Brown Lunation Number[5] by adding 953.

4.7 Saros Num (Saros Series Number)

Each lunar eclipse belongs to a Saros series using a numbering system first introduced by van den Bergh (1955). The eclipses with an even Saros number take place at the ascending node of the Moon's orbit; those with an odd Saros number take place at the descending node. This relationship is reversed for *solar* eclipses.

---

[5] The *Brown Lunation Number* defines lunation 1 as beginning at the first New Moon of 1923, the year when Ernest W. Brown's lunar theory was introduced in the major national astronomical almanacs.

The Saros is a period of 223 synodic months (~ 18 years, 11 days, and 8 hours). Eclipses separated by this period belong to the same Saros series and share similar geometry and characteristics.

4.8 Ecl Type (Lunar Eclipse Type)

The first character in this 2-character parameter gives the lunar eclipse type. The three basic types of lunar eclipses are:

1. Penumbral Lunar Eclipse (N) — The Moon passes through Earth's penumbral shadow
2. Partial Lunar Eclipse (P) — The Moon passes partly into Earth's umbral shadow
3. Total Lunar Eclipse (T) — The Moon passes completely into Earth's umbral shadow

The second character of the lunar eclipse type is a qualifier defined as follows.

1. m = Middle eclipse of Saros series
2. + = Central total eclipse (Moon's center passes north of shadow axis)
3. – = Central total eclipse (Moon's center passes south of shadow axis)
4. * = Total penumbral eclipse
5. b = Saros series begins (first penumbral eclipse in a Saros series)
6. e = Saros series ends (last penumbral eclipse in a Saros series)

Qualifiers 1 through 3 are used exclusively with total lunar eclipses while qualifiers 4 through 6 are exclusively used with penumbral eclipses.

4.9 QSE (Quincena Solar Eclipse Parameter)

A solar eclipse always occurs within 15 days of a lunar eclipse. The Quincena Solar Eclipse parameter (QSE) identifies the type of the solar eclipse (partial, annular, total, or hybrid) and whether it precedes or succeeds a particular lunar eclipse.

The QSE is a two-character string consisting of one or more of the above solar eclipse types (p, a, t, or h). The first character in the QSE identifies a solar eclipse preceding a lunar eclipse, while the second character identifies a solar eclipse succeeding a lunar eclipse. In most instances, one of the two characters is "–" indicating no solar eclipse occurs. On rare occasions, a double Quincena occurs in which a lunar eclipse is both preceded and succeeded by solar eclipses. See Section 3.6 for more information on the Quincena.

4.10 Gamma

Gamma is the minimum distance from the center of the Moon to the axis of Earth's umbral shadow cone in units of Earth's equatorial radius. This distance is positive or negative, depending on whether the Moon passes north or south, respectively, of the shadow cone axis.

4.11 Um Mag (Umbral Eclipse Magnitude)

The umbral magnitude is defined as the fraction of the Moon's diameter immersed in Earth's umbral shadow. During a partial lunar eclipse, some portion of the Moon's disk enters the umbral shadow. The umbral magnitude for partial eclipses in the *Canon* ranges from 0.0153 to 0.9760 (Sect. 3.3). Of course, the Moon also passes through the penumbra during a partial eclipse, so the penumbral magnitude is usually greater than 1.

In the case of a total lunar eclipse, the Moon's entire disk passes through Earth's umbral shadow, so the umbral magnitude is always equal to or greater than 1.0. During totality, the Moon can take on a range of colors from bright orange, to deep red, dark brown, or even very dark gray (Sect. 1.5). The only light reaching the Moon at this time is heavily filtered and attenuated by Earth's atmosphere. The umbral magnitude for total eclipses in the *Canon* ranges from 1.0019 to 1.8452 (Sect. 3.3).

4.12 Pen Mag (Penumbral Eclipse Magnitude)

The penumbral eclipse magnitude is defined as the fraction of the Moon's diameter immersed in Earth's penumbral shadow.

The penumbral eclipse magnitude of penumbral eclipses in the *Canon* ranges from 0.0032 to 1.0360 (Sect. 3.3). For most penumbral eclipses the penumbral magnitude is less than 1, meaning only a fraction of the Moon's disk enters the penumbra. When the penumbral magnitude is greater than or equal to 1, the Moon's entire disk is immersed in the penumbra and the event is termed a total penumbral eclipse. Penumbral eclipses are subtle events (Sect. 1.3). They cannot be detected visually (with or without optical aid) unless the eclipse magnitude is greater than ~0.6. The umbral eclipse magnitude of a penumbral eclipse is always negative. It is a measure of the distance of the Moon's limb to the edge of the umbral shadow in units of the Moon's diameter.

4.13 Phase Durations (Pen, Par & Total)

The duration of a penumbral eclipse "*Pen*" is the time between first and last external tangencies of the Moon with the penumbral shadow (= P4 – P1). Similarly, the duration of a partial eclipse "*Par*" is the time between first and last external tangencies of the Moon with the umbral shadow (= U4 – U1). Finally, the duration of a total eclipse "*Total*" is the time between first and last internal tangencies of the Moon with the umbral shadow (= U3 – U2). The position of the Moon at each contact is illustrated in Figure 2–1. The duration given for each eclipse phase is given to the nearest 0.1 of a minute.

4.14 Greatest in Zenith Lat Long (Latitude & Longitude)

The geographic latitude and longitude where the Moon appears in the zenith at greatest eclipse.

4.15 EclipseWise.com and the 21st Century Canon

Much of the content of the *21st Century Canon of Lunar Eclipses* is based on the eclipse predictions website *www.EclipseWise.com*. A plain text file containing the entire lunar eclipse catalog in *Appendix A* can be downloaded at: *www.EclipseWise.com/pubs/21CCLE.html.*

A web-based version of the catalog is available at: *www.eclipsewise.com/lunar/LEcatalog/21CCLEcat.html*

Section 5: Explanation of Lunar Eclipse Figures in Appendices B and C

5.1 Explanation of Small Eclipse Figures in Appendix B

Earth experiences 228 eclipses of the Moon during the 21st Century. An individual diagram and visibility map for each eclipse appears in *Appendix B*.

The figure for each eclipse consists of two diagrams. The top one depicts the Moon's path through Earth's penumbral and umbral shadows with Celestial North directed up. The second diagram is a map showing the global visibility of each eclipse phase. All salient features in these diagrams are identified in figure 5–1, which serves as a key.

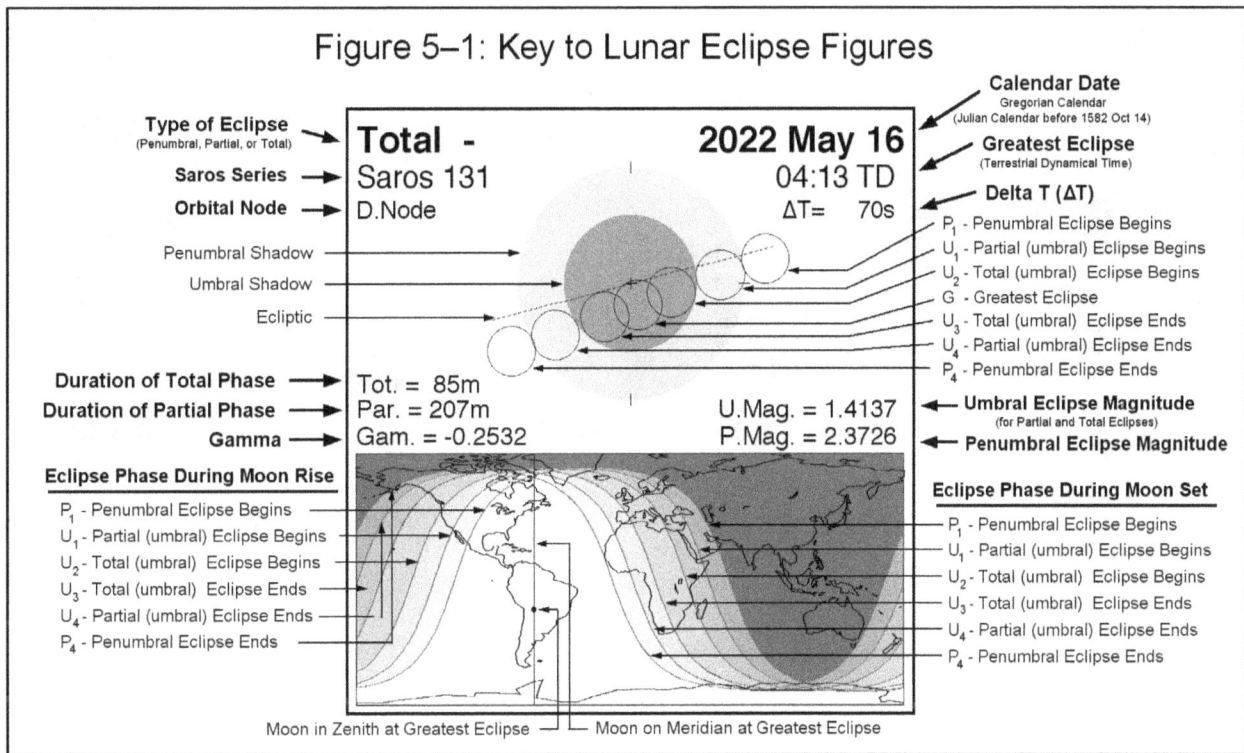

Figure 5–1: Key to Lunar Eclipse Figures

In the diagram, the Moon's orbital motion with respect to the shadows is from west to east (right to left). Each phase of the eclipse is defined by the instant when the Moon's limb is externally or internally tangent to the penumbra or umbra. The six primary contacts of the Moon with the penumbral and umbral shadows are defined as follows.

P1 — Instant of first exterior tangency of the Moon with the Penumbra
(Penumbral Eclipse Begins)

U1 — Instant of first exterior tangency of the Moon with the Umbra
(Partial Umbral Eclipse Begins)

U2 — Instant of first interior tangency of the Moon with the Umbra
(Total Umbral Eclipse Begins)

U3 — Instant of last interior tangency of the Moon with the Umbra
(Total Umbral Eclipse Ends)

U4 — Instant of last exterior tangency of the Moon with the Umbra
(Partial Umbral Eclipse Ends)

P4 — Instant of last exterior tangency of the Moon with the Penumbra
(Penumbral Eclipse Ends)

Penumbral lunar eclipses have two primary contacts: P1 and P4, but neither is observable because the edge of the penumbra is indistinct and extremely faint.

Photo 8 – The beginning, middle and end of the total lunar eclipse of 2015 September 28. ©2015 F. Espenak

In addition to the penumbral contacts, partial lunar eclipses have two more contacts as the Moon's limb enters and exits the umbral shadow: U1 and U4, respectively. These two contacts mark the instants when the partial phase of the eclipse begins and ends.

Total lunar eclipses undergo all six contacts. The two additional umbral contacts are the instants when the Moon's entire disk is first and last internally tangent to the umbra: U2 and U3, respectively. They mark the times when the total phase of the eclipse begins and ends.

The Moon passes closest to Earth's shadow axis at the instant of greatest eclipse. This corresponds to the maximum phase of the eclipse and the Moon's position at this instant is also depicted in the path diagrams.

The bottom diagram in each figure is an equidistant cylindrical projection map of Earth showing the geographic region of visibility at each phase of the eclipse. This is accomplished using a series of curves showing where Moonrise and Moonset occur at each eclipse contact. The map is shaded to indicate eclipse visibility. The entire eclipse is visible from the zone with no shading. Conversely, none of the eclipse can be seen from the zone with the darkest shading.

At greatest eclipse, the Moon is deepest in Earth's shadow. A vertical line running through the middle of the clear zone (complete eclipse visibility) indicates the meridian or line of longitude that the Moon is crossing. An observer positioned on this line would then see the Moon at its highest point in the sky due south or due north, depending on the observer's latitude and the Moon's declination. The geographic location where the Moon appears in the zenith at greatest eclipse is shown by a black dot on the meridian. All salient features of the eclipse figures are identified in Figure 5–1, which serves as the key to the figures in *Appendix B*.

Data relevant to the eclipse appear in the corners of each figure. To the top left are the eclipse type (penumbral, partial or total), the Saros series of the eclipse, and the node of the Moon's orbit where the eclipse occurs. To the top right are the Gregorian calendar date (Julian calendar dates are used prior to 1582 Oct 15), the time of greatest eclipse (Terrestrial Dynamical Time), and the value of ΔT (Delta T). The lower left corner lists the duration of the major phases of the eclipse in minutes. Depending on the eclipse type, the duration of the penumbral, partial or total phases are given. Beneath the eclipse durations is the quantity gamma, the minimum distance of the Moon's center from the axis of Earth's shadow cones at the instant of greatest eclipse. The umbral and penumbral eclipse magnitudes are given to the lower right. The list below summarizes the parameters appearing in each eclipse figure.

Lunar Eclipse Type – One of three basic types of lunar eclipses: Penumbral, Partial, or Total.

Saros Series – The Saros series to which the eclipse belongs. Eclipses with an odd number take place at the Moon's ascending node, while those with an even number are at the descending node.

Node – The orbital node near which the eclipse takes place. The ascending node (A. Node) is the point where the Moon travels from south to north through Earth's orbital plane. Similarly, the descending node (D. Node) is the point where the Moon travels from north to south.

Calendar Date – Gregorian calendar date at greatest eclipse is given in the ISO order of YEAR-MONTH-DAY.

Greatest Eclipse – The instant (Terrestrial Dynamical Time) when the distance between the Moon and the axis of Earth's shadow cone reaches a minimum.

ΔT (Delta T) – Difference, (seconds) between Terrestrial Dynamical Time (TD) and Universal Time (UT1).

Gamma – Minimum distance from the Moon's center to Earth's shadow axis (units of Earth's equatorial radius).

Duration – The duration of a penumbral eclipse (**Pen.**), a partial eclipse (**Par.**), and a total eclipse (**Tot.**).

Umbral Eclipse Magnitude (U. Mag) – The fraction of the Moon's diameter within the umbra at greatest eclipse.

Penumbral Eclipse Magnitude (P. Mag) – The fraction of the Moon's diameter within the penumbra at greatest eclipse.

Photo 9 – The setting Moon in total eclipse on 2018 January 31. ©2018 F. Espenak

5.2 Explanation of Full Page Lunar Eclipse Figures in Appendix C

The heart of the *21st Century Canon of Lunar Eclipses – Deluxe Edition* is the set of 228 full-page figures appearing in Appendix C — one map for every solar eclipse from 2001 through 2100. This feature is similar to the *Fifty Year Canon of Lunar Eclipses: 1986 – 2035* (Espenak 1989) and makes *21st Century Canon* its modern successor.

Each lunar eclipse has two diagrams associated with it along with data about the eclipse. The top figure shows the path of the Moon through Earth's penumbral and umbral shadows. Above this figure is listed the instant of greatest eclipse, expressed in Terrestrial Dynamical Time and Universal Time (UT1). The *Penumbral Magnitude* and *Umbral Magnitude* are defined as the fraction of the Moon's diameter immersed in each shadow at greatest eclipse. *Gamma* is the minimum distance of the Moon's center from Earth's shadow axis in Earth radii, and *Axis* is the same parameter expressed in degrees. The *Saros Series* of the eclipse is listed next. *Saros Member* consists of a pair of numbers. The first is the eclipse's relative position in the Saros series; the second is the total number of eclipses in the series.

In the upper left and right corners are the geocentric coordinates of the Sun and the Moon, respectively, at the instant of greatest eclipse. They are:

R.A. - Right Ascension
Dec. - Declination
S.D. - Apparent Semi-Diameter
H.P. - Horizontal Parallax

To the lower left are the durations of the penumbral, umbral (partial), and total eclipses. Below them are the solar/lunar ephemerides used in the predictions, the applied shadow *Rule,* and the extrapolated value of ΔT. To the lower right are the contact times of the Moon with Earth's penumbral and umbral shadows, as follows:

P1 - Instant of first exterior tangency of Moon with Penumbra.
(Penumbral Eclipse Begins)
U1 - Instant of first exterior tangency of Moon with Umbra.
(Partial Umbral Eclipse Begins)
U2 - Instant of first interior tangency of Moon with Umbra.
(Total Umbral Eclipse Begins)
U3 - Instant of last interior tangency of Moon with Umbra.
(Total Umbral Eclipse Ends)
U4 - Instant of last exterior tangency of Moon with Umbra
(Partial Umbral Eclipse Ends)
P4 - Instant of last exterior tangency of Moon with Penumbra.
(Penumbral Eclipse Ends)

The bottom figure is a cylindrical equidistant projection map of Earth that shows the regions of visibility for each stage of the eclipse. In particular, the moonrise/moonset terminator is plotted for each contact and is labeled accordingly. An asterisk indicates the point where the Moon is at the zenith at greatest eclipse. The region that is unshaded will observe the entire eclipse, while the darkest shaded area will witness no eclipse. The remaining lightly shaded areas will experience moonrise or moonset while the eclipse is in progress. The shaded zones east of the asterisk (*) will witness moonset before the eclipse ends, and the shaded zones west will witness moonrise after the eclipse has begun.

The salient features of these figures are identified in Figure 5–2, which serves as a key.

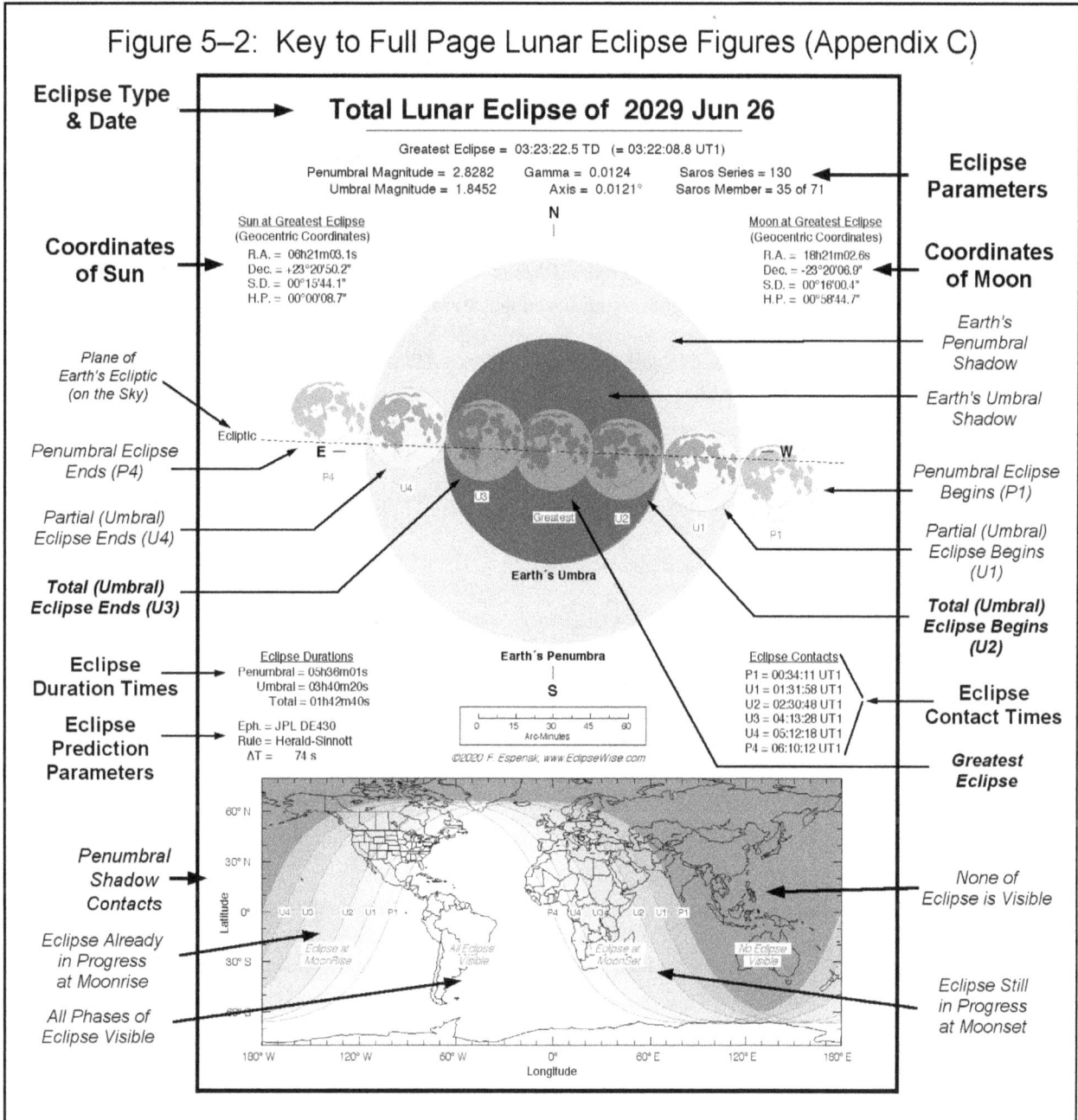

Figure 5–2: Key to Full Page Lunar Eclipse Figures (Appendix C)

References

Astronomical Almanac for 2011, Washington: US Government Printing Office; London: HM Stationery Office (2010).

Chauvenet, W.A., *Manual of Spherical and Practical Astronomy*, Vol.1, edition of 1891, (Dover reprint, New York, 1960).

Danjon, A., "Les éclipses de Lune par la pénombre en 1951," *L'Astronomie*, 65, 51–53 (1951).

Espenak, F., *Fifty Year Canon of Lunar Eclipses: 1986–2035*, Sky Publishing Corp., Cambridge, Massachusetts (1989).

Espenak, F., *Thousand Year Canon of Lunar Eclipses: 1501 to 2500*, Astropixels Publishing, Portal, Arizona (2014).

Espenak, F., and Meeus, J., *Five Millennium Canon of Lunar Eclipses: −1999 to +3000 (2000 BCE to 3000 CE)*, NASA Tech. Pub. 2006–214172, NASA Goddard Space Flight Center, Greenbelt, Maryland (2009a).

Espenak, F., and Meeus, J., *Five Millennium Catalog of Lunar Eclipses: −1999 to +3000 (2000 BCE to 3000 CE)*, NASA Tech. Pub. 2006–214173, NASA Goddard Space Flight Center, Greenbelt, Maryland (2009b).

Explanatory Supplement to the Ephemeris, H.M. Almanac Office, London (1974).

Herald, D., and Sinnott, R. W., "Analysis of Lunar Crater Timings, 1842–2011," *J. Br. Astron. Assoc.*, **124**, 5 (2014)

Keen, R. A., "Volcanic Aerosols and Lunar Eclipses", *Science*, vol. 222, p. 1011-1013, Dec. 2, 1983.

Lahire, P., *Tabulae Astronomicae* (Paris 1707).

Link, F., *Eclipse Phenomena in Astronomy*, Springer-Verlag, New York (1969).

Liu, Bao-Lin and Fiala, A. D., *Canon of Lunar Eclipses 1500 B.C.-A.D. 3000*, Willmann-Bell, p215 (1992).

Meeus, J. and Mucke, H., *Canon of Lunar Eclipse: -2002 to +2526*, Astronomisches Buro, Wein (1979).

Newcomb, S., "Tables of the Motion of the Earth on its Axis Around the Sun," *Astron. Papers Amer. Eph.*, Vol. 6, Part I (1895).

van den Bergh, *Periodicity and Variation of Solar (and Lunar) Eclipses*, Tjeenk Willink, and Haarlem, Netherlands (1955).

von Oppolzer, T.R., *Canon der Finsternisse*, Wien, (1887); Gingerich, O., (Translator) *Canon of Eclipses*, Dover Publications, New York (1962)

Appendix A

Lunar Eclipse Catalog: 2001 to 2100

Key to Lunar Eclipse Catalog

Cat Num — sequential Catalog Number assigned to each eclipse from 1 to 228

Calendar Date — Gregorian date of Greatest Eclipse

TD of Greatest Eclipse — Terrestrial Dynamical Time of Greatest Eclipse

ΔT — arithmetic difference between Terrestrial Dynamical Time Universal Time (seconds)

Luna Num — number of synodic months, or lunations, since New Moon on 2000 Jan 06

Saros Num — Saros Series Number of eclipse

Ecl Type — Lunar Eclipse Type

      N = Penumbral Lunar Eclipse
      P = Partial Lunar Eclipse
      T = Total Lunar Eclipse
      m = Middle eclipse of Saros series
      + = Central total eclipse (Moon's center passes north of shadow axis)
      – = Central total eclipse (Moon's center passes south of shadow axis)
      * = Total penumbral eclipse
      b = Saros series begins (first penumbral eclipse in a Saros series)
      e = Saros series ends (last penumbral eclipse in a Saros series)

QSE — Quincena Solar Eclipse Parameter

      p = Partial Solar Eclipse
      a = Annular Solar Eclipse
      t = Total Solar Eclipse
      h = Hybrid Solar Eclipse

Gamma — minimum distance from center of Moon to axis of Earth's umbral shadow cone

Pen Mag — Penumbral Magnitude; fraction of the Moon's diameter immersed in penumbra

Um Mag — Umbral Magnitude; fraction of the Moon's diameter immersed in umbra

Phase Durations
      Pen — elapsed time from contact P1 to P4 (minutes)
      Par — elapsed time from contact U1 to U4 (minutes)
      Total — elapsed time from contact U2 to U3 (minutes)

Greatest in Zenith
Lat & Long — latitude and longitude where Moon appears in zenith at greatest eclipse

Cat Num	Calendar Date	TD of Greatest Eclipse	ΔT s	Luna Num	Saros Num	Ecl Type	QSE	Gamma	Pen Mag	Um Mag	Pen m	Par m	Total m	Lat	Long
001	2001 Jan 09	20:21:40	64	12	134	T	p-	0.3720	2.1631	1.1902	311.9	197.1	61.7	22N	57E
002	2001 Jul 05	14:56:23	64	18	139	P	t-	-0.7287	1.5489	0.4961	326.0	160.0	-	23S	137E
003	2001 Dec 30	10:30:22	64	24	144	N	a-	1.0732	0.8948	-0.1141	244.3	-	-	24N	157W
004	2002 May 26	12:04:27	64	29	111	N	-a	1.1759	0.6910	-0.2871	217.3	-	-	20S	179E
005	2002 Jun 24	21:28:13	64	30	149	N	a-	-1.4440	0.2110	-0.7910	129.8	-	-	25S	39E
006	2002 Nov 20	01:47:41	64	35	116	N	-t	-1.1127	0.8618	-0.2246	265.1	-	-	19N	30W
007	2003 May 16	03:41:13	64	41	121	T	-a	0.4123	2.0765	1.1294	307.3	194.6	52.1	19S	56W
008	2003 Nov 09	01:19:38	64	47	126	T	-t	-0.4319	2.1157	1.0197	364.0	212.2	23.2	16N	24W
009	2004 May 04	20:31:17	65	53	131	T	p-	-0.3132	2.2645	1.3054	316.5	203.9	76.1	17S	51E
010	2004 Oct 28	03:05:11	65	59	136	T	p-	0.2846	2.3656	1.3100	354.6	219.4	81.1	13N	50W
011	2005 Apr 24	09:55:54	65	65	141	N	h-	-1.0885	0.8669	-0.1417	246.4	-	-	14S	150W
012	2005 Oct 17	12:04:27	65	71	146	P	a-	0.9796	1.0605	0.0645	260.5	57.0	-	10N	175E
013	2006 Mar 14	23:48:35	65	76	113	Nx	-t	1.0211	1.0320	-0.0583	288.2	-	-	3N	6E
014	2006 Sep 07	18:52:25	65	82	118	P	-a	-0.9262	1.1349	0.1857	255.0	91.8	-	7S	77E
015	2007 Mar 03	23:21:59	65	88	123	T	-p	0.3175	2.3208	1.2347	366.2	221.8	74.0	7N	13E
016	2007 Aug 28	10:38:27	65	94	128	T-	-p	-0.2146	2.4545	1.4777	328.0	212.8	90.6	10S	159W
017	2008 Feb 21	03:27:09	66	100	133	T	a-	-0.3992	2.1470	1.1081	339.7	206.1	50.5	10N	48W
018	2008 Aug 16	21:11:12	66	106	138	P	t-	0.5646	1.8385	0.8095	331.3	188.8	-	13S	43E
019	2009 Feb 09	14:39:21	66	112	143	N	a-	-1.0640	0.9013	-0.0863	239.5	-	-	14N	144E
020	2009 Jul 07	09:39:44	66	117	110	N	-t	-1.4916	0.1578	-0.9116	122.4	-	-	24S	143W
021	2009 Aug 06	00:40:16	66	118	148	N	t-	1.3572	0.4038	-0.6642	190.6	-	-	16S	9W
022	2009 Dec 31	19:23:46	66	123	115	P	-a	0.9766	1.0572	0.0779	251.8	60.8	-	24N	70E
023	2010 Jun 26	11:39:34	66	129	120	P	-t	-0.7091	1.5789	0.5383	323.0	163.6	-	24S	174W
024	2010 Dec 21	08:18:04	66	135	125	T	-p	0.3214	2.2821	1.2576	336.0	209.5	73.0	24N	125W
025	2011 Jun 15	20:13:43	67	141	130	T+	pp	0.0897	2.6883	1.7014	337.0	220.1	101.0	23S	57E
026	2011 Dec 10	14:32:57	67	147	135	T	p-	-0.3882	2.1875	1.1076	357.3	213.1	51.8	23N	140E
027	2012 Jun 04	11:04:20	67	153	140	P	a-	0.8248	1.3198	0.3718	270.8	127.2	-	22S	166W
028	2012 Nov 28	14:34:07	67	159	145	N	t-	-1.0869	0.9168	-0.1859	276.8	-	-	20N	139E
029	2013 Apr 25	20:08:37	67	164	112	P	-a	-1.0121	0.9878	0.0160	248.4	28.1	-	14S	57E
030	2013 May 25	04:11:07	67	165	150	Nb	a-	1.5351	0.0170	-0.9322	35.1	-	-	19S	63W
031	2013 Oct 18	23:51:25	67	170	117	N	-h	1.1508	0.7660	-0.2706	239.8	-	-	11N	2W
032	2014 Apr 15	07:46:48	67	176	122	T	-a	-0.3017	2.3193	1.2918	344.8	215.5	78.5	10S	116W
033	2014 Oct 08	10:55:44	67	182	127	T	-p	0.3827	2.1467	1.1670	318.9	200.2	59.4	6N	167W
034	2015 Apr 04	12:01:24	68	188	132	T	t-	0.4460	2.0802	1.0019	358.4	209.8	7.3	5S	179W
035	2015 Sep 28	02:48:17	68	194	137	T	p-	-0.3296	2.2307	1.2774	311.5	200.6	72.5	2N	44W
036	2016 Mar 23	11:48:22	68	200	142	N	t-	1.1592	0.7758	-0.3107	256.1	-	-	0S	175W
037	2016 Sep 16	18:55:27	68	206	147	N	a-	-1.0549	0.9091	-0.0624	239.9	-	-	3S	75E
038	2017 Feb 11	00:45:03	68	211	114	N	-a	-1.0255	0.9896	-0.0342	259.9	-	-	13N	8W
039	2017 Aug 07	18:21:38	69	217	119	P	-t	0.8669	1.2898	0.2477	301.7	115.9	-	15S	86E
040	2018 Jan 31	13:31:00	69	223	124	T	-p	-0.3014	2.2954	1.3167	318.0	203.5	76.7	17N	161E
041	2018 Jul 27	20:22:54	69	229	129	T+	pp	0.1168	2.6805	1.6100	374.8	235.4	103.8	19S	56E
042	2019 Jan 21	05:13:27	69	235	134	T	p-	0.3684	2.1697	1.1966	312.3	197.5	62.6	20N	75W
043	2019 Jul 16	21:31:55	69	241	139	P	t-	-0.6430	1.7050	0.6544	334.6	178.7	-	22S	39E
044	2020 Jan 10	19:11:11	69	247	144	N	a-	1.0727	0.8969	-0.1146	245.3	-	-	23N	74E
045	2020 Jun 05	19:26:14	70	252	111	N	-a	1.2406	0.5699	-0.4036	198.9	-	-	21S	69E
046	2020 Jul 05	04:31:12	70	253	149	N	a-	-1.3639	0.3560	-0.6422	165.7	-	-	24S	66W
047	2020 Nov 30	09:44:02	70	258	116	N	-t	-1.1309	0.8302	-0.2602	261.8	-	-	21N	148W
048	2021 May 26	11:19:53	70	264	121	T	-a	0.4774	1.9558	1.0112	302.8	188.1	15.9	21S	170W
049	2021 Nov 19	09:04:06	70	270	126	P	-t	-0.4552	2.0738	0.9760	362.4	209.2	-	19N	139W
050	2022 May 16	04:12:42	70	276	131	T-	p-	-0.2532	2.3743	1.4154	319.5	207.9	85.5	19S	64W
051	2022 Nov 08	11:00:22	71	282	136	T+	p-	0.2570	2.4161	1.3607	354.7	220.6	85.7	17N	169W
052	2023 May 05	17:24:04	71	288	141	N	h-	-1.0350	0.9655	-0.0438	258.3	-	-	17S	98E
053	2023 Oct 28	20:15:17	71	294	146	P	a-	0.9472	1.1200	0.1239	265.3	78.1	-	14N	52E
054	2024 Mar 25	07:14:00	71	299	113	N	-t	1.0610	0.9557	-0.1304	279.9	-	-	1S	106W
055	2024 Sep 18	02:45:26	71	305	118	P	-a	-0.9792	1.0392	0.0869	246.9	63.7	-	3S	42W
056	2025 Mar 14	06:59:56	72	311	123	T	-p	0.3485	2.2615	1.1804	363.4	218.9	66.1	3N	102W
057	2025 Sep 07	18:12:58	72	317	128	T	-p	-0.2752	2.3459	1.3638	327.4	210.0	82.7	6S	87E
058	2026 Mar 03	11:34:52	72	323	133	T	a-	-0.3765	2.1858	1.1526	339.4	207.8	59.0	6N	171W
059	2026 Aug 28	04:14:04	72	329	138	P	t-	0.4964	1.9664	0.9319	338.5	198.8	-	9S	63W
060	2027 Feb 20	23:14:05	73	335	143	N	a-	-1.0480	0.9286	-0.0549	241.7	-	-	10N	15E

Cat Num	Calendar Date	TD of Greatest Eclipse	ΔT s	Luna Num	Saros Num	Ecl Type	QSE	Gamma	Pen Mag	Um Mag	Phase ---- Durations ----			Greatest in Zenith	
											Pen m	Par m	Total m	Lat	Long
061	2027 Jul 18	16:04:10	73	340	110	Ne	-t	-1.5759	0.0032	-1.0662	17.7	–	–	22S	121E
062	2027 Aug 17	07:14:58	73	341	148	N	t-	1.2797	0.5476	-0.5234	219.4	–	–	12S	108W
063	2028 Jan 12	04:14:13	73	346	115	P	-a	0.9818	1.0485	0.0679	251.4	56.9	–	23N	61W
064	2028 Jul 06	18:20:57	73	352	120	P	-t	-0.7904	1.4282	0.3908	311.5	142.2	–	23S	86E
065	2028 Dec 31	16:53:15	73	358	125	T	-p	0.3258	2.2758	1.2478	337.1	209.6	72.0	23N	108E
066	2029 Jun 26	03:23:22	74	364	130	T+	pp	0.0124	2.8282	1.8452	336.0	220.3	102.7	23S	50W
067	2029 Dec 20	22:43:12	74	370	135	T	p-	-0.3811	2.2023	1.1190	358.9	214.1	54.4	23N	19E
068	2030 Jun 15	18:34:33	74	376	140	P	a-	0.7535	1.4495	0.5040	279.0	145.0	–	23S	82E
069	2030 Dec 09	22:28:51	74	382	145	N	t-	-1.0732	0.9430	-0.1613	280.1	–	–	22N	21E
070	2031 May 07	03:52:01	75	387	112	N	-a	-1.0695	0.8827	-0.0892	238.1	–	–	18S	59W
071	2031 Jun 05	11:45:17	75	388	150	N	a-	1.4732	0.1306	-0.8185	96.3	–	–	21S	176W
072	2031 Oct 30	07:46:44	75	393	117	N	-h	1.1774	0.7173	-0.3193	232.5	–	–	15N	121W
073	2032 Apr 25	15:14:51	75	399	122	T	-a	-0.3558	2.2204	1.1925	343.3	212.0	66.2	14S	131E
074	2032 Oct 18	19:03:40	75	405	127	T	-p	0.4169	2.0841	1.1039	316.2	196.6	47.7	10N	71E
075	2033 Apr 14	19:13:51	76	411	132	T	t-	0.3954	2.1722	1.0955	362.1	215.8	49.8	9S	72E
076	2033 Oct 08	10:56:23	76	417	137	T	p-	-0.2889	2.3068	1.3508	313.4	203.1	79.4	6N	167W
077	2034 Apr 03	19:07:00	76	423	142	N	t-	1.1144	0.8557	-0.2263	266.2	–	–	5S	75E
078	2034 Sep 28	02:47:37	76	429	147	P	a-	-1.0110	0.9922	0.0155	249.4	27.8	–	1N	44W
079	2035 Feb 22	09:06:11	76	434	114	N	-a	-1.0367	0.9663	-0.0523	256.4	–	–	9N	133W
080	2035 Aug 19	01:12:15	77	440	119	P	-a	0.9434	1.1519	0.1049	290.6	77.2	–	12S	17W
081	2036 Feb 11	22:13:06	77	446	124	T	-p	-0.3110	2.2762	1.3006	316.9	202.7	75.1	14N	31E
082	2036 Aug 07	02:52:32	77	452	129	T+	pp	0.2004	2.5279	1.4556	373.1	232.2	96.1	16S	41W
083	2037 Jan 31	14:01:38	77	458	134	T	p-	0.3619	2.1815	1.2086	312.9	198.2	64.3	18N	153E
084	2037 Jul 27	04:09:53	78	464	139	P	t-	-0.5582	1.8596	0.8108	341.7	193.2	–	20S	60W
085	2038 Jan 21	03:49:52	78	470	144	N	a-	1.0711	0.9009	-0.1127	246.5	–	–	21N	54W
086	2038 Jun 17	02:45:02	78	475	111	N	-a	1.3083	0.4438	-0.5259	177.0	–	–	22S	41W
087	2038 Jul 16	11:35:56	78	476	149	N	a-	-1.2838	0.5012	-0.4938	193.2	–	–	23S	172W
088	2038 Dec 11	17:45:00	79	481	116	N	-t	-1.1449	0.8062	-0.2876	259.3	–	–	22N	93E
089	2039 Jun 06	18:54:25	79	487	121	P	-a	0.5460	1.8288	0.8863	297.5	180.0	–	22S	77E
090	2039 Nov 30	16:56:28	79	493	126	P	-t	-0.4721	2.0435	0.9443	361.0	206.8	–	21N	103E
091	2040 May 26	11:46:22	79	499	131	T-	p-	-0.1872	2.4955	1.5365	322.2	211.5	92.9	21S	177W
092	2040 Nov 18	19:04:40	80	505	136	T+	p-	0.2361	2.4543	1.3991	354.4	221.2	88.6	20N	70E
093	2041 May 16	00:43:02	80	511	141	P	t-	-0.9747	1.0765	0.0663	270.5	59.4	–	20S	12W
094	2041 Nov 08	04:35:04	80	517	146	P	a-	0.9212	1.1675	0.1714	268.7	91.1	–	18N	73W
095	2042 Apr 05	14:30:12	80	522	113	N	-t	1.1080	0.8700	-0.2156	269.2	–	–	5S	144E
096	2042 Sep 29	10:45:47	81	528	118	N	-a	-1.0262	0.9548	-0.0010	239.2	–	–	2N	163W
097	2043 Mar 25	14:32:04	81	534	123	T	-t	0.3849	2.1920	1.1161	360.0	215.3	54.1	2S	144E
098	2043 Sep 19	01:51:50	81	540	128	T	-a	-0.3316	2.2452	1.2575	326.4	206.7	72.3	2S	29W
099	2044 Mar 13	19:38:33	82	546	133	T	a-	-0.3496	2.2322	1.2050	339.1	209.7	67.0	2N	68E
100	2044 Sep 07	11:20:44	82	552	138	T	t-	0.4318	2.0879	1.0476	344.8	206.9	34.8	5S	171W
101	2045 Mar 03	07:43:25	82	558	143	N	a-	-1.0274	0.9643	-0.0148	244.6	–	–	6N	113W
102	2045 Aug 27	13:54:49	82	564	148	N	t-	1.2061	0.6845	-0.3899	242.4	–	–	9S	151E
103	2046 Jan 22	13:02:37	83	569	115	P	-a	0.9886	1.0365	0.0550	250.7	51.4	–	21N	168E
104	2046 Jul 18	01:06:05	83	575	120	P	-t	-0.8692	1.2824	0.2478	299.0	115.3	–	22S	14W
105	2047 Jan 12	01:26:14	83	581	125	T	-p	0.3317	2.2665	1.2358	338.1	209.6	70.7	22N	19W
106	2047 Jul 07	10:35:45	84	587	130	T-	pp	-0.0636	2.7326	1.7529	334.3	219.3	101.6	23S	157W
107	2048 Jan 01	06:53:55	84	593	135	T	p-	-0.3746	2.2158	1.1297	360.4	215.1	56.7	23N	102W
108	2048 Jun 26	02:02:28	84	599	140	P	a-	0.6797	1.5841	0.6404	286.5	159.8	–	23S	30W
109	2048 Dec 20	06:27:48	84	605	145	N	t-	-1.0624	0.9632	-0.1420	282.4	–	–	22N	97W
110	2049 May 17	11:26:38	85	610	112	N	-a	-1.1337	0.7650	-0.2073	225.0	–	–	21S	172W
111	2049 Jun 15	19:14:12	85	611	150	N	a-	1.4069	0.2526	-0.6970	132.7	–	–	22S	72E
112	2049 Nov 09	15:52:11	85	616	117	N	-h	1.1965	0.6821	-0.3541	226.8	–	–	18N	118E
113	2050 May 06	22:32:02	85	622	122	T	-h	-0.4181	2.1064	1.0779	340.9	206.8	43.8	17S	21E
114	2050 Oct 30	03:21:47	86	628	127	T	-p	0.4435	2.0356	1.0549	313.9	193.6	35.1	14N	54W
115	2051 Apr 26	02:16:28	86	634	132	T	p-	0.3371	2.2785	1.2034	365.7	221.7	70.3	13S	34W
116	2051 Oct 19	19:11:50	86	640	137	T-	p-	-0.2542	2.3719	1.4130	315.0	205.0	84.2	10N	69E
117	2052 Apr 14	02:18:06	87	646	142	N	t-	1.0629	0.9478	-0.1294	276.8	–	–	9S	34W
118	2052 Oct 08	10:45:58	87	652	147	P	a-	-0.9727	1.0653	0.0832	257.3	63.9	–	5N	164W
119	2053 Mar 04	17:22:09	87	657	114	N	-a	-1.0531	0.9334	-0.0796	251.8	–	–	5N	102E
120	2053 Aug 29	08:05:50	87	663	119	Nx	-t	1.0165	1.0203	-0.0319	278.5	–	–	8S	121W

Cat Num	Calendar Date	TD of Greatest Eclipse	ΔT s	Luna Num	Saros Num	Ecl Type	QSE	Gamma	Pen Mag	Um Mag	Pen m	Par m	Total m	Lat	Long
121	2054 Feb 22	06:51:27	88	669	124	T	-p	-0.3242	2.2502	1.2780	315.5	201.6	72.7	10N	99W
122	2054 Aug 18	09:26:30	88	675	129	T	pp	0.2806	2.3817	1.3074	370.4	227.4	83.7	13S	140W
123	2055 Feb 11	22:46:17	88	681	134	T	p-	0.3526	2.1982	1.2258	313.7	199.1	66.6	14N	22E
124	2055 Aug 07	10:53:18	89	687	139	P	t-	-0.4769	2.0081	0.9606	347.2	204.2	-	17S	161W
125	2056 Feb 01	12:26:06	89	693	144	N	a-	1.0682	0.9069	-0.1084	248.0	-	-	18N	177E
126	2056 Jun 27	10:03:09	89	698	111	N	-a	1.3770	0.3158	-0.6504	150.6	-	-	22S	150W
127	2056 Jul 26	18:43:25	89	699	149	N	a-	-1.2048	0.6448	-0.3477	215.2	-	-	20S	81E
128	2056 Dec 22	01:48:56	90	704	116	N	-t	-1.1560	0.7872	-0.3093	257.3	-	-	22N	27W
129	2057 Jun 17	02:26:20	90	710	121	P	-a	0.6168	1.6982	0.7570	291.4	170.0	-	23S	36W
130	2057 Dec 11	00:53:38	90	716	126	P	-t	-0.4853	2.0194	0.9197	359.7	204.8	-	23N	15W
131	2058 Jun 06	19:15:48	91	722	131	T-	pp	-0.1181	2.6226	1.6628	324.5	214.1	98.0	23S	71E
132	2058 Nov 30	03:16:18	91	728	136	T+	p-	0.2208	2.4819	1.4277	353.9	221.5	90.4	22N	52W
133	2059 May 27	07:55:34	91	734	141	P	t-	-0.9098	1.1963	0.1846	282.5	97.9	-	22S	119W
134	2059 Nov 19	13:01:36	92	740	146	P	a-	0.9004	1.2055	0.2097	271.2	99.9	-	20N	161E
135	2060 Apr 15	21:37:05	92	745	113	N	-t	1.1622	0.7694	-0.3136	255.7	-	-	9S	37E
136	2060 Oct 09	18:53:33	92	751	118	N	-a	-1.0671	0.8816	-0.0779	232.0	-	-	6N	74E
137	2060 Nov 08	04:04:13	93	752	156	Nb	a-	1.5332	0.0286	-0.9356	45.3	-	-	18N	65W
138	2061 Apr 04	21:54:05	93	757	123	T	-t	0.4300	2.1064	1.0360	355.8	210.3	30.9	6S	33E
139	2061 Sep 29	09:38:13	93	763	128	T	-a	-0.3810	2.1576	1.1640	325.5	203.1	59.6	2N	146W
140	2062 Mar 25	03:33:50	94	769	133	T	p-	-0.3150	2.2925	1.2715	339.1	212.0	75.3	2S	52W
141	2062 Sep 18	18:34:02	94	775	138	T	p-	0.3736	2.1979	1.1515	350.0	211.1	60.2	1S	80E
142	2063 Mar 14	16:05:49	94	781	143	P	a-	-1.0008	1.0108	0.0363	248.5	41.9	-	1N	121E
143	2063 Sep 07	20:41:11	95	787	148	N	t-	1.1375	0.8121	-0.2657	261.2	-	-	5S	49E
144	2064 Feb 02	21:48:57	95	792	115	P	-a	0.9969	1.0215	0.0395	249.8	43.7	-	18N	37E
145	2064 Jul 28	07:52:48	95	798	120	P	-t	-0.9473	1.1378	0.1055	285.0	76.5	-	20S	116W
146	2065 Jan 22	09:58:58	96	804	125	T	-p	0.3371	2.2579	1.2248	339.0	209.7	69.4	20N	146W
147	2065 Jul 17	17:48:40	96	810	130	T-	pp	-0.1402	2.5907	1.6138	331.9	217.0	97.7	21S	95E
148	2066 Jan 11	15:04:47	96	816	135	T	p-	-0.3687	2.2276	1.1395	361.6	216.0	58.7	21N	136E
149	2066 Jul 07	09:30:29	97	822	140	P	a-	0.6056	1.7196	0.7770	293.1	172.0	-	22S	141W
150	2066 Dec 31	14:30:09	97	828	145	N	t-	-1.0540	0.9789	-0.1264	284.2	-	-	22N	143E
151	2067 May 28	18:56:07	97	833	112	N	-a	-1.2013	0.6416	-0.3316	209.2	-	-	23S	76E
152	2067 Jun 27	02:41:06	97	834	150	N	a-	1.3394	0.3770	-0.5736	160.5	-	-	22S	39W
153	2067 Nov 21	00:04:42	98	839	117	N	-h	1.2107	0.6557	-0.3798	222.2	-	-	21N	4W
154	2068 May 17	05:42:17	98	845	122	P	-t	-0.4852	1.9839	0.9545	337.5	199.8	-	20S	86W
155	2068 Nov 09	11:47:00	99	851	127	T	-p	0.4645	1.9974	1.0161	312.0	190.9	19.2	18N	180E
156	2069 May 06	09:09:57	99	857	132	T+	pp	0.2717	2.3977	1.3242	369.1	227.0	85.0	17S	138W
157	2069 Oct 30	03:35:06	99	863	137	T-	p-	-0.2263	2.4247	1.4628	316.3	206.4	87.4	14N	57W
158	2070 Apr 25	09:21:25	100	869	142	Nx	t-	1.0044	1.0527	-0.0197	287.7	-	-	12S	140W
159	2070 Oct 19	18:51:12	100	875	147	P	a-	-0.9406	1.1270	0.1395	263.9	82.3	-	9N	74E
160	2071 Mar 16	01:31:09	100	880	114	N	-a	-1.0757	0.8890	-0.1183	245.8	-	-	1N	21W
161	2071 Sep 09	15:05:40	101	886	119	N	-t	1.0835	0.9000	-0.1575	265.9	-	-	4S	133E
162	2072 Mar 04	15:23:07	101	892	124	T	-p	-0.3431	2.2137	1.2452	313.9	200.1	69.0	6N	132E
163	2072 Aug 28	16:05:42	102	898	129	T	-t	0.3563	2.2439	1.1673	366.9	221.1	64.8	9S	119E
164	2073 Feb 22	07:24:53	102	904	134	T	p-	0.3389	2.2230	1.2514	314.6	200.4	69.8	10N	107W
165	2073 Aug 17	17:42:41	102	910	139	T	t-	-0.3998	2.1490	1.1024	351.3	212.3	50.8	13S	96E
166	2074 Feb 11	20:55:58	103	916	144	N	a-	1.0612	0.9203	-0.0960	250.2	-	-	15N	50E
167	2074 Jul 08	17:21:37	103	921	111	N	-a	1.4457	0.1884	-0.7751	117.3	-	-	21S	101E
168	2074 Aug 07	01:56:04	103	922	149	N	a-	-1.1291	0.7826	-0.2079	232.9	-	-	17S	27W
169	2075 Jan 02	09:55:03	104	927	116	N	-t	-1.1643	0.7729	-0.3256	255.8	-	-	22N	147W
170	2075 Jun 28	09:55:36	104	933	121	P	-a	0.6897	1.5639	0.6235	284.2	157.6	-	23S	148W
171	2075 Dec 22	08:55:55	104	939	126	P	-t	-0.4945	2.0024	0.9028	358.6	203.4	-	23N	134W
172	2076 Jun 17	02:39:47	105	945	131	T-	pp	-0.0452	2.7570	1.7959	326.2	215.9	100.9	23S	39W
173	2076 Dec 10	11:34:51	105	951	136	T+	p-	0.2102	2.5006	1.4476	353.1	221.4	91.6	23N	175W
174	2077 Jun 06	14:59:52	106	957	141	P	t-	-0.8388	1.3274	0.3139	294.4	125.7	-	24S	135E
175	2077 Nov 29	21:35:53	106	963	146	P	a-	-0.8855	1.2326	0.2372	272.8	105.6	-	23N	33E
176	2078 Apr 27	04:35:45	106	968	113	N	-t	1.2223	0.6577	-0.4227	239.0	-	-	13S	69W
177	2078 Oct 21	03:08:04	107	974	118	N	-a	-1.1022	0.8191	-0.1442	225.5	-	-	10N	50W
178	2078 Nov 19	12:40:02	107	975	156	N	a-	1.5148	0.0633	-0.9028	67.2	-	-	21N	166E
179	2079 Apr 16	05:10:45	107	980	123	P	-t	0.4800	2.0119	0.9471	350.9	204.1	-	10S	77W
180	2079 Oct 10	17:30:30	108	986	128	T	-a	-0.4246	2.0806	1.0811	324.5	199.3	43.1	7N	95E

Cat Num	Calendar Date	TD of Greatest Eclipse	ΔT s	Luna Num	Saros Num	Ecl Type	QSE	Gamma	Pen Mag	Um Mag	Pen m	Par m	Total m	Greatest in Zenith Lat	Long
181	2080 Apr 04	11:23:38	108	992	133	T	p-	-0.2751	2.3626	1.3479	339.0	214.2	82.8	6S	170W
182	2080 Sep 29	01:52:42	109	998	138	T	p-	0.3203	2.2986	1.2462	354.5	218.0	74.4	3N	30W
183	2081 Mar 25	00:22:01	109	1004	143	P	a-	-0.9688	1.0673	0.0973	253.1	68.0	–	3S	4W
184	2081 Sep 18	03:35:25	109	1010	148	N	t-	1.0747	0.9291	-0.1524	276.4	–	–	1S	55W
185	2082 Feb 13	06:29:20	110	1015	115	P	-a	1.0101	0.9974	0.0153	247.9	27.3	–	14N	93W
186	2082 Aug 08	14:46:42	110	1021	120	Nx	-t	-1.0204	1.0030	-0.0275	270.5	–	–	17S	141E
187	2083 Feb 02	18:26:46	111	1027	125	T	-p	0.3464	2.2418	1.2070	339.7	209.5	67.2	17N	87E
188	2083 Jul 29	01:05:34	111	1033	130	T-	pp	-0.2143	2.4537	1.4791	328.8	213.6	91.1	19S	14W
189	2084 Jan 22	23:13:00	111	1039	135	T	p-	-0.3610	2.2425	1.1531	362.9	217.0	61.3	19N	15E
190	2084 Jul 17	16:58:51	112	1045	140	P	a-	0.5313	1.8557	0.9136	298.9	182.1	–	20S	107E
191	2085 Jan 10	22:32:29	112	1051	145	N	t-	-1.0453	0.9944	-0.1102	285.7	–	–	21N	24E
192	2085 Jun 08	02:17:36	113	1056	112	N	-a	-1.2746	0.5079	-0.4668	189.2	–	–	24S	34W
193	2085 Jul 07	10:04:39	113	1057	150	N	a-	1.2695	0.5064	-0.4461	184.2	–	–	21S	150W
194	2085 Dec 01	08:25:35	113	1062	117	N	-a	1.2190	0.6400	-0.3944	219.2	–	–	23N	129W
195	2086 May 28	12:43:47	114	1068	122	P	-t	-0.5585	1.8499	0.8193	332.9	190.2	–	22S	169E
196	2086 Nov 20	20:19:42	114	1074	127	P	-p	0.4800	1.9692	0.9877	310.4	188.9	–	20N	52E
197	2087 May 17	15:55:20	115	1080	132	T+	pp	0.1999	2.5289	1.4568	371.9	231.5	95.9	19S	121E
198	2087 Nov 10	12:05:33	115	1086	137	T-	p-	-0.2043	2.4667	1.5018	317.3	207.3	89.6	17N	175E
199	2088 May 05	16:16:50	115	1092	142	P	t-	0.9388	1.1708	0.1032	298.7	77.8	–	16S	116E
200	2088 Oct 30	03:03:20	116	1098	147	P	a-	-0.9147	1.1773	0.1843	269.3	94.2	–	13N	49W
201	2089 Mar 26	09:34:13	116	1103	114	N	-a	-1.1039	0.8343	-0.1670	238.4	–	–	4S	142W
202	2089 Sep 19	22:11:16	117	1109	119	N	-t	1.1448	0.7904	-0.2726	252.9	–	–	0N	26E
203	2090 Mar 15	23:48:31	117	1115	124	T	-p	-0.3675	2.1670	1.2023	312.0	198.1	63.6	1N	5E
204	2090 Sep 08	22:52:29	118	1121	129	T	-t	0.4257	2.1178	1.0387	362.9	213.9	32.5	5S	17E
205	2091 Mar 05	15:58:22	118	1127	134	T	p-	0.3212	2.2548	1.2843	315.8	202.0	73.5	6N	124E
206	2091 Aug 29	00:38:25	119	1133	139	T	t-	-0.3270	2.2821	1.2362	354.3	218.3	73.5	10S	9W
207	2092 Feb 23	05:21:00	119	1139	144	N	a-	1.0509	0.9395	-0.0777	253.1	–	–	11N	76W
208	2092 Jul 19	00:41:57	119	1144	111	Ne	-a	1.5132	0.0634	-0.8979	68.6	–	–	19S	9W
209	2092 Aug 17	09:14:00	120	1145	149	N	a-	-1.0569	0.9143	-0.0746	247.4	–	–	14S	137W
210	2093 Jan 12	18:00:02	120	1150	116	N	-t	-1.1734	0.7567	-0.3430	253.9	–	–	20N	93E
211	2093 Jul 08	17:24:18	120	1156	121	P	-a	0.7632	1.4289	0.4886	276.1	142.5	–	22S	101E
212	2094 Jan 01	17:00:06	121	1162	126	P	-t	-0.5025	1.9872	0.8886	357.4	202.0	–	22N	106E
213	2094 Jun 28	10:01:57	121	1168	131	T+	pp	0.0288	2.7879	1.8249	327.3	216.5	101.4	23S	149W
214	2094 Dec 21	19:56:32	122	1174	136	T+	p-	0.2016	2.5153	1.4642	352.1	221.3	92.4	24N	61E
215	2095 Jun 17	22:00:11	122	1180	141	P	t-	-0.7653	1.4632	0.4474	305.5	147.6	–	24S	31E
216	2095 Dec 11	06:15:02	123	1186	146	P	a-	0.8743	1.2526	0.2581	273.7	109.6	–	24N	95W
217	2096 May 07	11:24:44	123	1191	113	N	-t	1.2897	0.5327	-0.5451	217.7	–	–	16S	171W
218	2096 Jun 06	02:43:40	123	1192	151	Nb	t-	-1.5724	0.0064	-1.0567	24.7	–	–	24S	41W
219	2096 Oct 31	11:30:24	124	1197	118	N	-a	-1.1308	0.7685	-0.1987	220.0	–	–	13N	176W
220	2096 Nov 29	21:22:20	124	1198	156	N	a-	1.5018	0.0879	-0.8799	79.0	–	–	23N	37E
221	2097 Apr 26	12:18:17	124	1203	123	P	-t	0.5377	1.9032	0.8439	344.8	195.9	–	13S	176E
222	2097 Oct 21	01:30:55	125	1209	128	T	-a	-0.4608	2.0171	1.0116	323.8	195.9	16.7	11N	26W
223	2098 Apr 15	19:04:48	125	1215	133	T-	p-	-0.2272	2.4474	1.4389	339.0	216.5	89.7	10S	74E
224	2098 Oct 10	09:19:58	126	1221	138	T	pp	0.2749	2.3850	1.3266	358.2	221.7	83.4	7N	143W
225	2099 Apr 05	08:30:56	126	1227	143	P	a-	-0.9304	1.1353	0.1700	258.4	88.8	–	7S	127W
226	2099 Sep 29	10:36:37	127	1233	148	Nx	t-	1.0175	1.0360	-0.0491	289.1	–	–	3N	162W
227	2100 Feb 24	15:05:12	127	1238	115	N	-a	1.0267	0.9669	-0.0151	245.3	–	–	10N	138E
228	2100 Aug 19	21:44:59	128	1244	120	N	-t	-1.0906	0.8735	-0.1556	254.9	–	–	13S	36E

Appendix B

Small Lunar Eclipse Figures: 2001 to 2100

Key to Small Lunar Eclipse Figures

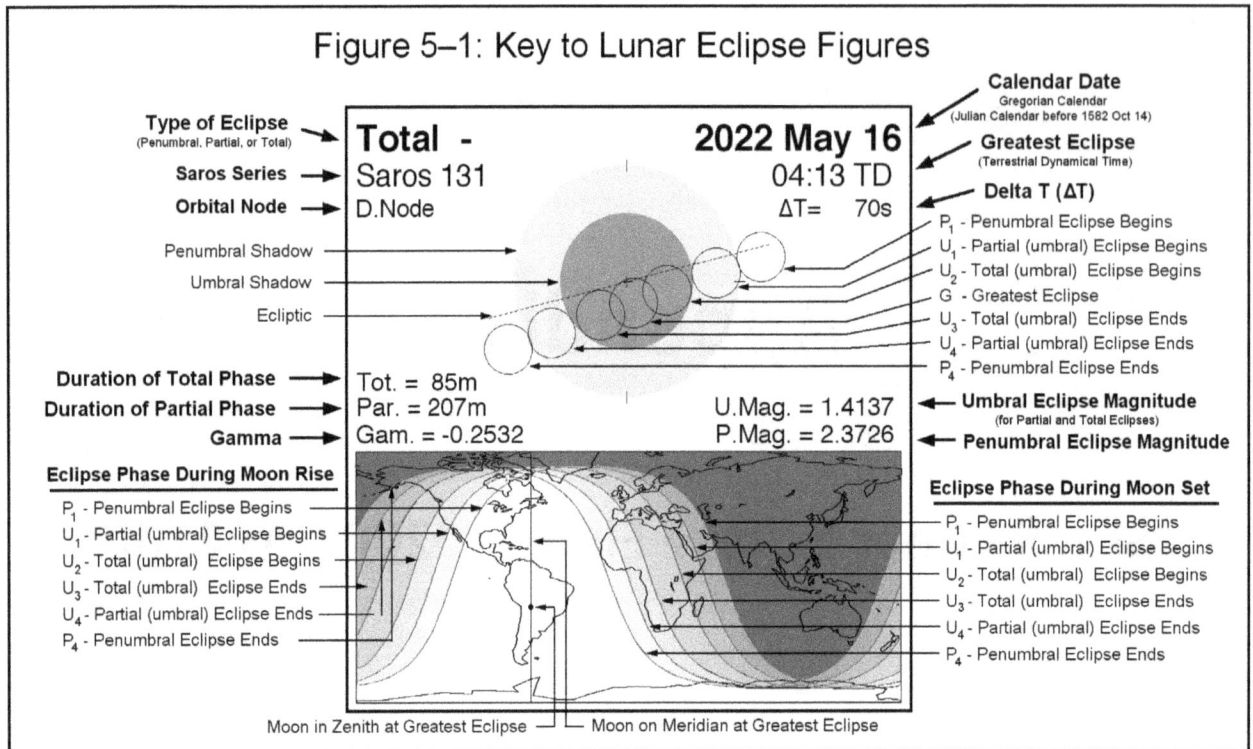

Figure 5–1: Key to Lunar Eclipse Figures

Lunar Eclipse Type – One of three basic types of lunar eclipses: Penumbral, Partial, or Total.

Saros Series – The Saros series that the eclipse belongs to. Eclipses with an even number take place at the Moon's ascending node, while those with an odd number are at the descending node.

Node – The orbital node near which the eclipse takes place. The ascending node (A. Node) is the point where the Moon travels from south to north through Earth's orbital plane. Similarly, the descending node (D. Node) is the point where the Moon travels from north to south.

Calendar Date – Gregorian calendar date at greatest eclipse is given in the ISO order of YEAR-MONTH-DAY.

Greatest Eclipse – The instant (Terrestrial Dynamical Time) when the distance between the Moon and the axis of Earth's shadow cone reaches a minimum.

ΔT (Delta T) – the arithmetic difference, in seconds, between Terrestrial Dynamical Time (TD) and Universal Time (UT1).

Gamma – The minimum distance from the center of the Moon to Earth's shadow axis, in units of Earth's equatorial radius.

Duration – The duration of a penumbral eclipse (**Pen.**), a partial eclipse (**Par.**) , and a total eclipse (**Tot.**).

Umbral Eclipse Magnitude (U. Mag) – The fraction of the Moon's diameter immersed in the umbra at greatest eclipse.

Penumbral Eclipse Magnitude (P. Mag) – The fraction of the Moon's diameter immersed in the penumbra at greatest eclipse.

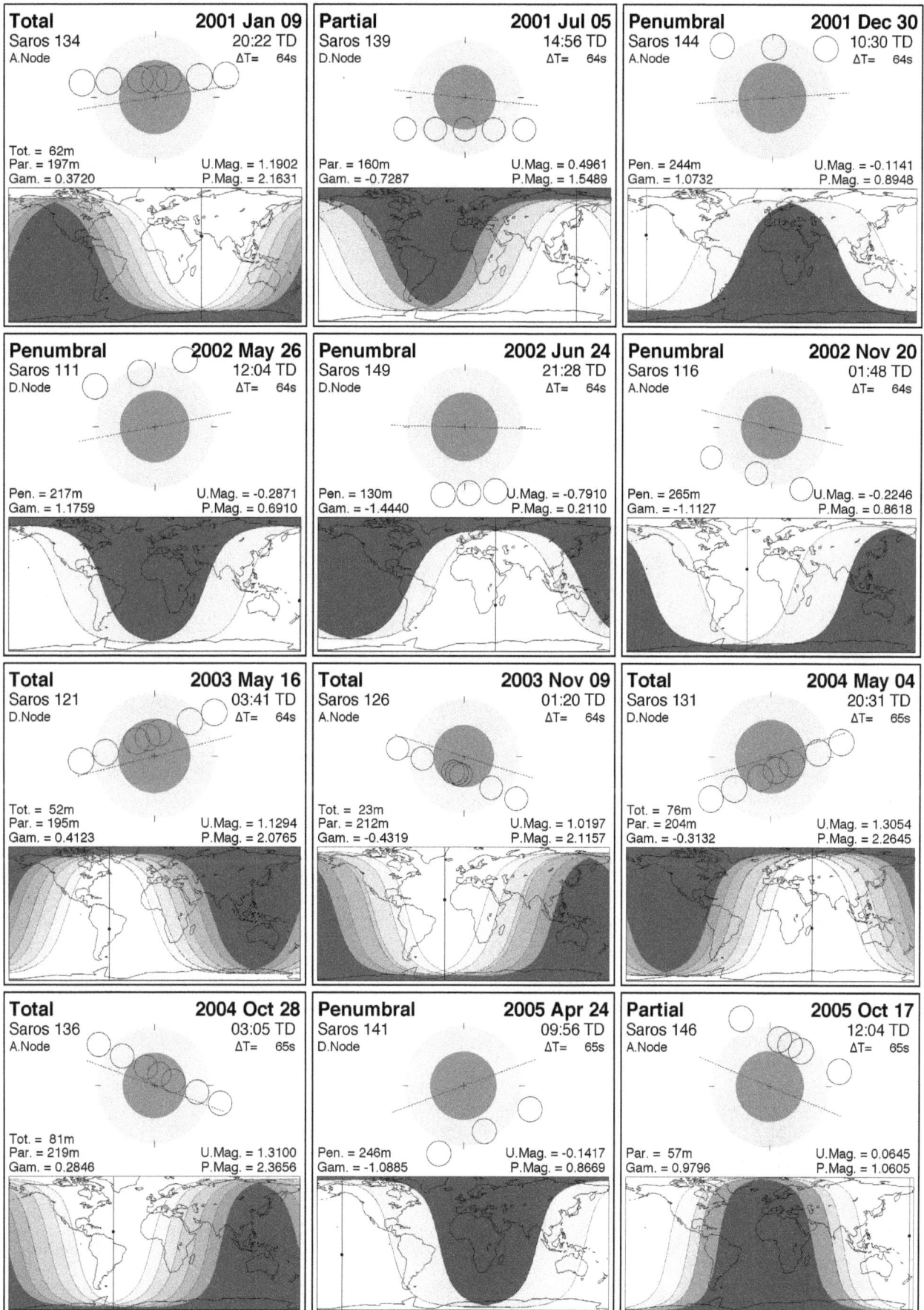

Total   **2001 Jan 09**
Saros 134   20:22 TD
A.Node   ΔT=   64s

Tot. = 62m
Par. = 197m
Gam. = 0.3720   U.Mag. = 1.1902
                  P.Mag. = 2.1631

Partial   **2001 Jul 05**
Saros 139   14:56 TD
D.Node   ΔT=   64s

Par. = 160m
Gam. = -0.7287   U.Mag. = 0.4961
                  P.Mag. = 1.5489

Penumbral   **2001 Dec 30**
Saros 144   10:30 TD
A.Node   ΔT=   64s

Pen. = 244m
Gam. = 1.0732   U.Mag. = -0.1141
                  P.Mag. = 0.8948

Penumbral   **2002 May 26**
Saros 111   12:04 TD
D.Node   ΔT=   64s

Pen. = 217m
Gam. = 1.1759   U.Mag. = -0.2871
                  P.Mag. = 0.6910

Penumbral   **2002 Jun 24**
Saros 149   21:28 TD
D.Node   ΔT=   64s

Pen. = 130m
Gam. = -1.4440   U.Mag. = -0.7910
                  P.Mag. = 0.2110

Penumbral   **2002 Nov 20**
Saros 116   01:48 TD
A.Node   ΔT=   64s

Pen. = 265m
Gam. = -1.1127   U.Mag. = -0.2246
                  P.Mag. = 0.8618

Total   **2003 May 16**
Saros 121   03:41 TD
D.Node   ΔT=   64s

Tot. = 52m
Par. = 195m
Gam. = 0.4123   U.Mag. = 1.1294
                  P.Mag. = 2.0765

Total   **2003 Nov 09**
Saros 126   01:20 TD
A.Node   ΔT=   64s

Tot. = 23m
Par. = 212m
Gam. = -0.4319   U.Mag. = 1.0197
                  P.Mag. = 2.1157

Total   **2004 May 04**
Saros 131   20:31 TD
D.Node   ΔT=   65s

Tot. = 76m
Par. = 204m
Gam. = -0.3132   U.Mag. = 1.3054
                  P.Mag. = 2.2645

Total   **2004 Oct 28**
Saros 136   03:05 TD
A.Node   ΔT=   65s

Tot. = 81m
Par. = 219m
Gam. = 0.2846   U.Mag. = 1.3100
                  P.Mag. = 2.3656

Penumbral   **2005 Apr 24**
Saros 141   09:56 TD
D.Node   ΔT=   65s

Pen. = 246m
Gam. = -1.0885   U.Mag. = -0.1417
                  P.Mag. = 0.8669

Partial   **2005 Oct 17**
Saros 146   12:04 TD
A.Node   ΔT=   65s

Par. = 57m
Gam. = 0.9796   U.Mag. = 0.0645
                  P.Mag. = 1.0605

Plate 001

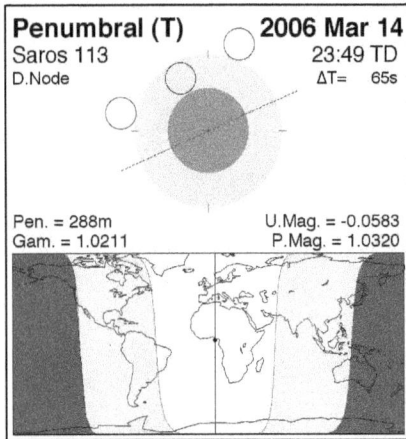

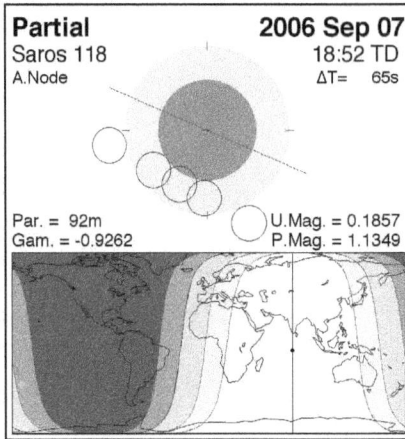

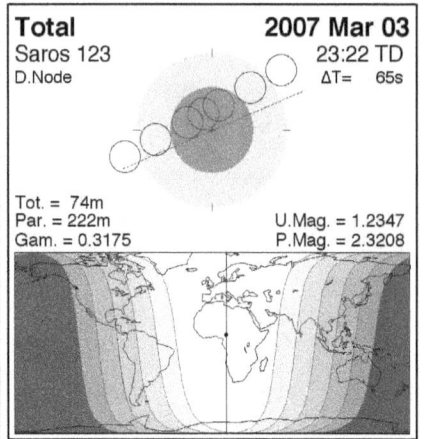

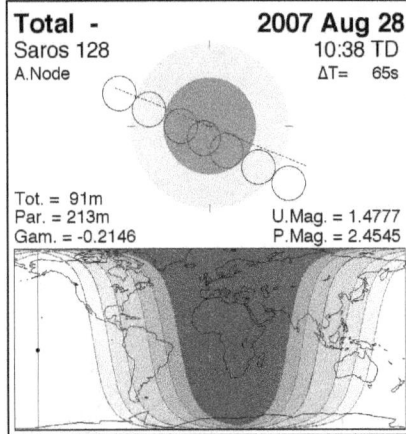

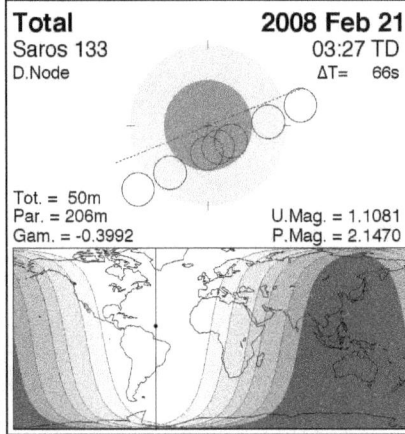

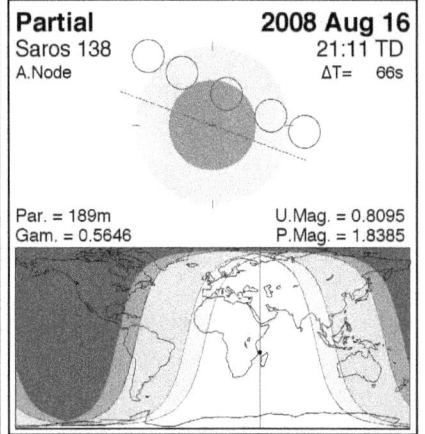

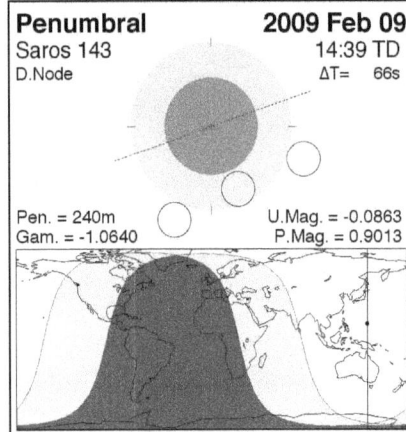

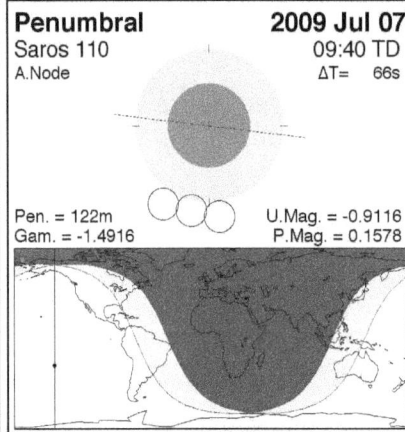

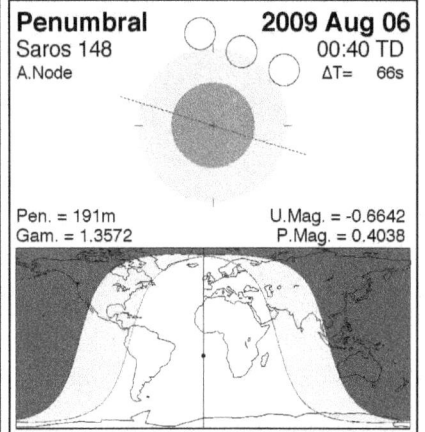

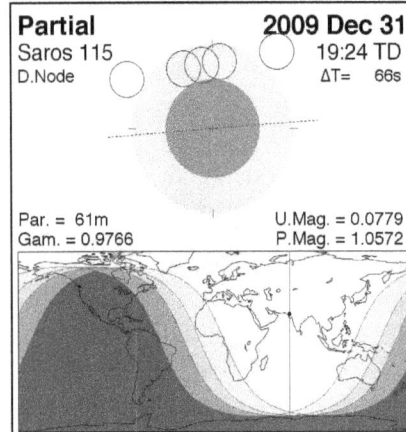

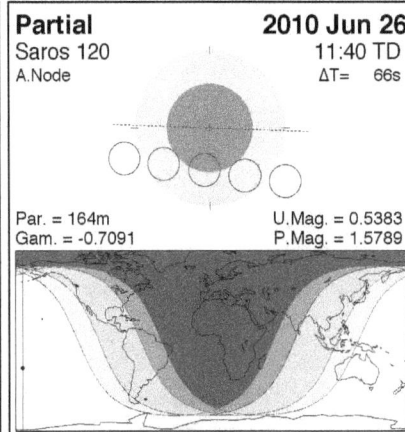

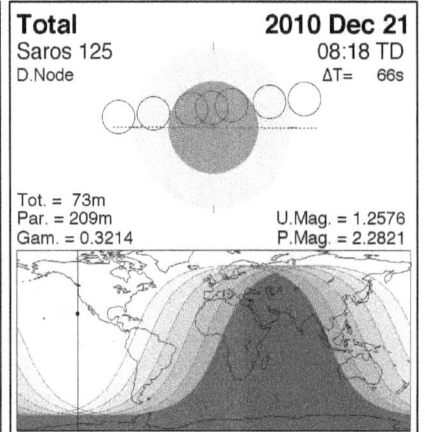

Penumbral (T)	2006 Mar 14
Saros 113	23:49 TD
D.Node	ΔT= 65s

Pen. = 288m U.Mag. = -0.0583
Gam. = 1.0211 P.Mag. = 1.0320

Partial	2006 Sep 07
Saros 118	18:52 TD
A.Node	ΔT= 65s

Par. = 92m U.Mag. = 0.1857
Gam. = -0.9262 P.Mag. = 1.1349

Total	2007 Mar 03
Saros 123	23:22 TD
D.Node	ΔT= 65s

Tot. = 74m
Par. = 222m U.Mag. = 1.2347
Gam. = 0.3175 P.Mag. = 2.3208

Total -	2007 Aug 28
Saros 128	10:38 TD
A.Node	ΔT= 65s

Tot. = 91m
Par. = 213m U.Mag. = 1.4777
Gam. = -0.2146 P.Mag. = 2.4545

Total	2008 Feb 21
Saros 133	03:27 TD
D.Node	ΔT= 66s

Tot. = 50m
Par. = 206m U.Mag. = 1.1081
Gam. = -0.3992 P.Mag. = 2.1470

Partial	2008 Aug 16
Saros 138	21:11 TD
A.Node	ΔT= 66s

Par. = 189m U.Mag. = 0.8095
Gam. = 0.5646 P.Mag. = 1.8385

Penumbral	2009 Feb 09
Saros 143	14:39 TD
D.Node	ΔT= 66s

Pen. = 240m U.Mag. = -0.0863
Gam. = -1.0640 P.Mag. = 0.9013

Penumbral	2009 Jul 07
Saros 110	09:40 TD
A.Node	ΔT= 66s

Pen. = 122m U.Mag. = -0.9116
Gam. = -1.4916 P.Mag. = 0.1578

Penumbral	2009 Aug 06
Saros 148	00:40 TD
A.Node	ΔT= 66s

Pen. = 191m U.Mag. = -0.6642
Gam. = 1.3572 P.Mag. = 0.4038

Partial	2009 Dec 31
Saros 115	19:24 TD
D.Node	ΔT= 66s

Par. = 61m U.Mag. = 0.0779
Gam. = 0.9766 P.Mag. = 1.0572

Partial	2010 Jun 26
Saros 120	11:40 TD
A.Node	ΔT= 66s

Par. = 164m U.Mag. = 0.5383
Gam. = -0.7091 P.Mag. = 1.5789

Total	2010 Dec 21
Saros 125	08:18 TD
D.Node	ΔT= 66s

Tot. = 73m
Par. = 209m U.Mag. = 1.2576
Gam. = 0.3214 P.Mag. = 2.2821

Plate 002

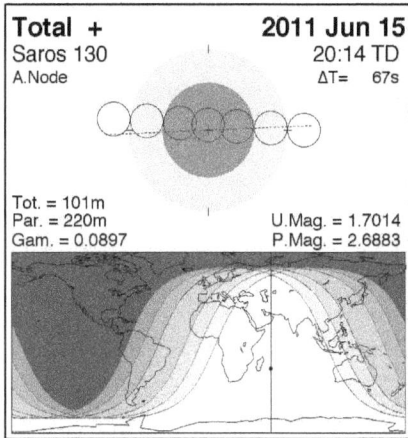

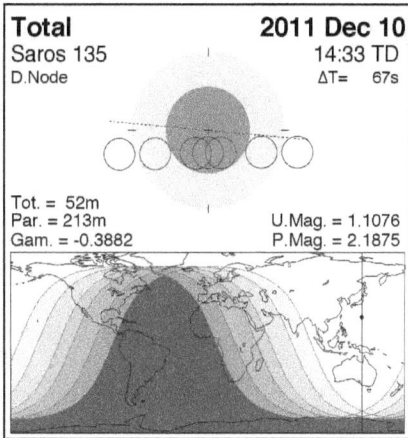

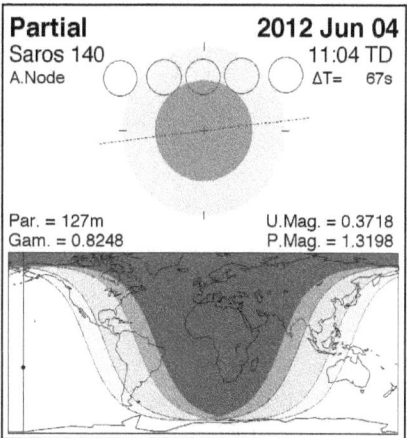

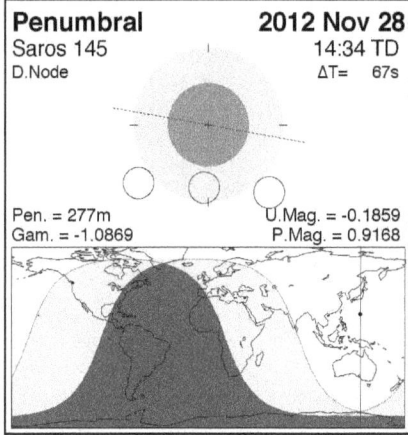

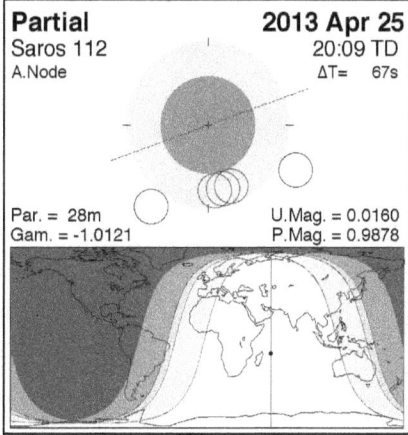

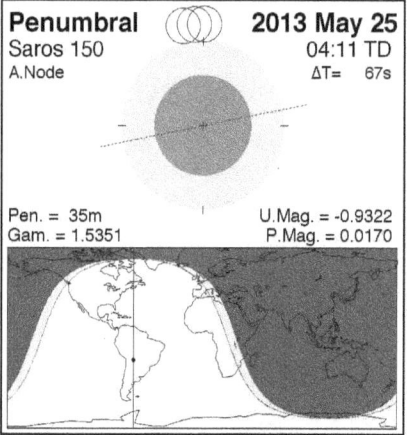

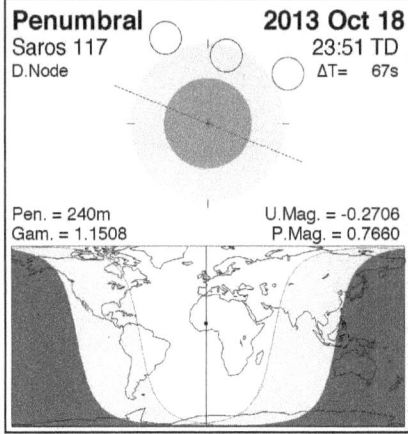

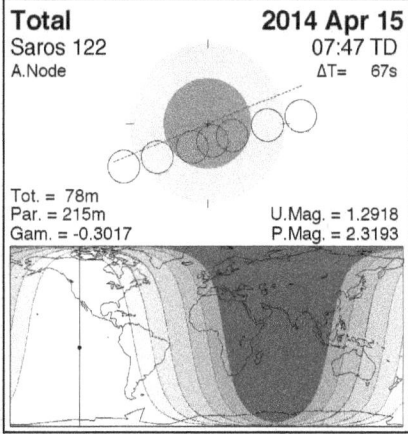

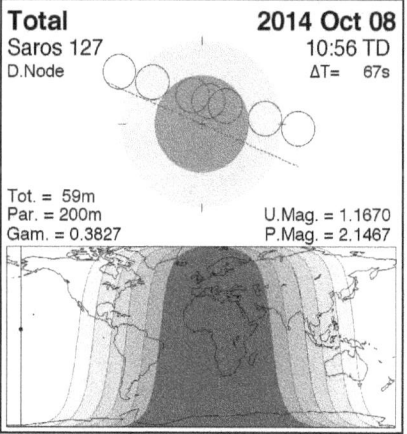

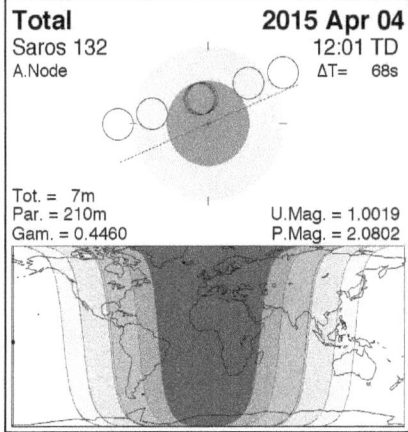

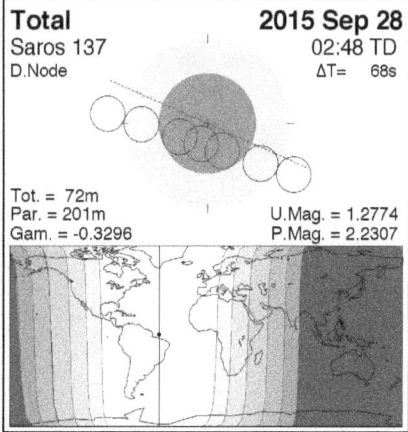

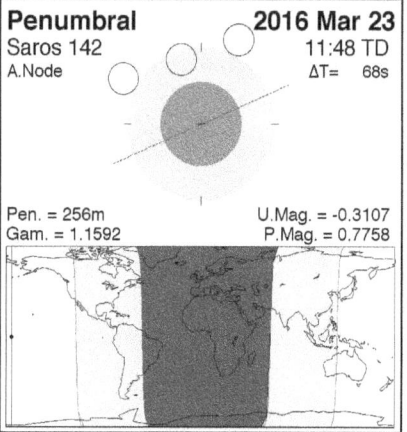

Total +	2011 Jun 15
Saros 130	20:14 TD
A.Node	ΔT= 67s

Tot. = 101m
Par. = 220m
Gam. = 0.0897
U.Mag. = 1.7014
P.Mag. = 2.6883

Total	2011 Dec 10
Saros 135	14:33 TD
D.Node	ΔT= 67s

Tot. = 52m
Par. = 213m
Gam. = -0.3882
U.Mag. = 1.1076
P.Mag. = 2.1875

Partial	2012 Jun 04
Saros 140	11:04 TD
A.Node	ΔT= 67s

Par. = 127m
Gam. = 0.8248
U.Mag. = 0.3718
P.Mag. = 1.3198

Penumbral	2012 Nov 28
Saros 145	14:34 TD
D.Node	ΔT= 67s

Pen. = 277m
Gam. = -1.0869
U.Mag. = -0.1859
P.Mag. = 0.9168

Partial	2013 Apr 25
Saros 112	20:09 TD
A.Node	ΔT= 67s

Par. = 28m
Gam. = -1.0121
U.Mag. = 0.0160
P.Mag. = 0.9878

Penumbral	2013 May 25
Saros 150	04:11 TD
A.Node	ΔT= 67s

Pen. = 35m
Gam. = 1.5351
U.Mag. = -0.9322
P.Mag. = 0.0170

Penumbral	2013 Oct 18
Saros 117	23:51 TD
D.Node	ΔT= 67s

Pen. = 240m
Gam. = 1.1508
U.Mag. = -0.2706
P.Mag. = 0.7660

Total	2014 Apr 15
Saros 122	07:47 TD
A.Node	ΔT= 67s

Tot. = 78m
Par. = 215m
Gam. = -0.3017
U.Mag. = 1.2918
P.Mag. = 2.3193

Total	2014 Oct 08
Saros 127	10:56 TD
D.Node	ΔT= 67s

Tot. = 59m
Par. = 200m
Gam. = 0.3827
U.Mag. = 1.1670
P.Mag. = 2.1467

Total	2015 Apr 04
Saros 132	12:01 TD
A.Node	ΔT= 68s

Tot. = 7m
Par. = 210m
Gam. = 0.4460
U.Mag. = 1.0019
P.Mag. = 2.0802

Total	2015 Sep 28
Saros 137	02:48 TD
D.Node	ΔT= 68s

Tot. = 72m
Par. = 201m
Gam. = -0.3296
U.Mag. = 1.2774
P.Mag. = 2.2307

Penumbral	2016 Mar 23
Saros 142	11:48 TD
A.Node	ΔT= 68s

Pen. = 256m
Gam. = 1.1592
U.Mag. = -0.3107
P.Mag. = 0.7758

Plate 003

Penumbral	2016 Sep 16
Saros 147	18:55 TD
D.Node	ΔT= 68s

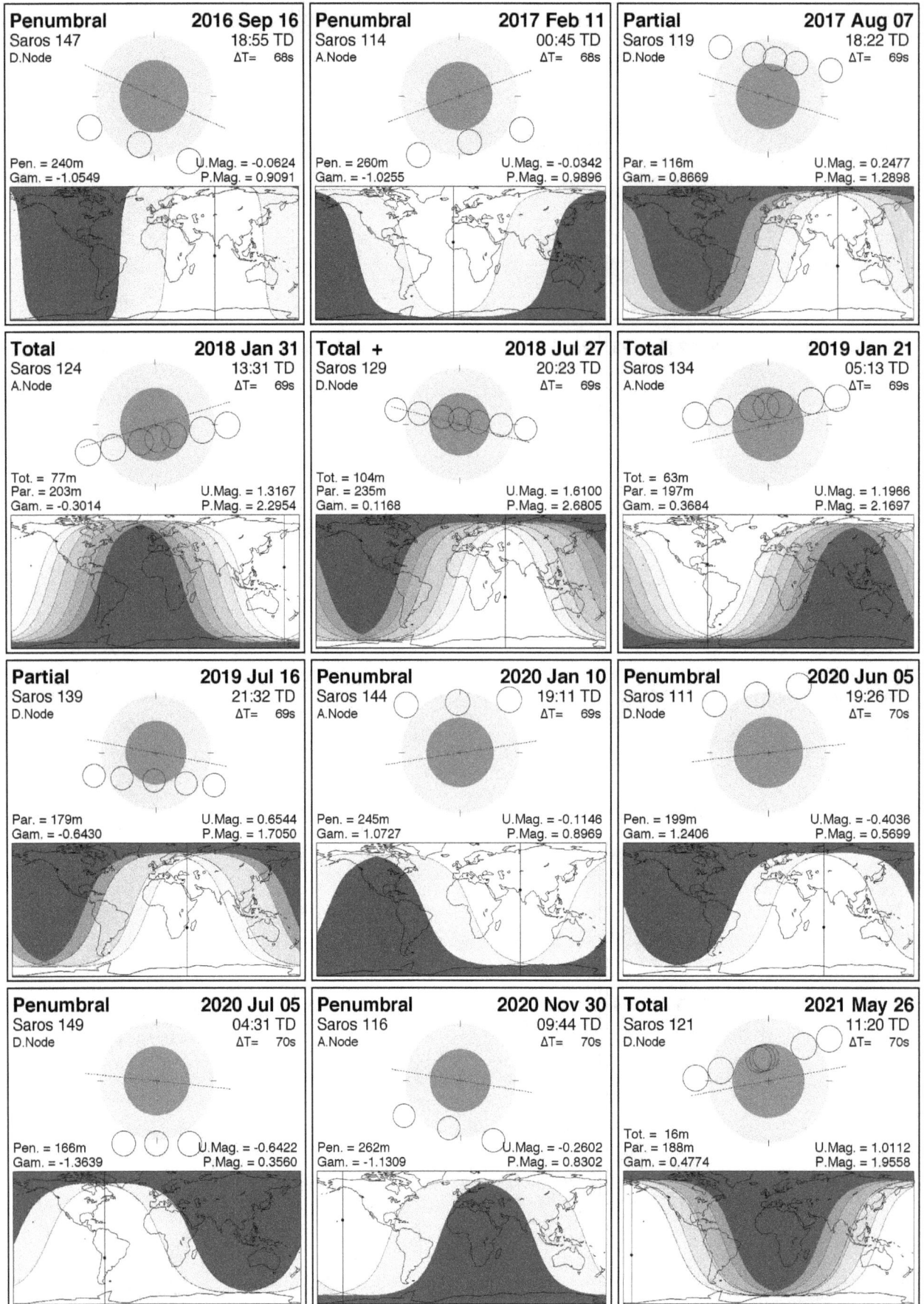

| Pen. = 240m | U.Mag. = -0.0624 |
| Gam. = -1.0549 | P.Mag. = 0.9091 |

Penumbral	2017 Feb 11
Saros 114	00:45 TD
A.Node	ΔT= 68s

| Pen. = 260m | U.Mag. = -0.0342 |
| Gam. = -1.0255 | P.Mag. = 0.9896 |

Partial	2017 Aug 07
Saros 119	18:22 TD
D.Node	ΔT= 69s

| Par. = 116m | U.Mag. = 0.2477 |
| Gam. = 0.8669 | P.Mag. = 1.2898 |

Total	2018 Jan 31
Saros 124	13:31 TD
A.Node	ΔT= 69s

Tot. = 77m	
Par. = 203m	U.Mag. = 1.3167
Gam. = -0.3014	P.Mag. = 2.2954

Total +	2018 Jul 27
Saros 129	20:23 TD
D.Node	ΔT= 69s

Tot. = 104m	
Par. = 235m	U.Mag. = 1.6100
Gam. = 0.1168	P.Mag. = 2.6805

Total	2019 Jan 21
Saros 134	05:13 TD
A.Node	ΔT= 69s

Tot. = 63m	
Par. = 197m	U.Mag. = 1.1966
Gam. = 0.3684	P.Mag. = 2.1697

Partial	2019 Jul 16
Saros 139	21:32 TD
D.Node	ΔT= 69s

| Par. = 179m | U.Mag. = 0.6544 |
| Gam. = -0.6430 | P.Mag. = 1.7050 |

Penumbral	2020 Jan 10
Saros 144	19:11 TD
A.Node	ΔT= 69s

| Pen. = 245m | U.Mag. = -0.1146 |
| Gam. = 1.0727 | P.Mag. = 0.8969 |

Penumbral	2020 Jun 05
Saros 111	19:26 TD
D.Node	ΔT= 70s

| Pen. = 199m | U.Mag. = -0.4036 |
| Gam. = 1.2406 | P.Mag. = 0.5699 |

Penumbral	2020 Jul 05
Saros 149	04:31 TD
D.Node	ΔT= 70s

| Pen. = 166m | U.Mag. = -0.6422 |
| Gam. = -1.3639 | P.Mag. = 0.3560 |

Penumbral	2020 Nov 30
Saros 116	09:44 TD
A.Node	ΔT= 70s

| Pen. = 262m | U.Mag. = -0.2602 |
| Gam. = -1.1309 | P.Mag. = 0.8302 |

Total	2021 May 26
Saros 121	11:20 TD
D.Node	ΔT= 70s

Tot. = 16m	
Par. = 188m	U.Mag. = 1.0112
Gam. = 0.4774	P.Mag. = 1.9558

Plate 004

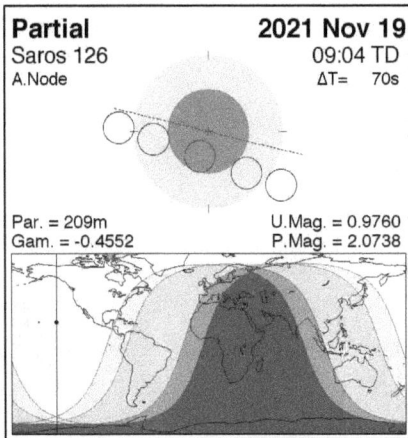

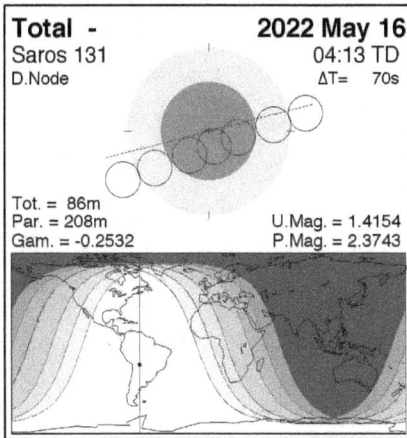

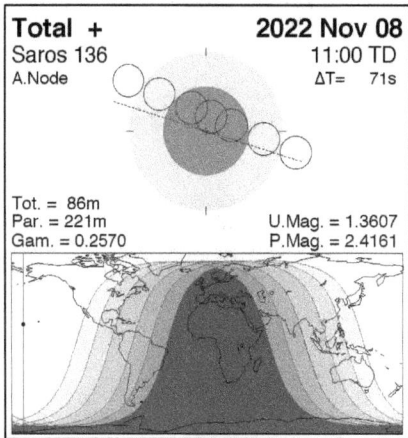

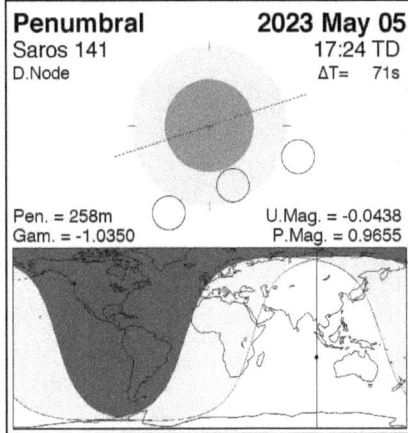

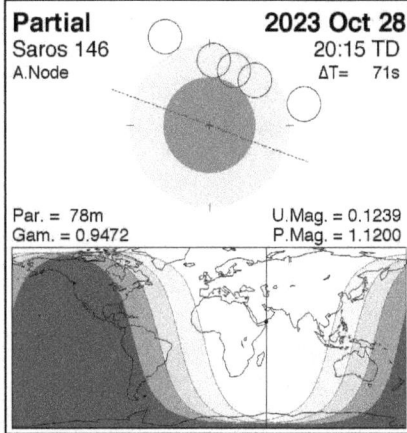

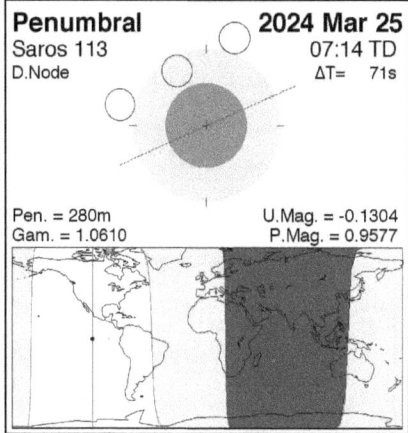

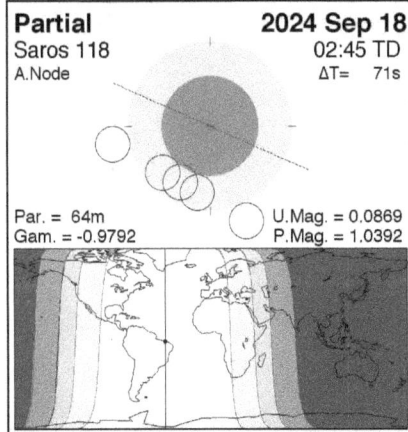

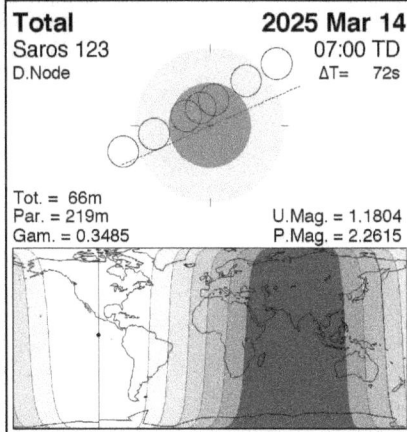

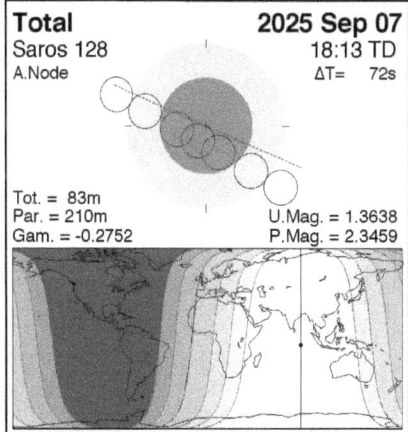

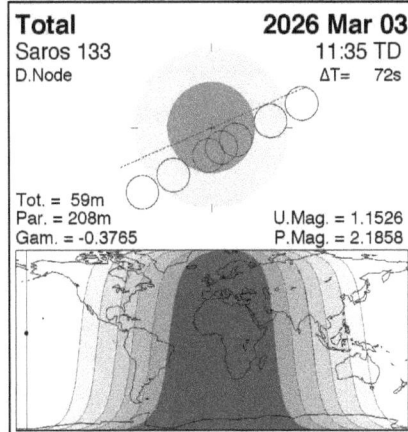

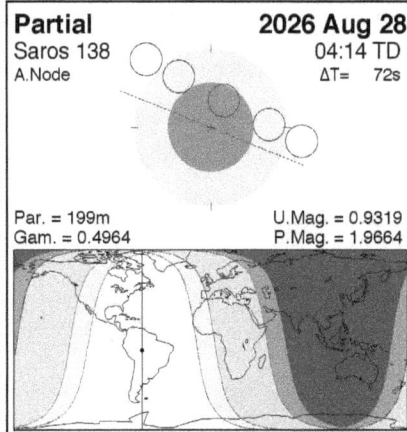

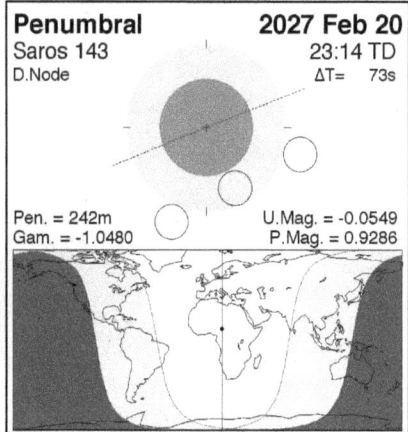

Partial	2021 Nov 19
Saros 126	09:04 TD
A.Node	ΔT= 70s

Par. = 209m U.Mag. = 0.9760
Gam. = -0.4552 P.Mag. = 2.0738

Total -	2022 May 16
Saros 131	04:13 TD
D.Node	ΔT= 70s

Tot. = 86m
Par. = 208m U.Mag. = 1.4154
Gam. = -0.2532 P.Mag. = 2.3743

Total +	2022 Nov 08
Saros 136	11:00 TD
A.Node	ΔT= 71s

Tot. = 86m
Par. = 221m U.Mag. = 1.3607
Gam. = 0.2570 P.Mag. = 2.4161

Penumbral	2023 May 05
Saros 141	17:24 TD
D.Node	ΔT= 71s

Pen. = 258m U.Mag. = -0.0438
Gam. = -1.0350 P.Mag. = 0.9655

Partial	2023 Oct 28
Saros 146	20:15 TD
A.Node	ΔT= 71s

Par. = 78m U.Mag. = 0.1239
Gam. = 0.9472 P.Mag. = 1.1200

Penumbral	2024 Mar 25
Saros 113	07:14 TD
D.Node	ΔT= 71s

Pen. = 280m U.Mag. = -0.1304
Gam. = 1.0610 P.Mag. = 0.9577

Partial	2024 Sep 18
Saros 118	02:45 TD
A.Node	ΔT= 71s

Par. = 64m U.Mag. = 0.0869
Gam. = -0.9792 P.Mag. = 1.0392

Total	2025 Mar 14
Saros 123	07:00 TD
D.Node	ΔT= 72s

Tot. = 66m
Par. = 219m U.Mag. = 1.1804
Gam. = 0.3485 P.Mag. = 2.2615

Total	2025 Sep 07
Saros 128	18:13 TD
A.Node	ΔT= 72s

Tot. = 83m
Par. = 210m U.Mag. = 1.3638
Gam. = -0.2752 P.Mag. = 2.3459

Total	2026 Mar 03
Saros 133	11:35 TD
D.Node	ΔT= 72s

Tot. = 59m
Par. = 208m U.Mag. = 1.1526
Gam. = -0.3765 P.Mag. = 2.1858

Partial	2026 Aug 28
Saros 138	04:14 TD
A.Node	ΔT= 72s

Par. = 199m U.Mag. = 0.9319
Gam. = 0.4964 P.Mag. = 1.9664

Penumbral	2027 Feb 20
Saros 143	23:14 TD
D.Node	ΔT= 73s

Pen. = 242m U.Mag. = -0.0549
Gam. = -1.0480 P.Mag. = 0.9286

Plate 005

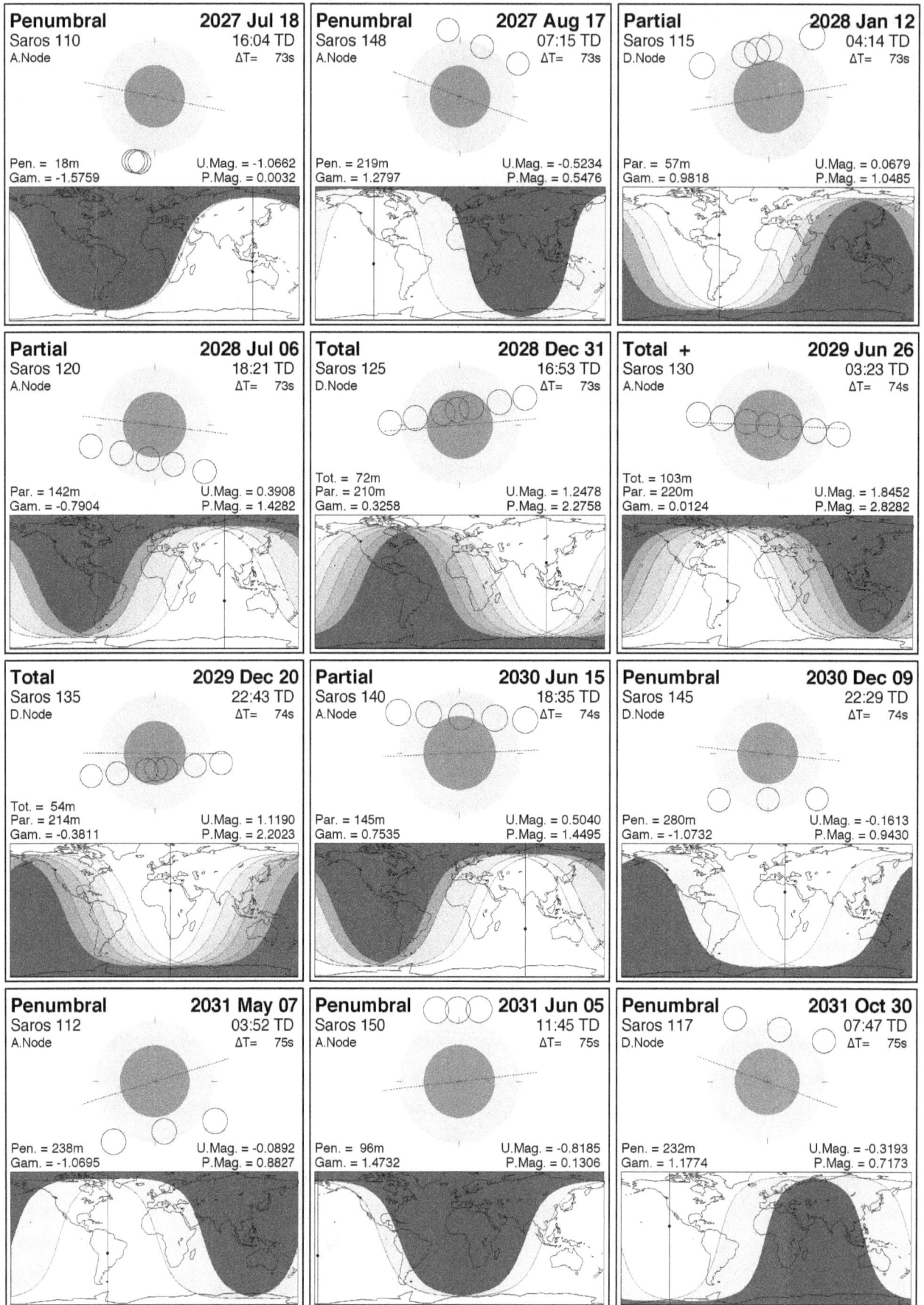

Penumbral	2027 Jul 18
Saros 110	16:04 TD
A.Node	ΔT= 73s

Pen. = 18m	U.Mag. = -1.0662
Gam. = -1.5759	P.Mag. = 0.0032

Penumbral	2027 Aug 17
Saros 148	07:15 TD
A.Node	ΔT= 73s

Pen. = 219m	U.Mag. = -0.5234
Gam. = 1.2797	P.Mag. = 0.5476

Partial	2028 Jan 12
Saros 115	04:14 TD
D.Node	ΔT= 73s

Par. = 57m	U.Mag. = 0.0679
Gam. = 0.9818	P.Mag. = 1.0485

Partial	2028 Jul 06
Saros 120	18:21 TD
A.Node	ΔT= 73s

Par. = 142m	U.Mag. = 0.3908
Gam. = -0.7904	P.Mag. = 1.4282

Total	2028 Dec 31
Saros 125	16:53 TD
D.Node	ΔT= 73s

Tot. = 72m	
Par. = 210m	U.Mag. = 1.2478
Gam. = 0.3258	P.Mag. = 2.2758

Total +	2029 Jun 26
Saros 130	03:23 TD
A.Node	ΔT= 74s

Tot. = 103m	
Par. = 220m	U.Mag. = 1.8452
Gam. = 0.0124	P.Mag. = 2.8282

Total	2029 Dec 20
Saros 135	22:43 TD
D.Node	ΔT= 74s

Tot. = 54m	
Par. = 214m	U.Mag. = 1.1190
Gam. = -0.3811	P.Mag. = 2.2023

Partial	2030 Jun 15
Saros 140	18:35 TD
A.Node	ΔT= 74s

Par. = 145m	U.Mag. = 0.5040
Gam. = 0.7535	P.Mag. = 1.4495

Penumbral	2030 Dec 09
Saros 145	22:29 TD
D.Node	ΔT= 74s

Pen. = 280m	U.Mag. = -0.1613
Gam. = -1.0732	P.Mag. = 0.9430

Penumbral	2031 May 07
Saros 112	03:52 TD
A.Node	ΔT= 75s

Pen. = 238m	U.Mag. = -0.0892
Gam. = -1.0695	P.Mag. = 0.8827

Penumbral	2031 Jun 05
Saros 150	11:45 TD
A.Node	ΔT= 75s

Pen. = 96m	U.Mag. = -0.8185
Gam. = 1.4732	P.Mag. = 0.1306

Penumbral	2031 Oct 30
Saros 117	07:47 TD
D.Node	ΔT= 75s

Pen. = 232m	U.Mag. = -0.3193
Gam. = 1.1774	P.Mag. = 0.7173

Plate 006

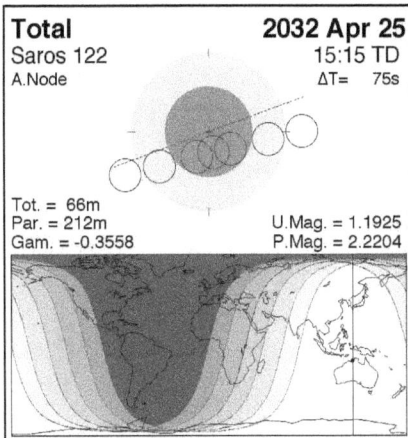

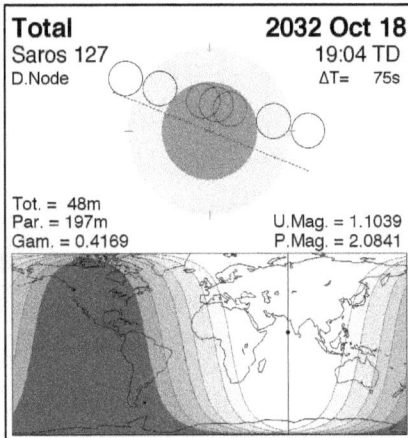

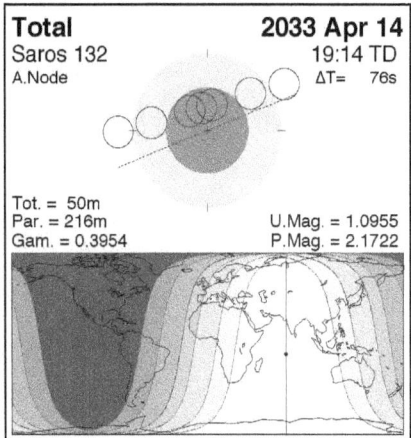

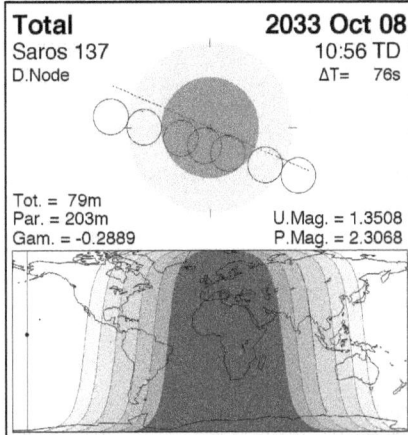

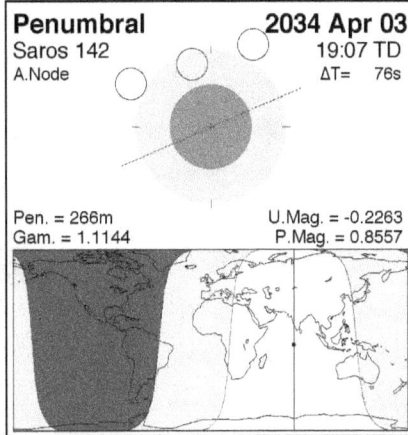

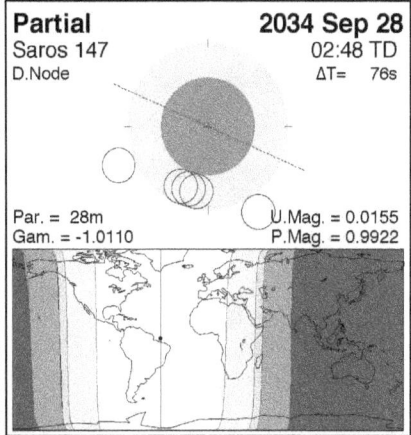

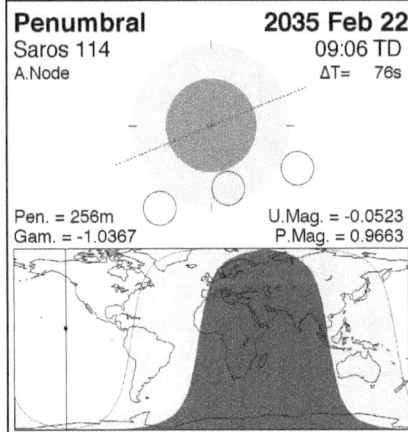

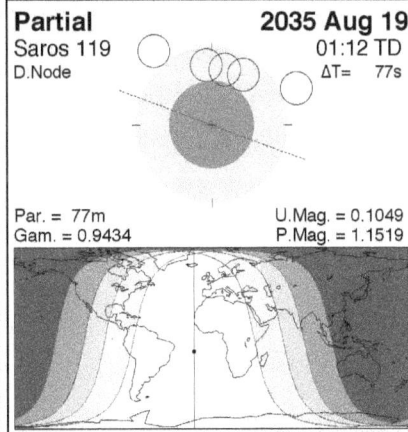

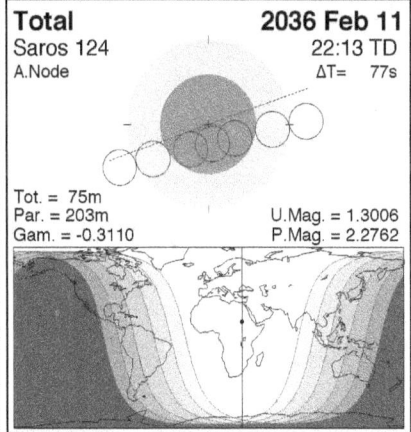

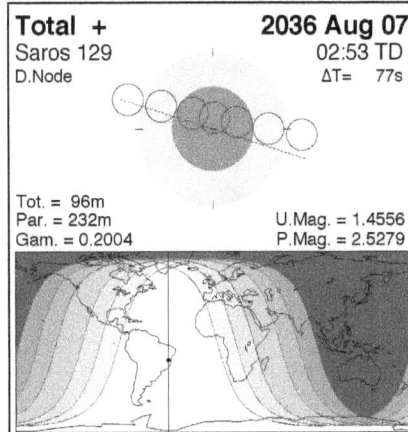

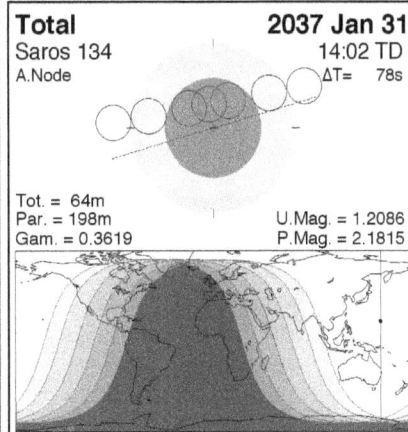

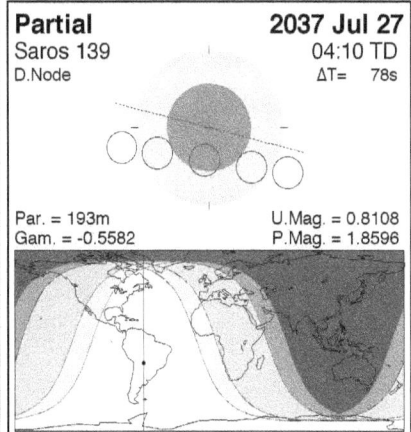

Total	2032 Apr 25
Saros 122	15:15 TD
A.Node	ΔT= 75s

Tot. = 66m
Par. = 212m
Gam. = -0.3558
U.Mag. = 1.1925
P.Mag. = 2.2204

Total	2032 Oct 18
Saros 127	19:04 TD
D.Node	ΔT= 75s

Tot. = 48m
Par. = 197m
Gam. = 0.4169
U.Mag. = 1.1039
P.Mag. = 2.0841

Total	2033 Apr 14
Saros 132	19:14 TD
A.Node	ΔT= 76s

Tot. = 50m
Par. = 216m
Gam. = 0.3954
U.Mag. = 1.0955
P.Mag. = 2.1722

Total	2033 Oct 08
Saros 137	10:56 TD
D.Node	ΔT= 76s

Tot. = 79m
Par. = 203m
Gam. = -0.2889
U.Mag. = 1.3508
P.Mag. = 2.3068

Penumbral	2034 Apr 03
Saros 142	19:07 TD
A.Node	ΔT= 76s

Pen. = 266m
Gam. = 1.1144
U.Mag. = -0.2263
P.Mag. = 0.8557

Partial	2034 Sep 28
Saros 147	02:48 TD
D.Node	ΔT= 76s

Par. = 28m
Gam. = -1.0110
U.Mag. = 0.0155
P.Mag. = 0.9922

Penumbral	2035 Feb 22
Saros 114	09:06 TD
A.Node	ΔT= 76s

Pen. = 256m
Gam. = -1.0367
U.Mag. = -0.0523
P.Mag. = 0.9663

Partial	2035 Aug 19
Saros 119	01:12 TD
D.Node	ΔT= 77s

Par. = 77m
Gam. = 0.9434
U.Mag. = 0.1049
P.Mag. = 1.1519

Total	2036 Feb 11
Saros 124	22:13 TD
A.Node	ΔT= 77s

Tot. = 75m
Par. = 203m
Gam. = -0.3110
U.Mag. = 1.3006
P.Mag. = 2.2762

Total +	2036 Aug 07
Saros 129	02:53 TD
D.Node	ΔT= 77s

Tot. = 96m
Par. = 232m
Gam. = 0.2004
U.Mag. = 1.4556
P.Mag. = 2.5279

Total	2037 Jan 31
Saros 134	14:02 TD
A.Node	ΔT= 78s

Tot. = 64m
Par. = 198m
Gam. = 0.3619
U.Mag. = 1.2086
P.Mag. = 2.1815

Partial	2037 Jul 27
Saros 139	04:10 TD
D.Node	ΔT= 78s

Par. = 193m
Gam. = -0.5582
U.Mag. = 0.8108
P.Mag. = 1.8596

Plate 007

43

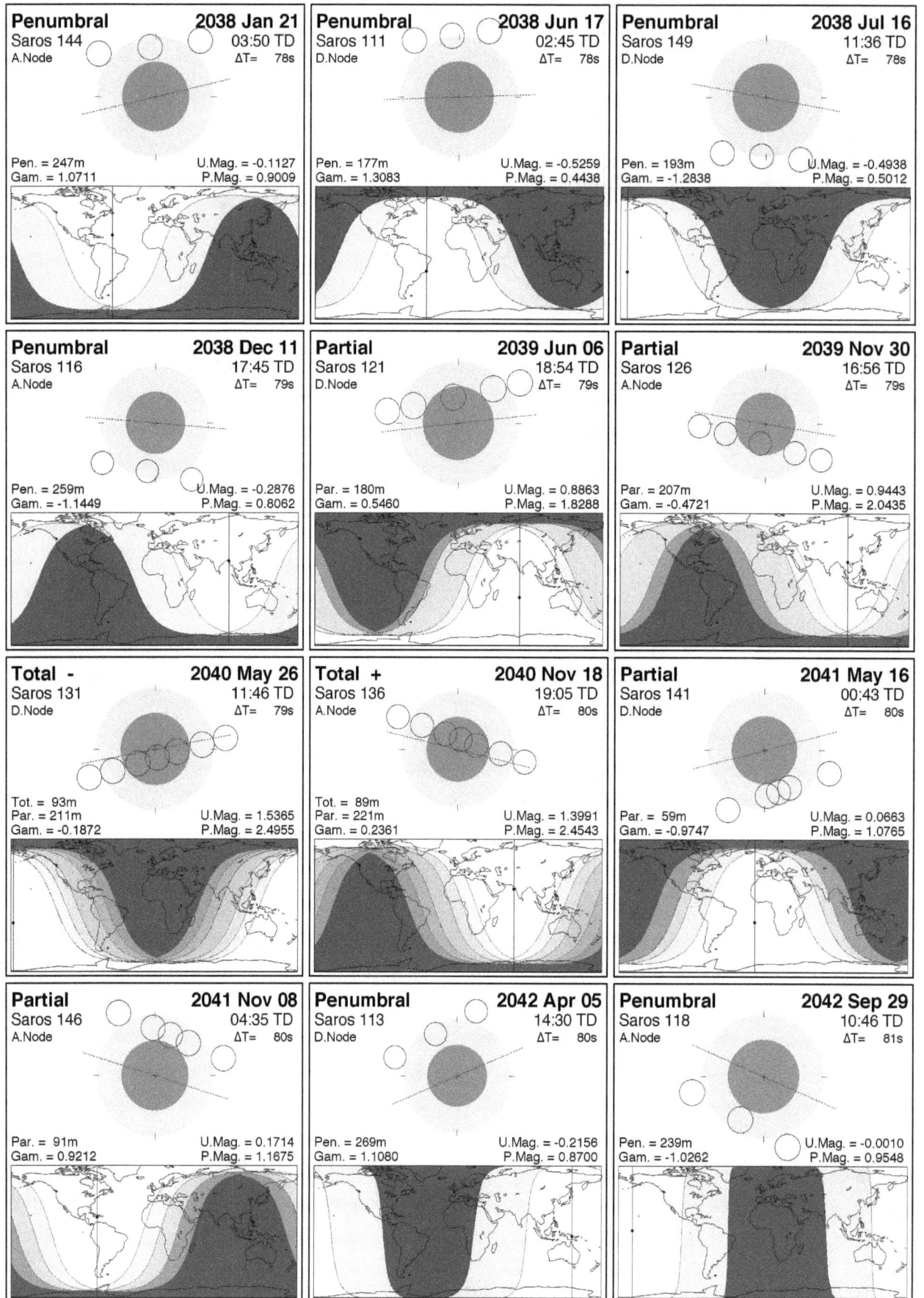

Penumbral	2038 Jan 21
Saros 144	03:50 TD
A.Node	ΔT= 78s

Pen. = 247m	U.Mag. = -0.1127
Gam. = 1.0711	P.Mag. = 0.9009

Penumbral	2038 Jun 17
Saros 111	02:45 TD
D.Node	ΔT= 78s

Pen. = 177m	U.Mag. = -0.5259
Gam. = 1.3083	P.Mag. = 0.4438

Penumbral	2038 Jul 16
Saros 149	11:36 TD
D.Node	ΔT= 78s

Pen. = 193m	U.Mag. = -0.4938
Gam. = -1.2838	P.Mag. = 0.5012

Penumbral	2038 Dec 11
Saros 116	17:45 TD
A.Node	ΔT= 79s

Pen. = 259m	U.Mag. = -0.2876
Gam. = -1.1449	P.Mag. = 0.8062

Partial	2039 Jun 06
Saros 121	18:54 TD
D.Node	ΔT= 79s

Par. = 180m	U.Mag. = 0.8863
Gam. = 0.5460	P.Mag. = 1.8288

Partial	2039 Nov 30
Saros 126	16:56 TD
A.Node	ΔT= 79s

Par. = 207m	U.Mag. = 0.9443
Gam. = -0.4721	P.Mag. = 2.0435

Total -	2040 May 26
Saros 131	11:46 TD
D.Node	ΔT= 79s

Tot. = 93m	
Par. = 211m	U.Mag. = 1.5365
Gam. = -0.1872	P.Mag. = 2.4955

Total +	2040 Nov 18
Saros 136	19:05 TD
A.Node	ΔT= 80s

Tot. = 89m	
Par. = 221m	U.Mag. = 1.3991
Gam. = 0.2361	P.Mag. = 2.4543

Partial	2041 May 16
Saros 141	00:43 TD
D.Node	ΔT= 80s

Par. = 59m	U.Mag. = 0.0663
Gam. = -0.9747	P.Mag. = 1.0765

Partial	2041 Nov 08
Saros 146	04:35 TD
A.Node	ΔT= 80s

Par. = 91m	U.Mag. = 0.1714
Gam. = 0.9212	P.Mag. = 1.1675

Penumbral	2042 Apr 05
Saros 113	14:30 TD
D.Node	ΔT= 80s

Pen. = 269m	U.Mag. = -0.2156
Gam. = 1.1080	P.Mag. = 0.8700

Penumbral	2042 Sep 29
Saros 118	10:46 TD
A.Node	ΔT= 81s

Pen. = 239m	U.Mag. = -0.0010
Gam. = -1.0262	P.Mag. = 0.9548

Plate 008

Total	2043 Mar 25
Saros 123	14:32 TD
D.Node	ΔT= 81s

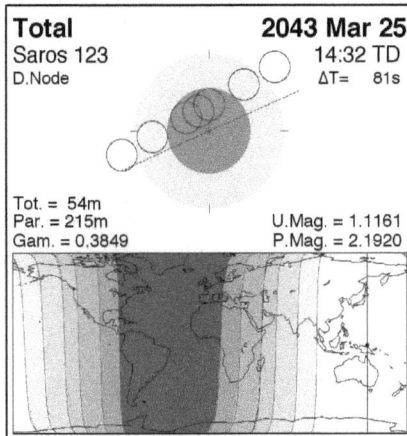

Tot. = 54m
Par. = 215m
Gam. = 0.3849
U.Mag. = 1.1161
P.Mag. = 2.1920

Total	2043 Sep 19
Saros 128	01:52 TD
A.Node	ΔT= 81s

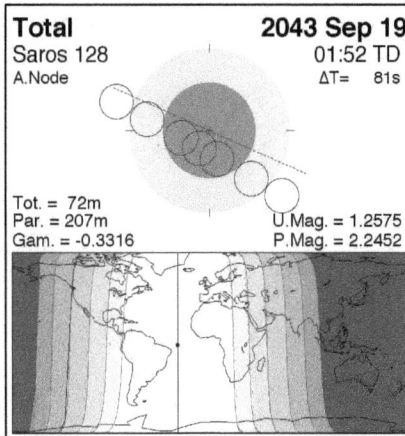

Tot. = 72m
Par. = 207m
Gam. = -0.3316
U.Mag. = 1.2575
P.Mag. = 2.2452

Total	2044 Mar 13
Saros 133	19:39 TD
D.Node	ΔT= 82s

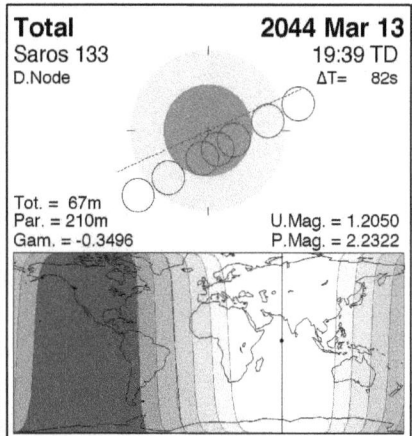

Tot. = 67m
Par. = 210m
Gam. = -0.3496
U.Mag. = 1.2050
P.Mag. = 2.2322

Total	2044 Sep 07
Saros 138	11:21 TD
A.Node	ΔT= 82s

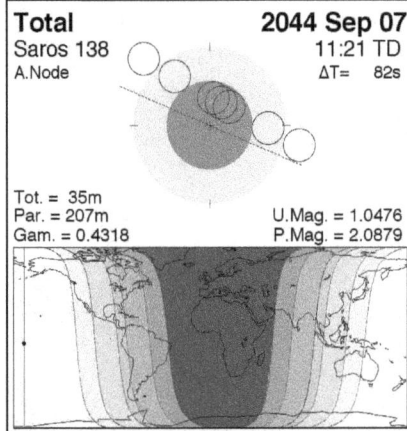

Tot. = 35m
Par. = 207m
Gam. = 0.4318
U.Mag. = 1.0476
P.Mag. = 2.0879

Penumbral	2045 Mar 03
Saros 143	07:43 TD
D.Node	ΔT= 82s

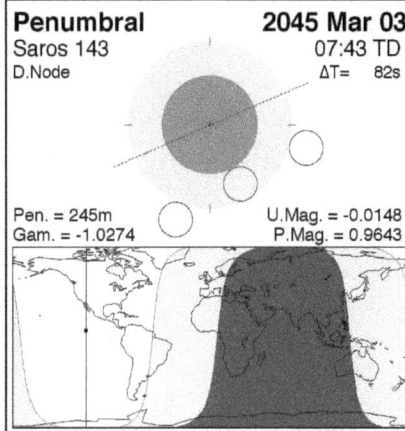

Pen. = 245m
Gam. = -1.0274
U.Mag. = -0.0148
P.Mag. = 0.9643

Penumbral	2045 Aug 27
Saros 148	13:55 TD
A.Node	ΔT= 82s

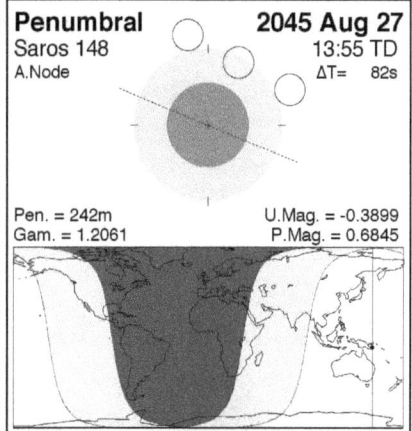

Pen. = 242m
Gam. = 1.2061
U.Mag. = -0.3899
P.Mag. = 0.6845

Partial	2046 Jan 22
Saros 115	13:03 TD
D.Node	ΔT= 83s

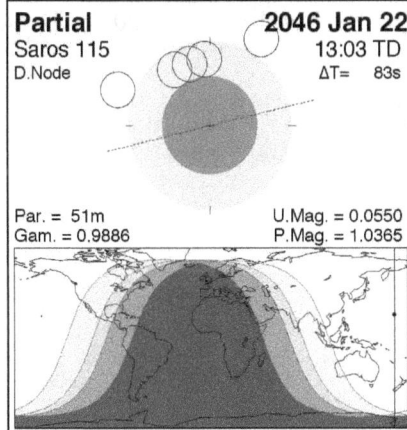

Par. = 51m
Gam. = 0.9886
U.Mag. = 0.0550
P.Mag. = 1.0365

Partial	2046 Jul 18
Saros 120	01:06 TD
A.Node	ΔT= 83s

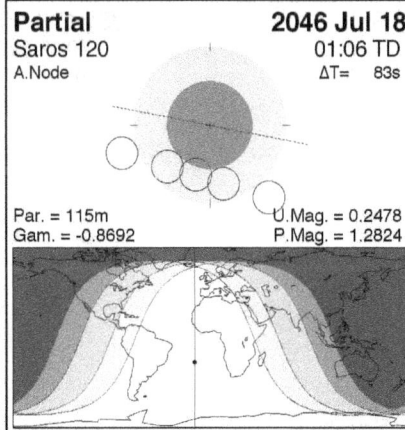

Par. = 115m
Gam. = -0.8692
U.Mag. = 0.2478
P.Mag. = 1.2824

Total	2047 Jan 12
Saros 125	01:26 TD
D.Node	ΔT= 83s

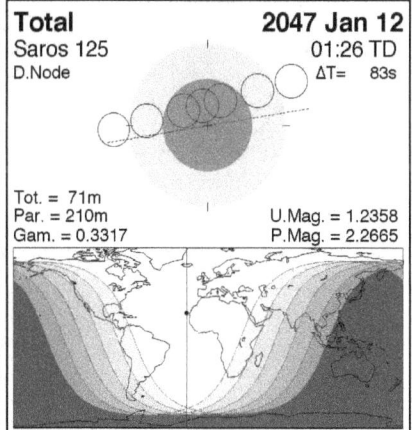

Tot. = 71m
Par. = 210m
Gam. = 0.3317
U.Mag. = 1.2358
P.Mag. = 2.2665

Total -	2047 Jul 07
Saros 130	10:36 TD
A.Node	ΔT= 84s

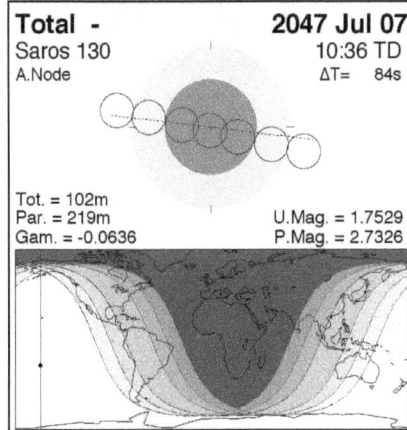

Tot. = 102m
Par. = 219m
Gam. = -0.0636
U.Mag. = 1.7529
P.Mag. = 2.7326

Total	2048 Jan 01
Saros 135	06:54 TD
D.Node	ΔT= 84s

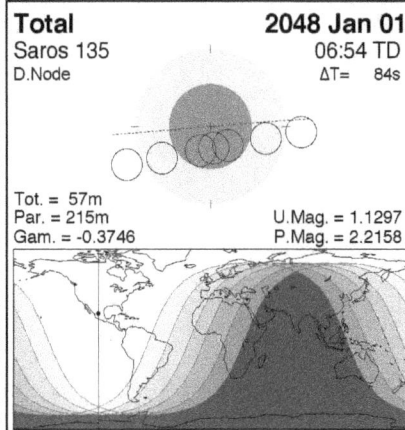

Tot. = 57m
Par. = 215m
Gam. = -0.3746
U.Mag. = 1.1297
P.Mag. = 2.2158

Partial	2048 Jun 26
Saros 140	02:02 TD
A.Node	ΔT= 84s

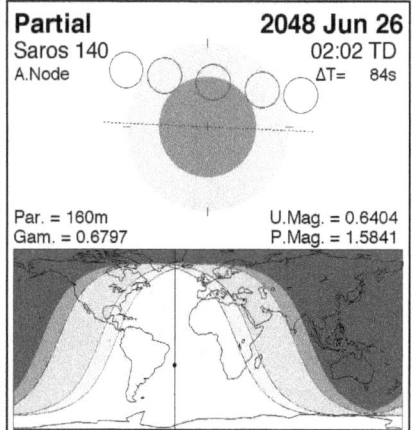

Par. = 160m
Gam. = 0.6797
U.Mag. = 0.6404
P.Mag. = 1.5841

Plate 009

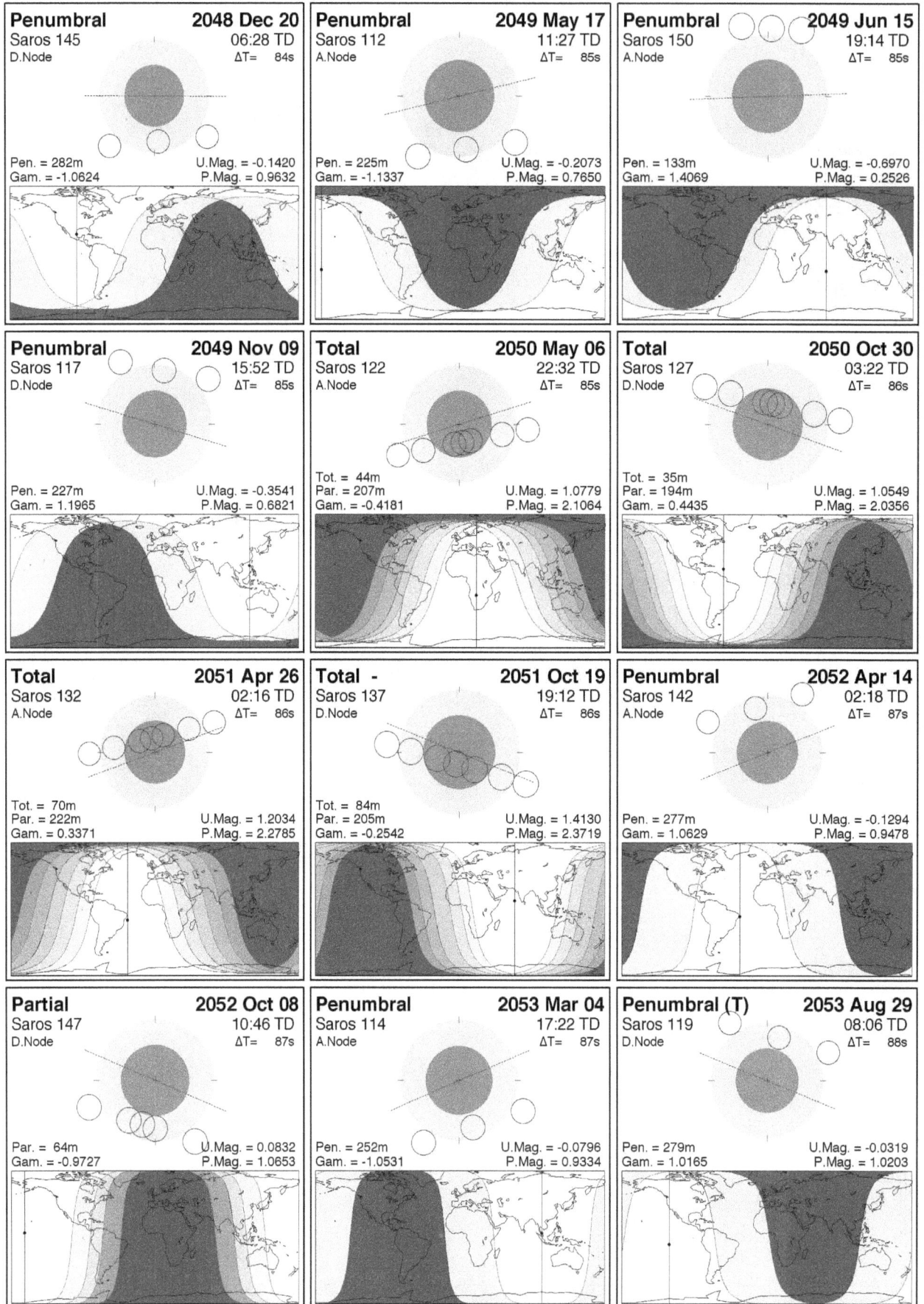

Penumbral	2048 Dec 20
Saros 145	06:28 TD
D.Node	ΔT= 84s

Pen. = 282m
Gam. = -1.0624
U.Mag. = -0.1420
P.Mag. = 0.9632

Penumbral	2049 May 17
Saros 112	11:27 TD
A.Node	ΔT= 85s

Pen. = 225m
Gam. = -1.1337
U.Mag. = -0.2073
P.Mag. = 0.7650

Penumbral	2049 Jun 15
Saros 150	19:14 TD
A.Node	ΔT= 85s

Pen. = 133m
Gam. = 1.4069
U.Mag. = -0.6970
P.Mag. = 0.2526

Penumbral	2049 Nov 09
Saros 117	15:52 TD
D.Node	ΔT= 85s

Pen. = 227m
Gam. = 1.1965
U.Mag. = -0.3541
P.Mag. = 0.6821

Total	2050 May 06
Saros 122	22:32 TD
A.Node	ΔT= 85s

Tot. = 44m
Par. = 207m
Gam. = -0.4181
U.Mag. = 1.0779
P.Mag. = 2.1064

Total	2050 Oct 30
Saros 127	03:22 TD
D.Node	ΔT= 86s

Tot. = 35m
Par. = 194m
Gam. = 0.4435
U.Mag. = 1.0549
P.Mag. = 2.0356

Total	2051 Apr 26
Saros 132	02:16 TD
A.Node	ΔT= 86s

Tot. = 70m
Par. = 222m
Gam. = 0.3371
U.Mag. = 1.2034
P.Mag. = 2.2785

Total -	2051 Oct 19
Saros 137	19:12 TD
D.Node	ΔT= 86s

Tot. = 84m
Par. = 205m
Gam. = -0.2542
U.Mag. = 1.4130
P.Mag. = 2.3719

Penumbral	2052 Apr 14
Saros 142	02:18 TD
A.Node	ΔT= 87s

Pen. = 277m
Gam. = 1.0629
U.Mag. = -0.1294
P.Mag. = 0.9478

Partial	2052 Oct 08
Saros 147	10:46 TD
D.Node	ΔT= 87s

Par. = 64m
Gam. = -0.9727
U.Mag. = 0.0832
P.Mag. = 1.0653

Penumbral	2053 Mar 04
Saros 114	17:22 TD
A.Node	ΔT= 87s

Pen. = 252m
Gam. = -1.0531
U.Mag. = -0.0796
P.Mag. = 0.9334

Penumbral (T)	2053 Aug 29
Saros 119	08:06 TD
D.Node	ΔT= 88s

Pen. = 279m
Gam. = 1.0165
U.Mag. = -0.0319
P.Mag. = 1.0203

Plate 010

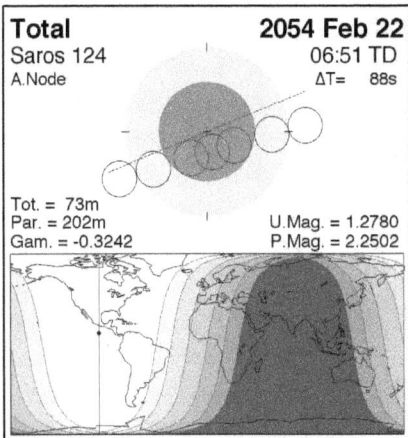

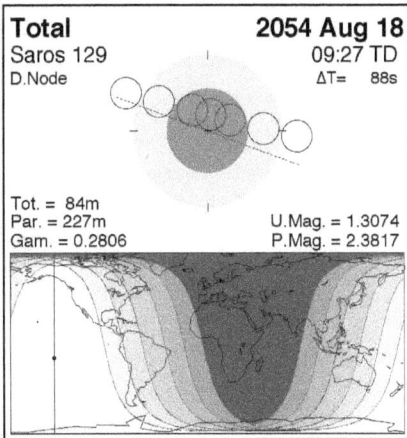

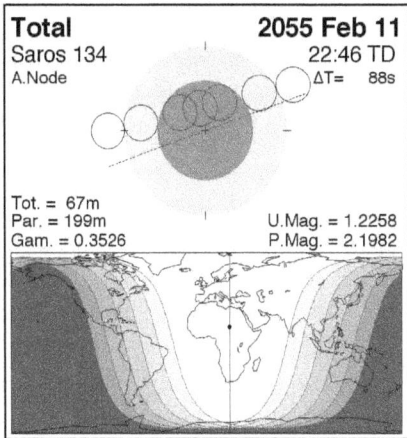

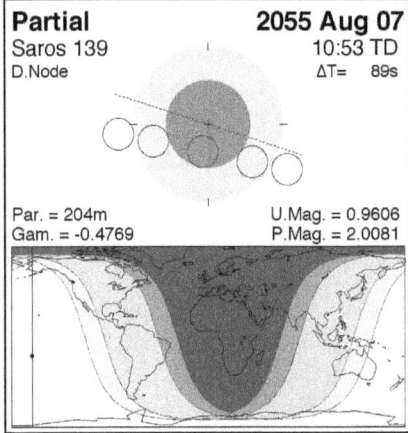

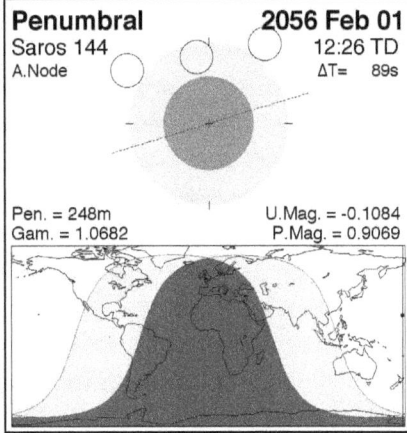

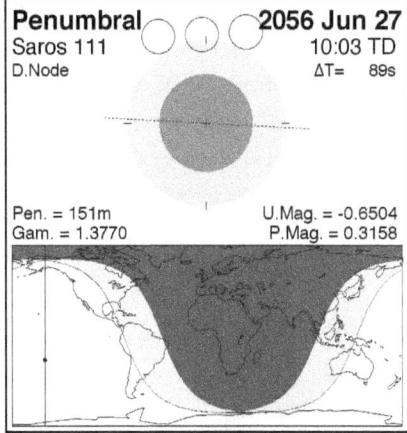

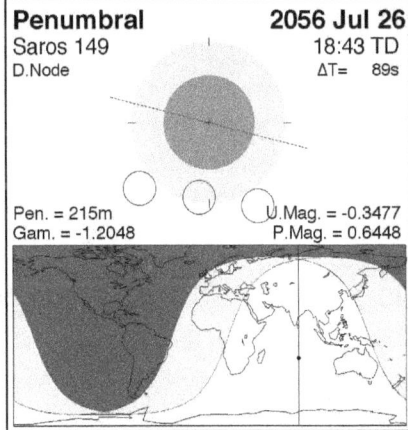

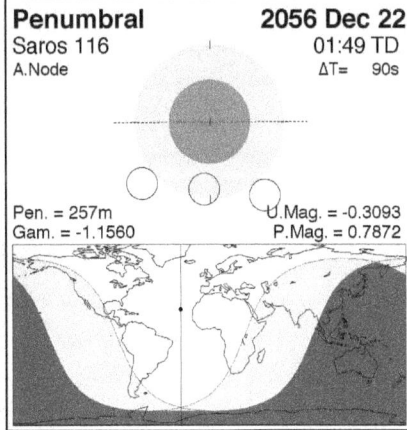

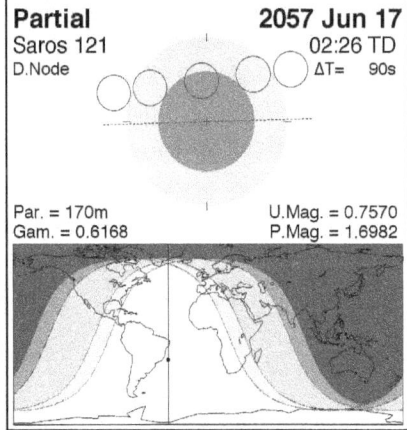

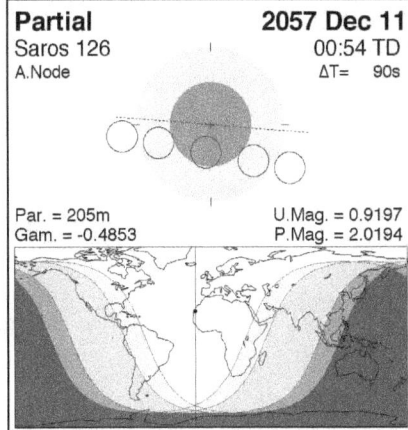

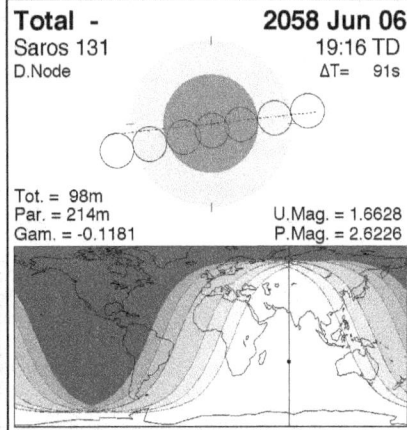

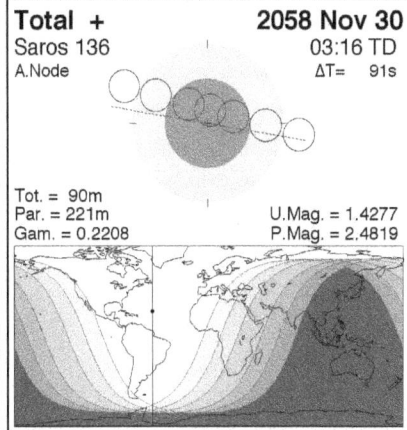

Total	2054 Feb 22
Saros 124	06:51 TD
A.Node	ΔT= 88s

Tot. = 73m
Par. = 202m
Gam. = -0.3242
U.Mag. = 1.2780
P.Mag. = 2.2502

Total	2054 Aug 18
Saros 129	09:27 TD
D.Node	ΔT= 88s

Tot. = 84m
Par. = 227m
Gam. = 0.2806
U.Mag. = 1.3074
P.Mag. = 2.3817

Total	2055 Feb 11
Saros 134	22:46 TD
A.Node	ΔT= 88s

Tot. = 67m
Par. = 199m
Gam. = 0.3526
U.Mag. = 1.2258
P.Mag. = 2.1982

Partial	2055 Aug 07
Saros 139	10:53 TD
D.Node	ΔT= 89s

Par. = 204m
Gam. = -0.4769
U.Mag. = 0.9606
P.Mag. = 2.0081

Penumbral	2056 Feb 01
Saros 144	12:26 TD
A.Node	ΔT= 89s

Pen. = 248m
Gam. = 1.0682
U.Mag. = -0.1084
P.Mag. = 0.9069

Penumbral	2056 Jun 27
Saros 111	10:03 TD
D.Node	ΔT= 89s

Pen. = 151m
Gam. = 1.3770
U.Mag. = -0.6504
P.Mag. = 0.3158

Penumbral	2056 Jul 26
Saros 149	18:43 TD
D.Node	ΔT= 89s

Pen. = 215m
Gam. = -1.2048
U.Mag. = -0.3477
P.Mag. = 0.6448

Penumbral	2056 Dec 22
Saros 116	01:49 TD
A.Node	ΔT= 90s

Pen. = 257m
Gam. = -1.1560
U.Mag. = -0.3093
P.Mag. = 0.7872

Partial	2057 Jun 17
Saros 121	02:26 TD
D.Node	ΔT= 90s

Par. = 170m
Gam. = 0.6168
U.Mag. = 0.7570
P.Mag. = 1.6982

Partial	2057 Dec 11
Saros 126	00:54 TD
A.Node	ΔT= 90s

Par. = 205m
Gam. = -0.4853
U.Mag. = 0.9197
P.Mag. = 2.0194

Total -	2058 Jun 06
Saros 131	19:16 TD
D.Node	ΔT= 91s

Tot. = 98m
Par. = 214m
Gam. = -0.1181
U.Mag. = 1.6628
P.Mag. = 2.6226

Total +	2058 Nov 30
Saros 136	03:16 TD
A.Node	ΔT= 91s

Tot. = 90m
Par. = 221m
Gam. = 0.2208
U.Mag. = 1.4277
P.Mag. = 2.4819

Plate 011

47

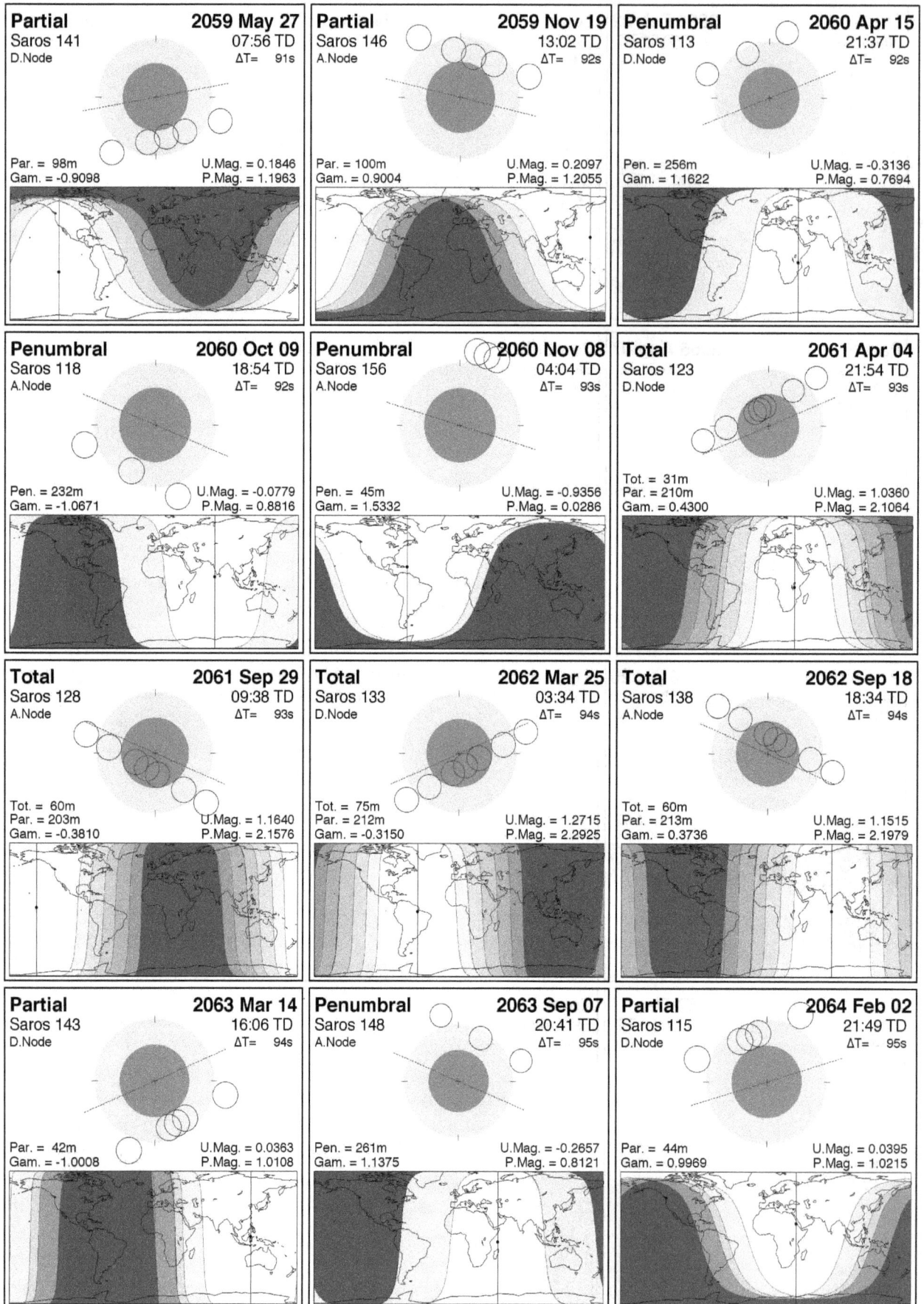

Partial	2059 May 27
Saros 141	07:56 TD
D.Node	ΔT= 91s
Par. = 98m	U.Mag. = 0.1846
Gam. = -0.9098	P.Mag. = 1.1963

Partial	2059 Nov 19
Saros 146	13:02 TD
A.Node	ΔT= 92s
Par. = 100m	U.Mag. = 0.2097
Gam. = 0.9004	P.Mag. = 1.2055

Penumbral	2060 Apr 15
Saros 113	21:37 TD
D.Node	ΔT= 92s
Pen. = 256m	U.Mag. = -0.3136
Gam. = 1.1622	P.Mag. = 0.7694

Penumbral	2060 Oct 09
Saros 118	18:54 TD
A.Node	ΔT= 92s
Pen. = 232m	U.Mag. = -0.0779
Gam. = -1.0671	P.Mag. = 0.8816

Penumbral	2060 Nov 08
Saros 156	04:04 TD
A.Node	ΔT= 93s
Pen. = 45m	U.Mag. = -0.9356
Gam. = 1.5332	P.Mag. = 0.0286

Total	2061 Apr 04
Saros 123	21:54 TD
D.Node	ΔT= 93s
Tot. = 31m	
Par. = 210m	U.Mag. = 1.0360
Gam. = 0.4300	P.Mag. = 2.1064

Total	2061 Sep 29
Saros 128	09:38 TD
A.Node	ΔT= 93s
Tot. = 60m	
Par. = 203m	U.Mag. = 1.1640
Gam. = -0.3810	P.Mag. = 2.1576

Total	2062 Mar 25
Saros 133	03:34 TD
D.Node	ΔT= 94s
Tot. = 75m	
Par. = 212m	U.Mag. = 1.2715
Gam. = -0.3150	P.Mag. = 2.2925

Total	2062 Sep 18
Saros 138	18:34 TD
A.Node	ΔT= 94s
Tot. = 60m	
Par. = 213m	U.Mag. = 1.1515
Gam. = 0.3736	P.Mag. = 2.1979

Partial	2063 Mar 14
Saros 143	16:06 TD
D.Node	ΔT= 94s
Par. = 42m	U.Mag. = 0.0363
Gam. = -1.0008	P.Mag. = 1.0108

Penumbral	2063 Sep 07
Saros 148	20:41 TD
A.Node	ΔT= 95s
Pen. = 261m	U.Mag. = -0.2657
Gam. = 1.1375	P.Mag. = 0.8121

Partial	2064 Feb 02
Saros 115	21:49 TD
D.Node	ΔT= 95s
Par. = 44m	U.Mag. = 0.0395
Gam. = 0.9969	P.Mag. = 1.0215

Plate 012

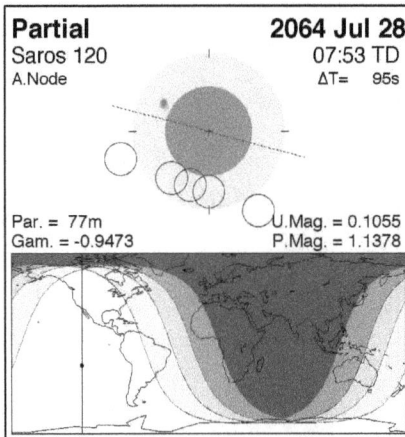

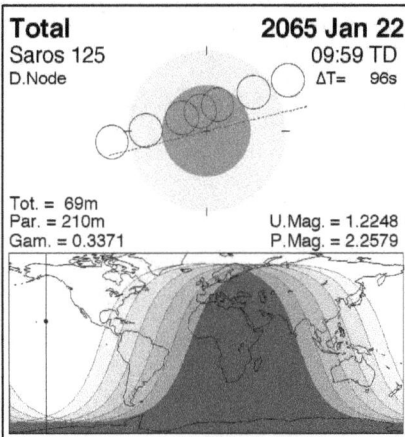

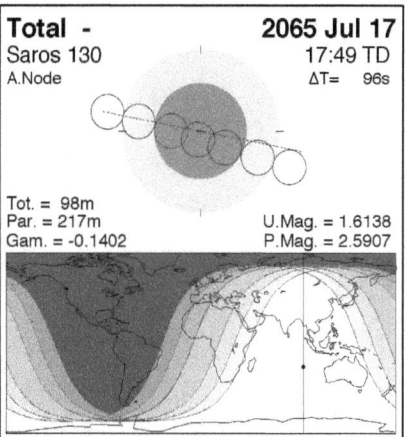

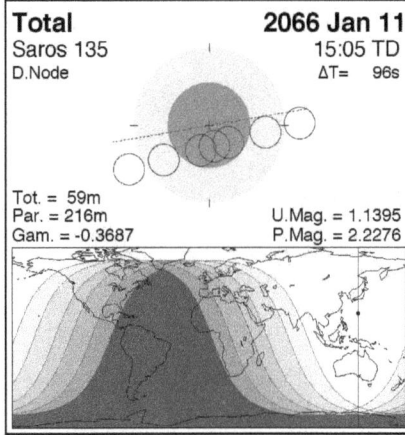

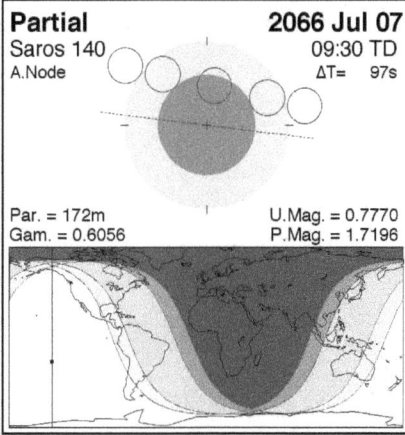

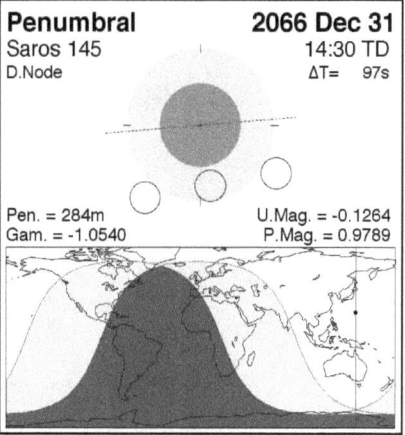

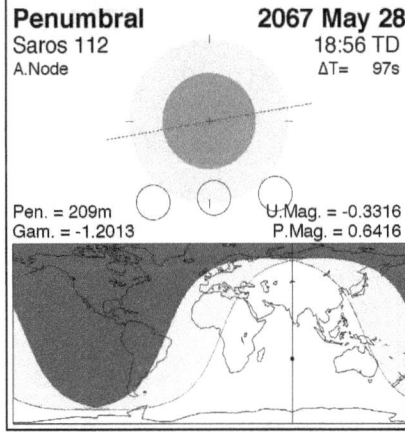

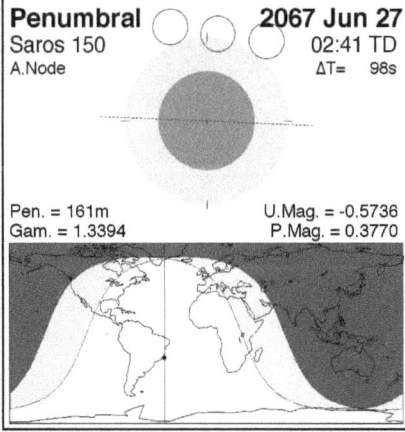

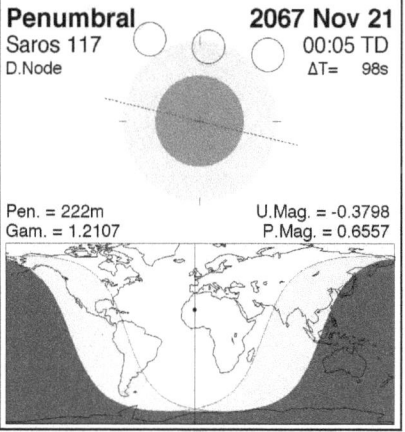

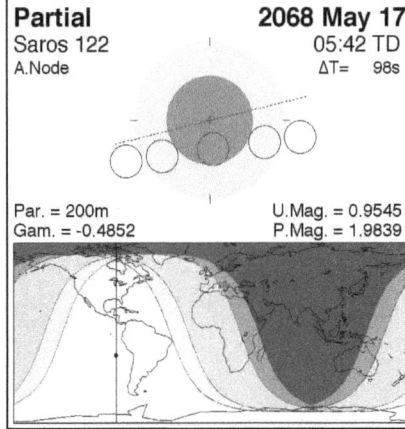

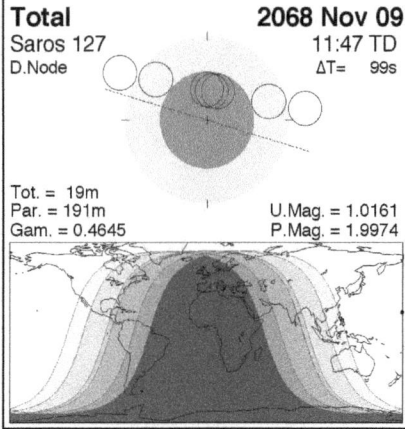

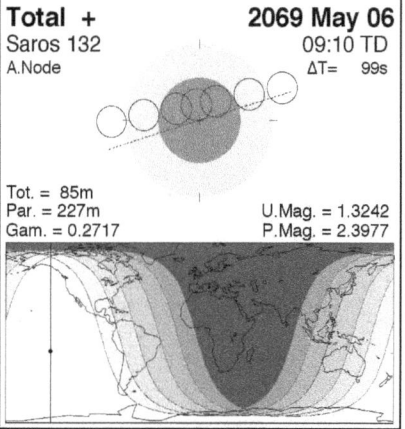

Partial	2064 Jul 28
Saros 120	07:53 TD
A.Node	ΔT= 95s

Par. = 77m
Gam. = -0.9473
U.Mag. = 0.1055
P.Mag. = 1.1378

Total	2065 Jan 22
Saros 125	09:59 TD
D.Node	ΔT= 96s

Tot. = 69m
Par. = 210m
Gam. = 0.3371
U.Mag. = 1.2248
P.Mag. = 2.2579

Total -	2065 Jul 17
Saros 130	17:49 TD
A.Node	ΔT= 96s

Tot. = 98m
Par. = 217m
Gam. = -0.1402
U.Mag. = 1.6138
P.Mag. = 2.5907

Total	2066 Jan 11
Saros 135	15:05 TD
D.Node	ΔT= 96s

Tot. = 59m
Par. = 216m
Gam. = -0.3687
U.Mag. = 1.1395
P.Mag. = 2.2276

Partial	2066 Jul 07
Saros 140	09:30 TD
A.Node	ΔT= 97s

Par. = 172m
Gam. = 0.6056
U.Mag. = 0.7770
P.Mag. = 1.7196

Penumbral	2066 Dec 31
Saros 145	14:30 TD
D.Node	ΔT= 97s

Pen. = 284m
Gam. = -1.0540
U.Mag. = -0.1264
P.Mag. = 0.9789

Penumbral	2067 May 28
Saros 112	18:56 TD
A.Node	ΔT= 97s

Pen. = 209m
Gam. = -1.2013
U.Mag. = -0.3316
P.Mag. = 0.6416

Penumbral	2067 Jun 27
Saros 150	02:41 TD
A.Node	ΔT= 98s

Pen. = 161m
Gam. = 1.3394
U.Mag. = -0.5736
P.Mag. = 0.3770

Penumbral	2067 Nov 21
Saros 117	00:05 TD
D.Node	ΔT= 98s

Pen. = 222m
Gam. = 1.2107
U.Mag. = -0.3798
P.Mag. = 0.6557

Partial	2068 May 17
Saros 122	05:42 TD
A.Node	ΔT= 98s

Par. = 200m
Gam. = -0.4852
U.Mag. = 0.9545
P.Mag. = 1.9839

Total	2068 Nov 09
Saros 127	11:47 TD
D.Node	ΔT= 99s

Tot. = 19m
Par. = 191m
Gam. = 0.4645
U.Mag. = 1.0161
P.Mag. = 1.9974

Total +	2069 May 06
Saros 132	09:10 TD
A.Node	ΔT= 99s

Tot. = 85m
Par. = 227m
Gam. = 0.2717
U.Mag. = 1.3242
P.Mag. = 2.3977

Plate 013

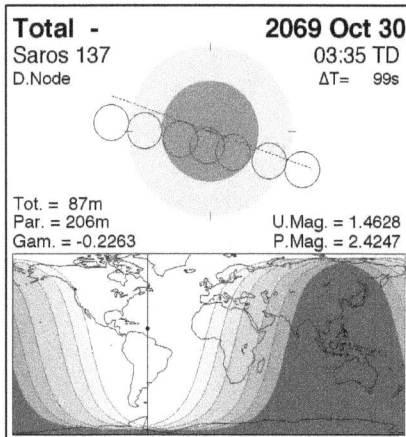

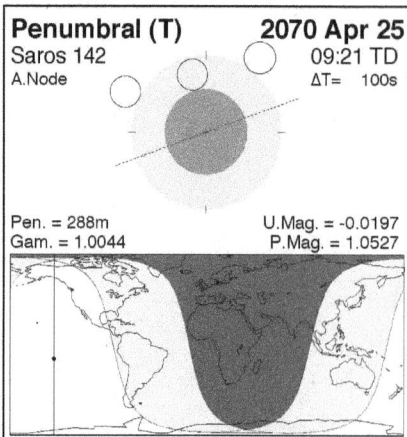

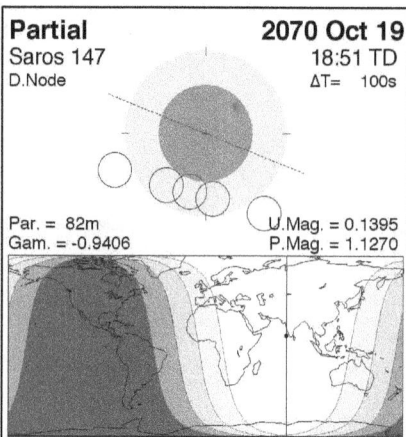

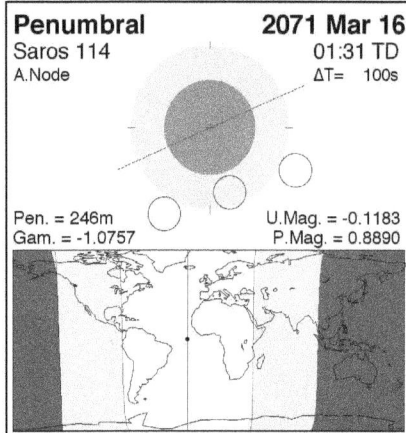

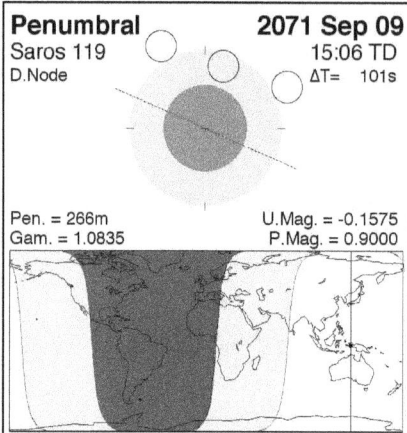

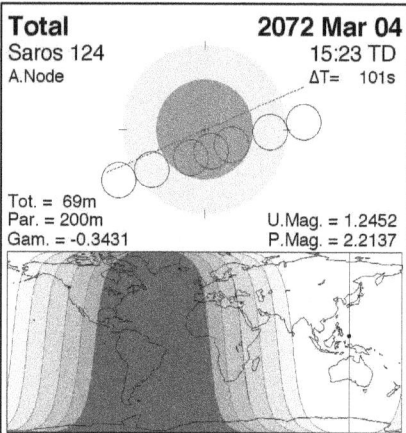

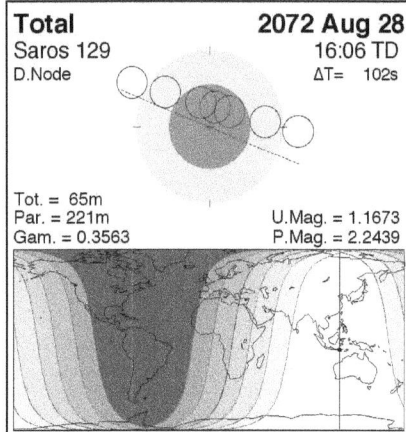

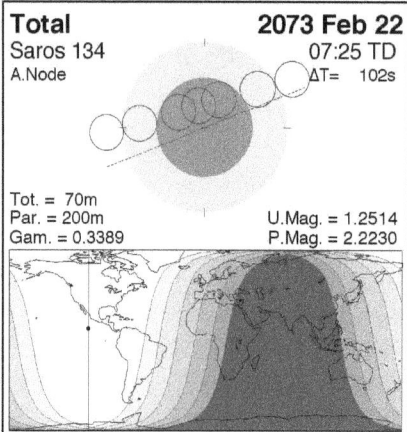

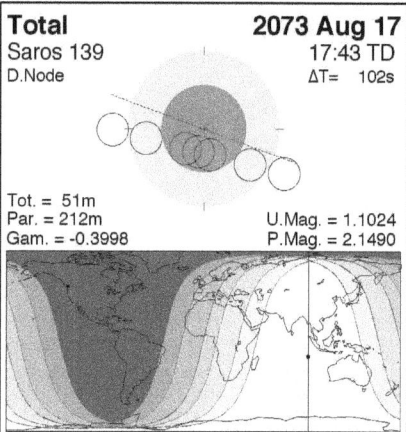

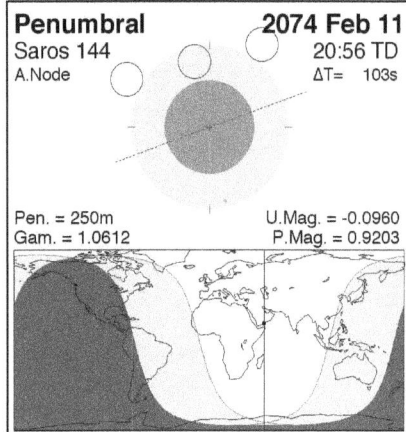

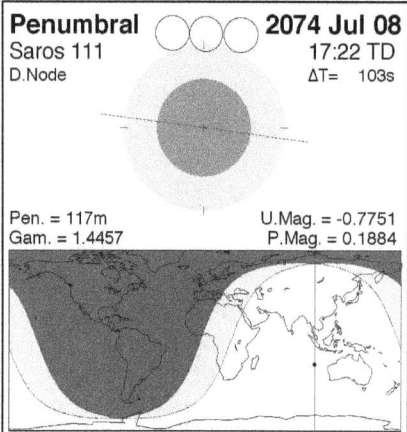

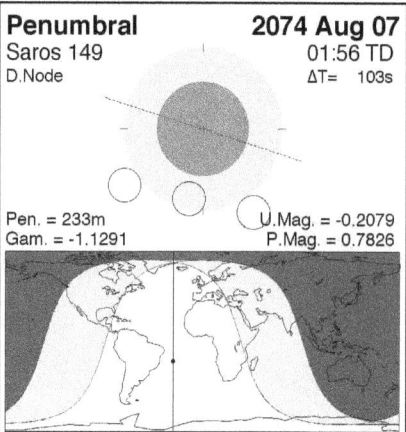

Total -	2069 Oct 30
Saros 137	03:35 TD
D.Node	ΔT= 99s

Tot. = 87m
Par. = 206m
Gam. = -0.2263
U.Mag. = 1.4628
P.Mag. = 2.4247

Penumbral (T)	2070 Apr 25
Saros 142	09:21 TD
A.Node	ΔT= 100s

Pen. = 288m
Gam. = 1.0044
U.Mag. = -0.0197
P.Mag. = 1.0527

Partial	2070 Oct 19
Saros 147	18:51 TD
D.Node	ΔT= 100s

Par. = 82m
Gam. = -0.9406
U.Mag. = 0.1395
P.Mag. = 1.1270

Penumbral	2071 Mar 16
Saros 114	01:31 TD
A.Node	ΔT= 100s

Pen. = 246m
Gam. = -1.0757
U.Mag. = -0.1183
P.Mag. = 0.8890

Penumbral	2071 Sep 09
Saros 119	15:06 TD
D.Node	ΔT= 101s

Pen. = 266m
Gam. = 1.0835
U.Mag. = -0.1575
P.Mag. = 0.9000

Total	2072 Mar 04
Saros 124	15:23 TD
A.Node	ΔT= 101s

Tot. = 69m
Par. = 200m
Gam. = -0.3431
U.Mag. = 1.2452
P.Mag. = 2.2137

Total	2072 Aug 28
Saros 129	16:06 TD
D.Node	ΔT= 102s

Tot. = 65m
Par. = 221m
Gam. = 0.3563
U.Mag. = 1.1673
P.Mag. = 2.2439

Total	2073 Feb 22
Saros 134	07:25 TD
A.Node	ΔT= 102s

Tot. = 70m
Par. = 200m
Gam. = 0.3389
U.Mag. = 1.2514
P.Mag. = 2.2230

Total	2073 Aug 17
Saros 139	17:43 TD
D.Node	ΔT= 102s

Tot. = 51m
Par. = 212m
Gam. = -0.3998
U.Mag. = 1.1024
P.Mag. = 2.1490

Penumbral	2074 Feb 11
Saros 144	20:56 TD
A.Node	ΔT= 103s

Pen. = 250m
Gam. = 1.0612
U.Mag. = -0.0960
P.Mag. = 0.9203

Penumbral	2074 Jul 08
Saros 111	17:22 TD
D.Node	ΔT= 103s

Pen. = 117m
Gam. = 1.4457
U.Mag. = -0.7751
P.Mag. = 0.1884

Penumbral	2074 Aug 07
Saros 149	01:56 TD
D.Node	ΔT= 103s

Pen. = 233m
Gam. = -1.1291
U.Mag. = -0.2079
P.Mag. = 0.7826

Plate 014

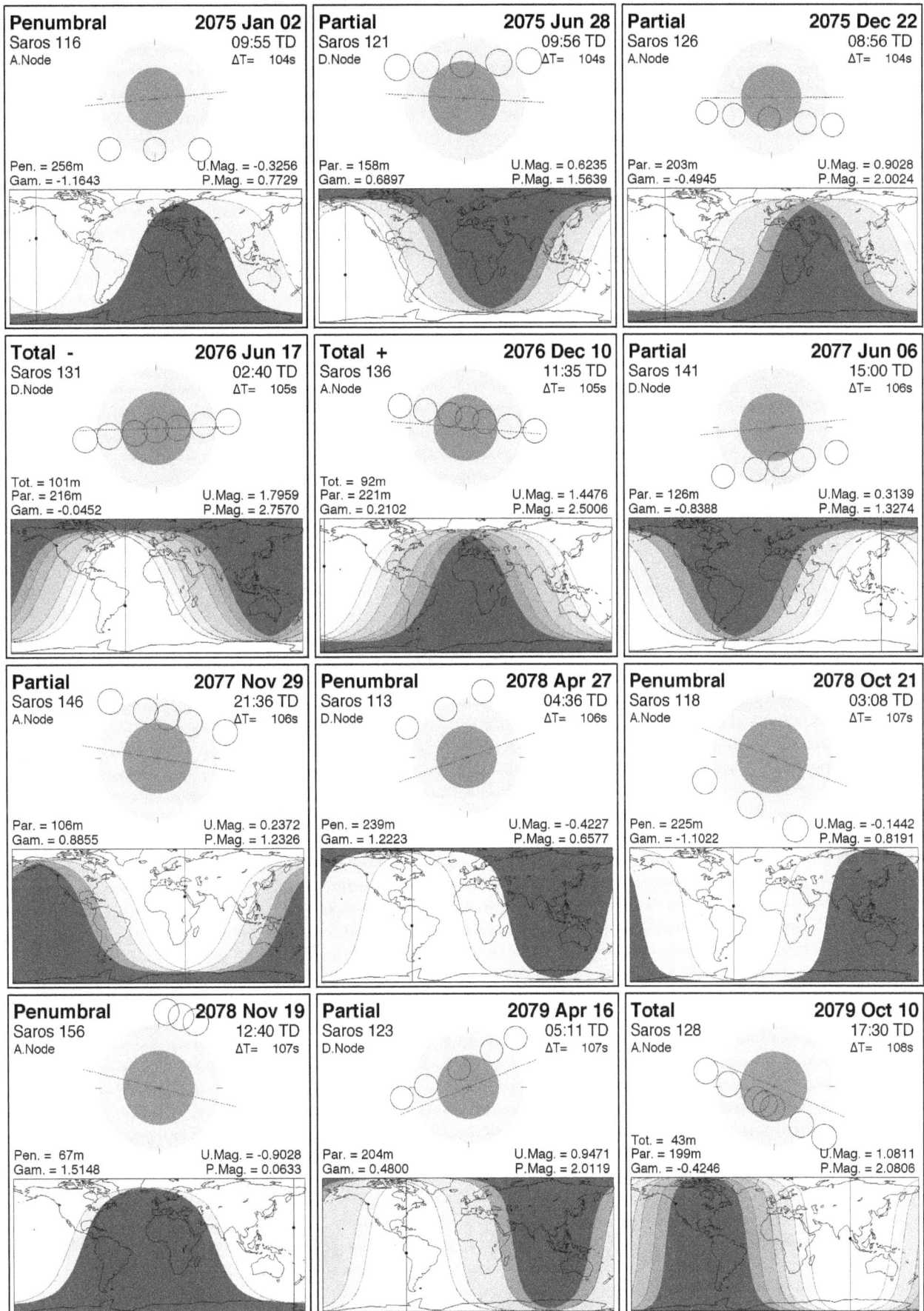

Penumbral	2075 Jan 02
Saros 116	09:55 TD
A.Node	ΔT= 104s

Pen. = 256m
Gam. = -1.1643
U.Mag. = -0.3256
P.Mag. = 0.7729

Partial	2075 Jun 28
Saros 121	09:56 TD
D.Node	ΔT= 104s

Par. = 158m
Gam. = 0.6897
U.Mag. = 0.6235
P.Mag. = 1.5639

Partial	2075 Dec 22
Saros 126	08:56 TD
A.Node	ΔT= 104s

Par. = 203m
Gam. = -0.4945
U.Mag. = 0.9028
P.Mag. = 2.0024

Total -	2076 Jun 17
Saros 131	02:40 TD
D.Node	ΔT= 105s

Tot. = 101m
Par. = 216m
Gam. = -0.0452
U.Mag. = 1.7959
P.Mag. = 2.7570

Total +	2076 Dec 10
Saros 136	11:35 TD
A.Node	ΔT= 105s

Tot. = 92m
Par. = 221m
Gam. = 0.2102
U.Mag. = 1.4476
P.Mag. = 2.5006

Partial	2077 Jun 06
Saros 141	15:00 TD
D.Node	ΔT= 106s

Par. = 126m
Gam. = -0.8388
U.Mag. = 0.3139
P.Mag. = 1.3274

Partial	2077 Nov 29
Saros 146	21:36 TD
A.Node	ΔT= 106s

Par. = 106m
Gam. = 0.8855
U.Mag. = 0.2372
P.Mag. = 1.2326

Penumbral	2078 Apr 27
Saros 113	04:36 TD
D.Node	ΔT= 106s

Pen. = 239m
Gam. = 1.2223
U.Mag. = -0.4227
P.Mag. = 0.6577

Penumbral	2078 Oct 21
Saros 118	03:08 TD
A.Node	ΔT= 107s

Pen. = 225m
Gam. = -1.1022
U.Mag. = -0.1442
P.Mag. = 0.8191

Penumbral	2078 Nov 19
Saros 156	12:40 TD
A.Node	ΔT= 107s

Pen. = 67m
Gam. = 1.5148
U.Mag. = -0.9028
P.Mag. = 0.0633

Partial	2079 Apr 16
Saros 123	05:11 TD
D.Node	ΔT= 107s

Par. = 204m
Gam. = 0.4800
U.Mag. = 0.9471
P.Mag. = 2.0119

Total	2079 Oct 10
Saros 128	17:30 TD
A.Node	ΔT= 108s

Tot. = 43m
Par. = 199m
Gam. = -0.4246
U.Mag. = 1.0811
P.Mag. = 2.0806

Plate 015

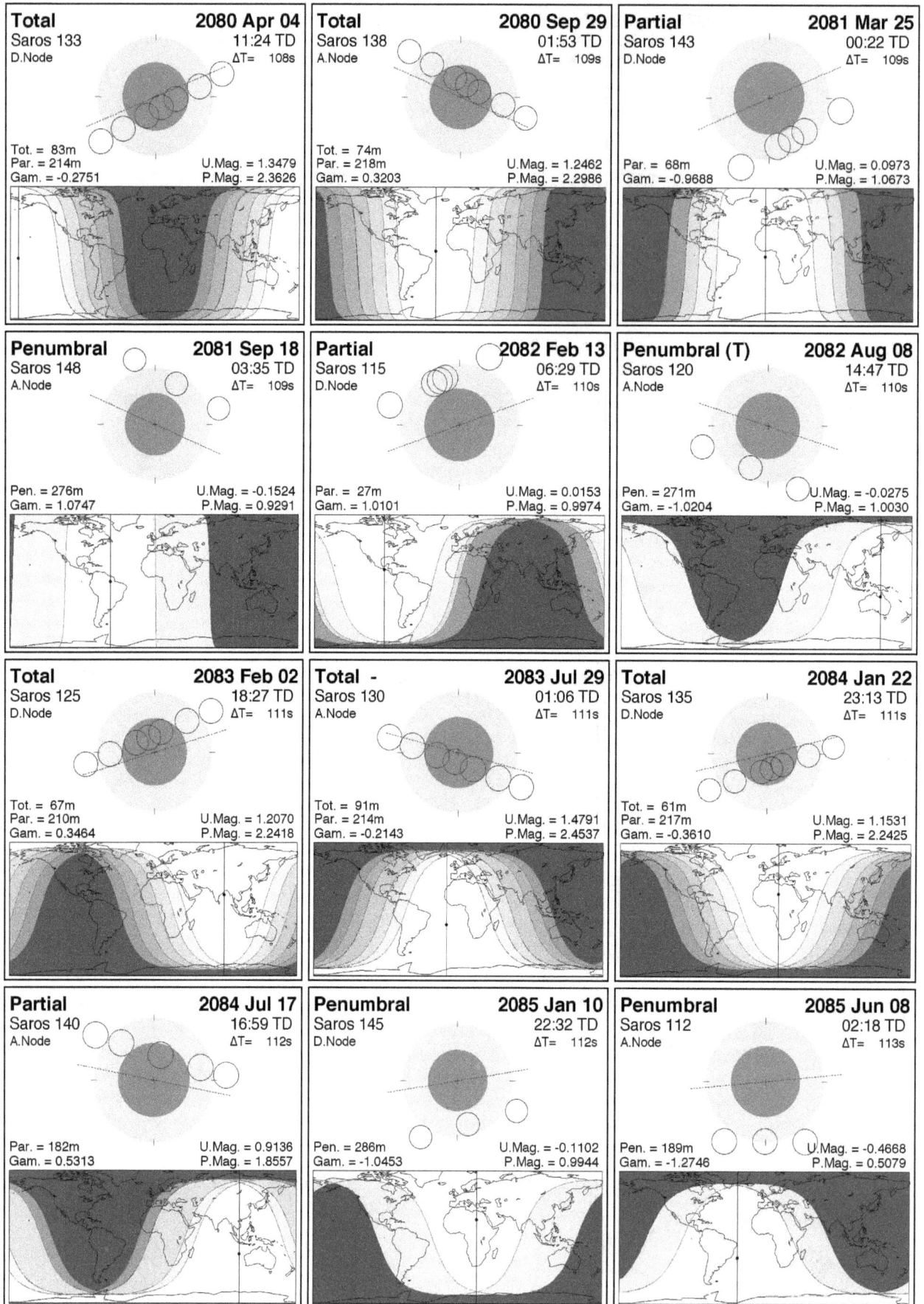

Total	2080 Apr 04
Saros 133	11:24 TD
D.Node	ΔT= 108s

Tot. = 83m
Par. = 214m U.Mag. = 1.3479
Gam. = -0.2751 P.Mag. = 2.3626

Total	2080 Sep 29
Saros 138	01:53 TD
A.Node	ΔT= 109s

Tot. = 74m
Par. = 218m U.Mag. = 1.2462
Gam. = 0.3203 P.Mag. = 2.2986

Partial	2081 Mar 25
Saros 143	00:22 TD
D.Node	ΔT= 109s

Par. = 68m U.Mag. = 0.0973
Gam. = -0.9688 P.Mag. = 1.0673

Penumbral	2081 Sep 18
Saros 148	03:35 TD
A.Node	ΔT= 109s

Pen. = 276m U.Mag. = -0.1524
Gam. = 1.0747 P.Mag. = 0.9291

Partial	2082 Feb 13
Saros 115	06:29 TD
D.Node	ΔT= 110s

Par. = 27m U.Mag. = 0.0153
Gam. = 1.0101 P.Mag. = 0.9974

Penumbral (T)	2082 Aug 08
Saros 120	14:47 TD
A.Node	ΔT= 110s

Pen. = 271m U.Mag. = -0.0275
Gam. = -1.0204 P.Mag. = 1.0030

Total	2083 Feb 02
Saros 125	18:27 TD
D.Node	ΔT= 111s

Tot. = 67m
Par. = 210m U.Mag. = 1.2070
Gam. = 0.3464 P.Mag. = 2.2418

Total -	2083 Jul 29
Saros 130	01:06 TD
A.Node	ΔT= 111s

Tot. = 91m
Par. = 214m U.Mag. = 1.4791
Gam. = -0.2143 P.Mag. = 2.4537

Total	2084 Jan 22
Saros 135	23:13 TD
D.Node	ΔT= 111s

Tot. = 61m
Par. = 217m U.Mag. = 1.1531
Gam. = -0.3610 P.Mag. = 2.2425

Partial	2084 Jul 17
Saros 140	16:59 TD
A.Node	ΔT= 112s

Par. = 182m U.Mag. = 0.9136
Gam. = 0.5313 P.Mag. = 1.8557

Penumbral	2085 Jan 10
Saros 145	22:32 TD
D.Node	ΔT= 112s

Pen. = 286m U.Mag. = -0.1102
Gam. = -1.0453 P.Mag. = 0.9944

Penumbral	2085 Jun 08
Saros 112	02:18 TD
A.Node	ΔT= 113s

Pen. = 189m U.Mag. = -0.4668
Gam. = -1.2746 P.Mag. = 0.5079

Plate 016

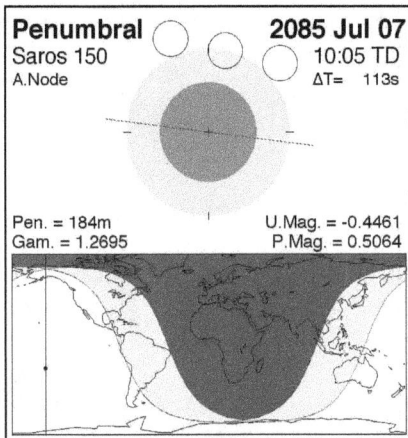

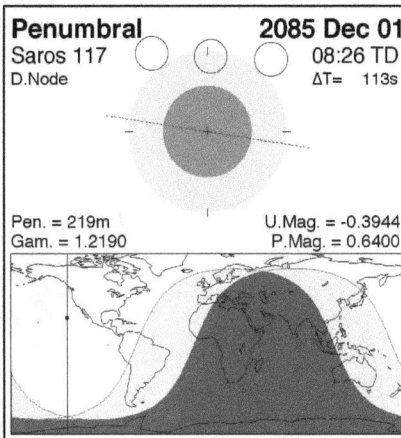

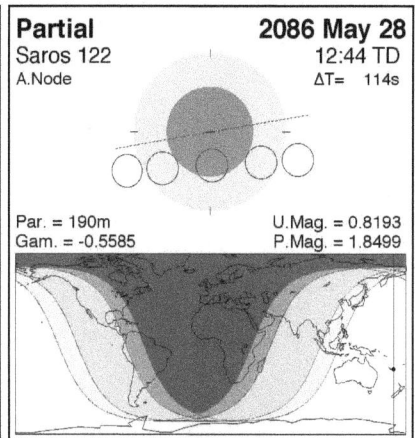

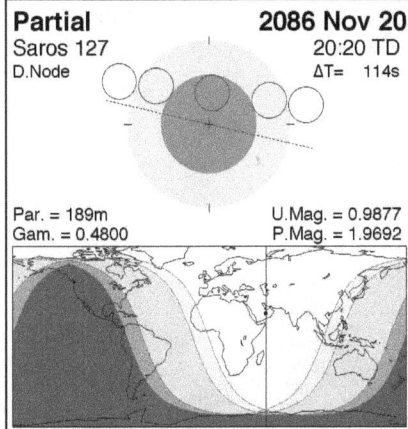

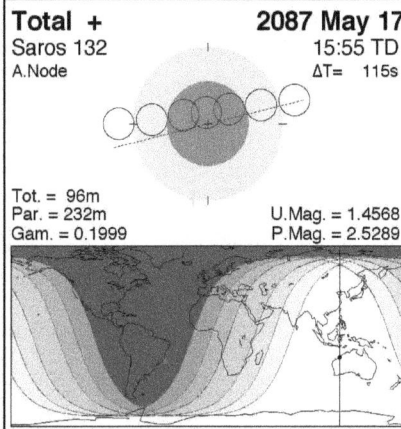

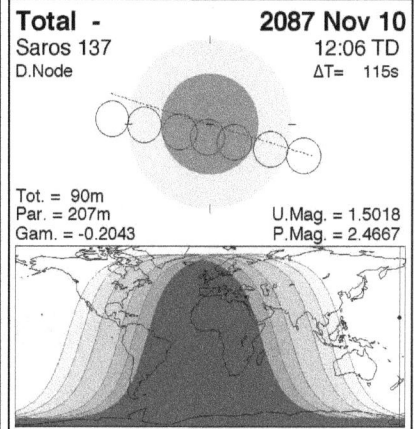

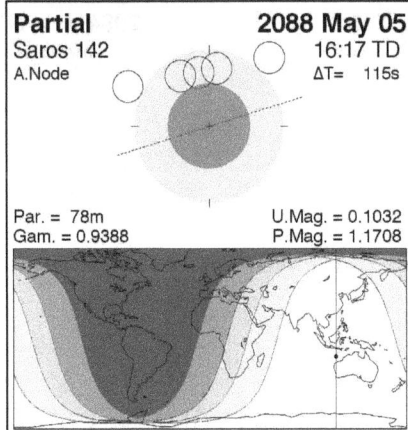

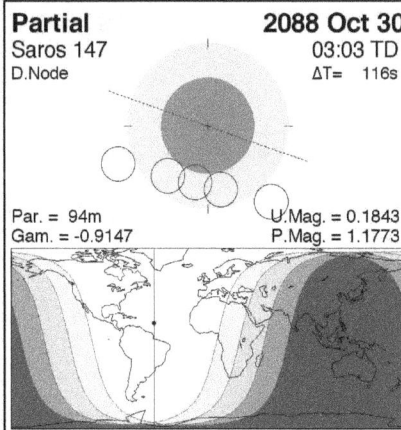

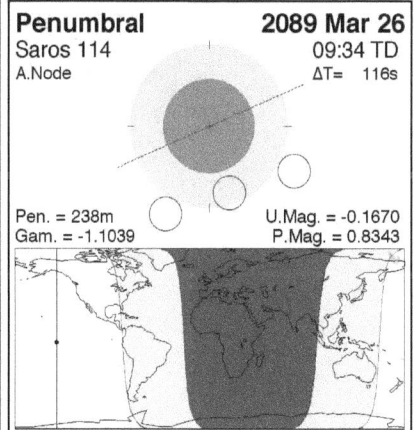

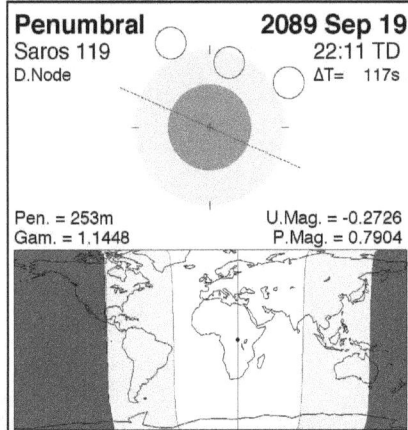

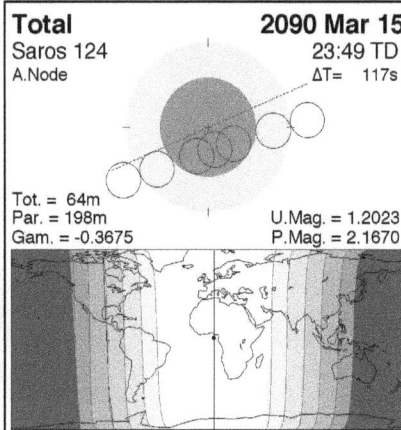

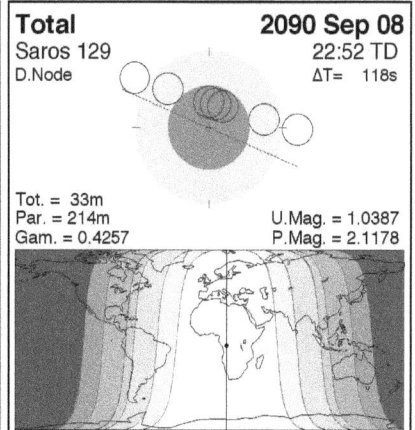

Penumbral	2085 Jul 07
Saros 150	10:05 TD
A.Node	ΔT= 113s

| Pen. = 184m | U.Mag. = -0.4461 |
| Gam. = 1.2695 | P.Mag. = 0.5064 |

Penumbral	2085 Dec 01
Saros 117	08:26 TD
D.Node	ΔT= 113s

| Pen. = 219m | U.Mag. = -0.3944 |
| Gam. = 1.2190 | P.Mag. = 0.6400 |

Partial	2086 May 28
Saros 122	12:44 TD
A.Node	ΔT= 114s

| Par. = 190m | U.Mag. = 0.8193 |
| Gam. = -0.5585 | P.Mag. = 1.8499 |

Partial	2086 Nov 20
Saros 127	20:20 TD
D.Node	ΔT= 114s

| Par. = 189m | U.Mag. = 0.9877 |
| Gam. = 0.4800 | P.Mag. = 1.9692 |

Total +	2087 May 17
Saros 132	15:55 TD
A.Node	ΔT= 115s

Tot. = 96m	
Par. = 232m	U.Mag. = 1.4568
Gam. = 0.1999	P.Mag. = 2.5289

Total -	2087 Nov 10
Saros 137	12:06 TD
D.Node	ΔT= 115s

Tot. = 90m	
Par. = 207m	U.Mag. = 1.5018
Gam. = -0.2043	P.Mag. = 2.4667

Partial	2088 May 05
Saros 142	16:17 TD
A.Node	ΔT= 115s

| Par. = 78m | U.Mag. = 0.1032 |
| Gam. = 0.9388 | P.Mag. = 1.1708 |

Partial	2088 Oct 30
Saros 147	03:03 TD
D.Node	ΔT= 116s

| Par. = 94m | U.Mag. = 0.1843 |
| Gam. = -0.9147 | P.Mag. = 1.1773 |

Penumbral	2089 Mar 26
Saros 114	09:34 TD
A.Node	ΔT= 116s

| Pen. = 238m | U.Mag. = -0.1670 |
| Gam. = -1.1039 | P.Mag. = 0.8343 |

Penumbral	2089 Sep 19
Saros 119	22:11 TD
D.Node	ΔT= 117s

| Pen. = 253m | U.Mag. = -0.2726 |
| Gam. = 1.1448 | P.Mag. = 0.7904 |

Total	2090 Mar 15
Saros 124	23:49 TD
A.Node	ΔT= 117s

Tot. = 64m	
Par. = 198m	U.Mag. = 1.2023
Gam. = -0.3675	P.Mag. = 2.1670

Total	2090 Sep 08
Saros 129	22:52 TD
D.Node	ΔT= 118s

Tot. = 33m	
Par. = 214m	U.Mag. = 1.0387
Gam. = 0.4257	P.Mag. = 2.1178

Plate 017

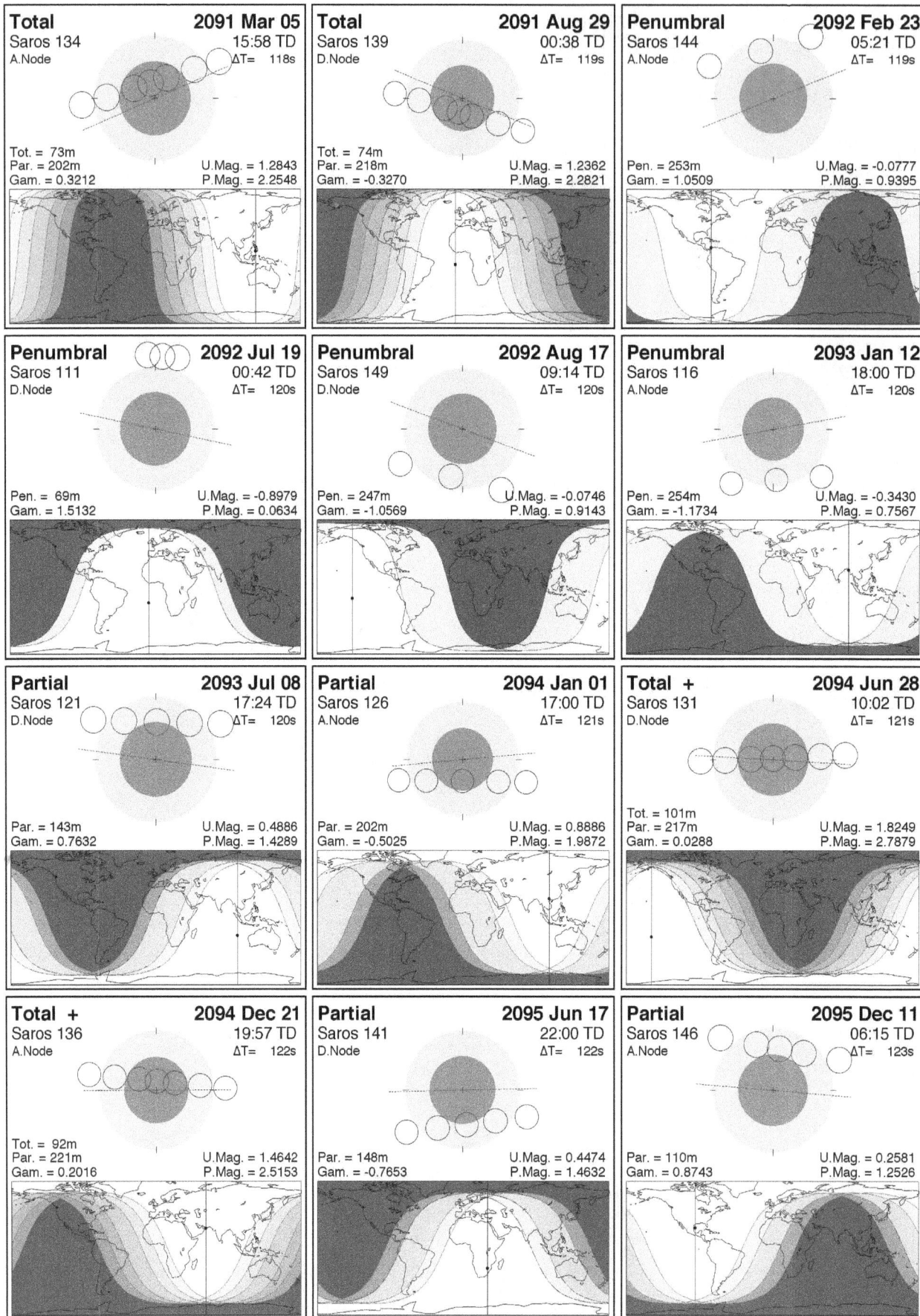

Total	2091 Mar 05
Saros 134	15:58 TD
A.Node	ΔT= 118s
Tot. = 73m	
Par. = 202m	U.Mag. = 1.2843
Gam. = 0.3212	P.Mag. = 2.2548

Total	2091 Aug 29
Saros 139	00:38 TD
D.Node	ΔT= 119s
Tot. = 74m	
Par. = 218m	U.Mag. = 1.2362
Gam. = -0.3270	P.Mag. = 2.2821

Penumbral	2092 Feb 23
Saros 144	05:21 TD
A.Node	ΔT= 119s
Pen. = 253m	U.Mag. = -0.0777
Gam. = 1.0509	P.Mag. = 0.9395

Penumbral	2092 Jul 19
Saros 111	00:42 TD
D.Node	ΔT= 120s
Pen. = 69m	U.Mag. = -0.8979
Gam. = 1.5132	P.Mag. = 0.0634

Penumbral	2092 Aug 17
Saros 149	09:14 TD
D.Node	ΔT= 120s
Pen. = 247m	U.Mag. = -0.0746
Gam. = -1.0569	P.Mag. = 0.9143

Penumbral	2093 Jan 12
Saros 116	18:00 TD
A.Node	ΔT= 120s
Pen. = 254m	U.Mag. = -0.3430
Gam. = -1.1734	P.Mag. = 0.7567

Partial	2093 Jul 08
Saros 121	17:24 TD
D.Node	ΔT= 120s
Par. = 143m	U.Mag. = 0.4886
Gam. = 0.7632	P.Mag. = 1.4289

Partial	2094 Jan 01
Saros 126	17:00 TD
A.Node	ΔT= 121s
Par. = 202m	U.Mag. = 0.8886
Gam. = -0.5025	P.Mag. = 1.9872

Total +	2094 Jun 28
Saros 131	10:02 TD
D.Node	ΔT= 121s
Tot. = 101m	
Par. = 217m	U.Mag. = 1.8249
Gam. = 0.0288	P.Mag. = 2.7879

Total +	2094 Dec 21
Saros 136	19:57 TD
A.Node	ΔT= 122s
Tot. = 92m	
Par. = 221m	U.Mag. = 1.4642
Gam. = 0.2016	P.Mag. = 2.5153

Partial	2095 Jun 17
Saros 141	22:00 TD
D.Node	ΔT= 122s
Par. = 148m	U.Mag. = 0.4474
Gam. = -0.7653	P.Mag. = 1.4632

Partial	2095 Dec 11
Saros 146	06:15 TD
A.Node	ΔT= 123s
Par. = 110m	U.Mag. = 0.2581
Gam. = 0.8743	P.Mag. = 1.2526

Plate 018

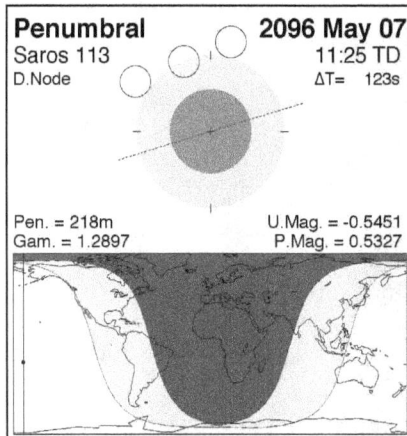

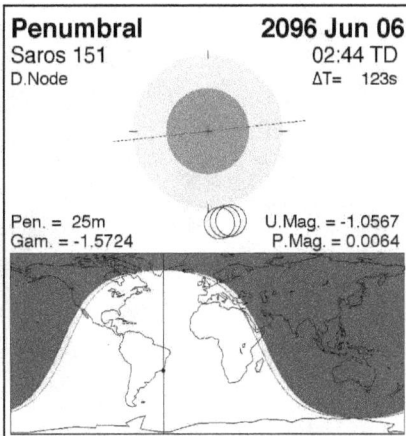

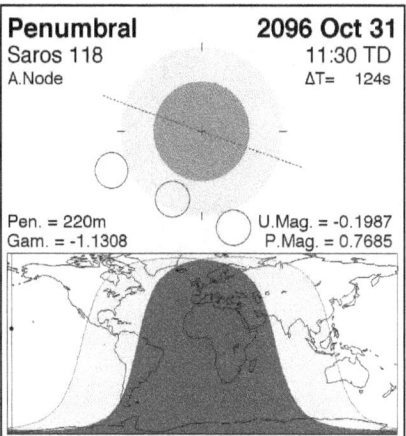

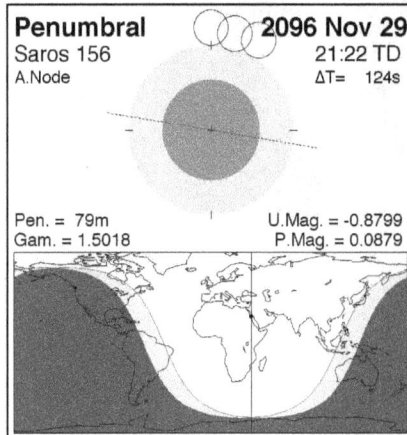

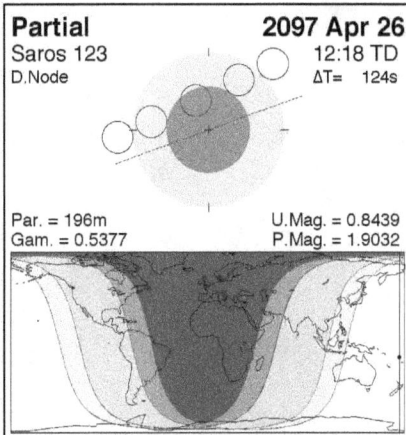

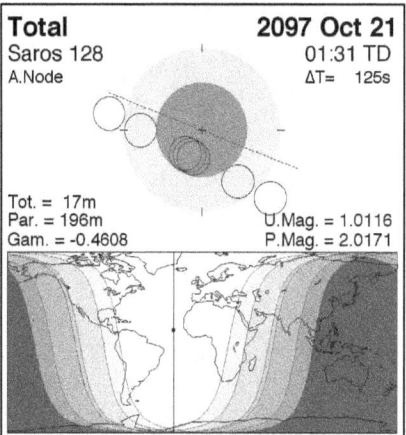

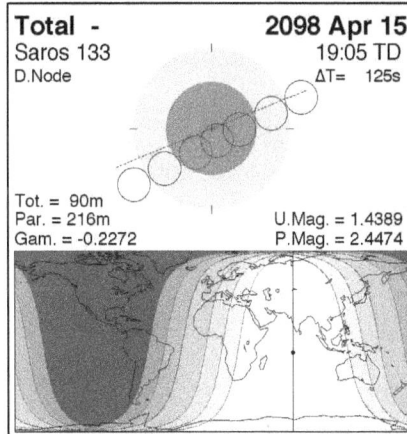

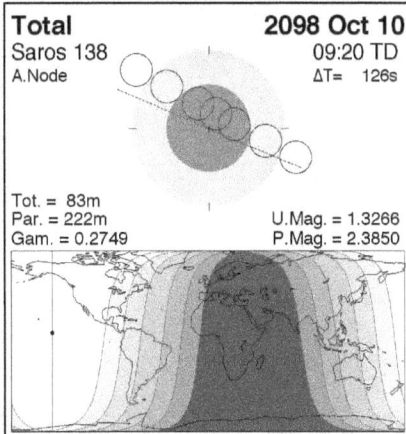

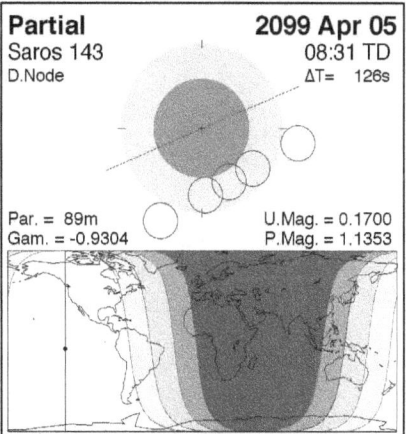

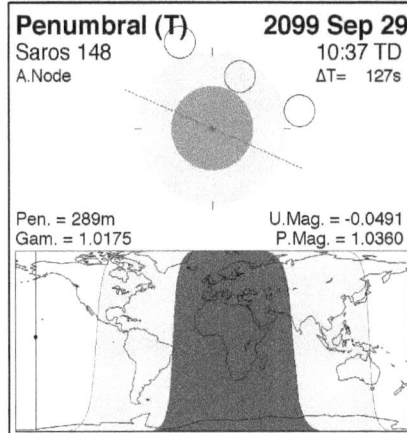

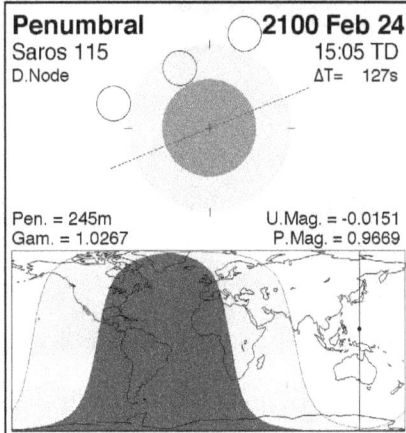

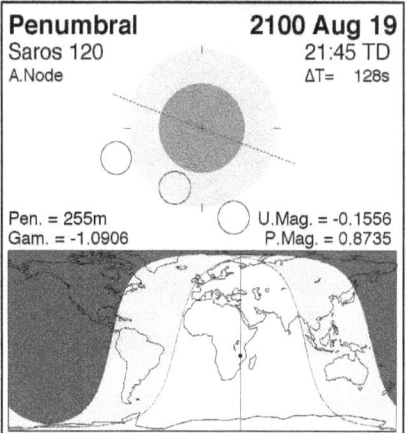

Penumbral **2096 May 07**	
Saros 113 11:25 TD	
D.Node ΔT= 123s	
Pen. = 218m U.Mag. = -0.5451	
Gam. = 1.2897 P.Mag. = 0.5327	

Penumbral   **2096 May 07**
Saros 113   11:25 TD
D.Node   ΔT= 123s
Pen. = 218m   U.Mag. = -0.5451
Gam. = 1.2897   P.Mag. = 0.5327

Penumbral   **2096 Jun 06**
Saros 151   02:44 TD
D.Node   ΔT= 123s
Pen. = 25m   U.Mag. = -1.0567
Gam. = -1.5724   P.Mag. = 0.0064

Penumbral   **2096 Oct 31**
Saros 118   11:30 TD
A.Node   ΔT= 124s
Pen. = 220m   U.Mag. = -0.1987
Gam. = -1.1308   P.Mag. = 0.7685

Penumbral   **2096 Nov 29**
Saros 156   21:22 TD
A.Node   ΔT= 124s
Pen. = 79m   U.Mag. = -0.8799
Gam. = 1.5018   P.Mag. = 0.0879

Partial   **2097 Apr 26**
Saros 123   12:18 TD
D.Node   ΔT= 124s
Par. = 196m   U.Mag. = 0.8439
Gam. = 0.5377   P.Mag. = 1.9032

Total   **2097 Oct 21**
Saros 128   01:31 TD
A.Node   ΔT= 125s
Tot. = 17m
Par. = 196m   U.Mag. = 1.0116
Gam. = -0.4608   P.Mag. = 2.0171

Total -   **2098 Apr 15**
Saros 133   19:05 TD
D.Node   ΔT= 125s
Tot. = 90m
Par. = 216m   U.Mag. = 1.4389
Gam. = -0.2272   P.Mag. = 2.4474

Total   **2098 Oct 10**
Saros 138   09:20 TD
A.Node   ΔT= 126s
Tot. = 83m
Par. = 222m   U.Mag. = 1.3266
Gam. = 0.2749   P.Mag. = 2.3850

Partial   **2099 Apr 05**
Saros 143   08:31 TD
D.Node   ΔT= 126s
Par. = 89m   U.Mag. = 0.1700
Gam. = -0.9304   P.Mag. = 1.1353

Penumbral (T)   **2099 Sep 29**
Saros 148   10:37 TD
A.Node   ΔT= 127s
Pen. = 289m   U.Mag. = -0.0491
Gam. = 1.0175   P.Mag. = 1.0360

Penumbral   **2100 Feb 24**
Saros 115   15:05 TD
D.Node   ΔT= 127s
Pen. = 245m   U.Mag. = -0.0151
Gam. = 1.0267   P.Mag. = 0.9669

Penumbral   **2100 Aug 19**
Saros 120   21:45 TD
A.Node   ΔT= 128s
Pen. = 255m   U.Mag. = -0.1556
Gam. = -1.0906   P.Mag. = 0.8735

Plate 019

55

Photo 10 – A time sequence shows the partial phases and totality during the total lunar eclipse of 2000 July 16 from Maui, Hawaii (Copyright ©2000 by Fred Espenak).

Appendix C

Full Page Lunar Eclipse Figures: 2001 to 2100

Key to Full Page Lunar Eclipse Figures

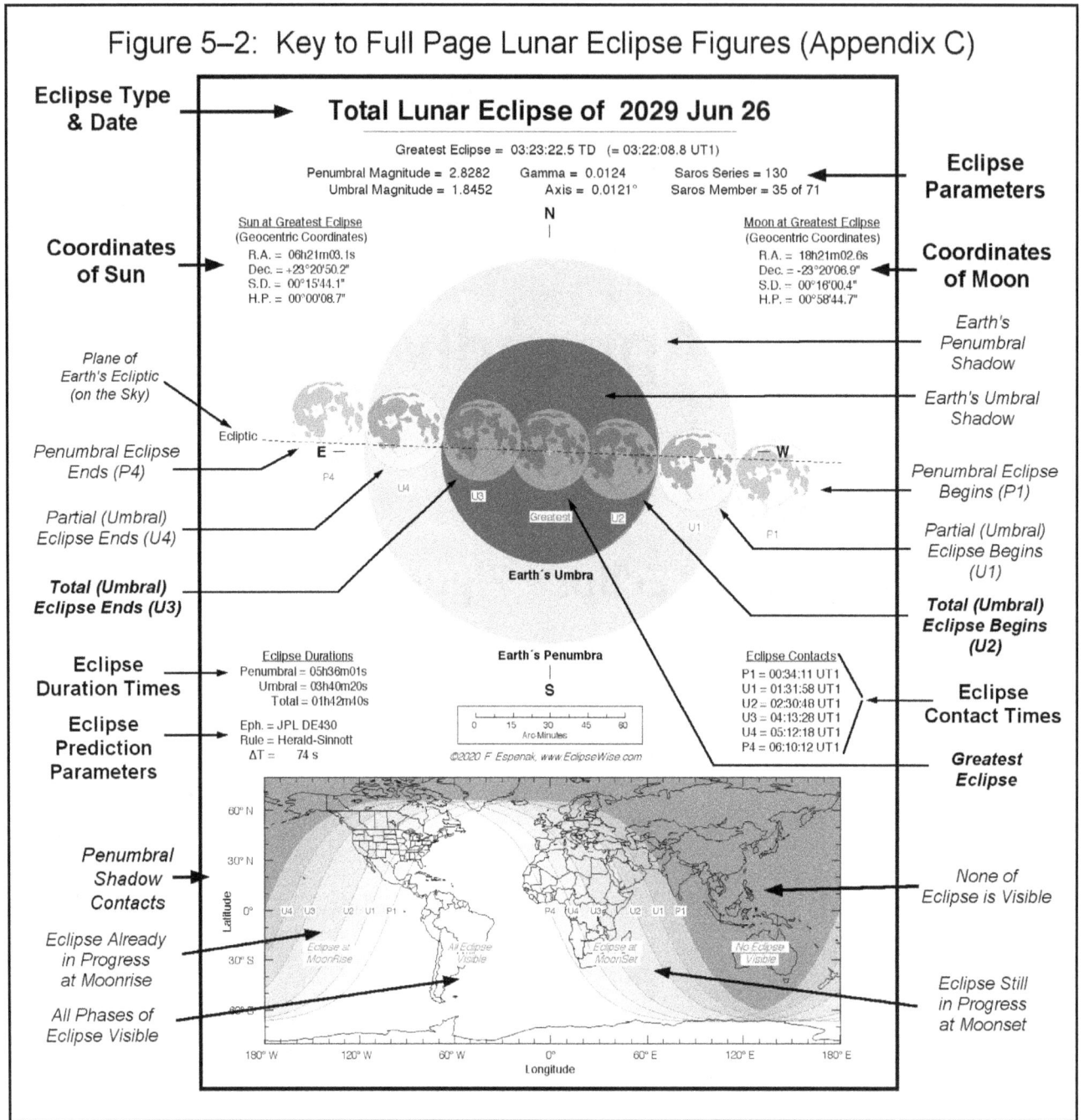

Figure 5–2: Key to Full Page Lunar Eclipse Figures (Appendix C)

Total Lunar Eclipse of 2001 Jan 09

Greatest Eclipse = 20:21:39.6 TD (= 20:20:35.5 UT1)

Penumbral Magnitude = 2.1631	Gamma = 0.3720	Saros Series = 134
Umbral Magnitude = 1.1902	Axis = 0.3803°	Saros Member = 26 of 72

Sun at Greatest Eclipse
(Geocentric Coordinates)

R.A. = 19h25m03.5s
Dec. = -21°59'58.3"
S.D. = 00°16'15.9"
H.P. = 00°00'08.9"

Moon at Greatest Eclipse
(Geocentric Coordinates)

R.A. = 07h25m08.0s
Dec. = +22°22'46.0"
S.D. = 00°16'43.0"
H.P. = 01°01'21.1"

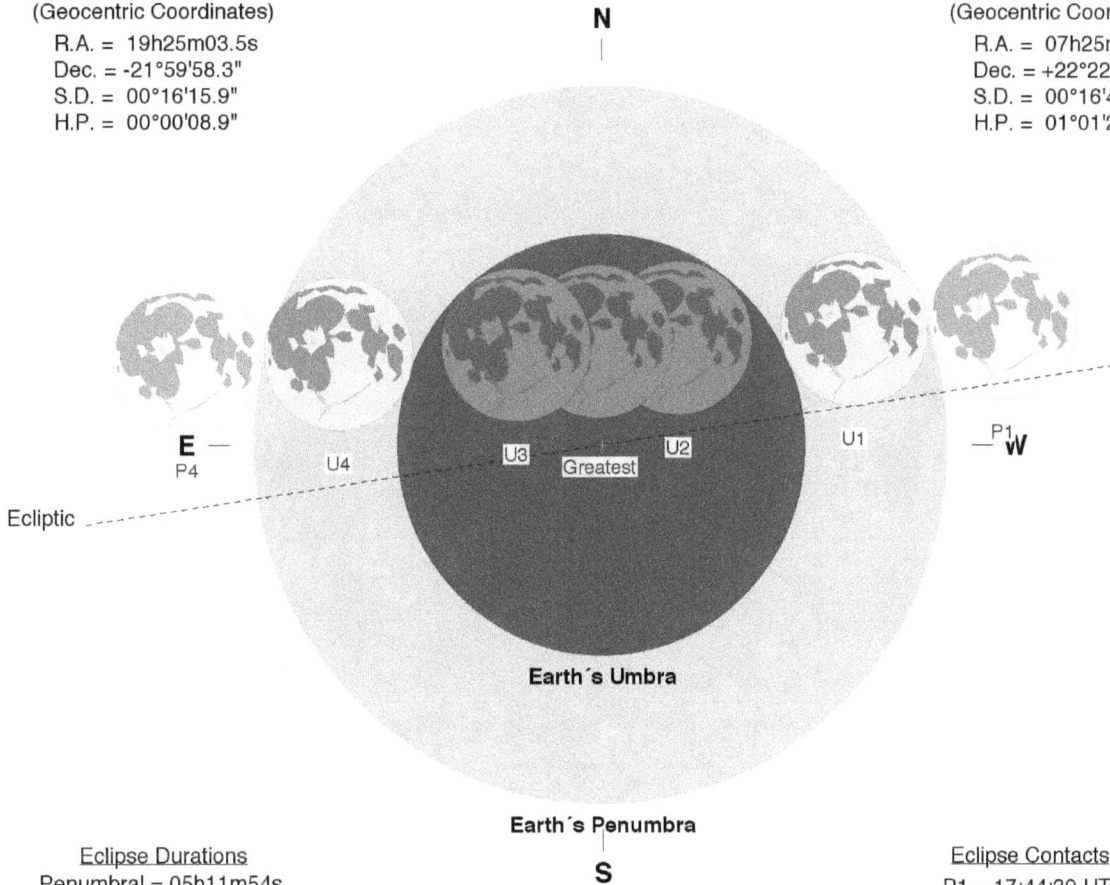

N

E —
P4

U4

U3

Greatest

U2

U1

P1
W

Ecliptic

Earth's Umbra

Earth's Penumbra

S

Eclipse Durations

Penumbral = 05h11m54s
Umbral = 03h17m04s
Total = 01h01m40s

Eph. = JPL DE430
Rule = Herald-Sinnott
ΔT =   64 s

0	15	30	45	60
Arc-Minutes

©2020 F. Espenak, www.EclipseWise.com

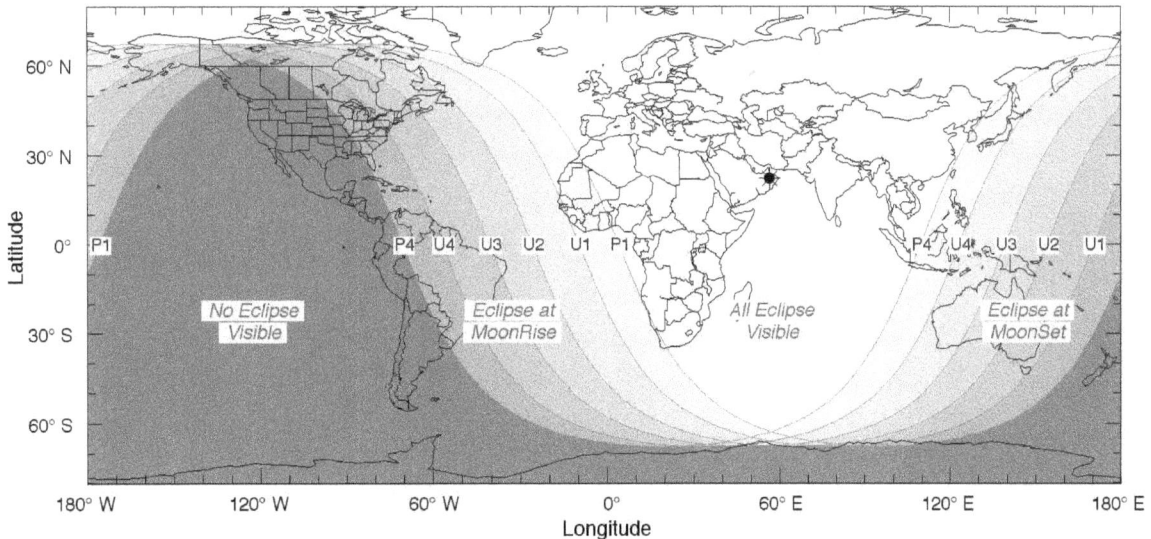

Eclipse Contacts

P1 = 17:44:39 UT1
U1 = 18:42:05 UT1
U2 = 19:49:47 UT1
U3 = 20:51:27 UT1
U4 = 21:59:08 UT1
P4 = 22:56:32 UT1

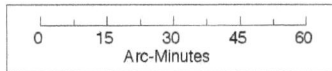

No Eclipse Visible

Eclipse at MoonRise

All Eclipse Visible

Eclipse at MoonSet

Partial Lunar Eclipse of 2001 Jul 05

Greatest Eclipse = 14:56:23.0 TD   (= 14:55:18.8 UT1)

Penumbral Magnitude = 1.5489	Gamma = -0.7287	Saros Series = 139
Umbral Magnitude = 0.4961	Axis = 0.6660°	Saros Member = 20 of 79

Sun at Greatest Eclipse
(Geocentric Coordinates)

R.A. = 06h59m16.1s
Dec. = +22°44'22.5"
S.D. = 00°15'43.9"
H.P. = 00°00'08.7"

N

Earth's Penumbra

Earth's Umbra

Moon at Greatest Eclipse
(Geocentric Coordinates)

R.A. = 18h59m16.6s
Dec. = -23°24'20.1"
S.D. = 00°14'56.6"
H.P. = 00°54'50.4"

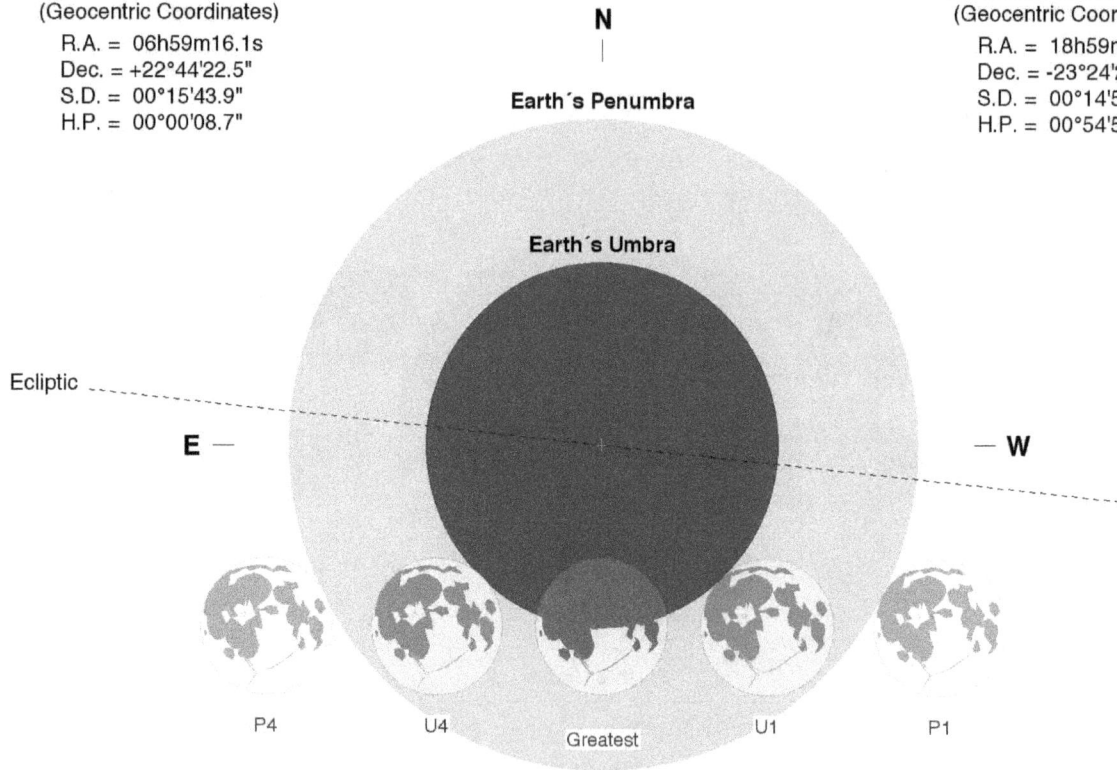

Ecliptic

E —

— **W**

P4   U4   Greatest   U1   P1

S

Eclipse Durations
Penumbral = 05h25m59s
Umbral = 02h40m01s

Eph. = JPL DE430
Rule = Herald-Sinnott
ΔT =   64 s

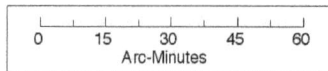

0   15   30   45   60
Arc-Minutes

©2020 F. Espenak, www.EclipseWise.com

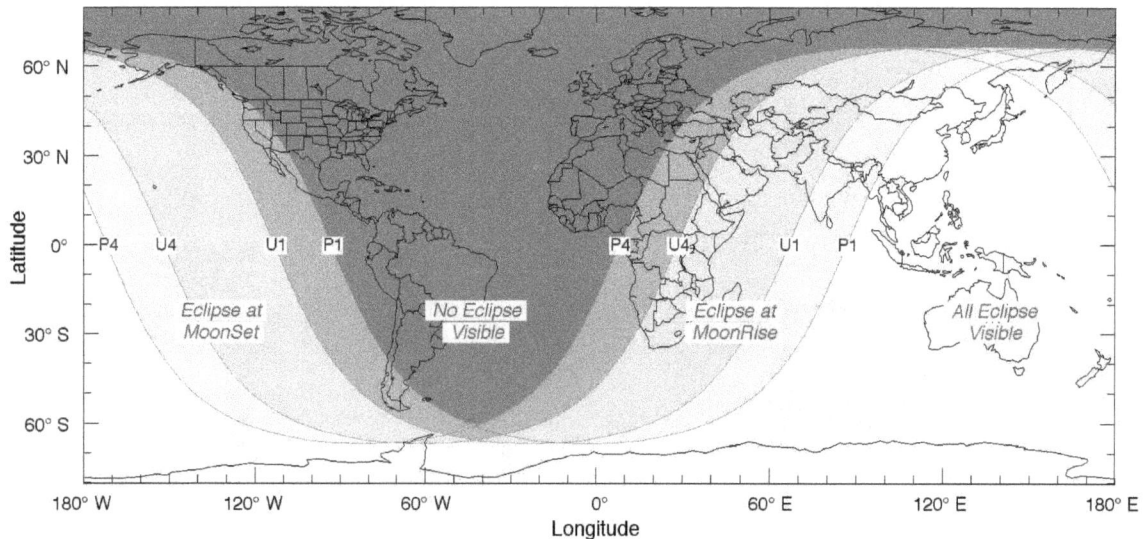

Eclipse Contacts
P1 = 12:12:20 UT1
U1 = 13:35:16 UT1
U4 = 16:15:16 UT1
P4 = 17:38:19 UT1

Eclipse at MoonSet   No Eclipse Visible   Eclipse at MoonRise   All Eclipse Visible

Penumbral Lunar Eclipse of 2001 Dec 30

Greatest Eclipse = 10:30:22.3 TD (= 10:29:18.0 UT1)

Penumbral Magnitude = 0.8948   Gamma = 1.0732   Saros Series = 144
Umbral Magnitude = -0.1141   Axis = 1.0582°   Saros Member = 15 of 71

Sun at Greatest Eclipse
(Geocentric Coordinates)

R.A. = 18h38m16.3s
Dec. = -23°08'50.7"
S.D. = 00°16'15.9"
H.P. = 00°00'08.9"

Moon at Greatest Eclipse
(Geocentric Coordinates)

R.A. = 06h38m07.7s
Dec. = +24°12'18.7"
S.D. = 00°16'07.4"
H.P. = 00°59'10.2"

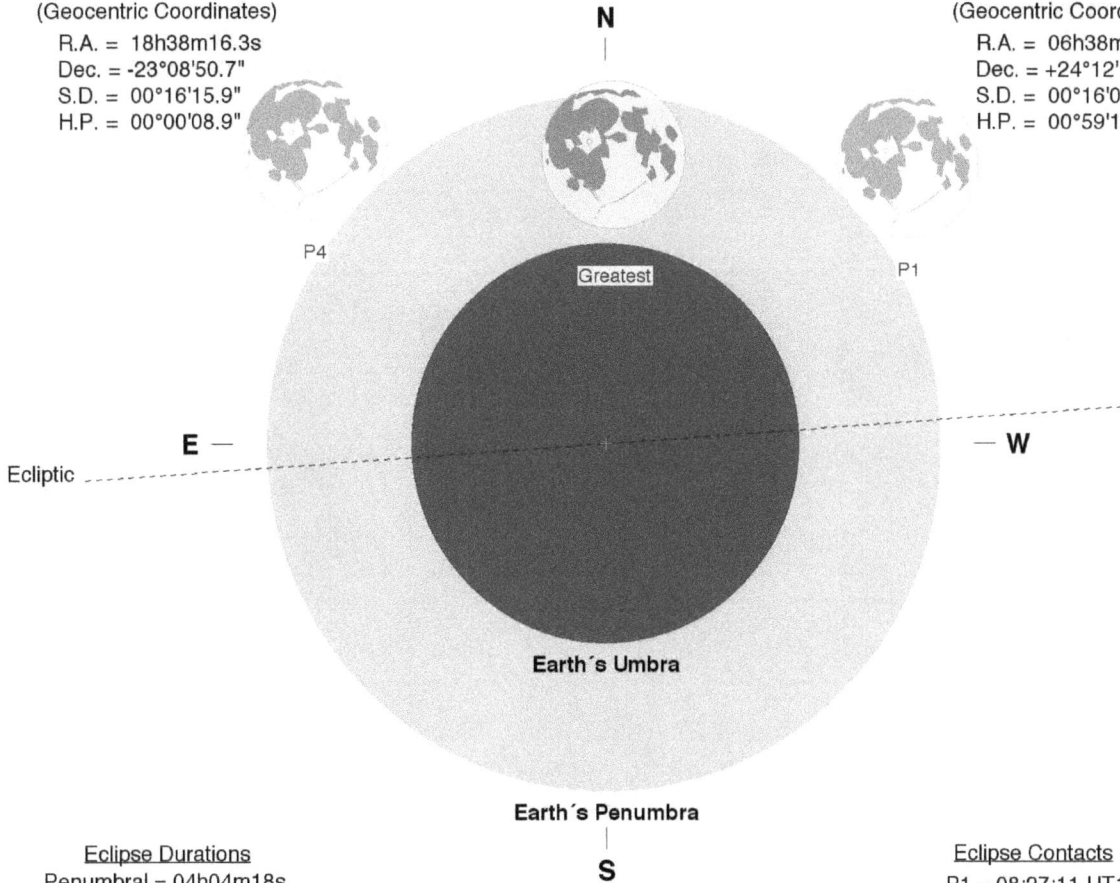

N

P4

P1

Greatest

E

W

Ecliptic

Earth's Umbra

Earth's Penumbra

S

Eclipse Durations
Penumbral = 04h04m18s

Eclipse Contacts
P1 = 08:27:11 UT1
P4 = 12:31:29 UT1

Eph. = JPL DE430
Rule = Herald-Sinnott
ΔT = 64 s

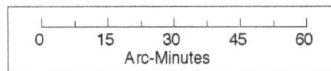

0   15   30   45   60
Arc-Minutes

©2020 F. Espenak, www.EclipseWise.com

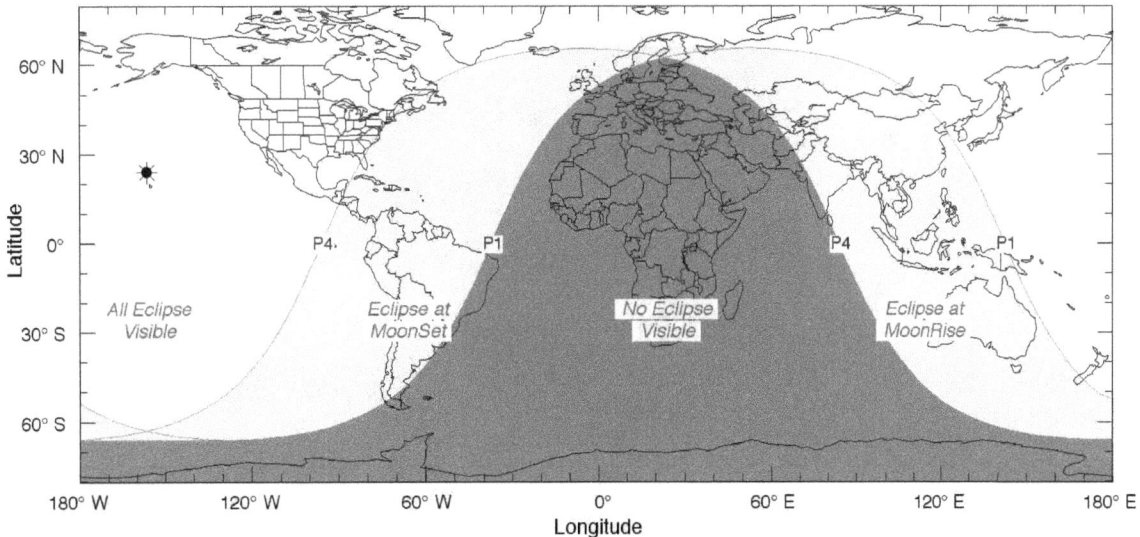

All Eclipse Visible   Eclipse at MoonSet   No Eclipse Visible   Eclipse at MoonRise

P4   P1   P4   P1

Latitude

Longitude

Penumbral Lunar Eclipse of 2002 May 26

Greatest Eclipse = 12:04:26.8 TD   (= 12:03:22.4 UT1)

Penumbral Magnitude = 0.6910	Gamma = 1.1759	Saros Series = 111
Umbral Magnitude = -0.2871	Axis = 1.1609°	Saros Member = 66 of 71

Sun at Greatest Eclipse
(Geocentric Coordinates)
R.A. = 04h12m31.0s
Dec. = +21°08'37.3"
S.D. = 00°15'47.3"
H.P. = 00°00'08.7"

Moon at Greatest Eclipse
(Geocentric Coordinates)
R.A. = 16h13m52.1s
Dec. = -20°01'35.7"
S.D. = 00°16'08.5"
H.P. = 00°59'14.5"

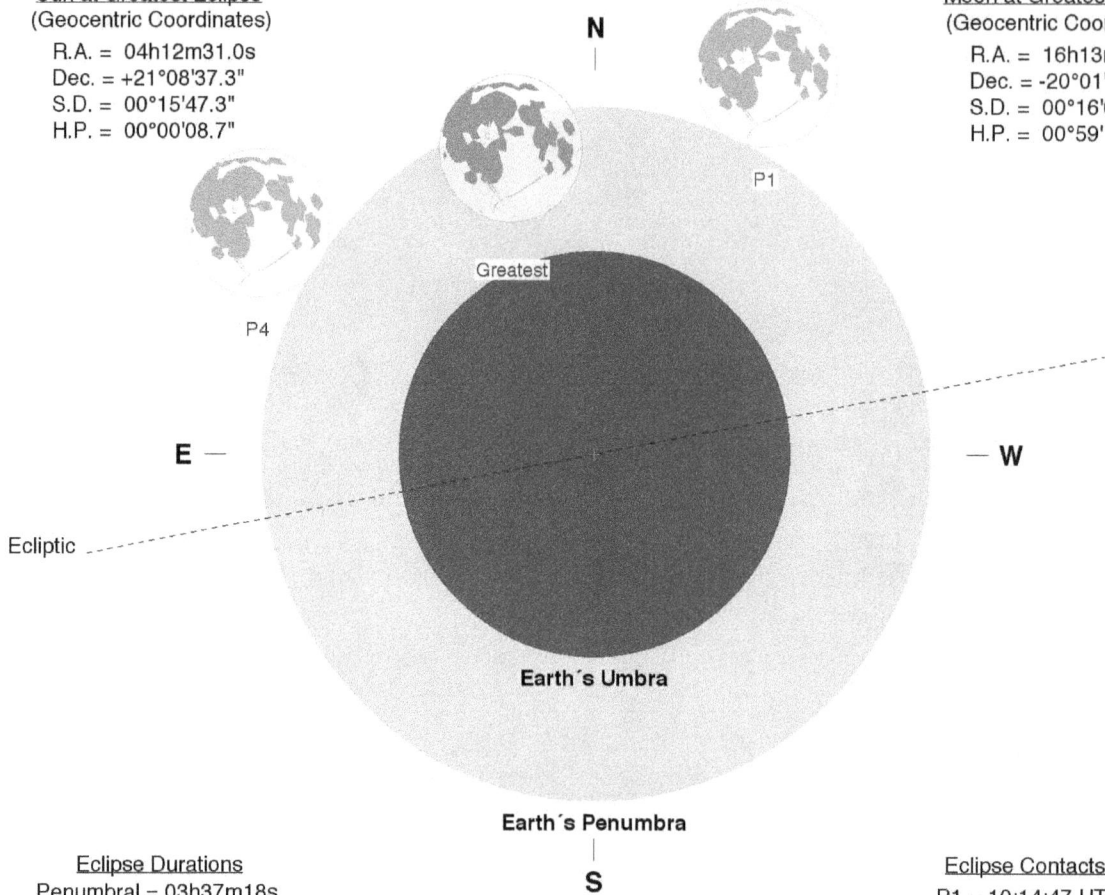

N

P1

P4

Greatest

Ecliptic

E —

— W

Earth's Umbra

Earth's Penumbra

S

Eclipse Durations
Penumbral = 03h37m18s

Eclipse Contacts
P1 = 10:14:47 UT1
P4 = 13:52:04 UT1

Eph. = JPL DE430
Rule = Herald-Sinnott
ΔT =   64 s

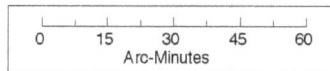

0	15	30	45	60

Arc-Minutes

©2020 F. Espenak, www.EclipseWise.com

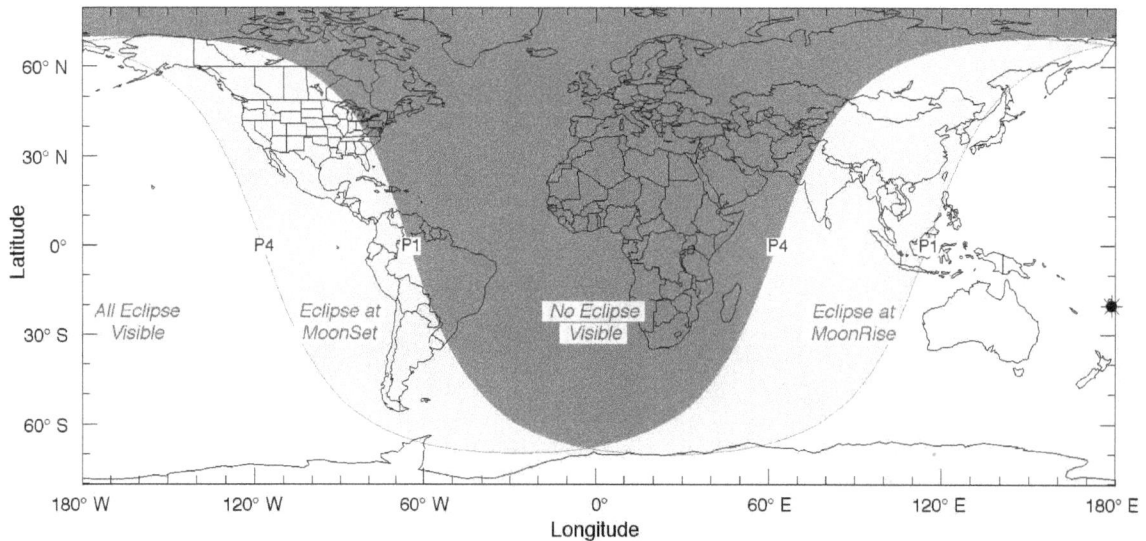

P4   P1   P4   P1

All Eclipse
Visible

Eclipse at
MoonSet

No Eclipse
Visible

Eclipse at
MoonRise

Latitude

Longitude

Penumbral Lunar Eclipse of 2002 Jun 24

Greatest Eclipse = 21:28:12.9 TD (= 21:27:08.5 UT1)

Penumbral Magnitude = 0.2110	Gamma = -1.4440	Saros Series = 149
Umbral Magnitude = -0.7910	Axis = 1.3870°	Saros Member = 2 of 71

Sun at Greatest Eclipse
(Geocentric Coordinates)

R.A. = 06h13m52.0s
Dec. = +23°24'03.8"
S.D. = 00°15'44.2"
H.P. = 00°00'08.7"

N

Earth's Penumbra

Earth's Umbra

Moon at Greatest Eclipse
(Geocentric Coordinates)

R.A. = 18h13m25.9s
Dec. = -24°47'04.8"
S.D. = 00°15'42.3"
H.P. = 00°57'38.4"

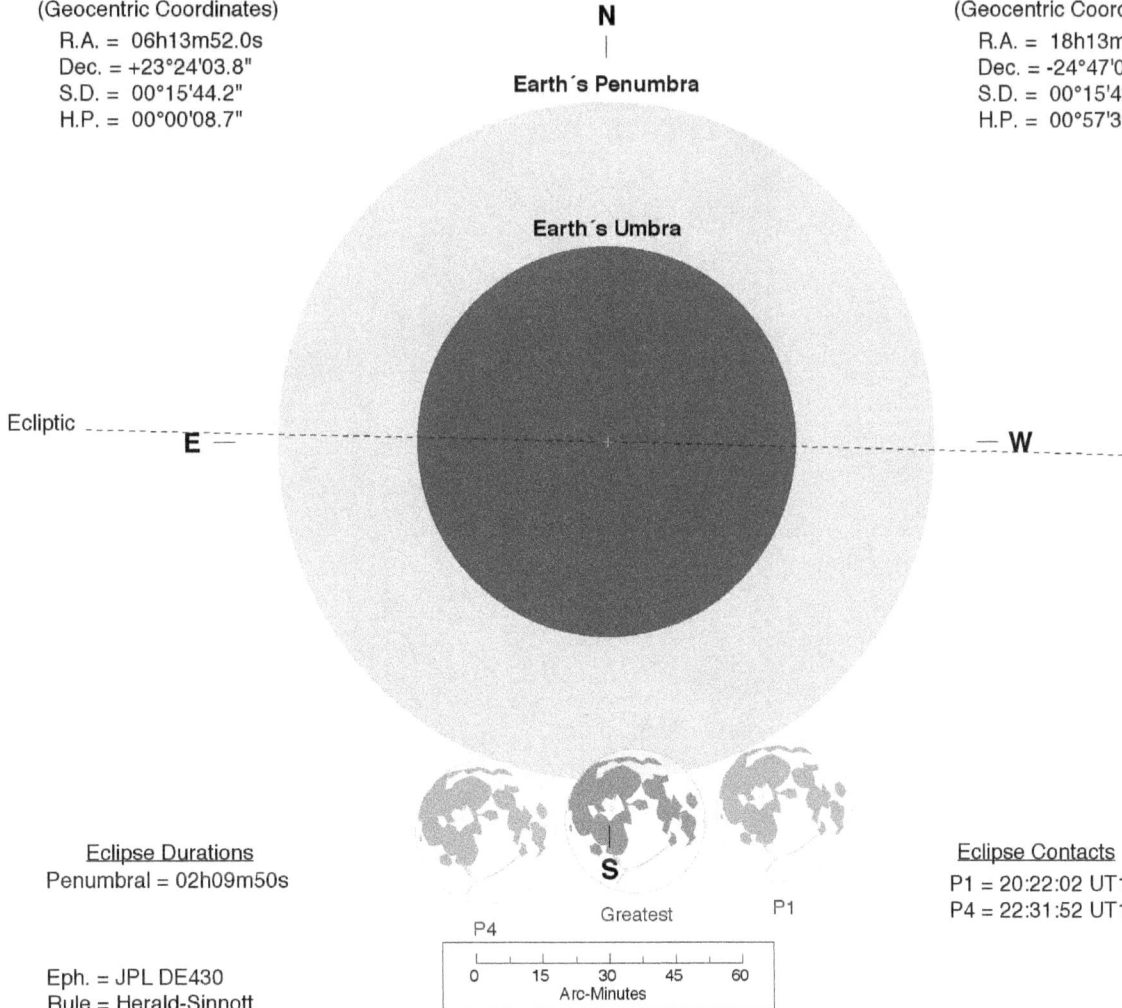

Ecliptic

E

W

S

P4   Greatest   P1

0   15   30   45   60
Arc-Minutes

Eclipse Durations
Penumbral = 02h09m50s

Eph. = JPL DE430
Rule = Herald-Sinnott
ΔT = 64 s

©2020 F. Espenak, www.EclipseWise.com

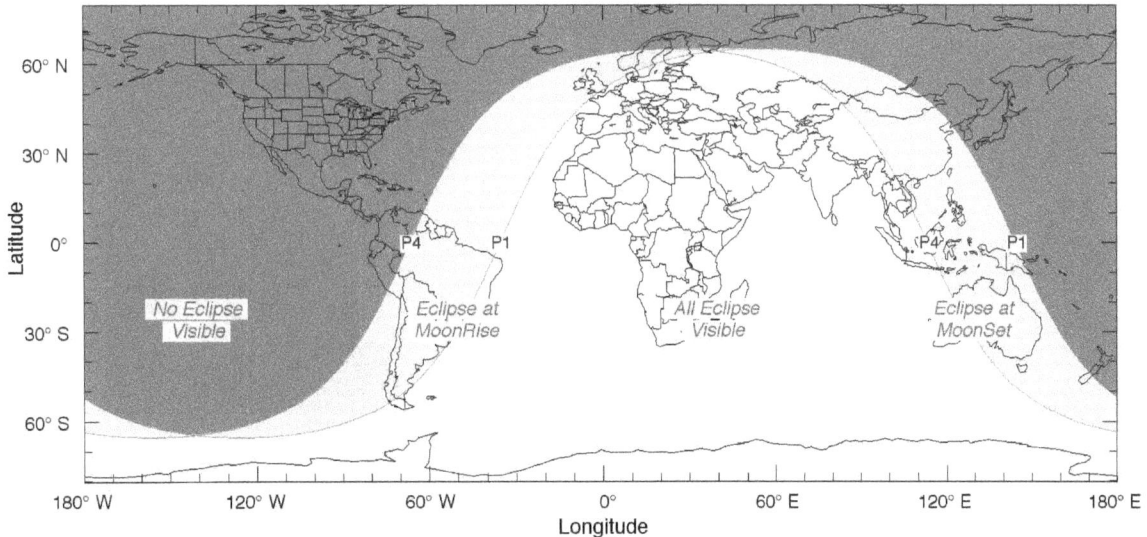

Eclipse Contacts
P1 = 20:22:02 UT1
P4 = 22:31:52 UT1

No Eclipse Visible

Eclipse at MoonRise

All Eclipse Visible

Eclipse at MoonSet

P4   P1   P4   P1

Penumbral Lunar Eclipse of 2002 Nov 20

Greatest Eclipse = 01:47:40.7 TD (= 01:46:36.3 UT1)

Penumbral Magnitude = 0.8618	Gamma = -1.1127	Saros Series = 116
Umbral Magnitude = -0.2246	Axis = 1.0139°	Saros Member = 57 of 73

Sun at Greatest Eclipse
(Geocentric Coordinates)
R.A. = 15h41m07.8s
Dec. = -19°36'53.3"
S.D. = 00°16'11.2"
H.P. = 00°00'08.9"

N

Earth´s Penumbra

Earth´s Umbra

Ecliptic

E

W

P4

Greatest

S

P1

Moon at Greatest Eclipse
(Geocentric Coordinates)
R.A. = 03h42m30.3s
Dec. = +18°39'15.4"
S.D. = 00°14'54.0"
H.P. = 00°54'40.9"

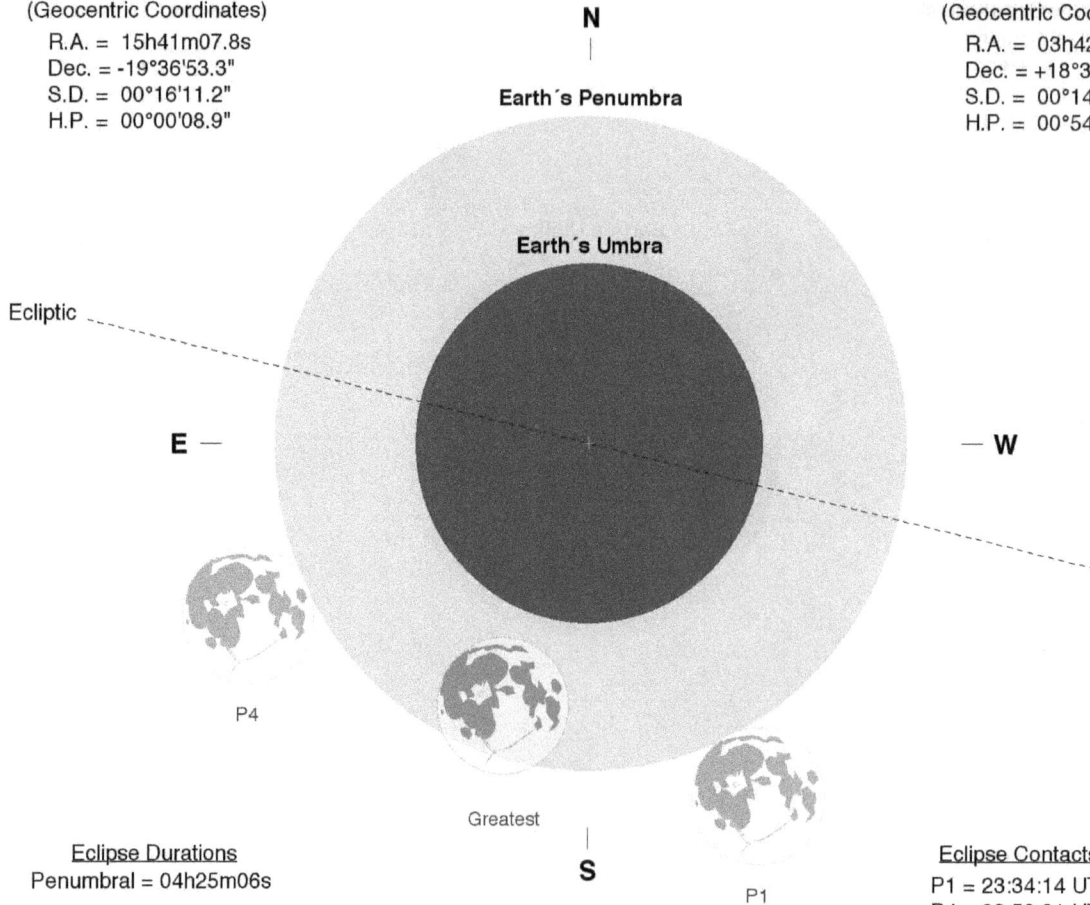

Eclipse Durations
Penumbral = 04h25m06s

Eclipse Contacts
P1 = 23:34:14 UT1
P4 = 03:59:21 UT1

0	15	30	45	60
Arc-Minutes

Eph. = JPL DE430
Rule = Herald-Sinnott
ΔT = 64 s

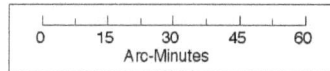

©2020 F. Espenak, www.EclipseWise.com

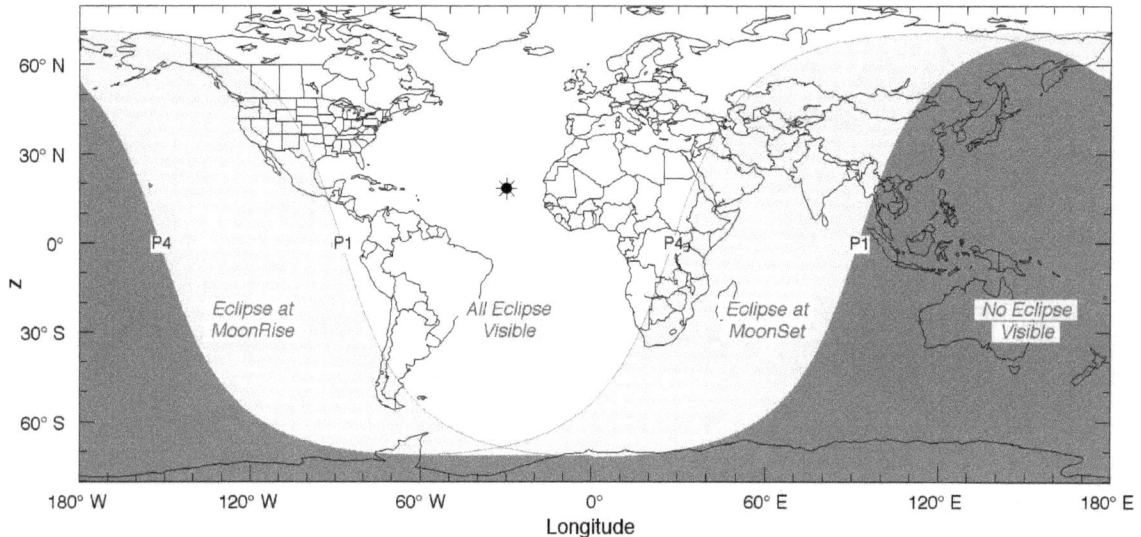

P4 P1 P4 P1

Eclipse at
MoonRise

All Eclipse
Visible

Eclipse at
MoonSet

No Eclipse
Visible

Longitude

Total Lunar Eclipse of 2003 May 16

Greatest Eclipse = 03:41:13.2 TD (= 03:40:08.8 UT1)

Penumbral Magnitude = 2.0765	Gamma = 0.4123	Saros Series = 121
Umbral Magnitude = 1.1294	Axis = 0.4212°	Saros Member = 54 of 82

Sun at Greatest Eclipse
(Geocentric Coordinates)

R.A. = 03h30m07.2s
Dec. = +18°59'20.2"
S.D. = 00°15'49.2"
H.P. = 00°00'08.7"

Moon at Greatest Eclipse
(Geocentric Coordinates)

R.A. = 15h30m43.0s
Dec. = -18°35'31.7"
S.D. = 00°16'42.2"
H.P. = 01°01'18.2"

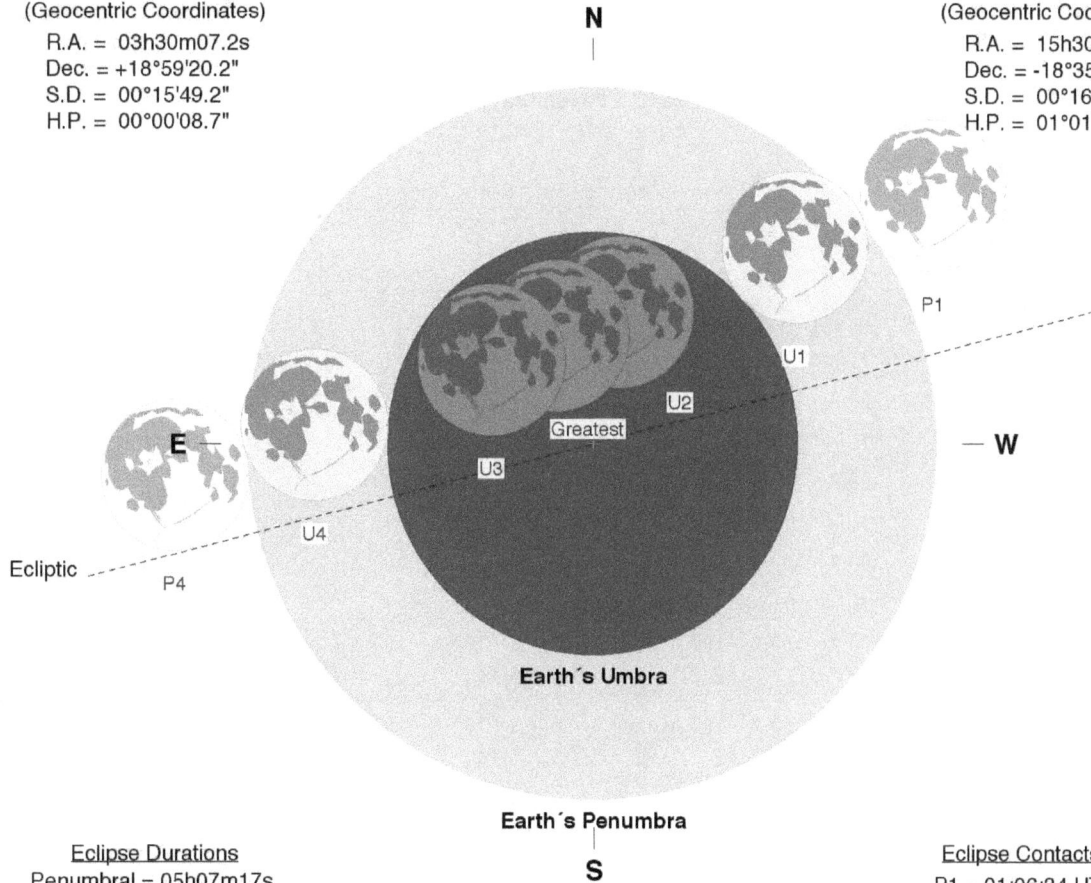

N

P1
U1
U2

E

Greatest

W

U3

U4

Ecliptic

P4

Earth's Umbra

Earth's Penumbra

S

Eclipse Durations
Penumbral = 05h07m17s
Umbral = 03h14m34s
Total = 00h52m03s

Eph. = JPL DE430
Rule = Herald-Sinnott
ΔT = 64 s

0   15   30   45   60
Arc-Minutes

©2020 F. Espenak, www.EclipseWise.com

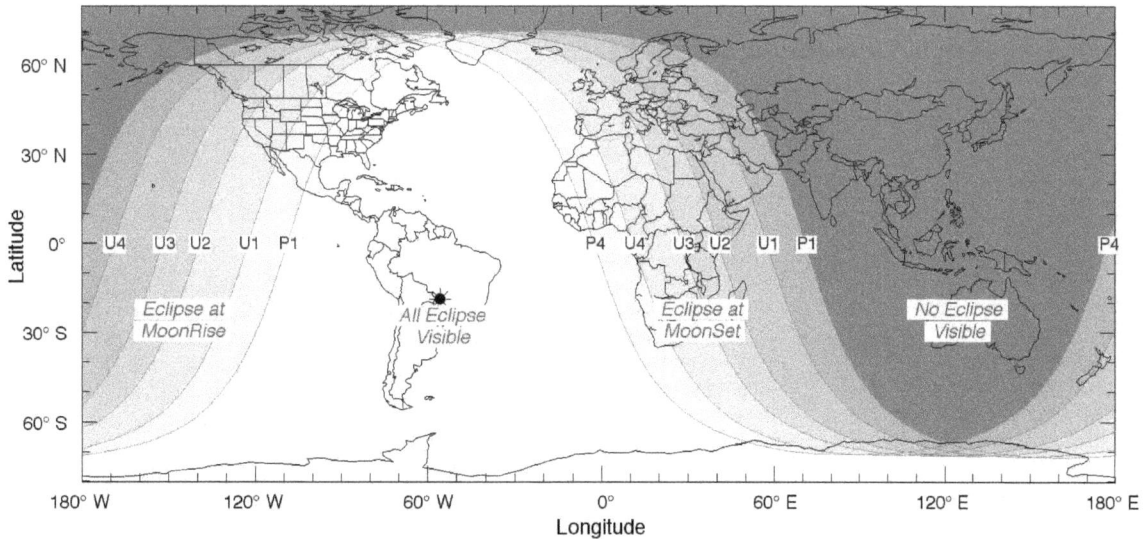

Eclipse Contacts
P1 = 01:06:34 UT1
U1 = 02:02:56 UT1
U2 = 03:14:17 UT1
U3 = 04:06:20 UT1
U4 = 05:17:30 UT1
P4 = 06:13:51 UT1

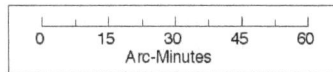

U4 U3 U2 U1 P1   P4 U4 U3 U2 U1 P1   P4

Eclipse at
MoonRise

All Eclipse
Visible

Eclipse at
MoonSet

No Eclipse
Visible

Total Lunar Eclipse of 2003 Nov 09

Greatest Eclipse = 01:19:38.0 TD (= 01:18:33.5 UT1)

Penumbral Magnitude = 2.1157	Gamma = -0.4319	Saros Series = 126
Umbral Magnitude = 1.0197	Axis = 0.3891°	Saros Member = 44 of 70

Sun at Greatest Eclipse
(Geocentric Coordinates)

R.A. = 14h54m59.9s
Dec. = -16°41'23.6"
S.D. = 00°16'08.7"
H.P. = 00°00'08.9"

N

Earth´s Penumbra

Earth´s Umbra

Ecliptic

E

P4

U4

U3

Greatest

U2

U1

P1

W

S

Moon at Greatest Eclipse
(Geocentric Coordinates)

R.A. = 02h55m37.1s
Dec. = +16°19'48.8"
S.D. = 00°14'43.8"
H.P. = 00°54'03.6"

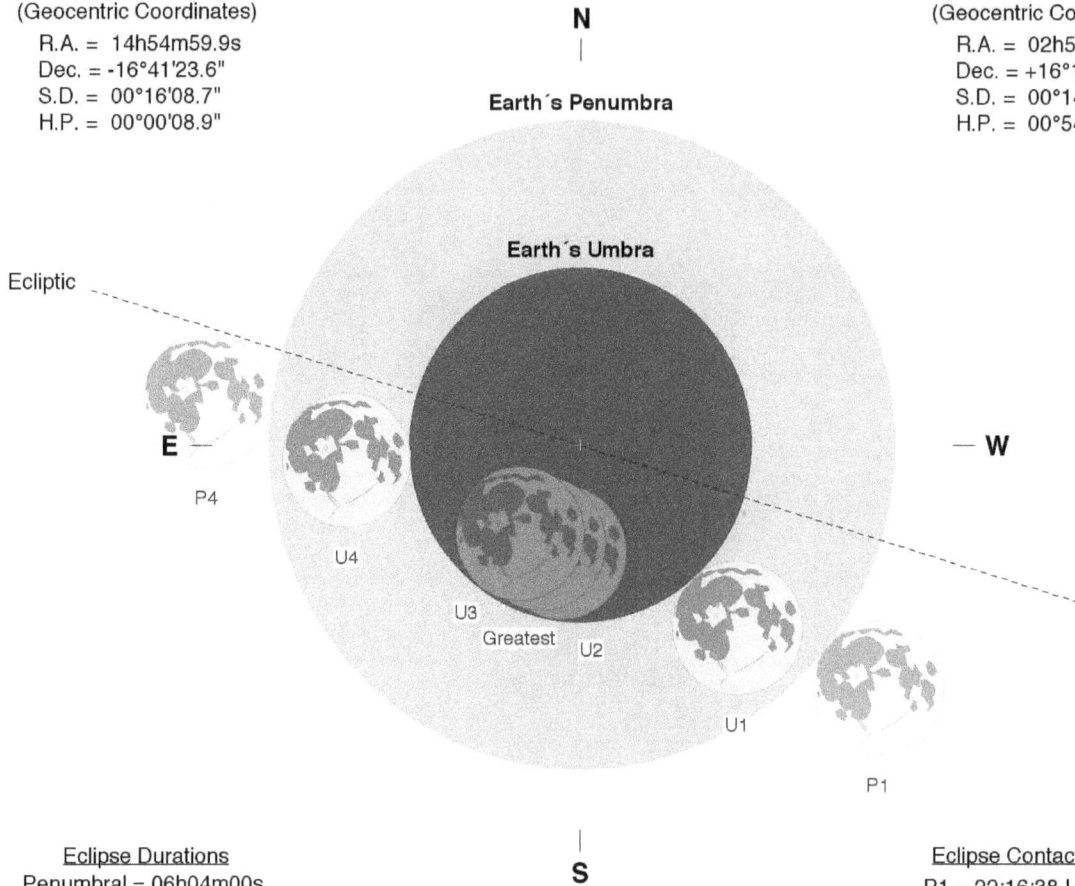

Eclipse Durations
Penumbral = 06h04m00s
Umbral = 03h32m10s
Total = 00h23m12s

Eph. = JPL DE430
Rule = Herald-Sinnott
ΔT =   64 s

0   15   30   45   60
Arc-Minutes

©2020 F. Espenak, www.EclipseWise.com

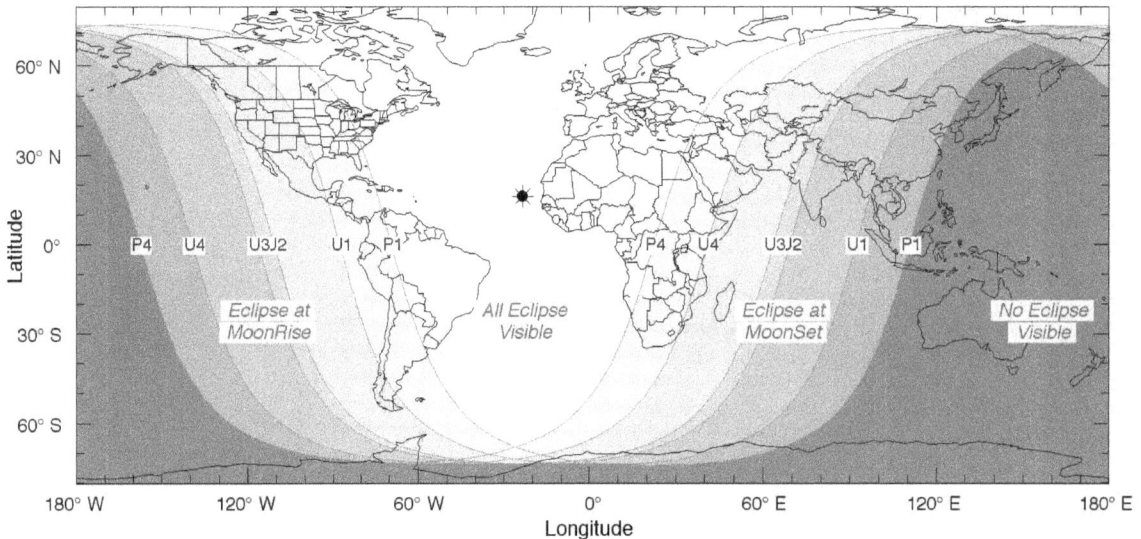

Eclipse Contacts
P1 = 22:16:38 UT1
U1 = 23:32:35 UT1
U2 = 01:07:12 UT1
U3 = 01:30:25 UT1
U4 = 03:04:45 UT1
P4 = 04:20:39 UT1

Total Lunar Eclipse of 2004 May 04

Greatest Eclipse = 20:31:17.1 TD (= 20:30:12.6 UT1)

Penumbral Magnitude = 2.2645	Gamma = -0.3132	Saros Series = 131
Umbral Magnitude = 1.3054	Axis = 0.3167°	Saros Member = 33 of 72

Sun at Greatest Eclipse
(Geocentric Coordinates)

R.A. = 02h48m55.8s
Dec. = +16°14'51.5"
S.D. = 00°15'51.5"
H.P. = 00°00'08.7"

N
|
Earth's Penumbra

Earth's Umbra

E —

— **W**
P1

U1

U2

Greatest

U3

U4

P4

Moon at Greatest Eclipse
(Geocentric Coordinates)

R.A. = 14h48m25.1s
Dec. = -16°32'22.6"
S.D. = 00°16'32.0"
H.P. = 01°00'40.8"

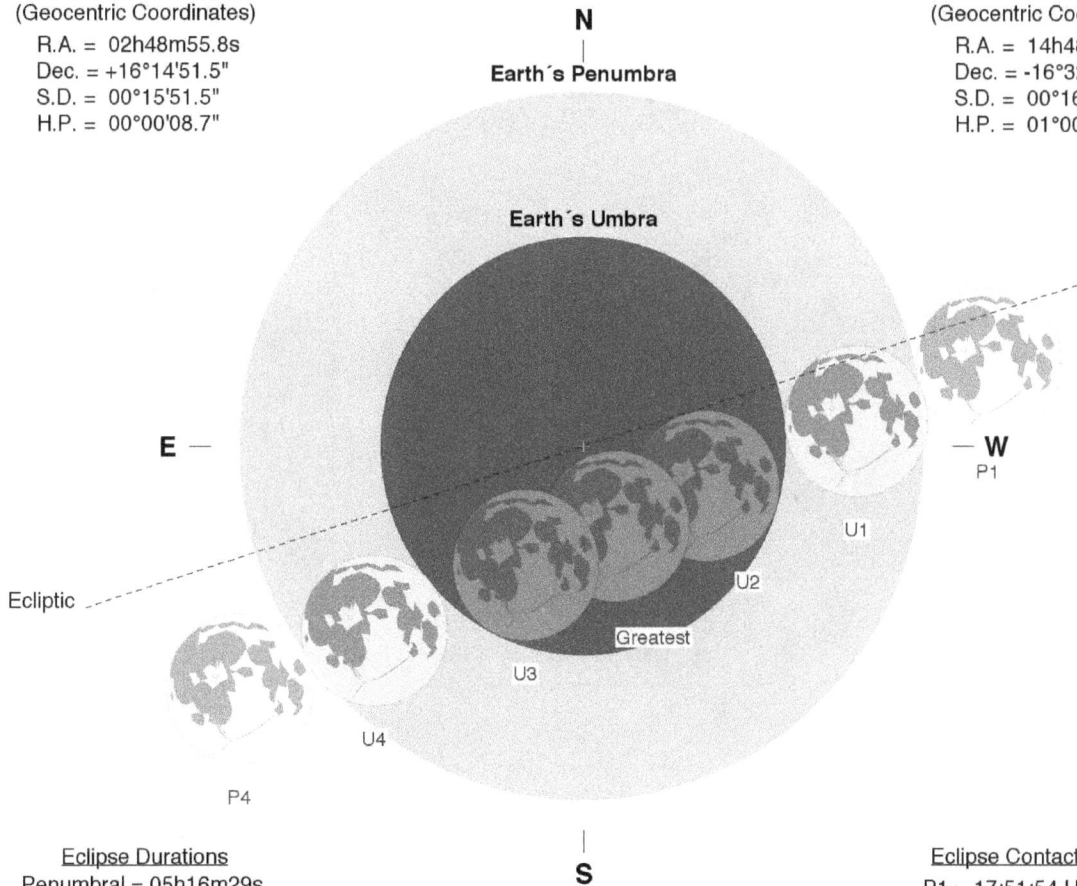

Ecliptic - - -

S

Eclipse Durations
Penumbral = 05h16m29s
Umbral = 03h23m53s
Total = 01h16m05s

Eph. = JPL DE430
Rule = Herald-Sinnott
ΔT =   65 s

0	15	30	45	60

Arc-Minutes

©2020 F. Espenak, www.EclipseWise.com

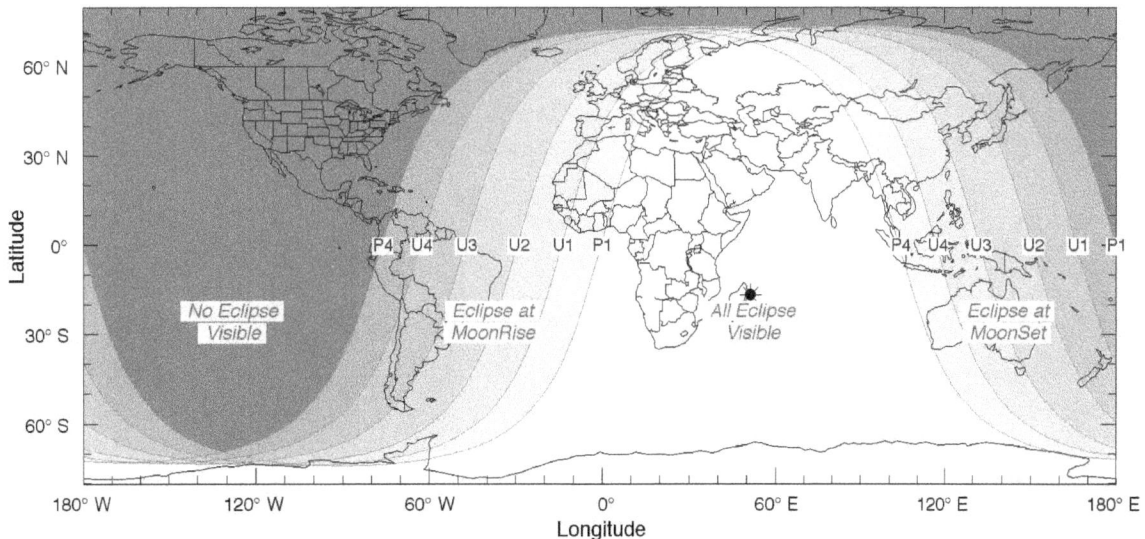

Eclipse Contacts
P1 = 17:51:54 UT1
U1 = 18:48:13 UT1
U2 = 19:52:01 UT1
U3 = 21:08:07 UT1
U4 = 22:12:06 UT1
P4 = 23:08:23 UT1

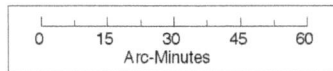

No Eclipse Visible

Eclipse at MoonRise

All Eclipse Visible

Eclipse at MoonSet

P4 U4 U3 U2 U1 P1   P4 U4 U3 U2 U1 P1

Total Lunar Eclipse of 2004 Oct 28

Greatest Eclipse = 03:05:11.3 TD (= 03:04:06.7 UT1)

Penumbral Magnitude = 2.3656	Gamma = 0.2846	Saros Series = 136
Umbral Magnitude = 1.3100	Axis = 0.2655°	Saros Member = 19 of 72

<u>Sun at Greatest Eclipse</u>
(Geocentric Coordinates)

R.A. = 14h11m00.6s
Dec. = -13°12'05.3"
S.D. = 00°16'06.0"
H.P. = 00°00'08.9"

<u>Moon at Greatest Eclipse</u>
(Geocentric Coordinates)

R.A. = 02h10m32.6s
Dec. = +13°26'29.6"
S.D. = 00°15'15.1"
H.P. = 00°55'58.4"

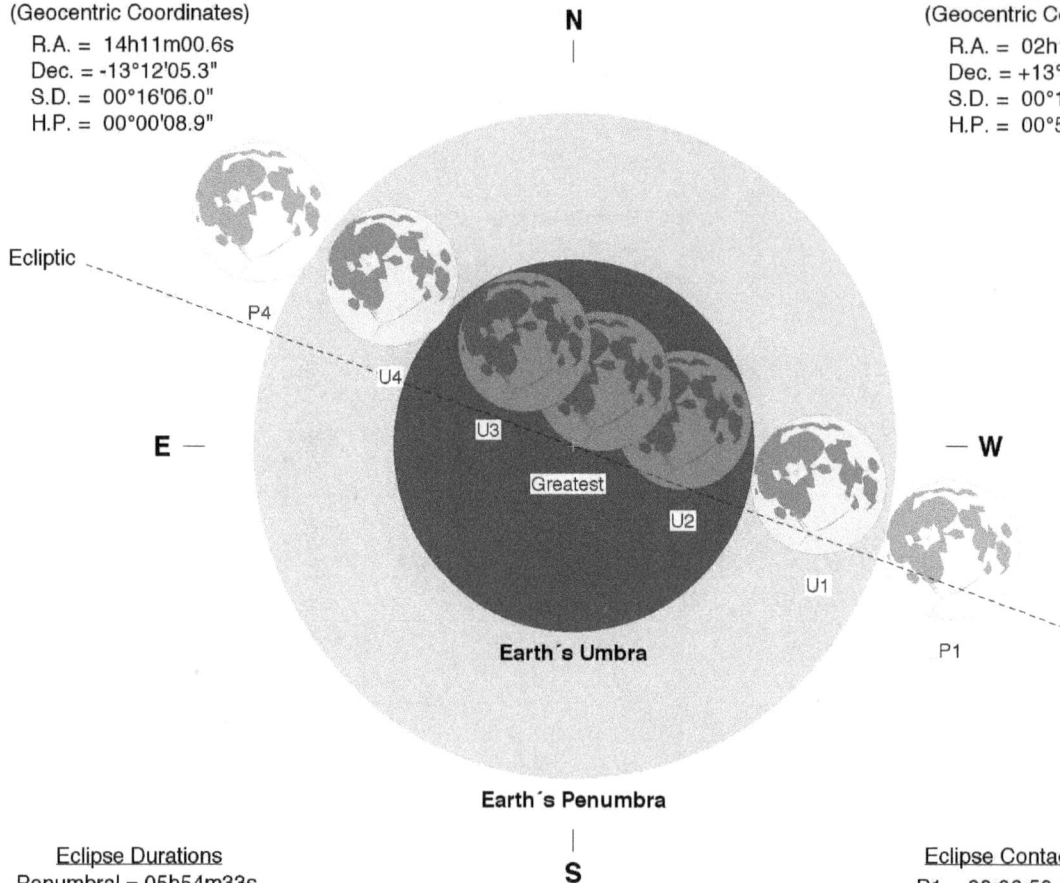

N

Ecliptic

P4
U4
U3
Greatest
U2
U1
P1

E

W

Earth's Umbra

Earth's Penumbra

S

<u>Eclipse Durations</u>
Penumbral = 05h54m33s
Umbral = 03h39m23s
Total = 01h21m08s

Eph. = JPL DE430
Rule = Herald-Sinnott
ΔT = 65 s

0	15	30	45	60

Arc-Minutes

©2020 F. Espenak, www.EclipseWise.com

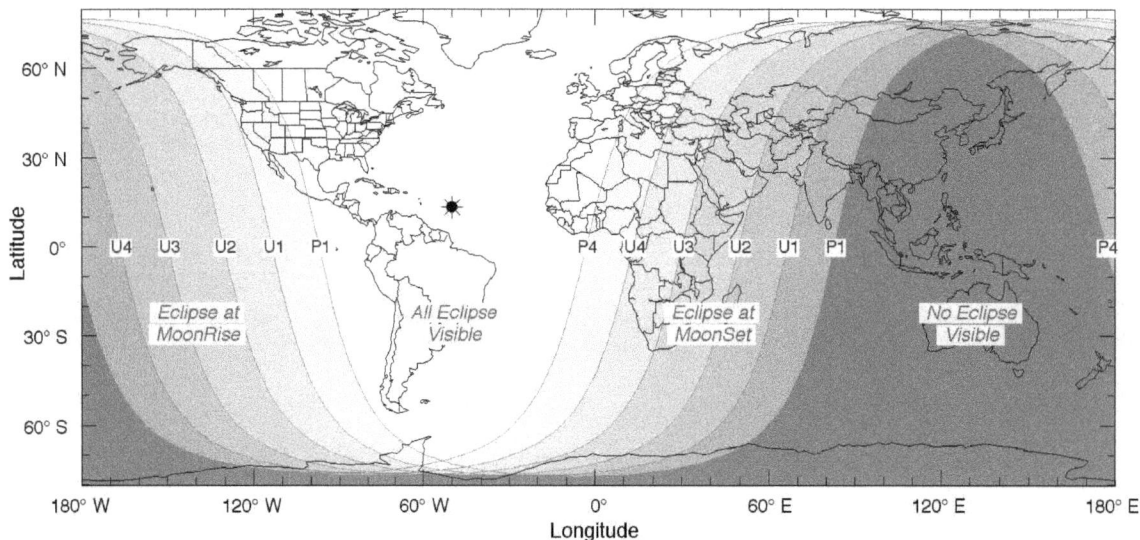

<u>Eclipse Contacts</u>
P1 = 00:06:50 UT1
U1 = 01:14:19 UT1
U2 = 02:23:21 UT1
U3 = 03:44:29 UT1
U4 = 04:53:42 UT1
P4 = 06:01:23 UT1

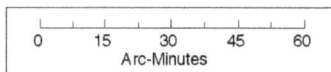

U4 U3 U2 U1 P1   P4 U4 U3 U2 U1 P1   P4

Eclipse at MoonRise   *All Eclipse Visible*   *Eclipse at MoonSet*   *No Eclipse Visible*

Penumbral Lunar Eclipse of 2005 Apr 24

Greatest Eclipse = 09:55:54.4 TD (= 09:54:49.7 UT1)

Penumbral Magnitude = 0.8669	Gamma = -1.0885	Saros Series = 141
Umbral Magnitude = -0.1417	Axis = 1.0496°	Saros Member = 23 of 72

Sun at Greatest Eclipse
(Geocentric Coordinates)

R.A. = 02h08m13.9s
Dec. = +12°57'36.8"
S.D. = 00°15'54.1"
H.P. = 00°00'08.7"

Moon at Greatest Eclipse
(Geocentric Coordinates)

R.A. = 14h06m23.1s
Dec. = -13°54'32.8"
S.D. = 00°15'46.0"
H.P. = 00°57'51.7"

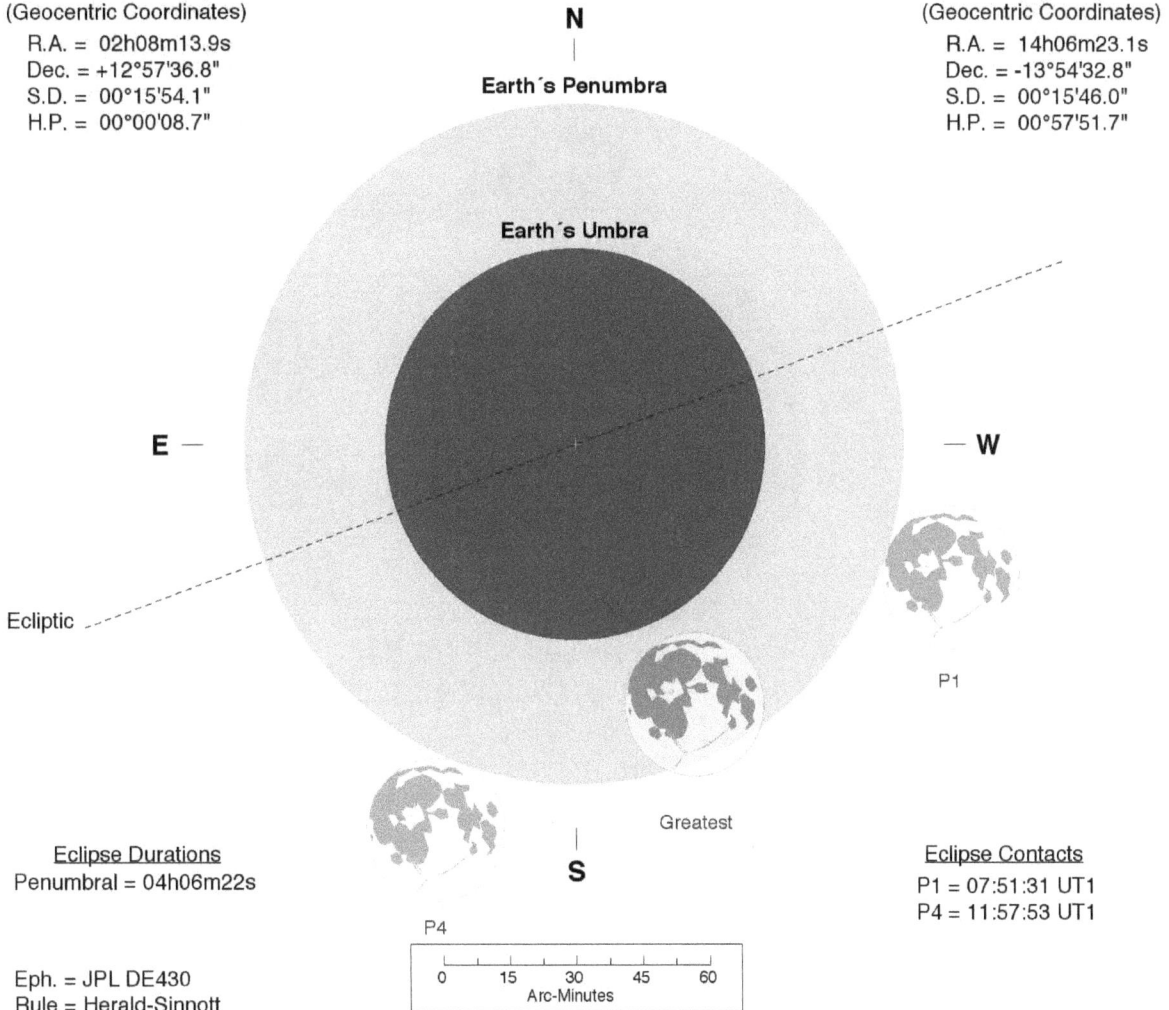

N

Earth's Penumbra

Earth's Umbra

E

W

Ecliptic

P1

Greatest

S

Eclipse Durations

Penumbral = 04h06m22s

P4

Eclipse Contacts

P1 = 07:51:31 UT1
P4 = 11:57:53 UT1

0   15   30   45   60
Arc-Minutes

Eph. = JPL DE430
Rule = Herald-Sinnott
ΔT =   65 s

©2020 F. Espenak, www.EclipseWise.com

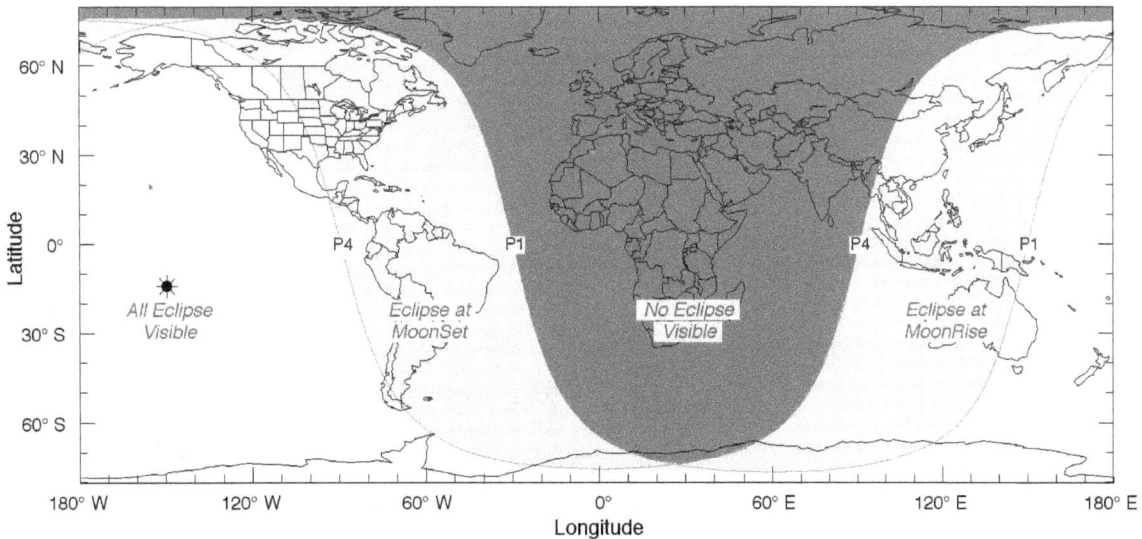

All Eclipse
Visible

Eclipse at
MoonSet

No Eclipse
Visible

Eclipse at
MoonRise

P4   P1   P4   P1

Latitude

Longitude

Partial Lunar Eclipse of 2005 Oct 17

Greatest Eclipse = 12:04:26.6 TD (= 12:03:21.8 UT1)

Penumbral Magnitude = 1.0605	Gamma = 0.9796	Saros Series = 146
Umbral Magnitude = 0.0645	Axis = 0.9655°	Saros Member = 10 of 72

Sun at Greatest Eclipse
(Geocentric Coordinates)

R.A. = 13h29m41.7s
Dec. = -09°23'29.0"
S.D. = 00°16'03.1"
H.P. = 00°00'08.8"

Moon at Greatest Eclipse
(Geocentric Coordinates)

R.A. = 01h27m54.2s
Dec. = +10°15'01.0"
S.D. = 00°16'06.9"
H.P. = 00°59'08.7"

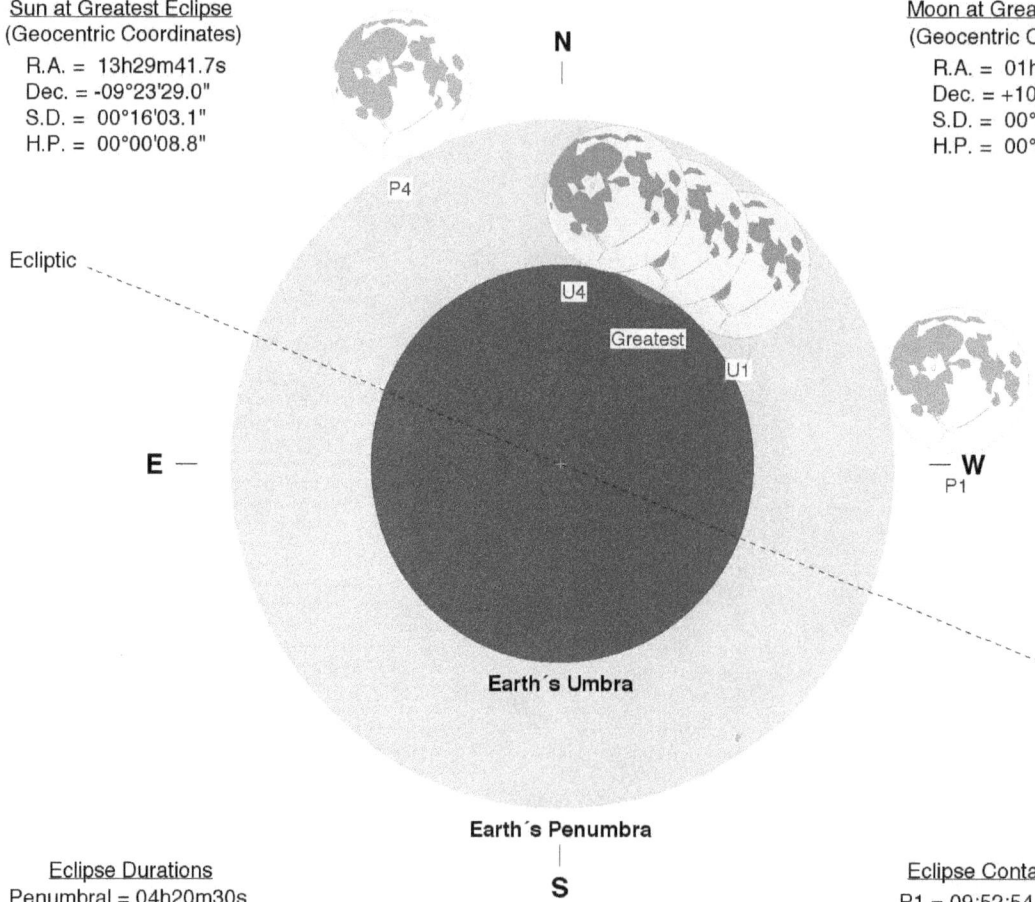

N

P4

Ecliptic

U4

Greatest

U1

E

W

P1

Earth's Umbra

Earth's Penumbra

S

Eclipse Durations
Penumbral = 04h20m30s
Umbral = 00h56m59s

Eph. = JPL DE430
Rule = Herald-Sinnott
ΔT = 65 s

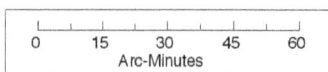

0 15 30 45 60
Arc-Minutes

©2020 F. Espenak, www.EclipseWise.com

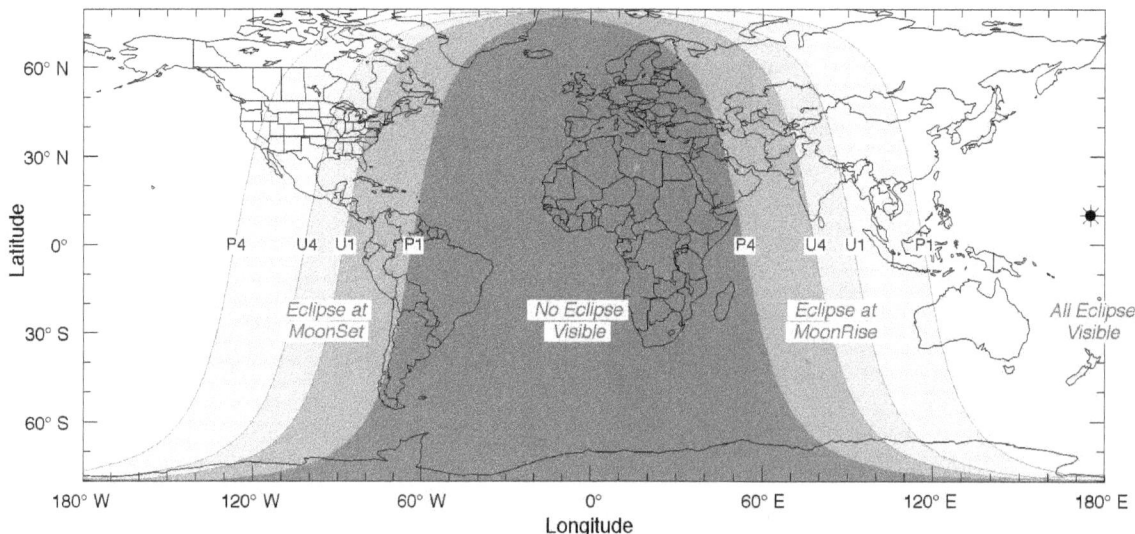

Eclipse Contacts
P1 = 09:52:54 UT1
U1 = 11:34:30 UT1
U4 = 12:31:29 UT1
P4 = 14:13:24 UT1

60° N

30° N

Latitude

0°

P4 U4 U1 P1

P4 U4 U1 P1

30° S

Eclipse at MoonSet

No Eclipse Visible

Eclipse at MoonRise

All Eclipse Visible

60° S

180° W   120° W   60° W   0°   60° E   120° E   180° E

Longitude

Penumbral Lunar Eclipse of 2006 Mar 14

Greatest Eclipse = 23:48:35.0 TD (= 23:47:30.1 UT1)

Penumbral Magnitude = 1.0320	Gamma = 1.0211	Saros Series = 113
Umbral Magnitude = -0.0583	Axis = 0.9212°	Saros Member = 63 of 71

Sun at Greatest Eclipse
(Geocentric Coordinates)

R.A. = 23h38m54.0s
Dec. = -02°16'57.9"
S.D. = 00°16'05.1"
H.P. = 00°00'08.8"

N

Moon at Greatest Eclipse
(Geocentric Coordinates)

R.A. = 11h40m41.4s
Dec. = +03°05'17.9"
S.D. = 00°14'45.1"
H.P. = 00°54'08.3"

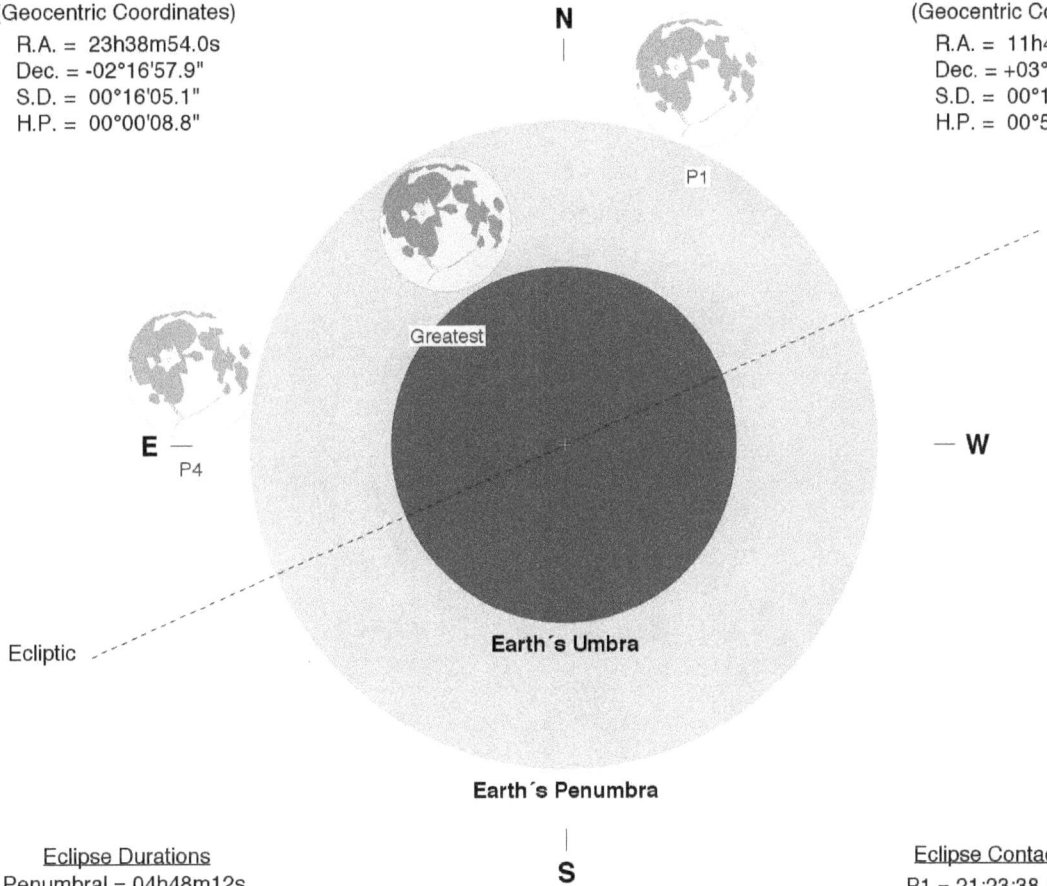

P1

Greatest

E
P4

— W

Ecliptic

Earth's Umbra

Earth's Penumbra

S

Eclipse Durations
Penumbral = 04h48m12s

Eclipse Contacts
P1 = 21:23:38 UT1
P4 = 02:11:50 UT1

Eph. = JPL DE430
Rule = Herald-Sinnott
ΔT =   65 s

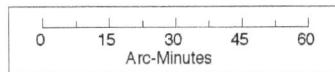

0	15	30	45	60

Arc-Minutes

©2020 F. Espenak, www.EclipseWise.com

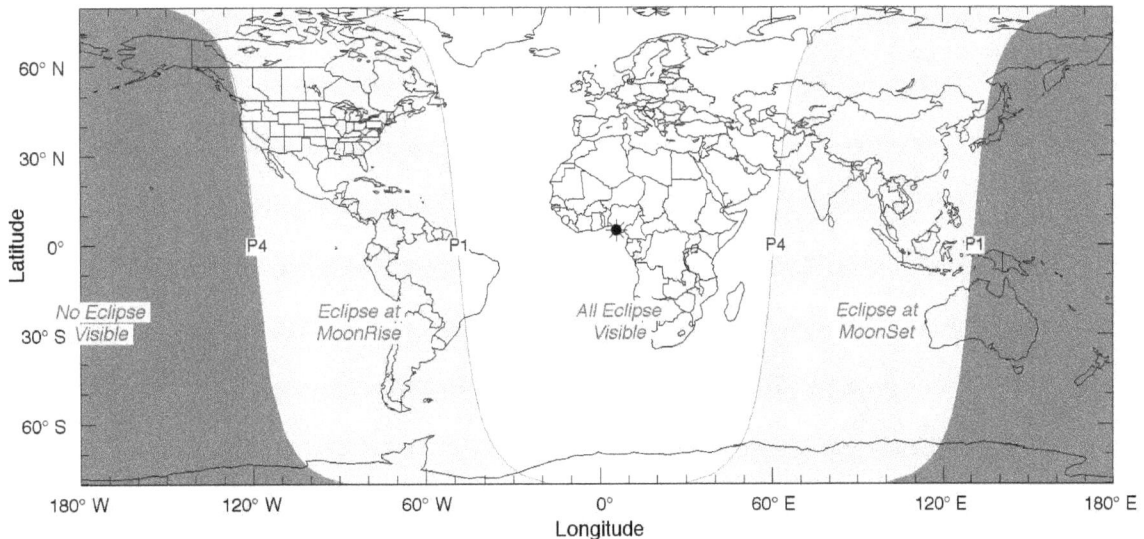

No Eclipse Visible

Eclipse at MoonRise

All Eclipse Visible

Eclipse at MoonSet

Partial Lunar Eclipse of 2006 Sep 07

Greatest Eclipse = 18:52:25.2 TD (= 18:51:20.1 UT1)

Penumbral Magnitude = 1.1349	Gamma = -0.9262	Saros Series = 118
Umbral Magnitude = 0.1857	Axis = 0.9473°	Saros Member = 51 of 73

Sun at Greatest Eclipse
(Geocentric Coordinates)
R.A. = 11h04m47.1s
Dec. = +05°54'23.1"
S.D. = 00°15'52.4"
H.P. = 00°00'08.7"

Moon at Greatest Eclipse
(Geocentric Coordinates)
R.A. = 23h06m35.6s
Dec. = -06°44'25.6"
S.D. = 00°16'43.3"
H.P. = 01°01'22.3"

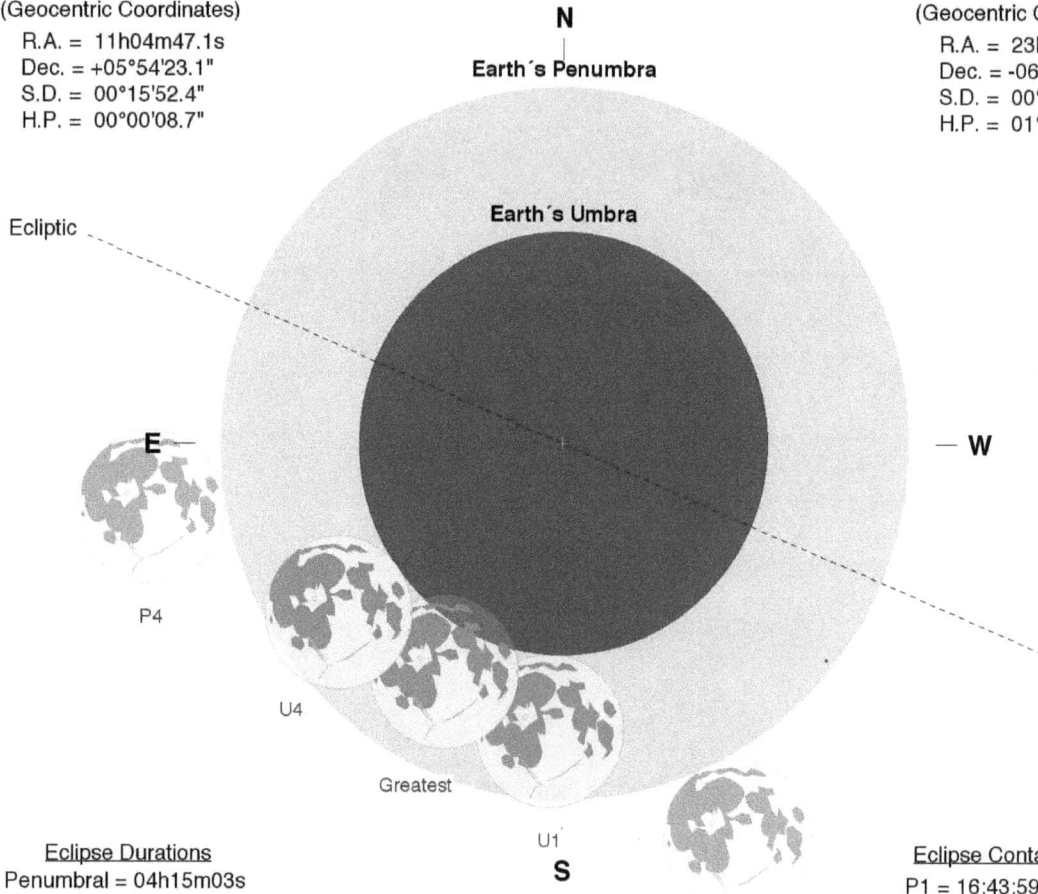

N

Earth's Penumbra

Ecliptic

Earth's Umbra

E

W

P4

U4

Greatest

U1

S

P1

Eclipse Durations
Penumbral = 04h15m03s
Umbral = 01h31m47s

Eph. = JPL DE430
Rule = Herald-Sinnott
ΔT = 65 s

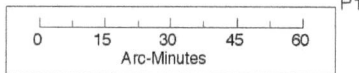

0   15   30   45   60
Arc-Minutes

©2020 F. Espenak, www.EclipseWise.com

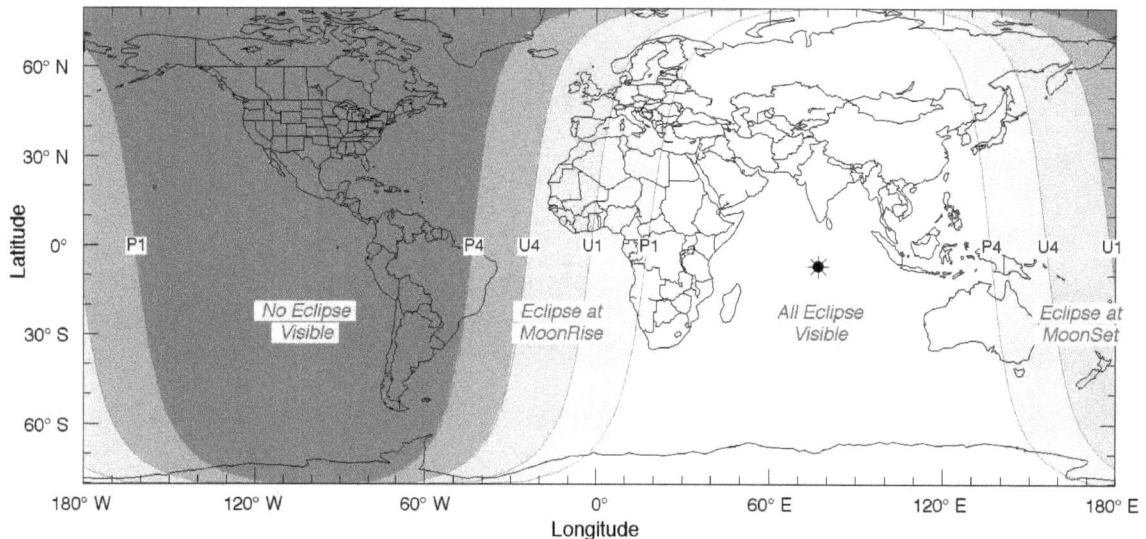

Eclipse Contacts
P1 = 16:43:59 UT1
U1 = 18:05:43 UT1
U4 = 19:37:30 UT1
P4 = 20:59:02 UT1

No Eclipse Visible

Eclipse at MoonRise

All Eclipse Visible

Eclipse at MoonSet

P1   P4   U4   U1   P1   P4   U4   U1

Latitude

Longitude

Total Lunar Eclipse of 2007 Mar 03

Greatest Eclipse = 23:21:58.7 TD   (= 23:20:53.5 UT1)

Penumbral Magnitude = 2.3208	Gamma = 0.3175	Saros Series = 123
Umbral Magnitude = 1.2347	Axis = 0.2884°	Saros Member = 52 of 72

Sun at Greatest Eclipse
(Geocentric Coordinates)

R.A. = 22h57m19.2s
Dec. = -06°40'46.3"
S.D. = 00°16'08.0"
H.P. = 00°00'08.9"

Moon at Greatest Eclipse
(Geocentric Coordinates)

R.A. = 10h57m52.2s
Dec. = +06°56'00.7"
S.D. = 00°14'51.3"
H.P. = 00°54'31.1"

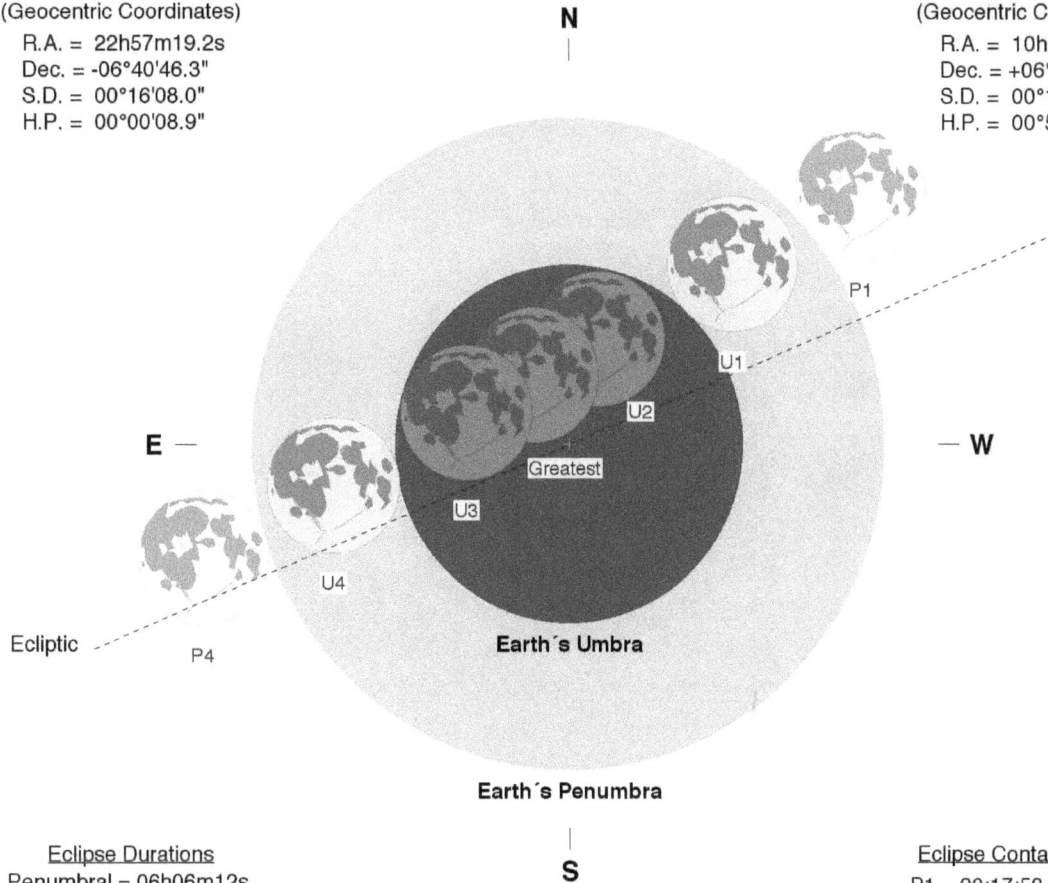

N

E — — W

Greatest

U1
U2
U3
U4
P1
P4

Earth's Umbra

Earth's Penumbra

Ecliptic

S

Eclipse Durations
Penumbral = 06h06m12s
Umbral = 03h41m46s
Total = 01h14m01s

Eph. = JPL DE430
Rule = Herald-Sinnott
ΔT = 65 s

0   15   30   45   60
Arc-Minutes

©2020 F. Espenak, www.EclipseWise.com

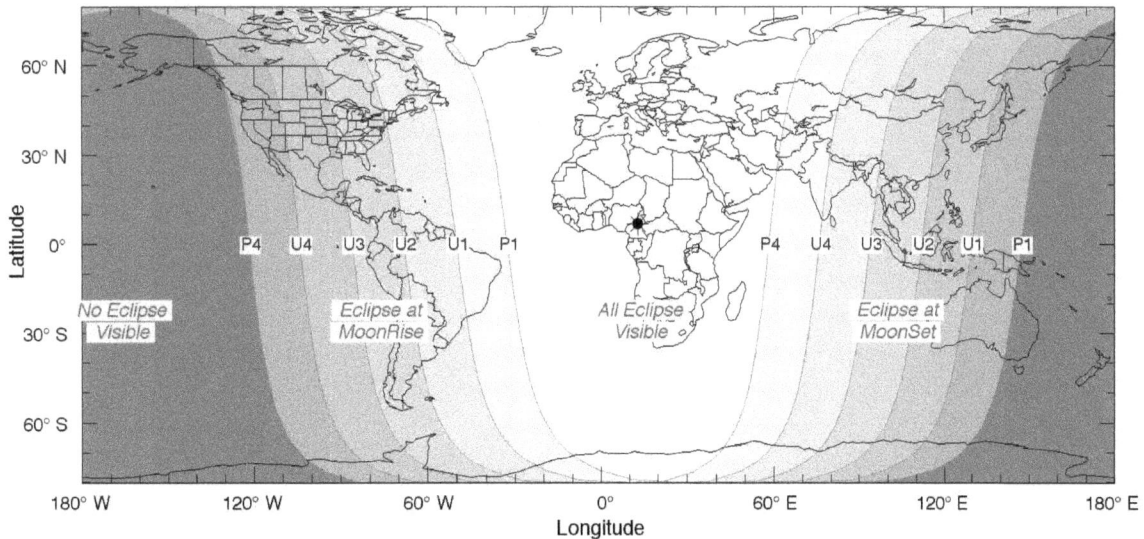

Eclipse Contacts
P1 = 20:17:53 UT1
U1 = 21:30:06 UT1
U2 = 22:44:06 UT1
U3 = 23:58:07 UT1
U4 = 01:11:52 UT1
P4 = 02:24:05 UT1

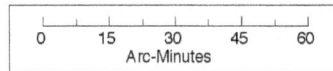

No Eclipse Visible

Eclipse at MoonRise

All Eclipse Visible

Eclipse at MoonSet

P4 U4 U3 U2 U1 P1   P4 U4 U3 U2 U1 P1

180° W   120° W   60° W   0°   60° E   120° E   180° E

Longitude

60° N
30° N
0°
30° S
60° S

Latitude

Total Lunar Eclipse of 2007 Aug 28

Greatest Eclipse = 10:38:26.8 TD   (= 10:37:21.4 UT1)

Penumbral Magnitude = 2.4545	Gamma = −0.2146	Saros Series = 128
Umbral Magnitude = 1.4777	Axis = 0.2127°	Saros Member = 40 of 71

Sun at Greatest Eclipse
(Geocentric Coordinates)

R.A. = 10h26m26.9s
Dec. = +09°45'56.7"
S.D. = 00°15'50.0"
H.P. = 00°00'08.7"

Moon at Greatest Eclipse
(Geocentric Coordinates)

R.A. = 22h26m50.4s
Dec. = −09°57'18.5"
S.D. = 00°16'12.5"
H.P. = 00°59'29.2"

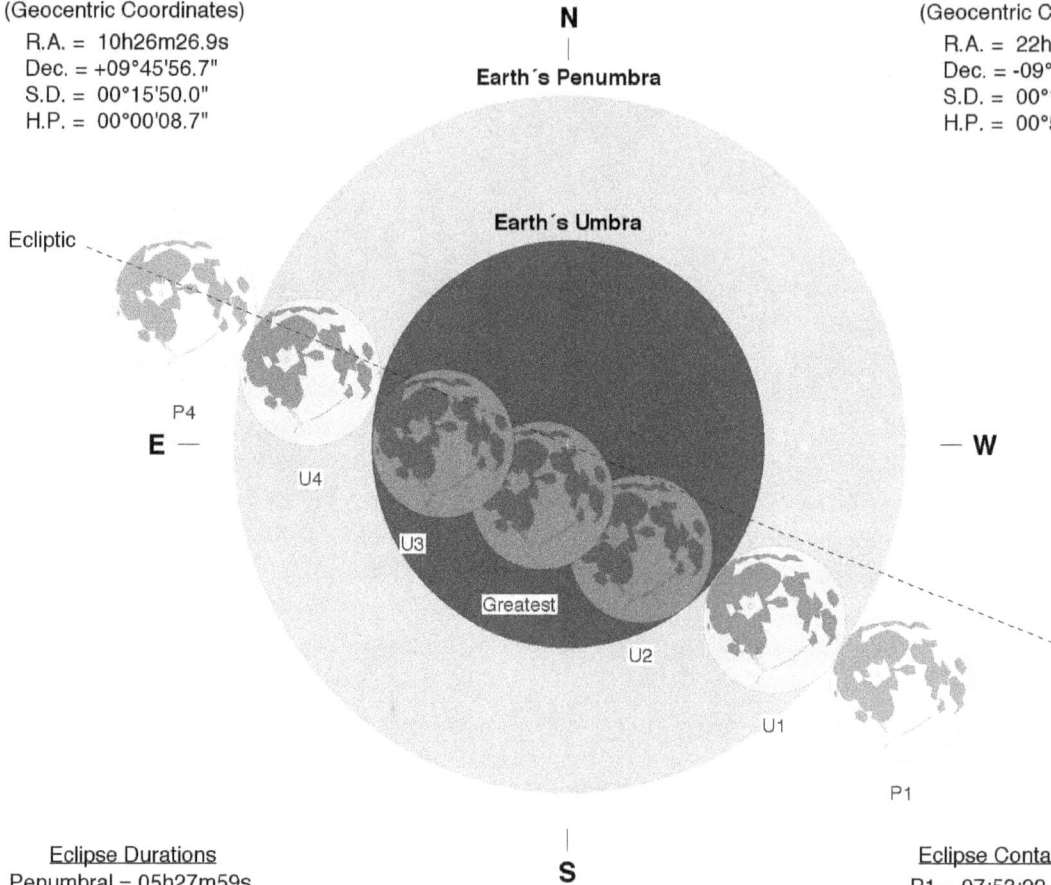

N

Earth's Penumbra

Earth's Umbra

Ecliptic

E

W

P4

U4

U3

Greatest

U2

U1

P1

S

Eclipse Durations
Penumbral = 05h27m59s
Umbral = 03h32m51s
Total = 01h30m38s

Eph. = JPL DE430
Rule = Herald-Sinnott
ΔT =   65 s

0   15   30   45   60
Arc-Minutes

©2020 F. Espenak, www.EclipseWise.com

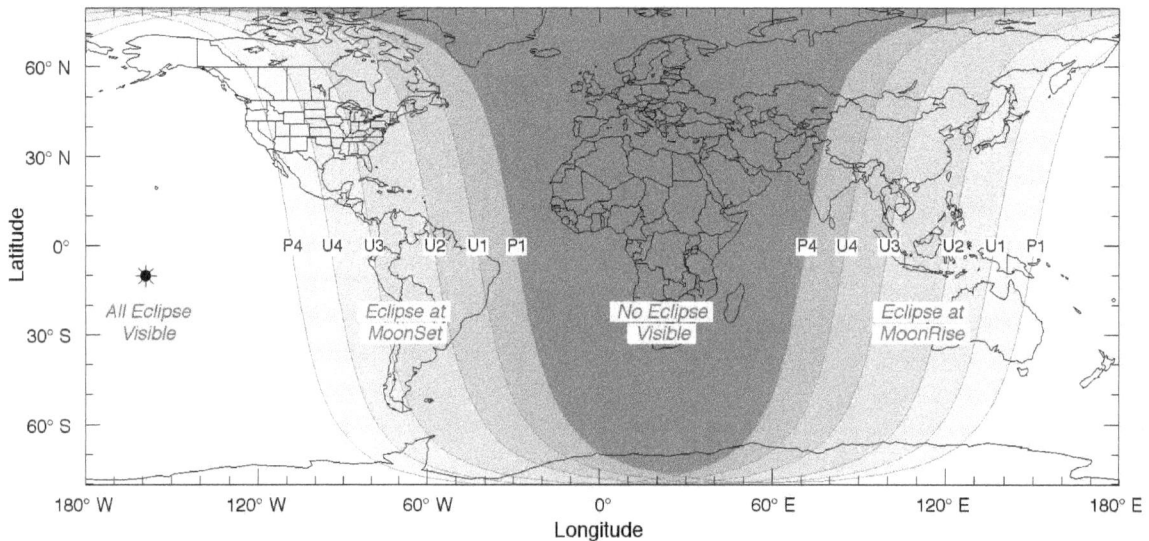

Eclipse Contacts
P1 = 07:53:22 UT1
U1 = 08:51:01 UT1
U2 = 09:52:11 UT1
U3 = 11:22:49 UT1
U4 = 12:23:51 UT1
P4 = 13:21:21 UT1

All Eclipse Visible

Eclipse at MoonSet

No Eclipse Visible

Eclipse at MoonRise

P4 U4 U3   U2 U1 P1

P4 U4 U3   U2 U1 P1

Longitude

Latitude

Total Lunar Eclipse of 2008 Feb 21

Greatest Eclipse = 03:27:08.8 TD   (= 03:26:03.3 UT1)

Penumbral Magnitude = 2.1470	Gamma = -0.3992	Saros Series = 133
Umbral Magnitude = 1.1081	Axis = 0.3802°	Saros Member = 26 of 71

Sun at Greatest Eclipse
(Geocentric Coordinates)

R.A. = 22h15m30.0s
Dec. = -10°48'31.3"
S.D. = 00°16'10.5"
H.P. = 00°00'08.9"

Moon at Greatest Eclipse
(Geocentric Coordinates)

R.A. = 10h14m48.5s
Dec. = +10°28'07.6"
S.D. = 00°15'34.2"
H.P. = 00°57'08.5"

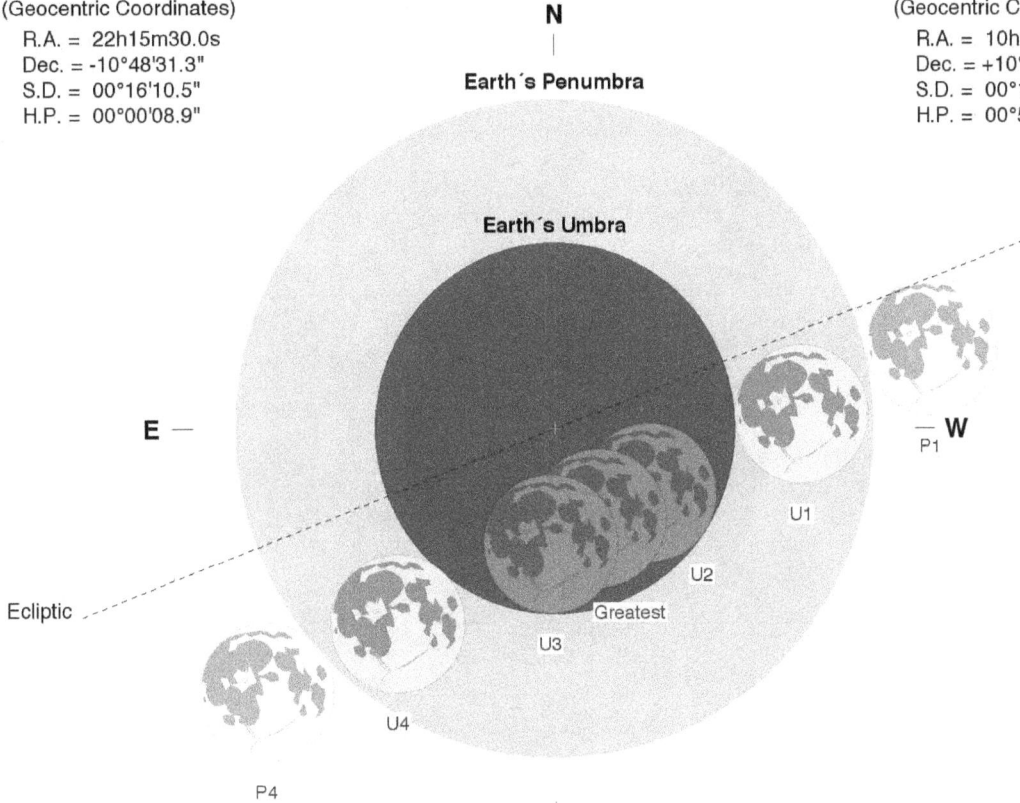

N

Earth's Penumbra

Earth's Umbra

E

W

P1

U1

U2

Greatest

U3

U4

P4

Ecliptic

S

Eclipse Durations
Penumbral = 05h39m43s
Umbral = 03h26m07s
Total = 00h50m27s

Eph. = JPL DE430
Rule = Herald-Sinnott
ΔT =   66 s

0   15   30   45   60
Arc-Minutes

©2020 F. Espenak, www.EclipseWise.com

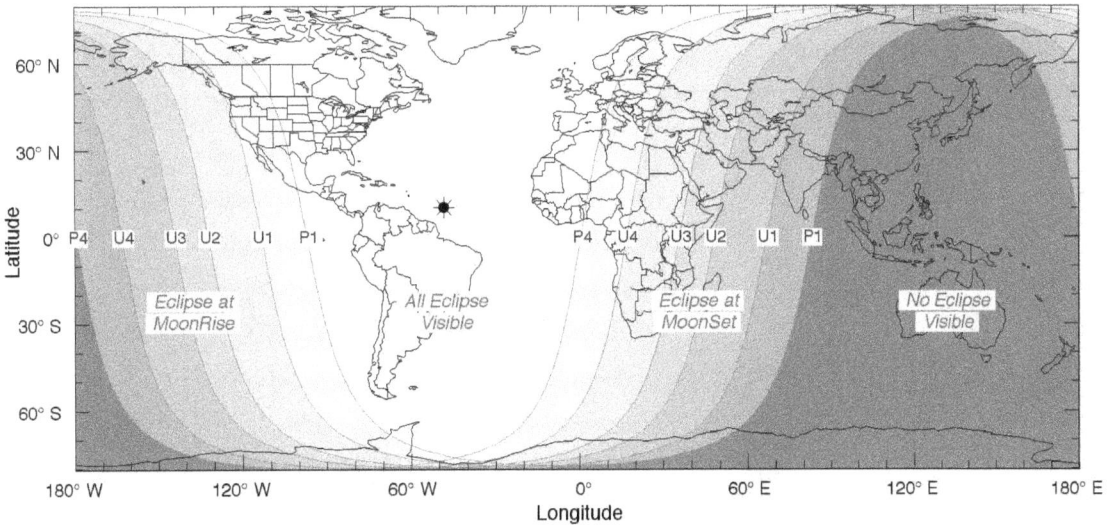

Eclipse Contacts
P1 = 00:36:10 UT1
U1 = 01:42:51 UT1
U2 = 03:00:33 UT1
U3 = 03:51:00 UT1
U4 = 05:08:58 UT1
P4 = 06:15:53 UT1

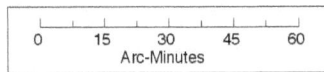

Partial Lunar Eclipse of 2008 Aug 16

Greatest Eclipse = 21:11:12.0 TD (= 21:10:06.4 UT1)

Penumbral Magnitude = 1.8385	Gamma = 0.5646	Saros Series = 138
Umbral Magnitude = 0.8095	Axis = 0.5302°	Saros Member = 28 of 82

Sun at Greatest Eclipse
(Geocentric Coordinates)

R.A. = 09h46m37.2s
Dec. = +13°24'18.2"
S.D. = 00°15'47.9"
H.P. = 00°00'08.7"

Moon at Greatest Eclipse
(Geocentric Coordinates)

R.A. = 21h45m41.8s
Dec. = -12°55'29.2"
S.D. = 00°15'21.1"
H.P. = 00°56'20.6"

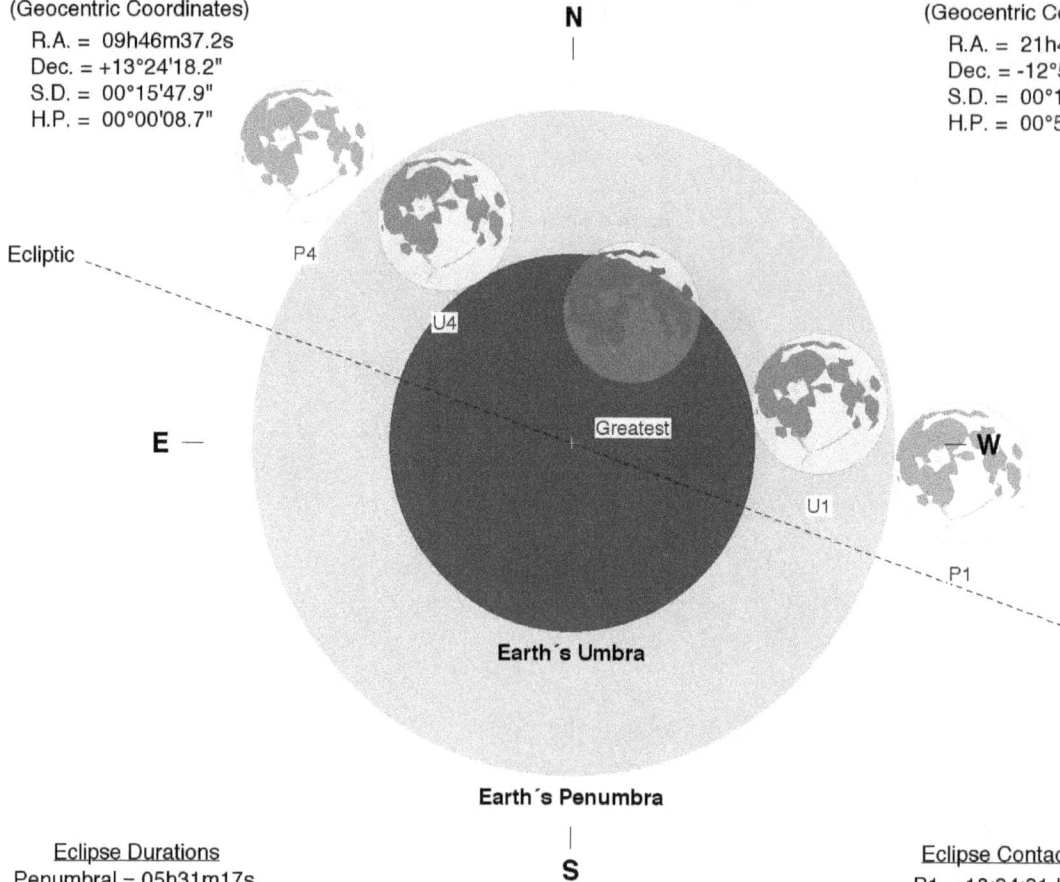

N

Ecliptic

P4

U4

Greatest

E

W

U1

P1

Earth's Umbra

Earth's Penumbra

S

Eclipse Durations
Penumbral = 05h31m17s
Umbral = 03h08m49s

Eph. = JPL DE430
Rule = Herald-Sinnott
ΔT =   66 s

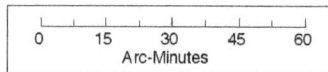

Eclipse Contacts
P1 = 18:24:21 UT1
U1 = 19:35:35 UT1
U4 = 22:44:24 UT1
P4 = 23:55:37 UT1

0   15   30   45   60
Arc-Minutes

©2020 F. Espenak, www.EclipseWise.com

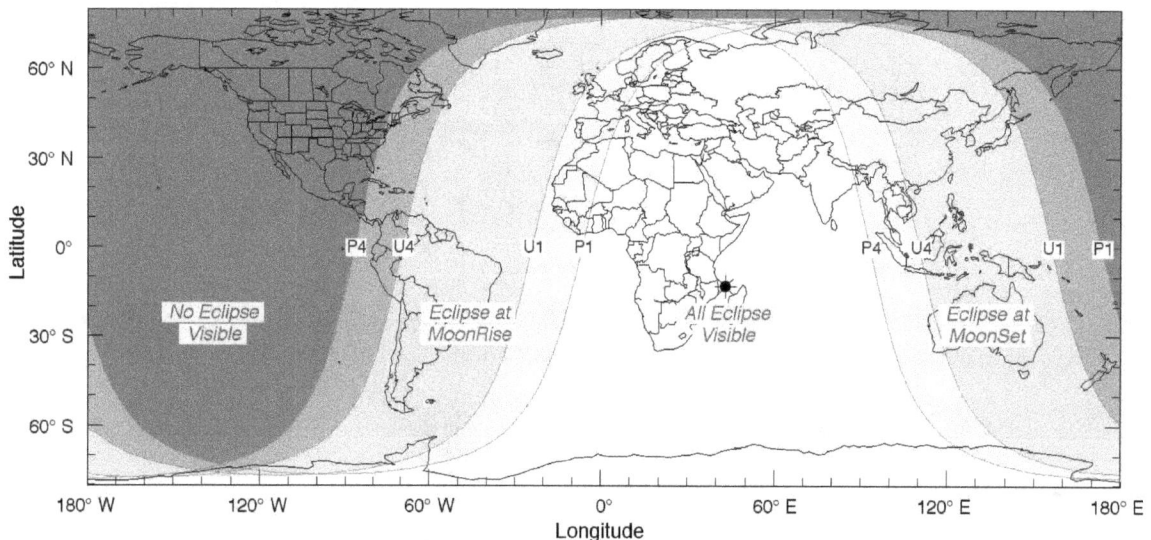

No Eclipse
Visible

Eclipse at
MoonRise

All Eclipse
Visible

Eclipse at
MoonSet

P4   U4'   U1   P1   P4   U4   U1   P1

Latitude

Longitude

Penumbral Lunar Eclipse of 2009 Feb 09

Greatest Eclipse = 14:39:21.2 TD (= 14:38:15.4 UT1)

Penumbral Magnitude = 0.9013	Gamma = -1.0640	Saros Series = 143
Umbral Magnitude = -0.0863	Axis = 1.0681°	Saros Member = 17 of 72

Sun at Greatest Eclipse
(Geocentric Coordinates)

R.A. = 21h33m30.0s
Dec. = -14°30'07.1"
S.D. = 00°16'12.6"
H.P. = 00°00'08.9"

Moon at Greatest Eclipse
(Geocentric Coordinates)

R.A. = 09h31m42.1s
Dec. = +13°31'37.5"
S.D. = 00°16'24.8"
H.P. = 01°00'14.2"

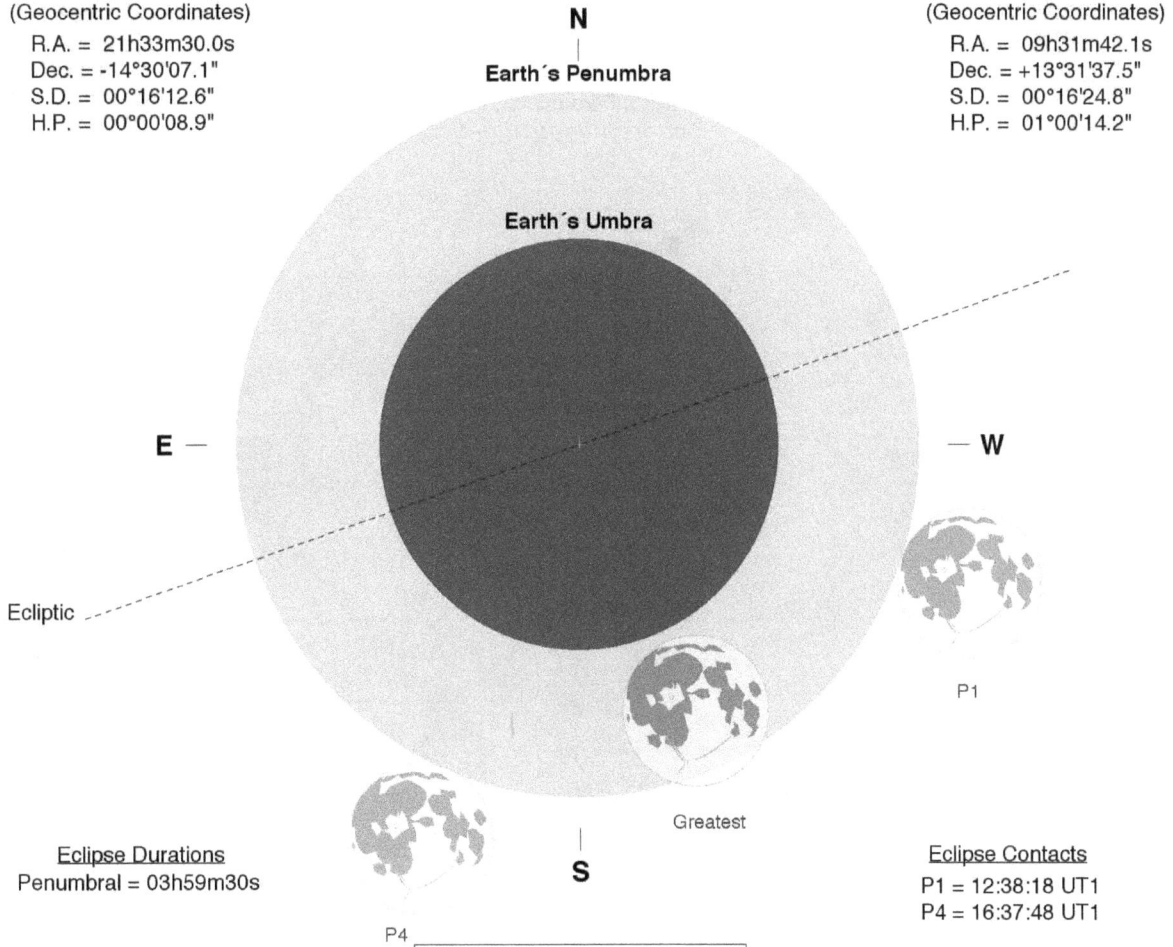

N

Earth's Penumbra

Earth's Umbra

E

W

Ecliptic

P1

Greatest

S

P4

Eclipse Durations
Penumbral = 03h59m30s

Eclipse Contacts
P1 = 12:38:18 UT1
P4 = 16:37:48 UT1

Eph. = JPL DE430
Rule = Herald-Sinnott
ΔT = 66 s

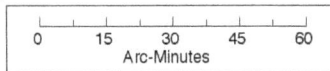

0   15   30   45   60
Arc-Minutes

©2020 F. Espenak, www.EclipseWise.com

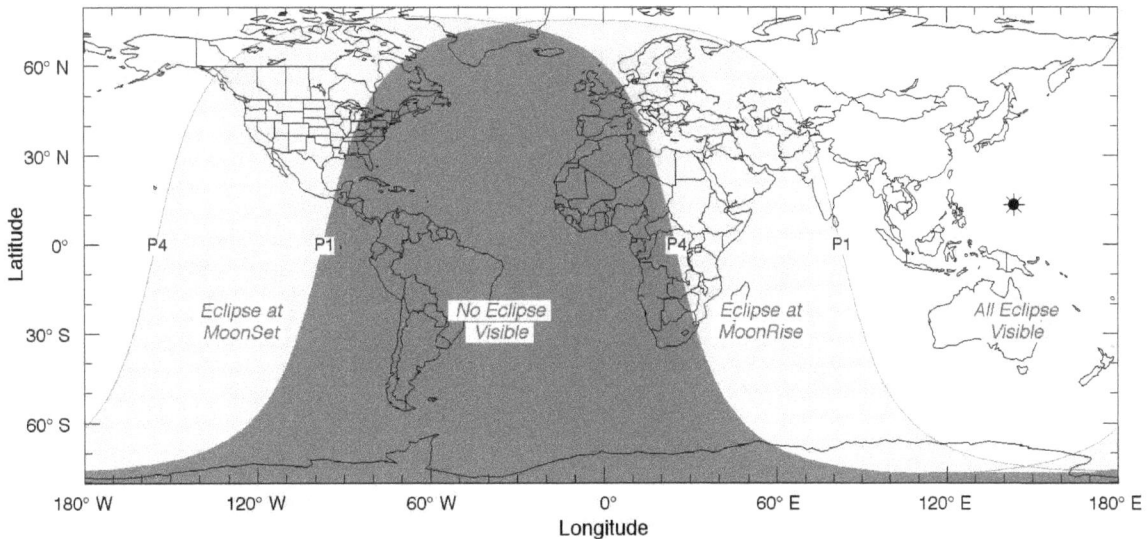

Penumbral Lunar Eclipse of 2009 Jul 07

Greatest Eclipse = 09:39:43.9 TD (= 09:38:38.0 UT1)

Penumbral Magnitude = 0.1578	Gamma = -1.4916	Saros Series = 110
Umbral Magnitude = -0.9116	Axis = 1.3420°	Saros Member = 71 of 72

Sun at Greatest Eclipse
(Geocentric Coordinates)

R.A. = 07h06m54.1s
Dec. = +22°32'55.2"
S.D. = 00°15'43.9"
H.P. = 00°00'08.7"

N

Earth's Penumbra

Earth's Umbra

Moon at Greatest Eclipse
(Geocentric Coordinates)

R.A. = 19h08m08.1s
Dec. = -23°51'38.0"
S.D. = 00°14'42.6"
H.P. = 00°53'59.3"

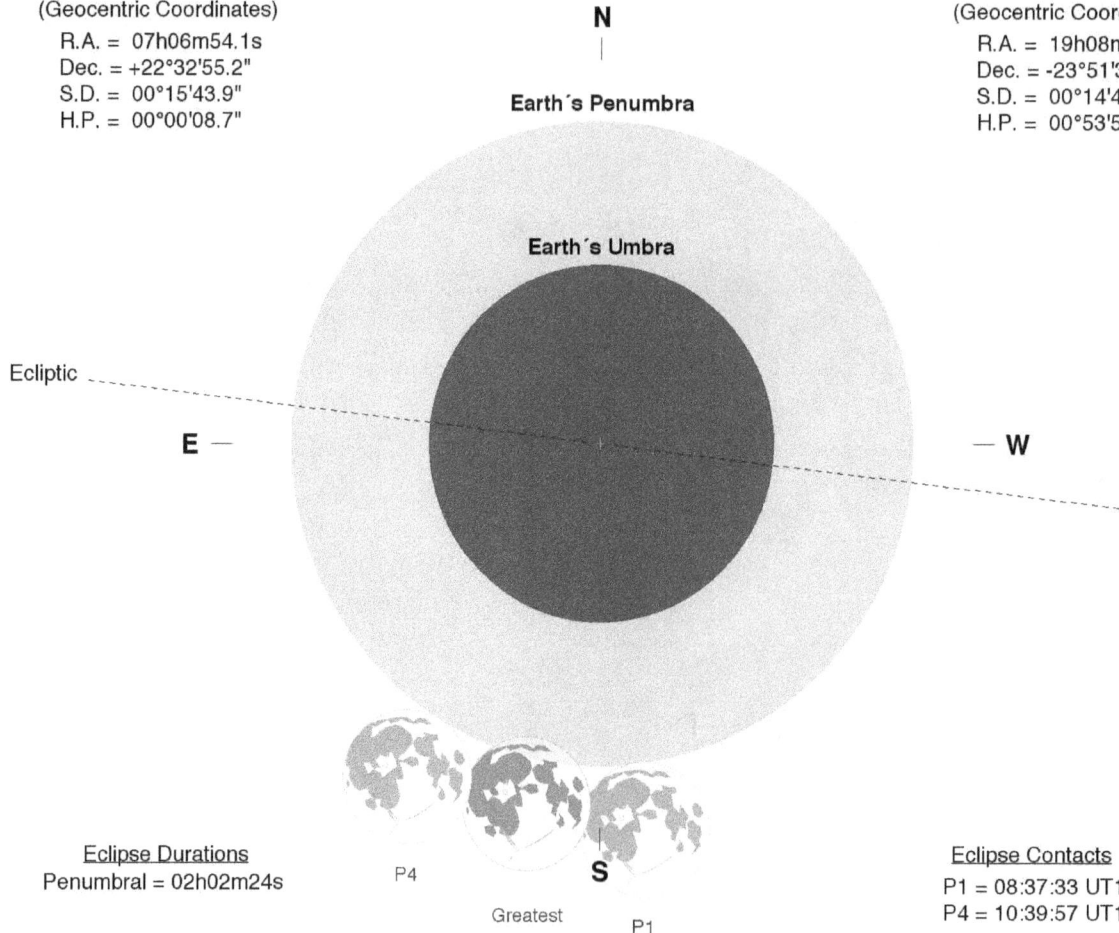

Ecliptic

E —

— **W**

S

P4

Greatest

P1

Eclipse Durations
Penumbral = 02h02m24s

Eph. = JPL DE430
Rule = Herald-Sinnott
ΔT =   66 s

0	15	30	45	60

Arc-Minutes

©2020 F. Espenak, www.EclipseWise.com

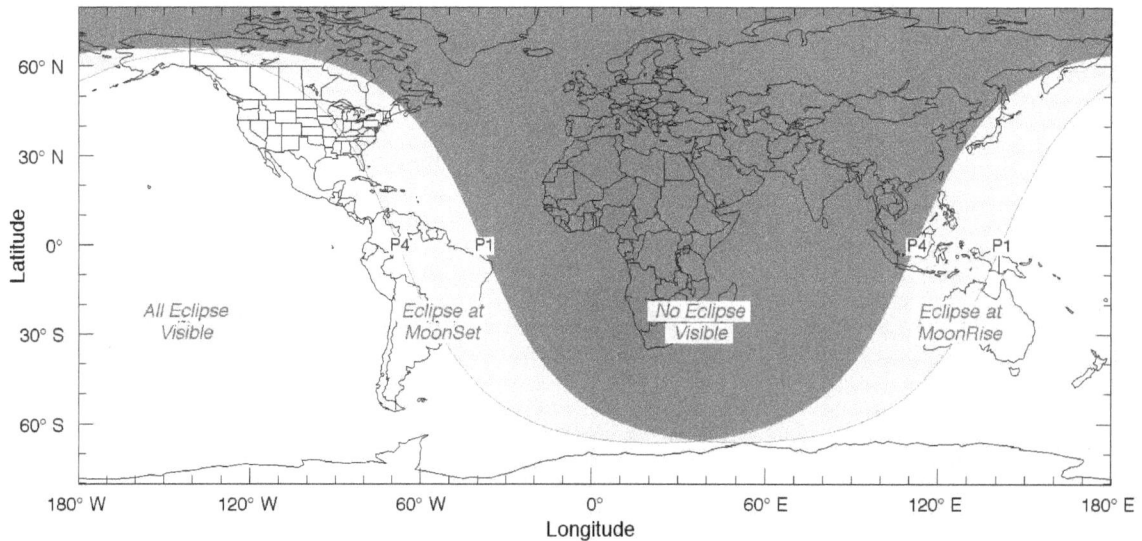

Eclipse Contacts
P1 = 08:37:33 UT1
P4 = 10:39:57 UT1

All Eclipse Visible

Eclipse at MoonSet

No Eclipse Visible

Eclipse at MoonRise

P4   P1

P4   P1

Penumbral Lunar Eclipse of 2009 Aug 06

Greatest Eclipse = 00:40:16.4 TD   (= 00:39:10.4 UT1)

Penumbral Magnitude = 0.4038	Gamma = 1.3572	Saros Series = 148
Umbral Magnitude = -0.6642	Axis = 1.2257°	Saros Member = 3 of 70

Sun at Greatest Eclipse
(Geocentric Coordinates)

R.A. = 09h04m42.0s
Dec. = +16°42'38.9"
S.D. = 00°15'46.1"
H.P. = 00°00'08.7"

Moon at Greatest Eclipse
(Geocentric Coordinates)

R.A. = 21h02m46.3s
Dec. = -15°34'32.9"
S.D. = 00°14'45.9"
H.P. = 00°54'11.4"

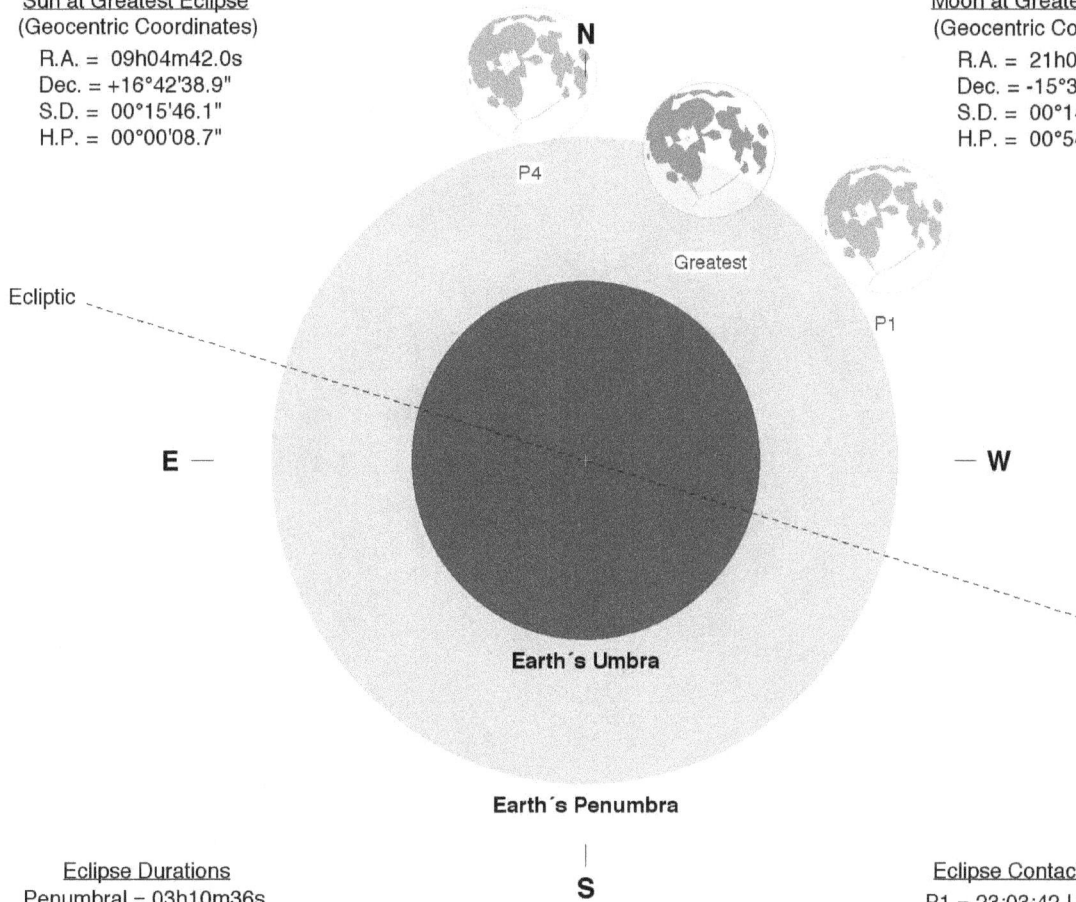

N

P4

Greatest

P1

Ecliptic

E

W

Earth's Umbra

Earth's Penumbra

S

Eclipse Durations
Penumbral = 03h10m36s

Eclipse Contacts
P1 = 23:03:42 UT1
P4 = 02:14:18 UT1

Eph. = JPL DE430
Rule = Herald-Sinnott
ΔT =   66 s

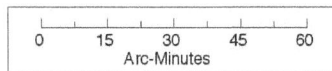

0   15   30   45   60
Arc-Minutes

©2020 F. Espenak, www.EclipseWise.com

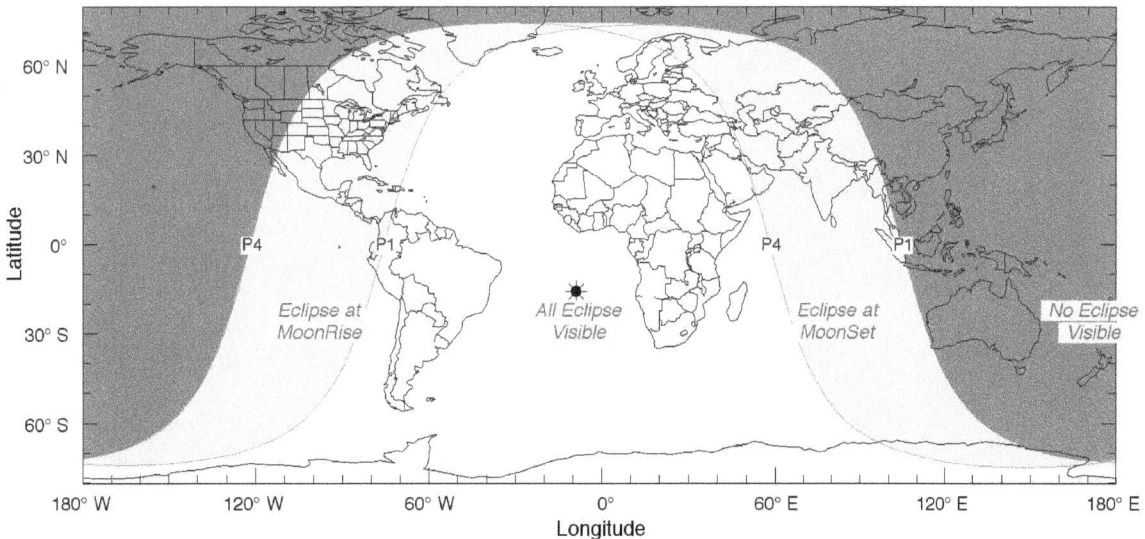

Eclipse at MoonRise

All Eclipse Visible

Eclipse at MoonSet

No Eclipse Visible

Partial Lunar Eclipse of 2009 Dec 31

Greatest Eclipse = 19:23:46.2 TD   (= 19:22:40.1 UT1)

Penumbral Magnitude = 1.0572	Gamma = 0.9766	Saros Series = 115
Umbral Magnitude = 0.0779	Axis = 0.9921°	Saros Member = 57 of 72

Sun at Greatest Eclipse
(Geocentric Coordinates)

R.A. = 18h44m37.2s
Dec. = -23°02'33.1"
S.D. = 00°16'15.9"
H.P. = 00°00'08.9"

Moon at Greatest Eclipse
(Geocentric Coordinates)

R.A. = 06h45m22.4s
Dec. = +24°01'10.3"
S.D. = 00°16'36.6"
H.P. = 01°00'57.6"

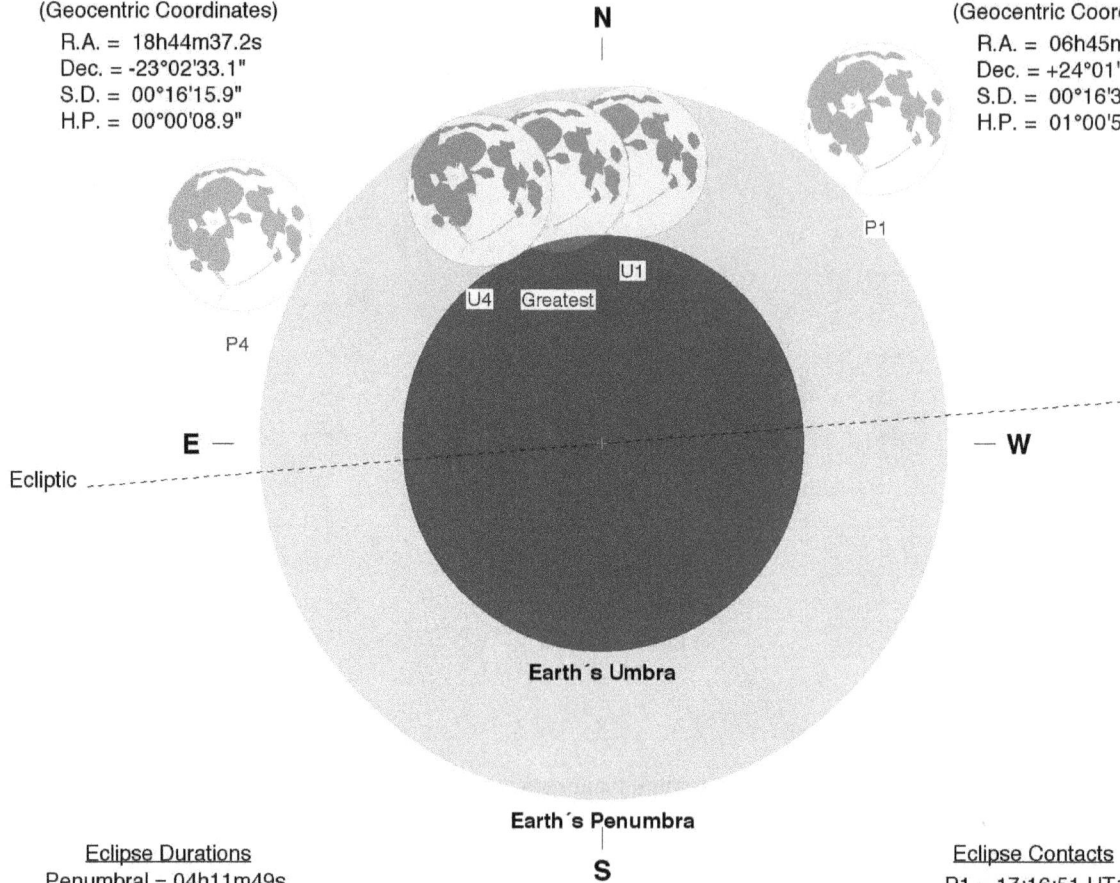

N

P1

U1

U4   Greatest

P4

E —

Ecliptic

— W

Earth's Umbra

Earth's Penumbra

S

Eclipse Durations
Penumbral = 04h11m49s
Umbral = 01h00m47s

Eph. = JPL DE430
Rule = Herald-Sinnott
ΔT =   66 s

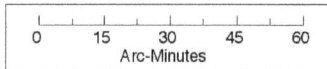

Eclipse Contacts
P1 = 17:16:51 UT1
U1 = 18:52:26 UT1
U4 = 19:53:12 UT1
P4 = 21:28:39 UT1

0	15	30	45	60

Arc-Minutes

©2020 F. Espenak, www.EclipseWise.com

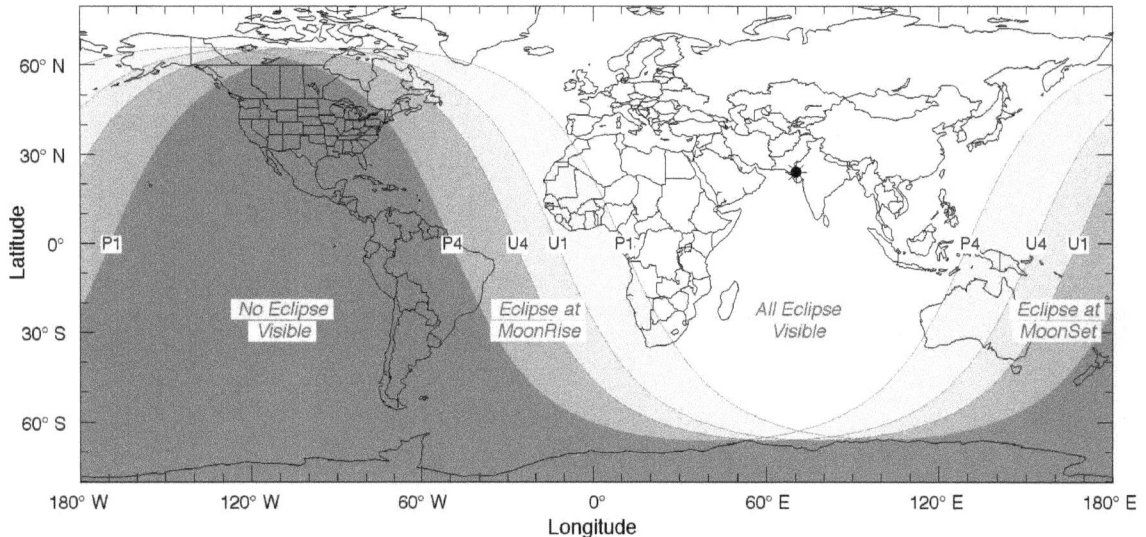

No Eclipse Visible

Eclipse at MoonRise

All Eclipse Visible

Eclipse at MoonSet

Partial Lunar Eclipse of 2010 Jun 26

Greatest Eclipse = 11:39:34.2 TD (= 11:38:27.9 UT1)

Penumbral Magnitude = 1.5789	Gamma = -0.7091	Saros Series = 120
Umbral Magnitude = 0.5383	Axis = 0.6558°	Saros Member = 57 of 83

Sun at Greatest Eclipse
(Geocentric Coordinates)

R.A. = 06h20m48.6s
Dec. = +23°21'07.6"
S.D. = 00°15'44.1"
H.P. = 00°00'08.7"

Moon at Greatest Eclipse
(Geocentric Coordinates)

R.A. = 18h21m11.8s
Dec. = -24°00'06.9"
S.D. = 00°15'07.3"
H.P. = 00°55'29.7"

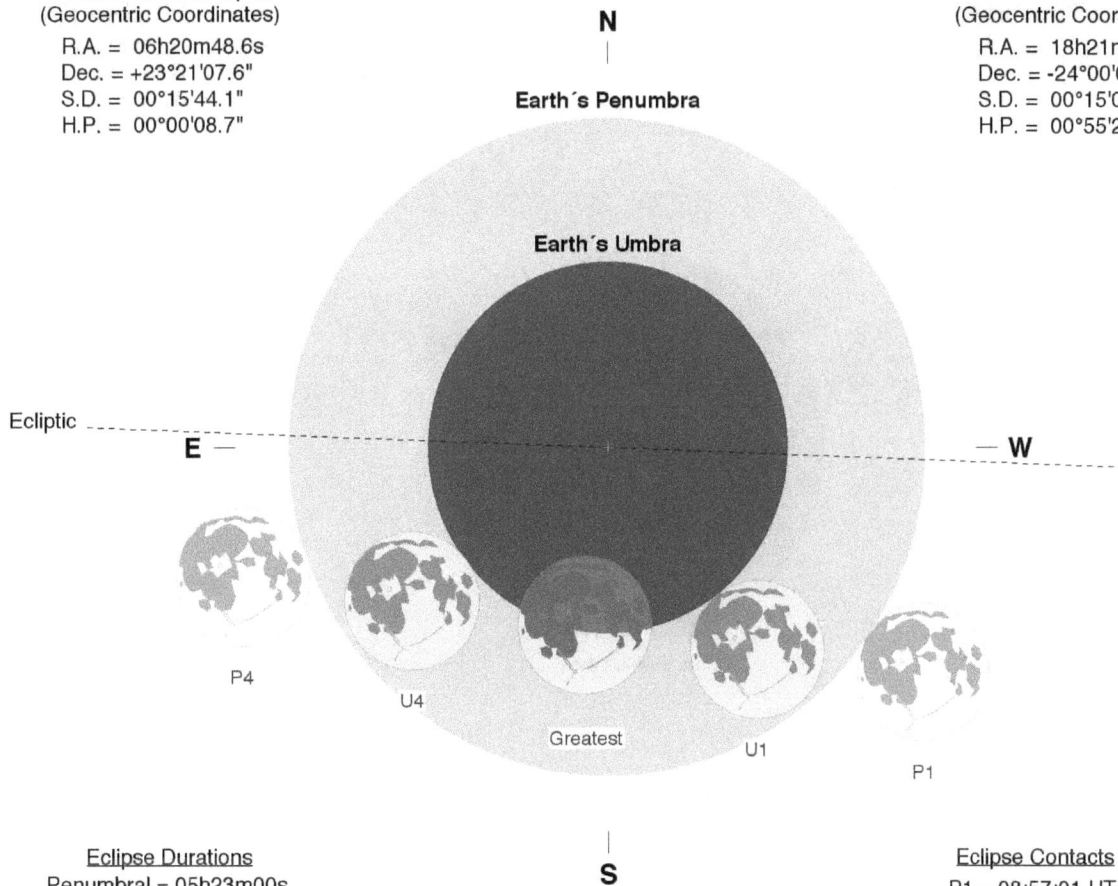

N

Earth's Penumbra

Earth's Umbra

Ecliptic

E

W

P4

U4

Greatest

U1

P1

S

Eclipse Durations
Penumbral = 05h23m00s
Umbral = 02h43m36s

Eph. = JPL DE430
Rule = Herald-Sinnott
ΔT =   66 s

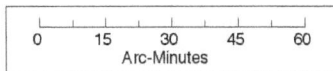

0	15	30	45	60

Arc-Minutes

©2020 F. Espenak, www.EclipseWise.com

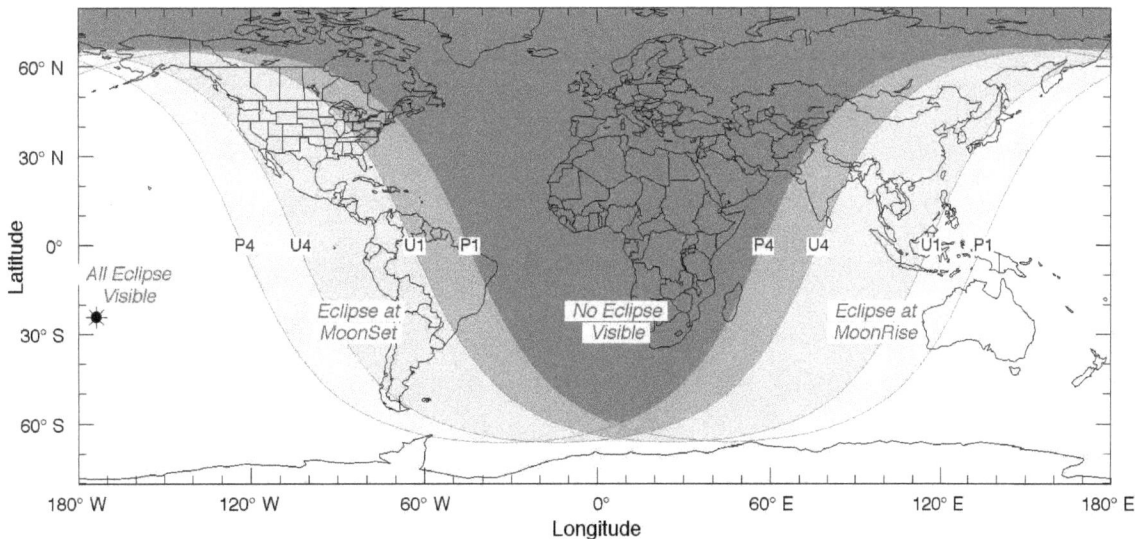

Eclipse Contacts
P1 = 08:57:01 UT1
U1 = 10:16:40 UT1
U4 = 13:00:16 UT1
P4 = 14:20:01 UT1

All Eclipse
Visible

Eclipse at
MoonSet

No Eclipse
Visible

Eclipse at
MoonRise

Latitude
60° N
30° N
0°
30° S
60° S

Longitude
180° W   120° W   60° W   0°   60° E   120° E   180° E

P4 U4 U1 P1   P4 U4 U1 P1

Total Lunar Eclipse of 2010 Dec 21

Greatest Eclipse = 08:18:04.2 TD   (= 08:16:57.8 UT1)

Penumbral Magnitude = 2.2821	Gamma = 0.3214	Saros Series = 125
Umbral Magnitude = 1.2576	Axis = 0.3119°	Saros Member = 48 of 72

Sun at Greatest Eclipse
(Geocentric Coordinates)

R.A. = 17h57m09.6s
Dec. = -23°26'09.9"
S.D. = 00°16'15.5"
H.P. = 00°00'08.9"

Moon at Greatest Eclipse
(Geocentric Coordinates)

R.A. = 05h57m17.3s
Dec. = +23°44'47.8"
S.D. = 00°15'52.1"
H.P. = 00°58'14.3"

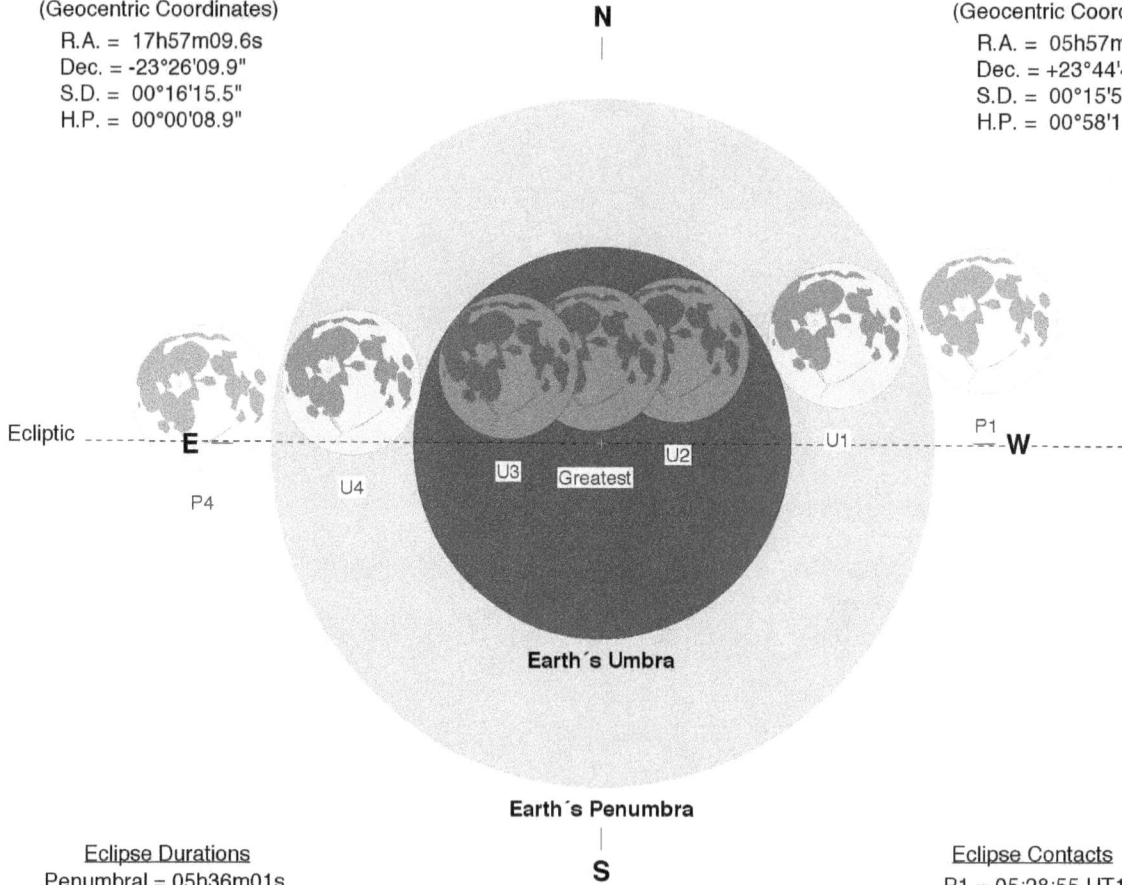

N

Ecliptic

E

W

P1

U1

U2

U3

Greatest

U4

P4

Earth's Umbra

Earth's Penumbra

S

Eclipse Durations

Penumbral = 05h36m01s
Umbral = 03h29m29s
Total = 01h13m02s

Eph. = JPL DE430
Rule = Herald-Sinnott
ΔT =   66 s

0   15   30   45   60
Arc-Minutes

©2020 F. Espenak, www.EclipseWise.com

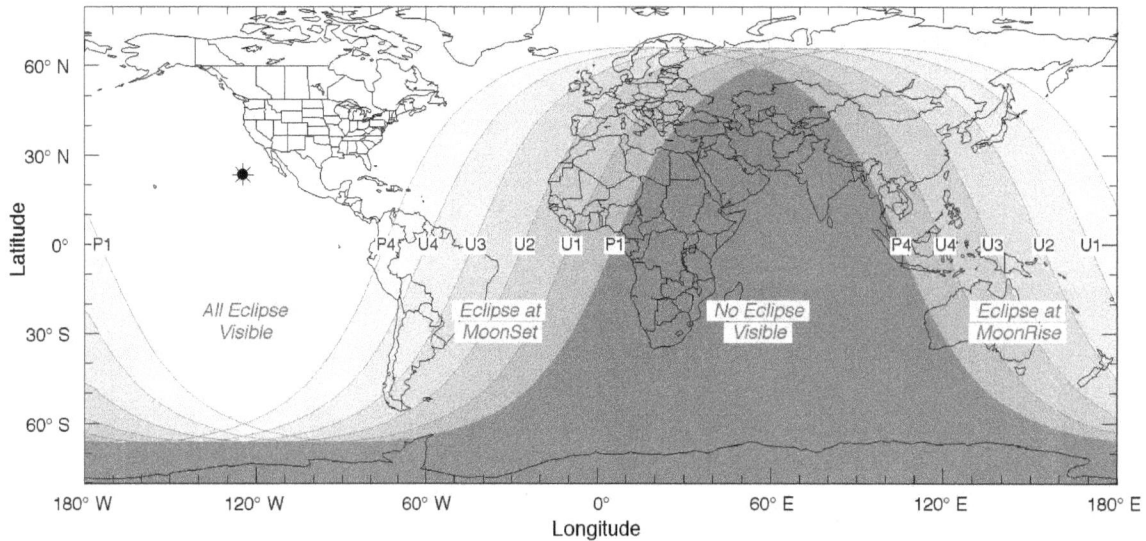

Eclipse Contacts

P1 = 05:28:55 UT1
U1 = 06:32:16 UT1
U2 = 07:40:30 UT1
U3 = 08:53:33 UT1
U4 = 10:01:45 UT1
P4 = 11:04:56 UT1

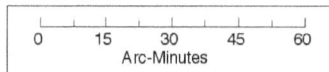

All Eclipse Visible

Eclipse at MoonSet

No Eclipse Visible

Eclipse at MoonRise

P1 P4 U4 U3 U2 U1 P1   P4 U4 U3 U2 U1

Total Lunar Eclipse of 2011 Jun 15

Greatest Eclipse = 20:13:43.4 TD (= 20:12:36.9 UT1)

Penumbral Magnitude = 2.6883	Gamma = 0.0897	Saros Series = 130
Umbral Magnitude = 1.7014	Axis = 0.0875°	Saros Member = 34 of 71

Sun at Greatest Eclipse
(Geocentric Coordinates)
R.A. = 05h35m33.6s
Dec. = +23°19'06.1"
S.D. = 00°15'44.7"
H.P. = 00°00'08.7"

N

Moon at Greatest Eclipse
(Geocentric Coordinates)
R.A. = 17h35m32.3s
Dec. = -23°13'51.6"
S.D. = 00°15'57.2"
H.P. = 00°58'33.0"

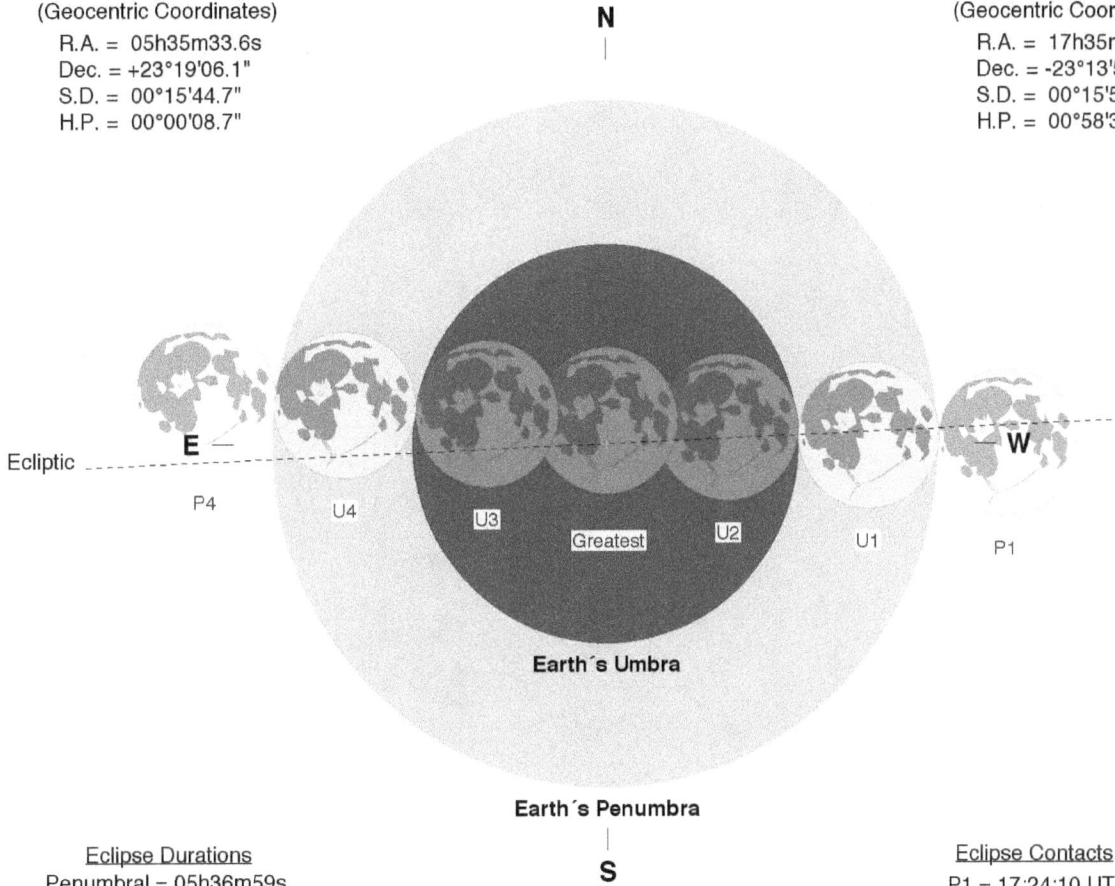

Ecliptic

E

P4 U4 U3 Greatest U2 U1 P1

W

Earth's Umbra

Earth's Penumbra

S

Eclipse Durations
Penumbral = 05h36m59s
Umbral = 03h40m06s
Total = 01h40m59s

Eph. = JPL DE430
Rule = Herald-Sinnott
ΔT =   67 s

0   15   30   45   60
Arc-Minutes

©2020 F. Espenak, www.EclipseWise.com

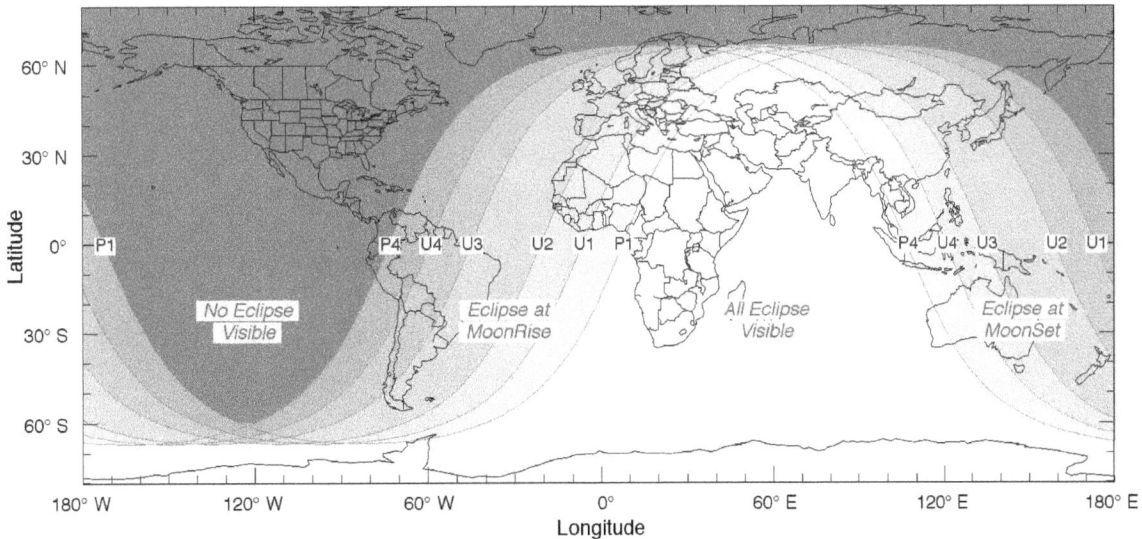

Eclipse Contacts
P1 = 17:24:10 UT1
U1 = 18:22:33 UT1
U2 = 19:22:06 UT1
U3 = 21:03:06 UT1
U4 = 22:02:39 UT1
P4 = 23:01:09 UT1

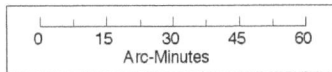

P1   P4 U4 U3   U2 U1 P1   P4 U4 U3   U2 U1

No Eclipse Visible   Eclipse at MoonRise   All Eclipse Visible   Eclipse at MoonSet

Total Lunar Eclipse of 2011 Dec 10

Greatest Eclipse = 14:32:56.5 TD (= 14:31:49.8 UT1)

Penumbral Magnitude = 2.1875	Gamma = -0.3882	Saros Series = 135
Umbral Magnitude = 1.1076	Axis = 0.3571°	Saros Member = 23 of 71

Sun at Greatest Eclipse
(Geocentric Coordinates)

R.A. = 17h08m35.0s
Dec. = -22°54'38.7"
S.D. = 00°16'14.5"
H.P. = 00°00'08.9"

N

Earth's Penumbra

Earth's Umbra

Ecliptic

E —

— **W**

P4   U4   U3   Greatest   U2   U1   P1

Moon at Greatest Eclipse
(Geocentric Coordinates)

R.A. = 05h08m33.9s
Dec. = +22°33'13.3"
S.D. = 00°15'02.4"
H.P. = 00°55'11.7"

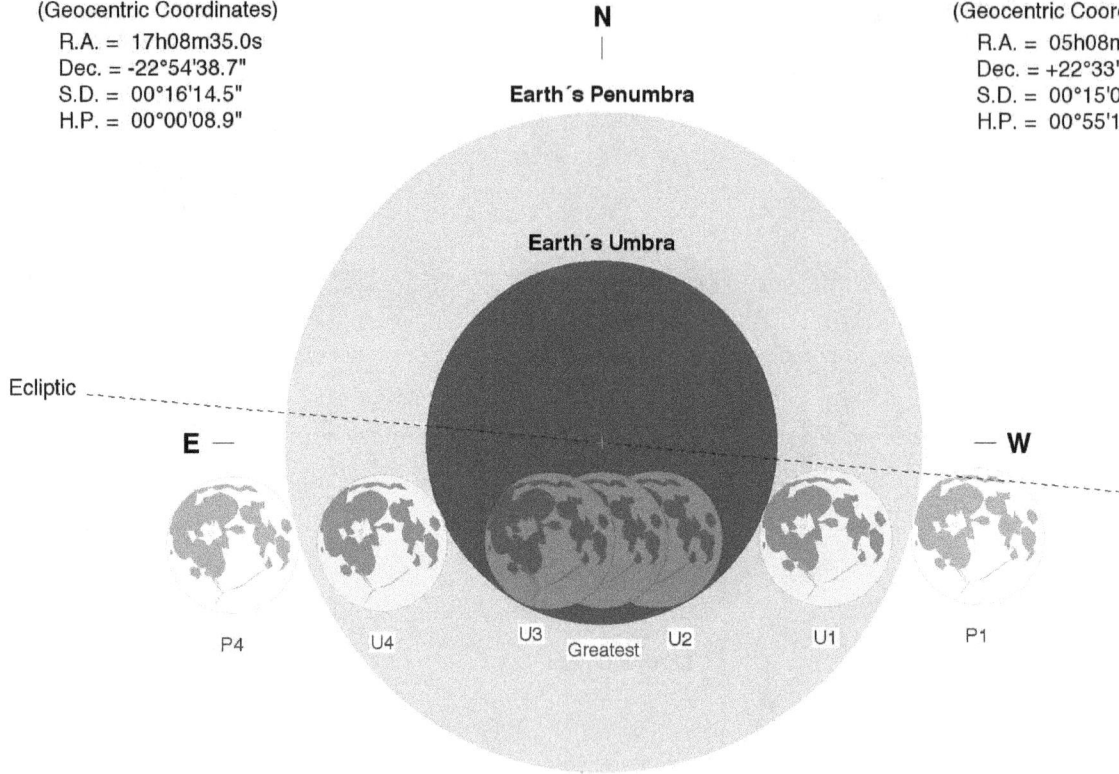

S

Eclipse Durations

Penumbral = 05h57m18s
Umbral = 03h33m05s
Total = 00h51m50s

Eph. = JPL DE430
Rule = Herald-Sinnott
ΔT = 67 s

0   15   30   45   60
Arc-Minutes

©2020 F. Espenak, www.EclipseWise.com

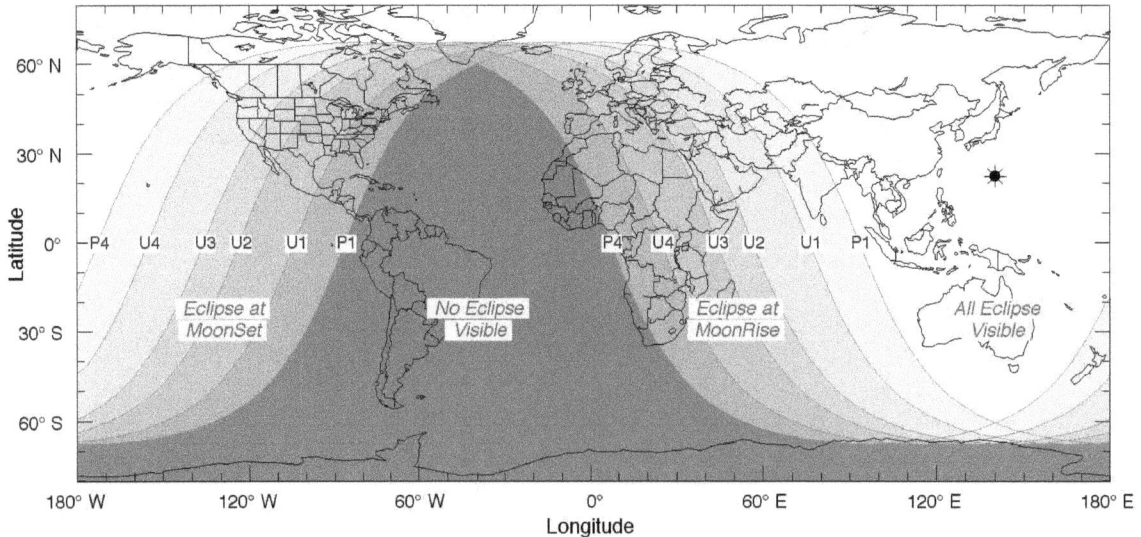

Eclipse Contacts

P1 = 11:33:09 UT1
U1 = 12:45:19 UT1
U2 = 14:05:56 UT1
U3 = 14:57:46 UT1
U4 = 16:18:24 UT1
P4 = 17:30:27 UT1

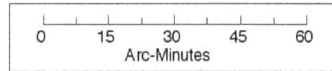

Eclipse at MoonSet   No Eclipse Visible   Eclipse at MoonRise   All Eclipse Visible

P4 U4 U3 U2 U1 P1   P4 U4 U3 U2 U1 P1

Partial Lunar Eclipse of 2012 Jun 04

Greatest Eclipse = 11:04:20.1 TD (= 11:03:13.3 UT1)

Penumbral Magnitude = 1.3198	Gamma = 0.8248	Saros Series = 140
Umbral Magnitude = 0.3718	Axis = 0.8390°	Saros Member = 24 of 77

Sun at Greatest Eclipse
(Geocentric Coordinates)
R.A. = 04h51m33.3s
Dec. = +22°30'16.0"
S.D. = 00°15'45.9"
H.P. = 00°00'08.7"

Moon at Greatest Eclipse
(Geocentric Coordinates)
R.A. = 16h51m37.6s
Dec. = -21°39'56.2"
S.D. = 00°16'37.9"
H.P. = 01°01'02.3"

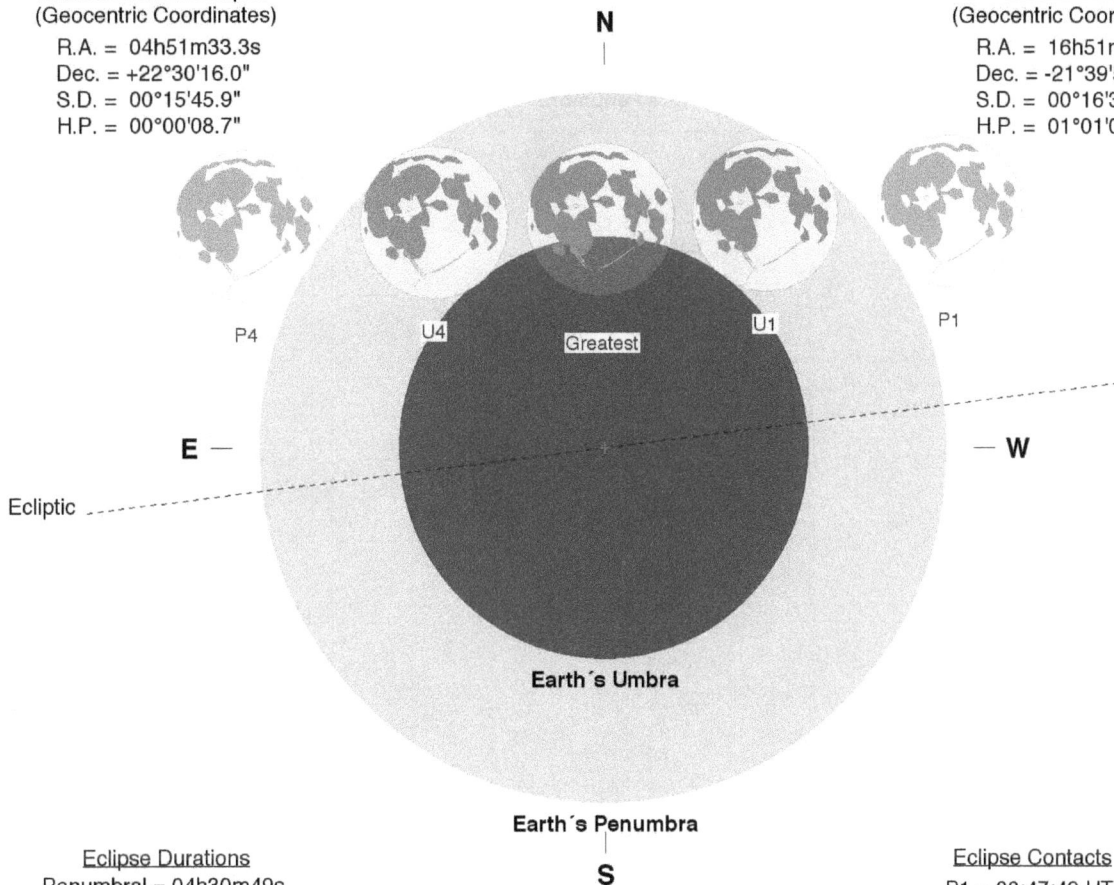

N

P4   U4   Greatest   U1   P1

E

W

Ecliptic

Earth's Umbra

Earth's Penumbra

S

Eclipse Durations
Penumbral = 04h30m49s
Umbral = 02h07m13s

Eph. = JPL DE430
Rule = Herald-Sinnott
ΔT = 67 s

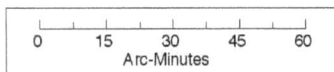

0   15   30   45   60
Arc-Minutes

©2020 F. Espenak, www.EclipseWise.com

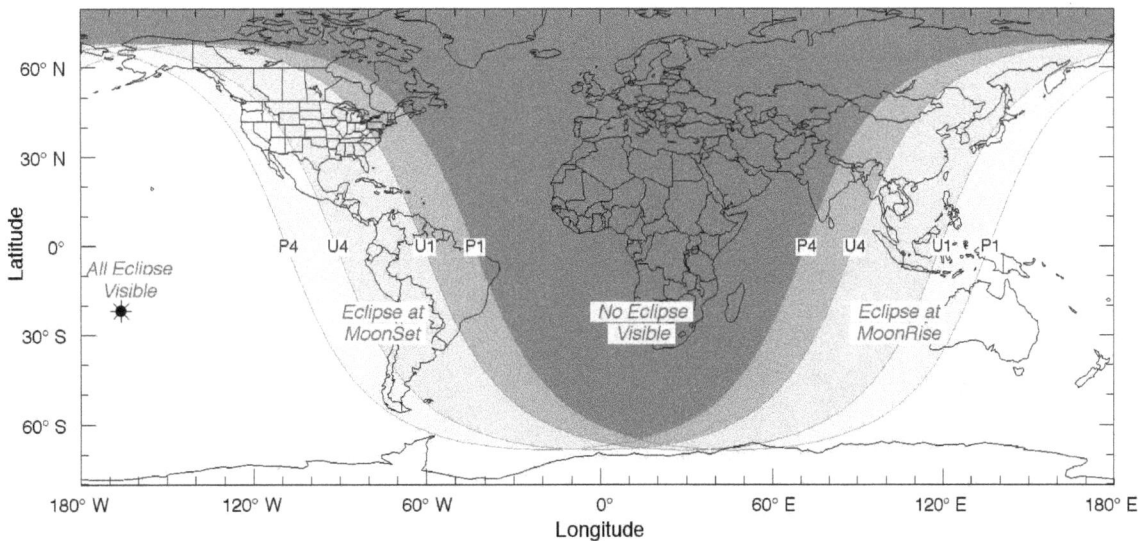

Eclipse Contacts
P1 = 08:47:49 UT1
U1 = 09:59:35 UT1
U4 = 12:06:49 UT1
P4 = 13:18:38 UT1

All Eclipse Visible

Eclipse at MoonSet

No Eclipse Visible

Eclipse at MoonRise

P4 U4 U1 P1   P4 U4 U1 P1

Penumbral Lunar Eclipse of 2012 Nov 28

Greatest Eclipse = 14:34:07.3 TD (= 14:33:00.3 UT1)

Penumbral Magnitude = 0.9168	Gamma = -1.0869	Saros Series = 145
Umbral Magnitude = -0.1859	Axis = 0.9774°	Saros Member = 11 of 71

Sun at Greatest Eclipse
(Geocentric Coordinates)

R.A. = 16h19m43.5s
Dec. = -21°26'15.1"
S.D. = 00°16'12.8"
H.P. = 00°00'08.9"

Moon at Greatest Eclipse
(Geocentric Coordinates)

R.A. = 04h20m01.1s
Dec. = +20°27'44.7"
S.D. = 00°14'42.2"
H.P. = 00°53'57.7"

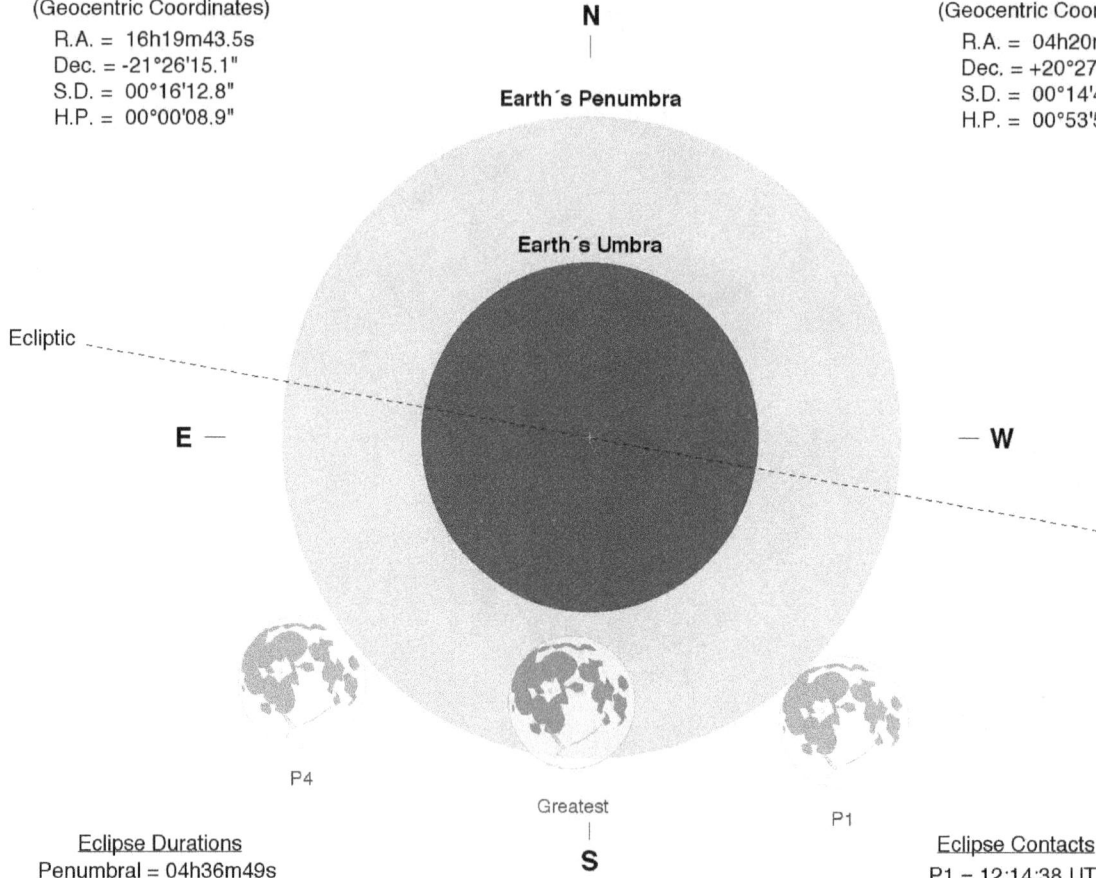

N

Earth's Penumbra

Earth's Umbra

Ecliptic

E

W

P4

Greatest

S

P1

Eclipse Durations
Penumbral = 04h36m49s

Eclipse Contacts
P1 = 12:14:38 UT1
P4 = 16:51:27 UT1

Eph. = JPL DE430
Rule = Herald-Sinnott
ΔT = 67 s

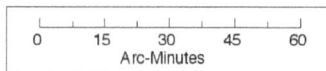

0   15   30   45   60
Arc-Minutes

©2020 F. Espenak, www.EclipseWise.com

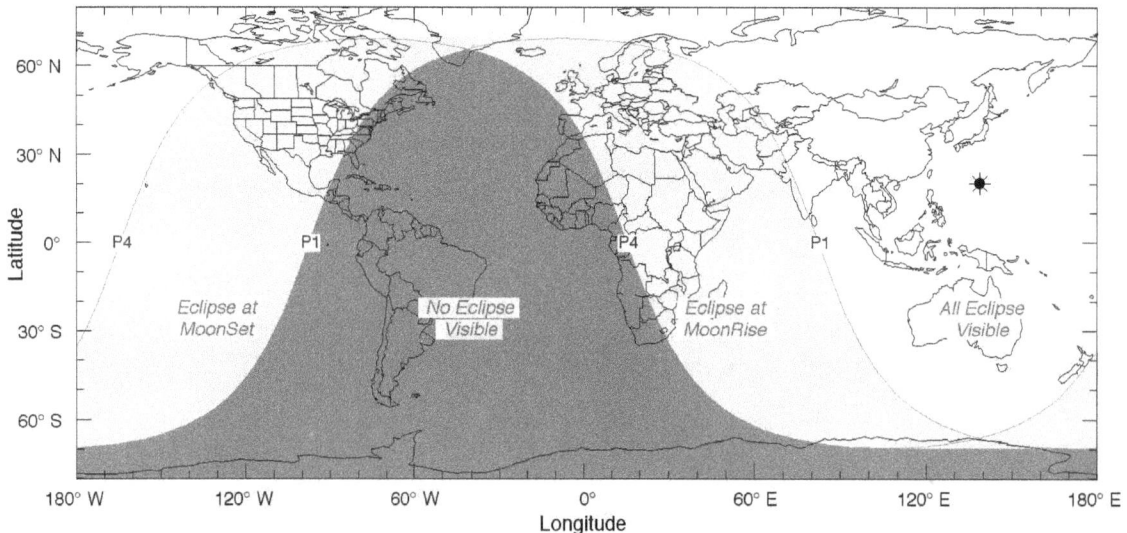

Partial Lunar Eclipse of 2013 Apr 25

Greatest Eclipse = 20:08:37.1 TD   (= 20:07:30.0 UT1)

Penumbral Magnitude = 0.9878	Gamma = -1.0121	Saros Series = 112
Umbral Magnitude = 0.0160	Axis = 1.0125°	Saros Member = 65 of 72

Sun at Greatest Eclipse
(Geocentric Coordinates)

R.A. = 02h13m51.3s
Dec. = +13°26'35.0"
S.D. = 00°15'53.7"
H.P. = 00°00'08.7"

Moon at Greatest Eclipse
(Geocentric Coordinates)

R.A. = 14h12m51.4s
Dec. = -14°25'34.1"
S.D. = 00°16'21.4"
H.P. = 01°00'01.6"

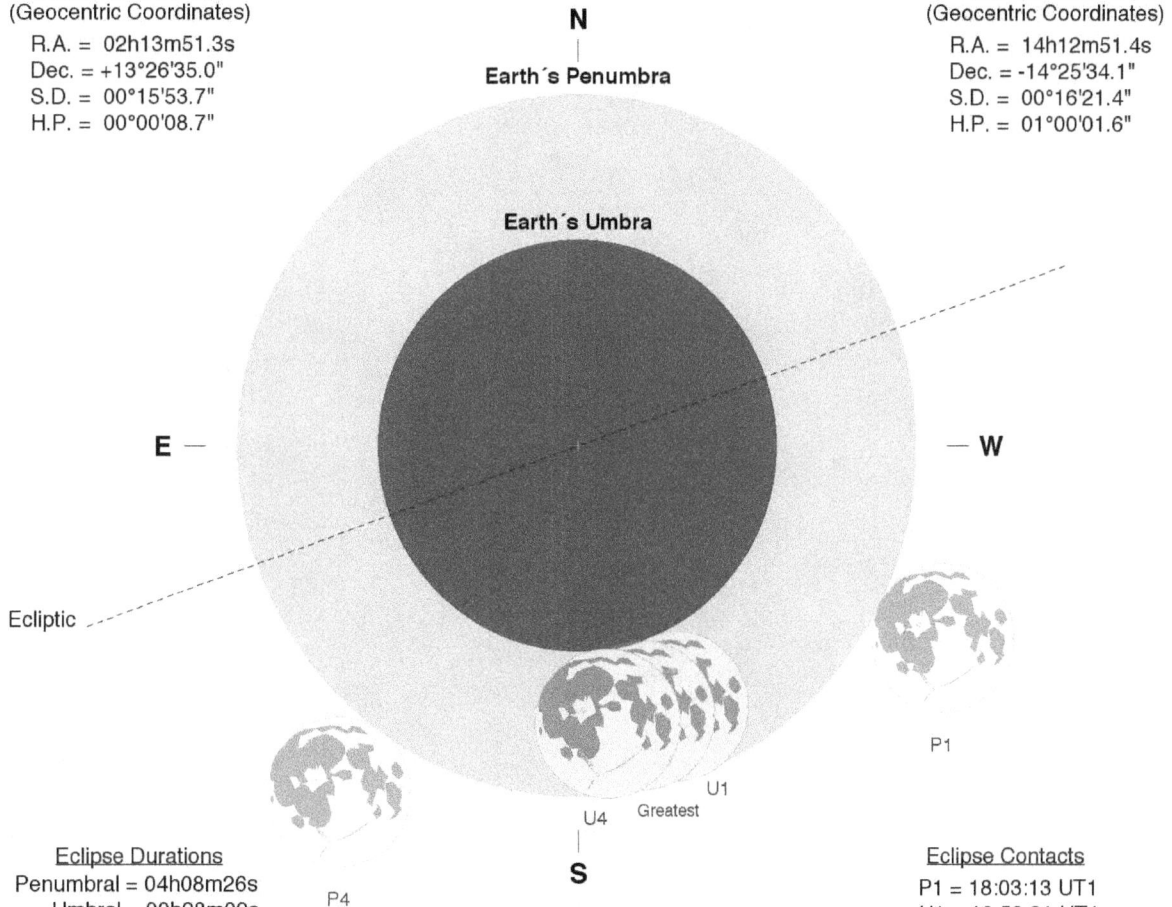

N

Earth's Penumbra

Earth's Umbra

E

W

Ecliptic

P1

U1

Greatest

U4

S

P4

Eclipse Durations
Penumbral = 04h08m26s
Umbral = 00h28m09s

Eph. = JPL DE430
Rule = Herald-Sinnott
ΔT =   67 s

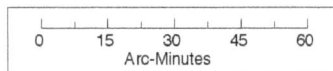

0   15   30   45   60
Arc-Minutes

©2020 F. Espenak, www.EclipseWise.com

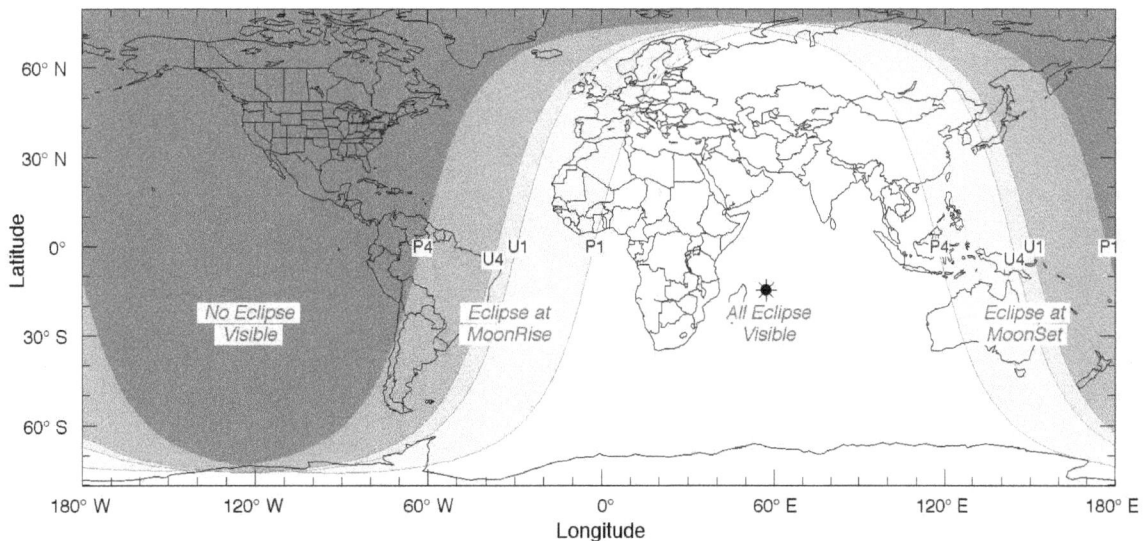

Eclipse Contacts
P1 = 18:03:13 UT1
U1 = 19:53:21 UT1
U4 = 20:21:30 UT1
P4 = 22:11:39 UT1

60° N

30° N

0°

Latitude

30° S

60° S

No Eclipse
Visible

P4   U1   P1
U4

Eclipse at
MoonRise

All Eclipse
Visible

P4   U1   P1
U4

Eclipse at
MoonSet

180° W   120° W   60° W   0°   60° E   120° E   180° E

Longitude

Penumbral Lunar Eclipse of 2013 May 25

Greatest Eclipse = 04:11:06.5 TD (= 04:09:59.4 UT1)

Penumbral Magnitude = 0.0170	Gamma = 1.5351	Saros Series = 150
Umbral Magnitude = -0.9322	Axis = 1.5620°	Saros Member = 1 of 71

N

Sun at Greatest Eclipse
(Geocentric Coordinates)

R.A. = 04h08m32.9s
Dec. = +20°58'05.1"
S.D. = 00°15'47.5"
H.P. = 00°00'08.7"

Moon at Greatest Eclipse
(Geocentric Coordinates)

R.A. = 16h09m09.9s
Dec. = -19°24'45.3"
S.D. = 00°16'38.2"
H.P. = 01°01'03.5"

P4   Greatest   P1

E —

— W

Ecliptic

Earth's Umbra

Earth's Penumbra

S

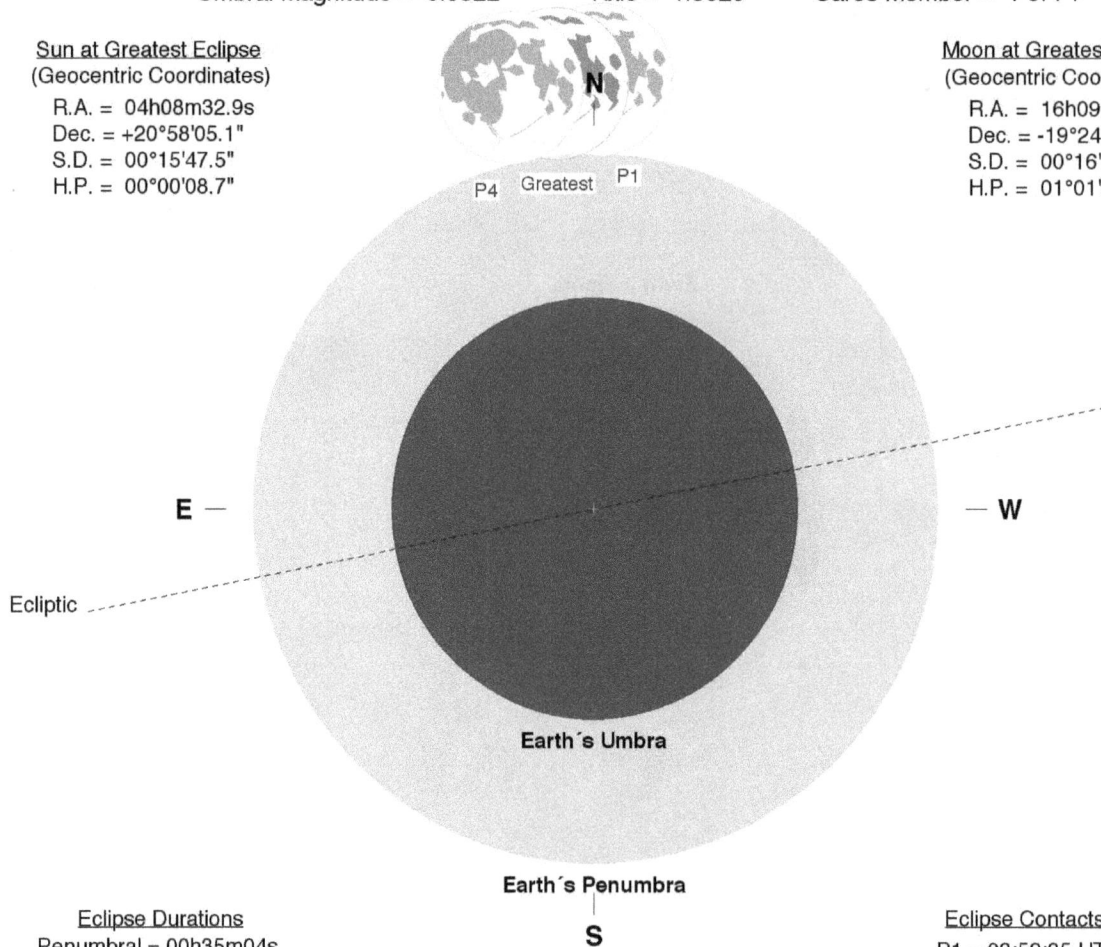

Eclipse Durations
Penumbral = 00h35m04s

Eclipse Contacts
P1 = 03:52:35 UT1
P4 = 04:27:39 UT1

Eph. = JPL DE430
Rule = Herald-Sinnott
ΔT = 67 s

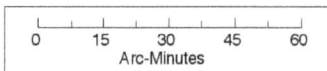

0	15	30	45	60

Arc-Minutes

©2020 F. Espenak, www.EclipseWise.com

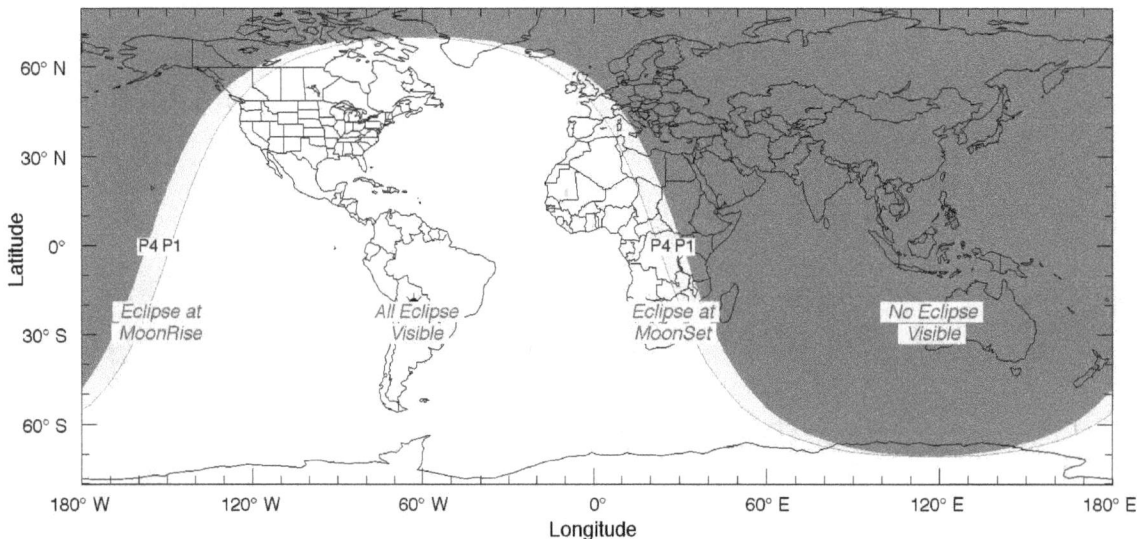

P4 P1

P4 P1

Eclipse at MoonRise

All Eclipse Visible

Eclipse at MoonSet

No Eclipse Visible

Penumbral Lunar Eclipse of 2013 Oct 18

Greatest Eclipse = 23:51:24.6 TD (= 23:50:17.4 UT1)

Penumbral Magnitude = 0.7660	Gamma = 1.1508	Saros Series = 117
Umbral Magnitude = -0.2706	Axis = 1.0902°	Saros Member = 52 of 71

Sun at Greatest Eclipse
(Geocentric Coordinates)

R.A. = 13h35m31.9s
Dec. = -09°57'14.9"
S.D. = 00°16'03.4"
H.P. = 00°00'08.8"

N

Moon at Greatest Eclipse
(Geocentric Coordinates)

R.A. = 01h34m19.6s
Dec. = +11°00'12.1"
S.D. = 00°15'29.3"
H.P. = 00°56'50.7"

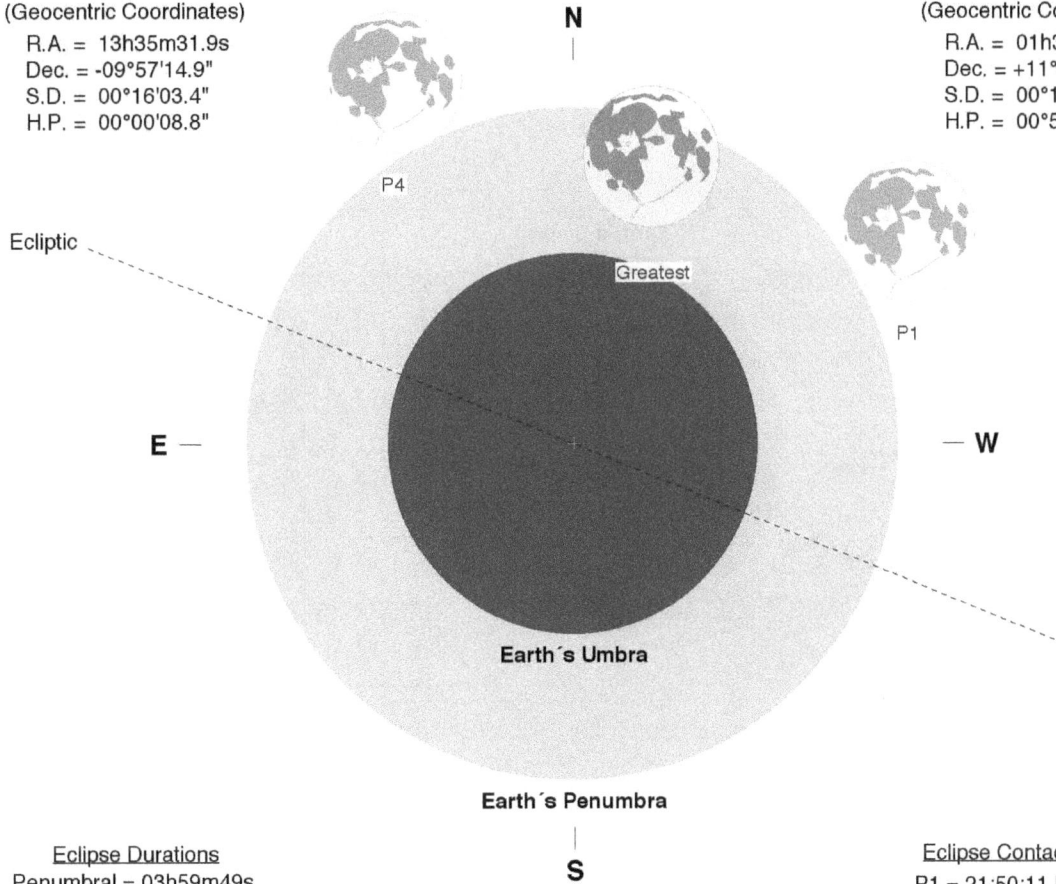

P4

Ecliptic

Greatest

P1

E —

— **W**

Earth's Umbra

Earth's Penumbra

Eclipse Durations
Penumbral = 03h59m49s

S

Eclipse Contacts
P1 = 21:50:11 UT1
P4 = 01:49:59 UT1

Eph. = JPL DE430
Rule = Herald-Sinnott
ΔT = 67 s

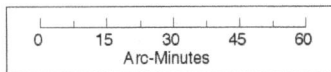

0	15	30	45	60

Arc-Minutes

©2020 F. Espenak, www.EclipseWise.com

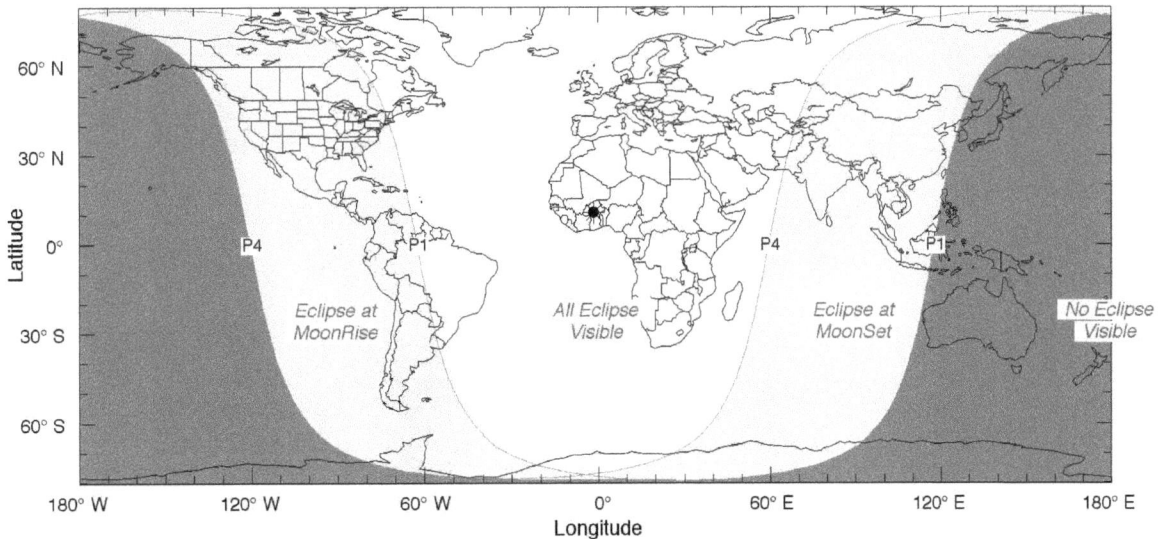

Eclipse at MoonRise — P4, P1

All Eclipse Visible

Eclipse at MoonSet — P4

No Eclipse Visible — P1

Total Lunar Eclipse of 2014 Apr 15

Greatest Eclipse = 07:46:47.6 TD (= 07:45:40.2 UT1)

Penumbral Magnitude = 2.3193	Gamma = -0.3017	Saros Series = 122
Umbral Magnitude = 1.2918	Axis = 0.2863°	Saros Member = 56 of 74

Sun at Greatest Eclipse
(Geocentric Coordinates)

R.A. = 01h33m40.0s
Dec. = +09°46'27.6"
S.D. = 00°15'56.6"
H.P. = 00°00'08.8"

Moon at Greatest Eclipse
(Geocentric Coordinates)

R.A. = 13h33m21.1s
Dec. = -10°02'59.8"
S.D. = 00°15'30.9"
H.P. = 00°56'56.4"

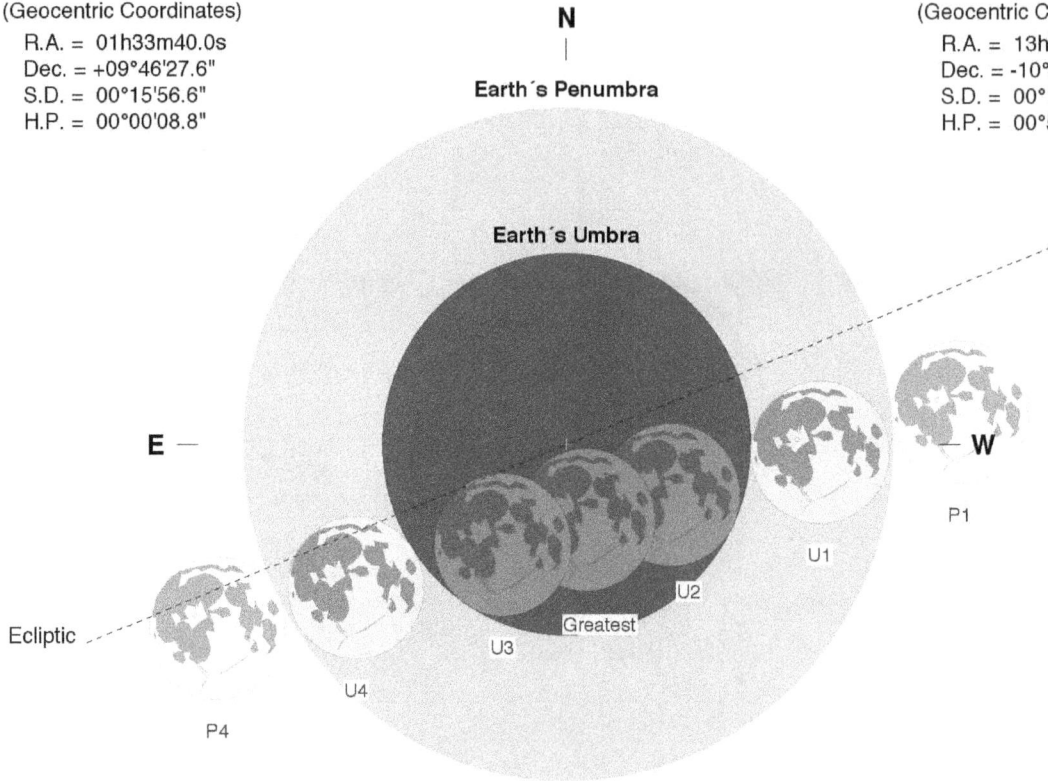

N

Earth's Penumbra

Earth's Umbra

E

W

Ecliptic

P1

U1

U2

Greatest

U3

U4

P4

S

Eclipse Durations
Penumbral = 05h44m45s
Umbral = 03h35m29s
Total = 01h18m27s

Eph. = JPL DE430
Rule = Herald-Sinnott
ΔT = 67 s

0   15   30   45   60
Arc-Minutes

©2020 F. Espenak, www.EclipseWise.com

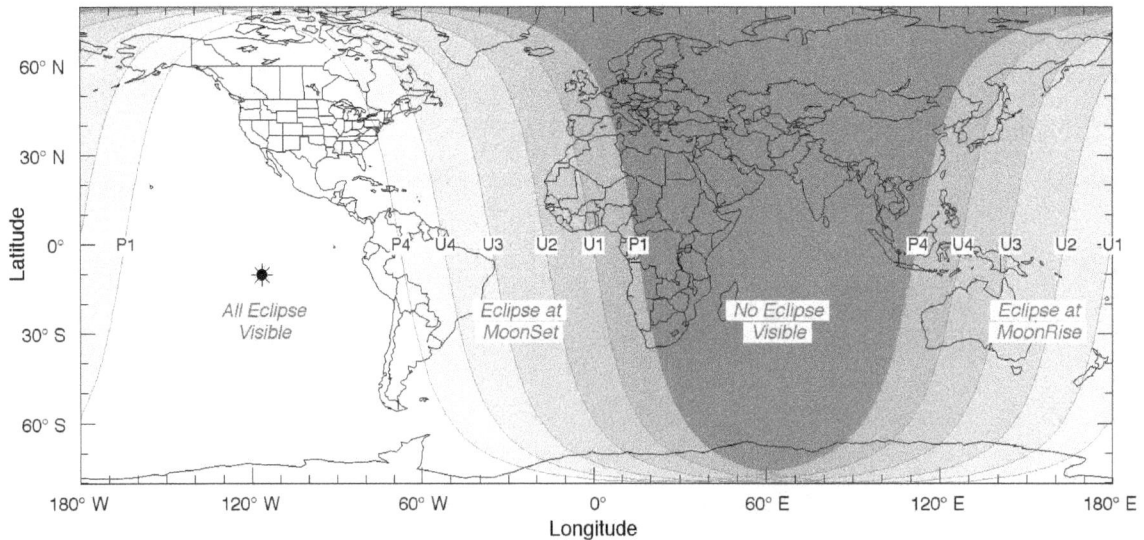

Eclipse Contacts
P1 = 04:53:13 UT1
U1 = 05:57:54 UT1
U2 = 07:06:20 UT1
U3 = 08:24:48 UT1
U4 = 09:33:23 UT1
P4 = 10:37:58 UT1

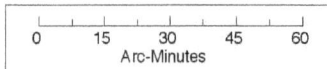

All Eclipse
Visible

Eclipse at
MoonSet

No Eclipse
Visible

Eclipse at
MoonRise

Total Lunar Eclipse of 2014 Oct 08

Greatest Eclipse = 10:55:43.9 TD (= 10:54:36.5 UT1)

Penumbral Magnitude = 2.1467	Gamma = 0.3827	Saros Series = 127
Umbral Magnitude = 1.1670	Axis = 0.3824°	Saros Member = 42 of 72

Sun at Greatest Eclipse
(Geocentric Coordinates)

R.A. = 12h55m34.3s
Dec. = -05°56'30.7"
S.D. = 00°16'00.4"
H.P. = 00°00'08.8"

Moon at Greatest Eclipse
(Geocentric Coordinates)

R.A. = 00h55m07.2s
Dec. = +06°18'26.7"
S.D. = 00°16'20.3"
H.P. = 00°59'57.9"

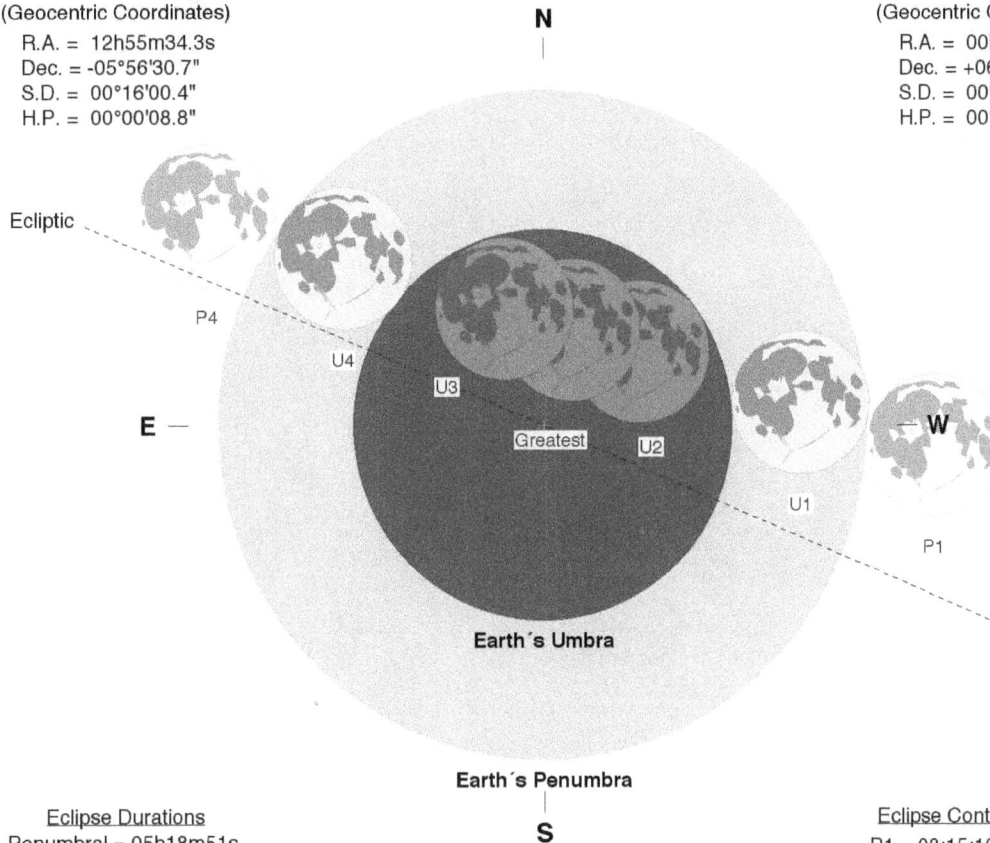

N

Ecliptic

P4

U4

E

U3

Greatest

U2

U1

P1

W

Earth's Umbra

Earth's Penumbra

S

Eclipse Durations
Penumbral = 05h18m51s
Umbral = 03h20m13s
Total = 00h59m23s

Eph. = JPL DE430
Rule = Herald-Sinnott
ΔT = 67 s

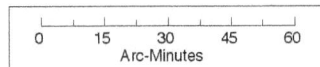

0 15 30 45 60
Arc-Minutes

©2020 F. Espenak, www.EclipseWise.com

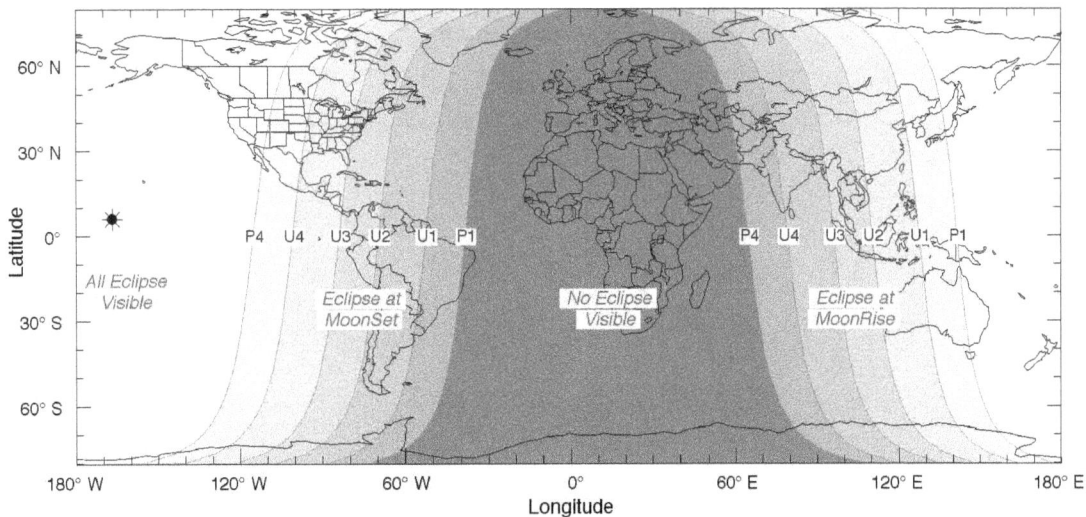

Eclipse Contacts
P1 = 08:15:10 UT1
U1 = 09:14:24 UT1
U2 = 10:24:44 UT1
U3 = 11:24:07 UT1
U4 = 12:34:37 UT1
P4 = 13:34:01 UT1

All Eclipse Visible

Eclipse at MoonSet

No Eclipse Visible

Eclipse at MoonRise

P4 U4 U3 U2 U1 P1

P4 U4 U3 U2 U1 P1

Longitude

Total Lunar Eclipse of 2015 Apr 04

Greatest Eclipse = 12:01:23.6 TD   (= 12:00:16.0 UT1)

Penumbral Magnitude = 2.0802	Gamma = 0.4460	Saros Series = 132
Umbral Magnitude = 1.0019	Axis = 0.4046°	Saros Member = 30 of 71

Sun at Greatest Eclipse
(Geocentric Coordinates)
R.A. = 00h53m01.2s
Dec. = +05°40'32.8"
S.D. = 00°15'59.6"
H.P. = 00°00'08.8"

Moon at Greatest Eclipse
(Geocentric Coordinates)
R.A. = 12h53m29.7s
Dec. = -05°17'20.2"
S.D. = 00°14'49.9"
H.P. = 00°54'25.9"

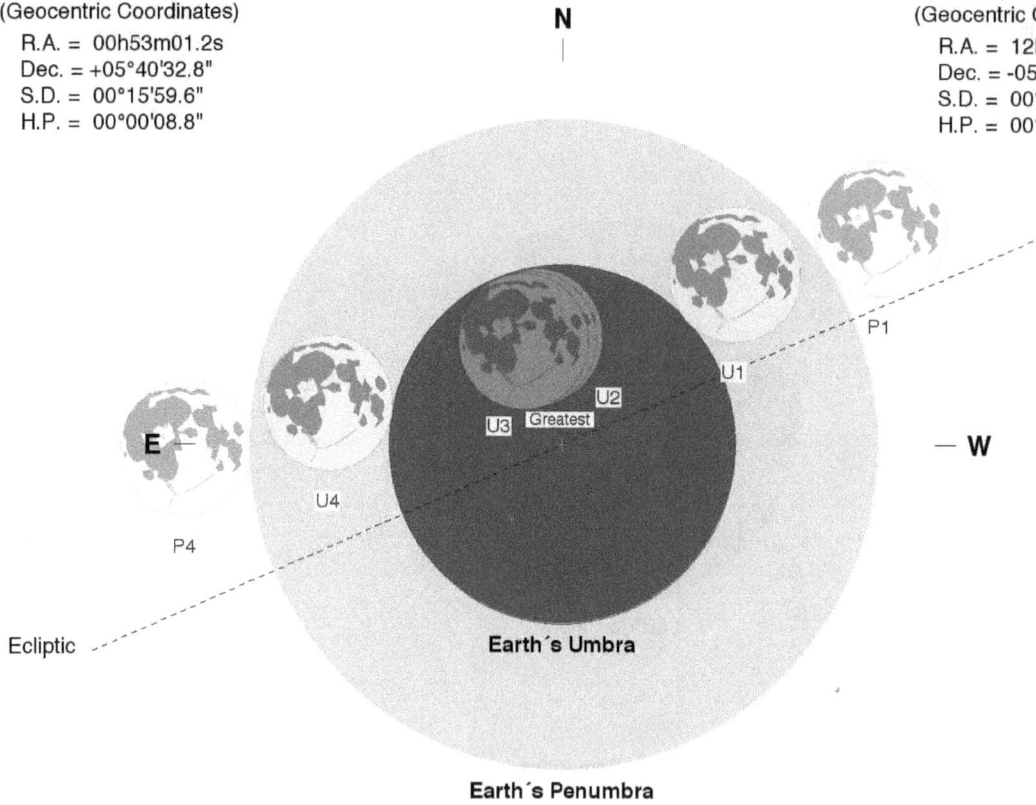

N

P1

U1

U2

U3 Greatest

E

U4

W

P4

Ecliptic

Earth's Umbra

Earth's Penumbra

S

Eclipse Durations
Penumbral = 05h58m24s
Umbral = 03h29m46s
Total = 00h07m16s

Eph. = JPL DE430
Rule = Herald-Sinnott
ΔT =   68 s

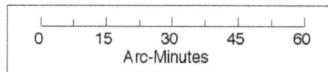

0	15	30	45	60
		Arc-Minutes		

©2020 F. Espenak, www.EclipseWise.com

Eclipse Contacts
P1 = 09:01:06 UT1
U1 = 10:15:30 UT1
U2 = 11:56:52 UT1
U3 = 12:04:08 UT1
U4 = 13:45:16 UT1
P4 = 14:59:30 UT1

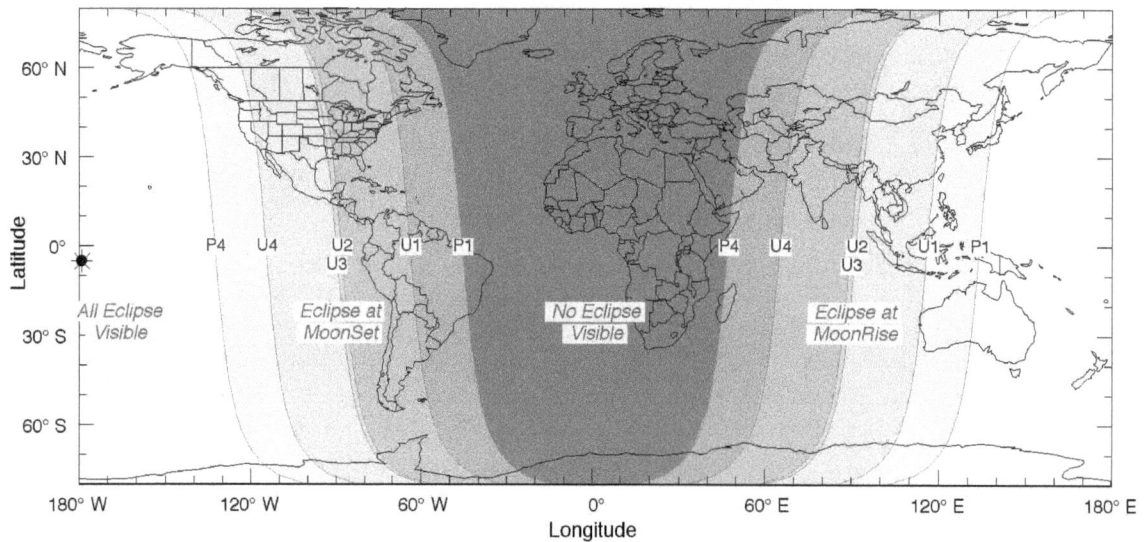

All Eclipse Visible

Eclipse at MoonSet

No Eclipse Visible

Eclipse at MoonRise

Total Lunar Eclipse of 2015 Sep 28

Greatest Eclipse = 02:48:16.8 TD   (= 02:47:09.0 UT1)

Penumbral Magnitude = 2.2307	Gamma = -0.3296	Saros Series = 137
Umbral Magnitude = 1.2774	Axis = 0.3375°	Saros Member = 26 of 78

Sun at Greatest Eclipse
(Geocentric Coordinates)

R.A. = 12h17m08.9s
Dec. = -01°51'20.9"
S.D. = 00°15'57.6"
H.P. = 00°00'08.8"

Moon at Greatest Eclipse
(Geocentric Coordinates)

R.A. = 00h17m33.6s
Dec. = +01°32'03.6"
S.D. = 00°16'44.5"
H.P. = 01°01'26.6"

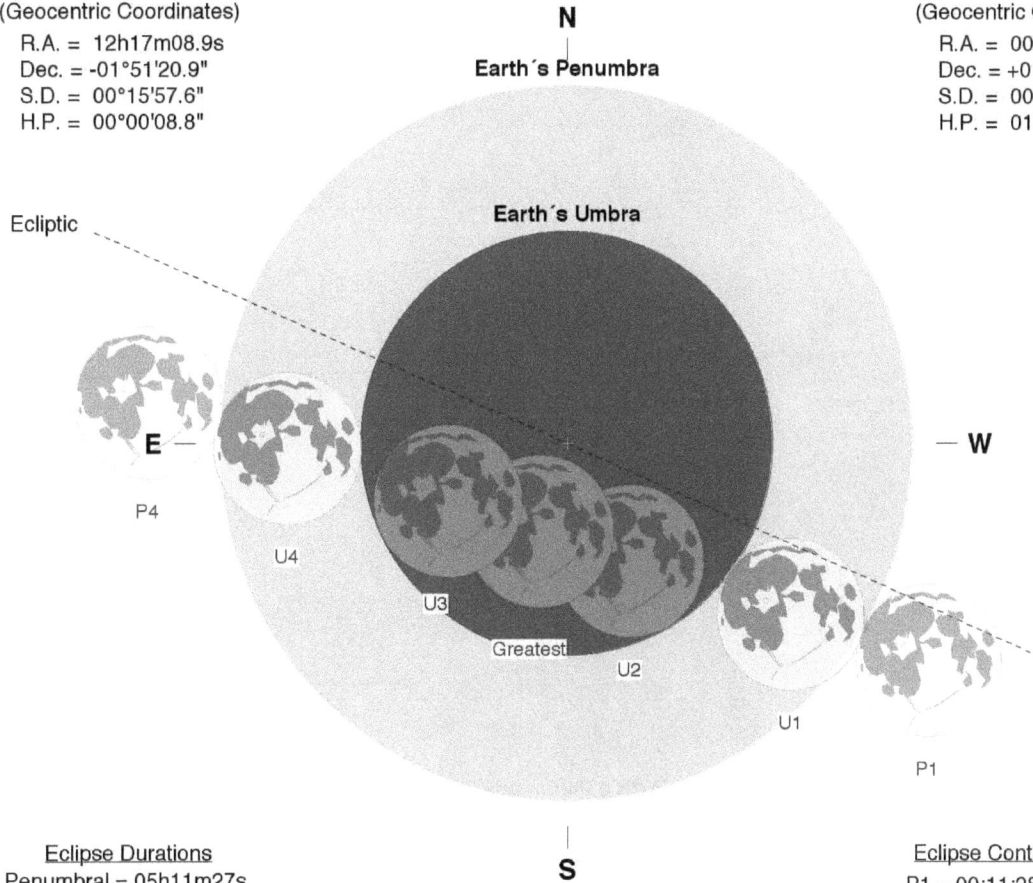

N

Earth's Penumbra

Earth's Umbra

Ecliptic

E

W

P4

U4

U3

Greatest

U2

U1

P1

S

Eclipse Durations
Penumbral = 05h11m27s
Umbral = 03h20m34s
Total = 01h12m29s

Eph. = JPL DE430
Rule = Herald-Sinnott
ΔT =   68 s

0   15   30   45   60
Arc-Minutes

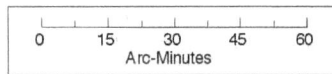

©2020 F. Espenak, www.EclipseWise.com

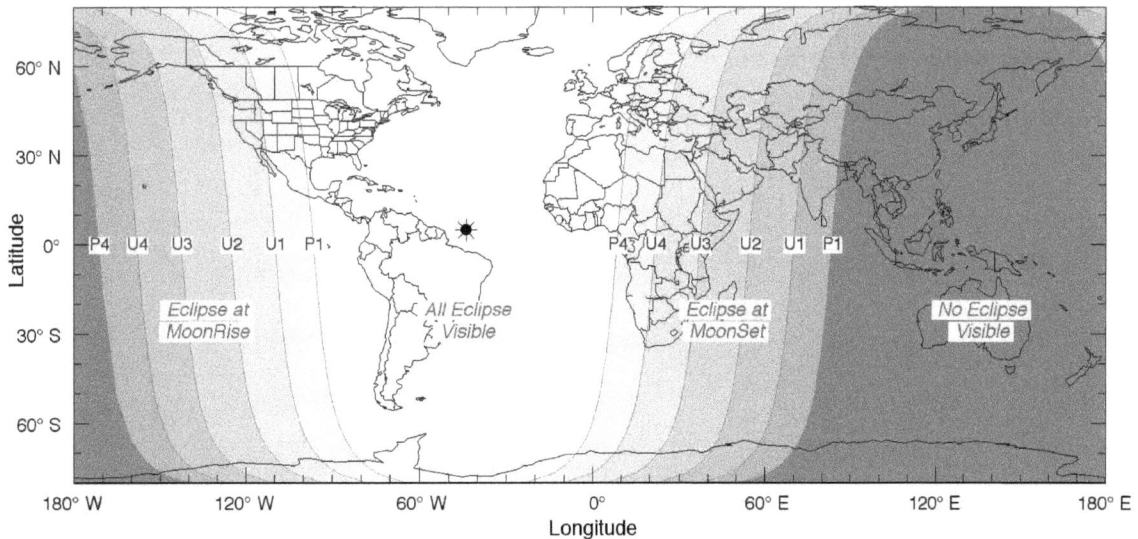

Eclipse Contacts
P1 = 00:11:28 UT1
U1 = 01:06:56 UT1
U2 = 02:11:03 UT1
U3 = 03:23:32 UT1
U4 = 04:27:30 UT1
P4 = 05:22:55 UT1

P4 U4 U3 U2 U1 P1   P4 U4 U3   U2 U1 P1

Eclipse at
MoonRise

All Eclipse
Visible

Eclipse at
MoonSet

No Eclipse
Visible

Longitude

Penumbral Lunar Eclipse of 2016 Mar 23

Greatest Eclipse = 11:48:22.1 TD (= 11:47:14.1 UT1)

Penumbral Magnitude = 0.7758	Gamma = 1.1592	Saros Series = 142
Umbral Magnitude = -0.3107	Axis = 1.0469°	Saros Member = 18 of 73

Sun at Greatest Eclipse
(Geocentric Coordinates)

R.A. = 00h12m02.0s
Dec. = +01°18'10.9"
S.D. = 00°16'02.7"
H.P. = 00°00'08.8"

Moon at Greatest Eclipse
(Geocentric Coordinates)

R.A. = 12h13m18.6s
Dec. = -00°18'21.4"
S.D. = 00°14'46.0"
H.P. = 00°54'11.6"

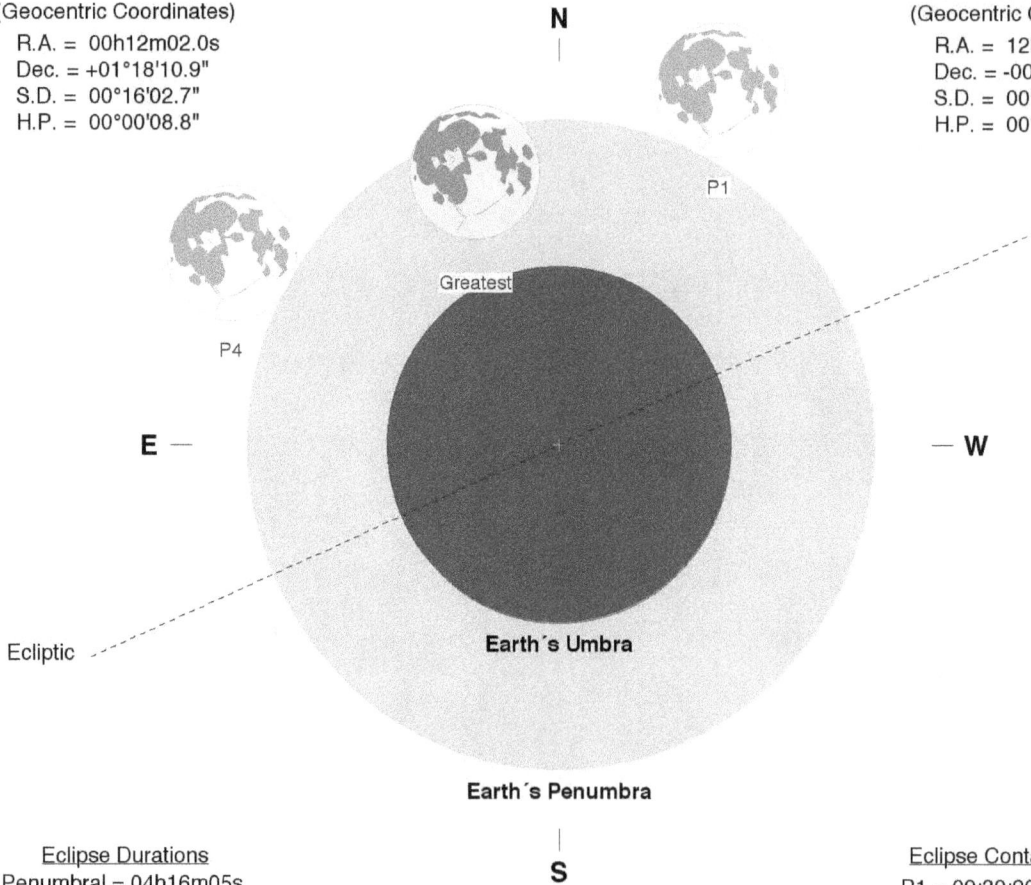

N

P1

Greatest

E

W

Earth's Umbra

Ecliptic

Earth's Penumbra

S

P4

Eclipse Durations
Penumbral = 04h16m05s

Eclipse Contacts
P1 = 09:39:20 UT1
P4 = 13:55:25 UT1

Eph. = JPL DE430
Rule = Herald-Sinnott
ΔT =   68 s

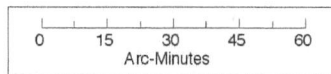

0	15	30	45	60

Arc-Minutes

©2020 F. Espenak, www.EclipseWise.com

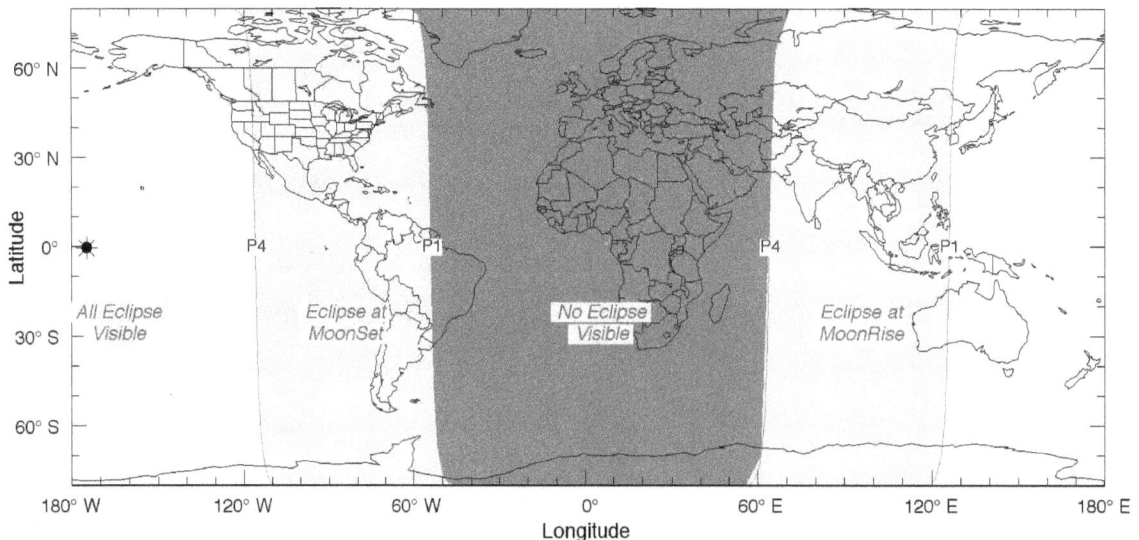

60° N

30° N

0°

30° S

60° S

P4   P1   P4   P1

All Eclipse
Visible

Eclipse at
MoonSet

No Eclipse
Visible

Eclipse at
MoonRise

180° W   120° W   60° W   0°   60° E   120° E   180° E

Longitude

Latitude

Penumbral Lunar Eclipse of 2016 Sep 16

Greatest Eclipse = 18:55:27.3 TD   (= 18:54:19.1 UT1)

Penumbral Magnitude = 0.9091	Gamma = -1.0549	Saros Series = 147
Umbral Magnitude = -0.0624	Axis = 1.0568°	Saros Member = 8 of 70

Sun at Greatest Eclipse
(Geocentric Coordinates)

R.A. = 11h39m09.7s
Dec. = +02°15'14.2"
S.D. = 00°15'54.8"
H.P. = 00°00'08.7"

Moon at Greatest Eclipse
(Geocentric Coordinates)

R.A. = 23h40m27.3s
Dec. = -03°15'36.5"
S.D. = 00°16'22.8"
H.P. = 01°00'06.8"

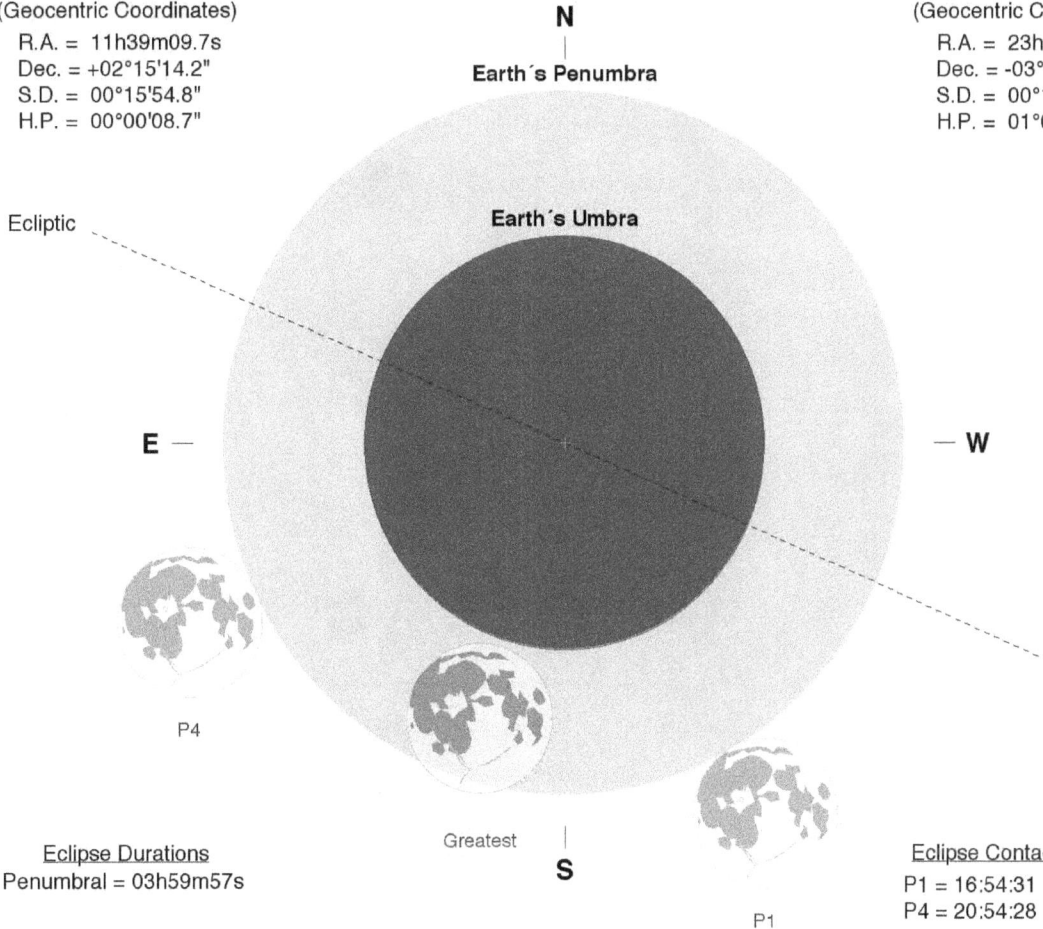

N

Earth's Penumbra

Ecliptic

Earth's Umbra

E —

— W

P4

Greatest

S

P1

Eclipse Durations
Penumbral = 03h59m57s

Eclipse Contacts
P1 = 16:54:31 UT1
P4 = 20:54:28 UT1

Eph. = JPL DE430
Rule = Herald-Sinnott
ΔT =   68 s

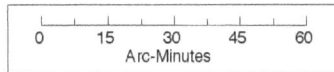

0	15	30	45	60

Arc-Minutes

©2020 F. Espenak, www.EclipseWise.com

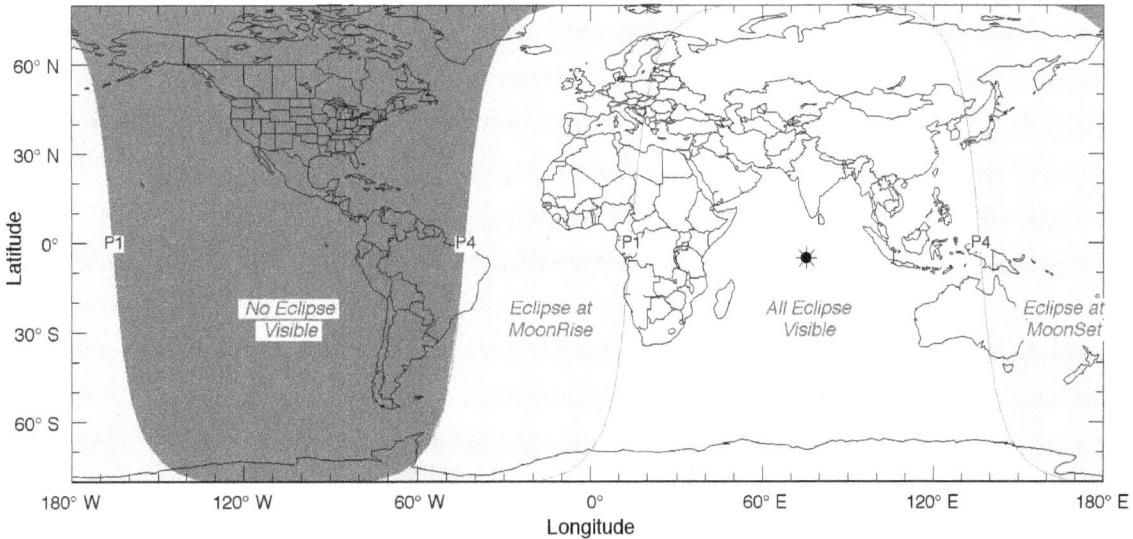

No Eclipse Visible

Eclipse at MoonRise

All Eclipse Visible

Eclipse at MoonSet

Penumbral Lunar Eclipse of 2017 Feb 11

Greatest Eclipse = 00:45:02.5 TD (= 00:43:54.2 UT1)

Penumbral Magnitude = 0.9896	Gamma = -1.0255	Saros Series = 114
Umbral Magnitude = -0.0342	Axis = 0.9928°	Saros Member = 59 of 71

Sun at Greatest Eclipse
(Geocentric Coordinates)

R.A. = 21h39m19.2s
Dec. = -14°01'07.8"
S.D. = 00°16'12.3"
H.P. = 00°00'08.9"

Moon at Greatest Eclipse
(Geocentric Coordinates)

R.A. = 09h38m22.6s
Dec. = +13°03'10.2"
S.D. = 00°15'49.8"
H.P. = 00°58'05.6"

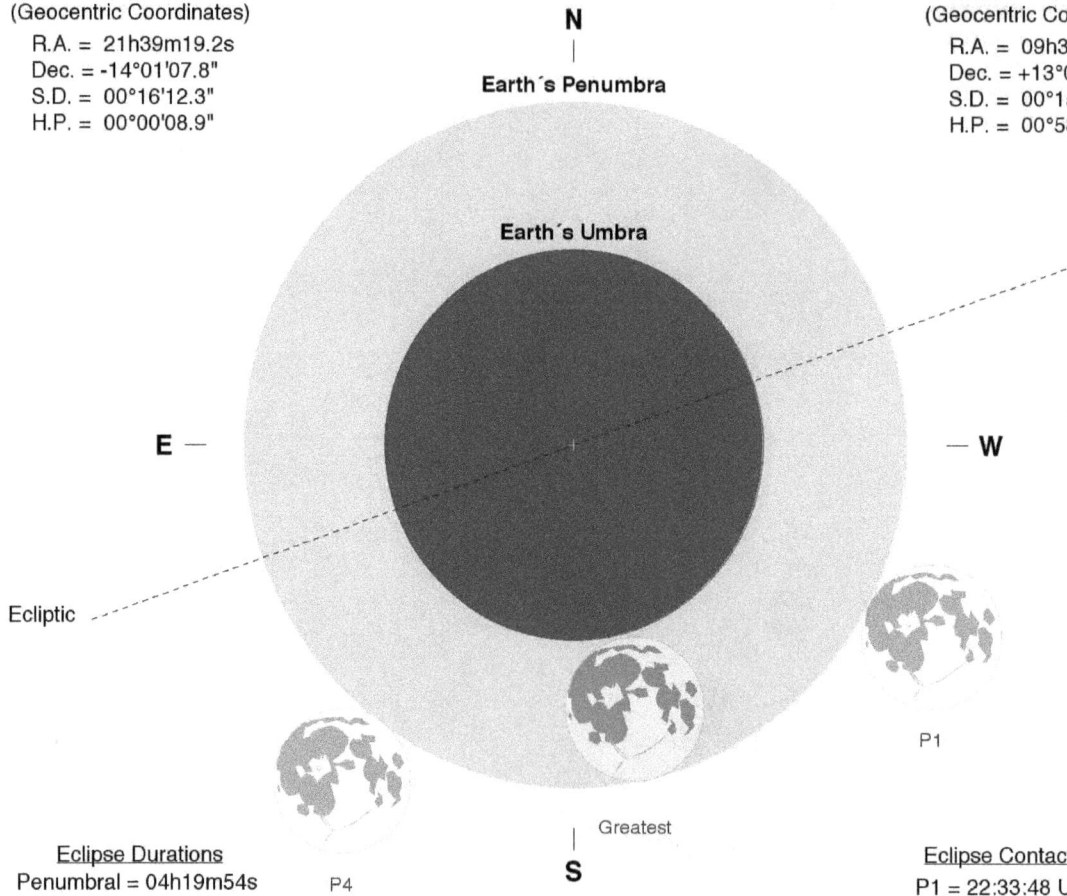

N

Earth´s Penumbra

Earth´s Umbra

E

W

Ecliptic

P1

Greatest

S

P4

Eclipse Durations
Penumbral = 04h19m54s

Eclipse Contacts
P1 = 22:33:48 UT1
P4 = 02:53:42 UT1

Eph. = JPL DE430
Rule = Herald-Sinnott
ΔT = 68 s

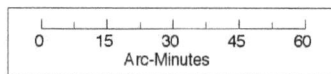

0	15	30	45	60

Arc-Minutes

©2020 F. Espenak, www.EclipseWise.com

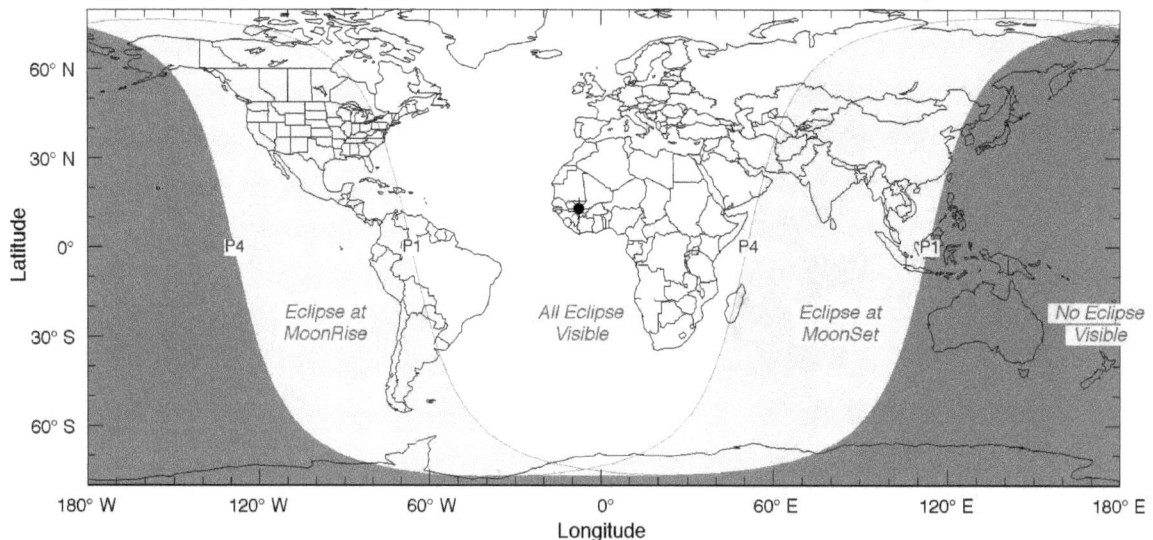

P4

P1

P4

P1

Eclipse at
MoonRise

All Eclipse
Visible

Eclipse at
MoonSet

No Eclipse
Visible

60° N

30° N

0°

Latitude

30° S

60° S

180° W   120° W   60° W   0°   60° E   120° E   180° E

Longitude

Partial Lunar Eclipse of 2017 Aug 07

Greatest Eclipse = 18:21:37.8 TD (= 18:20:29.2 UT1)

Penumbral Magnitude = 1.2898	Gamma = 0.8669	Saros Series = 119
Umbral Magnitude = 0.2477	Axis = 0.8024°	Saros Member = 61 of 82

Sun at Greatest Eclipse
(Geocentric Coordinates)
R.A. = 09h11m33.0s
Dec. = +16°12'28.1"
S.D. = 00°15'46.4"
H.P. = 00°00'08.7"

Moon at Greatest Eclipse
(Geocentric Coordinates)
R.A. = 21h10m53.1s
Dec. = -15°25'17.2"
S.D. = 00°15'08.1"
H.P. = 00°55'32.7"

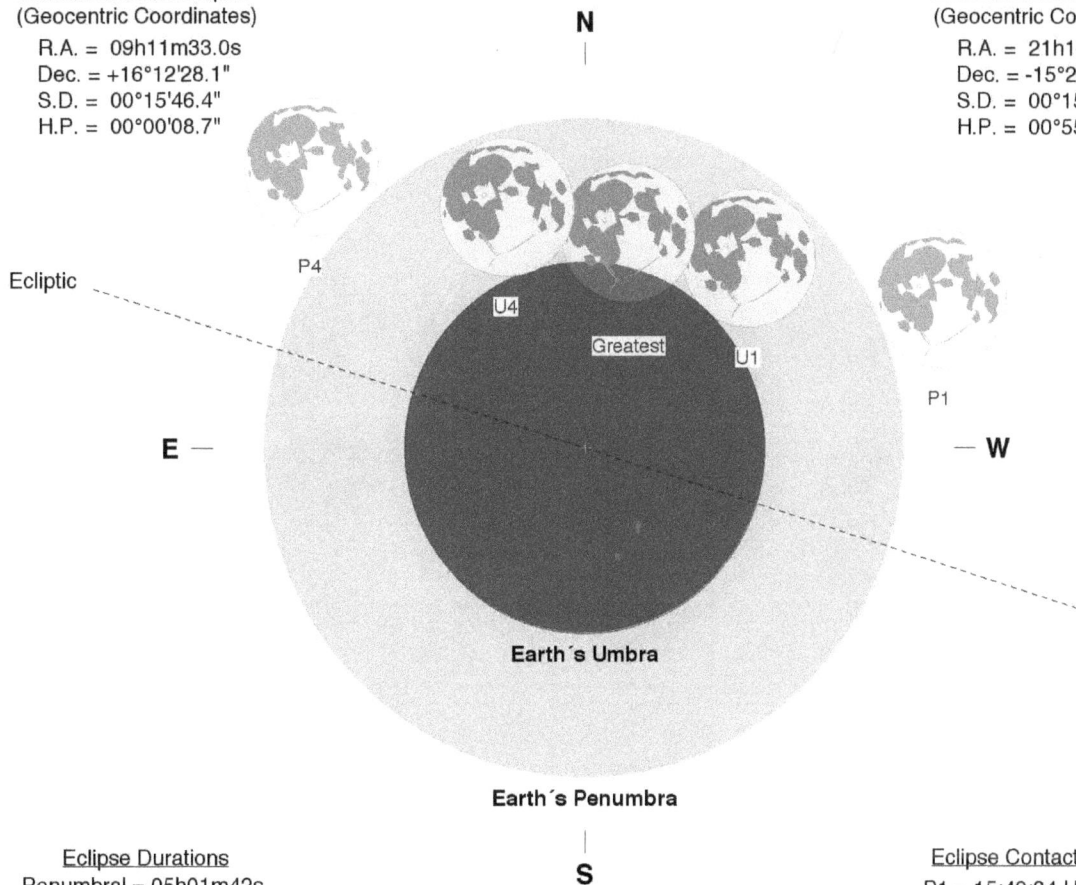

N

Ecliptic

P4

U4

Greatest

U1

E

W

P1

Earth´s Umbra

Earth´s Penumbra

S

Eclipse Durations
Penumbral = 05h01m42s
Umbral = 01h55m53s

Eph. = JPL DE430
Rule = Herald-Sinnott
ΔT =   69 s

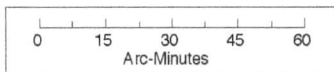

Eclipse Contacts
P1 = 15:49:34 UT1
U1 = 17:22:30 UT1
U4 = 19:18:23 UT1
P4 = 20:51:17 UT1

0   15   30   45   60
Arc-Minutes

©2020 F. Espenak, www.EclipseWise.com

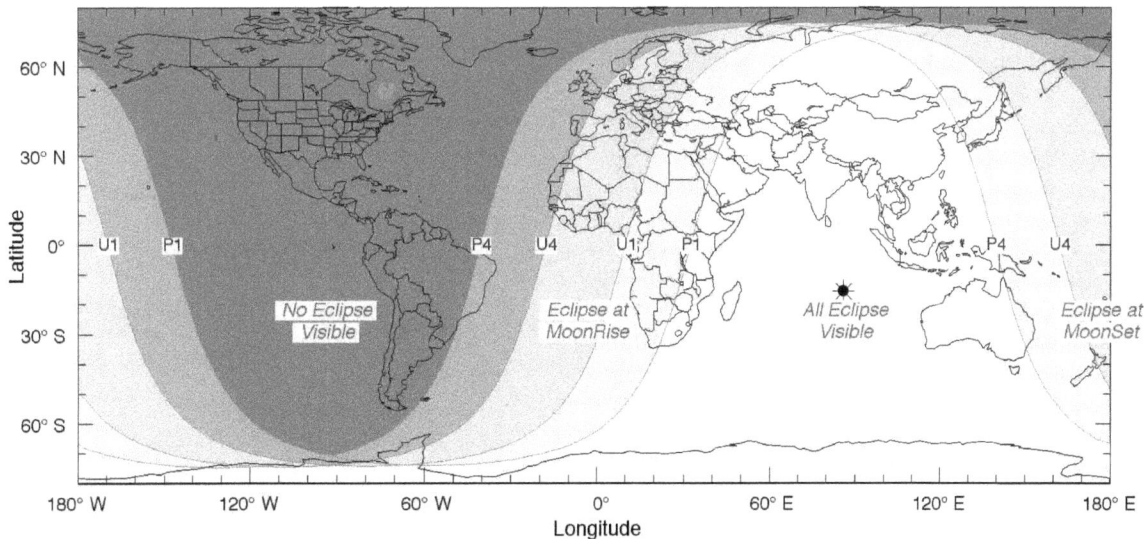

Total Lunar Eclipse of 2018 Jan 31

Greatest Eclipse = 13:31:00.1 TD   (= 13:29:51.4 UT1)

Penumbral Magnitude = 2.2954	Gamma = -0.3014	Saros Series = 124
Umbral Magnitude = 1.3167	Axis = 0.3058°	Saros Member = 49 of 73

Sun at Greatest Eclipse
(Geocentric Coordinates)

R.A. = 20h56m18.8s
Dec. = -17°17'47.0"
S.D. = 00°16'14.0"
H.P. = 00°00'08.9"

Moon at Greatest Eclipse
(Geocentric Coordinates)

R.A. = 08h56m05.0s
Dec. = +16°59'44.2"
S.D. = 00°16'35.2"
H.P. = 01°00'52.6"

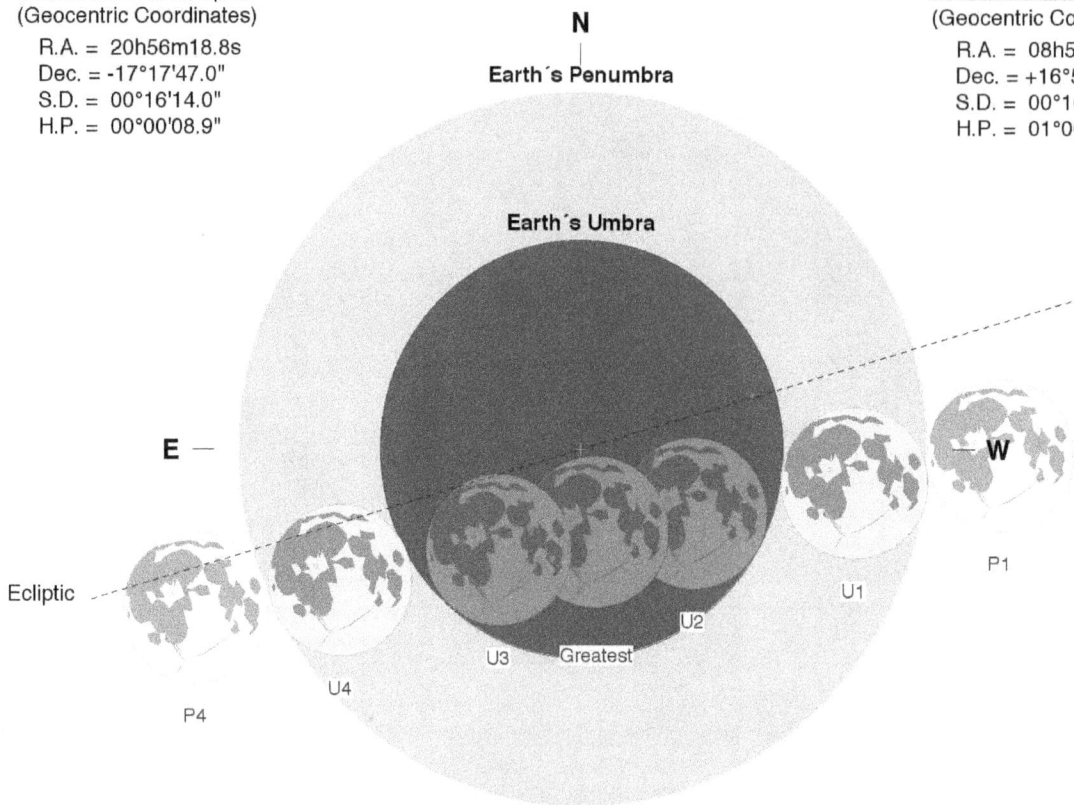

N

Earth's Penumbra

Earth's Umbra

E

W

Ecliptic

P1

U1

U2

Greatest

U3

U4

P4

S

Eclipse Durations
Penumbral = 05h18m03s
Umbral = 03h23m29s
Total = 01h16m43s

Eph. = JPL DE430
Rule = Herald-Sinnott
ΔT =   69 s

0	15	30	45	60

Arc-Minutes

©2020 F. Espenak, www.EclipseWise.com

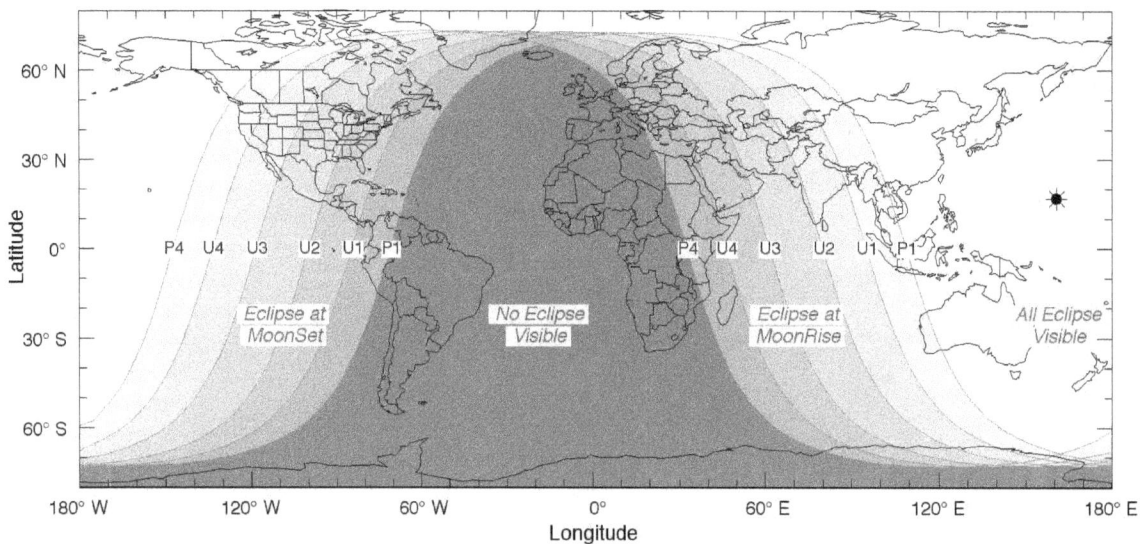

Eclipse Contacts
P1 = 10:50:50 UT1
U1 = 11:48:04 UT1
U2 = 12:51:25 UT1
U3 = 14:08:08 UT1
U4 = 15:11:33 UT1
P4 = 16:08:53 UT1

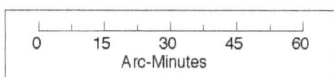

Total Lunar Eclipse of 2018 Jul 27

Greatest Eclipse = 20:22:54.3 TD   (= 20:21:45.4 UT1)

Penumbral Magnitude = 2.6805	Gamma = 0.1168	Saros Series = 129
Umbral Magnitude = 1.6100	Axis = 0.1051°	Saros Member = 38 of 71

Sun at Greatest Eclipse
(Geocentric Coordinates)

R.A. = 08h28m22.0s
Dec. = +19°04'25.2"
S.D. = 00°15'45.0"
H.P. = 00°00'08.7"

Moon at Greatest Eclipse
(Geocentric Coordinates)

R.A. = 20h28m18.2s
Dec. = -18°58'10.6"
S.D. = 00°14'42.7"
H.P. = 00°53'59.7"

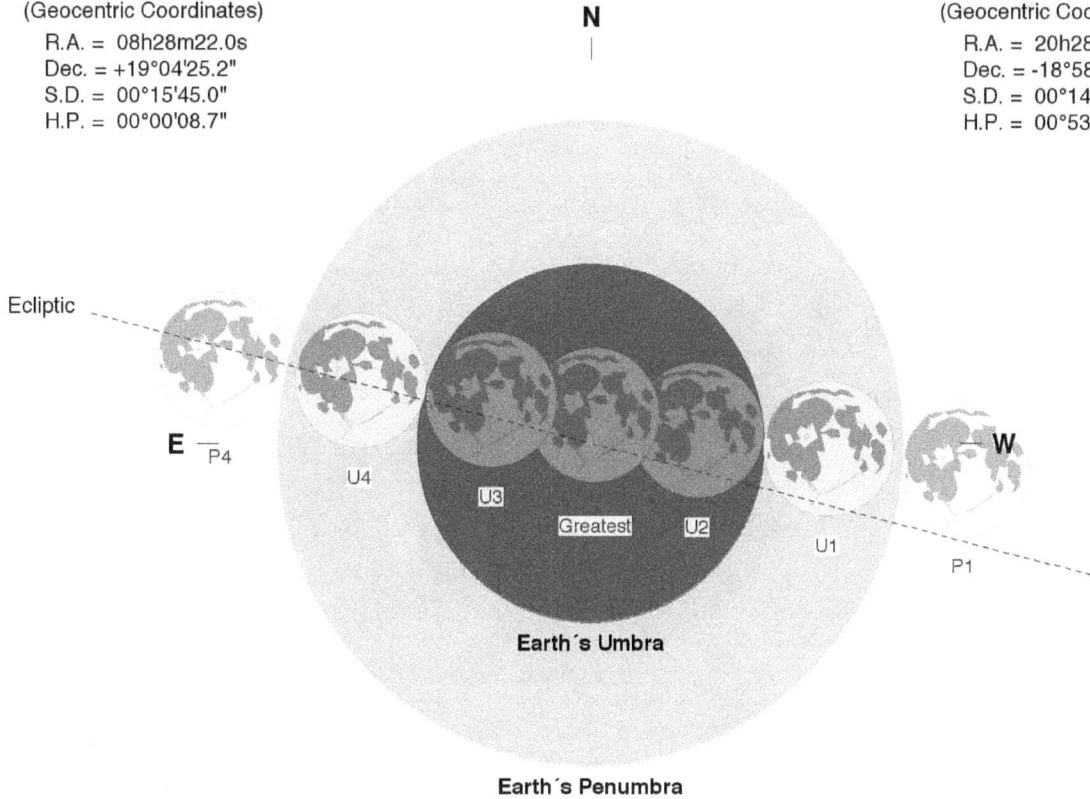

N

Ecliptic

E

P4

U4

U3

Greatest

U2

W

U1

P1

Earth´s Umbra

Earth´s Penumbra

S

Eclipse Durations
Penumbral = 06h14m45s
Umbral = 03h55m25s
Total = 01h43m47s

Eph. = JPL DE430
Rule = Herald-Sinnott
ΔT =   69 s

0   15   30   45   60
Arc-Minutes

©2020 F. Espenak, www.EclipseWise.com

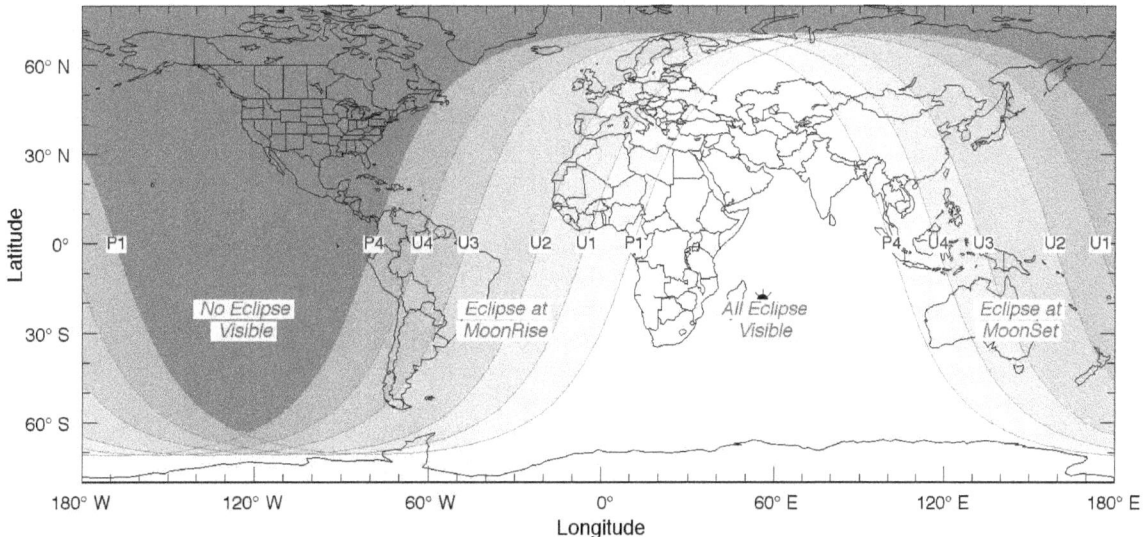

Eclipse Contacts
P1 = 17:14:22 UT1
U1 = 18:24:02 UT1
U2 = 19:29:50 UT1
U3 = 21:13:37 UT1
U4 = 22:19:27 UT1
P4 = 23:29:07 UT1

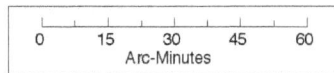

No Eclipse Visible

Eclipse at MoonRise

All Eclipse Visible

Eclipse at MoonSet

Total Lunar Eclipse of 2019 Jan 21

Greatest Eclipse = 05:13:27.1 TD   (= 05:12:18.0 UT1)

Penumbral Magnitude = 2.1697	Gamma = 0.3684	Saros Series = 134
Umbral Magnitude = 1.1966	Axis = 0.3763°	Saros Member = 27 of 72

Sun at Greatest Eclipse
(Geocentric Coordinates)

R.A. = 20h12m17.2s
Dec. = -19°57'48.1"
S.D. = 00°16'15.2"
H.P. = 00°00'08.9"

Moon at Greatest Eclipse
(Geocentric Coordinates)

R.A. = 08h12m28.7s
Dec. = +20°20'13.2"
S.D. = 00°16'42.1"
H.P. = 01°01'17.9"

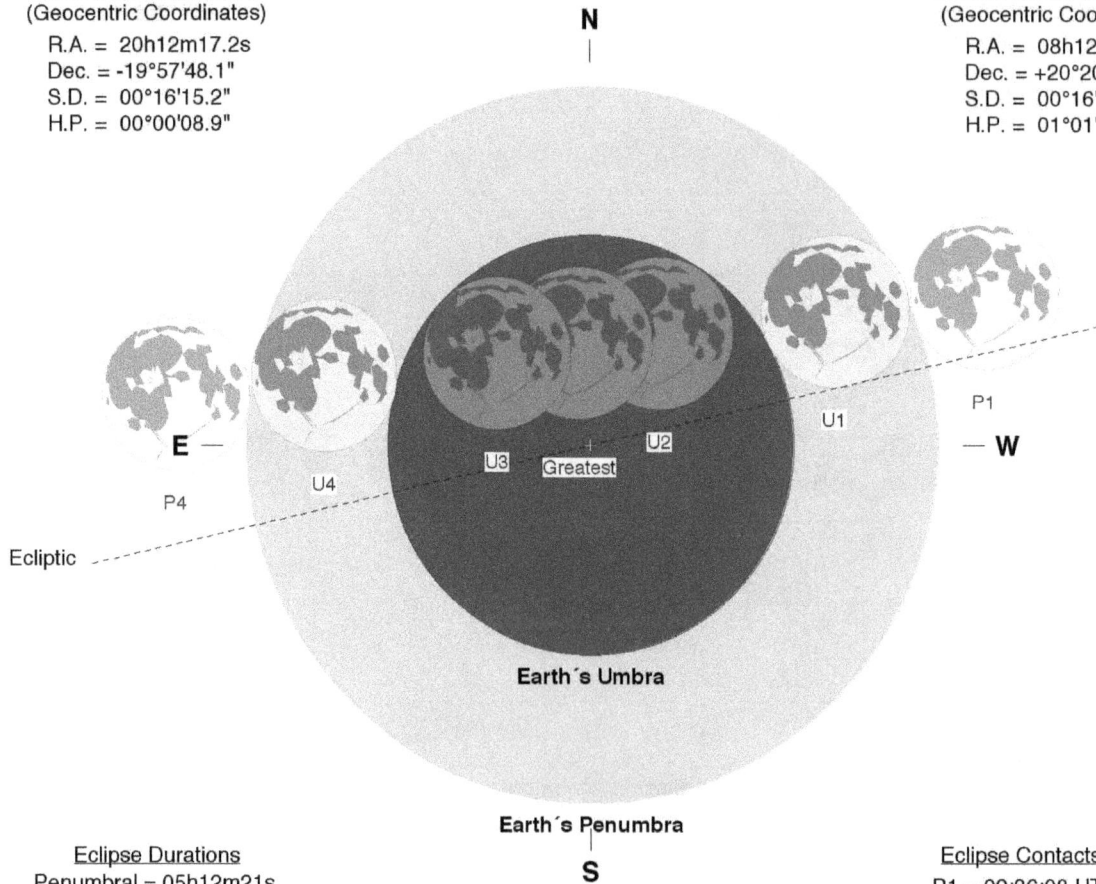

N

E — — W

P4

U4

U3   Greatest   U2

U1   P1

Ecliptic

Earth's Umbra

Earth's Penumbra

S

Eclipse Durations
Penumbral = 05h12m21s
Umbral = 03h17m30s
Total = 01h02m36s

Eph. = JPL DE430
Rule = Herald-Sinnott
ΔT =   69 s

Eclipse Contacts
P1 = 02:36:08 UT1
U1 = 03:33:35 UT1
U2 = 04:41:04 UT1
U3 = 05:43:40 UT1
U4 = 06:51:05 UT1
P4 = 07:48:29 UT1

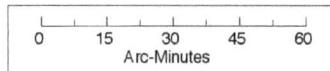

0   15   30   45   60
Arc-Minutes

©2020 F. Espenak, www.EclipseWise.com

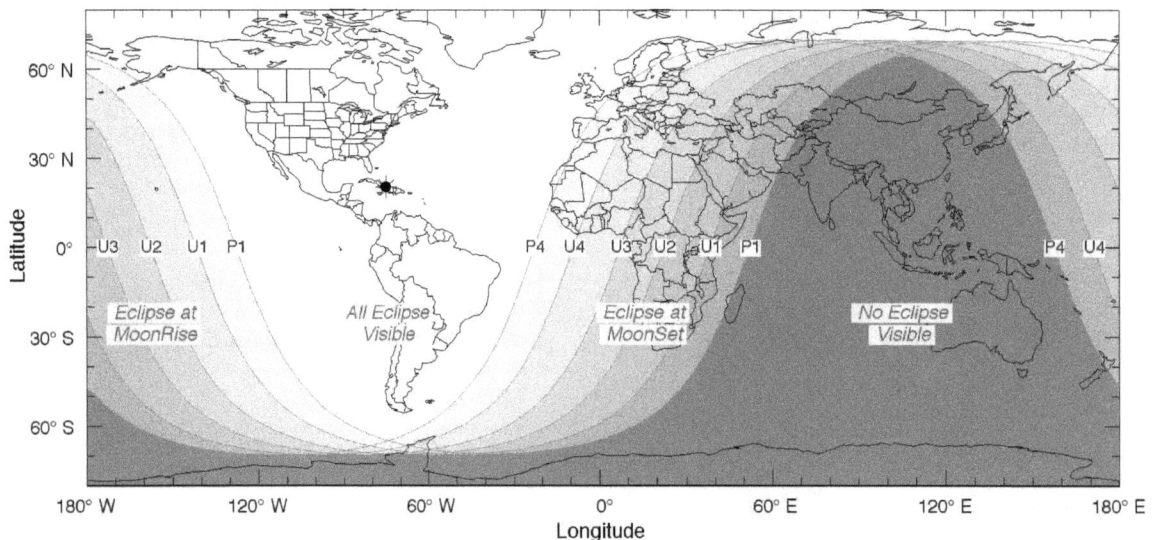

Partial Lunar Eclipse of 2019 Jul 16

Greatest Eclipse = 21:31:54.9 TD (= 21:30:45.6 UT1)

Penumbral Magnitude = 1.7050	Gamma = -0.6430	Saros Series = 139
Umbral Magnitude = 0.6544	Axis = 0.5890°	Saros Member = 21 of 79

Sun at Greatest Eclipse
(Geocentric Coordinates)

R.A. = 07h43m48.8s
Dec. = +21°17'38.5"
S.D. = 00°15'44.2"
H.P. = 00°00'08.7"

N

Earth's Penumbra

Earth's Umbra

Moon at Greatest Eclipse
(Geocentric Coordinates)

R.A. = 19h44m00.3s
Dec. = -21°52'53.0"
S.D. = 00°14'58.7"
H.P. = 00°54'58.2"

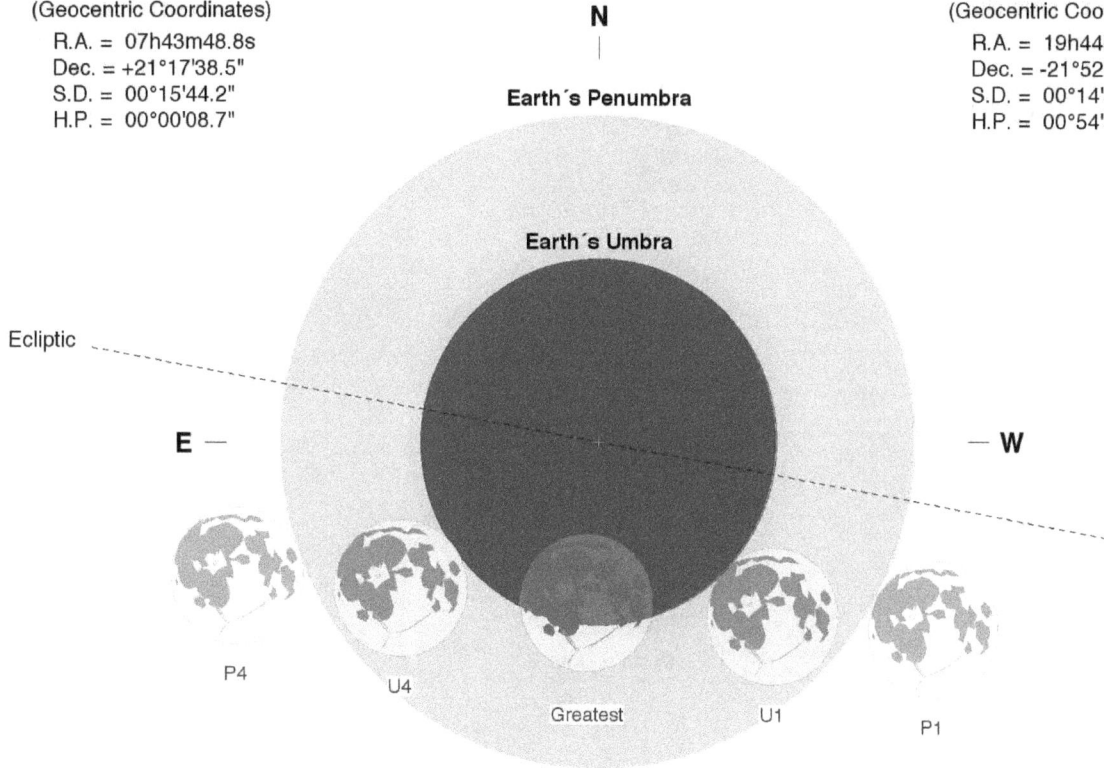

Ecliptic

E —

— **W**

P4   U4   Greatest   U1   P1

S

Eclipse Durations
Penumbral = 05h34m38s
Umbral = 02h58m42s

Eph. = JPL DE430
Rule = Herald-Sinnott
ΔT = 69 s

0	15	30	45	60

Arc-Minutes

©2020 F. Espenak, www.EclipseWise.com

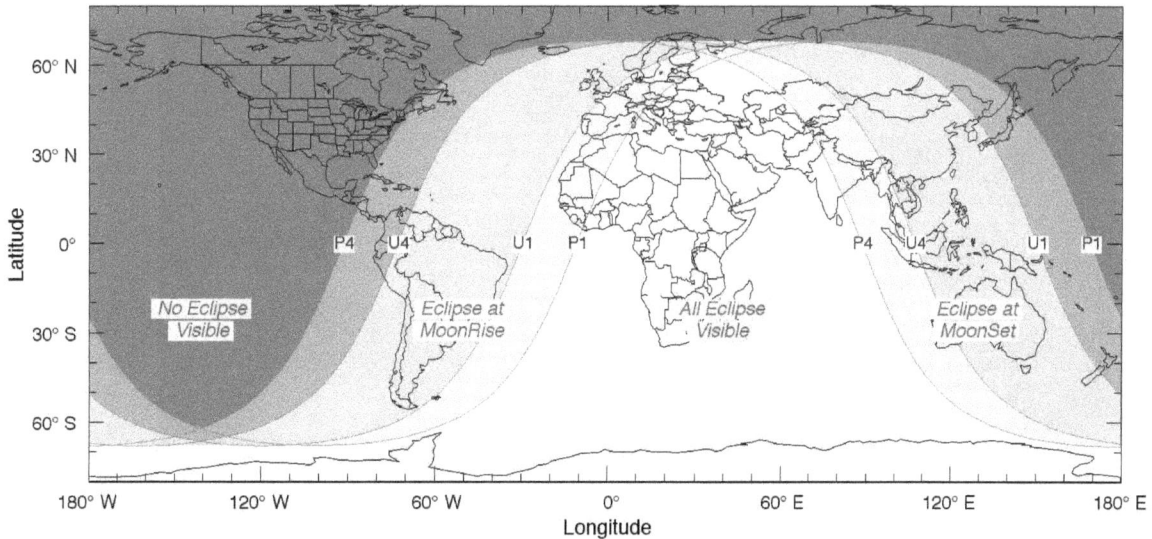

Eclipse Contacts
P1 = 18:43:29 UT1
U1 = 20:01:24 UT1
U4 = 23:00:06 UT1
P4 = 00:18:06 UT1

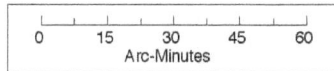

No Eclipse Visible

Eclipse at MoonRise

All Eclipse Visible

Eclipse at MoonSet

P4   U4   U1   P1   P4   U4   U1   P1

Penumbral Lunar Eclipse of 2020 Jan 10

Greatest Eclipse = 19:11:10.9 TD   (= 19:10:01.4 UT1)

Penumbral Magnitude = 0.8969	Gamma = 1.0727	Saros Series = 144
Umbral Magnitude = -0.1146	Axis = 1.0549°	Saros Member = 16 of 71

<u>Sun at Greatest Eclipse</u>
(Geocentric Coordinates)

R.A. = 19h26m32.0s
Dec. = -21°56'49.6"
S.D. = 00°16'15.9"
H.P. = 00°00'08.9"

<u>Moon at Greatest Eclipse</u>
(Geocentric Coordinates)

R.A. = 07h26m45.8s
Dec. = +23°00'02.8"
S.D. = 00°16'04.8"
H.P. = 00°59'00.8"

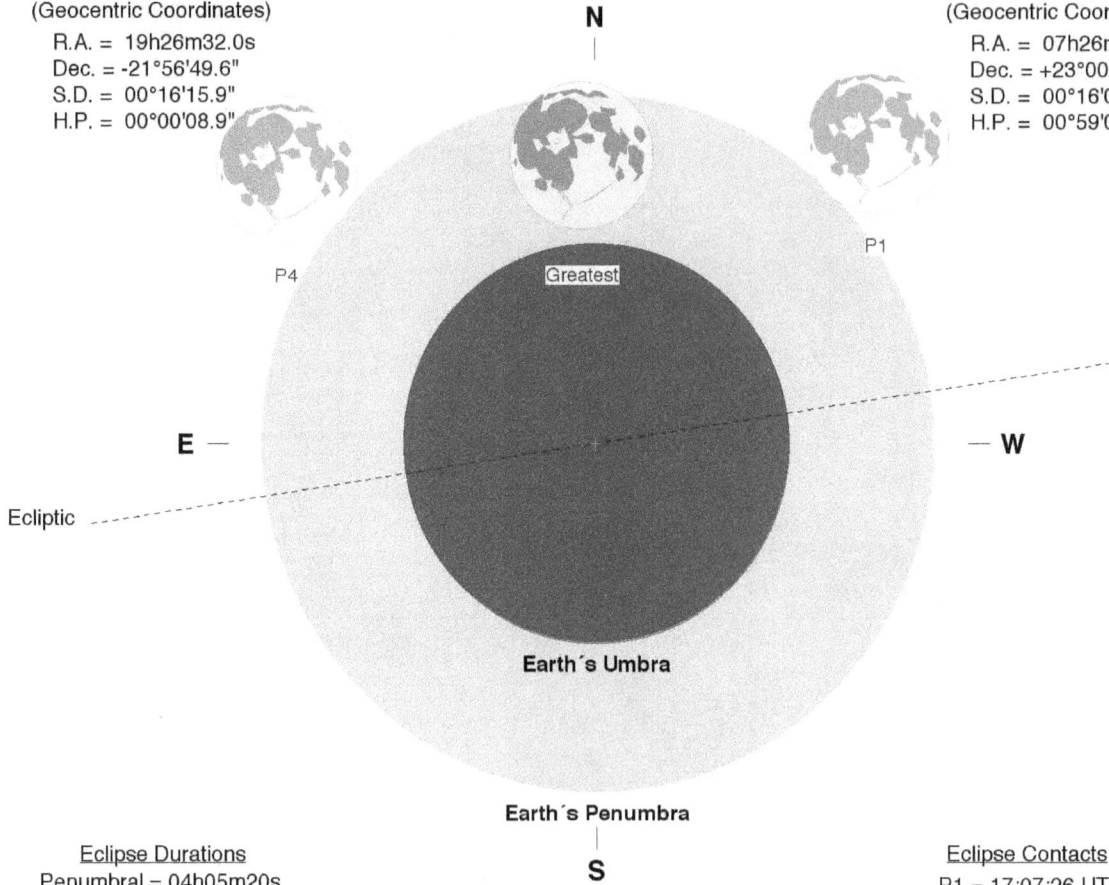

N

P4

P1

Greatest

E

W

Ecliptic

Earth's Umbra

Earth's Penumbra

S

<u>Eclipse Durations</u>
Penumbral = 04h05m20s

<u>Eclipse Contacts</u>
P1 = 17:07:26 UT1
P4 = 21:12:46 UT1

Eph. = JPL DE430
Rule = Herald-Sinnott
ΔT = 69 s

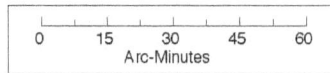

0	15	30	45	60

Arc-Minutes

©2020 F. Espenak, www.EclipseWise.com

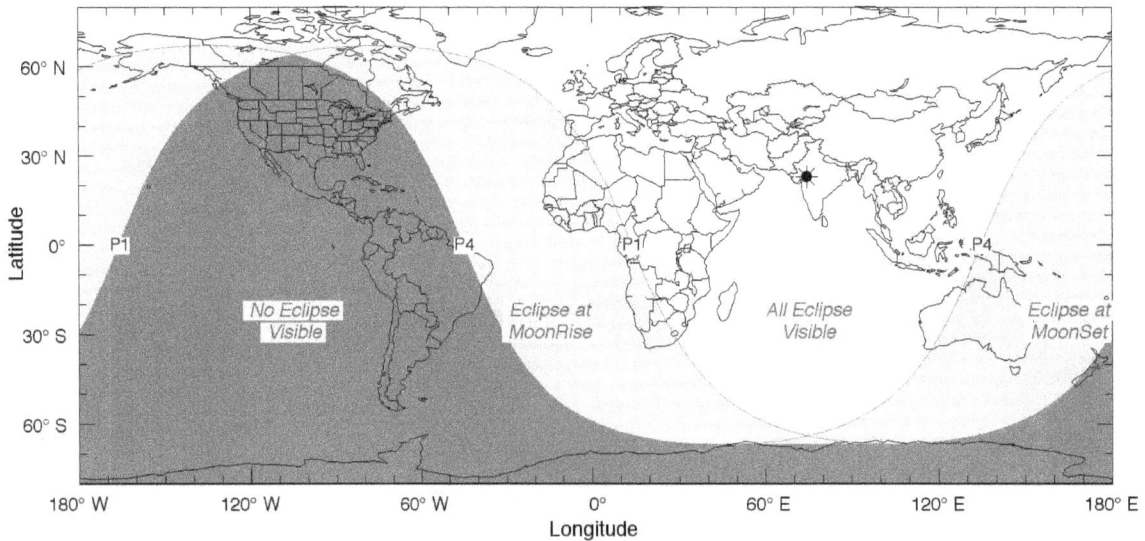

No Eclipse Visible

Eclipse at MoonRise

All Eclipse Visible

Eclipse at MoonSet

P1

P4

P1

P4

Penumbral Lunar Eclipse of 2020 Jun 05

Greatest Eclipse = 19:26:14.5 TD   (= 19:25:04.8 UT1)

Penumbral Magnitude = 0.5699	Gamma = 1.2406	Saros Series = 111
Umbral Magnitude = -0.4036	Axis = 1.2285°	Saros Member = 67 of 71

Sun at Greatest Eclipse
(Geocentric Coordinates)

R.A. = 04h57m21.6s
Dec. = +22°39'21.3"
S.D. = 00°15'45.7"
H.P. = 00°00'08.7"

N

Moon at Greatest Eclipse
(Geocentric Coordinates)

R.A. = 16h58m25.6s
Dec. = -21°27'08.8"
S.D. = 00°16'11.4"
H.P. = 00°59'25.1"

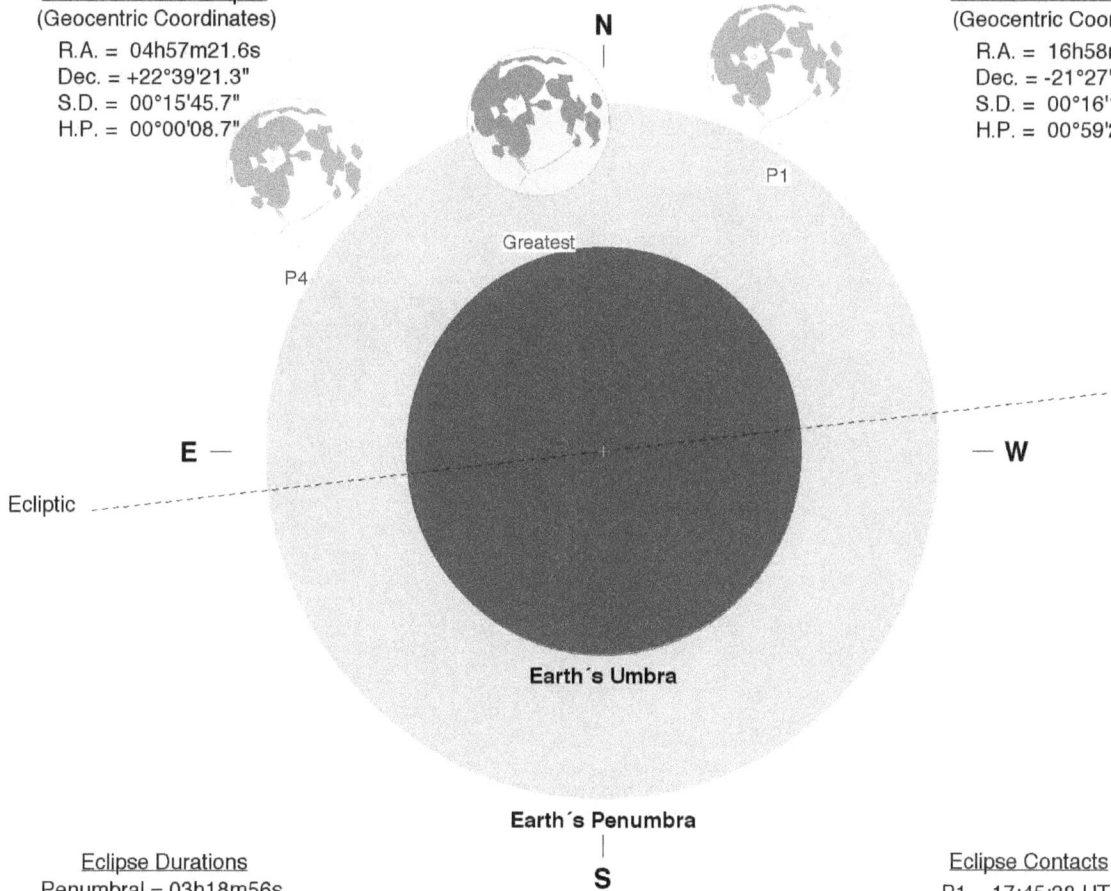

P1

Greatest

P4

E —

— **W**

Ecliptic

Earth´s Umbra

Earth´s Penumbra

S

Eclipse Durations
Penumbral = 03h18m56s

Eclipse Contacts
P1 = 17:45:38 UT1
P4 = 21:04:34 UT1

Eph. = JPL DE430
Rule = Herald-Sinnott
ΔT = 70 s

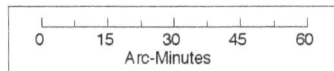

0	15	30	45	60

Arc-Minutes

©2020 F. Espenak, www.EclipseWise.com

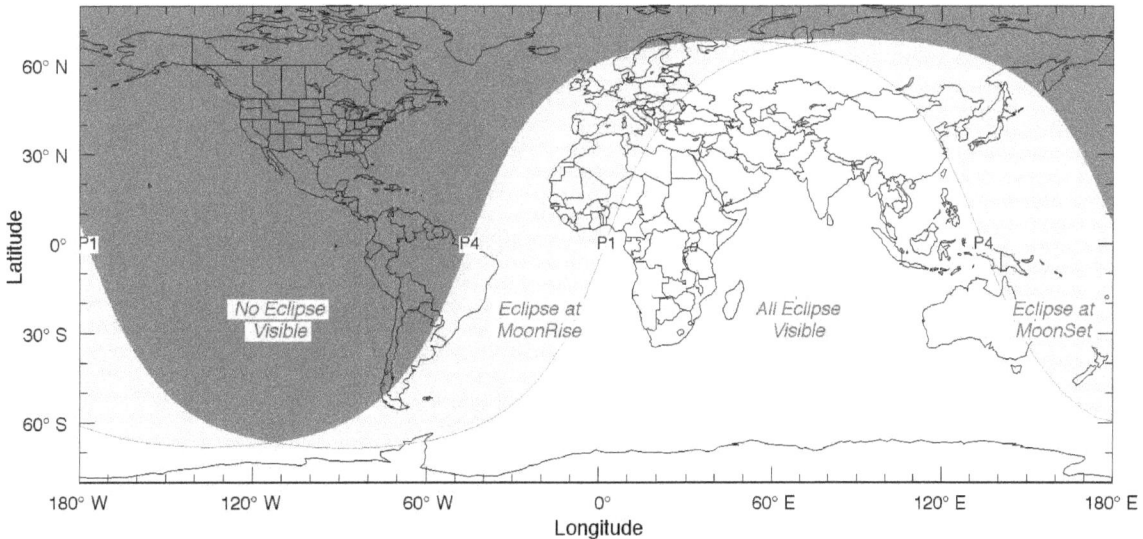

No Eclipse
Visible

Eclipse at
MoonRise

All Eclipse
Visible

Eclipse at
MoonSet

P1

P4

P1

P4

Penumbral Lunar Eclipse of 2020 Jul 05

Greatest Eclipse = 04:31:11.9 TD (= 04:30:02.2 UT1)

Penumbral Magnitude = 0.3560	Gamma = -1.3639	Saros Series = 149
Umbral Magnitude = -0.6422	Axis = 1.3146°	Saros Member = 3 of 71

Sun at Greatest Eclipse
(Geocentric Coordinates)
R.A. = 06h59m10.5s
Dec. = +22°44'23.3"
S.D. = 00°15'43.9"
H.P. = 00°00'08.6"

N

Earth´s Penumbra

Earth´s Umbra

Ecliptic

E —

— **W**

S

Moon at Greatest Eclipse
(Geocentric Coordinates)
R.A. = 18h59m12.6s
Dec. = -24°03'16.2"
S.D. = 00°15'45.6"
H.P. = 00°57'50.4"

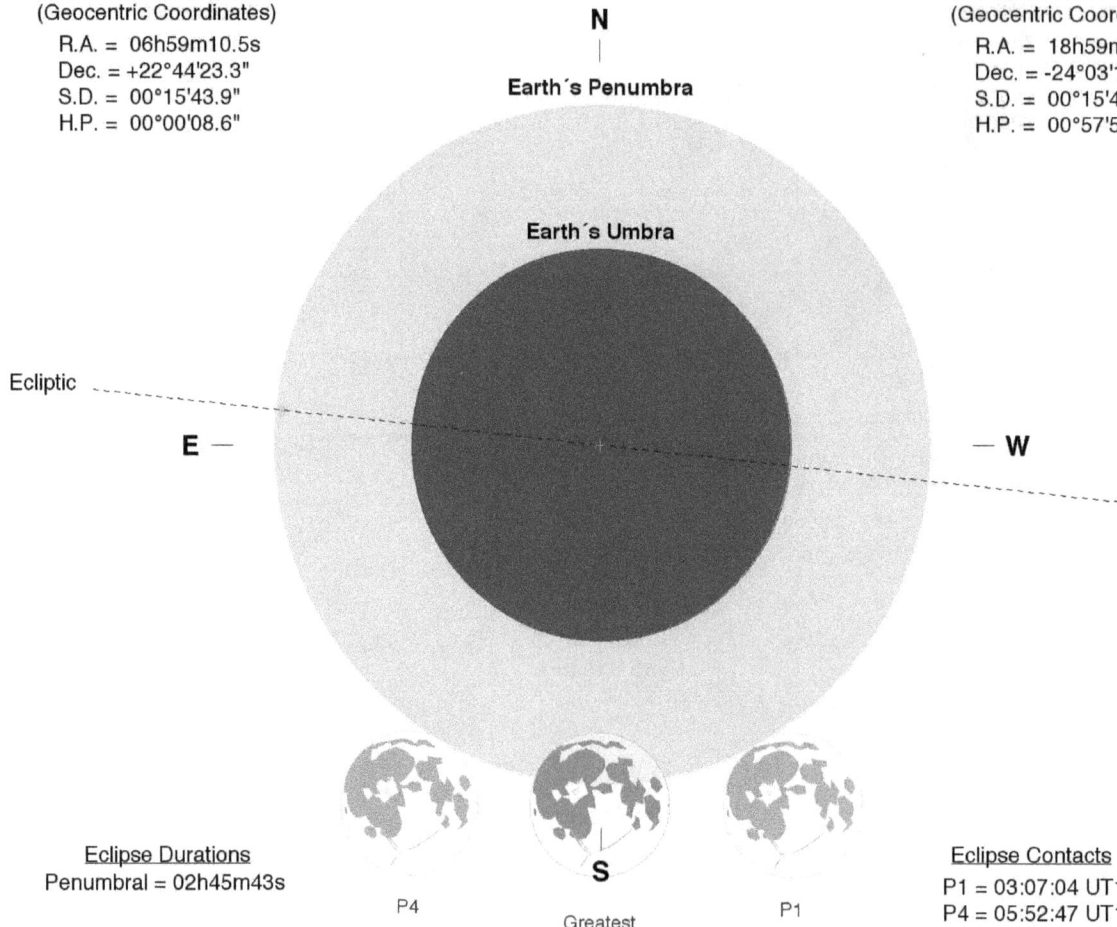

P4

Greatest

P1

Eclipse Durations
Penumbral = 02h45m43s

Eph. = JPL DE430
Rule = Herald-Sinnott
ΔT = 70 s

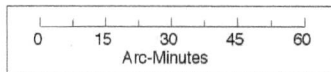

0	15	30	45	60
Arc-Minutes

Eclipse Contacts
P1 = 03:07:04 UT1
P4 = 05:52:47 UT1

©2020 F. Espenak, www.EclipseWise.com

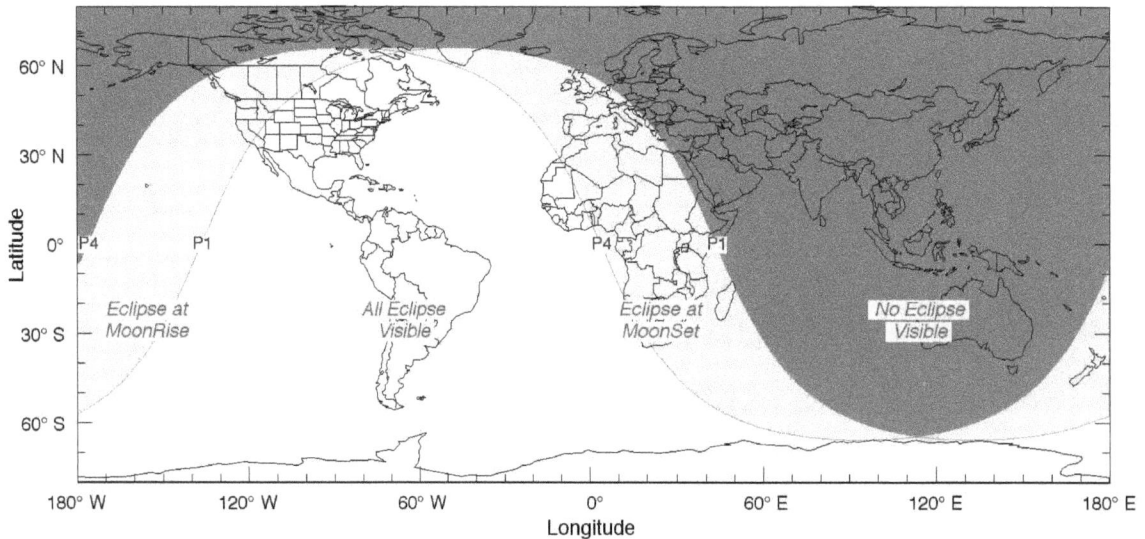

60° N

30° N

0°   P4   P1   P4   P1

Latitude

30° S   Eclipse at   All Eclipse   Eclipse at   No Eclipse
        MoonRise   Visible   MoonSet   Visible

60° S

180° W   120° W   60° W   0°   60° E   120° E   180° E

Longitude

Penumbral Lunar Eclipse of 2020 Nov 30

Greatest Eclipse = 09:44:01.7 TD (= 09:42:51.8 UT1)

Penumbral Magnitude = 0.8302	Gamma = -1.1309	Saros Series = 116
Umbral Magnitude = -0.2602	Axis = 1.0288°	Saros Member = 58 of 73

Sun at Greatest Eclipse
(Geocentric Coordinates)

R.A. = 16h27m40.0s
Dec. = -21°44'31.0"
S.D. = 00°16'13.1"
H.P. = 00°00'08.9"

N

Earth's Penumbra

Earth's Umbra

Ecliptic

E —

— **W**

P4

Greatest

S

P1

Moon at Greatest Eclipse
(Geocentric Coordinates)

R.A. = 04h28m46.7s
Dec. = +20°44'46.4"
S.D. = 00°14'52.4"
H.P. = 00°54'35.1"

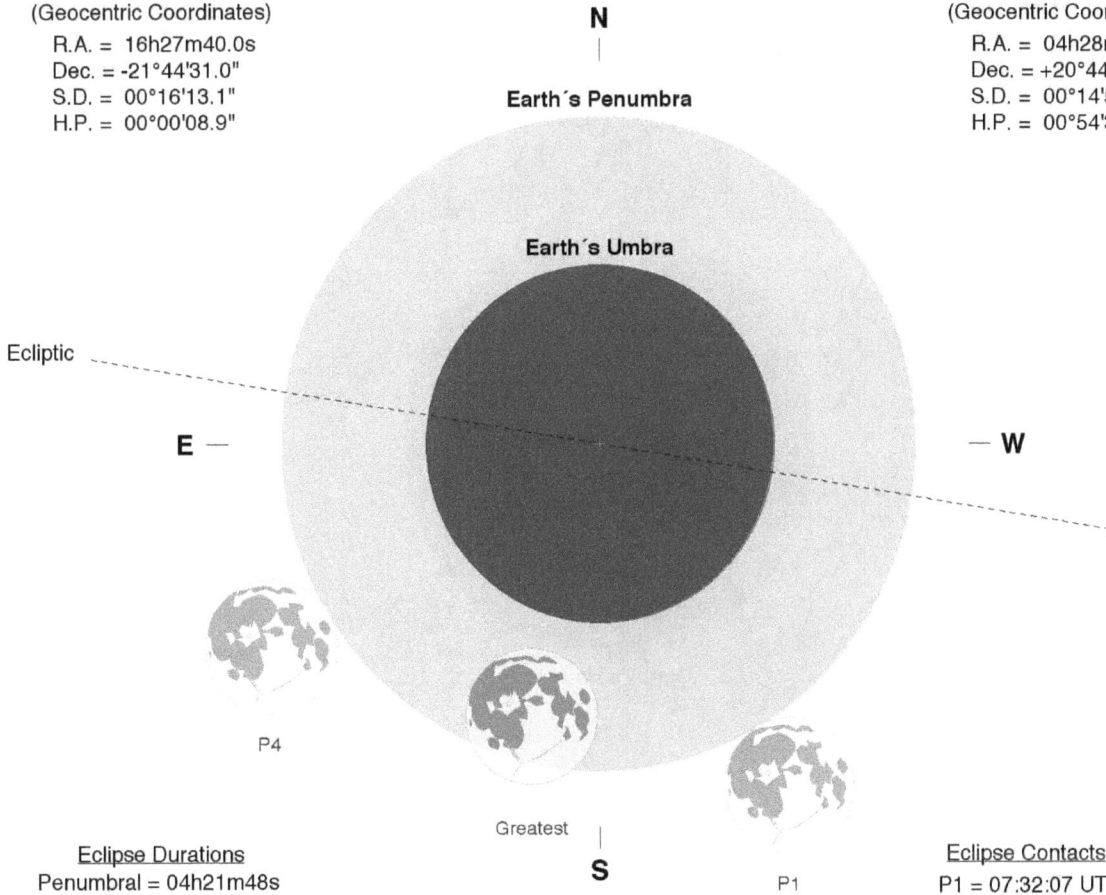

Eclipse Durations
Penumbral = 04h21m48s

Eclipse Contacts
P1 = 07:32:07 UT1
P4 = 11:53:55 UT1

Eph. = JPL DE430
Rule = Herald-Sinnott
ΔT = 70 s

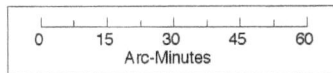

0   15   30   45   60
Arc-Minutes

©2020 F. Espenak, www.EclipseWise.com

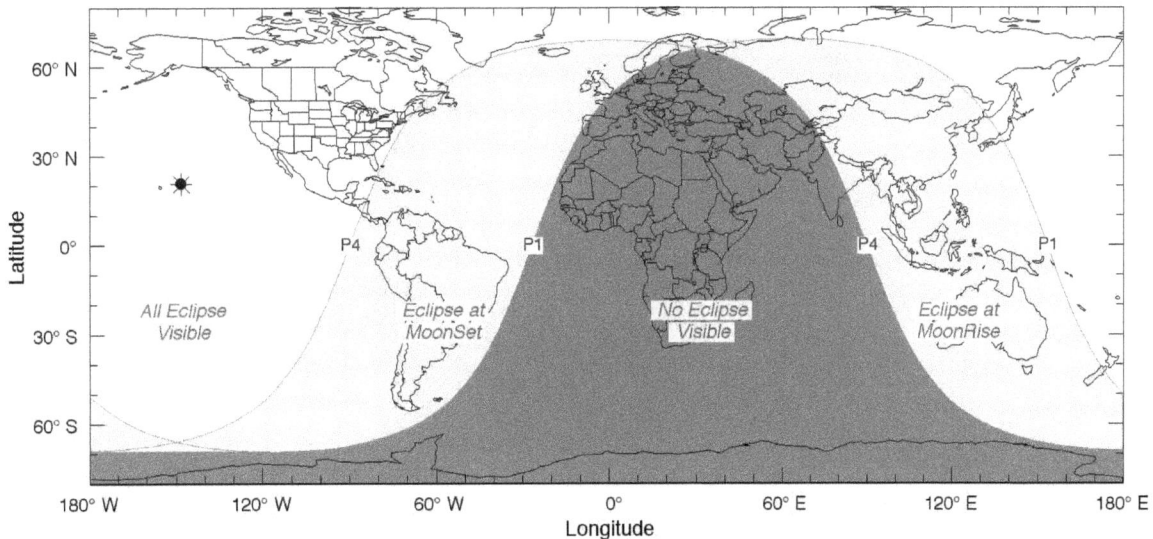

P4   P1   P4   P1

All Eclipse
Visible

Eclipse at
MoonSet

No Eclipse
Visible

Eclipse at
MoonRise

Total Lunar Eclipse of 2021 May 26

Greatest Eclipse = 11:19:52.7 TD   (= 11:18:42.7 UT1)

Penumbral Magnitude = 1.9558	Gamma = 0.4774	Saros Series = 121
Umbral Magnitude = 1.0112	Axis = 0.4880°	Saros Member = 55 of 82

Sun at Greatest Eclipse
(Geocentric Coordinates)
R.A. = 04h14m03.6s
Dec. = +21°12'25.4"
S.D. = 00°15'47.3"
H.P. = 00°00'08.7"

Moon at Greatest Eclipse
(Geocentric Coordinates)
R.A. = 16h14m37.8s
Dec. = -20°44'15.0"
S.D. = 00°16'42.9"
H.P. = 01°01'20.5"

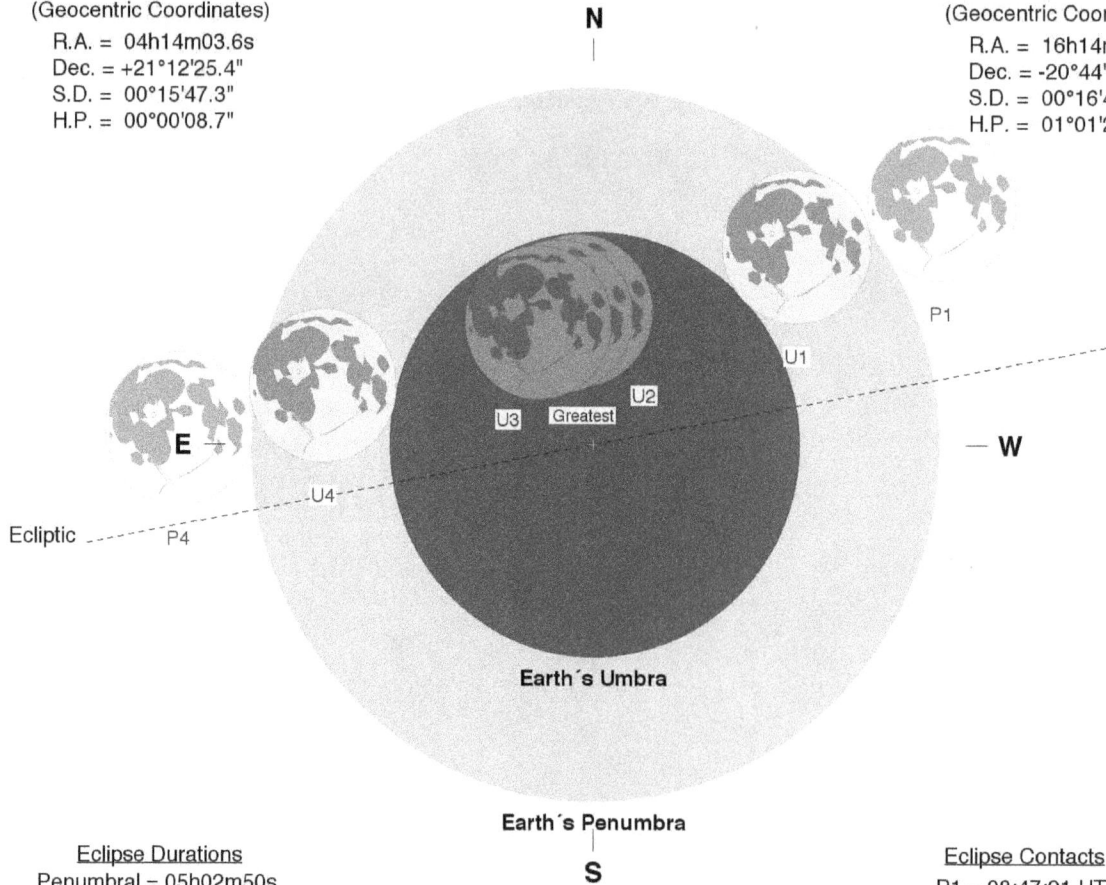

N

P1
U1
U2
U3 Greatest
E
W
U4
Ecliptic
P4

Earth's Umbra

Earth's Penumbra

S

Eclipse Durations
Penumbral = 05h02m50s
Umbral = 03h08m07s
Total = 00h15m52s

Eph. = JPL DE430
Rule = Herald-Sinnott
ΔT =   70 s

0   15   30   45   60
Arc-Minutes

©2020 F. Espenak, www.EclipseWise.com

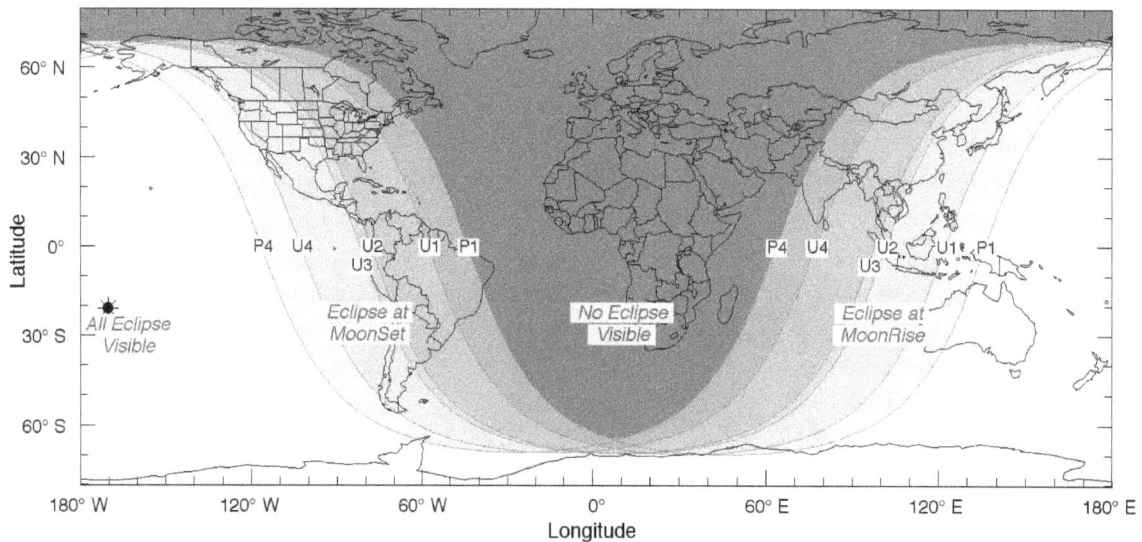

Eclipse Contacts
P1 = 08:47:21 UT1
U1 = 09:44:43 UT1
U2 = 11:10:56 UT1
U3 = 11:26:48 UT1
U4 = 12:52:50 UT1
P4 = 13:50:11 UT1

All Eclipse Visible

Eclipse at MoonSet

No Eclipse Visible

Eclipse at MoonRise

P4 U4 U2 U3 U1 P1   P4 U4 U2 U3 U1 P1

Partial Lunar Eclipse of 2021 Nov 19

Greatest Eclipse = 09:04:05.8 TD (= 09:02:55.5 UT1)

Penumbral Magnitude = 2.0738	Gamma = -0.4552	Saros Series = 126
Umbral Magnitude = 0.9760	Axis = 0.4104°	Saros Member = 45 of 70

Sun at Greatest Eclipse
(Geocentric Coordinates)
R.A. = 15h39m50.9s
Dec. = -19°32'33.1"
S.D. = 00°16'11.0"
H.P. = 00°00'08.9"

N

Earth's Penumbra

Earth's Umbra

Ecliptic

E

P4

U4

Greatest

U1

P1

— **W**

S

Moon at Greatest Eclipse
(Geocentric Coordinates)
R.A. = 03h40m24.8s
Dec. = +19°09'15.5"
S.D. = 00°14'44.5"
H.P. = 00°54'06.1"

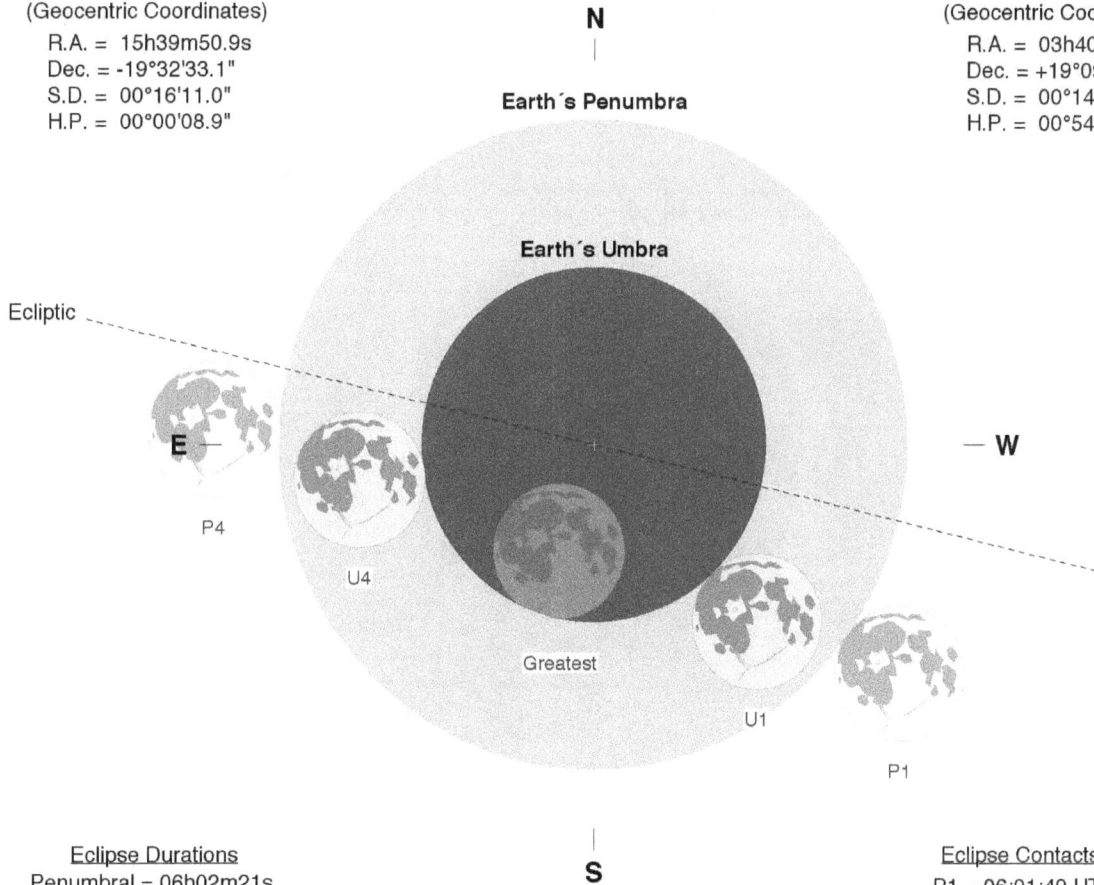

Eclipse Durations
Penumbral = 06h02m21s
Umbral = 03h29m09s

Eph. = JPL DE430
Rule = Herald-Sinnott
ΔT = 70 s

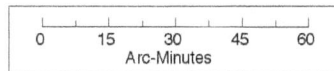

0 15 30 45 60
Arc-Minutes

©2020 F. Espenak, www.EclipseWise.com

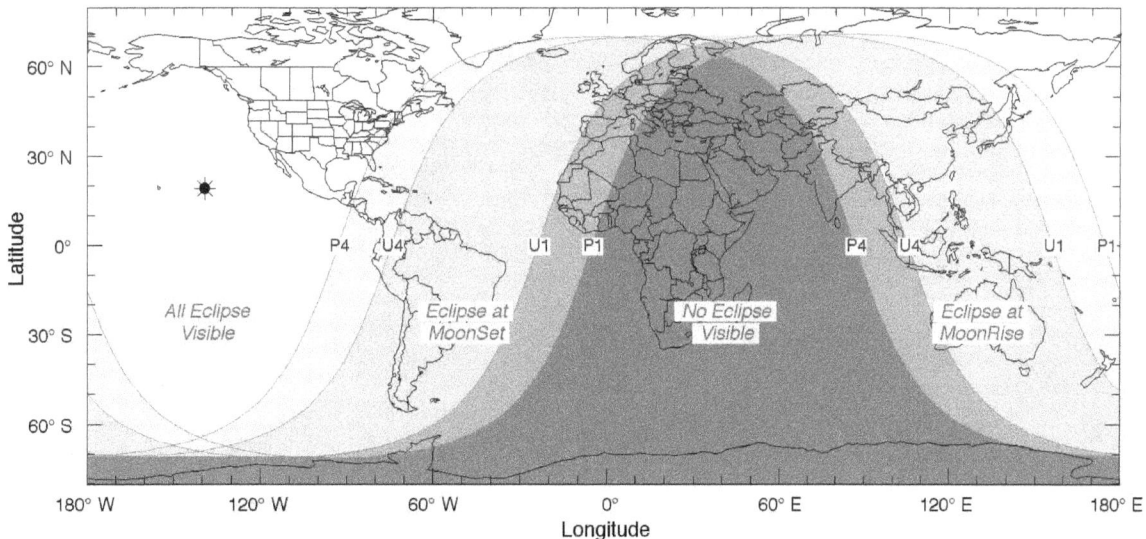

Eclipse Contacts
P1 = 06:01:49 UT1
U1 = 07:18:26 UT1
U4 = 10:47:36 UT1
P4 = 12:04:11 UT1

60° N

30° N

0°

30° S

60° S

Latitude

P4 U4   U1 P1   P4 U4   U1 P1

All Eclipse
Visible

Eclipse at
MoonSet

No Eclipse
Visible

Eclipse at
MoonRise

180° W   120° W   60° W   0°   60° E   120° E   180° E
Longitude

Total Lunar Eclipse of 2022 May 16

Greatest Eclipse = 04:12:41.7 TD (= 04:11:31.2 UT1)

Penumbral Magnitude = 2.3743	Gamma = -0.2532	Saros Series = 131
Umbral Magnitude = 1.4154	Axis = 0.2555°	Saros Member = 34 of 72

Sun at Greatest Eclipse
(Geocentric Coordinates)
R.A. = 03h31m49.5s
Dec. = +19°05'13.4"
S.D. = 00°15'49.2"
H.P. = 00°00'08.7"

Moon at Greatest Eclipse
(Geocentric Coordinates)
R.A. = 15h31m27.8s
Dec. = -19°19'40.4"
S.D. = 00°16'29.9"
H.P. = 01°00'33.1"

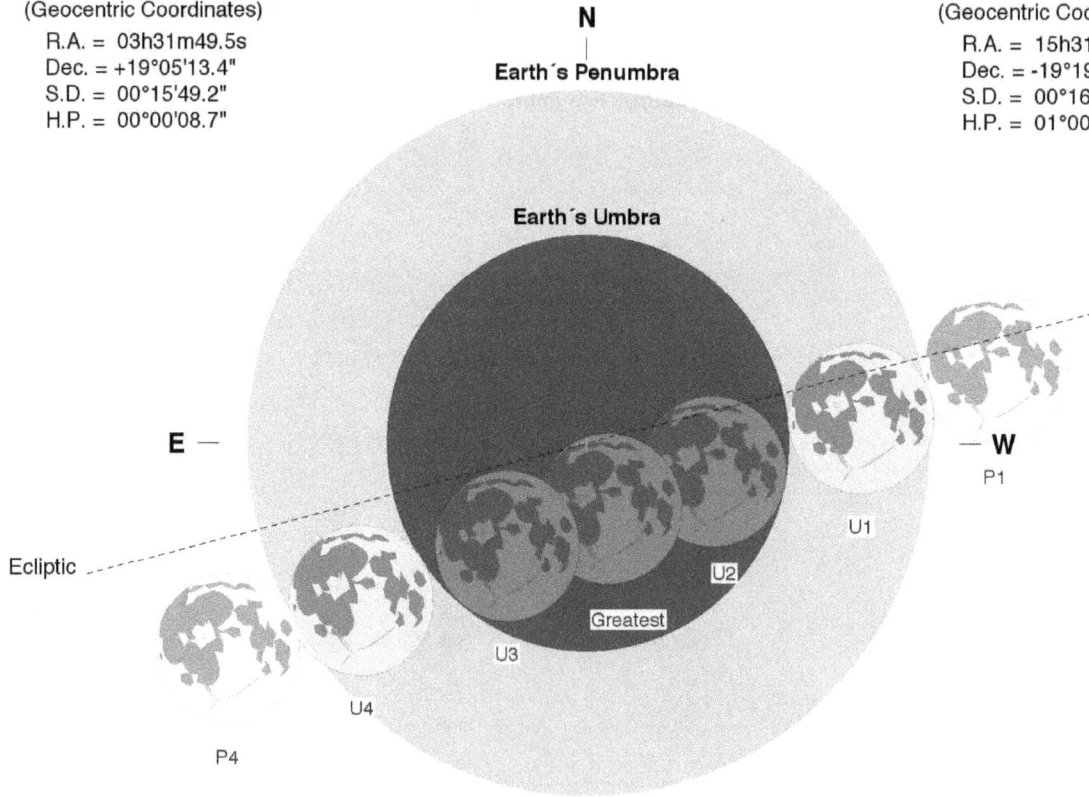

N

Earth's Penumbra

Earth's Umbra

E

W

P1

U1

U2

Greatest

U3

U4

P4

Ecliptic

S

Eclipse Durations
Penumbral = 05h19m28s
Umbral = 03h27m57s
Total = 01h25m32s

Eph. = JPL DE430
Rule = Herald-Sinnott
ΔT =   70 s

0   15   30   45   60
Arc-Minutes

©2020 F. Espenak, www.EclipseWise.com

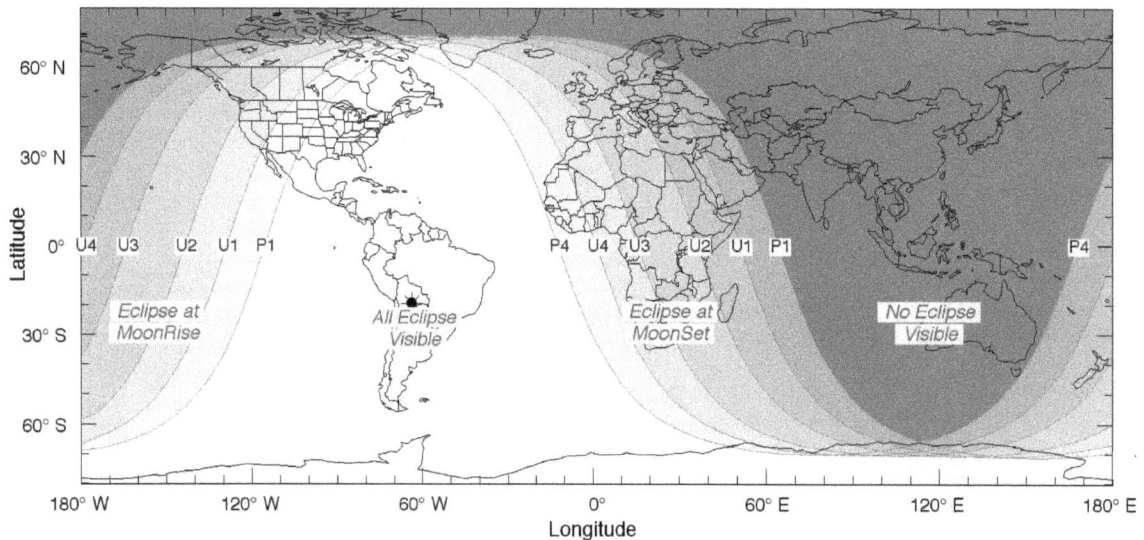

Eclipse Contacts
P1 = 01:31:44 UT1
U1 = 02:27:31 UT1
U2 = 03:28:40 UT1
U3 = 04:54:11 UT1
U4 = 05:55:27 UT1
P4 = 06:51:11 UT1

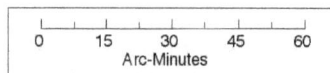

Eclipse at MoonRise

All Eclipse Visible

Eclipse at MoonSet

No Eclipse Visible

U4 U3 U2 U1 P1   P4 U4 U3 U2 U1 P1   P4

Total Lunar Eclipse of 2022 Nov 08

Greatest Eclipse = 11:00:22.0 TD (= 10:59:11.3 UT1)

Penumbral Magnitude = 2.4161	Gamma = 0.2570	Saros Series = 136
Umbral Magnitude = 1.3607	Axis = 0.2404°	Saros Member = 20 of 72

Sun at Greatest Eclipse
(Geocentric Coordinates)
R.A. = 14h54m11.2s
Dec. = -16°37'47.0"
S.D. = 00°16'08.5"
H.P. = 00°00'08.9"

Moon at Greatest Eclipse
(Geocentric Coordinates)
R.A. = 02h53m48.1s
Dec. = +16°51'06.7"
S.D. = 00°15'17.7"
H.P. = 00°56'07.8"

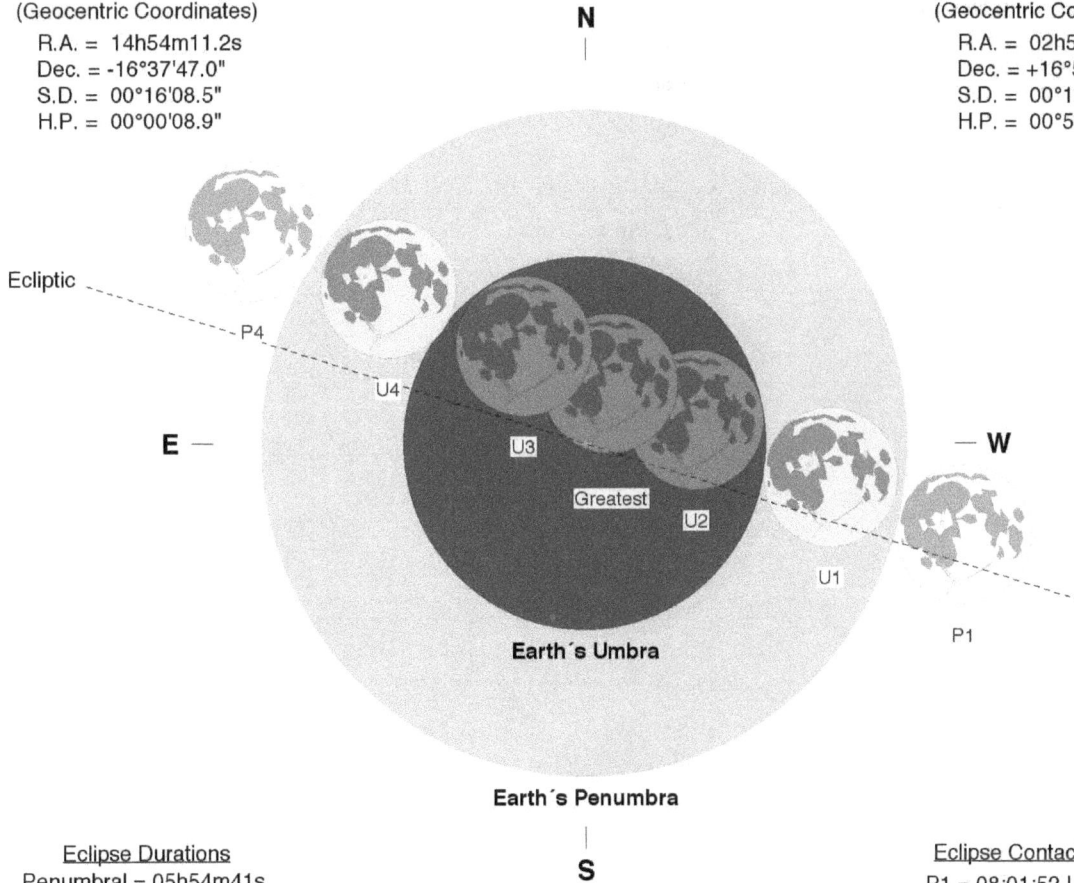

N

Ecliptic

P4

U4

E

U3

Greatest

U2

W

U1

P1

Earth´s Umbra

Earth´s Penumbra

S

Eclipse Durations
Penumbral = 05h54m41s
Umbral = 03h40m35s
Total = 01h25m40s

Eph. = JPL DE430
Rule = Herald-Sinnott
ΔT = 71 s

0   15   30   45   60
Arc-Minutes

©2020 F. Espenak, www.EclipseWise.com

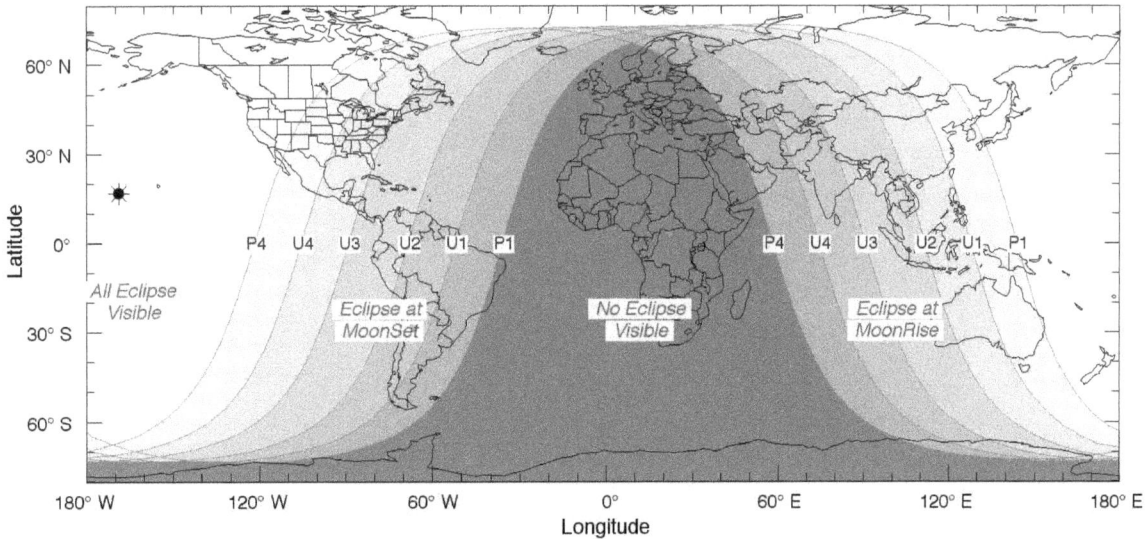

Eclipse Contacts
P1 = 08:01:52 UT1
U1 = 09:08:49 UT1
U2 = 10:16:12 UT1
U3 = 11:41:52 UT1
U4 = 12:49:24 UT1
P4 = 13:56:32 UT1

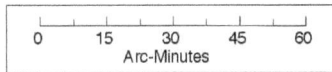

All Eclipse Visible

Eclipse at MoonSet

No Eclipse Visible

Eclipse at MoonRise

Latitude

Longitude

Penumbral Lunar Eclipse of 2023 May 05

Greatest Eclipse = 17:24:04.3 TD (= 17:22:53.5 UT1)

Penumbral Magnitude = 0.9655	Gamma = -1.0350	Saros Series = 141
Umbral Magnitude = -0.0438	Axis = 0.9946°	Saros Member = 24 of 72

Sun at Greatest Eclipse
(Geocentric Coordinates)
R.A. = 02h49m59.7s
Dec. = +16°19'27.9"
S.D. = 00°15'51.6"
H.P. = 00°00'08.7"

Moon at Greatest Eclipse
(Geocentric Coordinates)
R.A. = 14h48m23.5s
Dec. = -17°14'31.7"
S.D. = 00°15'42.8"
H.P. = 00°57'40.1"

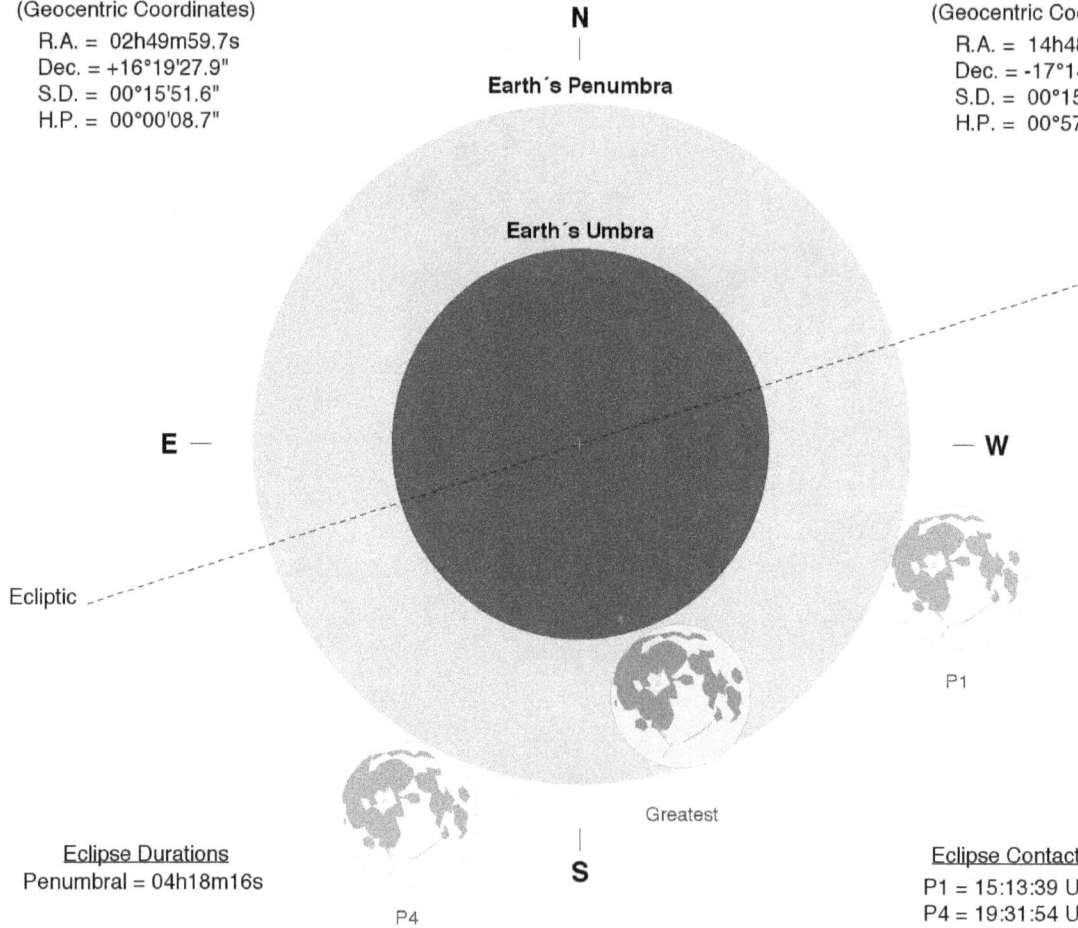

N

Earth's Penumbra

Earth's Umbra

E —

— **W**

Ecliptic

P1

Greatest

S

P4

Eclipse Durations
Penumbral = 04h18m16s

Eclipse Contacts
P1 = 15:13:39 UT1
P4 = 19:31:54 UT1

Eph. = JPL DE430
Rule = Herald-Sinnott
ΔT = 71 s

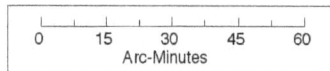

0	15	30	45	60

Arc-Minutes

©2020 F. Espenak, www.EclipseWise.com

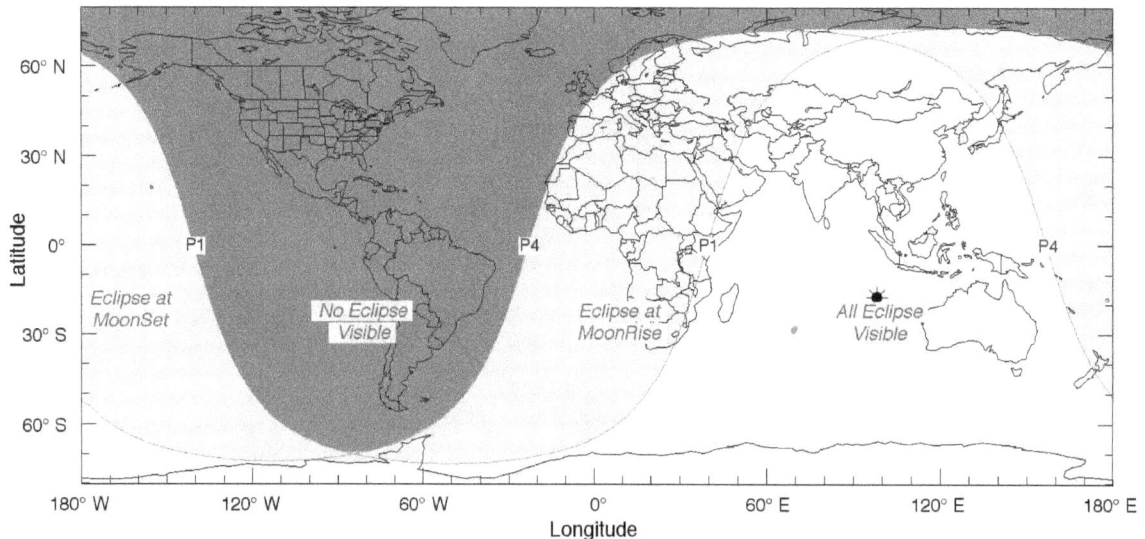

60° N

30° N

0°

Latitude

30° S

60° S

P1   P4   P1   P4

Eclipse at MoonSet   No Eclipse Visible   *Eclipse at MoonRise*   *All Eclipse Visible*

180° W   120° W   60° W   0°   60° E   120° E   180° E

Longitude

Partial Lunar Eclipse of 2023 Oct 28

Greatest Eclipse = 20:15:17.0 TD   (= 20:14:05.9 UT1)

Penumbral Magnitude = 1.1200	Gamma = 0.9472	Saros Series = 146
Umbral Magnitude = 0.1239	Axis = 0.9362°	Saros Member = 11 of 72

Sun at Greatest Eclipse
(Geocentric Coordinates)
R.A. = 14h11m25.9s
Dec. = -13°14'10.5"
S.D. = 00°16'05.9"
H.P. = 00°00'08.9"

Moon at Greatest Eclipse
(Geocentric Coordinates)
R.A. = 02h09m47.6s
Dec. = +14°05'01.6"
S.D. = 00°16'09.7"
H.P. = 00°59'18.9"

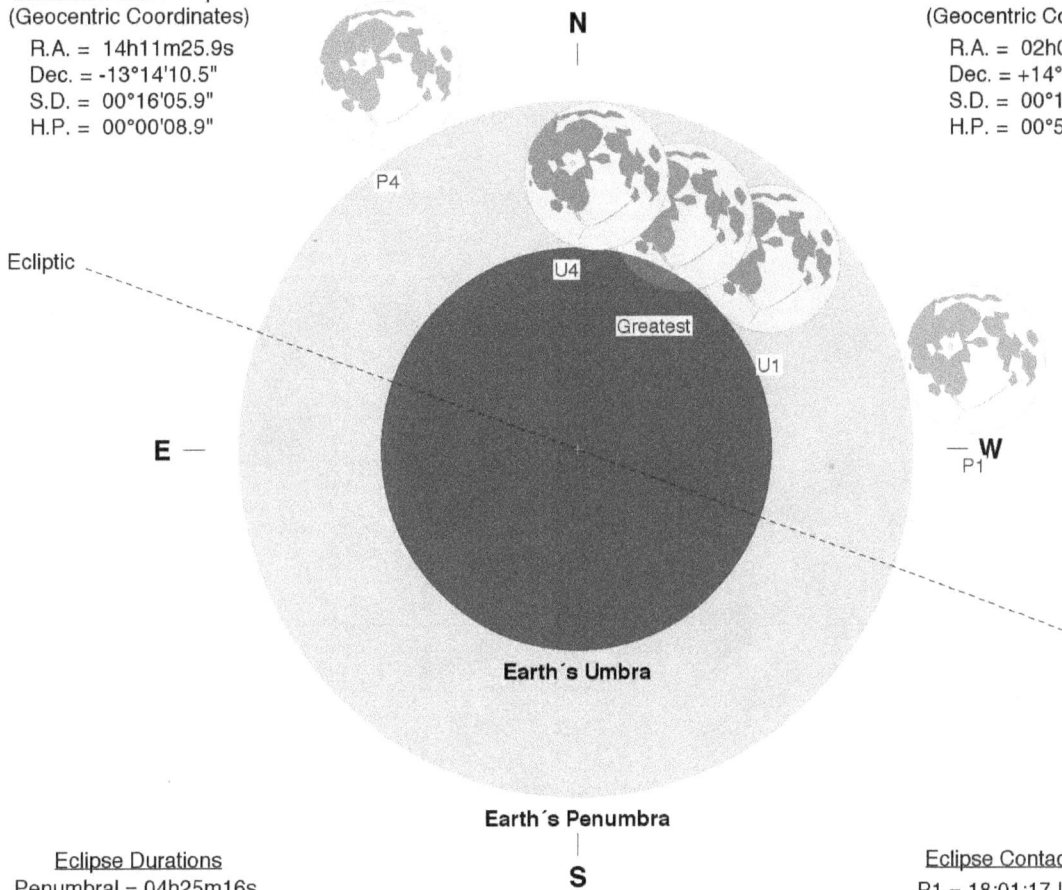

N

Ecliptic

P4

U4

Greatest

U1

E

W
P1

Earth's Umbra

Earth's Penumbra

S

Eclipse Durations
Penumbral = 04h25m16s
Umbral = 01h18m09s

Eph. = JPL DE430
Rule = Herald-Sinnott
ΔT =   71 s

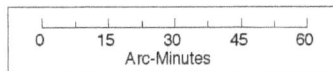

Eclipse Contacts
P1 = 18:01:17 UT1
U1 = 19:34:41 UT1
U4 = 20:52:50 UT1
P4 = 22:26:33 UT1

0   15   30   45   60
Arc-Minutes

©2020 F. Espenak, www.EclipseWise.com

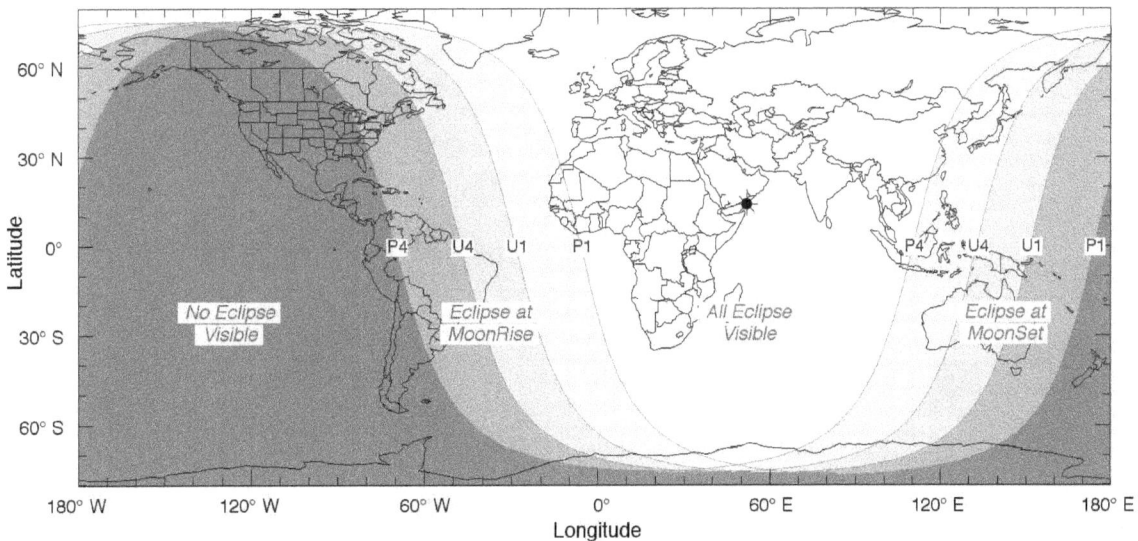

No Eclipse Visible

Eclipse at MoonRise

All Eclipse Visible

Eclipse at MoonSet

P4 U4 U1 P1   P4 U4 U1 P1

Penumbral Lunar Eclipse of 2024 Mar 25

Greatest Eclipse = 07:14:00.1 TD (= 07:12:48.8 UT1)

Penumbral Magnitude = 0.9577	Gamma = 1.0610	Saros Series = 113
Umbral Magnitude = -0.1304	Axis = 0.9564°	Saros Member = 64 of 71

Sun at Greatest Eclipse
(Geocentric Coordinates)
R.A. = 00h18m49.9s
Dec. = +02°02'16.6"
S.D. = 00°16'02.2"
H.P. = 00°00'08.8"

Moon at Greatest Eclipse
(Geocentric Coordinates)
R.A. = 12h20m41.3s
Dec. = -01°12'05.6"
S.D. = 00°14'44.3"
H.P. = 00°54'05.4"

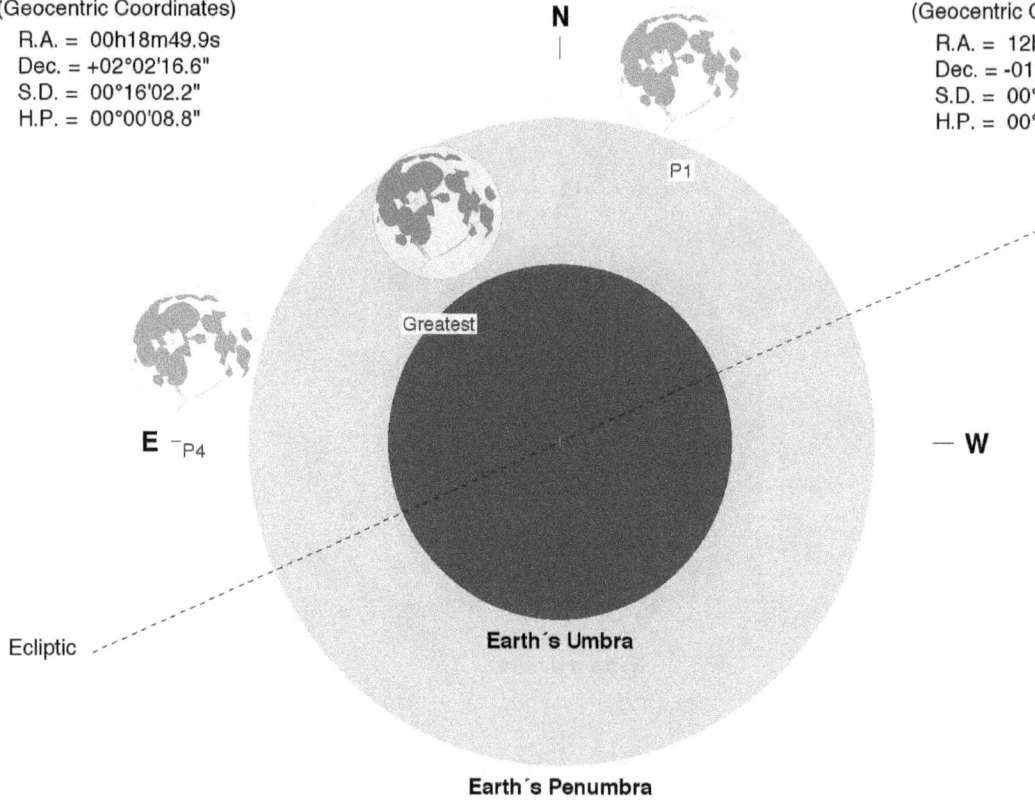

N

P1

Greatest

E P4

W

Ecliptic

Earth's Umbra

Earth's Penumbra

S

Eclipse Durations
Penumbral = 04h39m52s

Eclipse Contacts
P1 = 04:53:07 UT1
P4 = 09:32:59 UT1

Eph. = JPL DE430
Rule = Herald-Sinnott
ΔT = 71 s

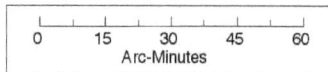

0	15	30	45	60

Arc-Minutes

©2020 F. Espenak, www.EclipseWise.com

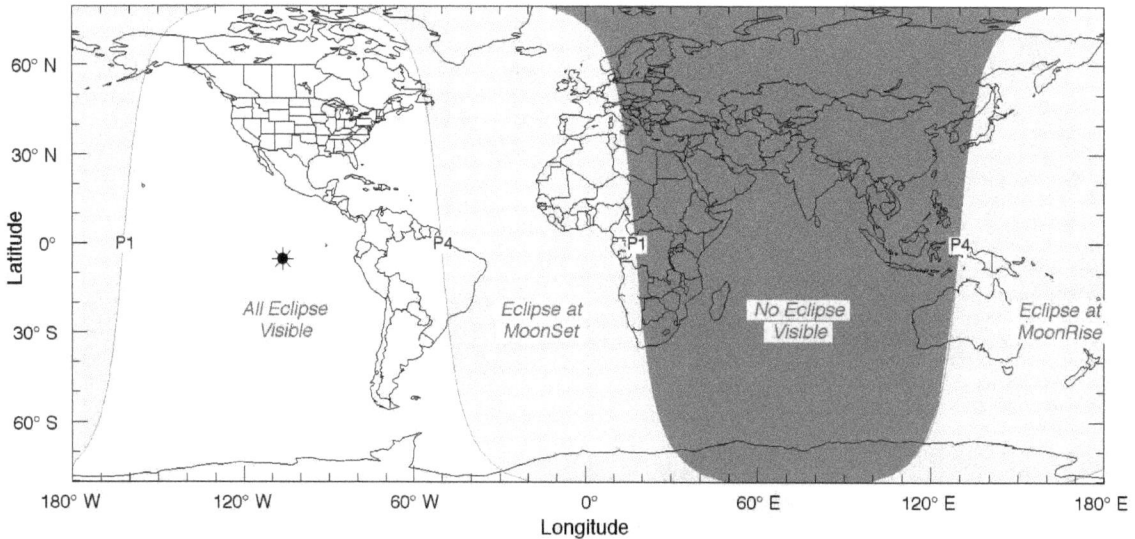

All Eclipse Visible

Eclipse at MoonSet

No Eclipse Visible

Eclipse at MoonRise

P1

P4

P1

P4

Partial Lunar Eclipse of 2024 Sep 18

Greatest Eclipse = 02:45:25.5 TD (= 02:44:14.1 UT1)

Penumbral Magnitude = 1.0392	Gamma = -0.9792	Saros Series = 118
Umbral Magnitude = 0.0869	Axis = 1.0009°	Saros Member = 52 of 73

Sun at Greatest Eclipse
(Geocentric Coordinates)

R.A. = 11h44m09.7s
Dec. = +01°42'52.9"
S.D. = 00°15'55.1"
H.P. = 00°00'08.8"

Moon at Greatest Eclipse
(Geocentric Coordinates)

R.A. = 23h46m06.1s
Dec. = -02°35'26.7"
S.D. = 00°16'42.8"
H.P. = 01°01'20.4"

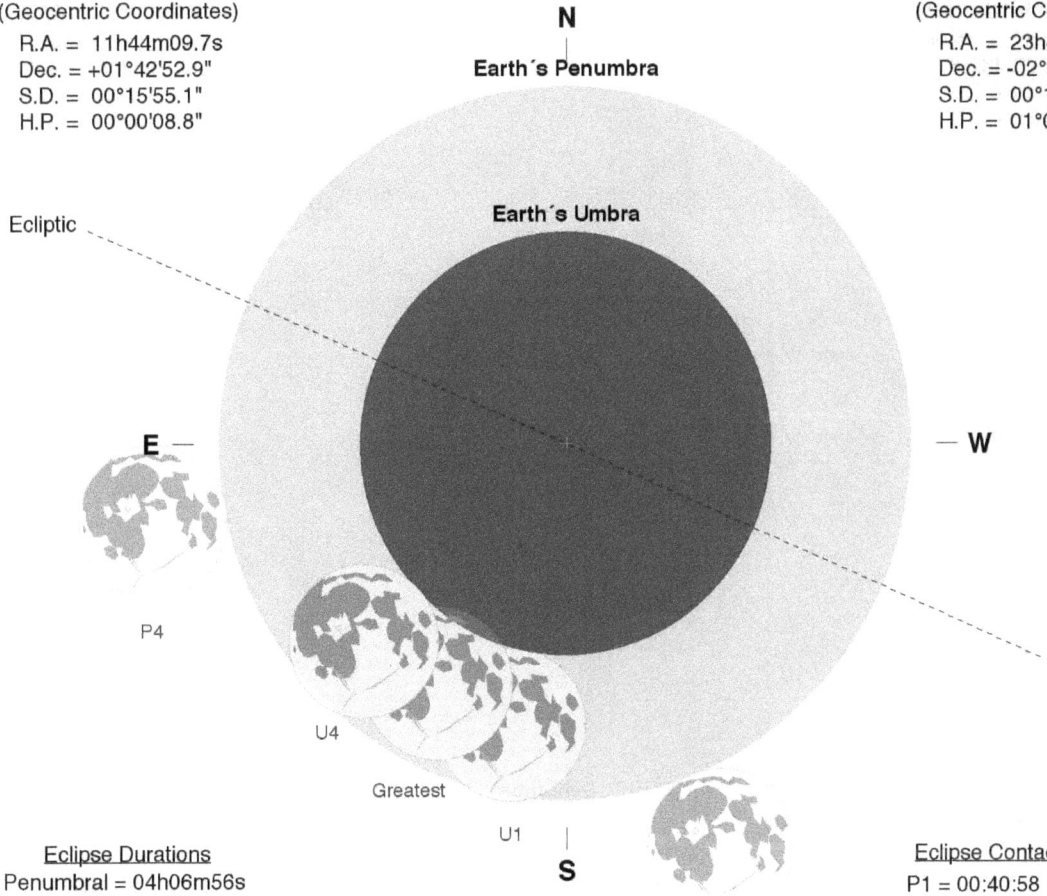

N

Earth's Penumbra

Earth's Umbra

Ecliptic

E

W

P4

U4

Greatest

U1

S

Eclipse Durations
Penumbral = 04h06m56s
Umbral = 01h03m40s

Eph. = JPL DE430
Rule = Herald-Sinnott
ΔT = 71 s

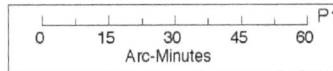

Eclipse Contacts
P1 = 00:40:58 UT1
U1 = 02:12:42 UT1
U4 = 03:16:22 UT1
P4 = 04:47:54 UT1

P1

0	15	30	45	60

Arc-Minutes

©2020 F. Espenak, www.EclipseWise.com

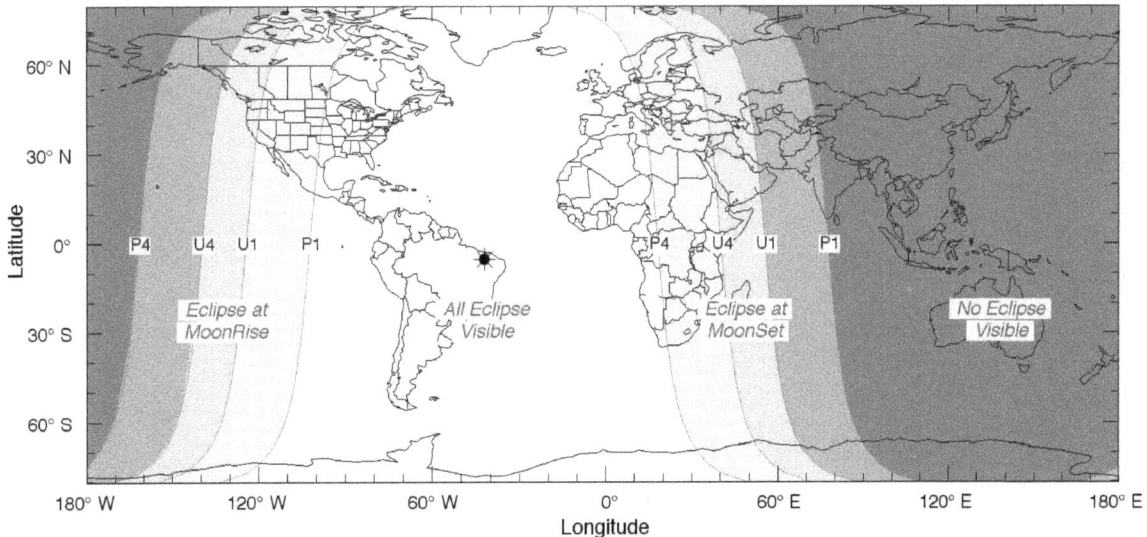

Total Lunar Eclipse of 2025 Mar 14

Greatest Eclipse = 06:59:56.2 TD (= 06:58:44.5 UT1)

Penumbral Magnitude = 2.2615	Gamma = 0.3485	Saros Series = 123
Umbral Magnitude = 1.1804	Axis = 0.3171°	Saros Member = 53 of 72

Sun at Greatest Eclipse
(Geocentric Coordinates)

R.A. = 23h37m46.0s
Dec. = -02°24'16.8"
S.D. = 00°16'05.2"
H.P. = 00°00'08.8"

Moon at Greatest Eclipse
(Geocentric Coordinates)

R.A. = 11h38m23.0s
Dec. = +02°40'54.6"
S.D. = 00°14'52.8"
H.P. = 00°54'36.8"

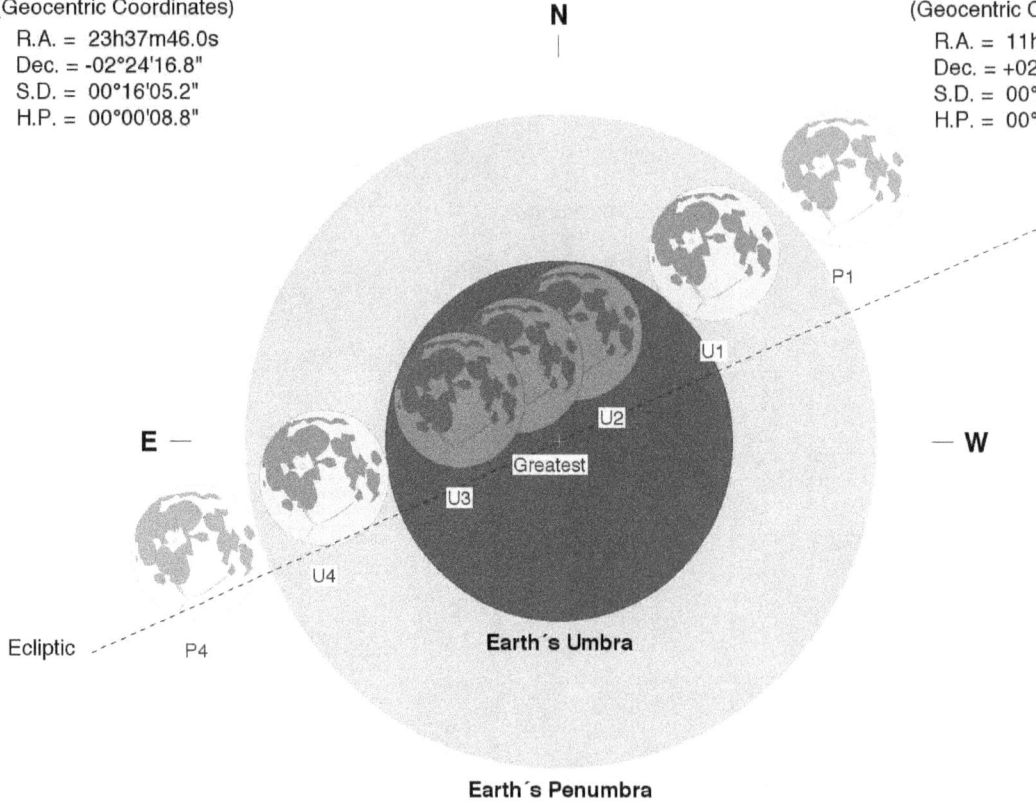

N

W

E

S

Greatest

U1
U2
U3
U4
P1
P4

Earth's Umbra

Earth's Penumbra

Ecliptic

Eclipse Durations
Penumbral = 06h03m22s
Umbral = 03h38m56s
Total = 01h06m04s

Eph. = JPL DE430
Rule = Herald-Sinnott
ΔT = 72 s

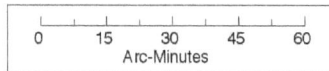

0   15   30   45   60
Arc-Minutes

©2020 F. Espenak, www.EclipseWise.com

Eclipse Contacts
P1 = 03:57:09 UT1
U1 = 05:09:23 UT1
U2 = 06:25:58 UT1
U3 = 07:32:01 UT1
U4 = 08:48:18 UT1
P4 = 10:00:32 UT1

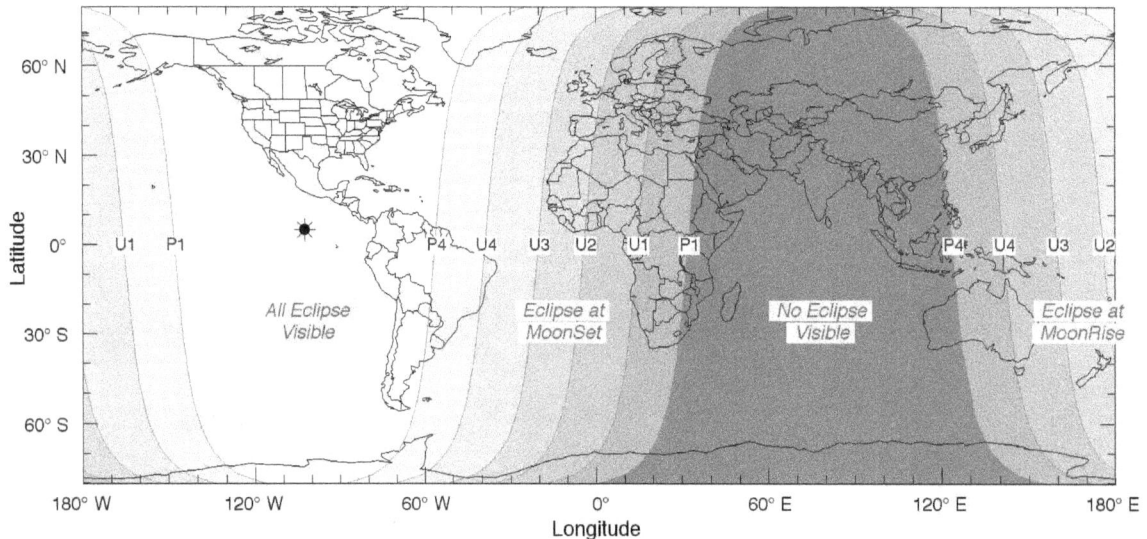

All Eclipse Visible

Eclipse at MoonSet

No Eclipse Visible

Eclipse at MoonRise

U1   P1   P4   U4   U3   U2   U1   P1   P4   U4   U3   U2

Total Lunar Eclipse of 2025 Sep 07

Greatest Eclipse = 18:12:58.0 TD   (= 18:11:46.1 UT1)

Penumbral Magnitude = 2.3459	Gamma = -0.2752	Saros Series = 128
Umbral Magnitude = 1.3638	Axis = 0.2721°	Saros Member = 41 of 71

Sun at Greatest Eclipse
(Geocentric Coordinates)

R.A. = 11h06m09.1s
Dec. = +05°45'47.6"
S.D. = 00°15'52.4"
H.P. = 00°00'08.7"

Moon at Greatest Eclipse
(Geocentric Coordinates)

R.A. = 23h06m40.4s
Dec. = -06°00'08.9"
S.D. = 00°16'09.8"
H.P. = 00°59'19.1"

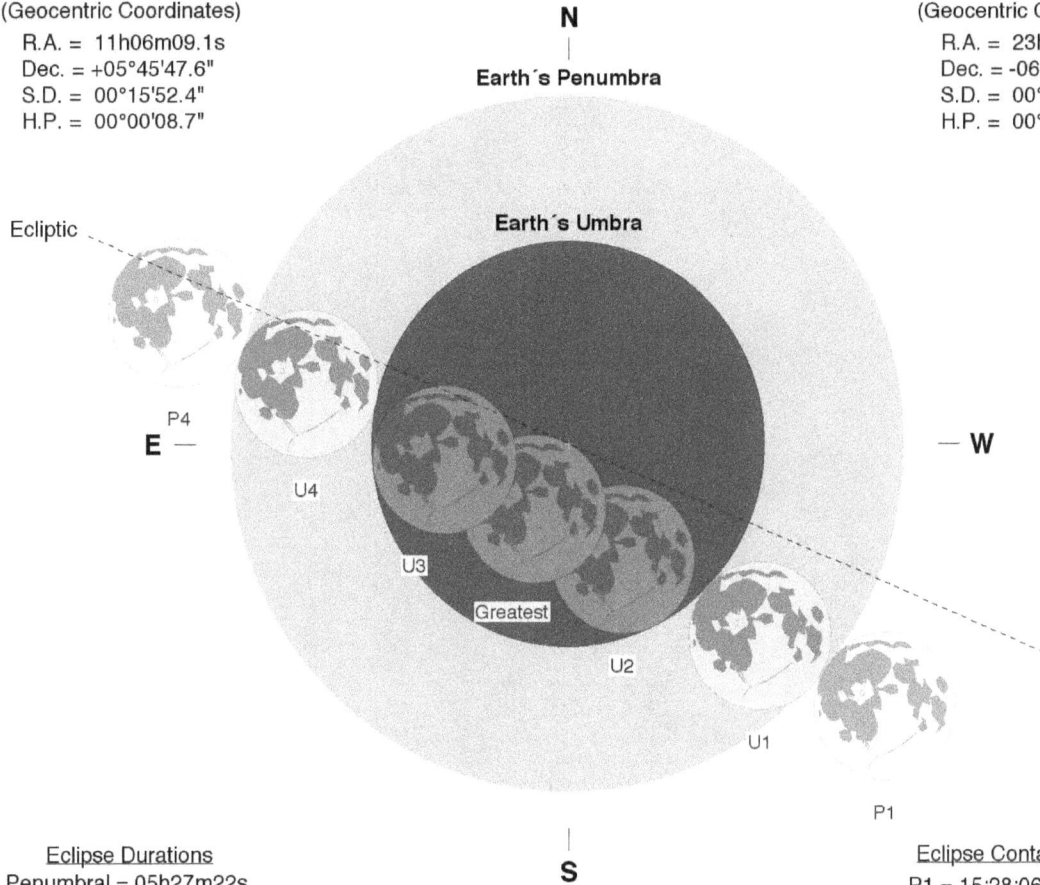

N

Earth's Penumbra

Earth's Umbra

Ecliptic

P4

E

U4

U3

Greatest

U2

U1

P1

W

S

Eclipse Durations
Penumbral = 05h27m22s
Umbral = 03h30m02s
Total = 01h22m41s

Eph. = JPL DE430
Rule = Herald-Sinnott
ΔT = 72 s

0	15	30	45	60
Arc-Minutes

©2020 F. Espenak, www.EclipseWise.com

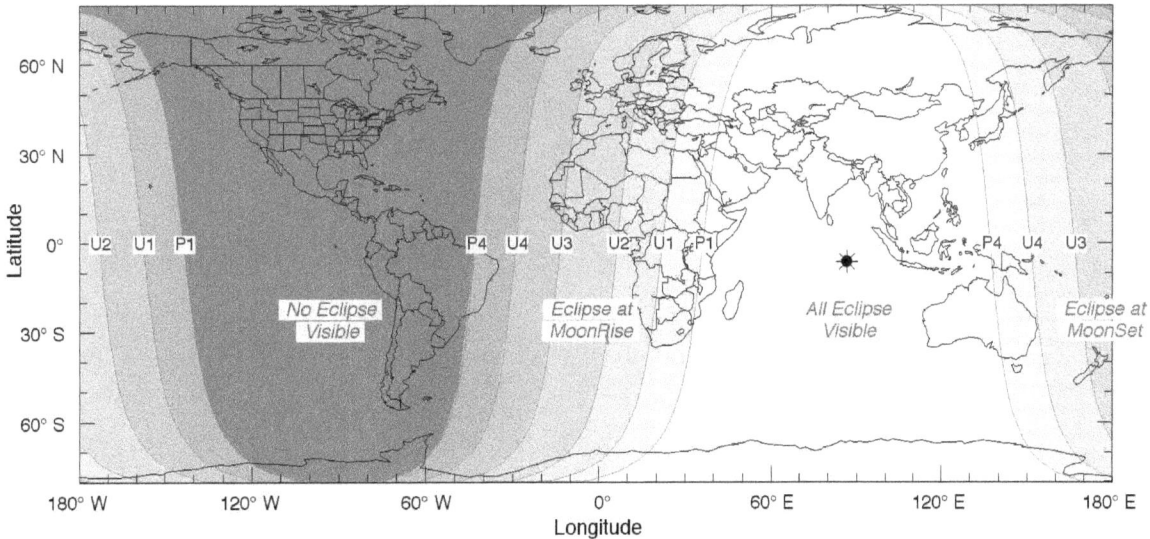

Eclipse Contacts
P1 = 15:28:06 UT1
U1 = 16:26:51 UT1
U2 = 17:30:37 UT1
U3 = 18:53:18 UT1
U4 = 19:56:53 UT1
P4 = 20:55:28 UT1

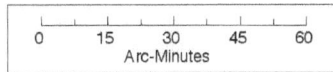

No Eclipse Visible

Eclipse at MoonRise

All Eclipse Visible

Eclipse at MoonSet

Total Lunar Eclipse of 2026 Mar 03

Greatest Eclipse = 11:34:52.1 TD   (= 11:33:40.0 UT1)

Penumbral Magnitude = 2.1858	Gamma = -0.3765	Saros Series = 133
Umbral Magnitude = 1.1526	Axis = 0.3596°	Saros Member = 27 of 71

Sun at Greatest Eclipse
(Geocentric Coordinates)
R.A. = 22h56m56.0s
Dec. = -06°43'06.4"
S.D. = 00°16'08.0"
H.P. = 00°00'08.9"

Moon at Greatest Eclipse
(Geocentric Coordinates)
R.A. = 10h56m15.0s
Dec. = +06°24'05.3"
S.D. = 00°15'37.0"
H.P. = 00°57'18.7"

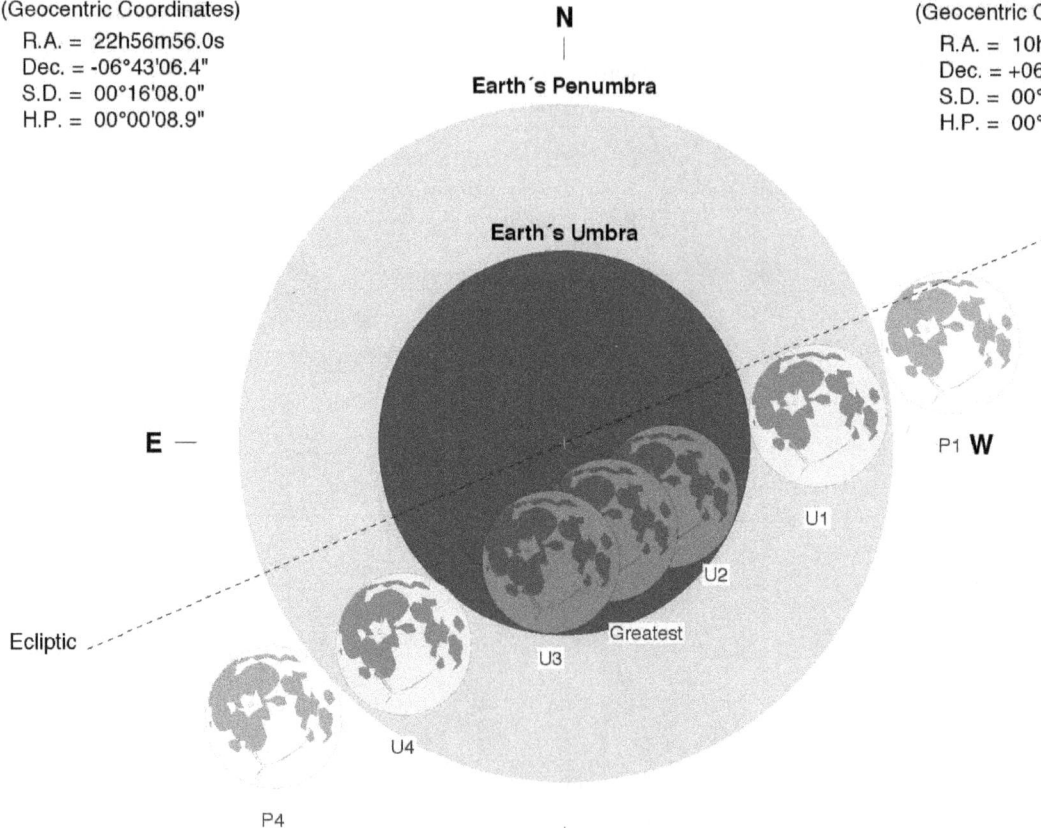

N

Earth's Penumbra

Earth's Umbra

E —

P1 **W**

U1

U2

Greatest

U3

Ecliptic

U4

P4

S

Eclipse Durations
Penumbral = 05h39m21s
Umbral = 03h27m49s
Total = 00h58m58s

Eph. = JPL DE430
Rule = Herald-Sinnott
ΔT =   72 s

0	15	30	45	60

Arc-Minutes

©2020 F. Espenak, www.EclipseWise.com

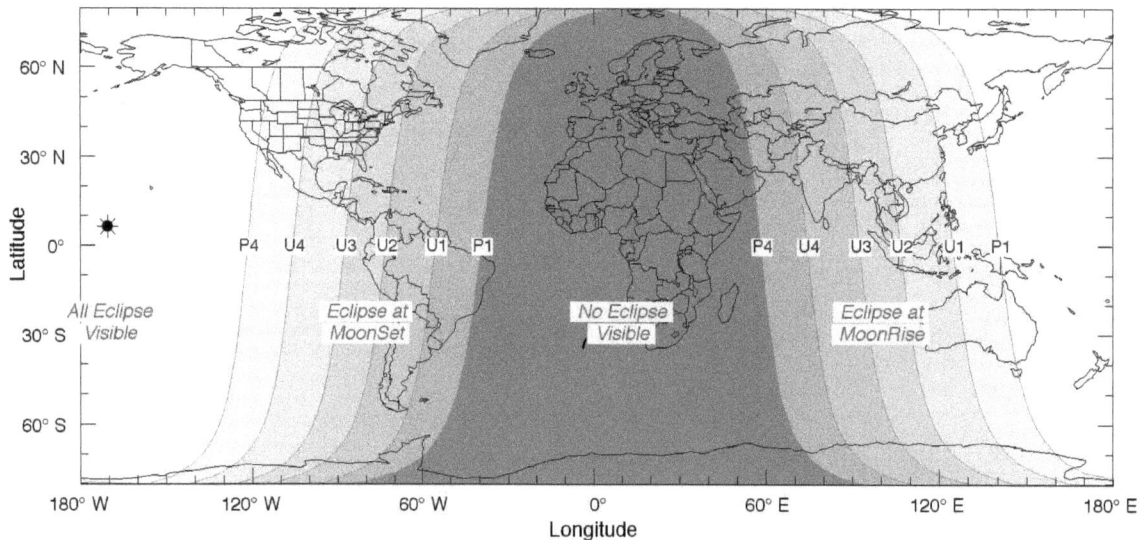

Eclipse Contacts
P1 = 08:43:58 UT1
U1 = 09:49:37 UT1
U2 = 11:03:54 UT1
U3 = 12:02:53 UT1
U4 = 13:17:26 UT1
P4 = 14:23:19 UT1

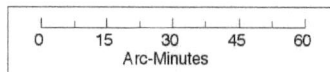

All Eclipse Visible

Eclipse at MoonSet

No Eclipse Visible

Eclipse at MoonRise

P4 U4 U3 U2 U1 P1   P4 U4 U3 U2 U1 P1

Longitude

Partial Lunar Eclipse of 2026 Aug 28

Greatest Eclipse = 04:14:04.3 TD (= 04:12:52.0 UT1)

Penumbral Magnitude = 1.9664	Gamma = 0.4964	Saros Series = 138
Umbral Magnitude = 0.9319	Axis = 0.4647°	Saros Member = 29 of 82

Sun at Greatest Eclipse
(Geocentric Coordinates)

R.A. = 10h26m57.9s
Dec. = +09°42'52.7"
S.D. = 00°15'50.0"
H.P. = 00°00'08.7"

Moon at Greatest Eclipse
(Geocentric Coordinates)

R.A. = 22h26m06.3s
Dec. = -09°18'03.6"
S.D. = 00°15'18.2"
H.P. = 00°56'09.9"

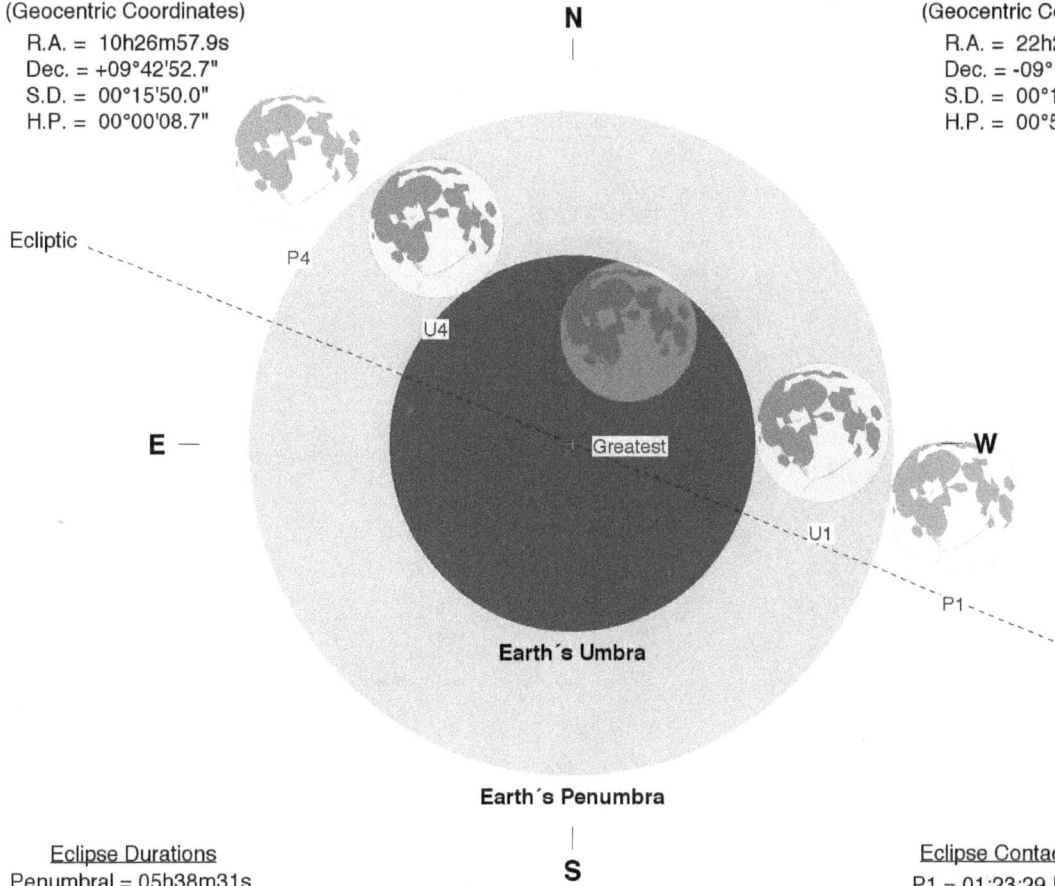

N

Ecliptic

P4

U4

Greatest

W

E

U1

P1

Earth's Umbra

Earth's Penumbra

S

Eclipse Durations
Penumbral = 05h38m31s
Umbral = 03h18m48s

Eph. = JPL DE430
Rule = Herald-Sinnott
ΔT = 72 s

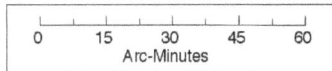

0   15   30   45   60
Arc-Minutes

©2020 F. Espenak, www.EclipseWise.com

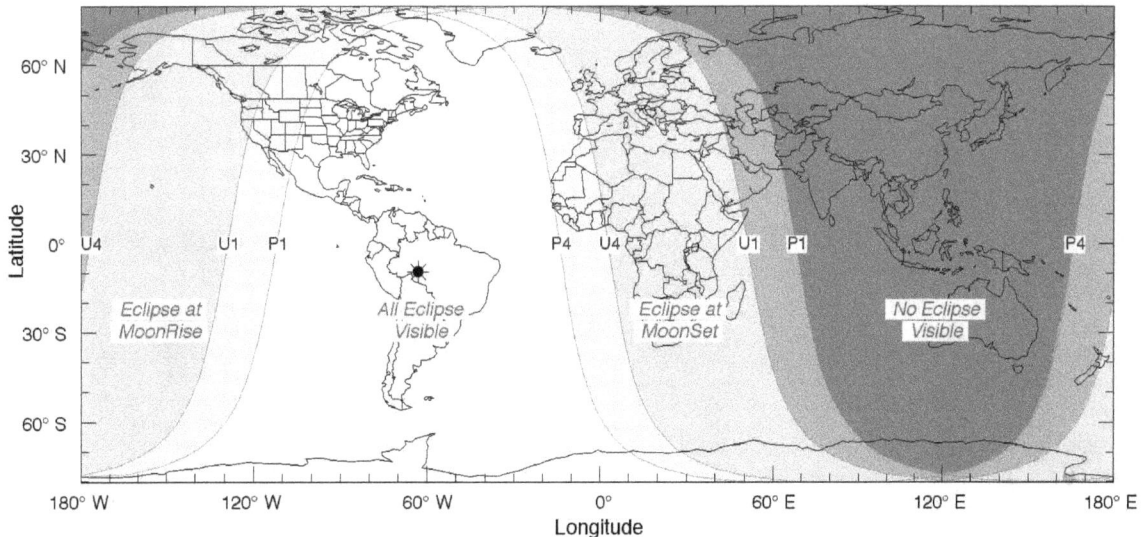

Eclipse Contacts
P1 = 01:23:29 UT1
U1 = 02:33:21 UT1
U4 = 05:52:09 UT1
P4 = 07:02:00 UT1

60° N

30° N

0°

30° S

60° S

Latitude

U4   U1   P1

P4   U4   U1   P1   P4

Eclipse at
MoonRise

All Eclipse
Visible

Eclipse at
MoonSet

No Eclipse
Visible

180° W   120° W   60° W   0°   60° E   120° E   180° E
Longitude

Penumbral Lunar Eclipse of 2027 Feb 20

Greatest Eclipse = 23:14:05.3 TD (= 23:12:52.7 UT1)

Penumbral Magnitude = 0.9286	Gamma = -1.0480	Saros Series = 143
Umbral Magnitude = -0.0549	Axis = 1.0542°	Saros Member = 18 of 72

Sun at Greatest Eclipse
(Geocentric Coordinates)
R.A. = 22h16m18.3s
Dec. = -10°43'53.9"
S.D. = 00°16'10.5"
H.P. = 00°00'08.9"

Moon at Greatest Eclipse
(Geocentric Coordinates)
R.A. = 10h14m23.7s
Dec. = +09°47'16.6"
S.D. = 00°16'26.8"
H.P. = 01°00'21.6"

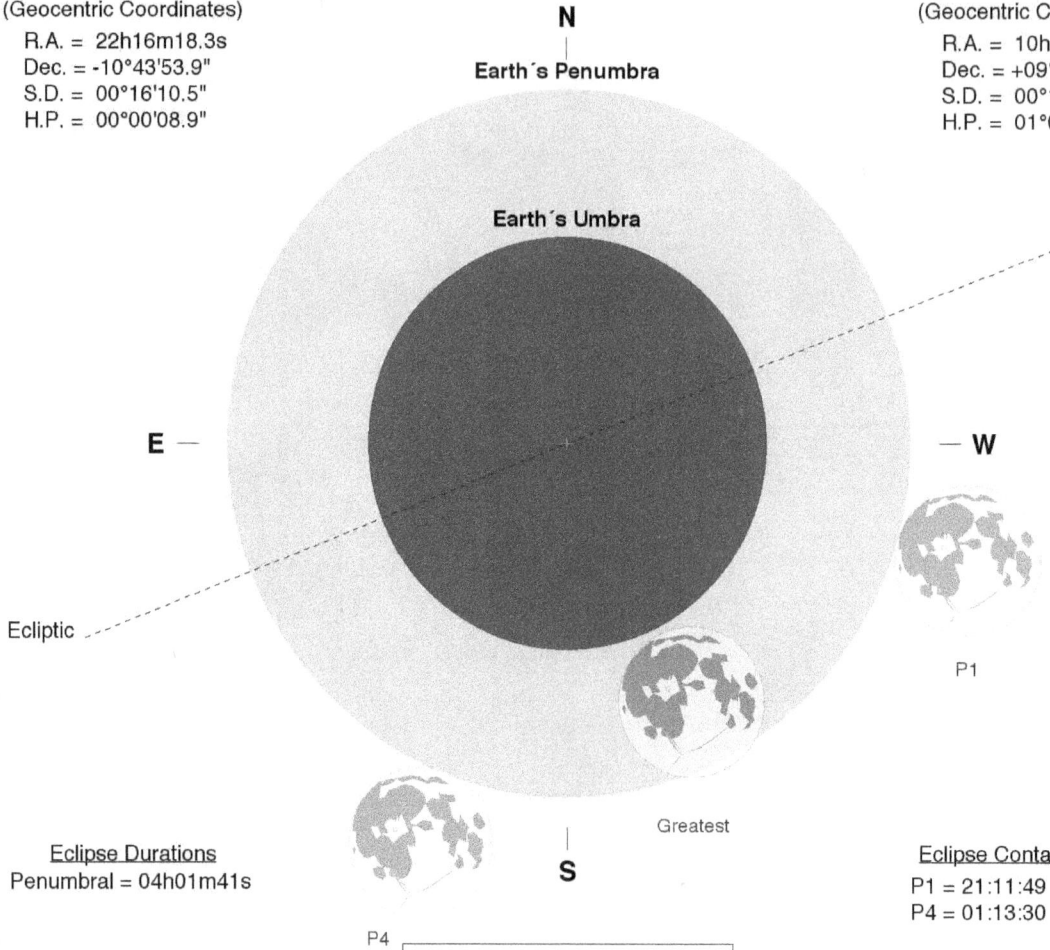

N
Earth's Penumbra

Earth's Umbra

E

W

Ecliptic

P1

Greatest

S

Eclipse Durations
Penumbral = 04h01m41s

P4

Eclipse Contacts
P1 = 21:11:49 UT1
P4 = 01:13:30 UT1

0	15	30	45	60

Arc-Minutes

Eph. = JPL DE430
Rule = Herald-Sinnott
ΔT =   73 s

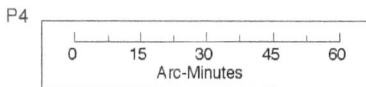

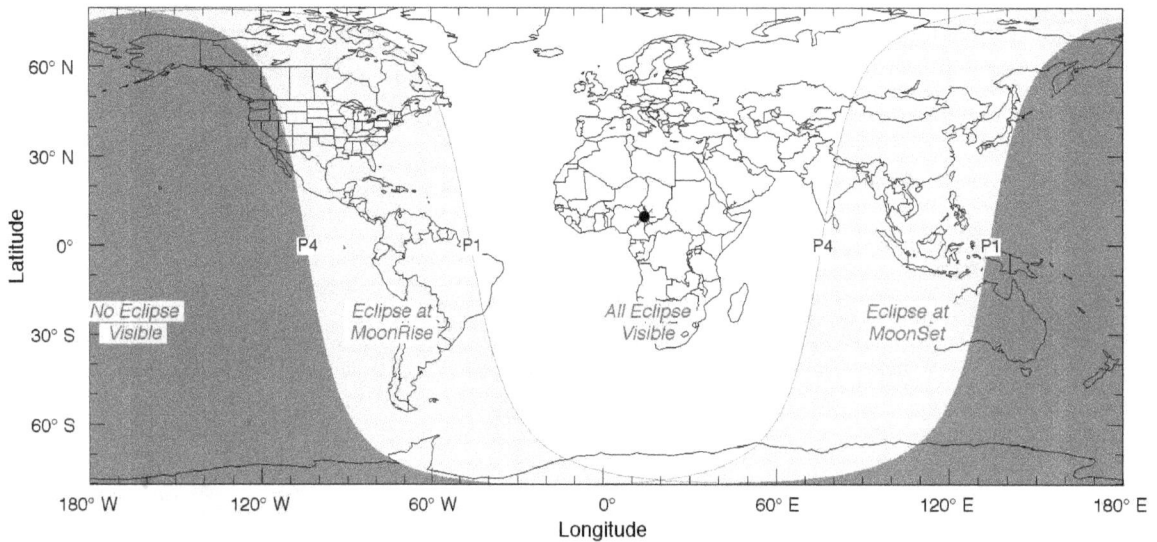

60° N

30° N

0°

Latitude

30° S

60° S

P4

P1

No Eclipse
Visible

Eclipse at
MoonRise

All Eclipse
Visible

P4

P1

Eclipse at
MoonSet

180° W   120° W   60° W   0°   60° E   120° E   180° E

Longitude

Penumbral Lunar Eclipse of 2027 Jul 18

Greatest Eclipse = 16:04:10.5 TD (= 16:02:57.8 UT1)

Penumbral Magnitude = 0.0032	Gamma = -1.5759	Saros Series = 110
Umbral Magnitude = -1.0662	Axis = 1.4184°	Saros Member = 72 of 72

Sun at Greatest Eclipse
(Geocentric Coordinates)

R.A. = 07h51m14.4s
Dec. = +20°58'43.6"
S.D. = 00°15'44.3"
H.P. = 00°00'08.7"

N

Earth's Penumbra

Earth's Umbra

Ecliptic

E —

— **W**

Moon at Greatest Eclipse
(Geocentric Coordinates)

R.A. = 19h52m57.2s
Dec. = -22°20'25.3"
S.D. = 00°14'43.0"
H.P. = 00°54'00.6"

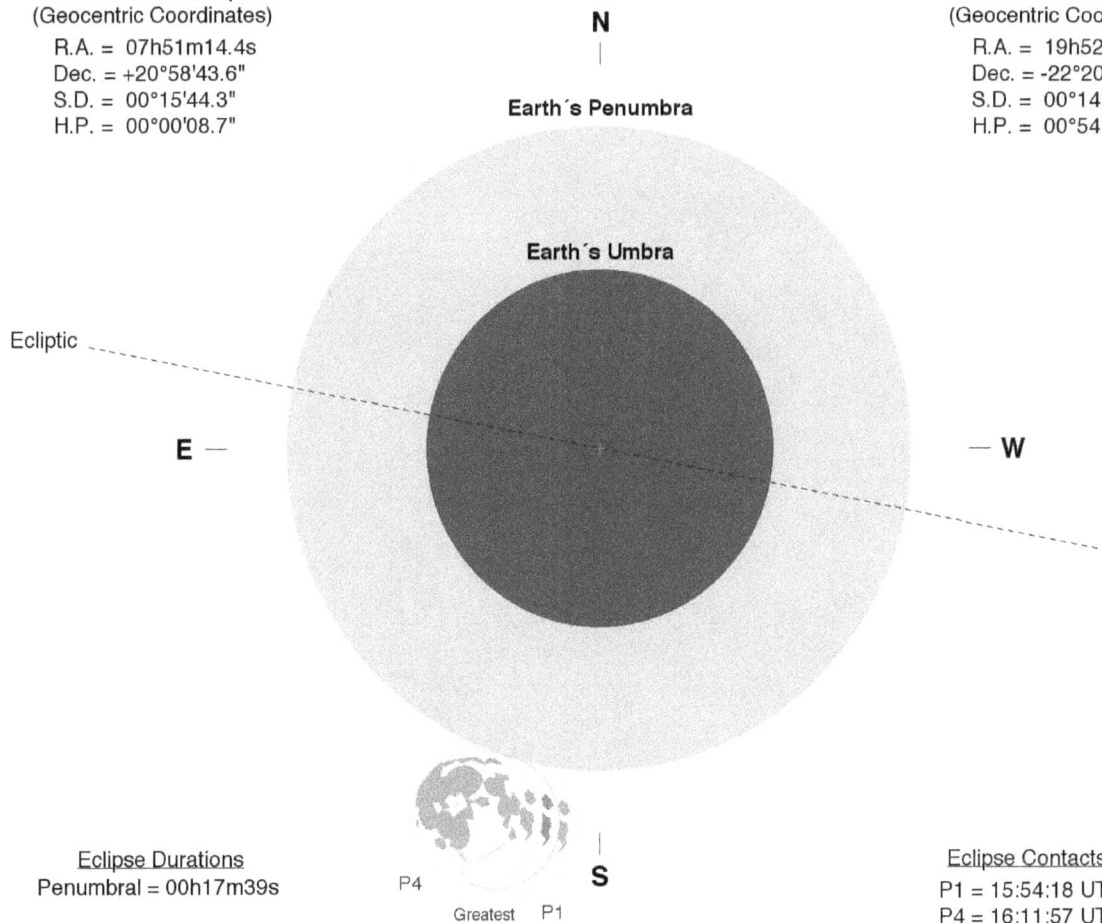

Eclipse Durations
Penumbral = 00h17m39s

P4

Greatest   P1

S

Eclipse Contacts
P1 = 15:54:18 UT1
P4 = 16:11:57 UT1

Eph. = JPL DE430
Rule = Herald-Sinnott
ΔT = 73 s

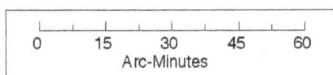

0   15   30   45   60
Arc-Minutes

©2020 F. Espenak, www.EclipseWise.com

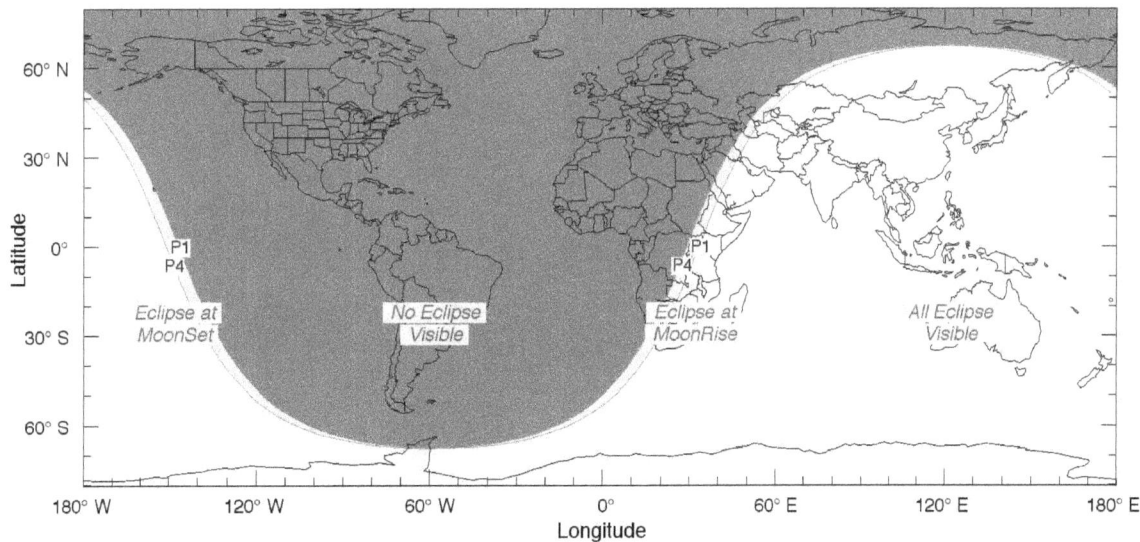

Penumbral Lunar Eclipse of 2027 Aug 17

Greatest Eclipse = 07:14:57.9 TD   (= 07:13:45.1 UT1)

Penumbral Magnitude = 0.5476	Gamma = 1.2797	Saros Series = 148
Umbral Magnitude = -0.5234	Axis = 1.1544°	Saros Member = 4 of 70

Sun at Greatest Eclipse
(Geocentric Coordinates)
R.A. = 09h45m58.6s
Dec. = +13°27'30.2"
S.D. = 00°15'47.8"
H.P. = 00°00'08.7"

Moon at Greatest Eclipse
(Geocentric Coordinates)
R.A. = 21h43m58.8s
Dec. = -12°24'40.9"
S.D. = 00°14'44.9"
H.P. = 00°54'07.8"

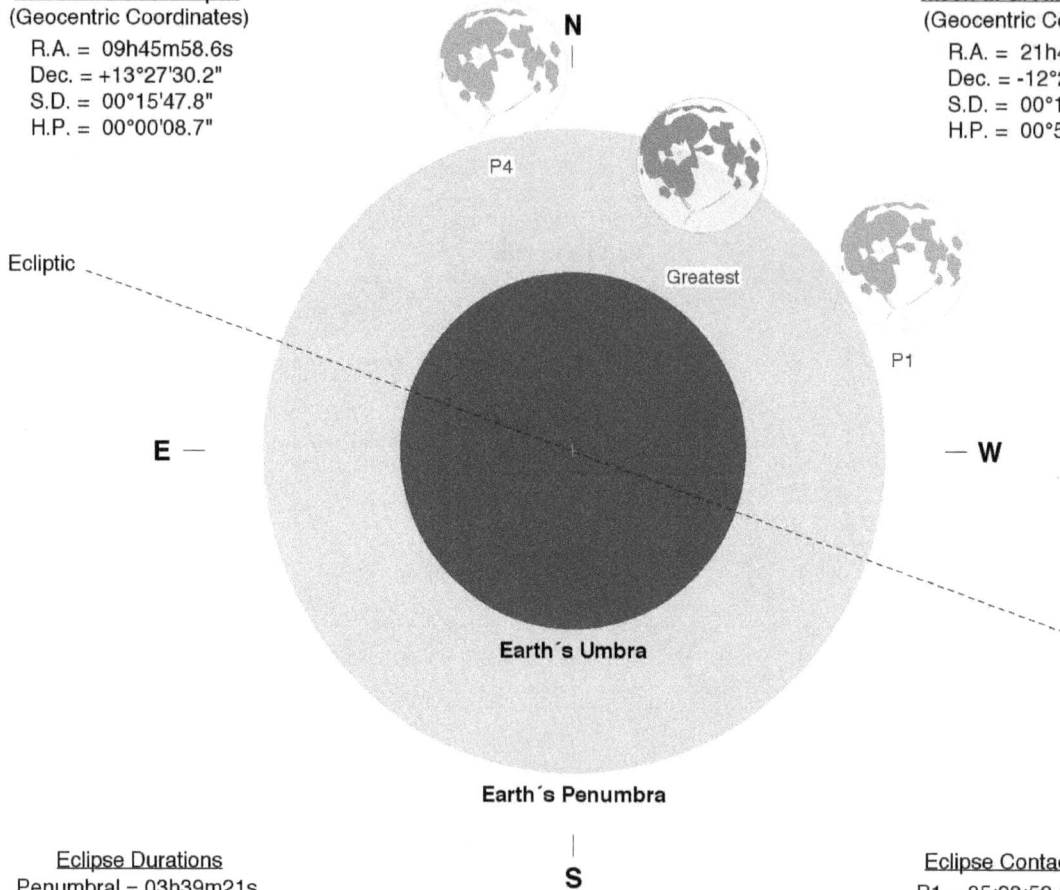

N

P4

Greatest

P1

Ecliptic

E —

— W

Earth's Umbra

Earth's Penumbra

S

Eclipse Durations
Penumbral = 03h39m21s

Eclipse Contacts
P1 = 05:23:52 UT1
P4 = 09:03:14 UT1

Eph. = JPL DE430
Rule = Herald-Sinnott
ΔT =   73 s

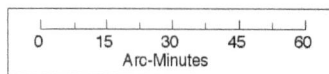

0	15	30	45	60

Arc-Minutes

©2020 F. Espenak, www.EclipseWise.com

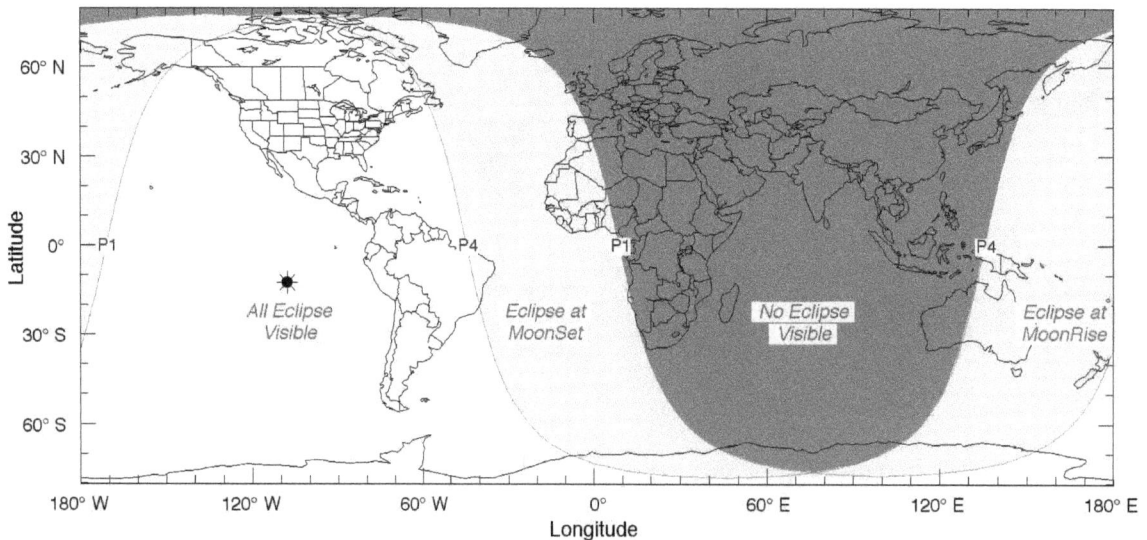

All Eclipse
Visible

Eclipse at
MoonSet

No Eclipse
Visible

Eclipse at
MoonRise

P1

P4

P1

P4

Partial Lunar Eclipse of 2028 Jan 12

Greatest Eclipse = 04:14:13.4 TD   (= 04:13:00.5 UT1)

Penumbral Magnitude = 1.0485	Gamma = 0.9818	Saros Series = 115
Umbral Magnitude = 0.0679	Axis = 0.9958°	Saros Member = 58 of 72

Sun at Greatest Eclipse
(Geocentric Coordinates)

R.A. = 19h32m47.8s
Dec. = -21°43'29.4"
S.D. = 00°16'15.8"
H.P. = 00°00'08.9"

Moon at Greatest Eclipse
(Geocentric Coordinates)

R.A. = 07h33m53.0s
Dec. = +22°41'18.2"
S.D. = 00°16'35.1"
H.P. = 01°00'52.0"

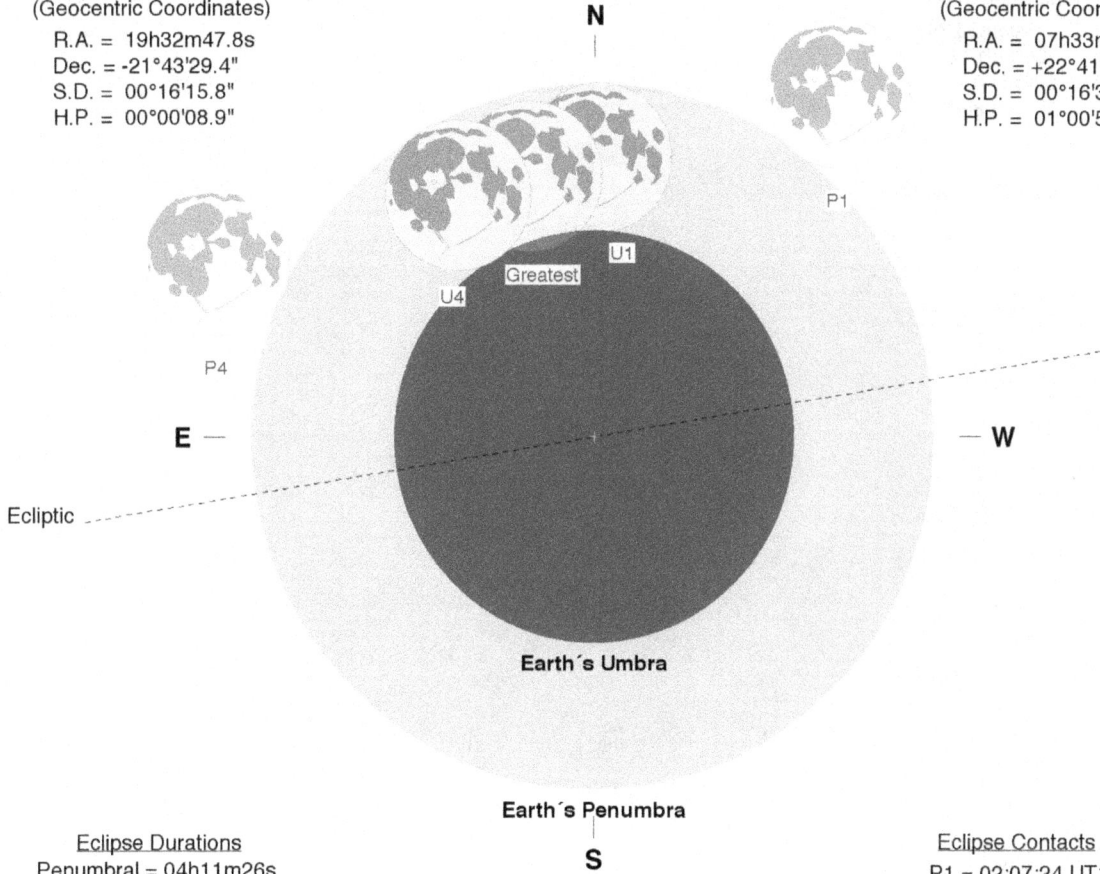

N

P1

U1

Greatest

U4

P4

E —

Ecliptic

— W

Earth´s Umbra

Earth´s Penumbra

S

Eclipse Durations
Penumbral = 04h11m26s
Umbral = 00h56m53s

Eph. = JPL DE430
Rule = Herald-Sinnott
ΔT =   73 s

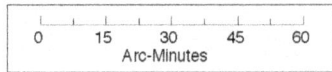

0	15	30	45	60

Arc-Minutes

©2020 F. Espenak, www.EclipseWise.com

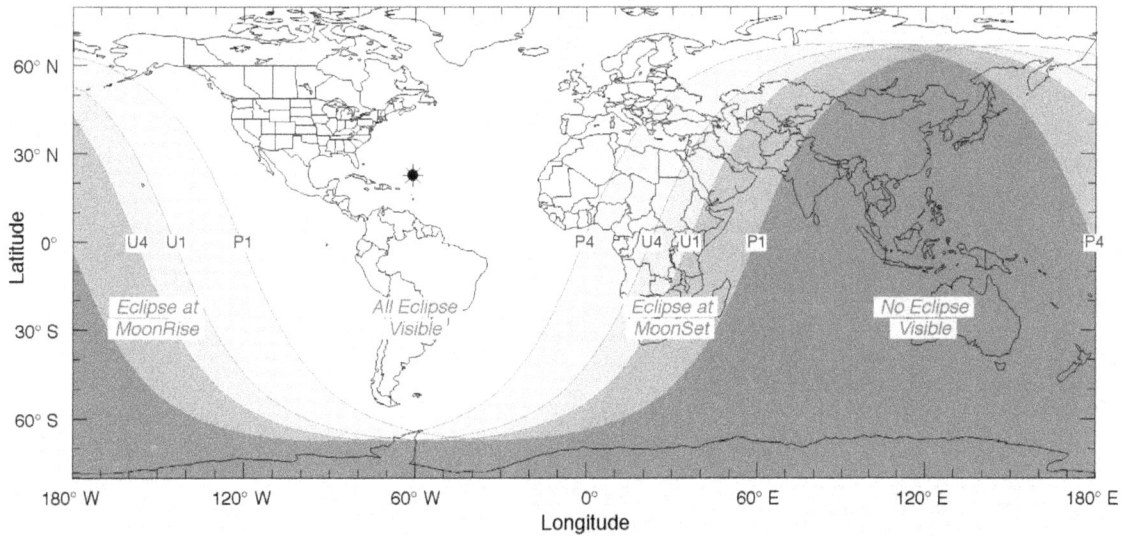

Eclipse Contacts
P1 = 02:07:24 UT1
U1 = 03:44:46 UT1
U4 = 04:41:39 UT1
P4 = 06:18:50 UT1

Partial Lunar Eclipse of 2028 Jul 06

Greatest Eclipse = 18:20:57.2 TD (= 18:19:44.0 UT1)

Penumbral Magnitude = 1.4282	Gamma = -0.7904	Saros Series = 120
Umbral Magnitude = 0.3908	Axis = 0.7331°	Saros Member = 58 of 83

Sun at Greatest Eclipse
(Geocentric Coordinates)
R.A. = 07h05m56.7s
Dec. = +22°34'16.5"
S.D. = 00°15'43.9"
H.P. = 00°00'08.6"

N

Earth´s Penumbra

Earth´s Umbra

Ecliptic

E —

— **W**

P4

U4

Greatest

U1

P1

S

Moon at Greatest Eclipse
(Geocentric Coordinates)
R.A. = 19h06m37.0s
Dec. = -23°17'16.4"
S.D. = 00°15'09.9"
H.P. = 00°55'39.4"

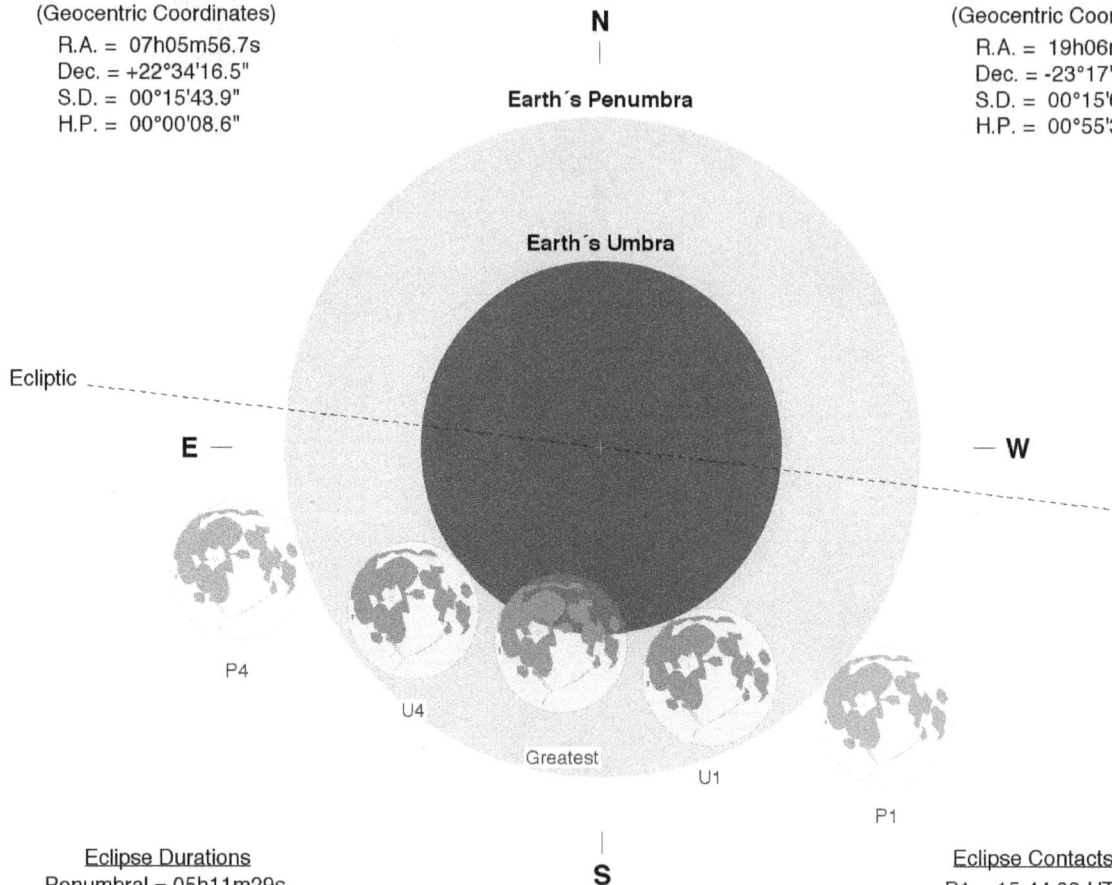

Eclipse Durations
Penumbral = 05h11m29s
Umbral = 02h22m13s

Eph. = JPL DE430
Rule = Herald-Sinnott
ΔT =   73 s

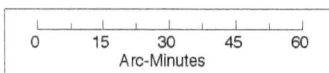

0   15   30   45   60
Arc-Minutes

©2020 F. Espenak, www.EclipseWise.com

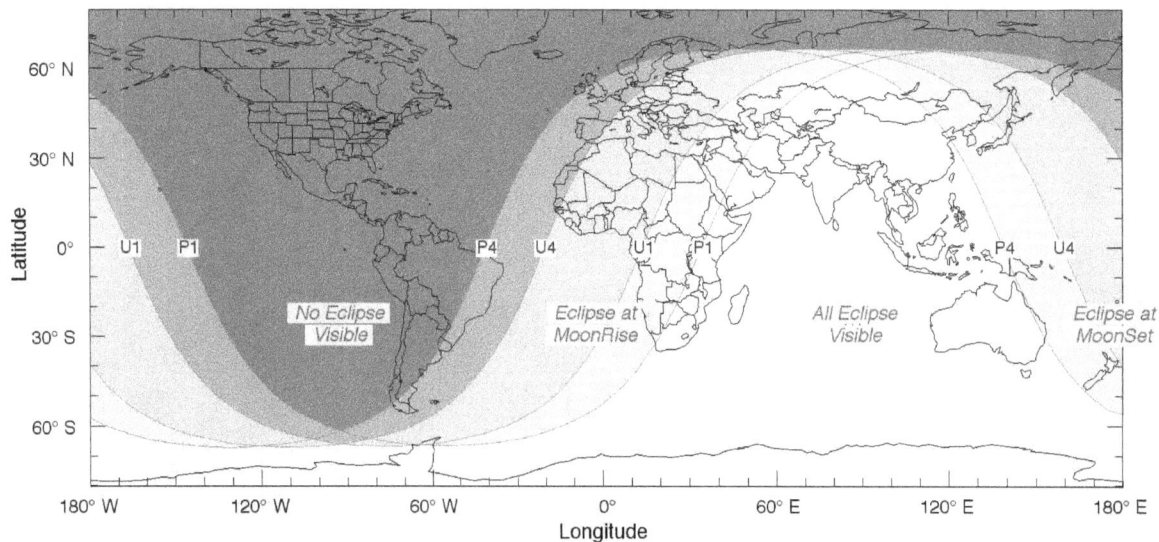

Eclipse Contacts
P1 = 15:44:03 UT1
U1 = 17:08:40 UT1
U4 = 19:30:53 UT1
P4 = 20:55:33 UT1

No Eclipse
Visible

Eclipse at
MoonRise

All Eclipse
Visible

Eclipse at
MoonSet

Total Lunar Eclipse of 2028 Dec 31

Greatest Eclipse = 16:53:15.1 TD (= 16:52:01.7 UT1)

Penumbral Magnitude = 2.2758	Gamma = 0.3258	Saros Series = 125
Umbral Magnitude = 1.2478	Axis = 0.3153°	Saros Member = 49 of 72

Sun at Greatest Eclipse
(Geocentric Coordinates)

R.A. = 18h45m53.7s
Dec. = -23°01'00.5"
S.D. = 00°16'15.9"
H.P. = 00°00'08.9"

N

Moon at Greatest Eclipse
(Geocentric Coordinates)

R.A. = 06h46m08.4s
Dec. = +23°19'37.5"
S.D. = 00°15'49.4"
H.P. = 00°58'04.3"

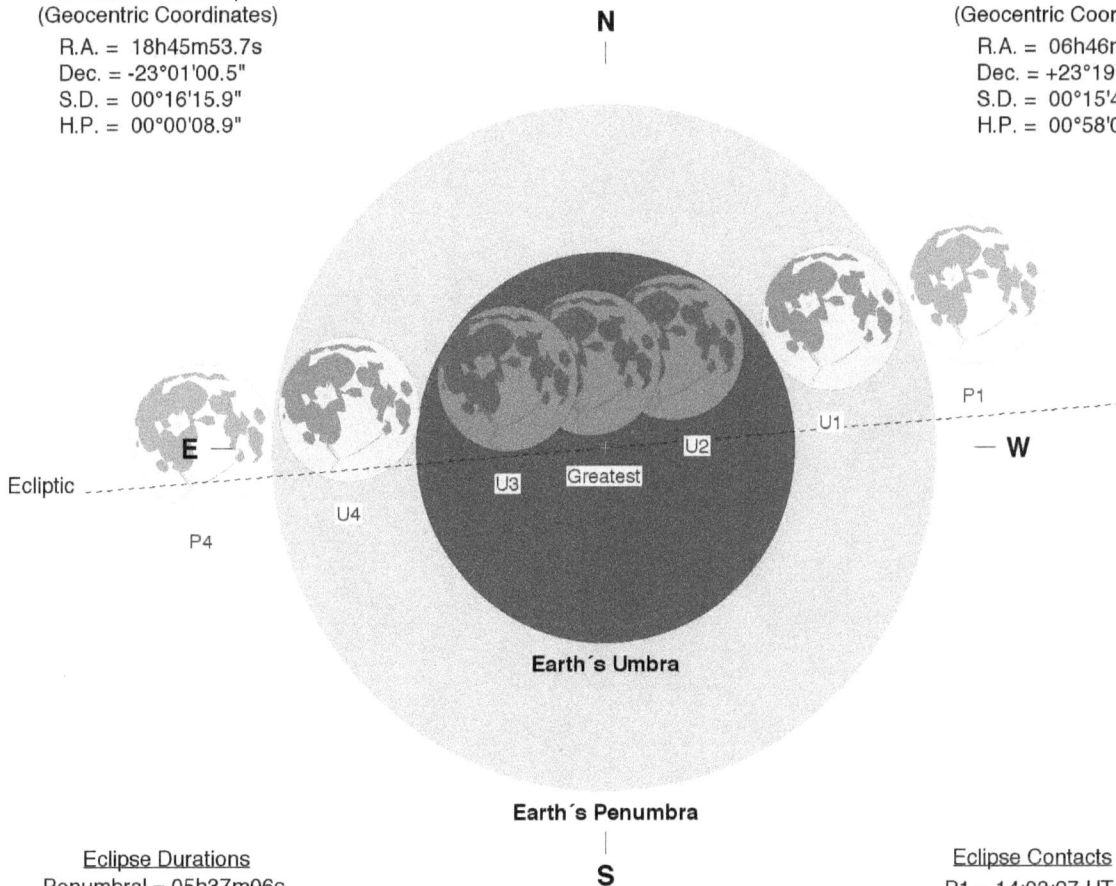

P1

U1

Ecliptic

E

W

U3 Greatest U2

U4

P4

Earth's Umbra

Earth's Penumbra

S

Eclipse Durations
Penumbral = 05h37m06s
Umbral = 03h29m37s
Total = 01h12m02s

Eph. = JPL DE430
Rule = Herald-Sinnott
ΔT =   73 s

0	15	30	45	60

Arc-Minutes

©2020 F. Espenak, www.EclipseWise.com

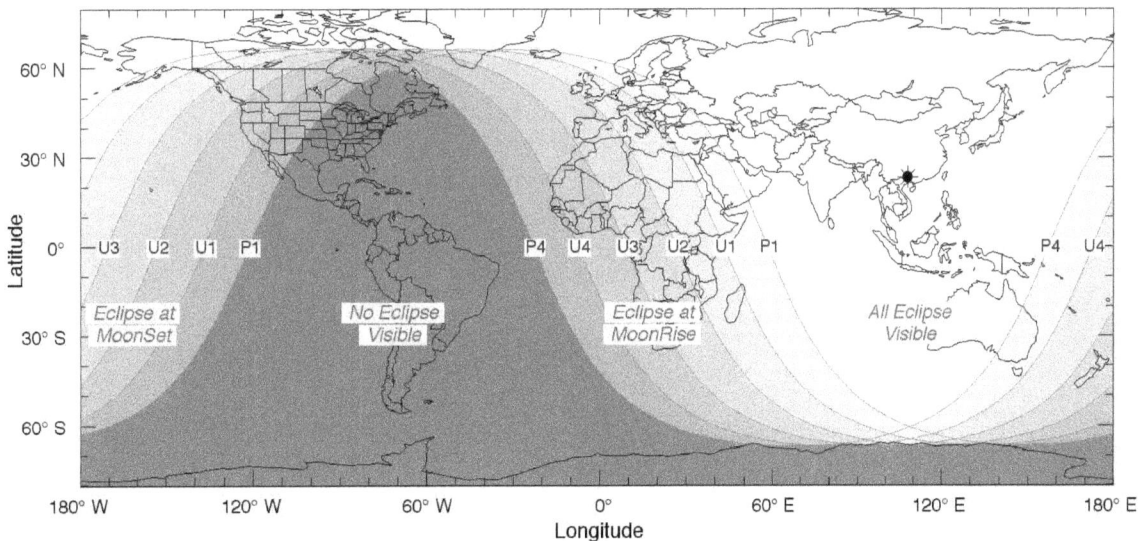

Eclipse Contacts
P1 = 14:03:27 UT1
U1 = 15:07:17 UT1
U2 = 16:16:07 UT1
U3 = 17:28:08 UT1
U4 = 18:36:54 UT1
P4 = 19:40:34 UT1

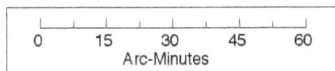

U3 U2 U1 P1

P4 U4 U3 U2 U1 P1

P4 U4

Eclipse at
MoonSet

No Eclipse
Visible

Eclipse at
MoonRise

All Eclipse
Visible

Latitude

Longitude

Total Lunar Eclipse of 2029 Jun 26

Greatest Eclipse = 03:23:22.5 TD (= 03:22:08.8 UT1)

Penumbral Magnitude = 2.8282	Gamma = 0.0124	Saros Series = 130
Umbral Magnitude = 1.8452	Axis = 0.0121°	Saros Member = 35 of 71

Sun at Greatest Eclipse
(Geocentric Coordinates)

R.A. = 06h21m03.1s
Dec. = +23°20'50.2"
S.D. = 00°15'44.1"
H.P. = 00°00'08.7"

Moon at Greatest Eclipse
(Geocentric Coordinates)

R.A. = 18h21m02.6s
Dec. = -23°20'06.9"
S.D. = 00°16'00.4"
H.P. = 00°58'44.7"

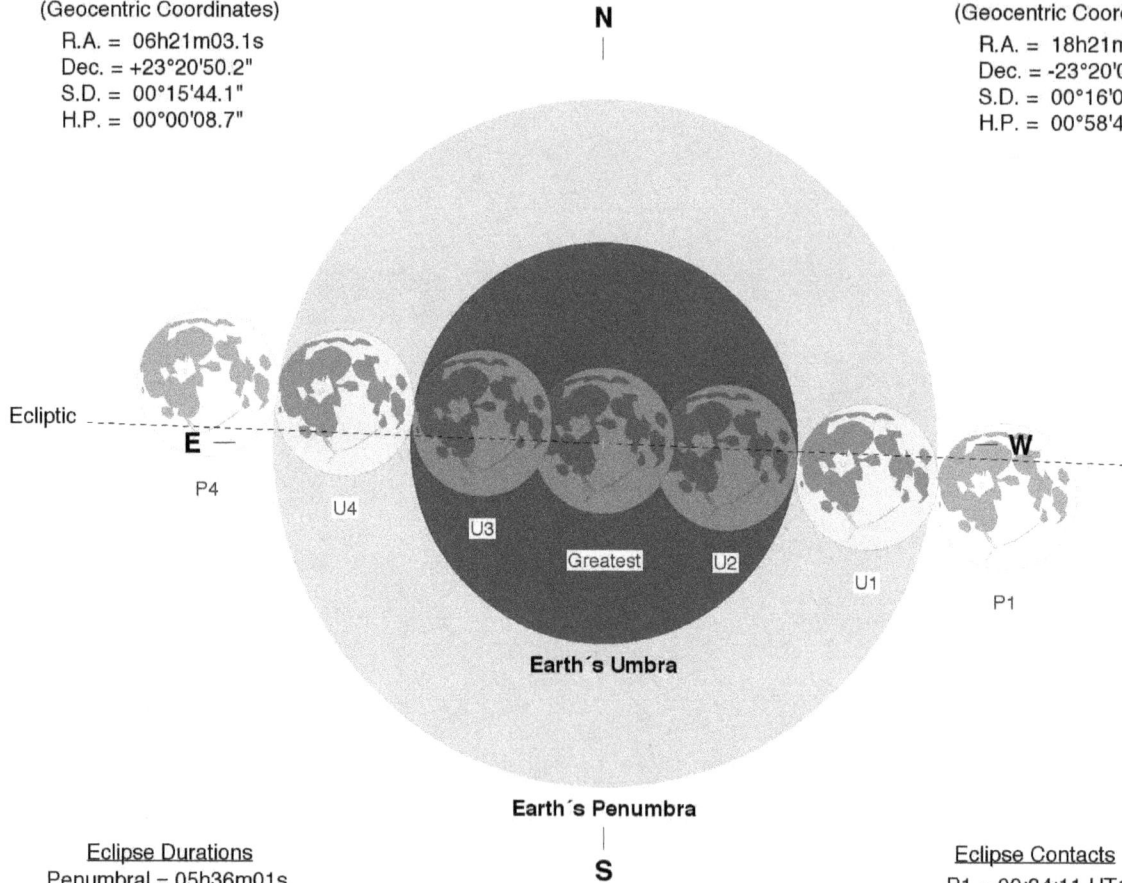

N

Ecliptic

E

W

P4

U4

U3

Greatest

U2

U1

P1

Earth's Umbra

Earth's Penumbra

S

Eclipse Durations

Penumbral = 05h36m01s
Umbral = 03h40m20s
Total = 01h42m40s

Eph. = JPL DE430
Rule = Herald-Sinnott
ΔT =   74 s

0   15   30   45   60
Arc-Minutes

©2020 F. Espenak, www.EclipseWise.com

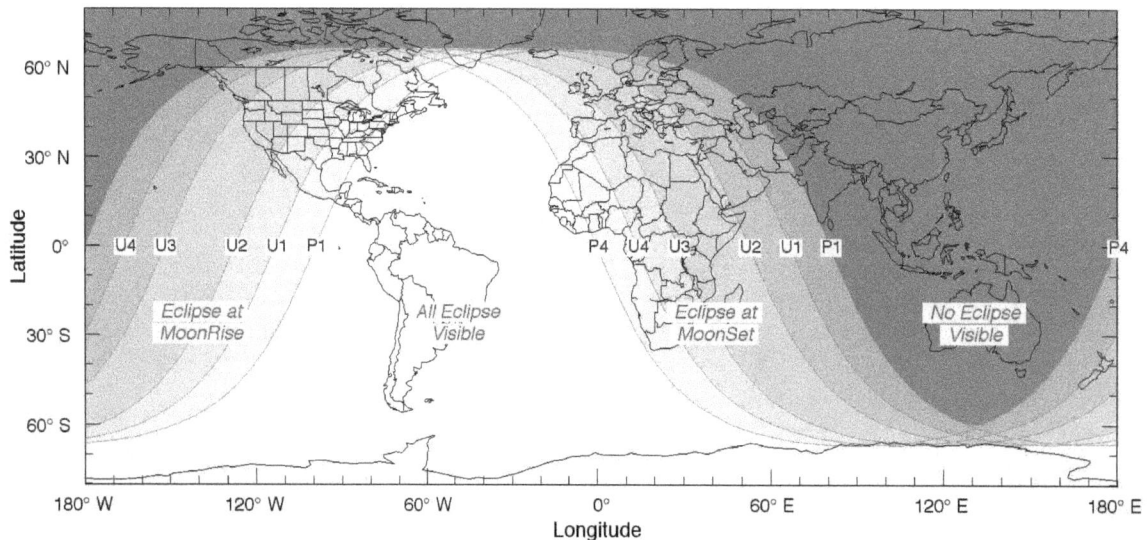

Eclipse Contacts

P1 = 00:34:11 UT1
U1 = 01:31:58 UT1
U2 = 02:30:48 UT1
U3 = 04:13:28 UT1
U4 = 05:12:18 UT1
P4 = 06:10:12 UT1

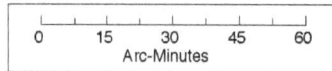

Total Lunar Eclipse of 2029 Dec 20

Greatest Eclipse = 22:43:11.8 TD (= 22:41:57.9 UT1)

Penumbral Magnitude = 2.2023	Gamma = -0.3811	Saros Series = 135
Umbral Magnitude = 1.1190	Axis = 0.3498°	Saros Member = 24 of 71

Sun at Greatest Eclipse
(Geocentric Coordinates)

R.A. = 17h57m07.6s
Dec. = -23°26'00.2"
S.D. = 00°16'15.5"
H.P. = 00°00'08.9"

Moon at Greatest Eclipse
(Geocentric Coordinates)

R.A. = 05h56m59.0s
Dec. = +23°05'06.7"
S.D. = 00°15'00.4"
H.P. = 00°55'04.6"

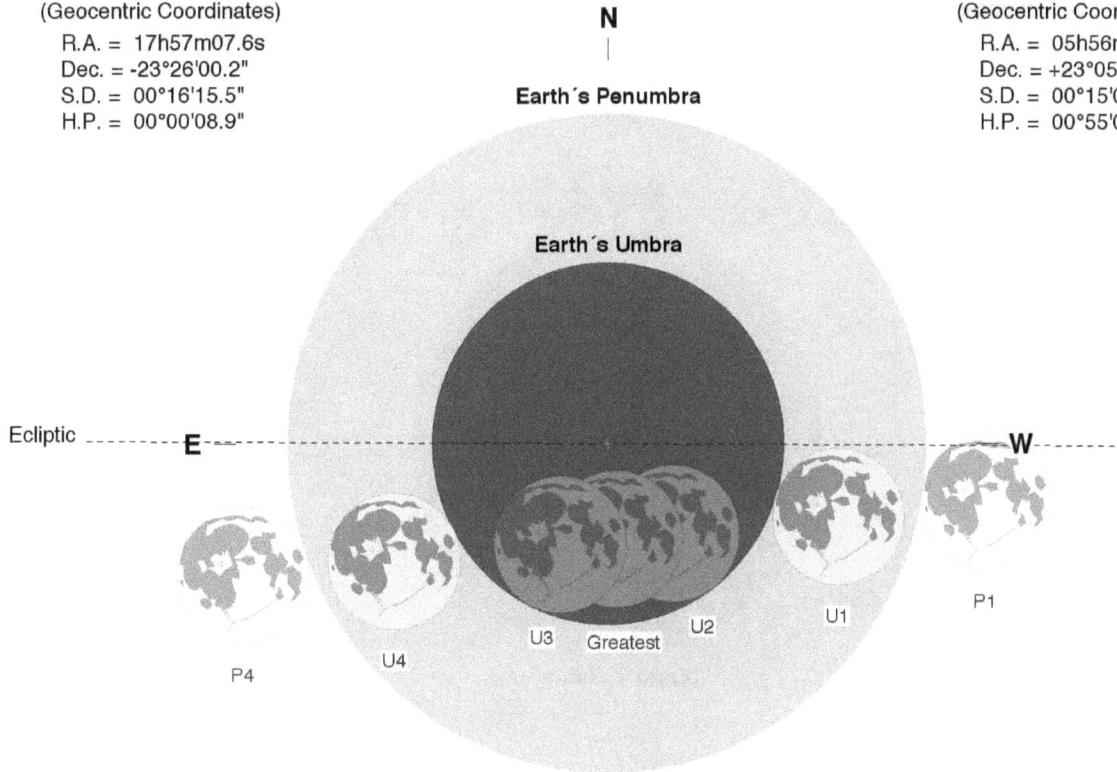

N

Earth's Penumbra

Earth's Umbra

Ecliptic

E

W

U3   Greatest   U2

U4   U1

P4   P1

S

Eclipse Durations

Penumbral = 05h58m54s
Umbral = 03h34m07s
Total = 00h54m23s

Eph. = JPL DE430
Rule = Herald-Sinnott
ΔT =   74 s

0   15   30   45   60
Arc-Minutes

©2020 F. Espenak, www.EclipseWise.com

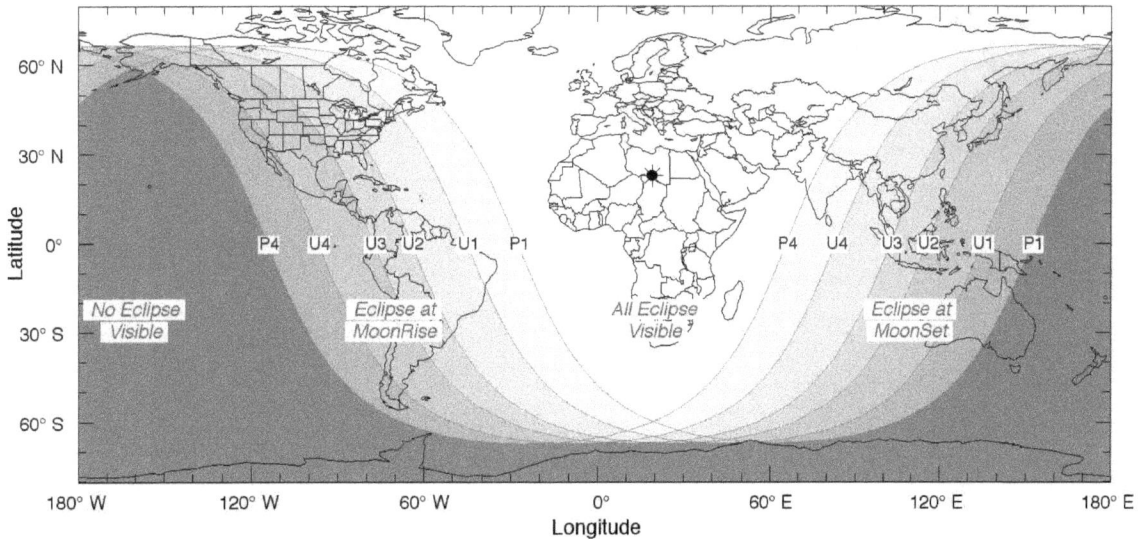

Eclipse Contacts

P1 = 19:42:28 UT1
U1 = 20:54:54 UT1
U2 = 22:14:44 UT1
U3 = 23:09:07 UT1
U4 = 00:29:01 UT1
P4 = 01:41:22 UT1

P4   U4   U3 U2   U1   P1   P4   U4   U3 U2   U1   P1

No Eclipse
Visible

Eclipse at
MoonRise

All Eclipse
Visible

Eclipse at
MoonSet

125

Partial Lunar Eclipse of 2030 Jun 15

Greatest Eclipse = 18:34:33.5 TD   (= 18:33:19.4 UT1)

Penumbral Magnitude = 1.4495	Gamma = 0.7535	Saros Series = 140
Umbral Magnitude = 0.5040	Axis = 0.7674°	Saros Member = 25 of 77

Sun at Greatest Eclipse
(Geocentric Coordinates)
R.A. = 05h36m57.6s
Dec. = +23°19'44.0"
S.D. = 00°15'44.7"
H.P. = 00°00'08.7"

Moon at Greatest Eclipse
(Geocentric Coordinates)
R.A. = 17h36m46.1s
Dec. = -22°33'45.8"
S.D. = 00°16'39.2"
H.P. = 01°01'07.1"

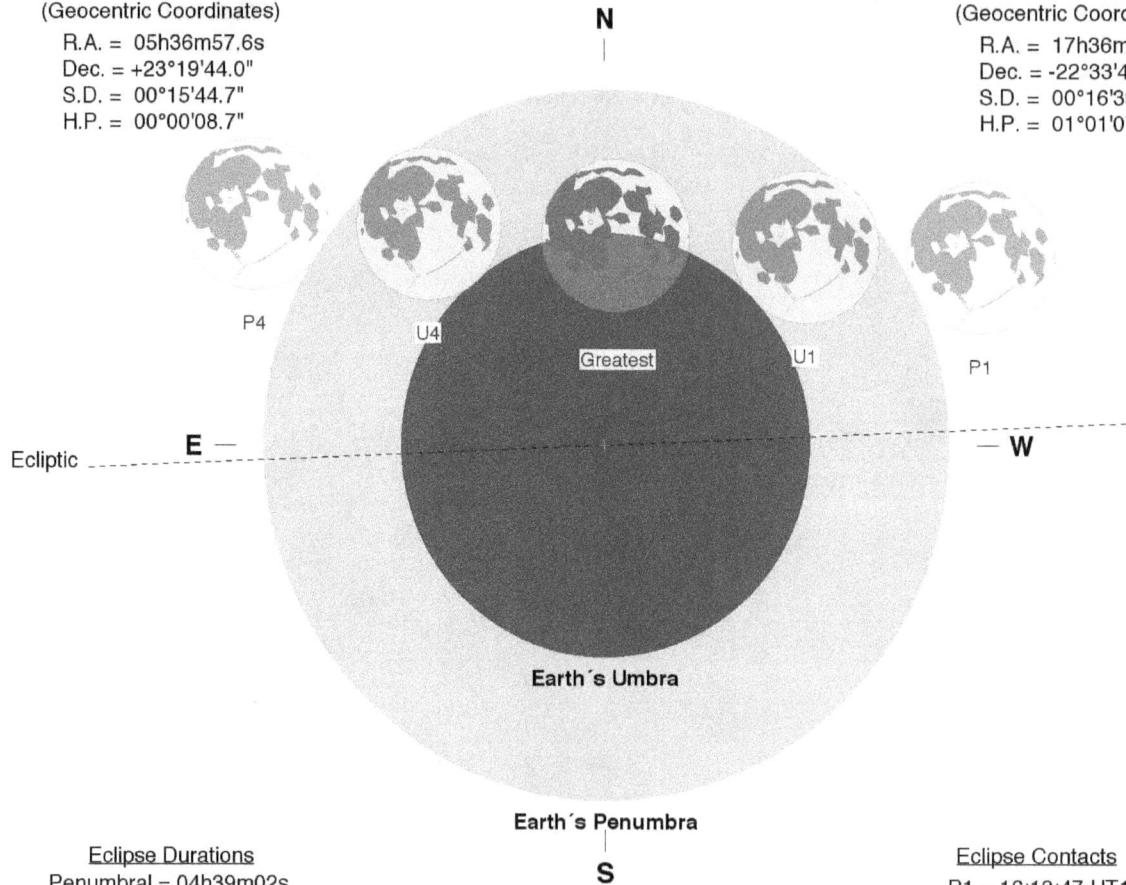

N

P4   U4   Greatest   U1   P1

Ecliptic

E

W

Earth's Umbra

Earth's Penumbra

S

Eclipse Durations
Penumbral = 04h39m02s
Umbral = 02h25m02s

Eph. = JPL DE430
Rule = Herald-Sinnott
ΔT =   74 s

Eclipse Contacts
P1 = 16:13:47 UT1
U1 = 17:20:46 UT1
U4 = 19:45:47 UT1
P4 = 20:52:49 UT1

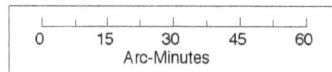

0   15   30   45   60
Arc-Minutes

©2020 F. Espenak, www.EclipseWise.com

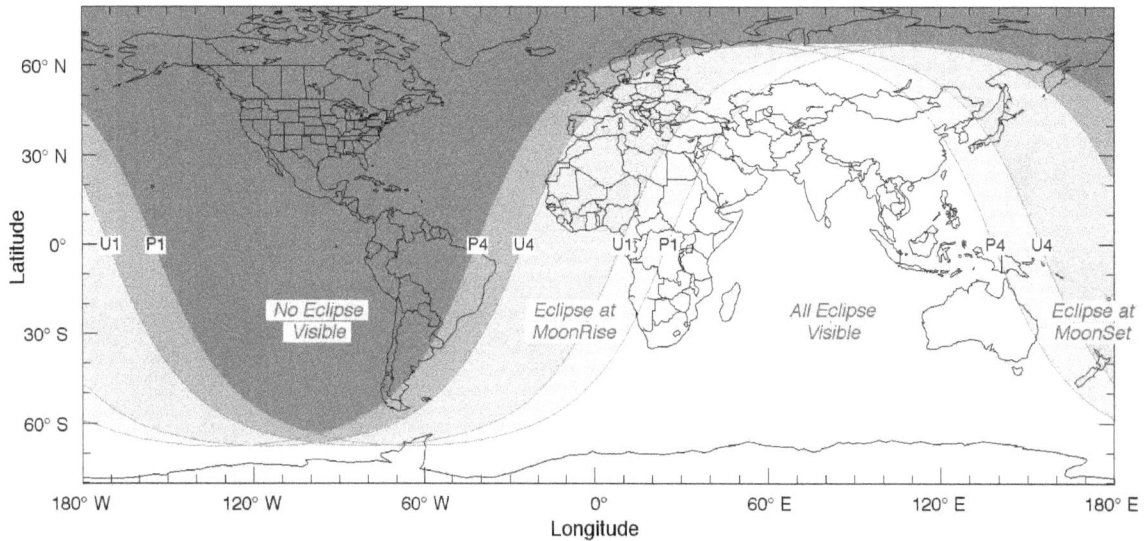

No Eclipse Visible

Eclipse at MoonRise

All Eclipse Visible

Eclipse at MoonSet

U1   P1   P4   U4   U1   P1   P4   U4

Penumbral Lunar Eclipse of 2030 Dec 09

Greatest Eclipse = 22:28:51.1 TD (= 22:27:36.7 UT1)

Penumbral Magnitude = 0.9430	Gamma = -1.0732	Saros Series = 145
Umbral Magnitude = -0.1613	Axis = 0.9652°	Saros Member = 12 of 71

Sun at Greatest Eclipse
(Geocentric Coordinates)
R.A. = 17h07m21.3s
Dec. = -22°52'57.8"
S.D. = 00°16'14.4"
H.P. = 00°00'08.9"

N

Earth´s Penumbra

Earth´s Umbra

Ecliptic

E —

— **W**

Moon at Greatest Eclipse
(Geocentric Coordinates)
R.A. = 05h07m19.1s
Dec. = +21°55'03.1"
S.D. = 00°14'42.3"
H.P. = 00°53'58.2"

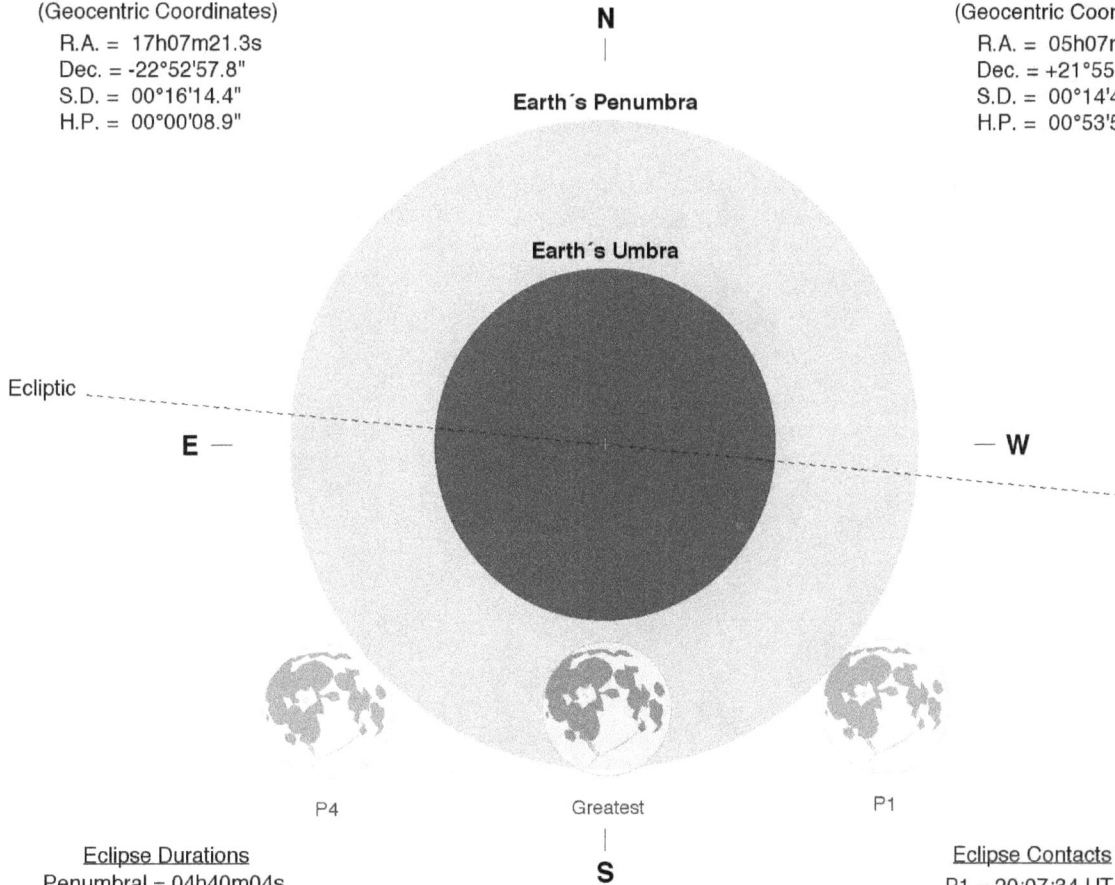

P4   Greatest   P1

S

Eclipse Durations
Penumbral = 04h40m04s

Eclipse Contacts
P1 = 20:07:34 UT1
P4 = 00:47:38 UT1

Eph. = JPL DE430
Rule = Herald-Sinnott
ΔT =   74 s

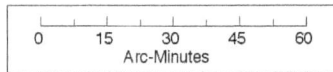

0   15   30   45   60
Arc-Minutes

©2020 F. Espenak, www.EclipseWise.com

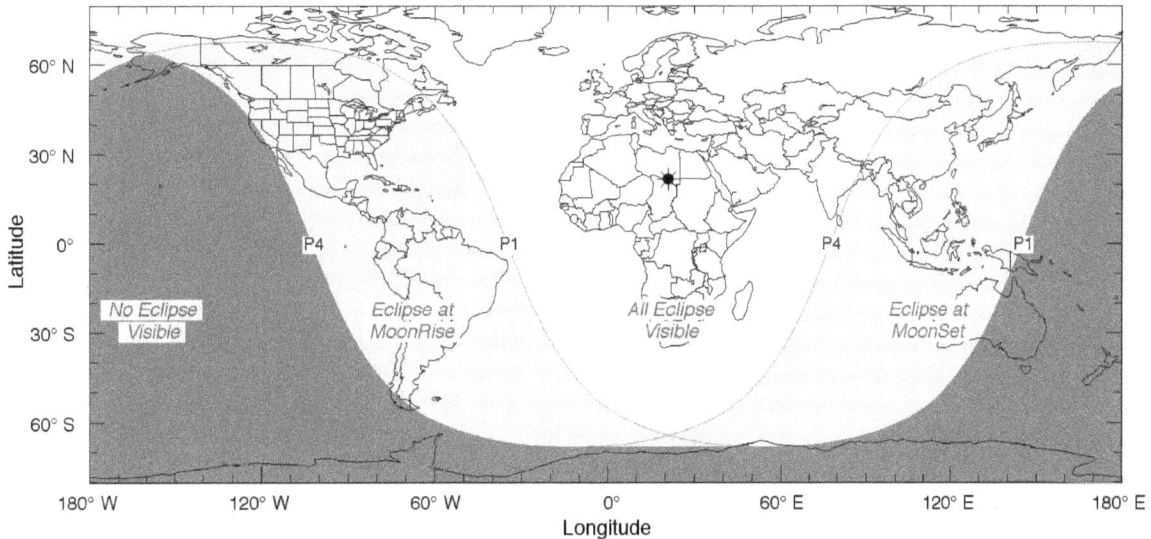

Penumbral Lunar Eclipse of 2031 May 07

Greatest Eclipse = 03:52:01.3 TD   (= 03:50:46.7 UT1)

Penumbral Magnitude = 0.8827	Gamma = -1.0695	Saros Series = 112
Umbral Magnitude = -0.0892	Axis = 1.0670°	Saros Member = 66 of 72

Sun at Greatest Eclipse
(Geocentric Coordinates)
R.A. = 02h55m49.7s
Dec. = +16°44'40.2"
S.D. = 00°15'51.2"
H.P. = 00°00'08.7"

N

Earth´s Penumbra

Moon at Greatest Eclipse
(Geocentric Coordinates)
R.A. = 14h54m58.0s
Dec. = -17°47'29.4"
S.D. = 00°16'18.7"
H.P. = 00°59'52.0"

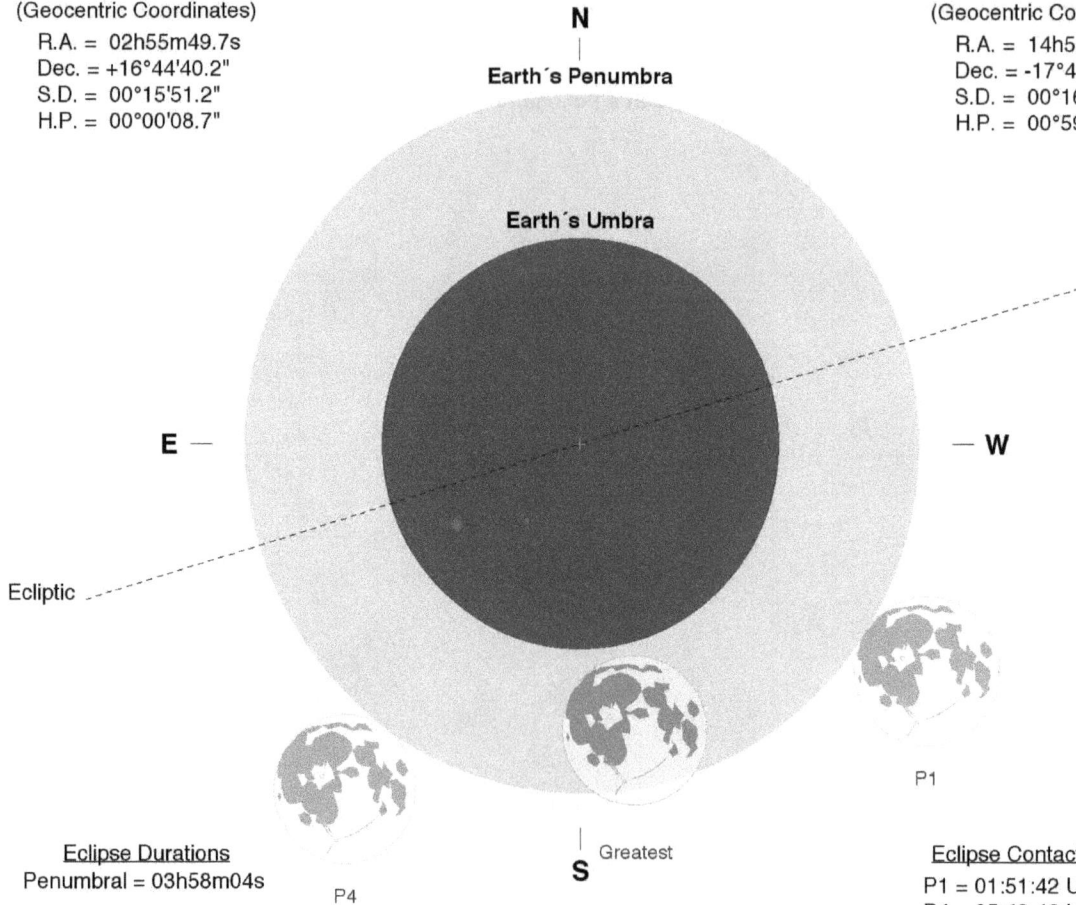

Earth´s Umbra

E

W

Ecliptic

P1

S Greatest

Eclipse Durations
Penumbral = 03h58m04s

P4

Eclipse Contacts
P1 = 01:51:42 UT1
P4 = 05:49:46 UT1

Eph. = JPL DE430
Rule = Herald-Sinnott
ΔT =   75 s

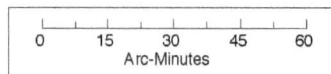

0	15	30	45	60

Arc-Minutes

©2020 F. Espenak, www.EclipseWise.com

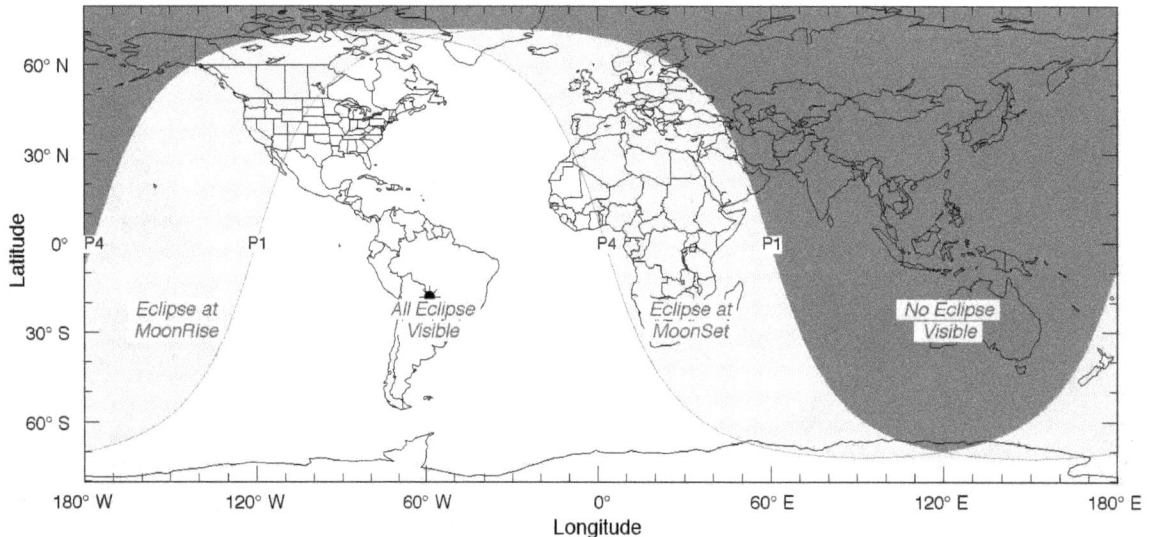

Eclipse at MoonRise

All Eclipse Visible

Eclipse at MoonSet

No Eclipse Visible

Latitude

Longitude

Penumbral Lunar Eclipse of 2031 Jun 05

Greatest Eclipse = 11:45:17.4 TD (= 11:44:02.8 UT1)

Penumbral Magnitude = 0.1306	Gamma = 1.4732	Saros Series = 150
Umbral Magnitude = -0.8185	Axis = 1.4967°	Saros Member = 2 of 71

Sun at Greatest Eclipse
(Geocentric Coordinates)
R.A. = 04h53m21.6s
Dec. = +22°33'01.5"
S.D. = 00°15'45.9"
H.P. = 00°00'08.7"

Moon at Greatest Eclipse
(Geocentric Coordinates)
R.A. = 16h53m29.4s
Dec. = -21°03'14.0"
S.D. = 00°16'36.6"
H.P. = 01°00'57.7"

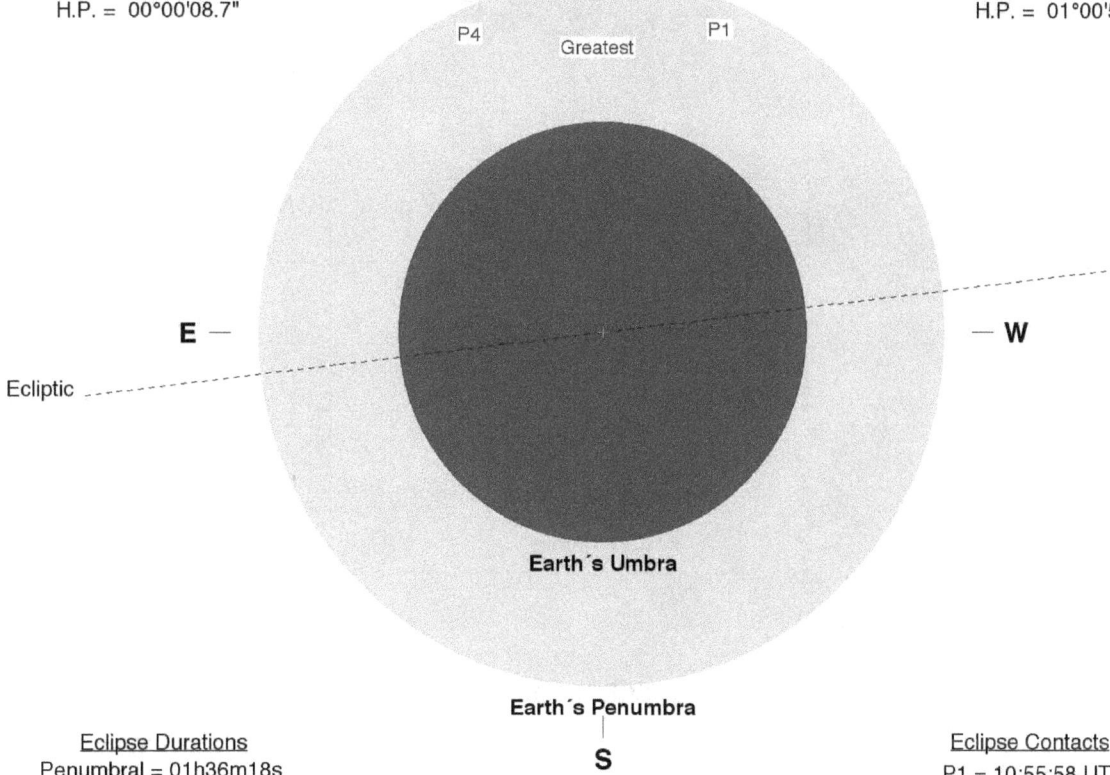

N

P4　Greatest　P1

E

W

Ecliptic

Earth's Umbra

Earth's Penumbra

S

Eclipse Durations
Penumbral = 01h36m18s

Eclipse Contacts
P1 = 10:55:58 UT1
P4 = 12:32:17 UT1

Eph. = JPL DE430
Rule = Herald-Sinnott
ΔT = 75 s

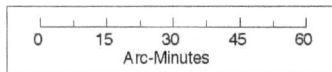

0　15　30　45　60
Arc-Minutes

©2020 F. Espenak, www.EclipseWise.com

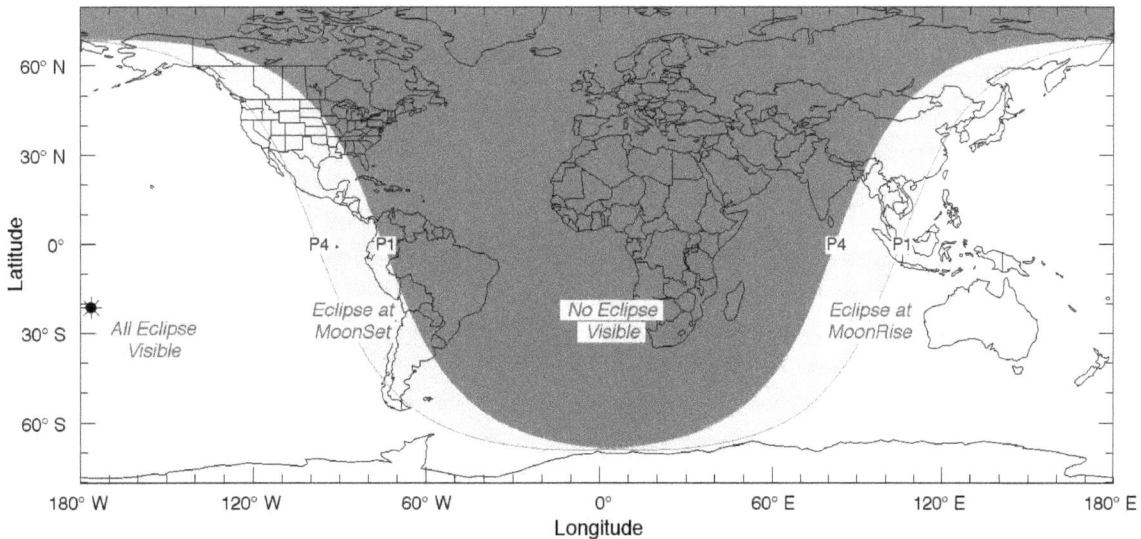

All Eclipse
Visible

P4　P1

Eclipse at
MoonSet

No Eclipse
Visible

P4　P1

Eclipse at
MoonRise

Penumbral Lunar Eclipse of 2031 Oct 30

Greatest Eclipse = 07:46:44.2 TD (= 07:45:29.4 UT1)

Penumbral Magnitude = 0.7173	Gamma = 1.1774	Saros Series = 117
Umbral Magnitude = -0.3193	Axis = 1.1188°	Saros Member = 53 of 71

Sun at Greatest Eclipse
(Geocentric Coordinates)
R.A. = 14h17m25.0s
Dec. = -13°44'38.7"
S.D. = 00°16'06.3"
H.P. = 00°00'08.9"

Moon at Greatest Eclipse
(Geocentric Coordinates)
R.A. = 02h16m19.7s
Dec. = +14°49'53.3"
S.D. = 00°15'32.2"
H.P. = 00°57'01.3"

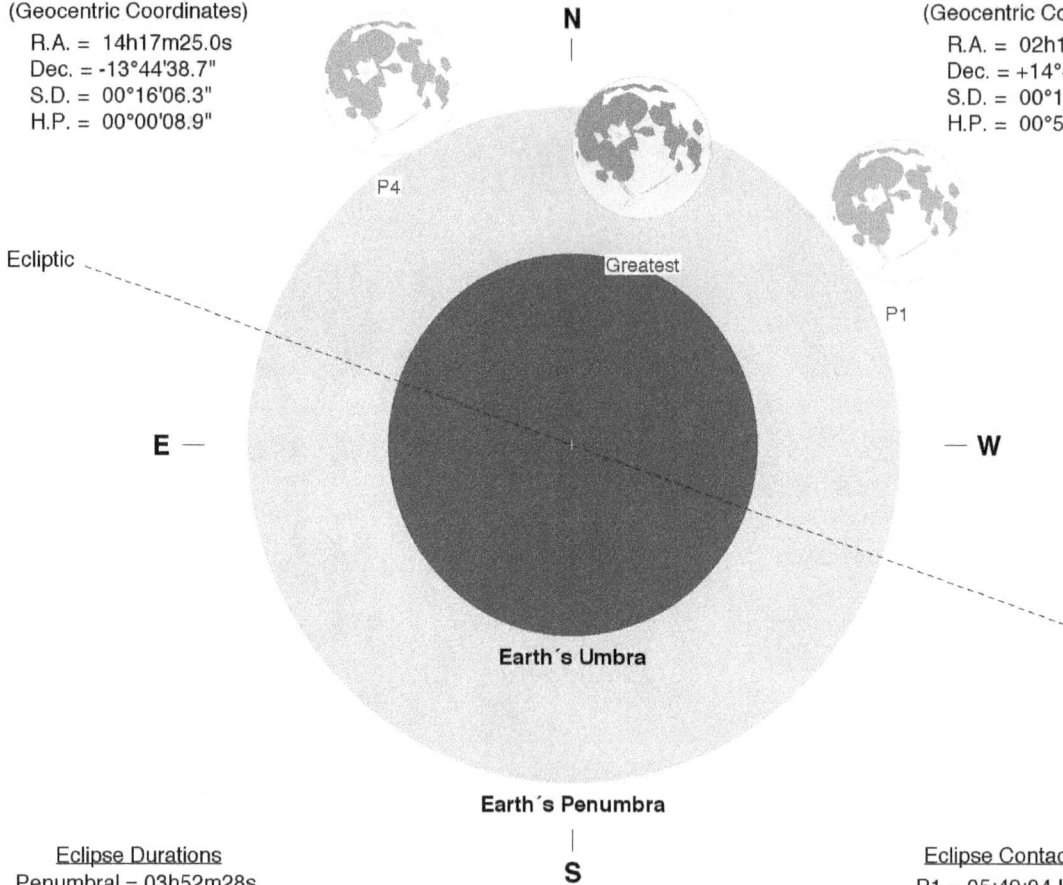

N

P4

Ecliptic

Greatest

P1

E

W

Earth's Umbra

Earth's Penumbra

S

Eclipse Durations
Penumbral = 03h52m28s

Eclipse Contacts
P1 = 05:49:04 UT1
P4 = 09:41:32 UT1

Eph. = JPL DE430
Rule = Herald-Sinnott
ΔT = 75 s

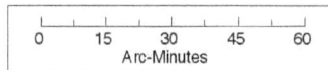

0	15	30	45	60

Arc-Minutes

©2020 F. Espenak, www.EclipseWise.com

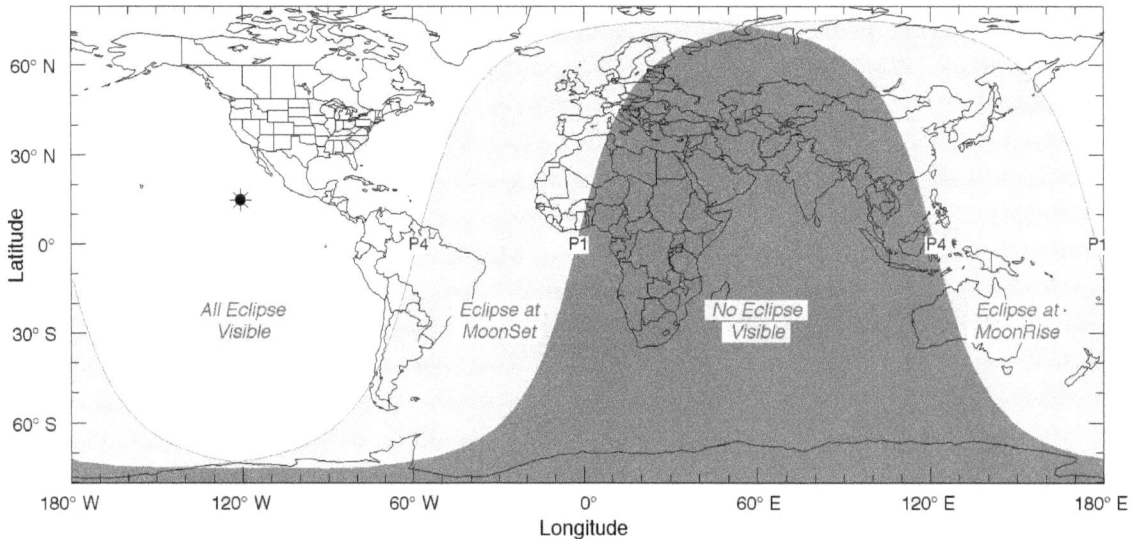

All Eclipse
Visible

Eclipse at
MoonSet

No Eclipse
Visible

Eclipse at
MoonRise

Total Lunar Eclipse of 2032 Apr 25

Greatest Eclipse = 15:14:50.6 TD (= 15:13:35.5 UT1)

Penumbral Magnitude = 2.2204	Gamma = -0.3558	Saros Series = 122
Umbral Magnitude = 1.1925	Axis = 0.3365°	Saros Member = 57 of 74

Sun at Greatest Eclipse
(Geocentric Coordinates)

R.A. = 02h14m38.2s
Dec. = +13°30'28.8"
S.D. = 00°15'53.8"
H.P. = 00°00'08.7"

N

Earth´s Penumbra

Earth´s Umbra

E —

— **W**

P1

U1

U2

Greatest

Ecliptic

U3

U4

P4

Moon at Greatest Eclipse
(Geocentric Coordinates)

R.A. = 14h14m18.6s
Dec. = -13°50'06.1"
S.D. = 00°15'27.9"
H.P. = 00°56'45.4"

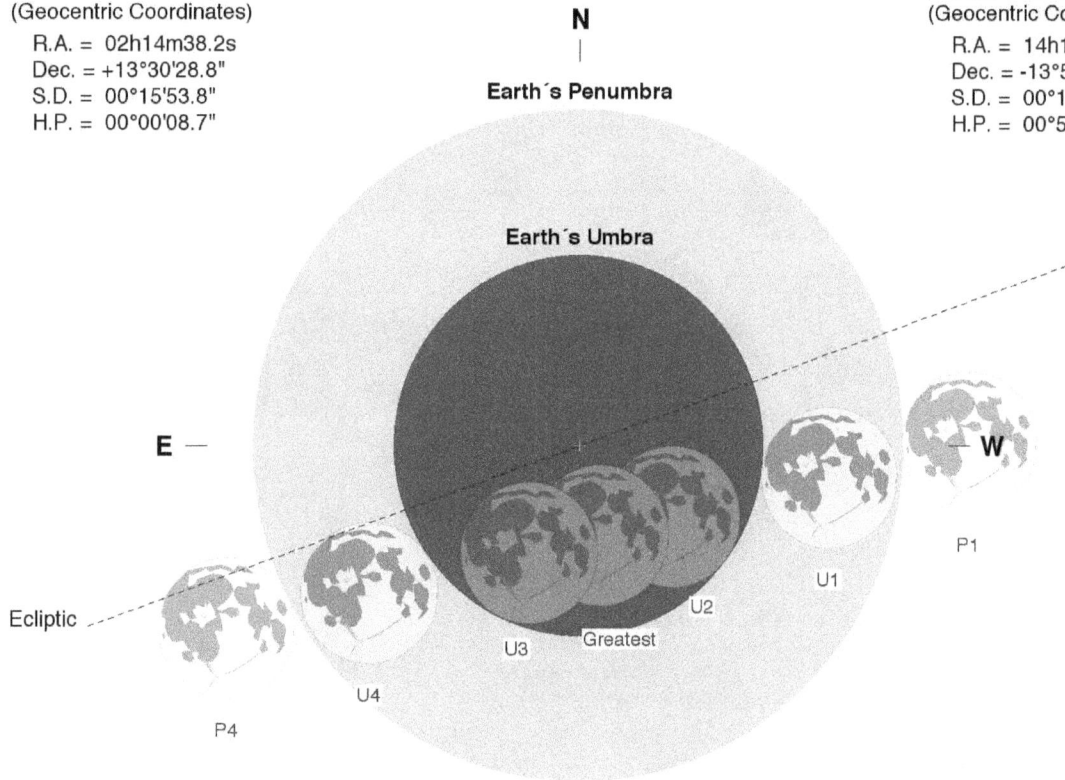

S

Eclipse Durations
Penumbral = 05h43m18s
Umbral = 03h31m58s
Total = 01h06m10s

Eph. = JPL DE430
Rule = Herald-Sinnott
ΔT = 75 s

0	15	30	45	60

Arc-Minutes

©2020 F. Espenak, www.EclipseWise.com

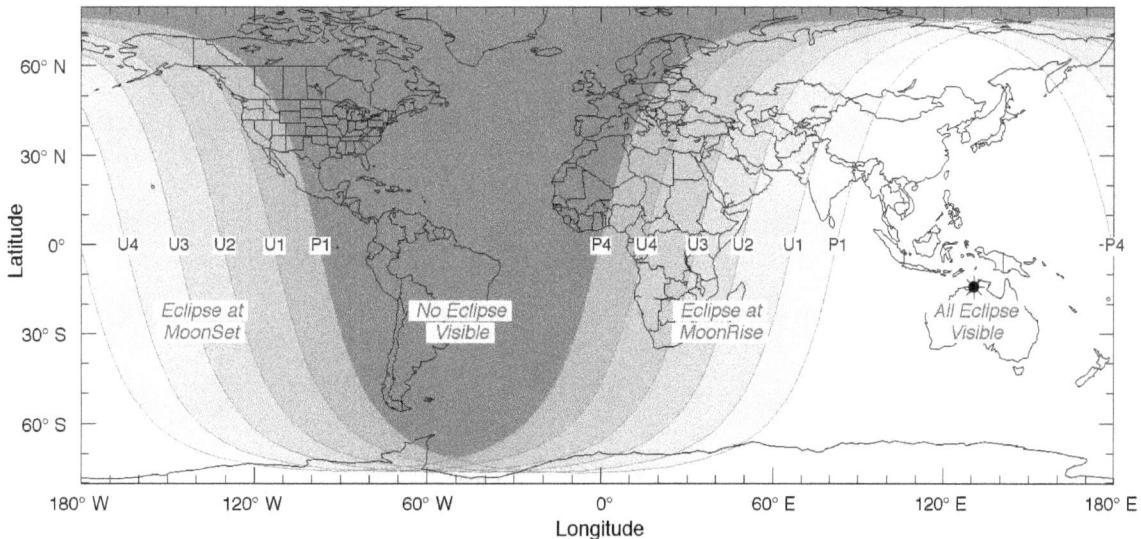

Eclipse Contacts
P1 = 12:21:52 UT1
U1 = 13:27:35 UT1
U2 = 14:40:24 UT1
U3 = 15:46:34 UT1
U4 = 16:59:33 UT1
P4 = 18:05:10 UT1

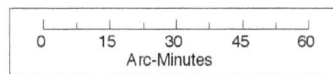

Total Lunar Eclipse of 2032 Oct 18

Greatest Eclipse = 19:03:40.2 TD (= 19:02:24.9 UT1)

Penumbral Magnitude = 2.0841	Gamma = 0.4169	Saros Series = 127
Umbral Magnitude = 1.1039	Axis = 0.4177°	Saros Member = 43 of 72

Sun at Greatest Eclipse
(Geocentric Coordinates)
R.A. = 13h36m15.4s
Dec. = -10°01'20.8"
S.D. = 00°16'03.4"
H.P. = 00°00'08.8"

Moon at Greatest Eclipse
(Geocentric Coordinates)
R.A. = 01h35m47.9s
Dec. = +10°25'28.7"
S.D. = 00°16'22.8"
H.P. = 01°00'07.0"

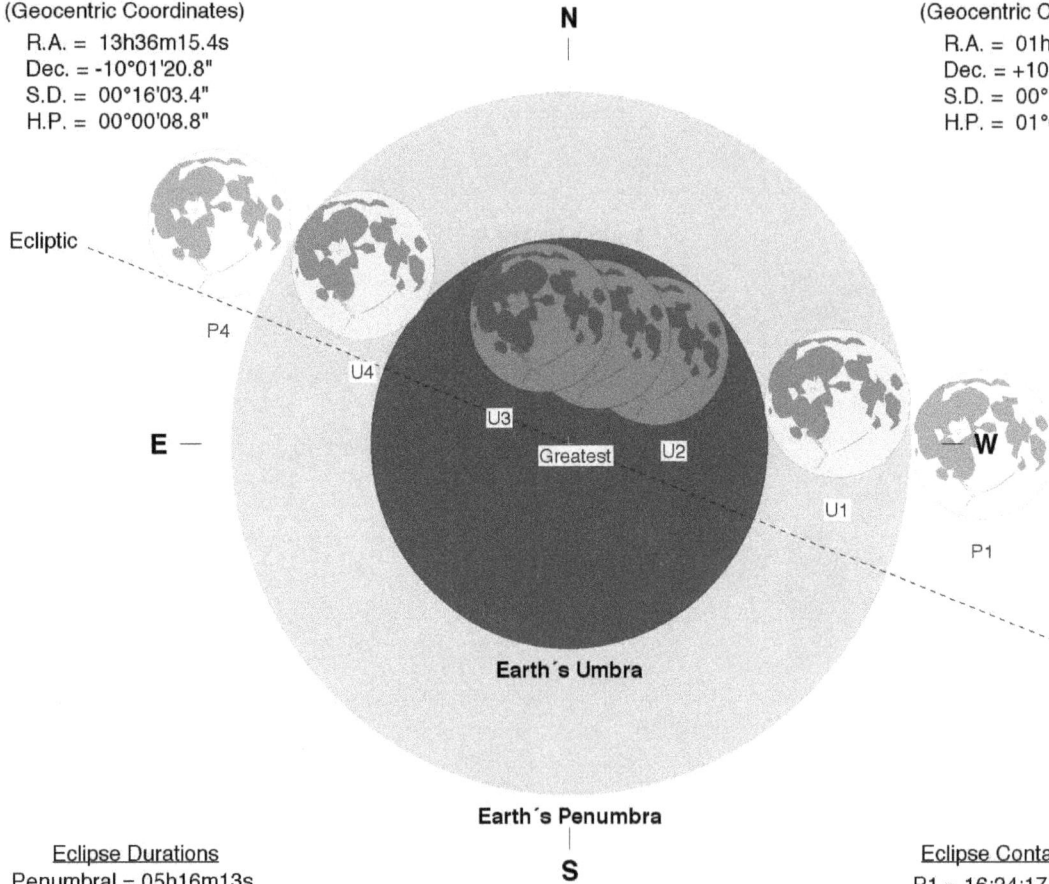

N

Ecliptic

P4

U4

U3

E

Greatest

U2

W

U1

P1

Earth´s Umbra

Earth´s Penumbra

S

Eclipse Durations
Penumbral = 05h16m13s
Umbral = 03h16m38s
Total = 00h47m40s

Eph. = JPL DE430
Rule = Herald-Sinnott
ΔT =   75 s

0	15	30	45	60

Arc-Minutes

©2020 F. Espenak, www.EclipseWise.com

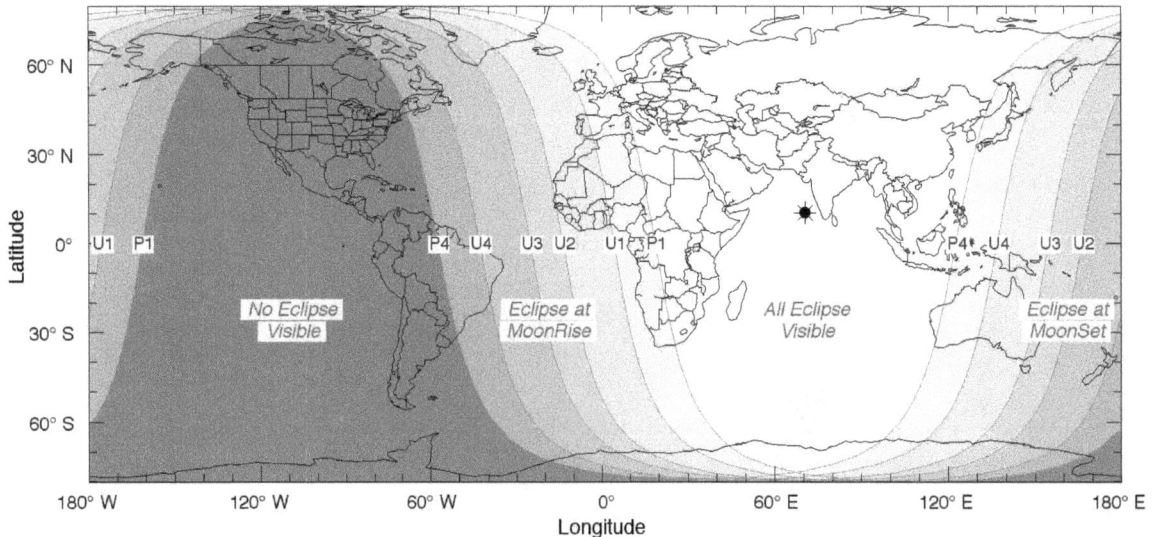

Eclipse Contacts
P1 = 16:24:17 UT1
U1 = 17:24:00 UT1
U2 = 18:38:24 UT1
U3 = 19:26:04 UT1
U4 = 20:40:38 UT1
P4 = 21:40:30 UT1

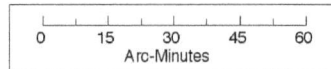

Total Lunar Eclipse of 2033 Apr 14

Greatest Eclipse = 19:13:51.4 TD   (= 19:12:35.9 UT1)

Penumbral Magnitude = 2.1722	Gamma = 0.3954	Saros Series = 132
Umbral Magnitude = 1.0955	Axis = 0.3581°	Saros Member = 31 of 71

Sun at Greatest Eclipse
(Geocentric Coordinates)

R.A. = 01h33m13.7s
Dec. = +09°43'50.2"
S.D. = 00°15'56.7"
H.P. = 00°00'08.8"

N

Moon at Greatest Eclipse
(Geocentric Coordinates)

R.A. = 13h33m37.2s
Dec. = -09°23'08.7"
S.D. = 00°14'48.5"
H.P. = 00°54'21.0"

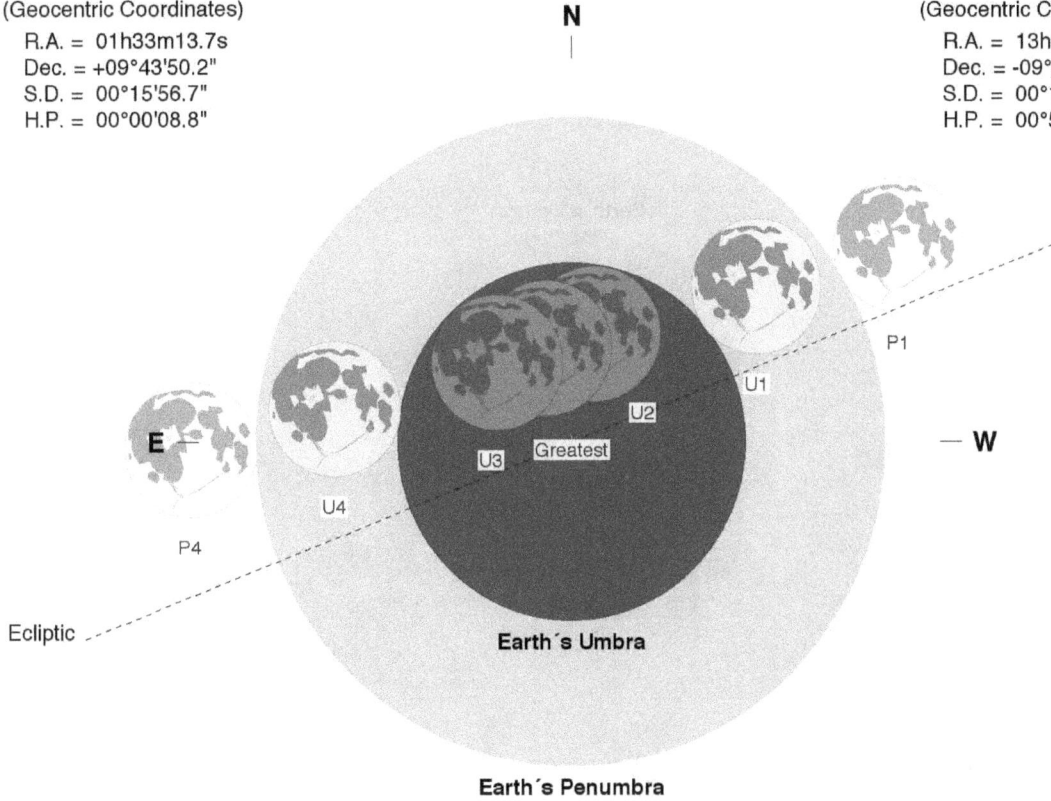

E — — W

P1
U1
U2
U3 Greatest
U4
P4

Earth's Umbra

Earth's Penumbra

Ecliptic

S

Eclipse Durations
Penumbral = 06h02m05s
Umbral = 03h35m47s
Total = 00h49m50s

Eph. = JPL DE430
Rule = Herald-Sinnott
ΔT = 76 s

0   15   30   45   60
Arc-Minutes

©2020 F. Espenak, www.EclipseWise.com

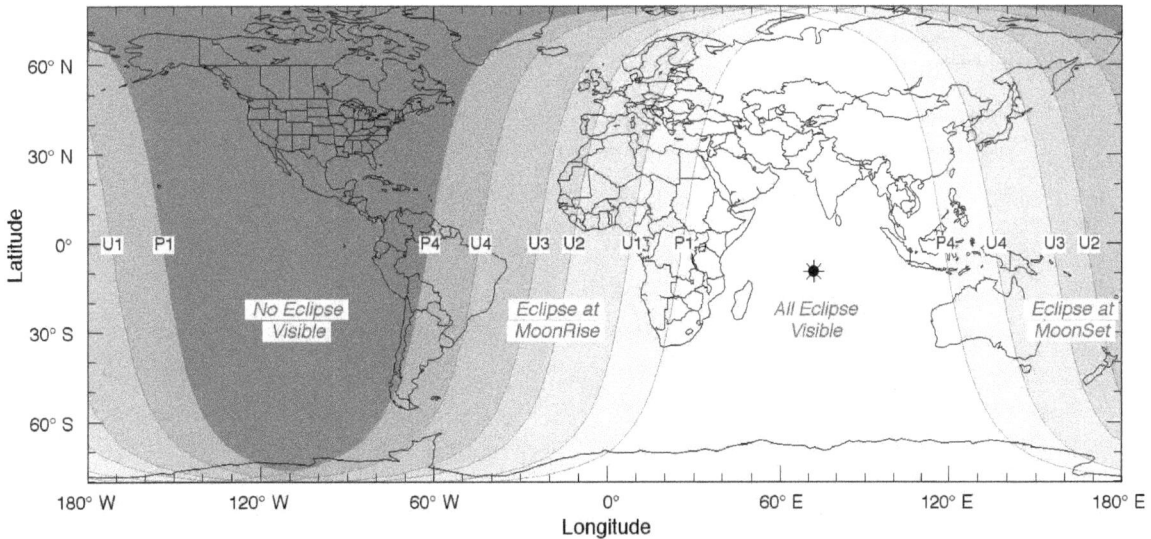

Eclipse Contacts
P1 = 16:11:35 UT1
U1 = 17:24:48 UT1
U2 = 18:47:53 UT1
U3 = 19:37:42 UT1
U4 = 21:00:35 UT1
P4 = 22:13:40 UT1

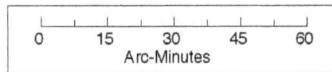

No Eclipse Visible

Eclipse at MoonRise

All Eclipse Visible

Eclipse at MoonSet

133

Total Lunar Eclipse of 2033 Oct 08

Greatest Eclipse = 10:56:22.6 TD   (= 10:55:06.9 UT1)

Penumbral Magnitude = 2.3068	Gamma = -0.2889	Saros Series = 137
Umbral Magnitude = 1.3508	Axis = 0.2958°	Saros Member = 27 of 78

Sun at Greatest Eclipse
(Geocentric Coordinates)

R.A. = 12h57m01.9s
Dec. = -06°05'34.4"
S.D. = 00°16'00.5"
H.P. = 00°00'08.8"

Moon at Greatest Eclipse
(Geocentric Coordinates)

R.A. = 00h57m22.8s
Dec. = +05°48'36.0"
S.D. = 00°16'44.6"
H.P. = 01°01'27.1"

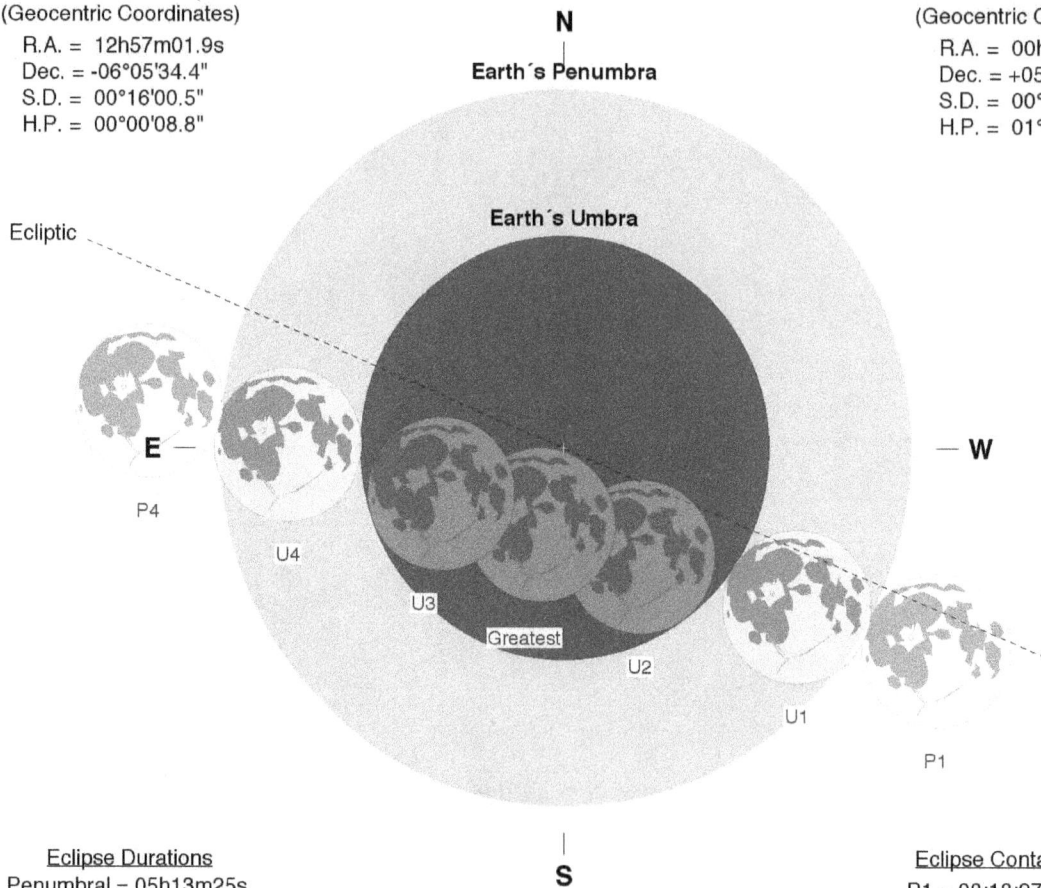

N

Earth's Penumbra

Earth's Umbra

Ecliptic

E

W

P4

U4

U3

Greatest

U2

U1

P1

S

Eclipse Durations
Penumbral = 05h13m25s
Umbral = 03h23m07s
Total = 01h19m25s

Eph. = JPL DE430
Rule = Herald-Sinnott
ΔT = 76 s

0   15   30   45   60
Arc-Minutes

©2020 F. Espenak, www.EclipseWise.com

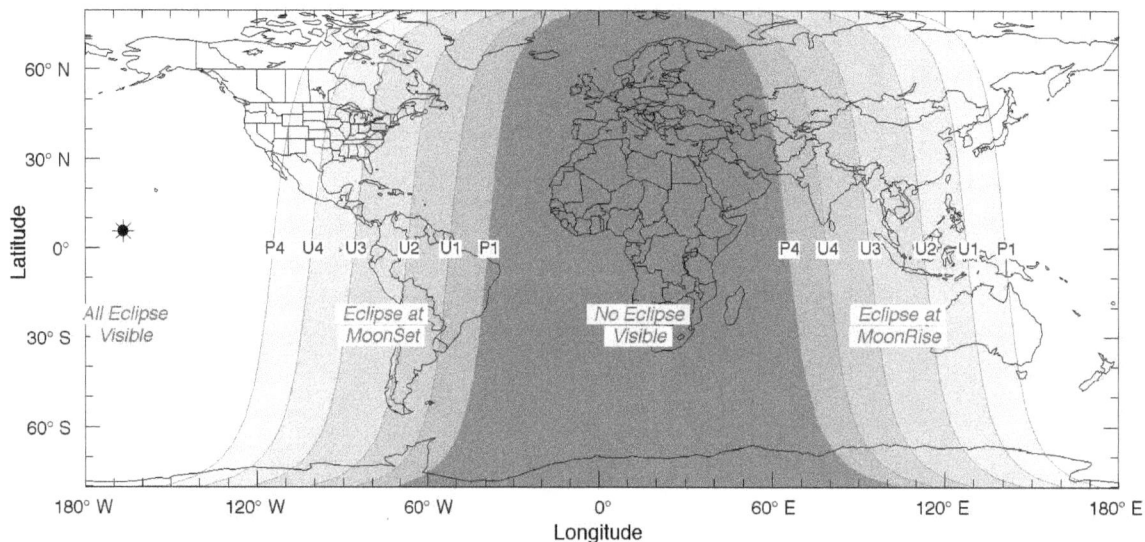

Eclipse Contacts
P1 = 08:18:27 UT1
U1 = 09:13:37 UT1
U2 = 10:15:31 UT1
U3 = 11:34:56 UT1
U4 = 12:36:43 UT1
P4 = 13:31:51 UT1

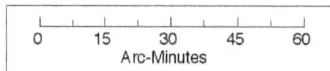

P4 U4 U3 U2 U1 P1   P4 U4 U3 U2 U1 P1

All Eclipse Visible

Eclipse at MoonSet

No Eclipse Visible

Eclipse at MoonRise

Penumbral Lunar Eclipse of 2034 Apr 03

Greatest Eclipse = 19:06:59.9 TD   (= 19:05:43.9 UT1)

Penumbral Magnitude = 0.8557	Gamma = 1.1144	Saros Series = 142
Umbral Magnitude = -0.2263	Axis = 1.0077°	Saros Member = 19 of 73

Sun at Greatest Eclipse
(Geocentric Coordinates)

R.A. = 00h51m54.0s
Dec. = +05°33'29.1"
S.D. = 00°15'59.8"
H.P. = 00°00'08.8"

Moon at Greatest Eclipse
(Geocentric Coordinates)

R.A. = 12h53m05.6s
Dec. = -04°35'42.2"
S.D. = 00°14'47.1"
H.P. = 00°54'15.6"

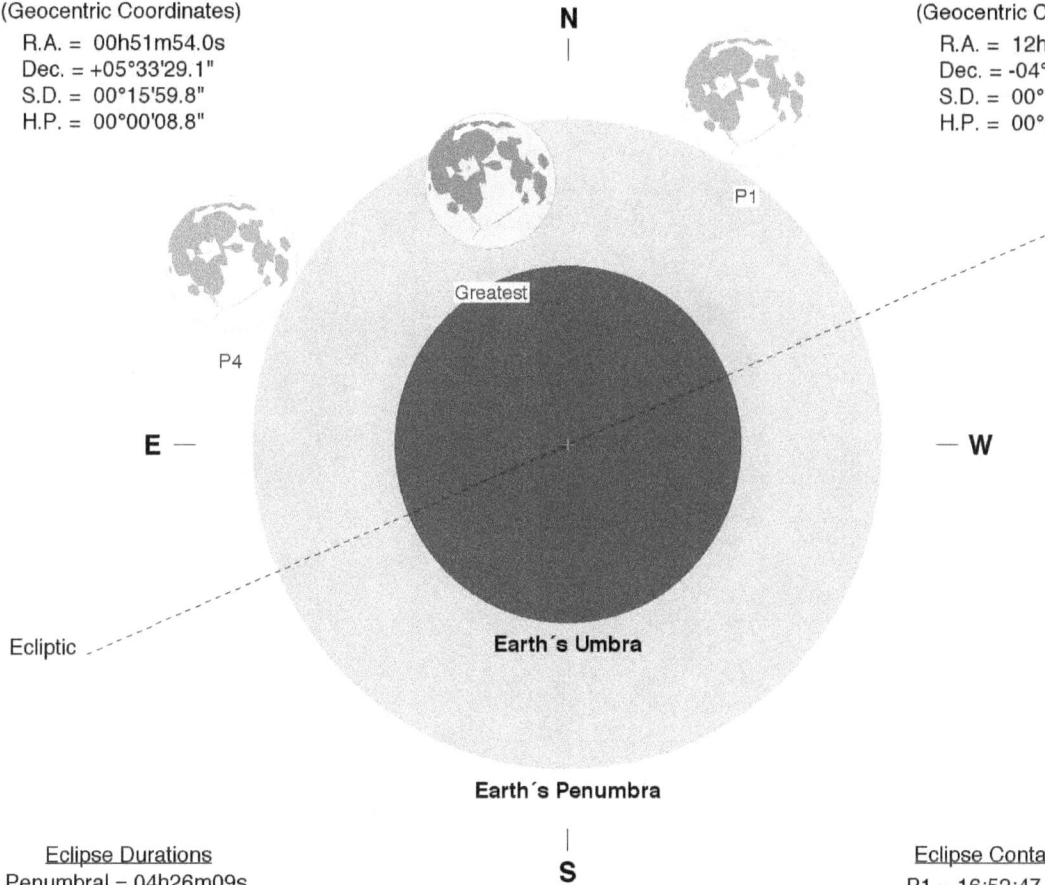

N

P1

Greatest

P4

E —

— W

Earth's Umbra

Earth's Penumbra

S

Ecliptic

Eclipse Durations
Penumbral = 04h26m09s

Eclipse Contacts
P1 = 16:52:47 UT1
P4 = 21:18:56 UT1

Eph. = JPL DE430
Rule = Herald-Sinnott
ΔT =   76 s

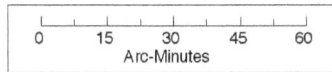

0	15	30	45	60

Arc-Minutes

©2020 F. Espenak, www.EclipseWise.com

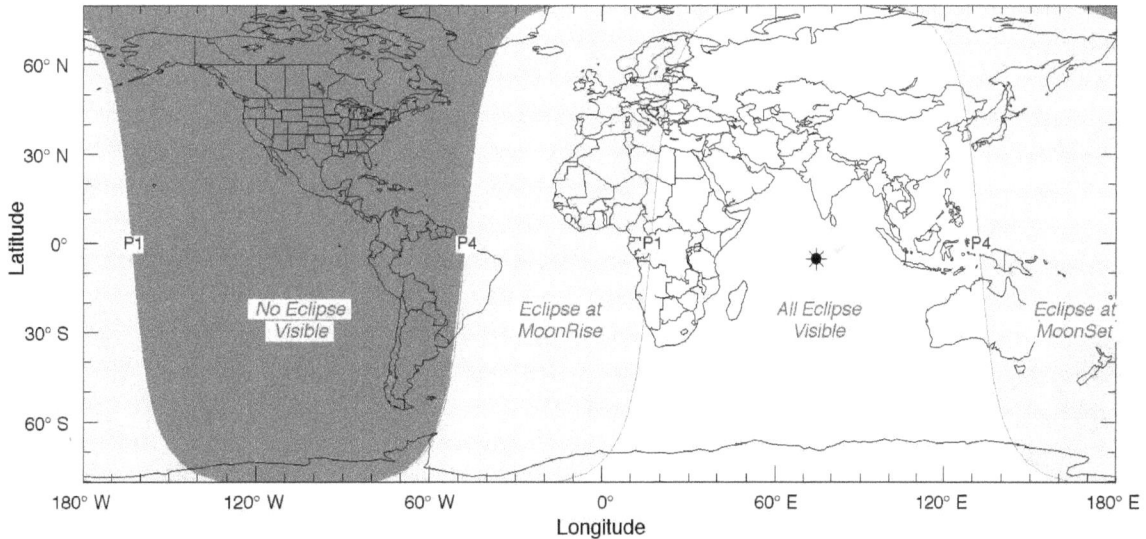

No Eclipse Visible

P1

P4

Eclipse at MoonRise

P1

All Eclipse Visible

P4

Eclipse at MoonSet

Latitude

Longitude

135

Partial Lunar Eclipse of 2034 Sep 28

Greatest Eclipse = 02:47:37.3 TD   (= 02:46:21.1 UT1)

Penumbral Magnitude = 0.9922	Gamma = -1.0110	Saros Series = 147
Umbral Magnitude = 0.0155	Axis = 1.0104°	Saros Member = 9 of 70

Sun at Greatest Eclipse
(Geocentric Coordinates)

R.A. = 12h18m35.8s
Dec. = -02°00'43.0"
S.D. = 00°15'57.6"
H.P. = 00°00'08.8"

Moon at Greatest Eclipse
(Geocentric Coordinates)

R.A. = 00h19m50.0s
Dec. = +01°02'59.4"
S.D. = 00°16'20.4"
H.P. = 00°59'58.2"

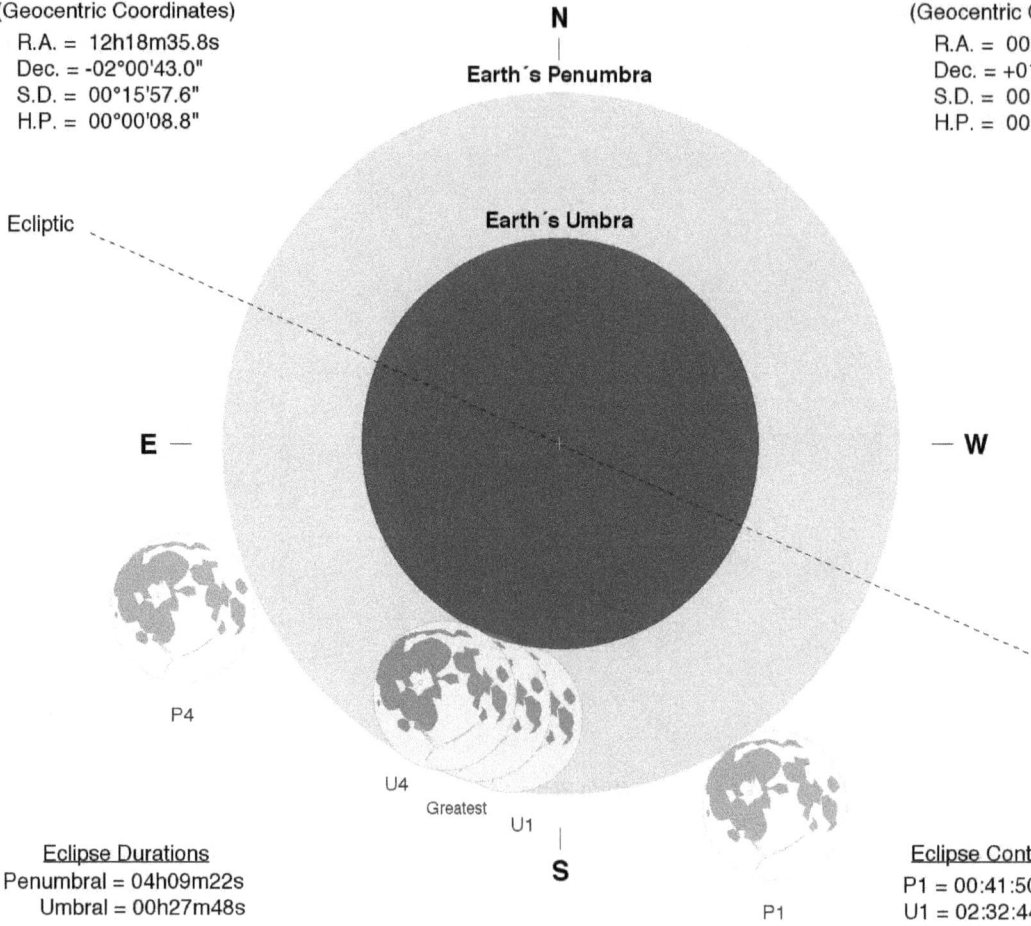

N

Earth's Penumbra

Ecliptic

Earth's Umbra

E —

— W

P4

U4

Greatest

U1

S

P1

Eclipse Durations
Penumbral = 04h09m22s
Umbral = 00h27m48s

Eph. = JPL DE430
Rule = Herald-Sinnott
ΔT = 76 s

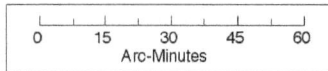

0	15	30	45	60

Arc-Minutes

©2020 F. Espenak, www.EclipseWise.com

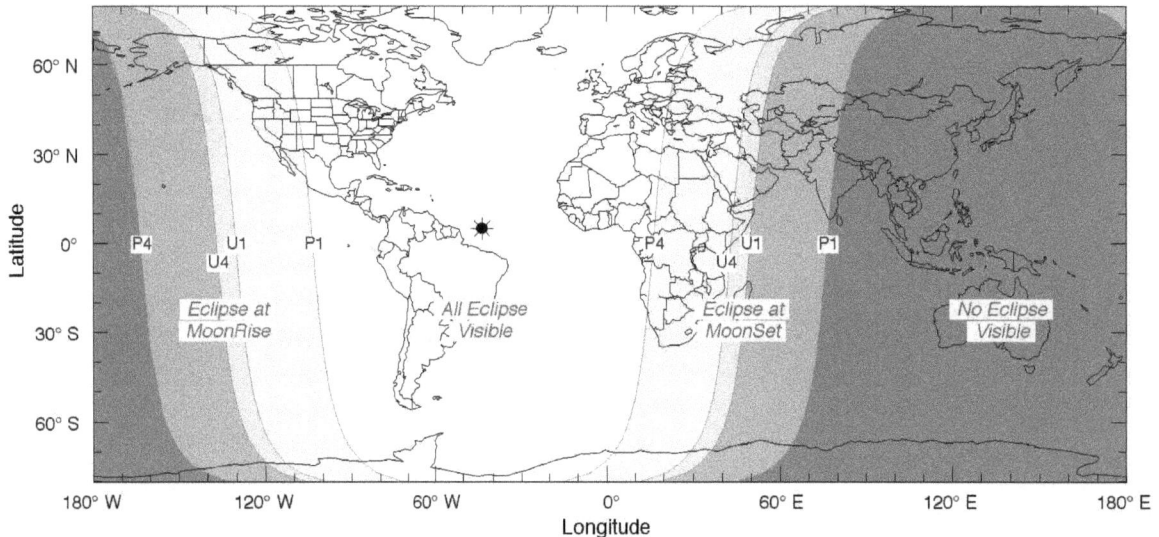

Eclipse Contacts
P1 = 00:41:50 UT1
U1 = 02:32:44 UT1
U4 = 03:00:33 UT1
P4 = 04:51:12 UT1

Penumbral Lunar Eclipse of 2035 Feb 22

Greatest Eclipse = 09:06:11.4 TD (= 09:04:55.0 UT1)

Penumbral Magnitude = 0.9663	Gamma = -1.0367	Saros Series = 114
Umbral Magnitude = -0.0523	Axis = 1.0066°	Saros Member = 60 of 71

Sun at Greatest Eclipse
(Geocentric Coordinates)

R.A. = 22h21m54.2s
Dec. = -10°11'53.9"
S.D. = 00°16'10.2"
H.P. = 00°00'08.9"

Moon at Greatest Eclipse
(Geocentric Coordinates)

R.A. = 10h20m48.3s
Dec. = +09°13'43.5"
S.D. = 00°15'52.5"
H.P. = 00°58'15.8"

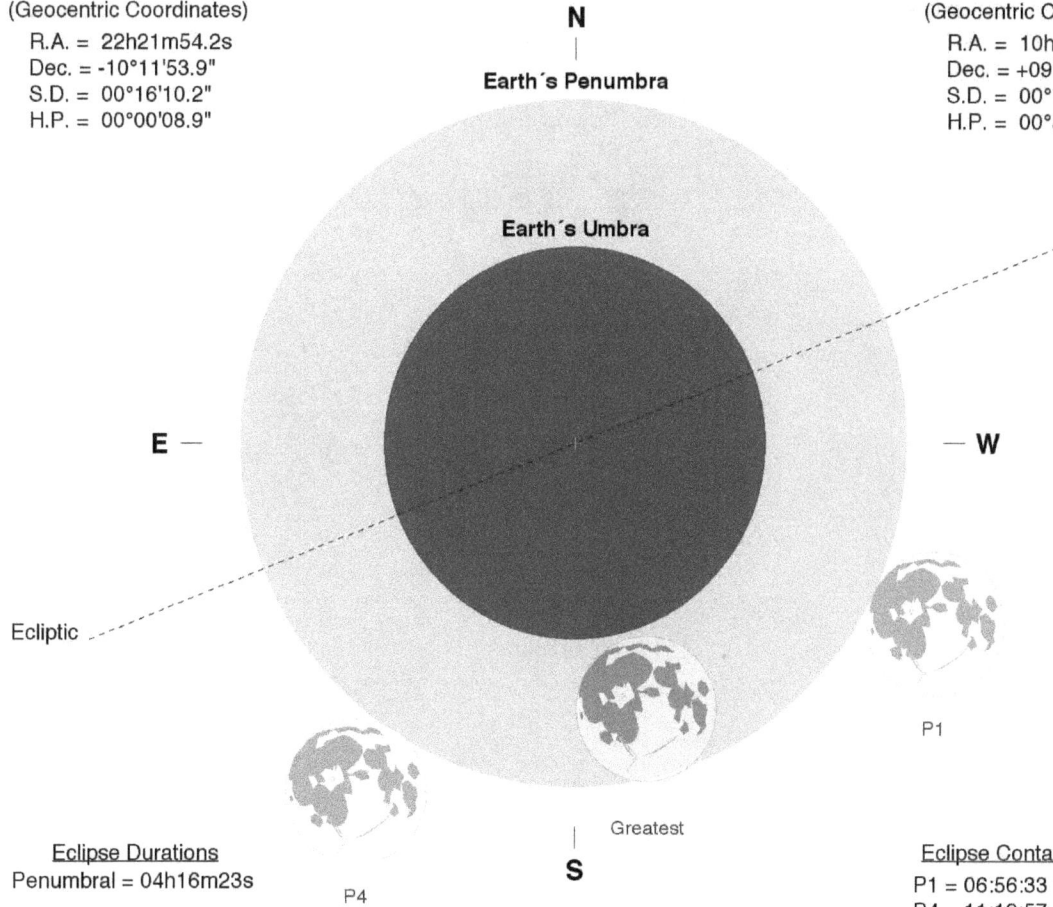

N

Earth's Penumbra

Earth's Umbra

E

W

Ecliptic

P1

Greatest

S

P4

Eclipse Durations
Penumbral = 04h16m23s

Eclipse Contacts
P1 = 06:56:33 UT1
P4 = 11:12:57 UT1

Eph. = JPL DE430
Rule = Herald-Sinnott
ΔT = 76 s

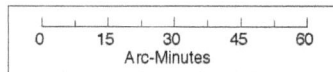

0	15	30	45	60

Arc-Minutes

©2020 F. Espenak, www.EclipseWise.com

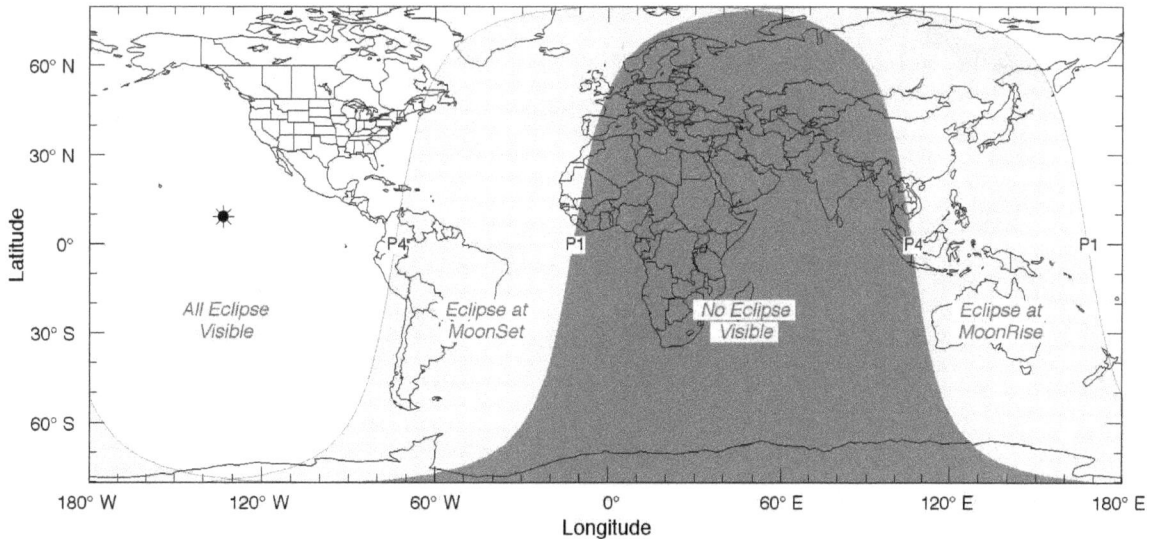

All Eclipse Visible

Eclipse at MoonSet

No Eclipse Visible

Eclipse at MoonRise

P4   P1   P4   P1

Longitude

Partial Lunar Eclipse of 2035 Aug 19

Greatest Eclipse = 01:12:15.0 TD   (= 01:10:58.2 UT1)

Penumbral Magnitude = 1.1519	Gamma = 0.9434	Saros Series = 119
Umbral Magnitude = 0.1049	Axis = 0.8708°	Saros Member = 62 of 82

Sun at Greatest Eclipse
(Geocentric Coordinates)
R.A. = 09h52m42.8s
Dec. = +12°52'21.3"
S.D. = 00°15'48.1"
H.P. = 00°00'08.7"

Moon at Greatest Eclipse
(Geocentric Coordinates)
R.A. = 21h51m50.7s
Dec. = -12°01'40.6"
S.D. = 00°15'05.5"
H.P. = 00°55'23.4"

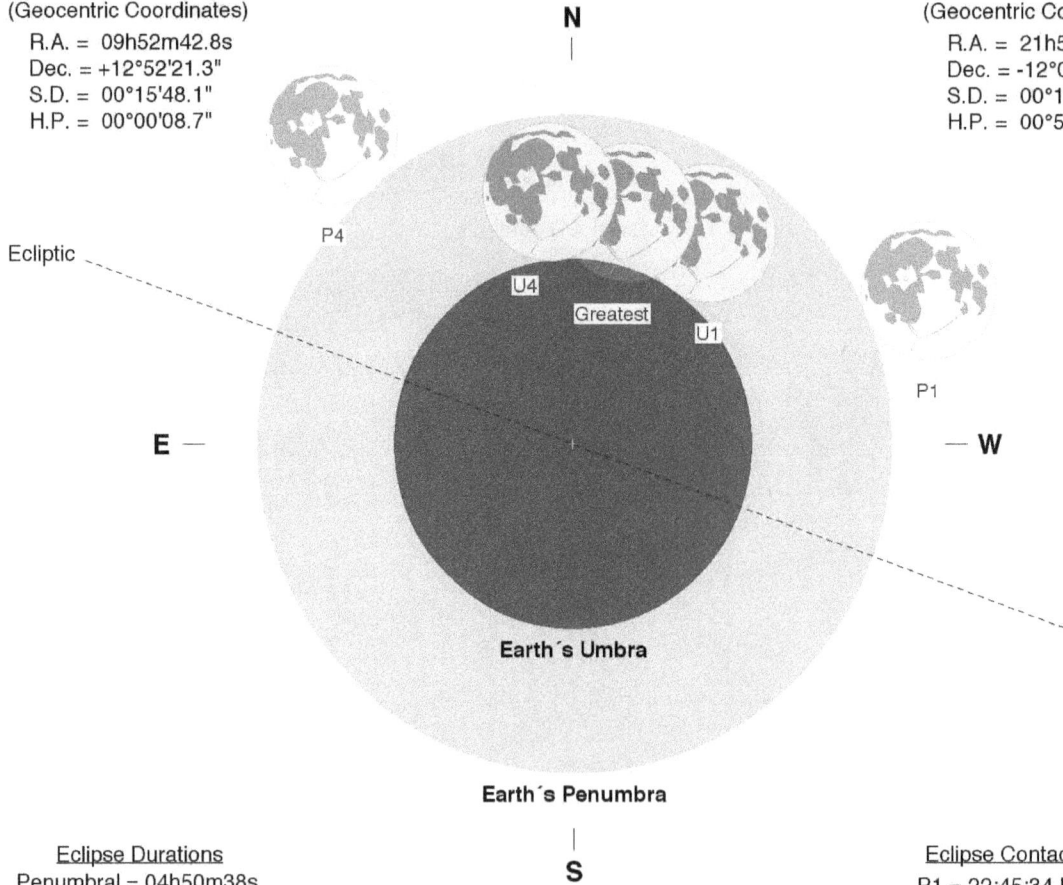

N

Ecliptic

P4

U4

Greatest

U1

P1

E

W

Earth's Umbra

Earth's Penumbra

S

Eclipse Durations
Penumbral = 04h50m38s
Umbral = 01h17m11s

Eph. = JPL DE430
Rule = Herald-Sinnott
ΔT =   77 s

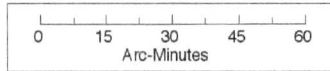

Arc-Minutes
0   15   30   45   60

©2020 F. Espenak, www.EclipseWise.com

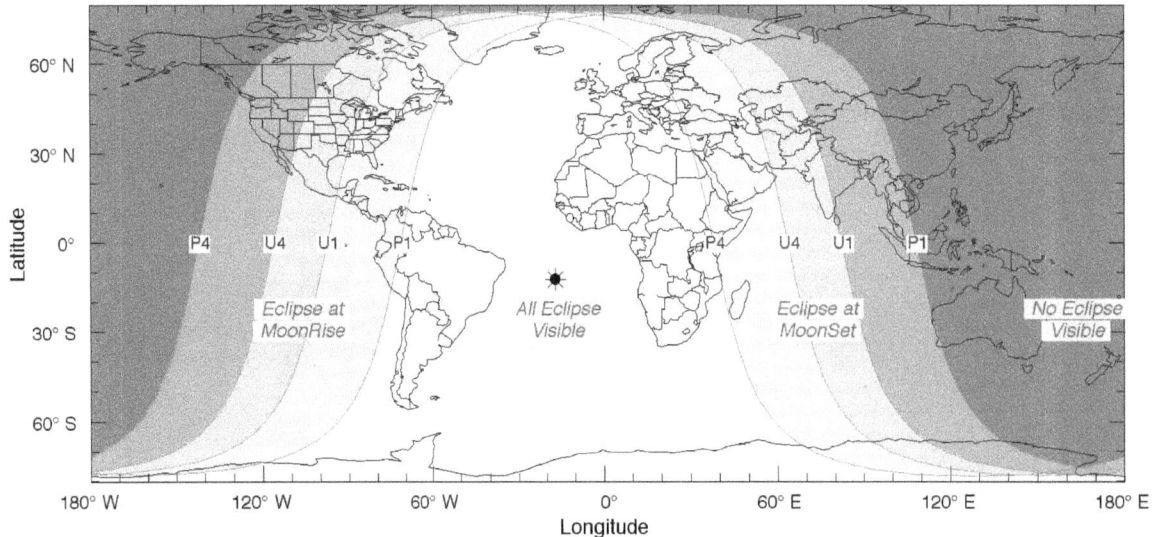

Eclipse Contacts
P1 = 22:45:34 UT1
U1 = 00:32:18 UT1
U4 = 01:49:29 UT1
P4 = 03:36:13 UT1

Total Lunar Eclipse of 2036 Feb 11

Greatest Eclipse = 22:13:06.2 TD   (= 22:11:49.2 UT1)

Penumbral Magnitude = 2.2762	Gamma = -0.3110	Saros Series = 124
Umbral Magnitude = 1.3006	Axis = 0.3159°	Saros Member = 50 of 73

Sun at Greatest Eclipse
(Geocentric Coordinates)

R.A. = 21h40m25.4s
Dec. = -13°55'30.0"
S.D. = 00°16'12.3"
H.P. = 00°00'08.9"

N

Earth's Penumbra

Moon at Greatest Eclipse
(Geocentric Coordinates)

R.A. = 09h40m07.3s
Dec. = +13°37'03.4"
S.D. = 00°16'36.7"
H.P. = 01°00'57.8"

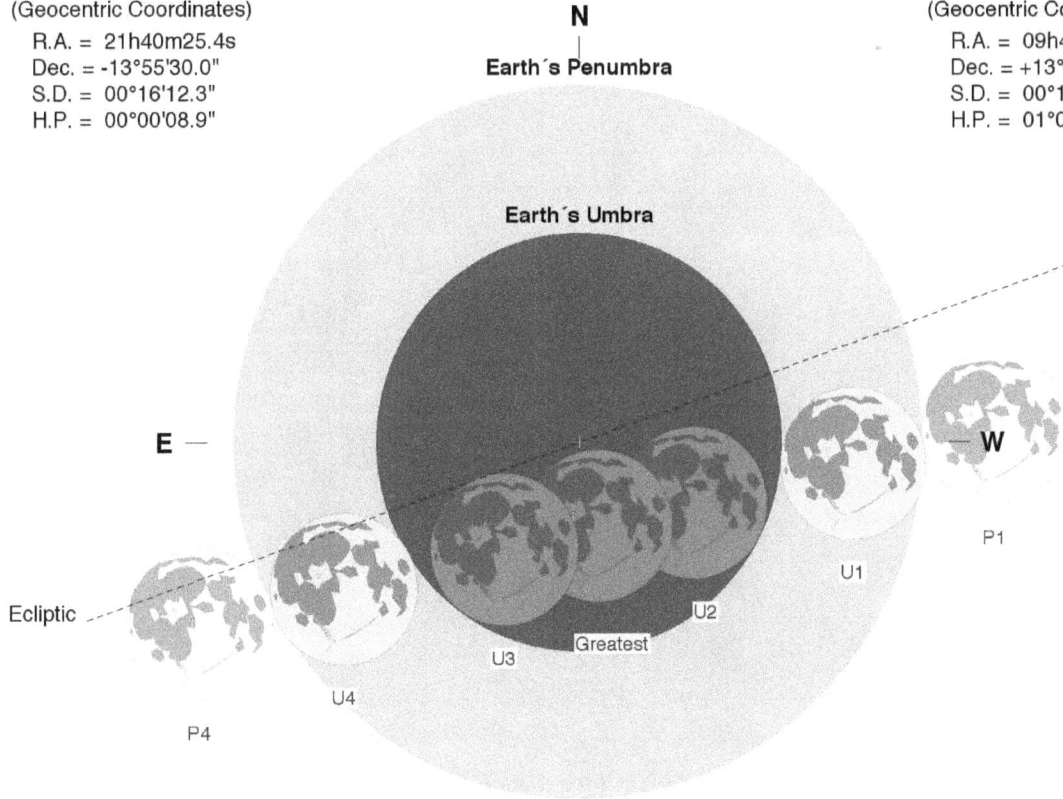

Earth's Umbra

E —

W

P1

Ecliptic

U1

U2

Greatest

U3

U4

P4

S

Eclipse Durations

Penumbral = 05h16m53s
Umbral = 03h22m39s
Total = 01h15m06s

Eph. = JPL DE430
Rule = Herald-Sinnott
ΔT =   77 s

0	15	30	45	60

Arc-Minutes

©2020 F. Espenak, www.EclipseWise.com

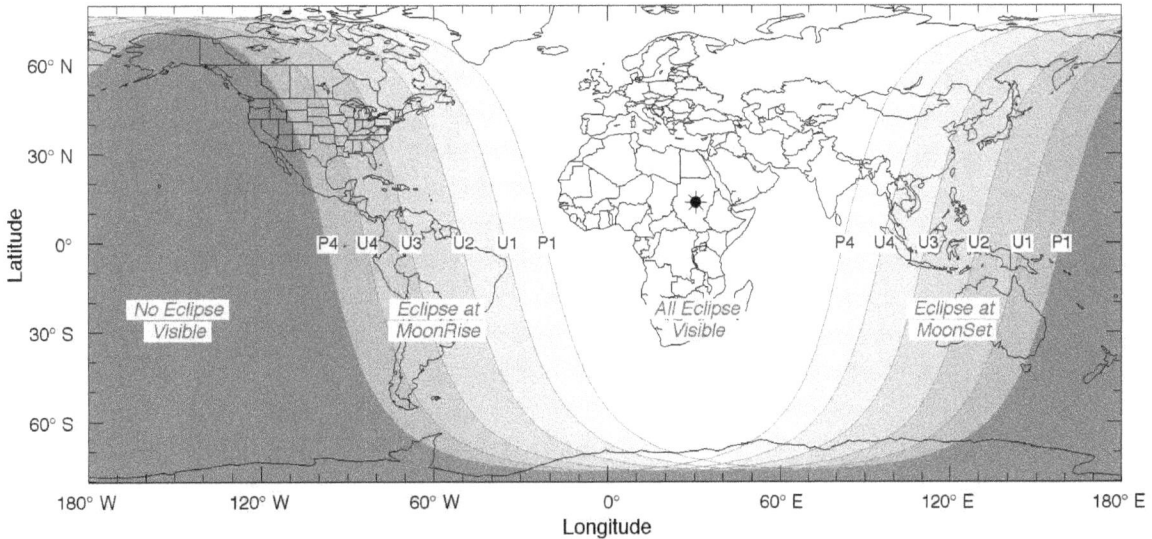

Eclipse Contacts

P1 = 19:33:22 UT1
U1 = 20:30:26 UT1
U2 = 21:34:10 UT1
U3 = 22:49:16 UT1
U4 = 23:53:05 UT1
P4 = 00:50:15 UT1

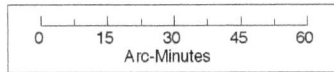

No Eclipse
Visible

Eclipse at
MoonRise

All Eclipse
Visible

Eclipse at
MoonSet

P4 U4 U3 U2 U1 P1   P4 U4 U3 U2 U1 P1

Total Lunar Eclipse of 2036 Aug 07

Greatest Eclipse = 02:52:32.4 TD   (= 02:51:15.2 UT1)

Penumbral Magnitude = 2.5279	Gamma = 0.2004	Saros Series = 129
Umbral Magnitude = 1.4556	Axis = 0.1803°	Saros Member = 39 of 71

Sun at Greatest Eclipse
(Geocentric Coordinates)
R.A. = 09h10m39.1s
Dec. = +16°16'20.8"
S.D. = 00°15'46.3"
H.P. = 00°00'08.7"

Moon at Greatest Eclipse
(Geocentric Coordinates)
R.A. = 21h10m30.3s
Dec. = -16°05'44.3"
S.D. = 00°14'42.5"
H.P. = 00°53'58.8"

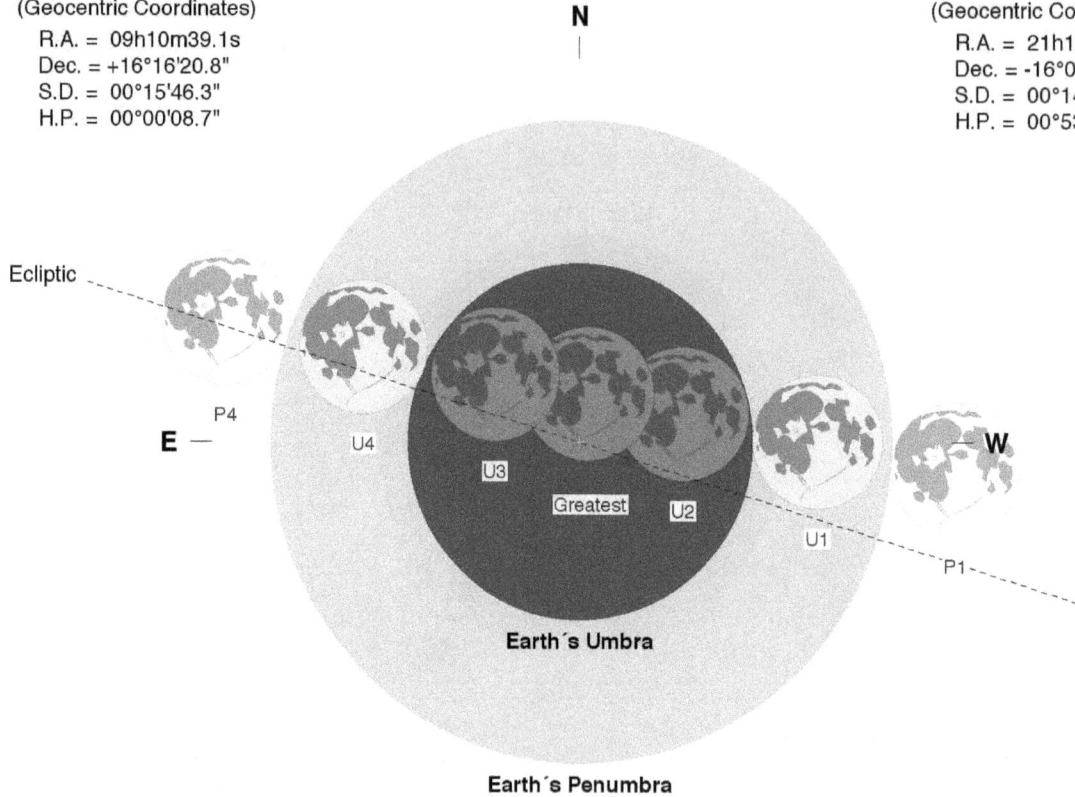

N

Ecliptic

P4

E

U4

U3

Greatest

U2

U1

P1

W

Earth's Umbra

Earth's Penumbra

S

Eclipse Durations
Penumbral = 06h13m03s
Umbral = 03h52m11s
Total = 01h36m06s

Eph. = JPL DE430
Rule = Herald-Sinnott
ΔT = 77 s

0   15   30   45   60
Arc-Minutes

©2020 F. Espenak, www.EclipseWise.com

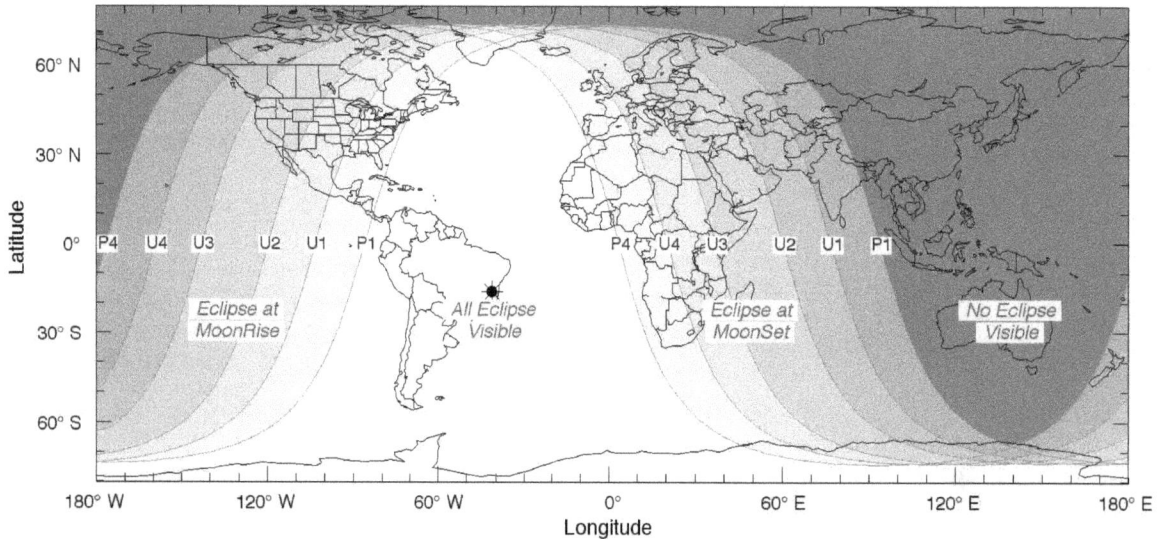

Eclipse Contacts
P1 = 23:44:42 UT1
U1 = 00:55:08 UT1
U2 = 02:03:08 UT1
U3 = 03:39:15 UT1
U4 = 04:47:19 UT1
P4 = 05:57:45 UT1

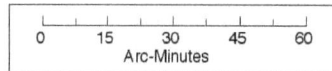

Total Lunar Eclipse of 2037 Jan 31

Greatest Eclipse = 14:01:38.4 TD (= 14:00:20.9 UT1)

Penumbral Magnitude = 2.1815	Gamma = 0.3619	Saros Series = 134
Umbral Magnitude = 1.2086	Axis = 0.3693°	Saros Member = 28 of 72

Sun at Greatest Eclipse
(Geocentric Coordinates)

R.A. = 20h57m58.6s
Dec. = -17°10'47.4"
S.D. = 00°16'14.0"
H.P. = 00°00'08.9"

Moon at Greatest Eclipse
(Geocentric Coordinates)

R.A. = 08h58m15.6s
Dec. = +17°32'34.5"
S.D. = 00°16'41.1"
H.P. = 01°01'14.2"

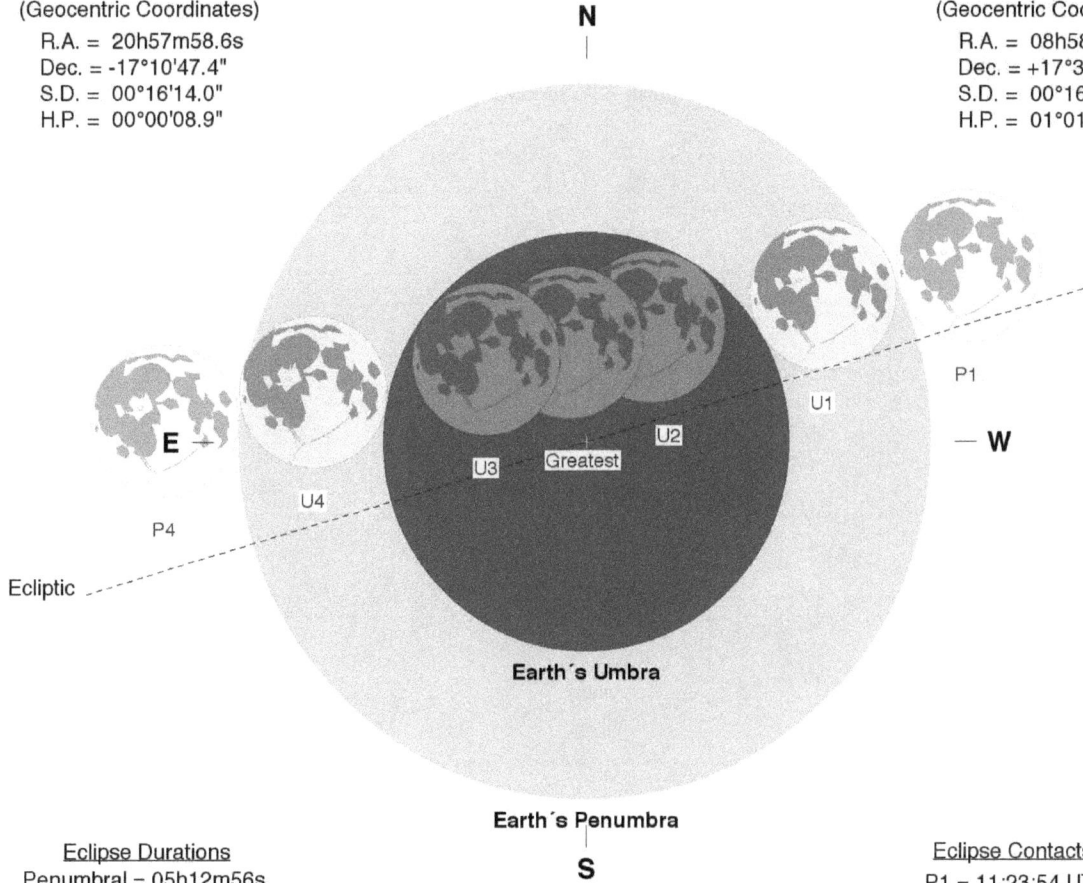

N

E

W

P1

U1

U2

Greatest

U3

U4

P4

Ecliptic

Earth's Umbra

Earth's Penumbra

S

Eclipse Durations

Penumbral = 05h12m56s
Umbral = 03h18m12s
Total = 01h04m17s

Eph. = JPL DE430
Rule = Herald-Sinnott
ΔT =   77 s

0   15   30   45   60
Arc-Minutes

©2020 F. Espenak, www.EclipseWise.com

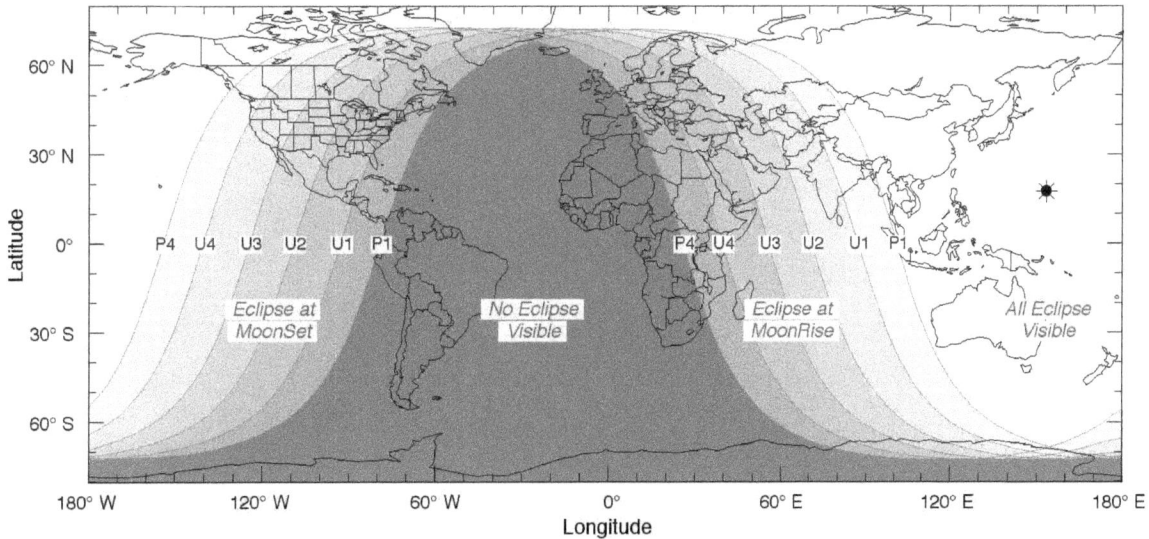

Eclipse Contacts

P1 = 11:23:54 UT1
U1 = 12:21:18 UT1
U2 = 13:28:18 UT1
U3 = 14:32:35 UT1
U4 = 15:39:30 UT1
P4 = 16:36:50 UT1

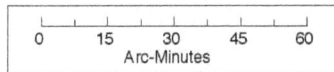

P4 U4 U3 U2 U1 P1

P4 U4   U3 U2 U1 P1

Eclipse at
MoonSet

No Eclipse
Visible

Eclipse at
MoonRise

All Eclipse
Visible

Partial Lunar Eclipse of 2037 Jul 27

Greatest Eclipse = 04:09:53.3 TD   (= 04:08:35.6 UT1)

Penumbral Magnitude = 1.8596	Gamma = -0.5582	Saros Series = 139
Umbral Magnitude = 0.8108	Axis = 0.5126°	Saros Member = 22 of 79

Sun at Greatest Eclipse
(Geocentric Coordinates)
R.A. = 08h27m18.9s
Dec. = +19°07'58.8"
S.D. = 00°15'45.0"
H.P. = 00°00'08.7"

Moon at Greatest Eclipse
(Geocentric Coordinates)
R.A. = 20h27m37.3s
Dec. = -19°38'25.9"
S.D. = 00°15'00.9"
H.P. = 00°55'06.5"

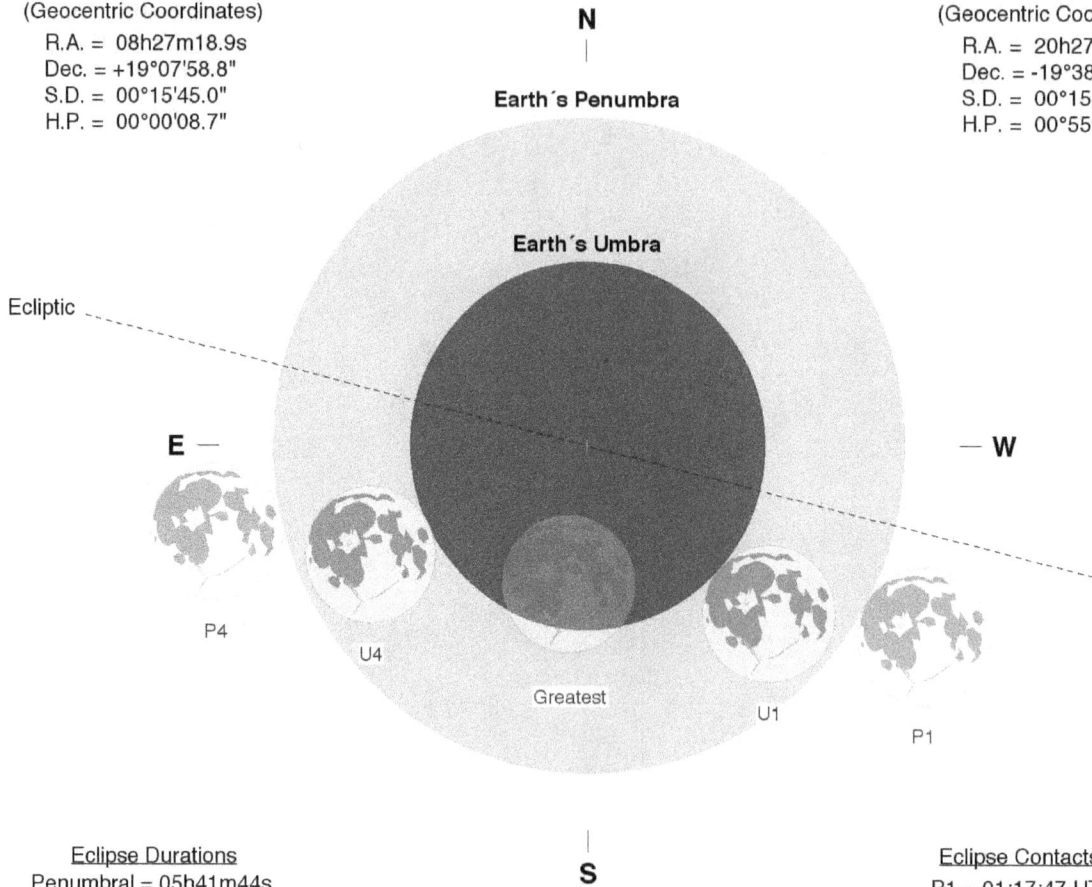

N

Earth's Penumbra

Earth's Umbra

Ecliptic

E

W

P4

U4

Greatest

U1

P1

S

Eclipse Durations
Penumbral = 05h41m44s
Umbral = 03h13m11s

Eph. = JPL DE430
Rule = Herald-Sinnott
ΔT =   78 s

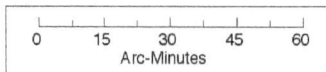

0	15	30	45	60
		Arc-Minutes		

©2020 F. Espenak, www.EclipseWise.com

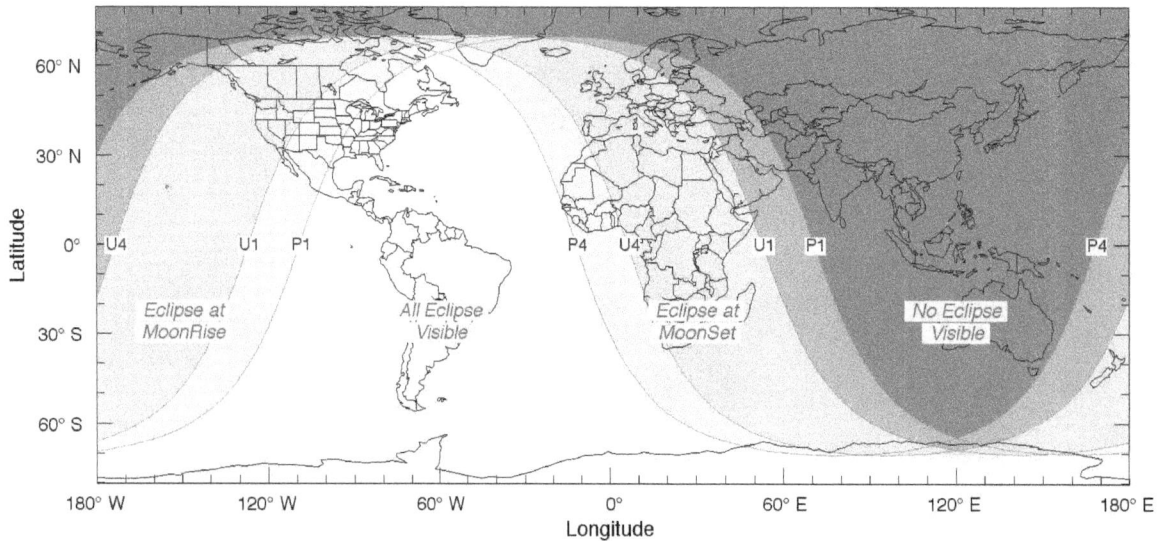

Eclipse Contacts
P1 = 01:17:47 UT1
U1 = 02:32:01 UT1
U4 = 05:45:12 UT1
P4 = 06:59:31 UT1

Penumbral Lunar Eclipse of 2038 Jan 21

Greatest Eclipse = 03:49:51.9 TD (= 03:48:33.9 UT1)

Penumbral Magnitude = 0.9009	Gamma = 1.0711	Saros Series = 144
Umbral Magnitude = -0.1127	Axis = 1.0505°	Saros Member = 17 of 71

Sun at Greatest Eclipse
(Geocentric Coordinates)

R.A. = 20h13m39.3s
Dec. = -19°53'23.0"
S.D. = 00°16'15.2"
H.P. = 00°00'08.9"

Moon at Greatest Eclipse
(Geocentric Coordinates)

R.A. = 08h14m12.5s
Dec. = +20°55'55.8"
S.D. = 00°16'02.1"
H.P. = 00°58'51.1"

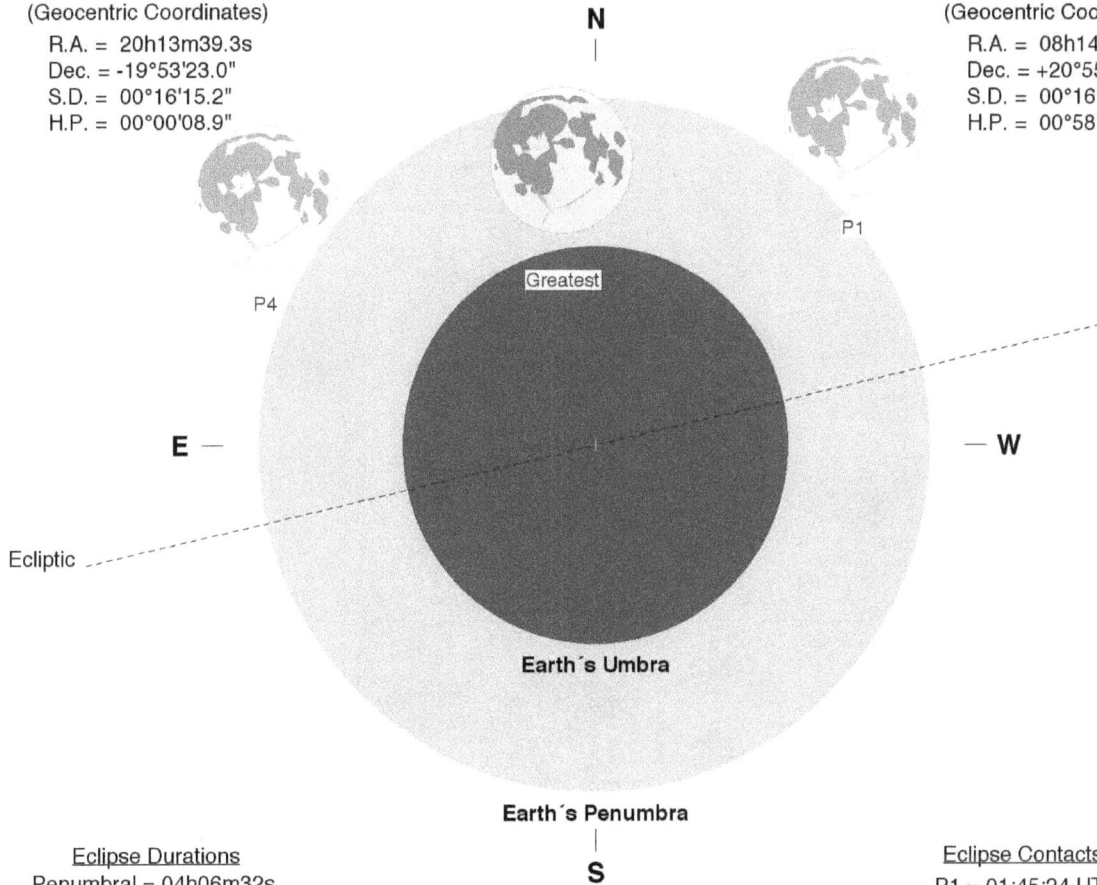

N

P1

P4

Greatest

E —

— W

Ecliptic

Earth´s Umbra

Earth´s Penumbra

S

Eclipse Durations
Penumbral = 04h06m32s

Eclipse Contacts
P1 = 01:45:24 UT1
P4 = 05:51:56 UT1

Eph. = JPL DE430
Rule = Herald-Sinnott
ΔT = 78 s

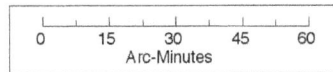

0	15	30	45	60

Arc-Minutes

©2020 F. Espenak, www.EclipseWise.com

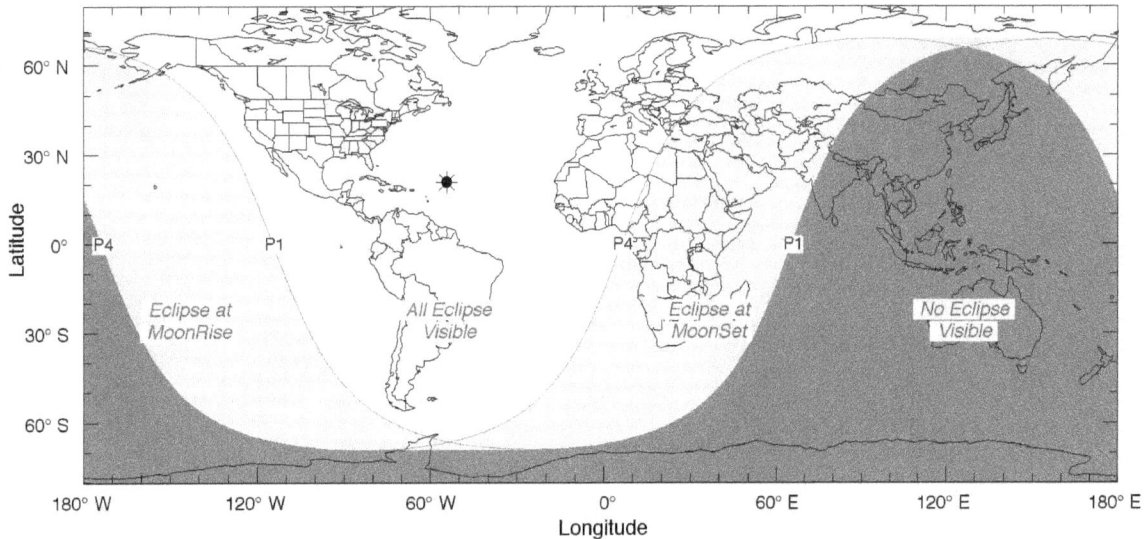

60° N

30° N

0° — P4 — P1 — P4 — P1

30° S

60° S

Eclipse at
MoonRise

All Eclipse
Visible

Eclipse at
MoonSet

No Eclipse
Visible

Latitude

180° W   120° W   60° W   0°   60° E   120° E   180° E

Longitude

Penumbral Lunar Eclipse of 2038 Jun 17

Greatest Eclipse = 02:45:02.2 TD   (= 02:43:44.0 UT1)

Penumbral Magnitude = 0.4438	Gamma = 1.3083	Saros Series = 111
Umbral Magnitude = -0.5259	Axis = 1.2993°	Saros Member = 68 of 71

Sun at Greatest Eclipse
(Geocentric Coordinates)

R.A. = 05h42m46.1s
Dec. = +23°22'28.6"
S.D. = 00°15'44.7"
H.P. = 00°00'08.7"

Moon at Greatest Eclipse
(Geocentric Coordinates)

R.A. = 17h43m28.2s
Dec. = -22°05'07.2"
S.D. = 00°16'14.3"
H.P. = 00°59'35.6"

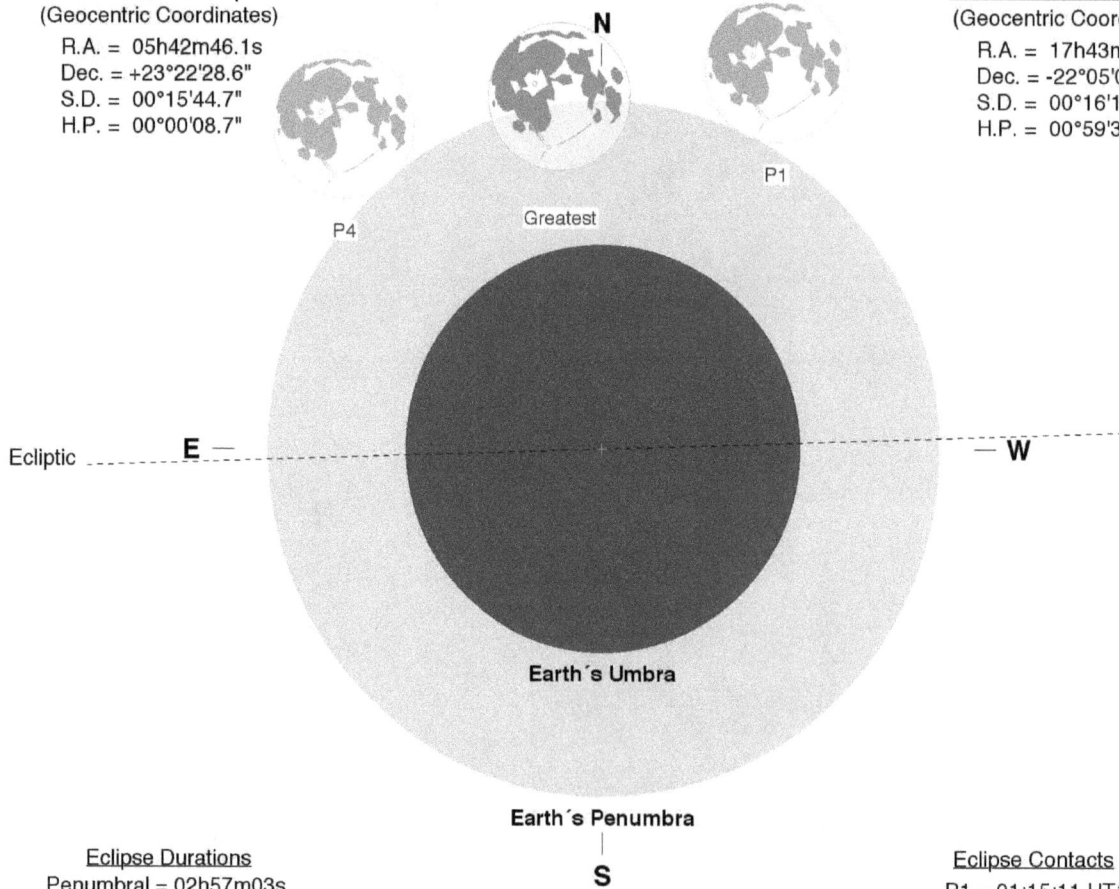

N

P1

P4

Greatest

Ecliptic

E

W

Earth's Umbra

Earth's Penumbra

S

Eclipse Durations
Penumbral = 02h57m03s

Eclipse Contacts
P1 = 01:15:11 UT1
P4 = 04:12:14 UT1

Eph. = JPL DE430
Rule = Herald-Sinnott
ΔT = 78 s

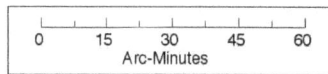

0   15   30   45   60
Arc-Minutes

©2020 F. Espenak, www.EclipseWise.com

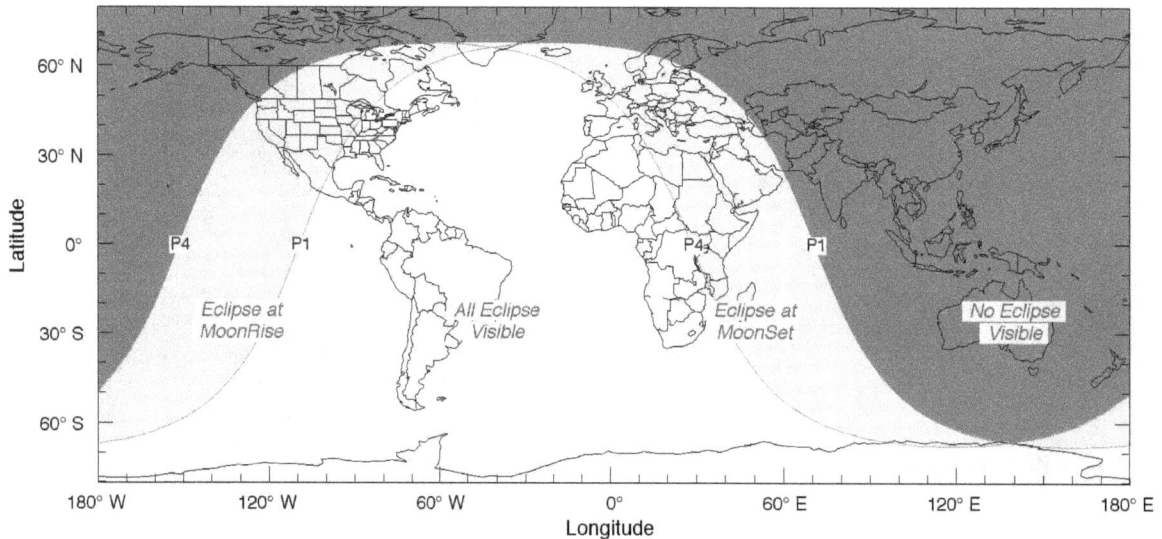

P4   P1

P4   P1

Eclipse at MoonRise

All Eclipse Visible

Eclipse at MoonSet

No Eclipse Visible

Latitude

60° N
30° N
0°
30° S
60° S

180° W   120° W   60° W   0°   60° E   120° E   180° E
Longitude

Penumbral Lunar Eclipse of 2038 Jul 16

Greatest Eclipse = 11:35:56.3 TD (= 11:34:38.0 UT1)

Penumbral Magnitude = 0.5012	Gamma = -1.2838	Saros Series = 149
Umbral Magnitude = -0.4938	Axis = 1.2417°	Saros Member = 4 of 71

Sun at Greatest Eclipse
(Geocentric Coordinates)

R.A. = 07h43m47.7s
Dec. = +21°17'34.6"
S.D. = 00°15'44.2"
H.P. = 00°00'08.7"

N

Earth's Penumbra

Earth's Umbra

Ecliptic

E —

— **W**

Moon at Greatest Eclipse
(Geocentric Coordinates)

R.A. = 19h44m13.1s
Dec. = -22°31'51.1"
S.D. = 00°15'48.9"
H.P. = 00°58'02.4"

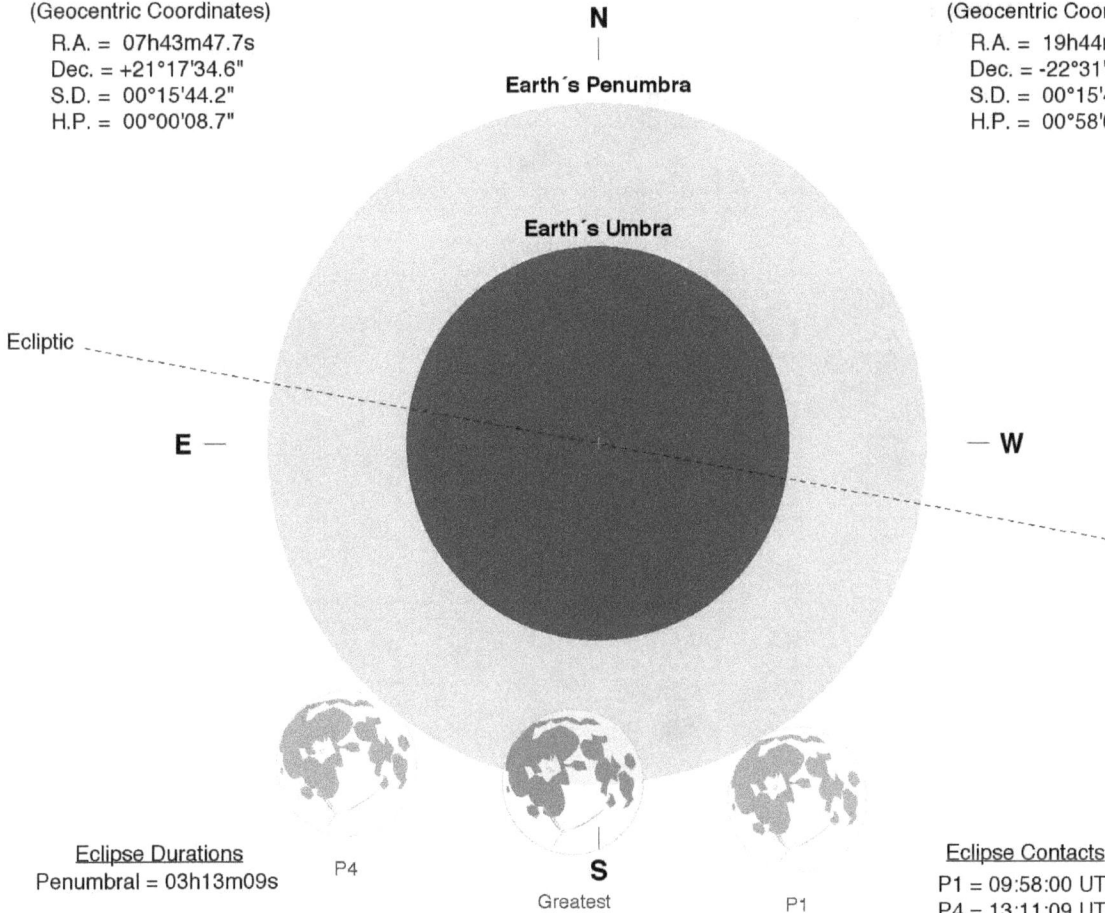

P4

S

Greatest

P1

Eclipse Durations
Penumbral = 03h13m09s

Eclipse Contacts
P1 = 09:58:00 UT1
P4 = 13:11:09 UT1

0	15	30	45	60

Arc-Minutes

Eph. = JPL DE430
Rule = Herald-Sinnott
ΔT = 78 s

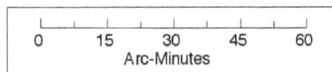

©2020 F. Espenak, www.EclipseWise.com

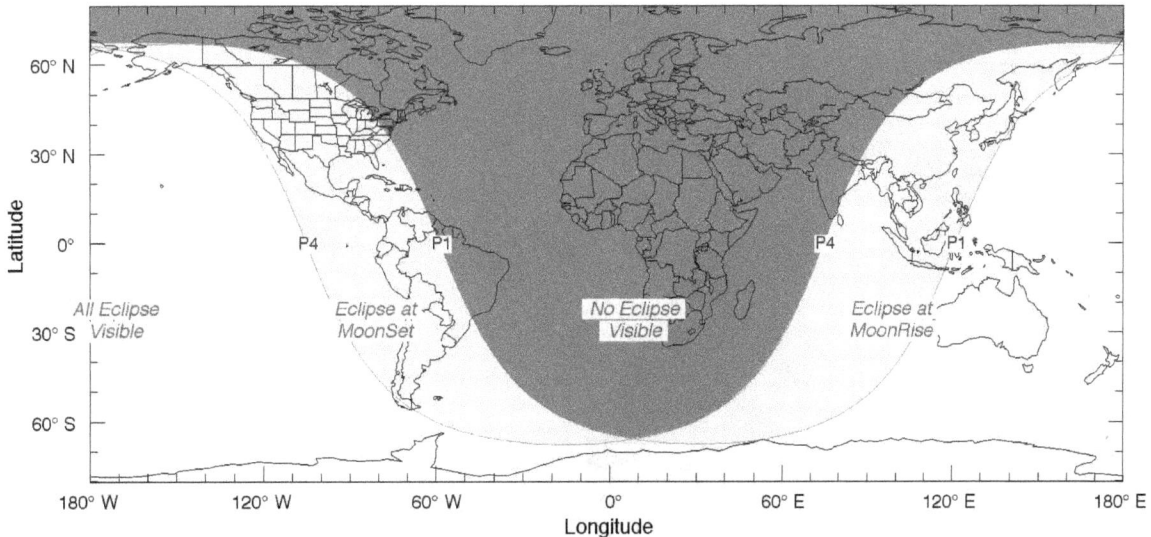

All Eclipse
Visible

Eclipse at
MoonSet

No Eclipse
Visible

Eclipse at
MoonRise

P4

P1

P4

P1

Penumbral Lunar Eclipse of 2038 Dec 11

Greatest Eclipse = 17:45:00.4 TD   (= 17:43:41.9 UT1)

Penumbral Magnitude = 0.8062	Gamma = -1.1449	Saros Series = 116
Umbral Magnitude = -0.2876	Axis = 1.0398°	Saros Member = 59 of 73

Sun at Greatest Eclipse
(Geocentric Coordinates)
R.A. = 17h15m29.9s
Dec. = -23°02'24.2"
S.D. = 00°16'14.6"
H.P. = 00°00'08.9"

Moon at Greatest Eclipse
(Geocentric Coordinates)
R.A. = 05h16m16.9s
Dec. = +22°00'57.8"
S.D. = 00°14'51.0"
H.P. = 00°54'29.8"

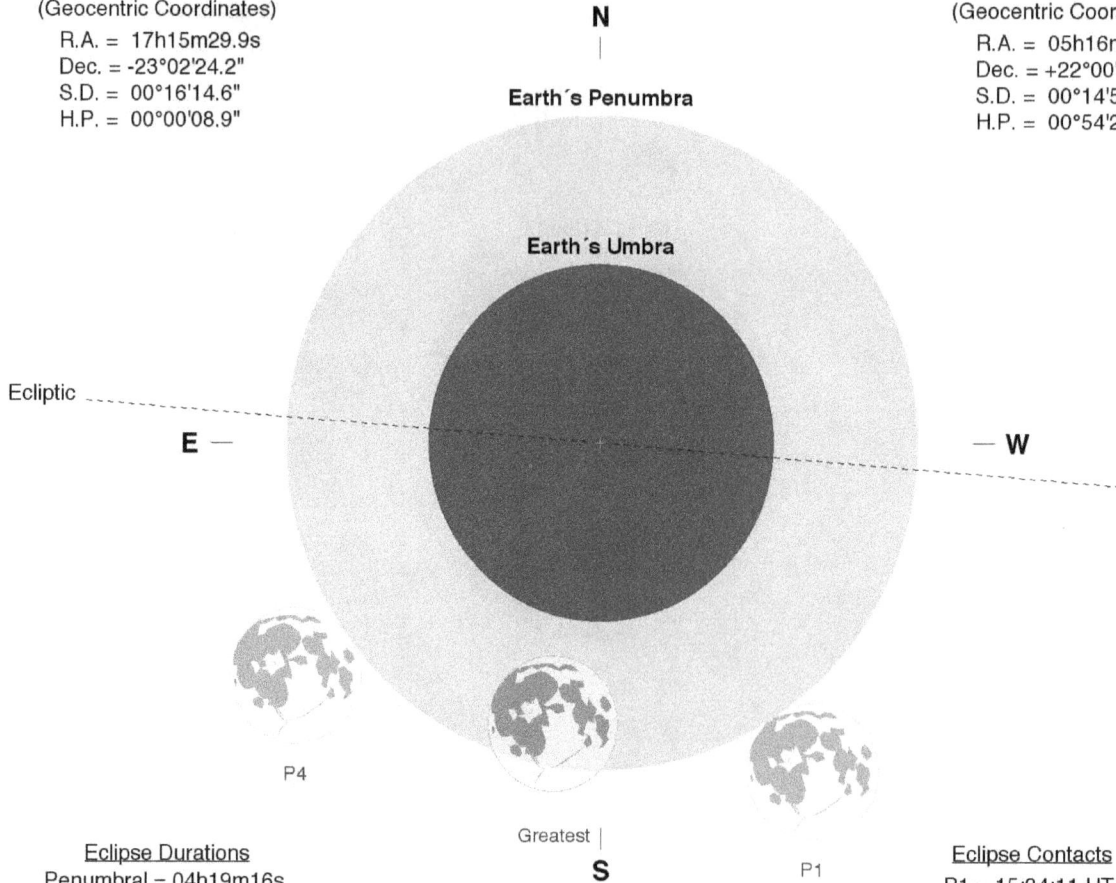

N

Earth's Penumbra

Earth's Umbra

Ecliptic

E

W

P4

Greatest

S

P1

Eclipse Durations
Penumbral = 04h19m16s

Eclipse Contacts
P1 = 15:34:11 UT1
P4 = 19:53:27 UT1

Eph. = JPL DE430
Rule = Herald-Sinnott
ΔT =   79 s

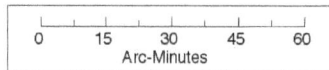

0   15   30   45   60
Arc-Minutes

©2020 F. Espenak, www.EclipseWise.com

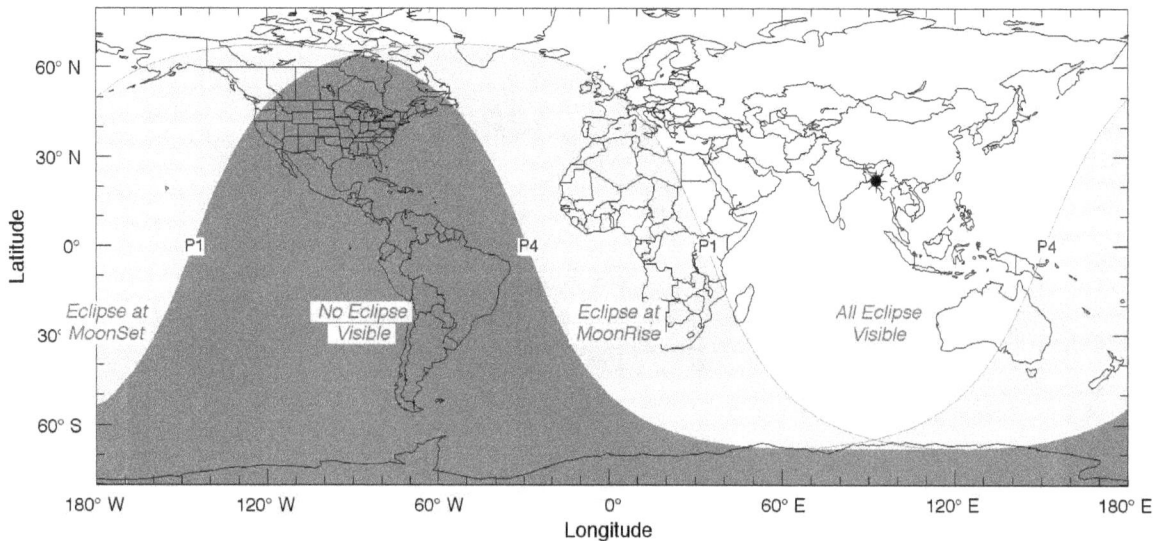

Latitude / Longitude map:

60° N

30° N

0°

30° S

60° S

180° W   120° W   60° W   0°   60° E   120° E   180° E

Longitude

Eclipse at MoonSet

No Eclipse Visible

Eclipse at MoonRise

All Eclipse Visible

P1   P4   P1   P4

Partial Lunar Eclipse of 2039 Jun 06

Greatest Eclipse = 18:54:25.4 TD (= 18:53:06.7 UT1)

Penumbral Magnitude = 1.8288	Gamma = 0.5460	Saros Series = 121
Umbral Magnitude = 0.8863	Axis = 0.5584°	Saros Member = 56 of 82

<u>Sun at Greatest Eclipse</u>
(Geocentric Coordinates)

R.A. = 04h58m56.4s
Dec. = +22°41'33.8"
S.D. = 00°15'45.7"
H.P. = 00°00'08.7"

<u>Moon at Greatest Eclipse</u>
(Geocentric Coordinates)

R.A. = 16h59m25.6s
Dec. = -22°08'44.6"
S.D. = 00°16'43.4"
H.P. = 01°01'22.4"

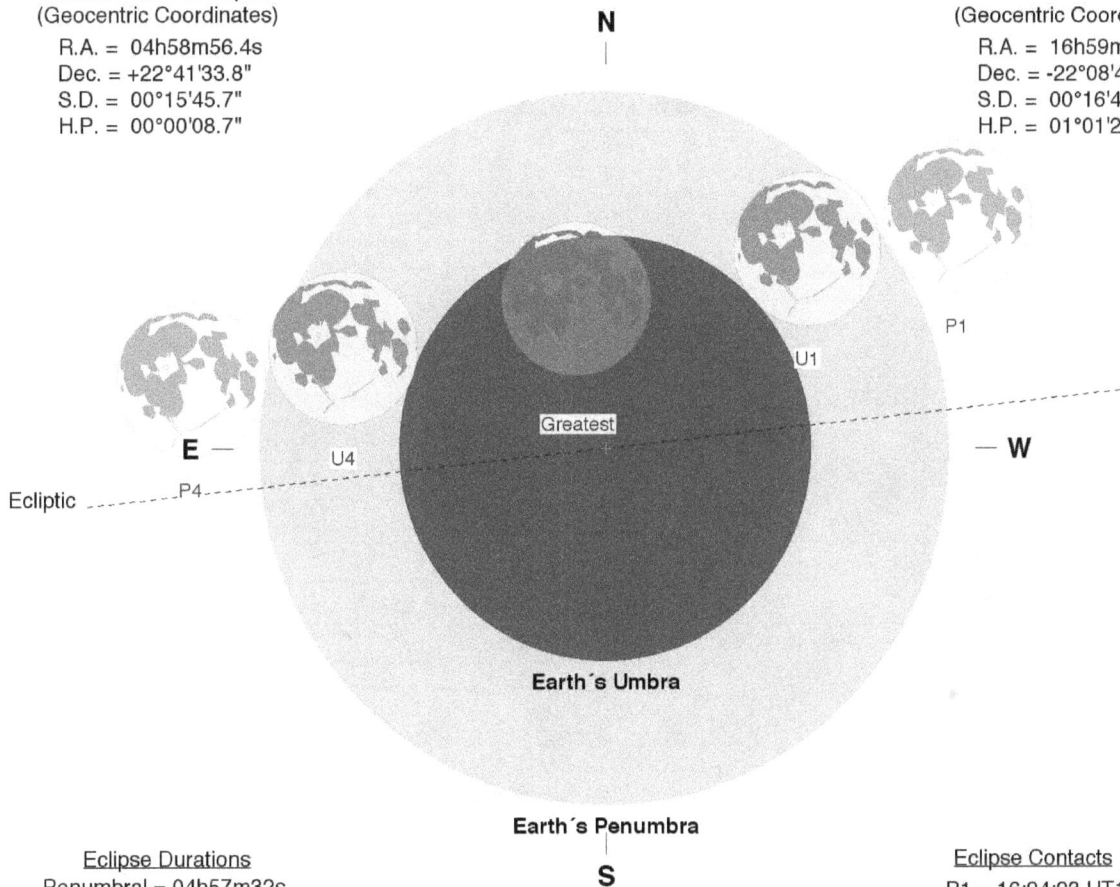

N

E — — W

Ecliptic

P1

U1

Greatest

U4

P4

Earth's Umbra

Earth's Penumbra

S

<u>Eclipse Durations</u>
Penumbral = 04h57m32s
Umbral = 03h00m00s

Eph. = JPL DE430
Rule = Herald-Sinnott
ΔT = 79 s

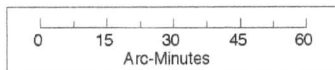

0   15   30   45   60
Arc-Minutes

©2020 F. Espenak, www.EclipseWise.com

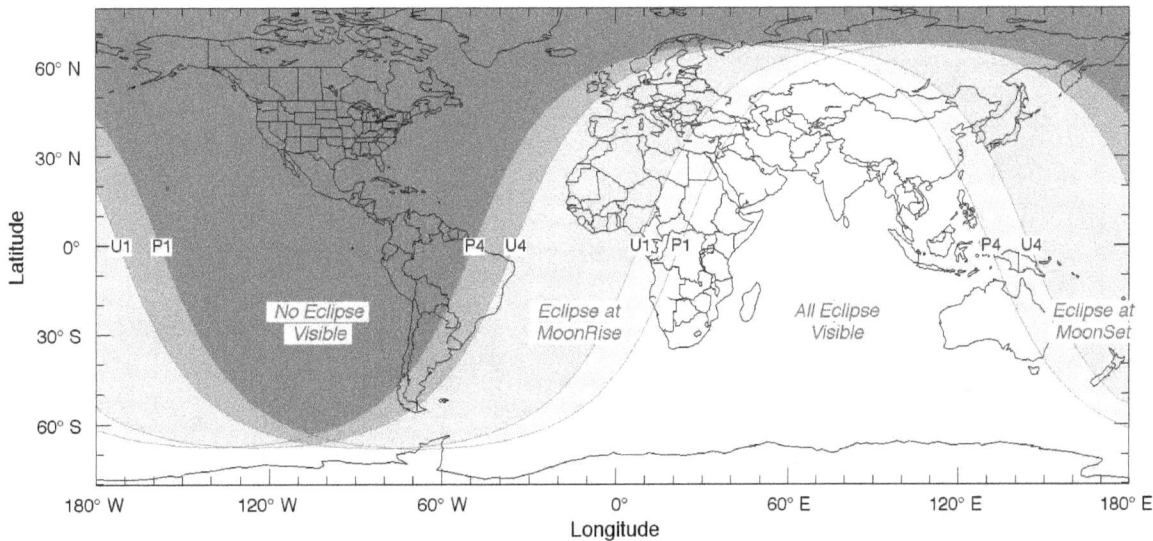

<u>Eclipse Contacts</u>
P1 = 16:24:23 UT1
U1 = 17:23:10 UT1
U4 = 20:23:10 UT1
P4 = 21:21:55 UT1

U1 P1   P4 U4   U1 P1   P4 U4

No Eclipse Visible   *Eclipse at MoonRise*   *All Eclipse Visible*   *Eclipse at MoonSet*

Partial Lunar Eclipse of 2039 Nov 30

Greatest Eclipse = 16:56:27.6 TD   (= 16:55:08.5 UT1)

Penumbral Magnitude = 2.0435	Gamma = -0.4721	Saros Series = 126
Umbral Magnitude = 0.9443	Axis = 0.4260°	Saros Member = 46 of 70

Sun at Greatest Eclipse
(Geocentric Coordinates)

R.A. = 16h26m20.8s
Dec. = -21°41'27.9"
S.D. = 00°16'13.0"
H.P. = 00°00'08.9"

Moon at Greatest Eclipse
(Geocentric Coordinates)

R.A. = 04h26m48.9s
Dec. = +21°16'45.4"
S.D. = 00°14'45.3"
H.P. = 00°54'08.9"

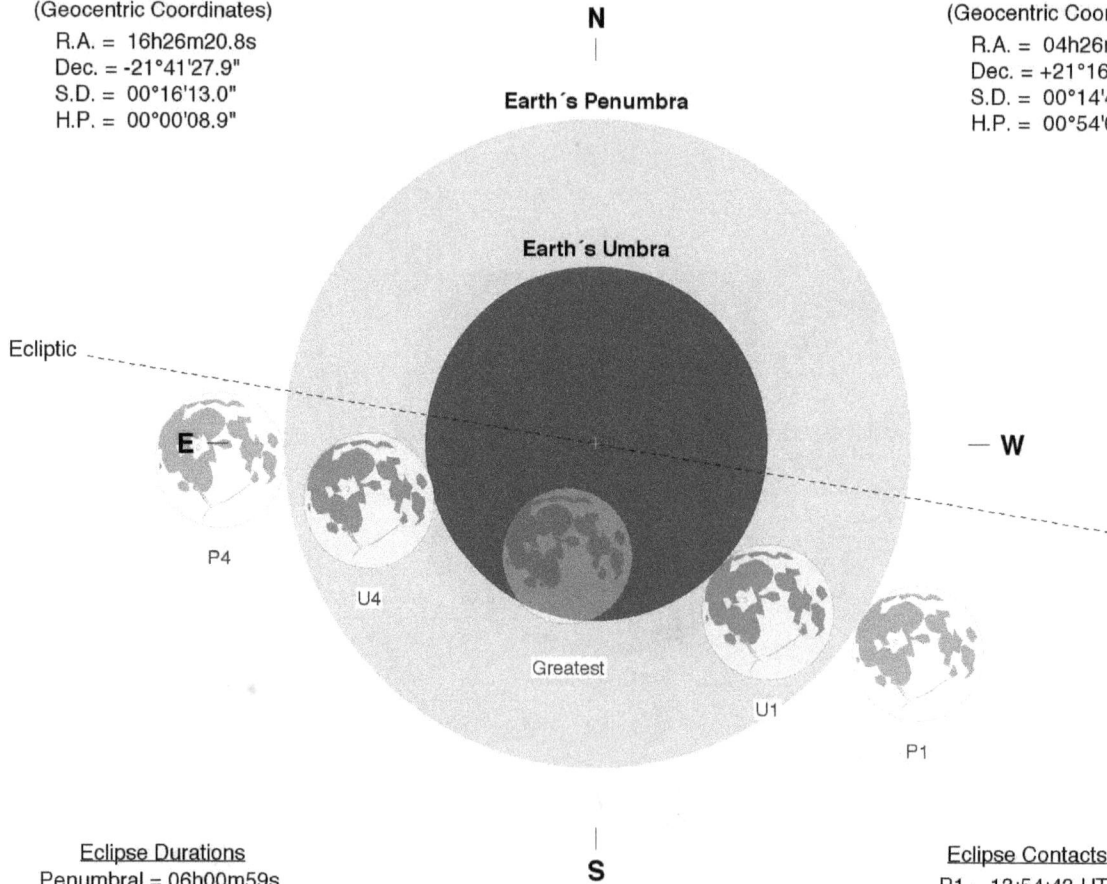

N

Earth's Penumbra

Earth's Umbra

Ecliptic

E

P4

U4

Greatest

U1

P1

W

S

Eclipse Durations
Penumbral = 06h00m59s
Umbral = 03h26m48s

Eph. = JPL DE430
Rule = Herald-Sinnott
ΔT = 79 s

0   15   30   45   60
Arc-Minutes

©2020 F. Espenak, www.EclipseWise.com

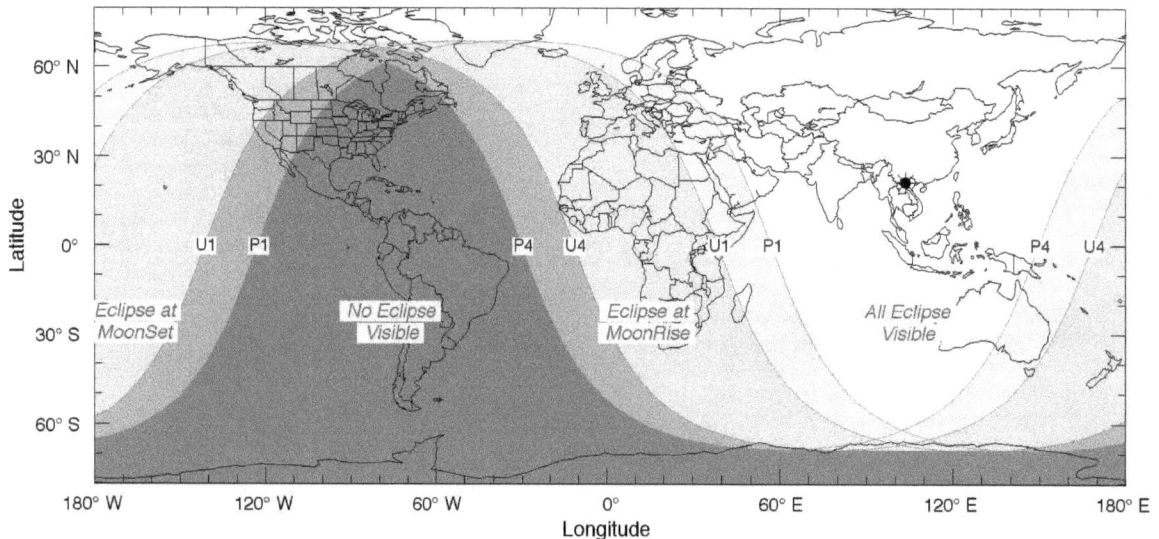

Eclipse Contacts
P1 = 13:54:43 UT1
U1 = 15:11:49 UT1
U4 = 18:38:37 UT1
P4 = 19:55:42 UT1

Eclipse at
MoonSet

No Eclipse
Visible

Eclipse at
MoonRise

All Eclipse
Visible

U1   P1   P4   U4   U1   P1   P4   U4

Total Lunar Eclipse of 2040 May 26

Greatest Eclipse = 11:46:21.6 TD (= 11:45:02.2 UT1)

Penumbral Magnitude = 2.4955	Gamma = -0.1872	Saros Series = 131
Umbral Magnitude = 1.5365	Axis = 0.1885°	Saros Member = 35 of 72

Sun at Greatest Eclipse
(Geocentric Coordinates)

R.A. = 04h15m46.6s
Dec. = +21°16'35.1"
S.D. = 00°15'47.2"
H.P. = 00°00'08.7"

Moon at Greatest Eclipse
(Geocentric Coordinates)

R.A. = 16h15m33.4s
Dec. = -21°27'28.2"
S.D. = 00°16'27.7"
H.P. = 01°00'24.9"

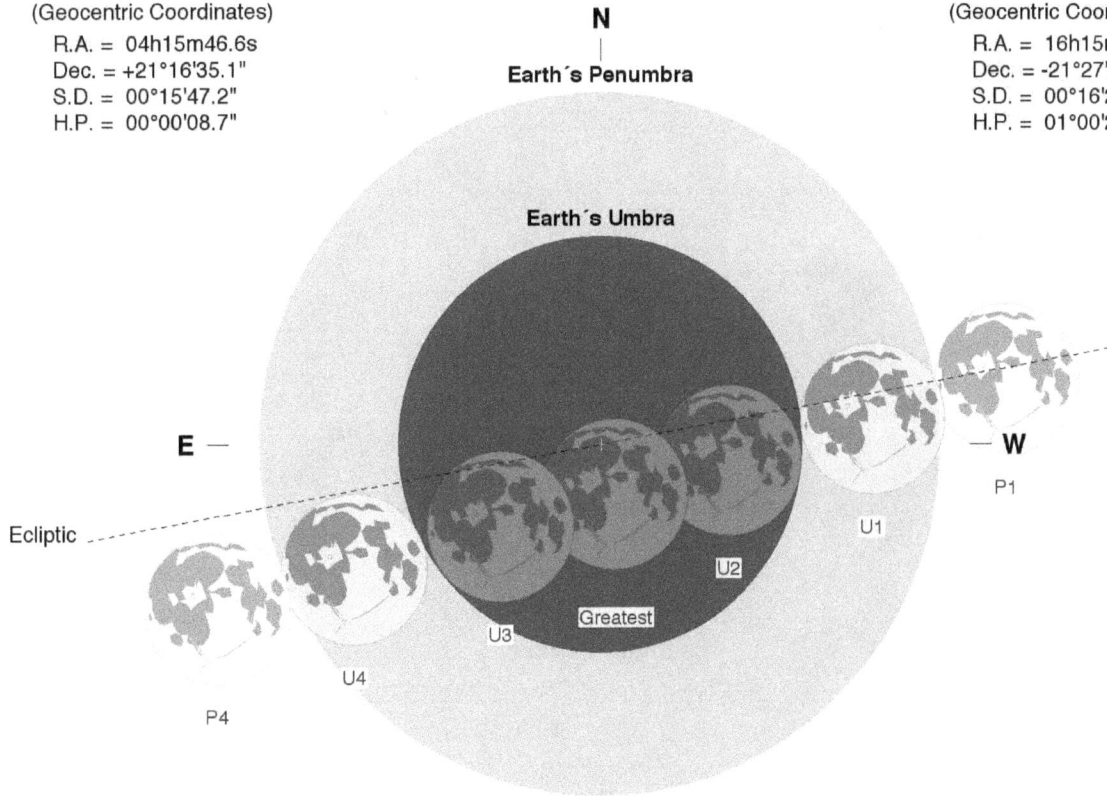

N

Earth's Penumbra

Earth's Umbra

E

Ecliptic

W

P1

U1

U2

Greatest

U3

U4

P4

S

Eclipse Durations
Penumbral = 05h22m11s
Umbral = 03h31m28s
Total = 01h32m56s

Eph. = JPL DE430
Rule = Herald-Sinnott
ΔT = 79 s

0   15   30   45   60
Arc-Minutes

©2020 F. Espenak, www.EclipseWise.com

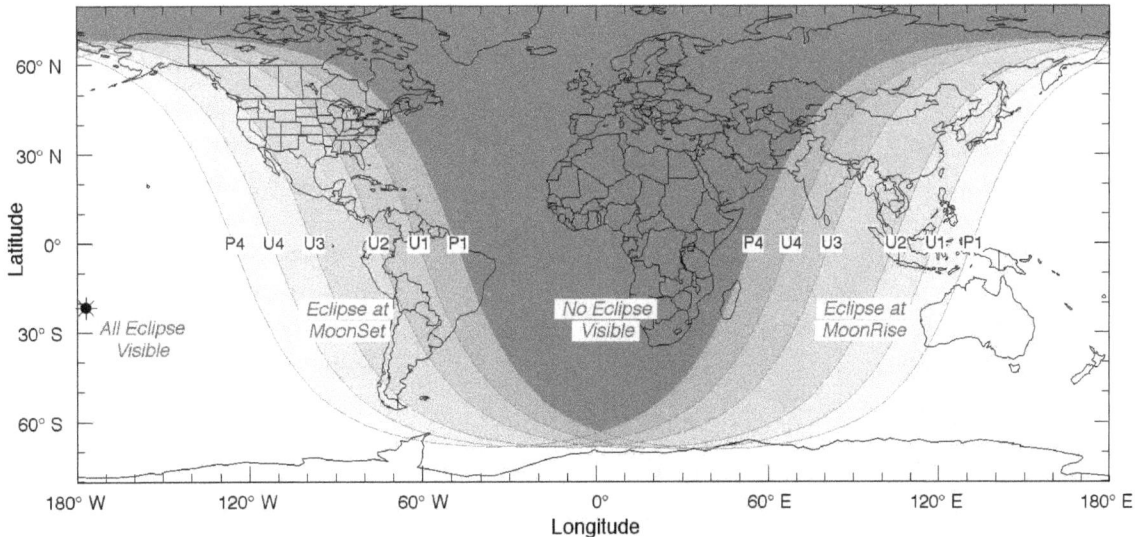

Eclipse Contacts
P1 = 09:03:54 UT1
U1 = 09:59:17 UT1
U2 = 10:58:31 UT1
U3 = 12:31:27 UT1
U4 = 13:30:45 UT1
P4 = 14:26:05 UT1

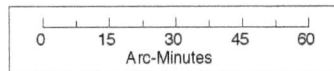

All Eclipse Visible

Eclipse at MoonSet

No Eclipse Visible

Eclipse at MoonRise

P4 U4 U3 U2 U1 P1

P4 U4 U3 U2 U1 P1

Total Lunar Eclipse of 2040 Nov 18

Greatest Eclipse = 19:04:40.5 TD   (= 19:03:20.9 UT1)

Penumbral Magnitude = 2.4543	Gamma = 0.2361	Saros Series = 136
Umbral Magnitude = 1.3991	Axis = 0.2215°	Saros Member = 21 of 72

Sun at Greatest Eclipse
(Geocentric Coordinates)

R.A. = 15h39m03.9s
Dec. = -19°29'49.7"
S.D. = 00°16'11.0"
H.P. = 00°00'08.9"

Moon at Greatest Eclipse
(Geocentric Coordinates)

R.A. = 03h38m45.6s
Dec. = +19°42'23.6"
S.D. = 00°15'20.2"
H.P. = 00°56'17.3"

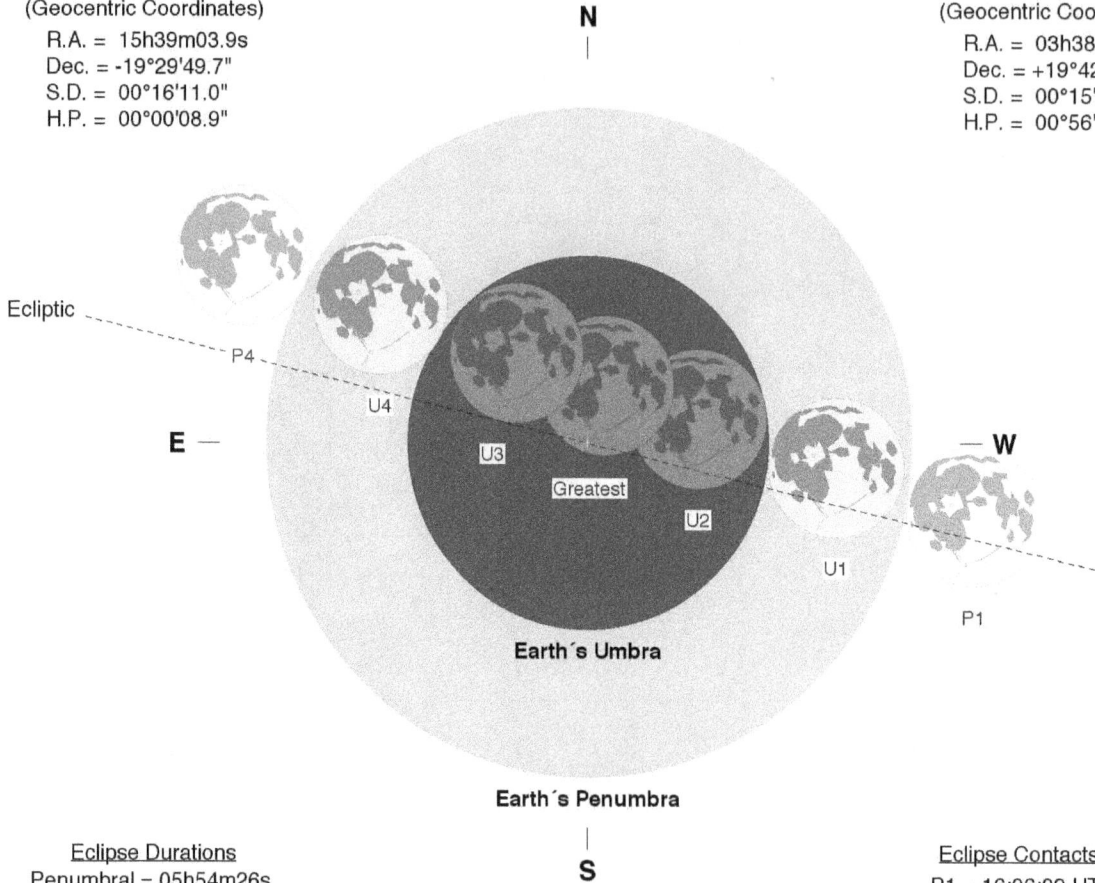

N

Ecliptic

P4
U4
U3
Greatest
U2
U1
P1

Earth's Umbra

Earth's Penumbra

E

W

S

Eclipse Durations
Penumbral = 05h54m26s
Umbral = 03h41m13s
Total = 01h28m33s

Eph. = JPL DE430
Rule = Herald-Sinnott
ΔT =   80 s

0	15	30	45	60

Arc-Minutes

©2020 F. Espenak, www.EclipseWise.com

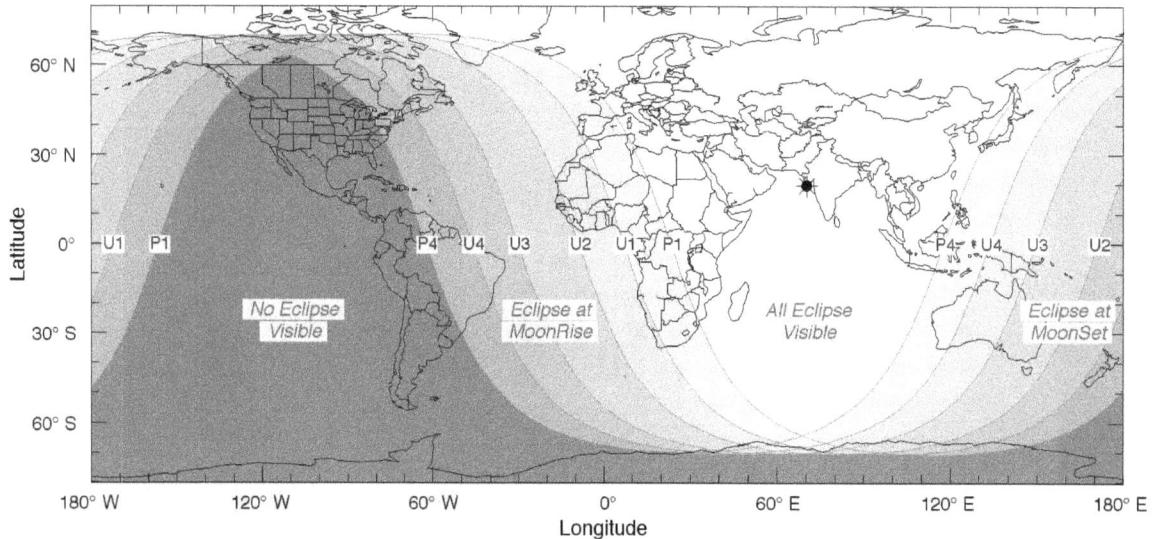

Eclipse Contacts
P1 = 16:06:09 UT1
U1 = 17:12:40 UT1
U2 = 18:18:57 UT1
U3 = 19:47:30 UT1
U4 = 20:53:53 UT1
P4 = 22:00:35 UT1

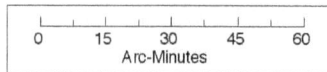

Partial Lunar Eclipse of 2041 May 16

Greatest Eclipse = 00:43:02.1 TD (= 00:41:42.2 UT1)

Penumbral Magnitude = 1.0765	Gamma = -0.9747	Saros Series = 141
Umbral Magnitude = 0.0663	Axis = 0.9335°	Saros Member = 25 of 72

Sun at Greatest Eclipse
(Geocentric Coordinates)

R.A. = 03h32m49.6s
Dec. = +19°08'35.5"
S.D. = 00°15'49.2"
H.P. = 00°00'08.7"

Moon at Greatest Eclipse
(Geocentric Coordinates)

R.A. = 15h31m30.5s
Dec. = -20°01'25.1"
S.D. = 00°15'39.6"
H.P. = 00°57'28.4"

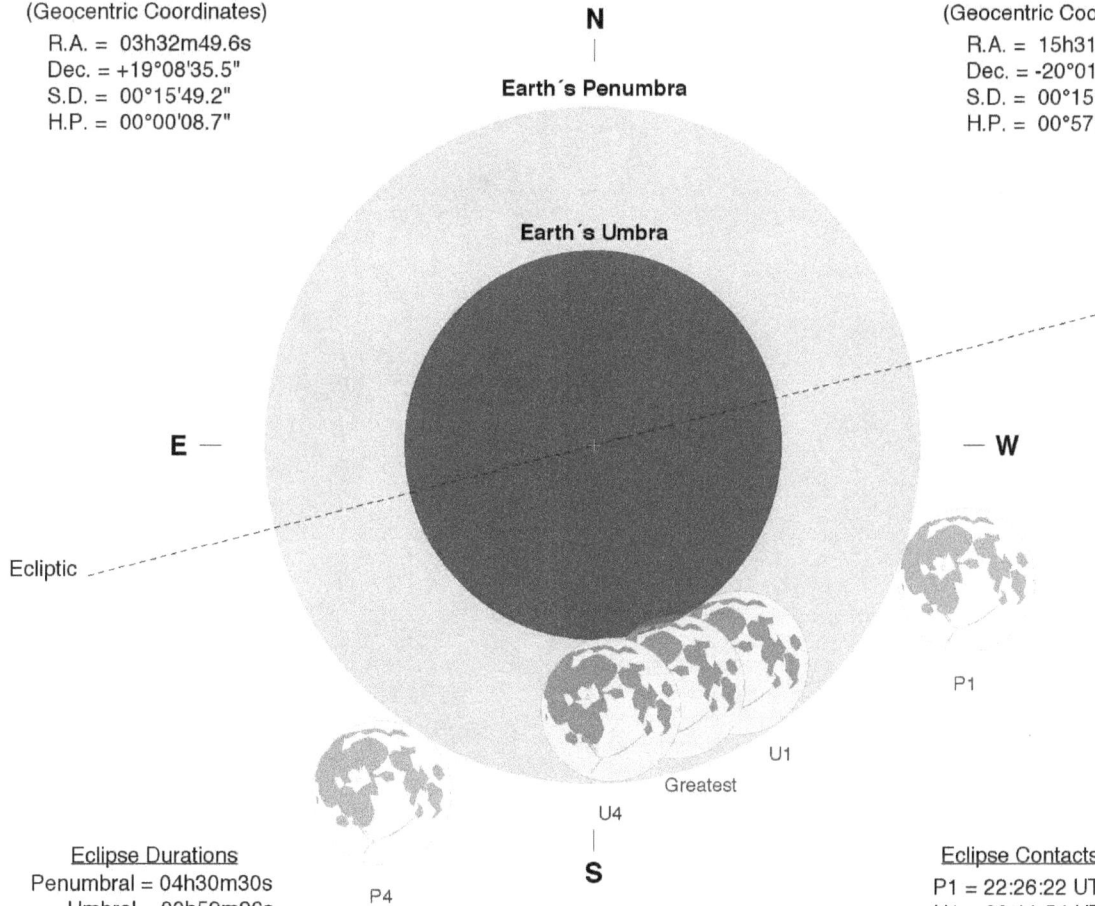

N

Earth's Penumbra

Earth's Umbra

E

W

Ecliptic

P1

U1

Greatest

U4

S

P4

Eclipse Durations
Penumbral = 04h30m30s
Umbral = 00h59m26s

Eph. = JPL DE430
Rule = Herald-Sinnott
ΔT = 80 s

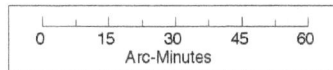

0   15   30   45   60
Arc-Minutes

©2020 F. Espenak, www.EclipseWise.com

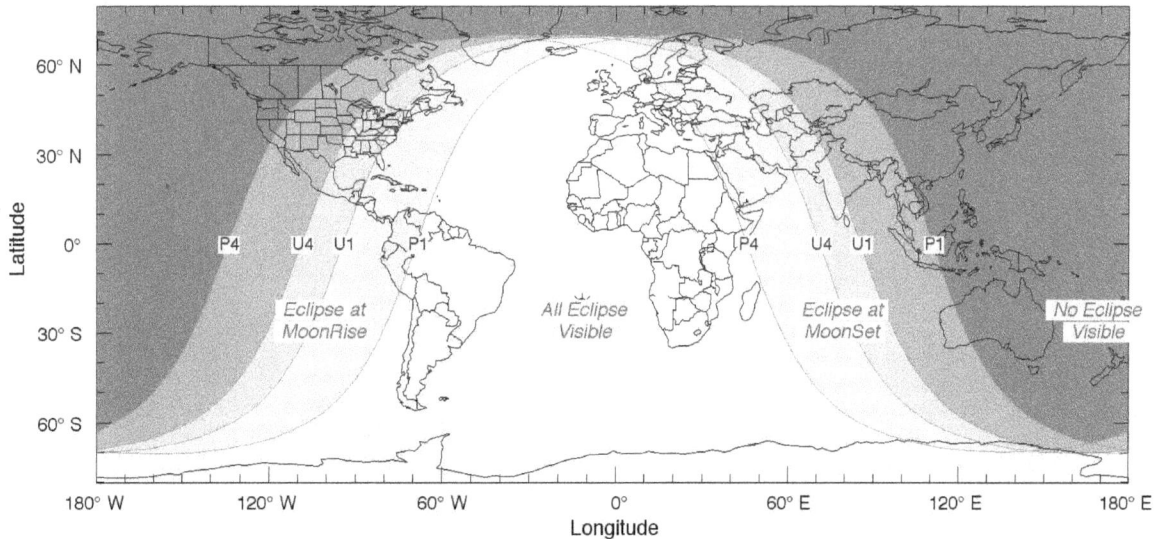

Eclipse Contacts
P1 = 22:26:22 UT1
U1 = 00:11:54 UT1
U4 = 01:11:20 UT1
P4 = 02:56:51 UT1

Eclipse at MoonRise

All Eclipse Visible

Eclipse at MoonSet

No Eclipse Visible

151

Partial Lunar Eclipse of 2041 Nov 08

Greatest Eclipse = 04:35:04.2 TD   (= 04:33:44.0 UT1)

Penumbral Magnitude = 1.1675	Gamma = 0.9212	Saros Series = 146
Umbral Magnitude = 0.1714	Axis = 0.9131°	Saros Member = 12 of 72

Sun at Greatest Eclipse
(Geocentric Coordinates)
R.A. = 14h54m42.6s
Dec. = -16°39'56.0"
S.D. = 00°16'08.5"
H.P. = 00°00'08.9"

Moon at Greatest Eclipse
(Geocentric Coordinates)
R.A. = 02h53m15.3s
Dec. = +17°30'36.2"
S.D. = 00°16'12.4"
H.P. = 00°59'28.8"

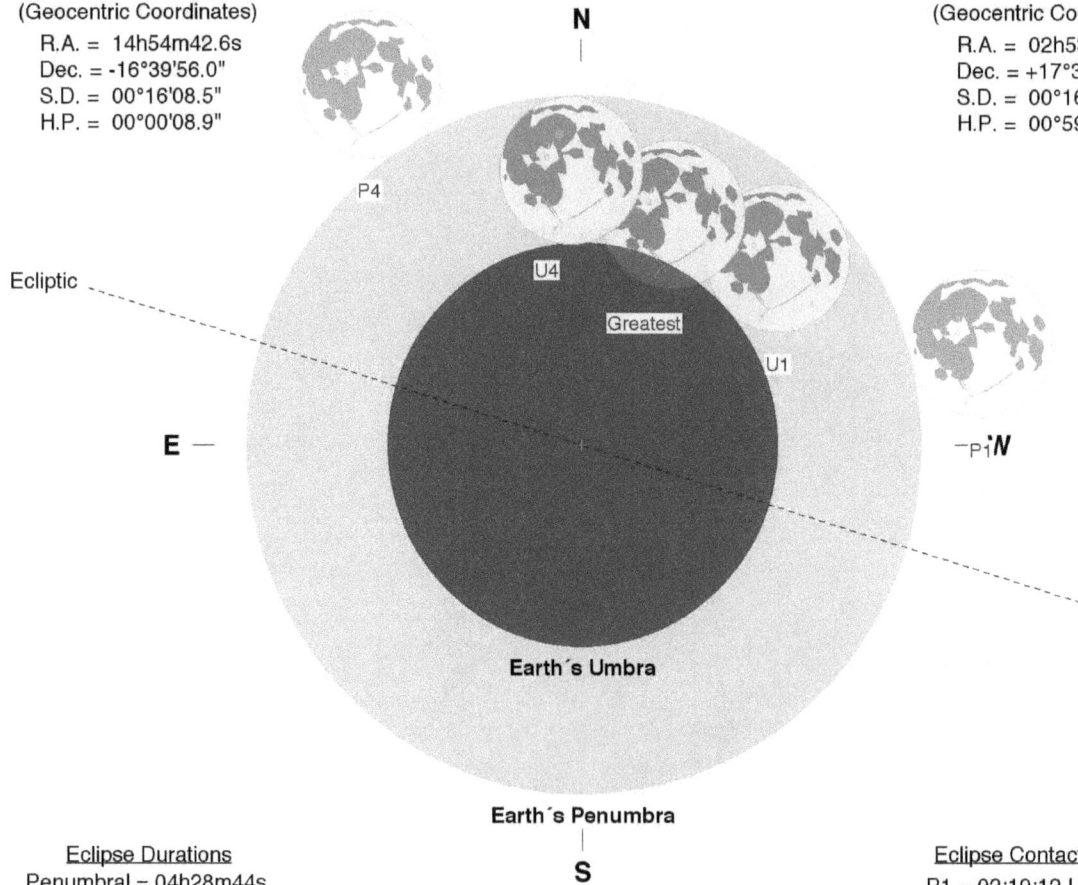

N

P4

Ecliptic

U4

Greatest

U1

E

P1 N

Earth's Umbra

Earth's Penumbra

S

Eclipse Durations
Penumbral = 04h28m44s
Umbral = 01h31m04s

Eph. = JPL DE430
Rule = Herald-Sinnott
ΔT =   80 s

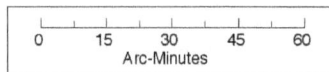

0   15   30   45   60
Arc-Minutes

©2020 F. Espenak, www.EclipseWise.com

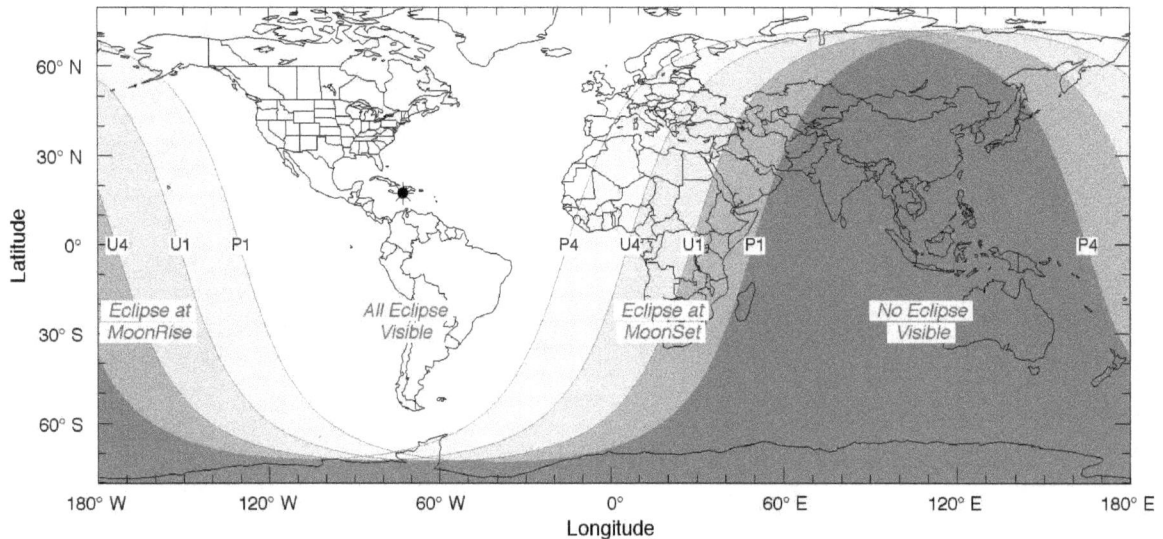

Eclipse Contacts
P1 = 02:19:12 UT1
U1 = 03:47:54 UT1
U4 = 05:18:59 UT1
P4 = 06:47:57 UT1

Eclipse at MoonRise

All Eclipse Visible

Eclipse at MoonSet

No Eclipse Visible

Penumbral Lunar Eclipse of 2042 Apr 05

Greatest Eclipse = 14:30:12.0 TD (= 14:28:51.6 UT1)

Penumbral Magnitude = 0.8700	Gamma = 1.1080	Saros Series = 113
Umbral Magnitude = -0.2156	Axis = 0.9981°	Saros Member = 65 of 71

Sun at Greatest Eclipse
(Geocentric Coordinates)
R.A. = 00h58m43.2s
Dec. = +06°16'08.8"
S.D. = 00°15'59.3"
H.P. = 00°00'08.8"

N

Moon at Greatest Eclipse
(Geocentric Coordinates)
R.A. = 13h00m37.2s
Dec. = -05°23'23.8"
S.D. = 00°14'43.6"
H.P. = 00°54'03.0"

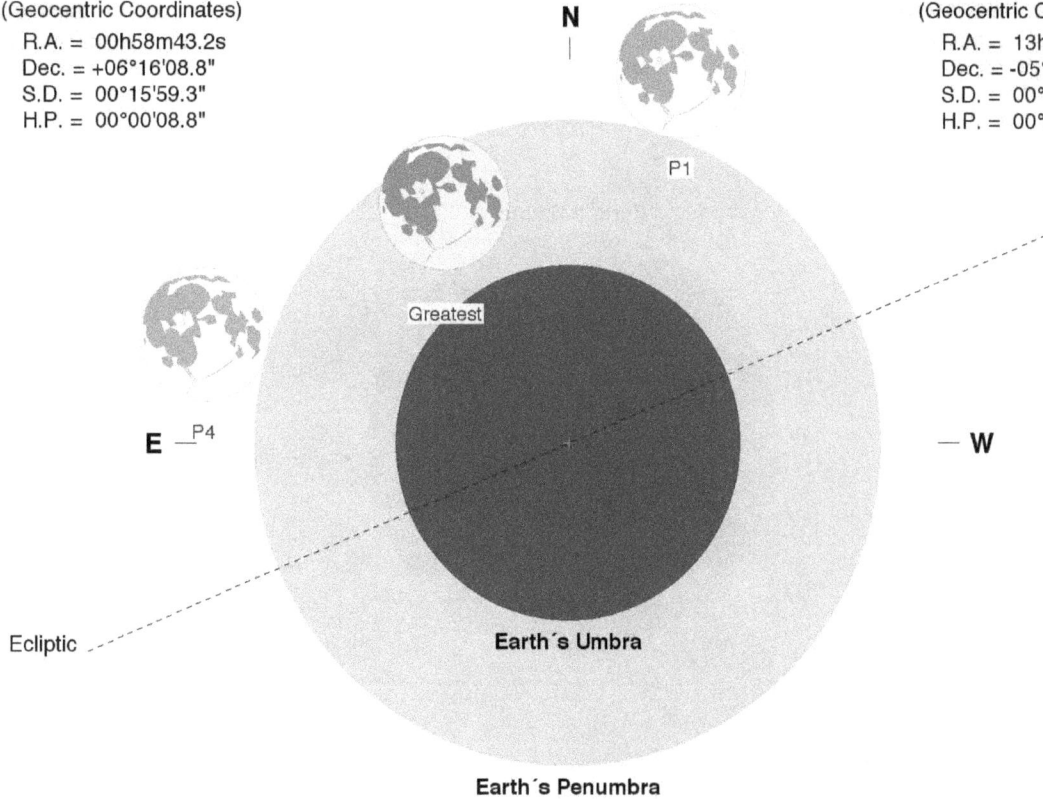

P1

Greatest

E P4

W

Ecliptic

Earth's Umbra

Earth's Penumbra

S

Eclipse Durations
Penumbral = 04h29m11s

Eclipse Contacts
P1 = 12:14:30 UT1
P4 = 16:43:42 UT1

Eph. = JPL DE430
Rule = Herald-Sinnott
ΔT = 80 s

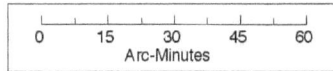

0   15   30   45   60
Arc-Minutes

©2020 F. Espenak, www.EclipseWise.com

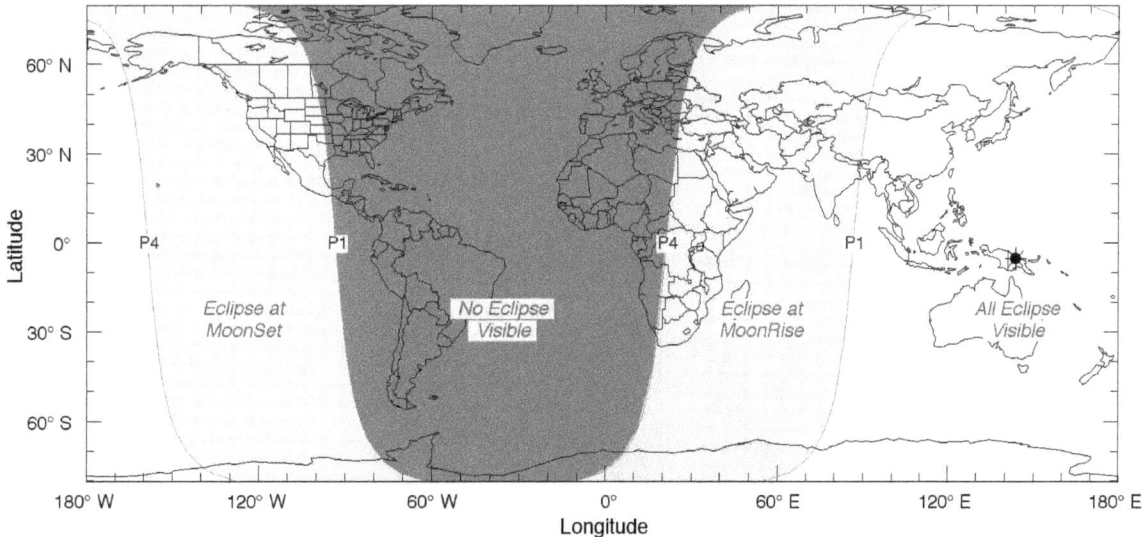

Penumbral Lunar Eclipse of 2042 Sep 29

Greatest Eclipse = 10:45:47.4 TD   (= 10:44:26.8 UT1)

Penumbral Magnitude = 0.9548	Gamma = -1.0262	Saros Series = 118
Umbral Magnitude = -0.0010	Axis = 1.0483°	Saros Member = 53 of 73

Sun at Greatest Eclipse
(Geocentric Coordinates)

R.A. = 12h23m37.3s
Dec. = -02°33'13.4"
S.D. = 00°15'57.9"
H.P. = 00°00'08.8"

Moon at Greatest Eclipse
(Geocentric Coordinates)

R.A. = 00h25m38.7s
Dec. = +01°38'07.3"
S.D. = 00°16'42.1"
H.P. = 01°01'18.0"

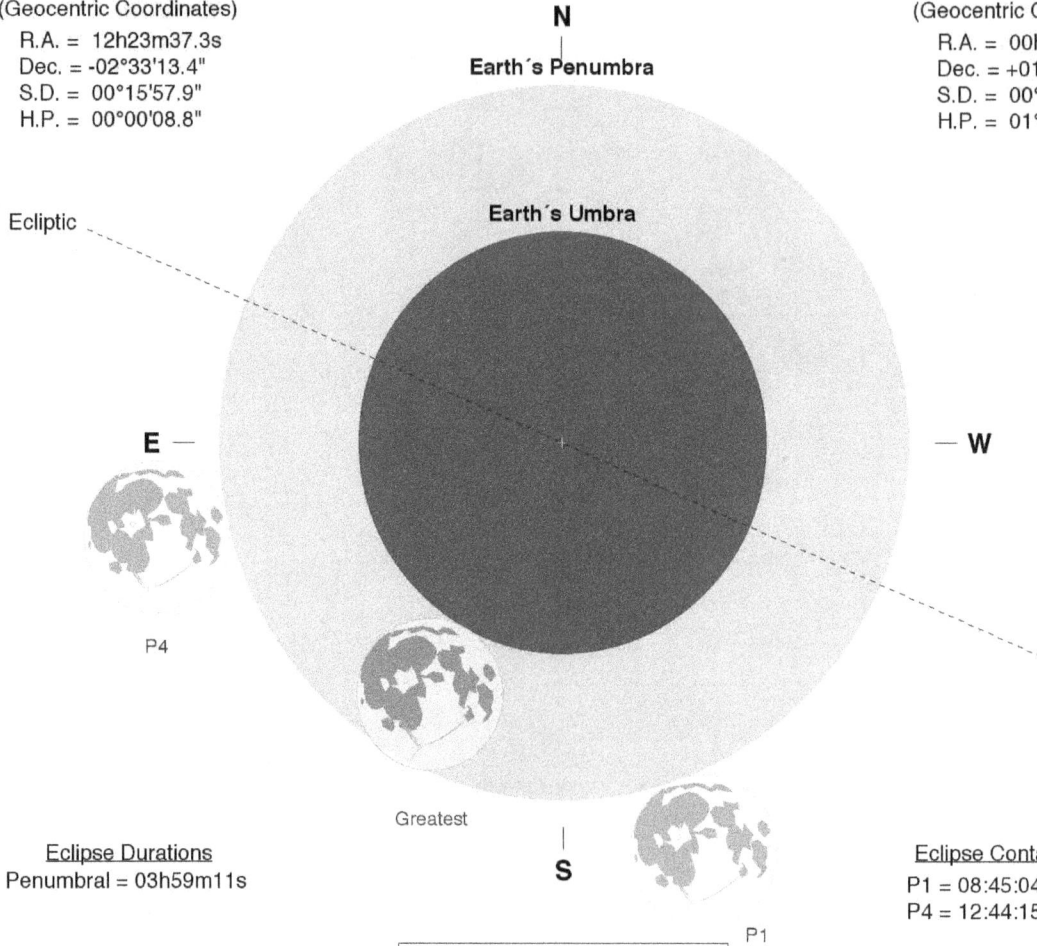

N

Earth's Penumbra

Ecliptic

Earth's Umbra

E

W

P4

Greatest

S

P1

Eclipse Durations
Penumbral = 03h59m11s

Eclipse Contacts
P1 = 08:45:04 UT1
P4 = 12:44:15 UT1

0	15	30	45	60

Arc-Minutes

Eph. = JPL DE430
Rule = Herald-Sinnott
ΔT =   81 s

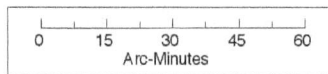

©2020 F. Espenak, www.EclipseWise.com

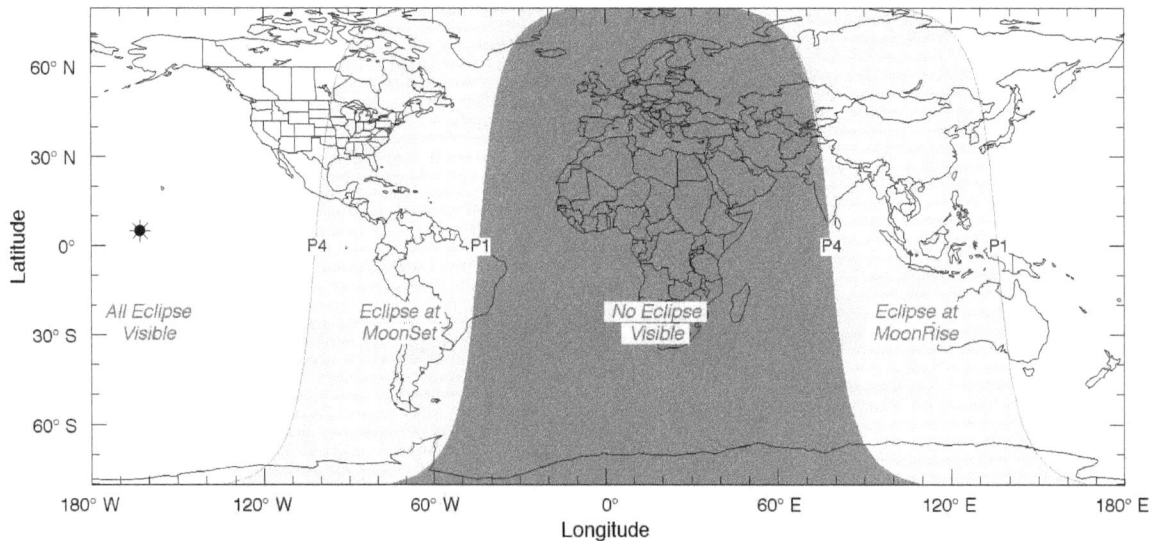

Total Lunar Eclipse of 2043 Mar 25

Greatest Eclipse = 14:32:03.6 TD (= 14:30:42.6 UT1)

Penumbral Magnitude = 2.1920	Gamma = 0.3849	Saros Series = 123
Umbral Magnitude = 1.1161	Axis = 0.3510°	Saros Member = 54 of 72

Sun at Greatest Eclipse
(Geocentric Coordinates)

R.A. = 00h17m45.9s
Dec. = +01°55'21.5"
S.D. = 00°16'02.4"
H.P. = 00°00'08.8"

Moon at Greatest Eclipse
(Geocentric Coordinates)

R.A. = 12h18m26.9s
Dec. = -01°36'57.6"
S.D. = 00°14'54.5"
H.P. = 00°54'42.9"

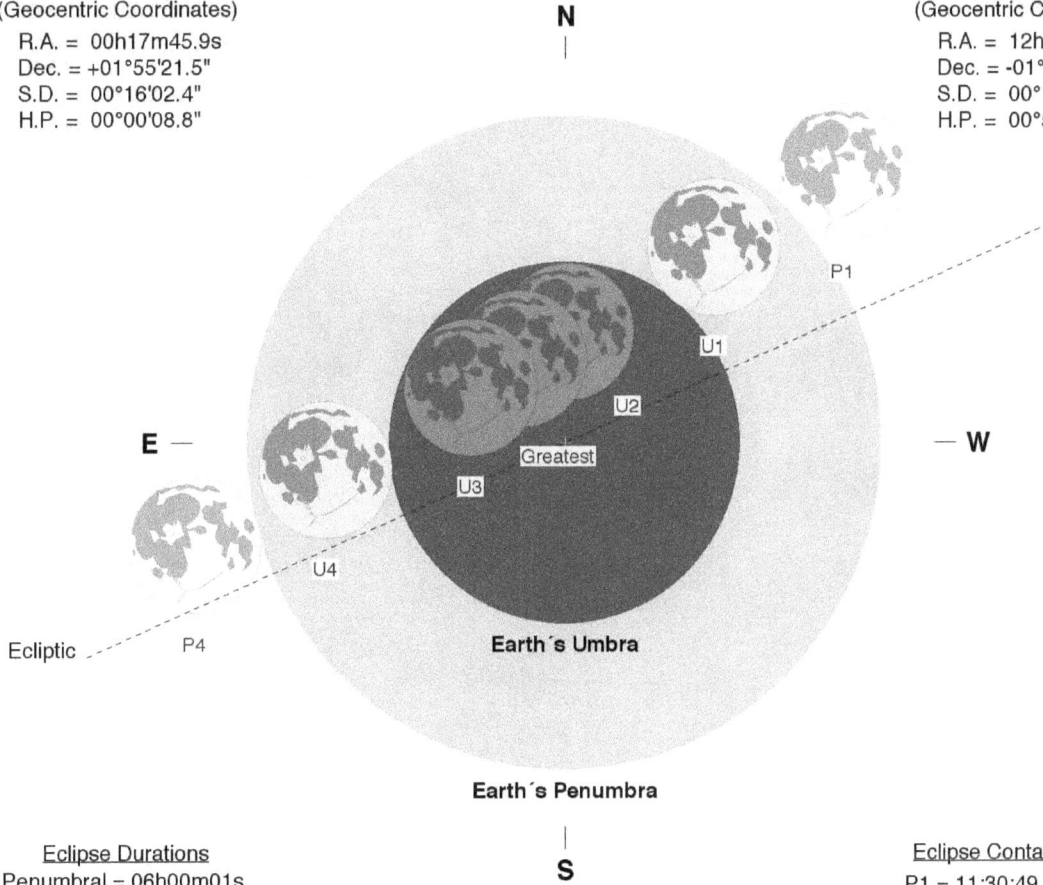

N

E — — W

Greatest

Earth's Umbra

Earth's Penumbra

S

Eclipse Durations
Penumbral = 06h00m01s
Umbral = 03h35m18s
Total = 00h54m06s

Eph. = JPL DE430
Rule = Herald-Sinnott
ΔT = 81 s

0   15   30   45   60
Arc-Minutes

©2020 F. Espenak, www.EclipseWise.com

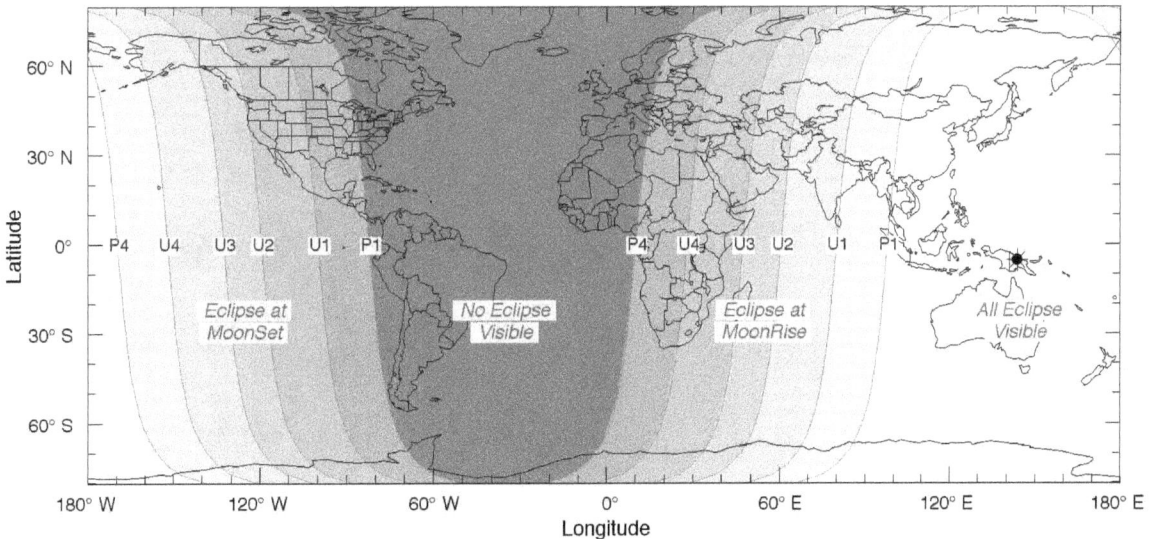

Eclipse Contacts
P1 = 11:30:49 UT1
U1 = 12:43:10 UT1
U2 = 14:03:56 UT1
U3 = 14:58:02 UT1
U4 = 16:18:28 UT1
P4 = 17:30:50 UT1

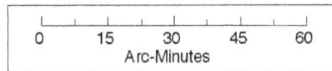

Total Lunar Eclipse of 2043 Sep 19

Greatest Eclipse = 01:51:49.8 TD   (= 01:50:28.6 UT1)

Penumbral Magnitude = 2.2452	Gamma = -0.3316	Saros Series = 128
Umbral Magnitude = 1.2575	Axis = 0.3269°	Saros Member = 42 of 71

Sun at Greatest Eclipse
(Geocentric Coordinates)

R.A. = 11h45m28.0s
Dec. = +01°34'24.4"
S.D. = 00°15'55.1"
H.P. = 00°00'08.8"

Moon at Greatest Eclipse
(Geocentric Coordinates)

R.A. = 23h46m06.1s
Dec. = -01°51'33.2"
S.D. = 00°16'07.0"
H.P. = 00°59'08.8"

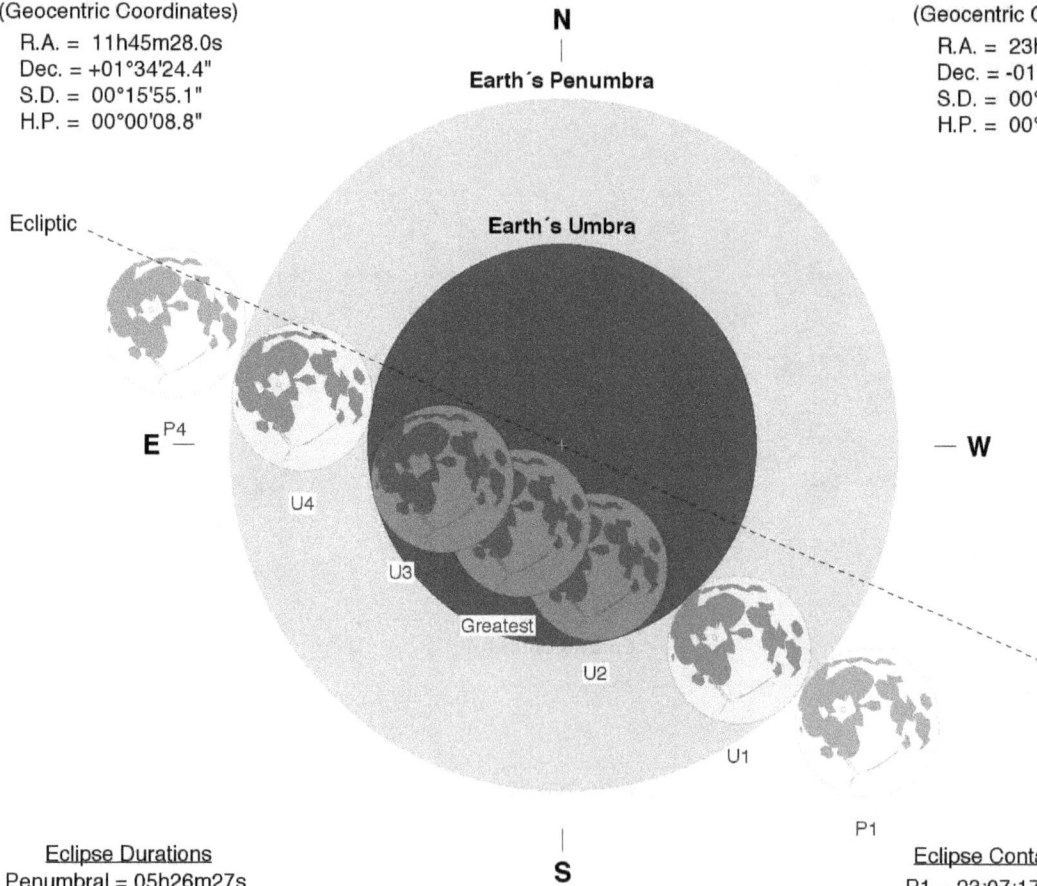

N

Earth's Penumbra

Ecliptic

Earth's Umbra

E

P4

U4

U3

Greatest

U2

U1

W

P1

S

Eclipse Durations

Penumbral = 05h26m27s
Umbral = 03h26m39s
Total = 01h12m20s

Eph. = JPL DE430
Rule = Herald-Sinnott
ΔT =   81 s

0   15   30   45   60
Arc-Minutes

©2020 F. Espenak, www.EclipseWise.com

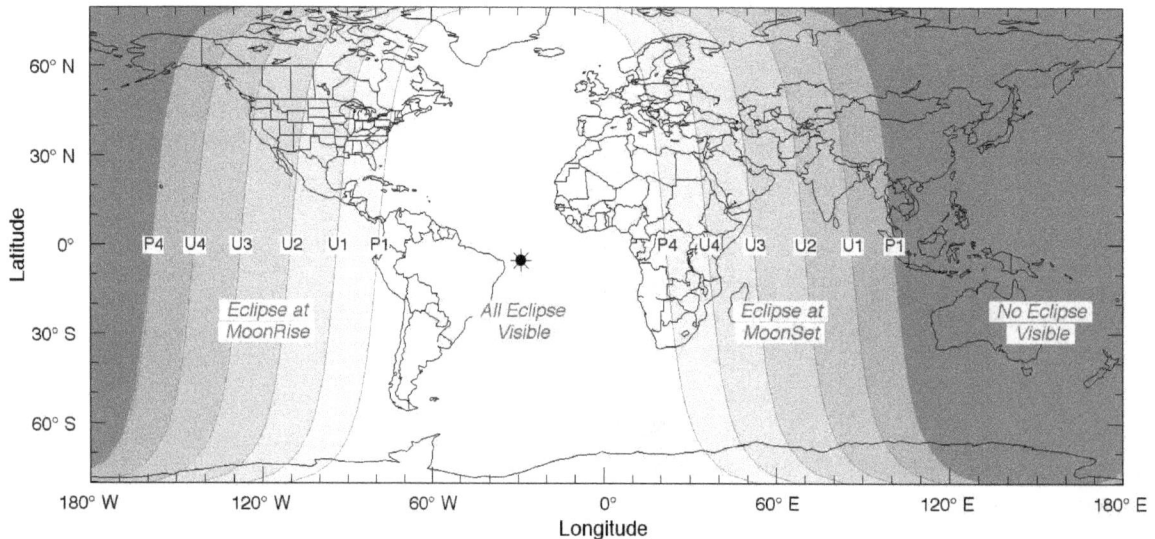

Eclipse Contacts

P1 = 23:07:17 UT1
U1 = 00:07:16 UT1
U2 = 01:14:33 UT1
U3 = 02:26:52 UT1
U4 = 03:33:56 UT1
P4 = 04:33:44 UT1

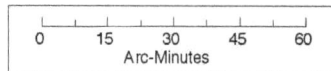

P4 U4 U3 U2 U1 P1   P4 U4   U3   U2   U1   P1

Eclipse at
MoonRise

All Eclipse
Visible

Eclipse at
MoonSet

No Eclipse
Visible

Total Lunar Eclipse of 2044 Mar 13

Greatest Eclipse = 19:38:32.5 TD (= 19:37:11.0 UT1)

Penumbral Magnitude = 2.2322	Gamma = -0.3496	Saros Series = 133
Umbral Magnitude = 1.2050	Axis = 0.3349°	Saros Member = 28 of 71

<u>Sun at Greatest Eclipse</u>
(Geocentric Coordinates)

R.A. = 23h37m30.3s
Dec. = -02°25'56.9"
S.D. = 00°16'05.4"
H.P. = 00°00'08.8"

N

Earth´s Penumbra

Earth´s Umbra

E —

P1 W

U1

U2

Greatest

U3

U4

P4

Ecliptic

<u>Moon at Greatest Eclipse</u>
(Geocentric Coordinates)

R.A. = 11h36m51.3s
Dec. = +02°08'22.5"
S.D. = 00°15'39.8"
H.P. = 00°57'29.1"

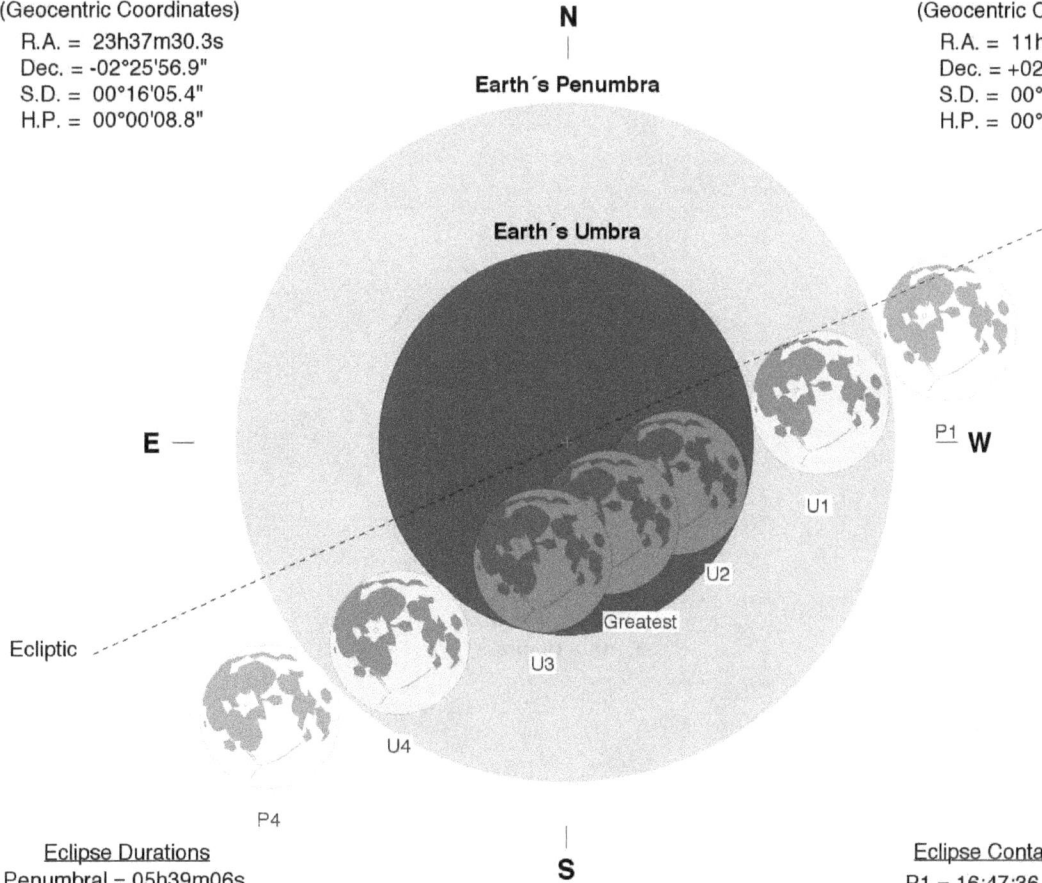

S

<u>Eclipse Durations</u>
Penumbral = 05h39m06s
Umbral = 03h29m43s
Total = 01h07m02s

Eph. = JPL DE430
Rule = Herald-Sinnott
ΔT = 82 s

0   15   30   45   60
Arc-Minutes

©2020 F. Espenak, www.EclipseWise.com

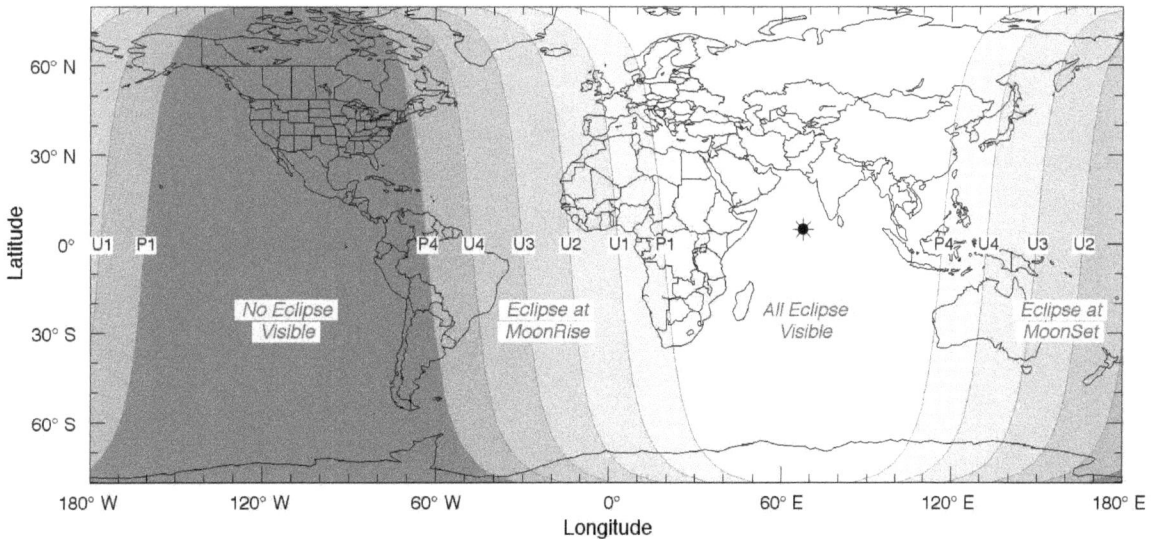

<u>Eclipse Contacts</u>
P1 = 16:47:36 UT1
U1 = 17:52:11 UT1
U2 = 19:03:24 UT1
U3 = 20:10:26 UT1
U4 = 21:21:54 UT1
P4 = 22:26:42 UT1

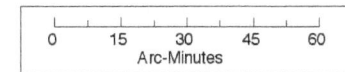

Total Lunar Eclipse of 2044 Sep 07

Greatest Eclipse = 11:20:44.4 TD   (= 11:19:22.6 UT1)

Penumbral Magnitude = 2.0879	Gamma = 0.4318	Saros Series = 138
Umbral Magnitude = 1.0476	Axis = 0.4030°	Saros Member = 30 of 82

Sun at Greatest Eclipse
(Geocentric Coordinates)
R.A. = 11h06m33.5s
Dec. = +05°43'12.4"
S.D. = 00°15'52.4"
H.P. = 00°00'08.7"

Moon at Greatest Eclipse
(Geocentric Coordinates)
R.A. = 23h05m47.2s
Dec. = -05°21'56.9"
S.D. = 00°15'15.4"
H.P. = 00°55'59.6"

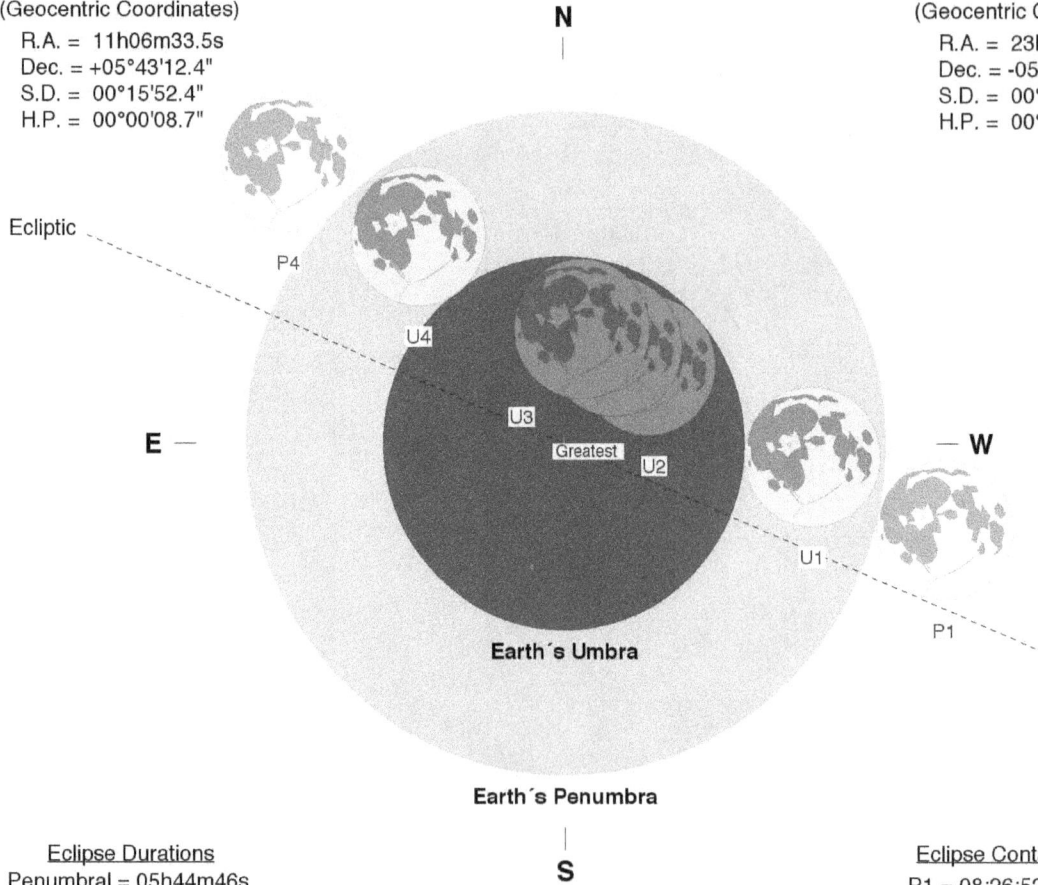

N

Ecliptic

P4

U4

U3

Greatest

U2

Greatest

E —

— W

U1

P1

Earth's Umbra

Earth's Penumbra

S

Eclipse Durations
Penumbral = 05h44m46s
Umbral = 03h26m52s
Total = 00h34m47s

Eph. = JPL DE430
Rule = Herald-Sinnott
ΔT =   82 s

0	15	30	45	60

Arc-Minutes

©2020 F. Espenak, www.EclipseWise.com

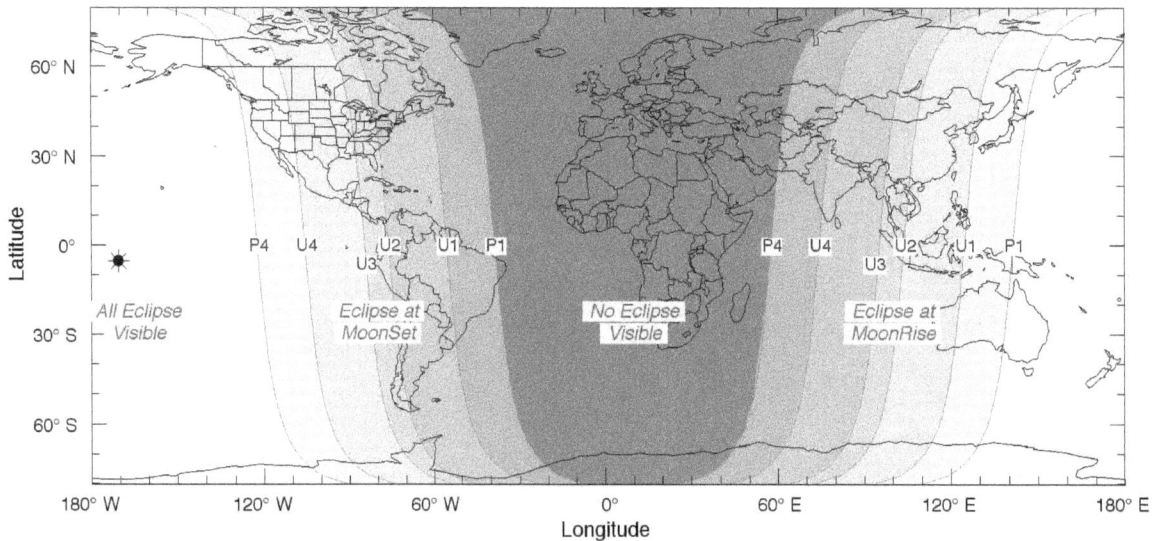

Eclipse Contacts
P1 = 08:26:52 UT1
U1 = 09:35:50 UT1
U2 = 11:01:42 UT1
U3 = 11:36:30 UT1
U4 = 13:02:42 UT1
P4 = 14:11:38 UT1

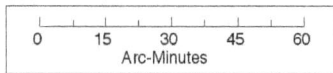

All Eclipse Visible

Eclipse at MoonSet

No Eclipse Visible

Eclipse at MoonRise

Penumbral Lunar Eclipse of 2045 Mar 03

Greatest Eclipse = 07:43:24.9 TD (= 07:42:02.8 UT1)

Penumbral Magnitude = 0.9643	Gamma = -1.0274	Saros Series = 143
Umbral Magnitude = -0.0148	Axis = 1.0354°	Saros Member = 19 of 72

Sun at Greatest Eclipse
(Geocentric Coordinates)

R.A. = 22h57m49.1s
Dec. = -06°37'35.6"
S.D. = 00°16'08.1"
H.P. = 00°00'08.9"

Moon at Greatest Eclipse
(Geocentric Coordinates)

R.A. = 10h55m51.5s
Dec. = +05°42'46.0"
S.D. = 00°16'28.7"
H.P. = 01°00'28.6"

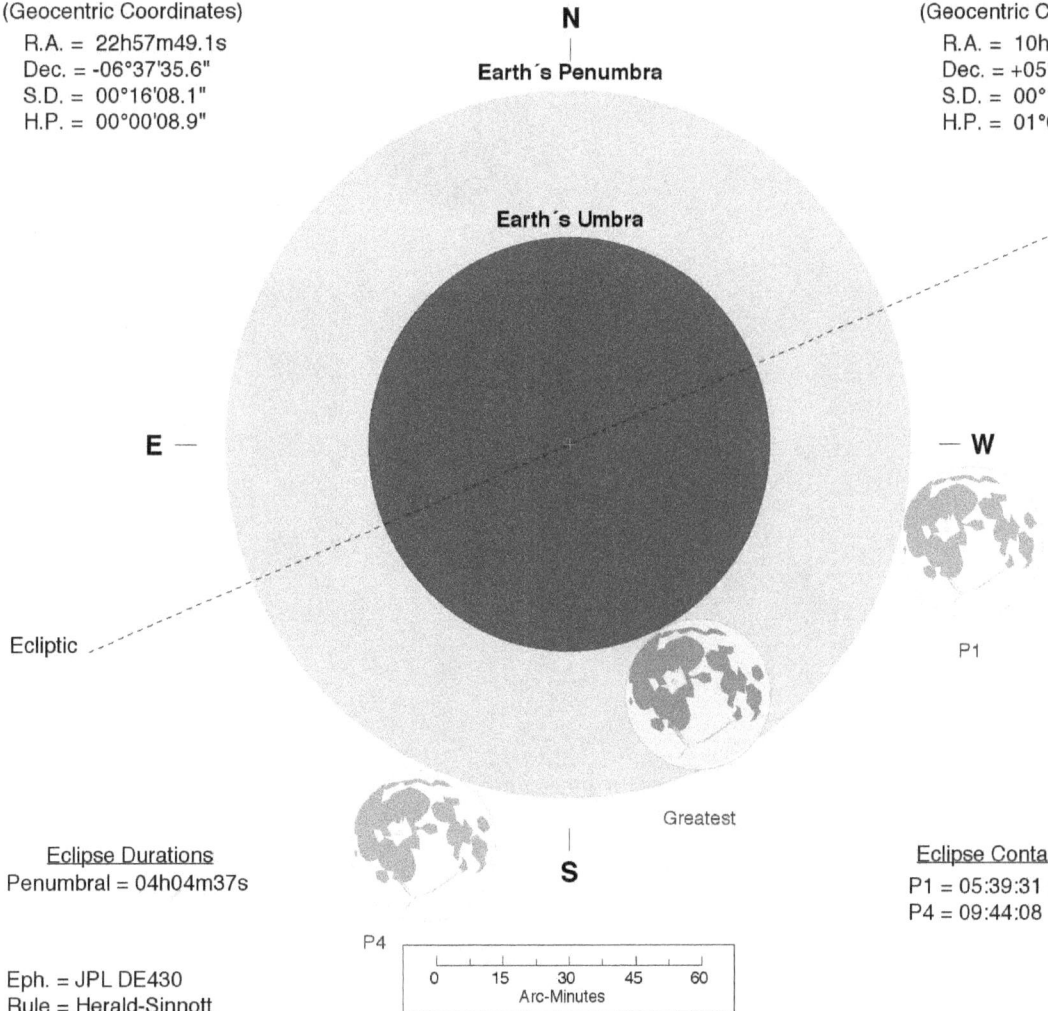

N

Earth's Penumbra

Earth's Umbra

E

W

Ecliptic

P1

Greatest

S

Eclipse Durations
Penumbral = 04h04m37s

P4

Eclipse Contacts
P1 = 05:39:31 UT1
P4 = 09:44:08 UT1

Eph. = JPL DE430
Rule = Herald-Sinnott
ΔT =   82 s

0	15	30	45	60

Arc-Minutes

©2020 F. Espenak, www.EclipseWise.com

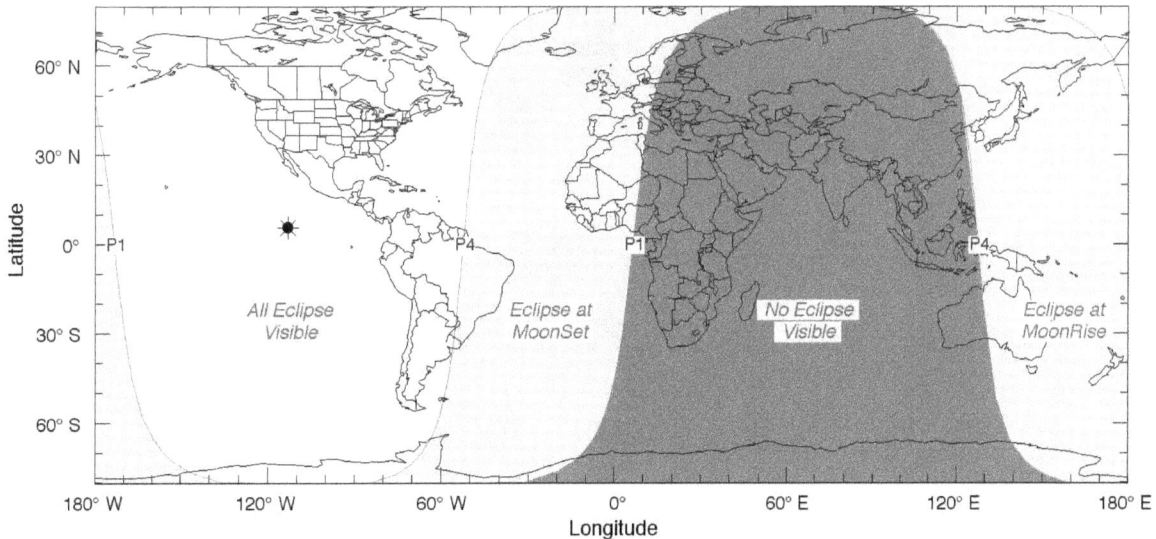

All Eclipse
Visible

Eclipse at
MoonSet

No Eclipse
Visible

Eclipse at
MoonRise

Latitude

Longitude

Penumbral Lunar Eclipse of 2045 Aug 27

Greatest Eclipse = 13:54:48.8 TD (= 13:53:26.4 UT1)

Penumbral Magnitude = 0.6845	Gamma = 1.2061	Saros Series = 148
Umbral Magnitude = -0.3899	Axis = 1.0869°	Saros Member = 5 of 70

Sun at Greatest Eclipse
(Geocentric Coordinates)

R.A. = 10h26m15.1s
Dec. = +09°46'56.3"
S.D. = 00°15'49.9"
H.P. = 00°00'08.7"

Moon at Greatest Eclipse
(Geocentric Coordinates)

R.A. = 22h24m15.1s
Dec. = -08°48'49.2"
S.D. = 00°14'44.1"
H.P. = 00°54'04.7"

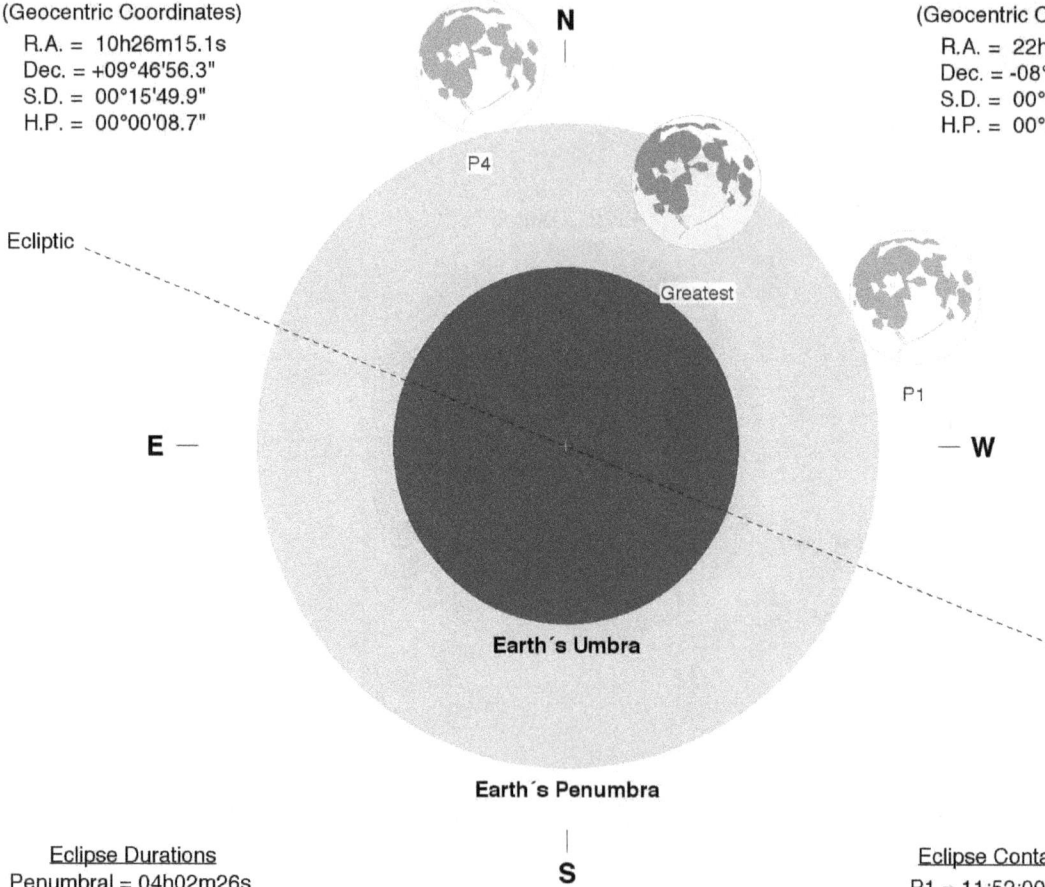

N

P4

Ecliptic

Greatest

Earth's Umbra

E

W

P1

Earth's Penumbra

S

Eclipse Durations
Penumbral = 04h02m26s

Eclipse Contacts
P1 = 11:52:00 UT1
P4 = 15:54:26 UT1

Eph. = JPL DE430
Rule = Herald-Sinnott
ΔT =   82 s

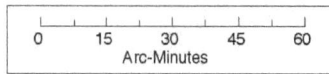

0   15   30   45   60
Arc-Minutes

©2020 F. Espenak, www.EclipseWise.com

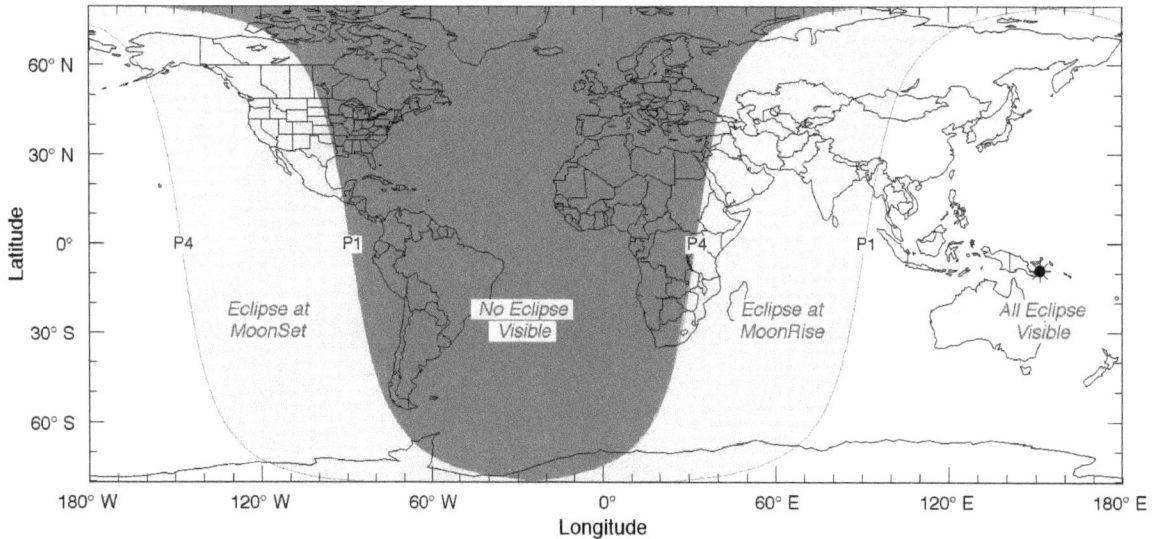

Eclipse at MoonSet

No Eclipse Visible

Eclipse at MoonRise

All Eclipse Visible

P4   P1   P4   P1

Partial Lunar Eclipse of 2046 Jan 22

Greatest Eclipse = 13:02:37.2 TD (= 13:01:14.5 UT1)

Penumbral Magnitude = 1.0365	Gamma = 0.9886	Saros Series = 115
Umbral Magnitude = 0.0550	Axis = 1.0011°	Saros Member = 59 of 72

Sun at Greatest Eclipse
(Geocentric Coordinates)
  R.A. = 20h19m45.5s
  Dec. = -19°33'42.8"
  S.D. = 00°16'15.1"
  H.P. = 00°00'08.9"

Moon at Greatest Eclipse
(Geocentric Coordinates)
  R.A. = 08h21m07.9s
  Dec. = +20°30'34.8"
  S.D. = 00°16'33.4"
  H.P. = 01°00'46.0"

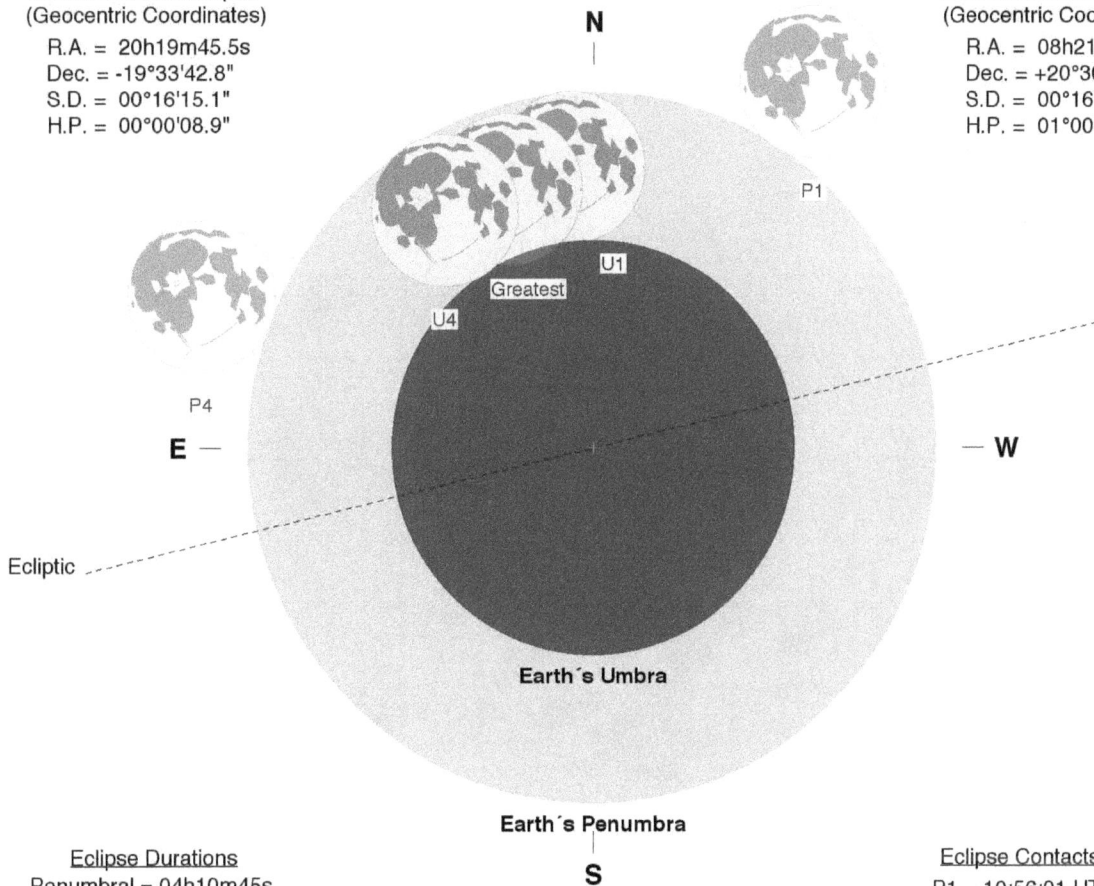

N

P1

U1

Greatest

U4

P4

E —

— W

Ecliptic

Earth's Umbra

Earth's Penumbra

S

Eclipse Durations
Penumbral = 04h10m45s
Umbral = 00h51m22s

Eph. = JPL DE430
Rule = Herald-Sinnott
ΔT = 83 s

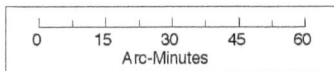

0 15 30 45 60
Arc-Minutes

©2020 F. Espenak, www.EclipseWise.com

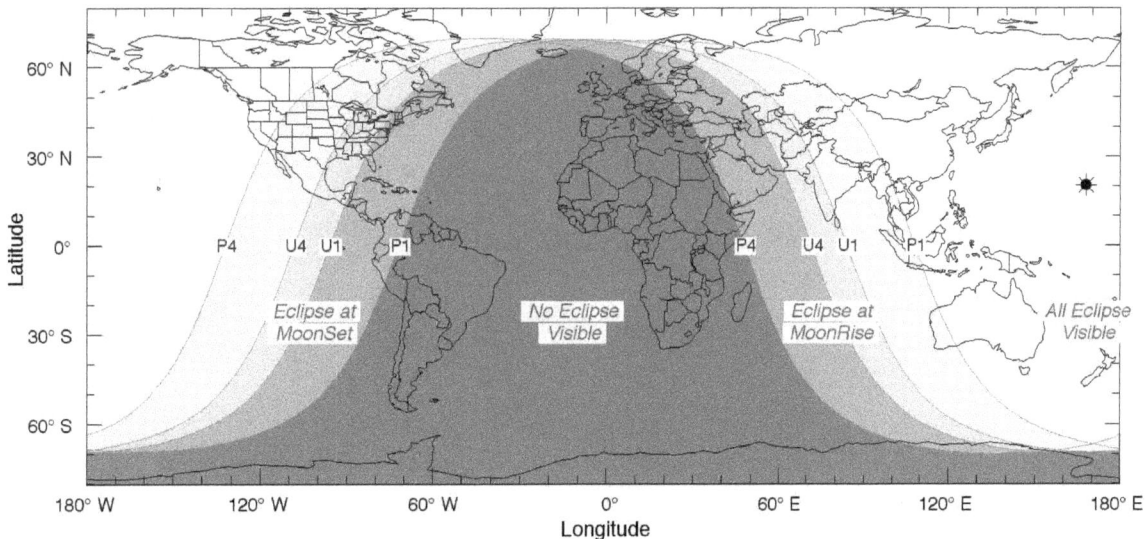

Eclipse Contacts
P1 = 10:56:01 UT1
U1 = 12:35:48 UT1
U4 = 13:27:10 UT1
P4 = 15:06:45 UT1

P4 U4 U1 P1

Eclipse at
MoonSet

No Eclipse
Visible

P4 U4 U1 P1

Eclipse at
MoonRise

All Eclipse
Visible

Partial Lunar Eclipse of 2046 Jul 18

Greatest Eclipse = 01:06:05.5 TD   (= 01:04:42.5 UT1)

Penumbral Magnitude = 1.2824	Gamma = -0.8692	Saros Series = 120
Umbral Magnitude = 0.2478	Axis = 0.8086°	Saros Member = 59 of 83

Sun at Greatest Eclipse
(Geocentric Coordinates)

R.A. = 07h50m23.8s
Dec. = +21°00'48.3"
S.D. = 00°15'44.2"
H.P. = 00°00'08.7"

Moon at Greatest Eclipse
(Geocentric Coordinates)

R.A. = 19h51m22.3s
Dec. = -21°47'22.3"
S.D. = 00°15'12.7"
H.P. = 00°55'49.5"

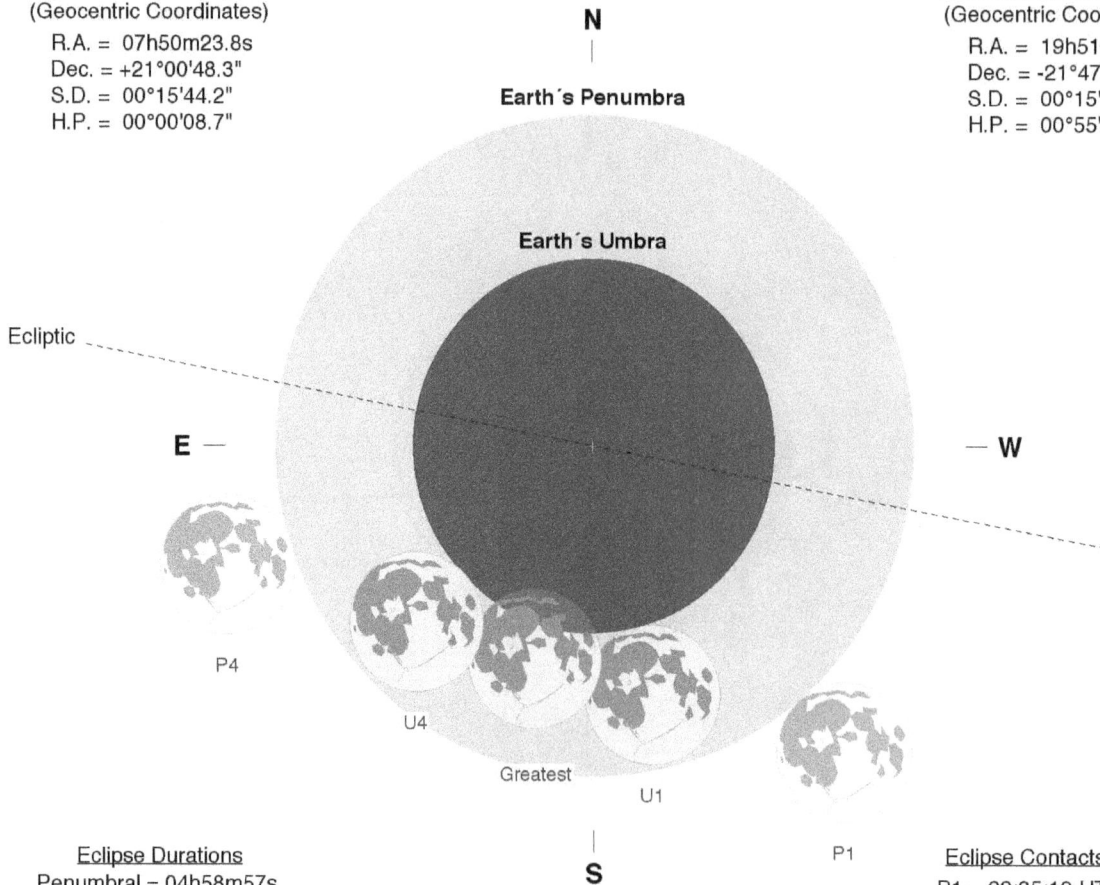

N

Earth's Penumbra

Earth's Umbra

Ecliptic

E

W

P4

U4

Greatest

U1

P1

S

Eclipse Durations
Penumbral = 04h58m57s
Umbral = 01h55m18s

Eph. = JPL DE430
Rule = Herald-Sinnott
ΔT = 83 s

0	15	30	45	60

Arc-Minutes

©2020 F. Espenak, www.EclipseWise.com

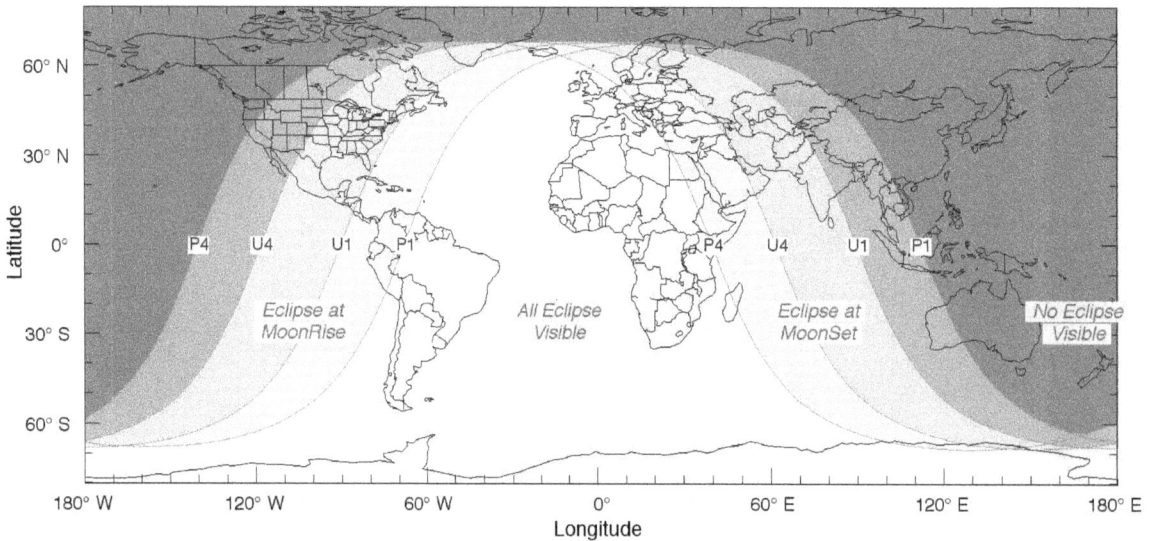

Eclipse Contacts
P1 = 22:35:19 UT1
U1 = 00:07:08 UT1
U4 = 02:02:26 UT1
P4 = 03:34:16 UT1

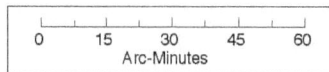

Eclipse at MoonRise

All Eclipse Visible

Eclipse at MoonSet

No Eclipse Visible

P4   U4   U1   P1   P4   U4   U1   P1

Total Lunar Eclipse of 2047 Jan 12

Greatest Eclipse = 01:26:14.4 TD   (= 01:24:51.2 UT1)

Penumbral Magnitude = 2.2665	Gamma = 0.3317	Saros Series = 125
Umbral Magnitude = 1.2358	Axis = 0.3201°	Saros Member = 50 of 72

Sun at Greatest Eclipse
(Geocentric Coordinates)

R.A. = 19h33m56.9s
Dec. = -21°40'46.3"
S.D. = 00°16'15.8"
H.P. = 00°00'08.9"

Moon at Greatest Eclipse
(Geocentric Coordinates)

R.A. = 07h34m18.1s
Dec. = +21°59'20.2"
S.D. = 00°15'46.6"
H.P. = 00°57'54.2"

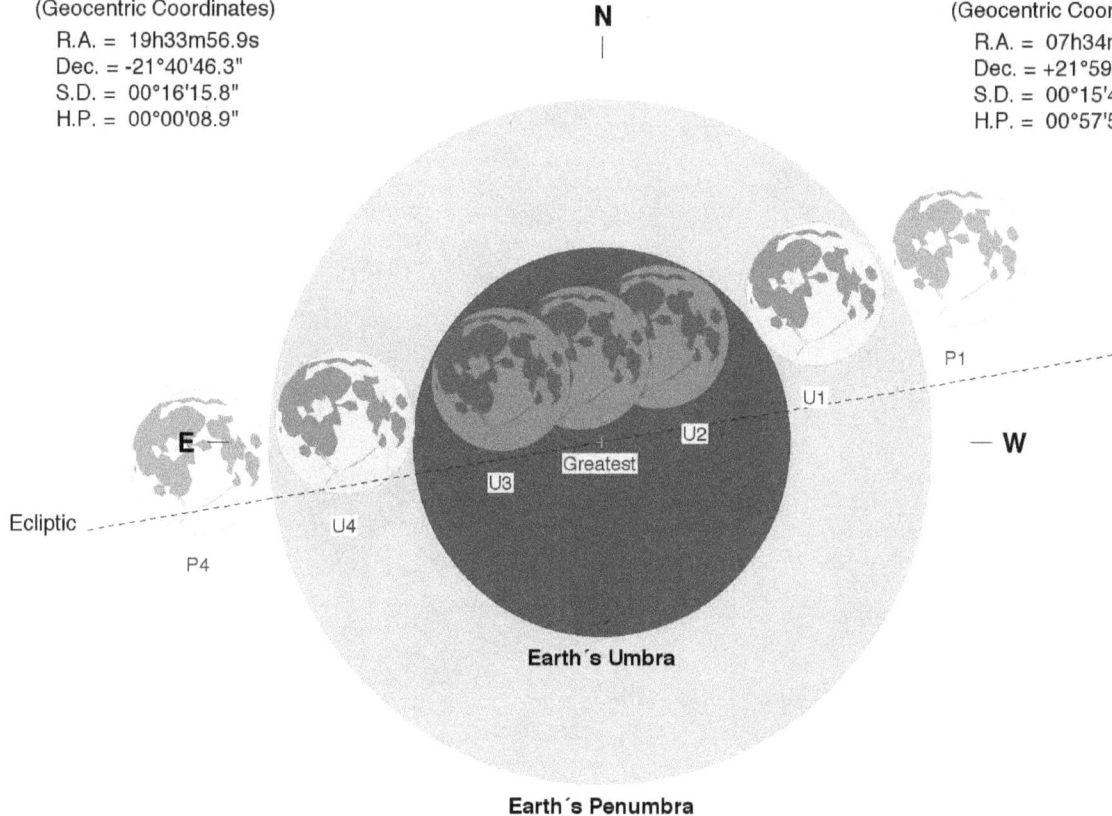

N

P1

U1

E

U2

W

Ecliptic

Greatest

U3

U4

P4

Earth´s Umbra

Earth´s Penumbra

S

Eclipse Durations

Penumbral = 05h38m05s
Umbral = 03h29m39s
Total = 01h10m41s

Eph. = JPL DE430
Rule = Herald-Sinnott
ΔT =   83 s

0	15	30	45	60

Arc-Minutes

©2020 F. Espenak, www.EclipseWise.com

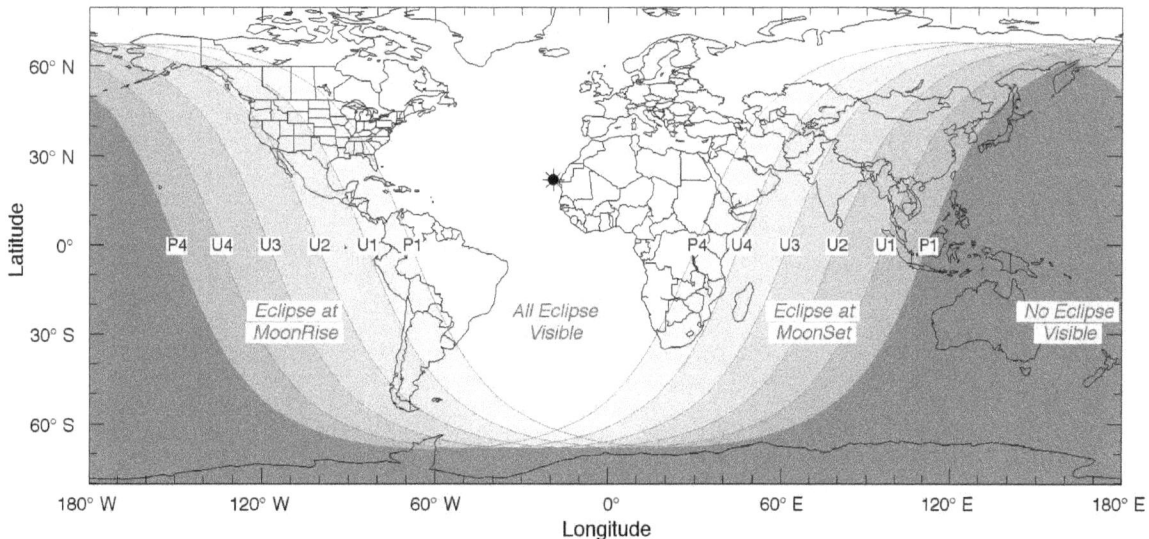

Eclipse Contacts

P1 = 22:35:48 UT1
U1 = 23:40:07 UT1
U2 = 00:49:39 UT1
U3 = 02:00:20 UT1
U4 = 03:09:45 UT1
P4 = 04:13:53 UT1

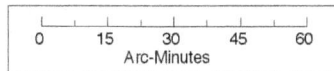

P4 U4 U3 U2 U1 P1   P4 U4 U3 U2 U1 P1

Eclipse at MoonRise   *All Eclipse Visible*   *Eclipse at MoonSet*   *No Eclipse Visible*

163

Total Lunar Eclipse of 2047 Jul 07

Greatest Eclipse = 10:35:45.4 TD (= 10:34:21.9 UT1)

Penumbral Magnitude = 2.7326	Gamma = -0.0636	Saros Series = 130
Umbral Magnitude = 1.7529	Axis = 0.0625°	Saros Member = 36 of 71

Sun at Greatest Eclipse
(Geocentric Coordinates)

R.A. = 07h06m19.6s
Dec. = +22°33'30.9"
S.D. = 00°15'43.9"
H.P. = 00°00'08.7"

Moon at Greatest Eclipse
(Geocentric Coordinates)

R.A. = 19h06m23.0s
Dec. = -22°37'10.8"
S.D. = 00°16'03.5"
H.P. = 00°58'56.1"

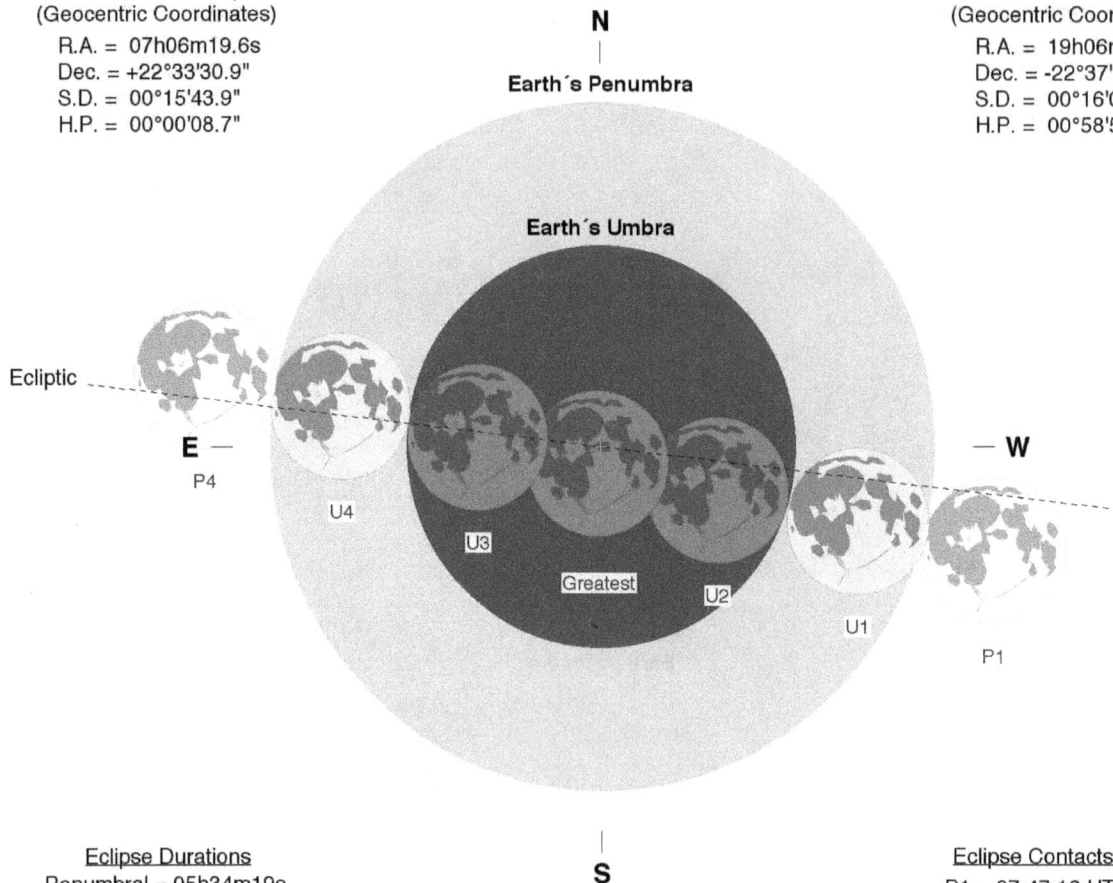

N

Earth´s Penumbra

Earth´s Umbra

Ecliptic

E
P4

U4

U3

Greatest

U2

U1

P1

W

S

Eclipse Durations
Penumbral = 05h34m19s
Umbral = 03h39m18s
Total = 01h41m35s

Eph. = JPL DE430
Rule = Herald-Sinnott
ΔT = 84 s

Arc-Minutes
0   15   30   45   60

©2020 F. Espenak, www.EclipseWise.com

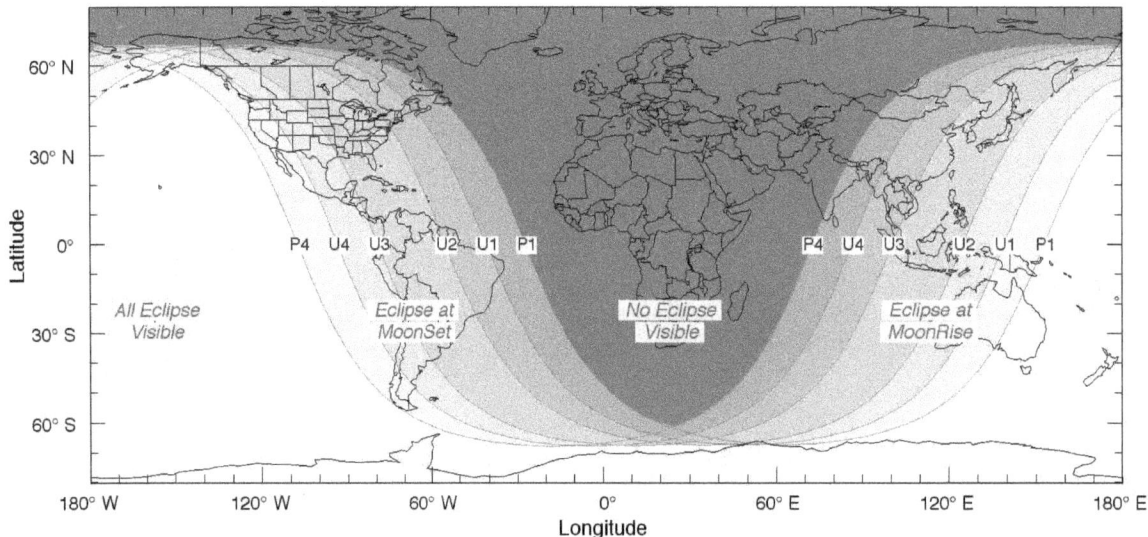

Eclipse Contacts
P1 = 07:47:16 UT1
U1 = 08:44:42 UT1
U2 = 09:43:35 UT1
U3 = 11:25:10 UT1
U4 = 12:24:01 UT1
P4 = 13:21:34 UT1

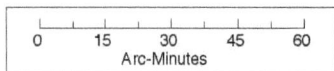

All Eclipse
Visible

Eclipse at
MoonSet

No Eclipse
Visible

Eclipse at
MoonRise

Total Lunar Eclipse of 2048 Jan 01

Greatest Eclipse = 06:53:54.8 TD   (= 06:52:30.9 UT1)

Penumbral Magnitude = 2.2158	Gamma = -0.3746	Saros Series = 135
Umbral Magnitude = 1.1297	Axis = 0.3431°	Saros Member = 25 of 71

Sun at Greatest Eclipse
(Geocentric Coordinates)

R.A. = 18h45m45.0s
Dec. = -23°01'00.1"
S.D. = 00°16'15.9"
H.P. = 00°00'08.9"

N

Earth's Penumbra

Earth's Umbra

Moon at Greatest Eclipse
(Geocentric Coordinates)

R.A. = 06h45m29.1s
Dec. = +22°40'44.8"
S.D. = 00°14'58.6"
H.P. = 00°54'57.7"

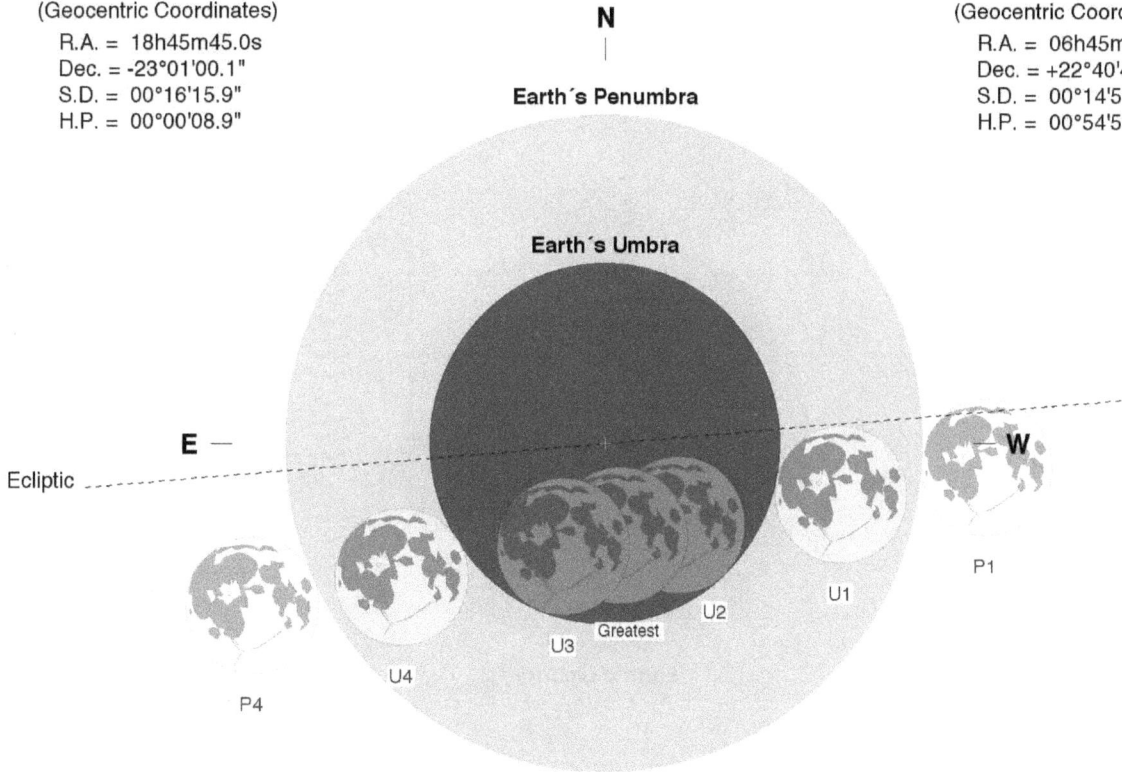

E —

Ecliptic

W

P1

U1

U2

Greatest

U3

U4

P4

S

Eclipse Durations
Penumbral = 06h00m21s
Umbral = 03h35m05s
Total = 00h56m40s

Eph. = JPL DE430
Rule = Herald-Sinnott
ΔT =   84 s

0	15	30	45	60

Arc-Minutes

©2020 F. Espenak, www.EclipseWise.com

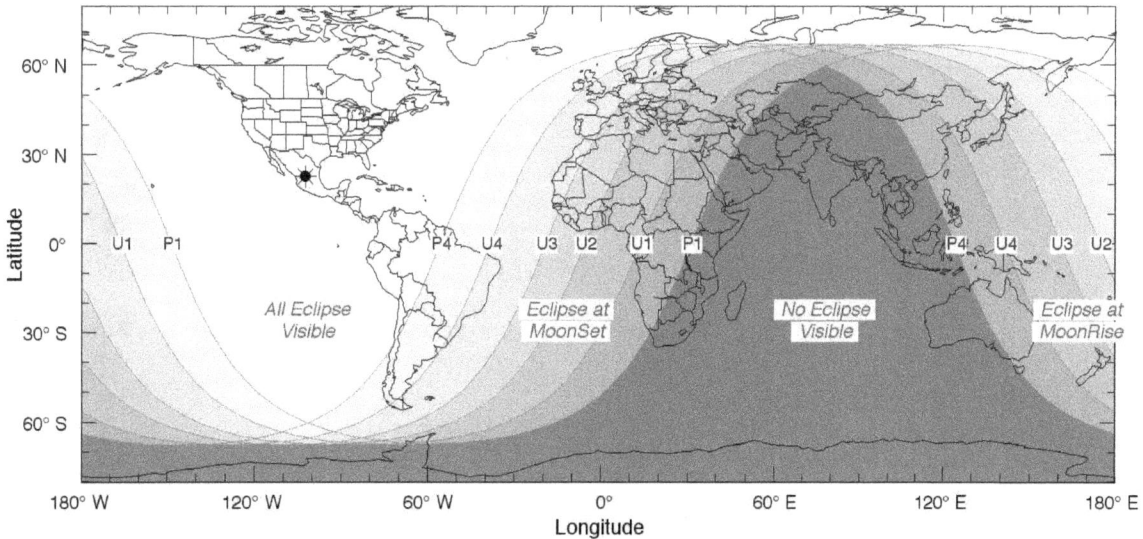

Eclipse Contacts
P1 = 03:52:17 UT1
U1 = 05:04:57 UT1
U2 = 06:24:06 UT1
U3 = 07:20:46 UT1
U4 = 08:40:02 UT1
P4 = 09:52:38 UT1

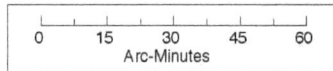

All Eclipse
Visible

U1   P1

P4   U4   U3   U2   U1   P1

Eclipse at
MoonSet

No Eclipse
Visible

P4   U4   U3   U2

Eclipse at
MoonRise

Partial Lunar Eclipse of 2048 Jun 26

Greatest Eclipse = 02:02:28.0 TD   (= 02:01:03.9 UT1)

Penumbral Magnitude = 1.5841	Gamma = 0.6797	Saros Series = 140
Umbral Magnitude = 0.6404	Axis = 0.6931°	Saros Member = 26 of 77

Sun at Greatest Eclipse
(Geocentric Coordinates)

R.A. = 06h22m31.9s
Dec. = +23°19'54.0"
S.D. = 00°15'44.1"
H.P. = 00°00'08.7"

Moon at Greatest Eclipse
(Geocentric Coordinates)

R.A. = 18h22m07.4s
Dec. = -22°38'42.2"
S.D. = 00°16'40.4"
H.P. = 01°01'11.5"

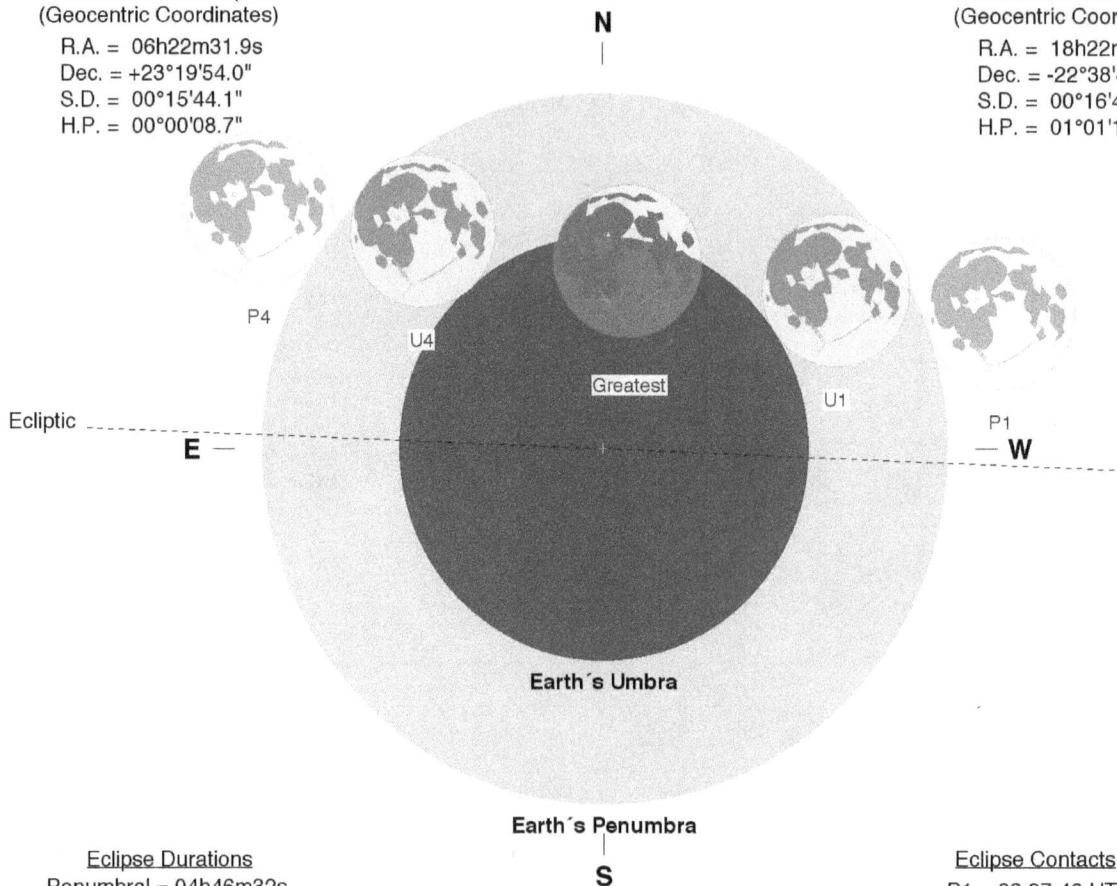

N

P4

U4

Greatest

U1

P1

Ecliptic

E — — W

Earth's Umbra

Earth's Penumbra

S

Eclipse Durations
Penumbral = 04h46m32s
Umbral = 02h39m51s

Eph. = JPL DE430
Rule = Herald-Sinnott
ΔT =   84 s

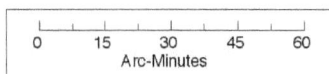

Eclipse Contacts
P1 = 23:37:46 UT1
U1 = 00:41:04 UT1
U4 = 03:20:55 UT1
P4 = 04:24:18 UT1

Arc-Minutes
0   15   30   45   60

©2020 F. Espenak, www.EclipseWise.com

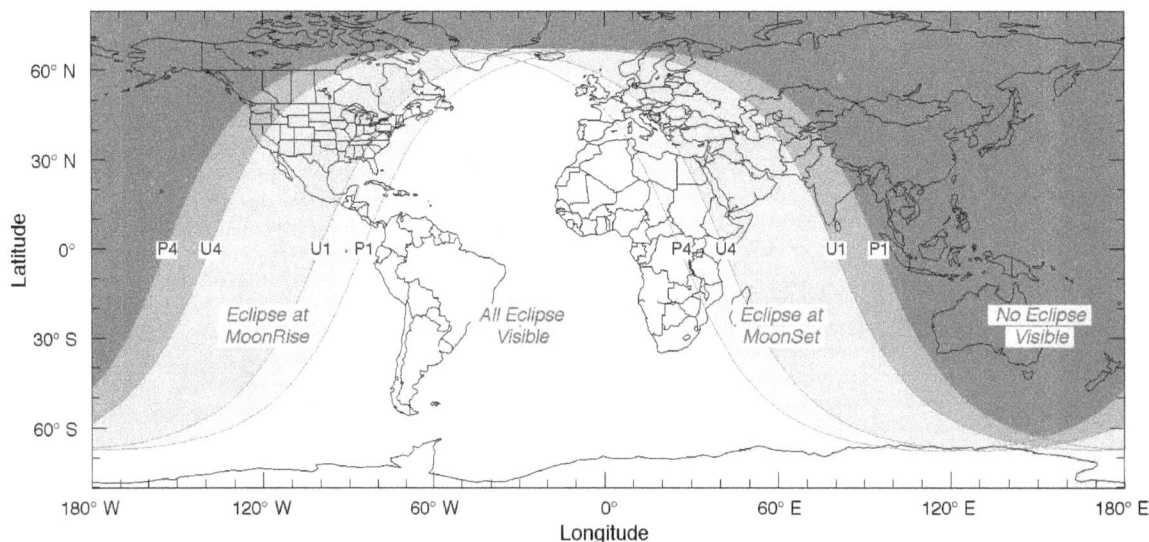

Eclipse at
MoonRise

All Eclipse
Visible

Eclipse at
MoonSet

No Eclipse
Visible

P4 U4   U1 P1   P4 U4   U1 P1

Penumbral Lunar Eclipse of 2048 Dec 20

Greatest Eclipse = 06:27:47.7 TD   (= 06:26:23.3 UT1)

Penumbral Magnitude = 0.9632	Gamma = -1.0624	Saros Series = 145
Umbral Magnitude = -0.1420	Axis = 0.9558°	Saros Member = 13 of 71

Sun at Greatest Eclipse
(Geocentric Coordinates)

R.A. = 17h55m49.3s
Dec. = -23°25'43.8"
S.D. = 00°16'15.4"
H.P. = 00°00'08.9"

Moon at Greatest Eclipse
(Geocentric Coordinates)

R.A. = 05h55m26.5s
Dec. = +22°28'37.2"
S.D. = 00°14'42.6"
H.P. = 00°53'59.0"

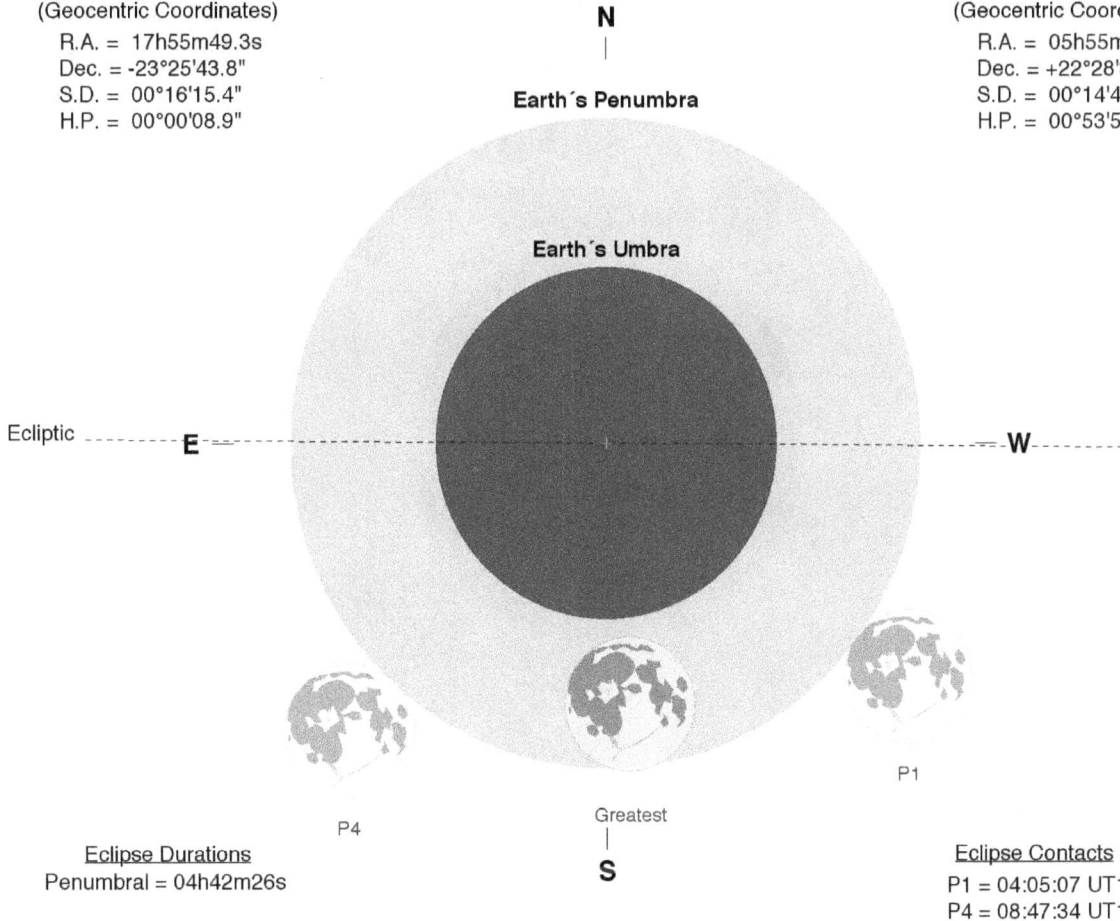

N

Earth's Penumbra

Earth's Umbra

Ecliptic E W

P1

P4

Greatest

S

Eclipse Durations
Penumbral = 04h42m26s

Eclipse Contacts
P1 = 04:05:07 UT1
P4 = 08:47:34 UT1

Eph. = JPL DE430
Rule = Herald-Sinnott
ΔT = 84 s

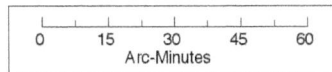

0   15   30   45   60
Arc-Minutes

©2020 F. Espenak, www.EclipseWise.com

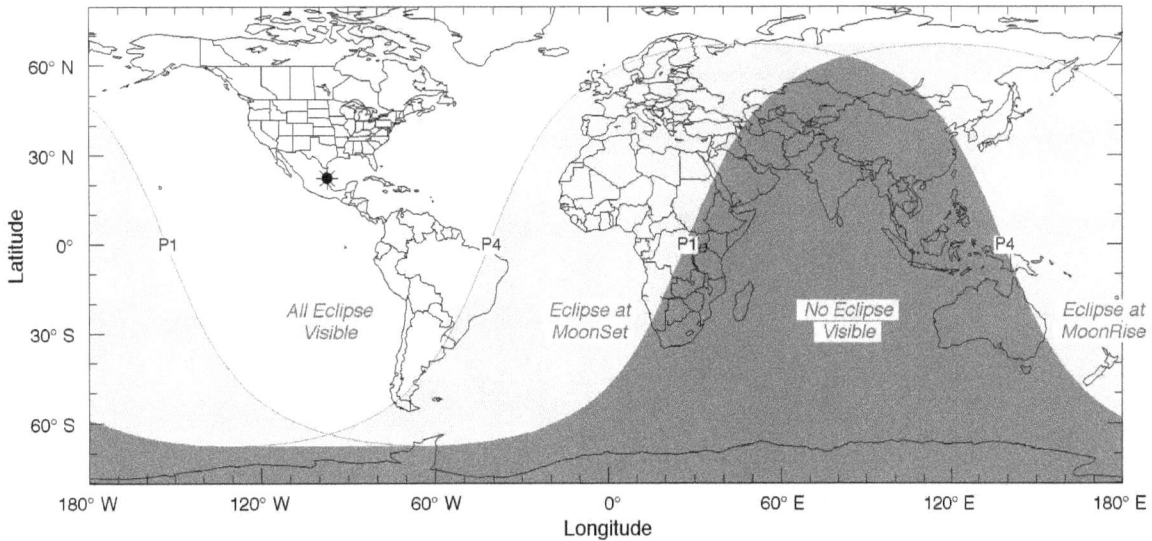

All Eclipse Visible

Eclipse at MoonSet

No Eclipse Visible

Eclipse at MoonRise

P1

P4

P1

P4

Penumbral Lunar Eclipse of 2049 May 17

Greatest Eclipse = 11:26:38.4 TD   (= 11:25:13.7 UT1)

Penumbral Magnitude = 0.7650	Gamma = -1.1337	Saros Series = 112
Umbral Magnitude = -0.2073	Axis = 1.1279°	Saros Member = 67 of 72

Sun at Greatest Eclipse
(Geocentric Coordinates)

R.A. = 03h38m51.9s
Dec. = +19°28'58.4"
S.D. = 00°15'49.0"
H.P. = 00°00'08.7"

Moon at Greatest Eclipse
(Geocentric Coordinates)

R.A. = 15h38m12.8s
Dec. = -20°36'01.8"
S.D. = 00°16'16.0"
H.P. = 00°59'41.9"

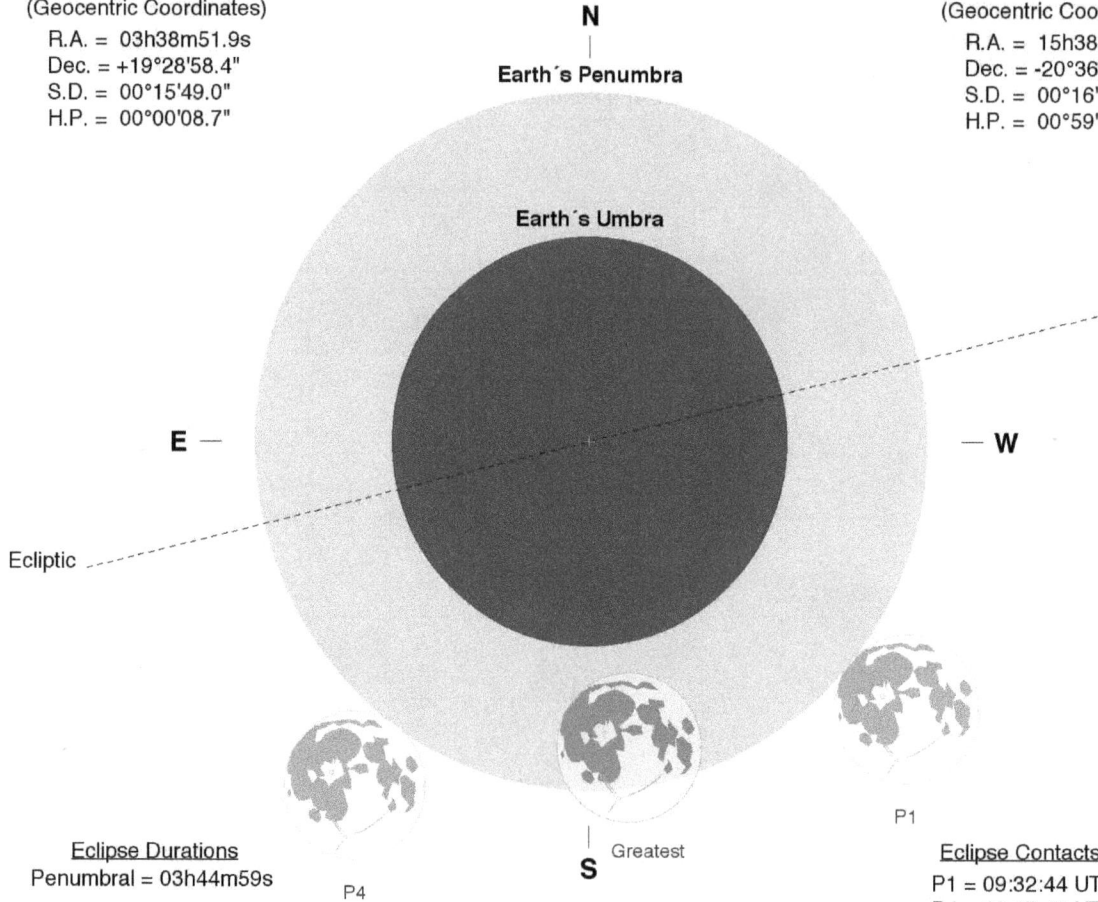

N

Earth's Penumbra

Earth's Umbra

E

W

Ecliptic

P1

S   Greatest

P4

Eclipse Durations
Penumbral = 03h44m59s

Eclipse Contacts
P1 = 09:32:44 UT1
P4 = 13:17:43 UT1

Eph. = JPL DE430
Rule = Herald-Sinnott
ΔT =   85 s

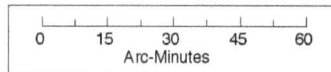

0	15	30	45	60

Arc-Minutes

©2020 F. Espenak, www.EclipseWise.com

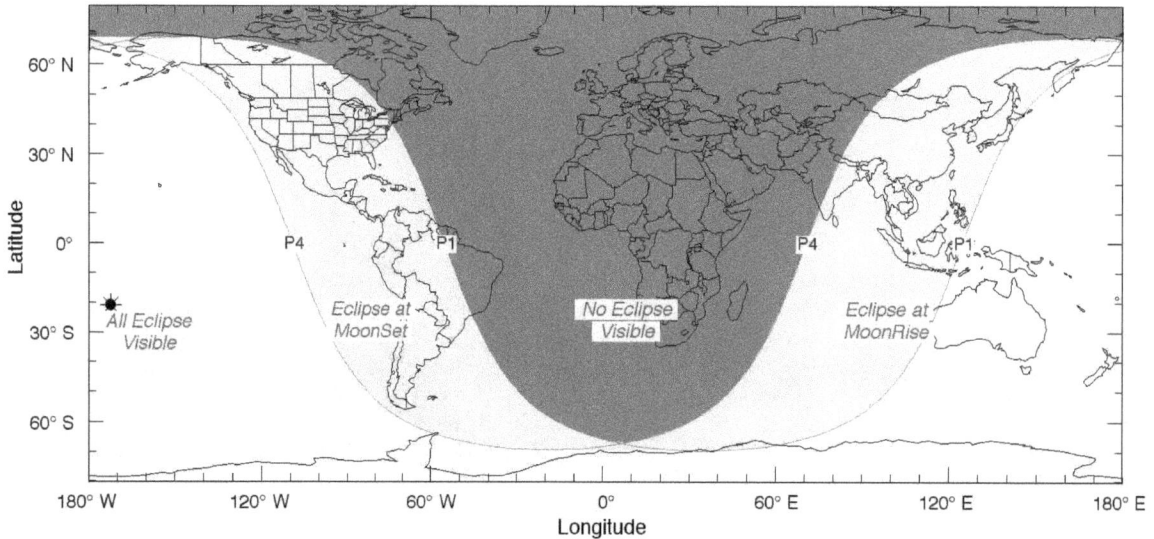

All Eclipse
Visible

Eclipse at
MoonSet

No Eclipse
Visible

Eclipse at
MoonRise

P4   P1

P4   P1

Penumbral Lunar Eclipse of 2049 Jun 15

Greatest Eclipse = 19:14:11.5 TD (= 19:12:46.8 UT1)

Penumbral Magnitude = 0.2526	Gamma = 1.4069	Saros Series = 150
Umbral Magnitude = -0.6970	Axis = 1.4268°	Saros Member = 3 of 71

Sun at Greatest Eclipse
(Geocentric Coordinates)

R.A. = 05h38m45.5s
Dec. = +23°20'31.0"
S.D. = 00°15'44.8"
H.P. = 00°00'08.7"

Moon at Greatest Eclipse
(Geocentric Coordinates)

R.A. = 17h38m24.2s
Dec. = -21°55'02.3"
S.D. = 00°16'34.9"
H.P. = 01°00'51.4"

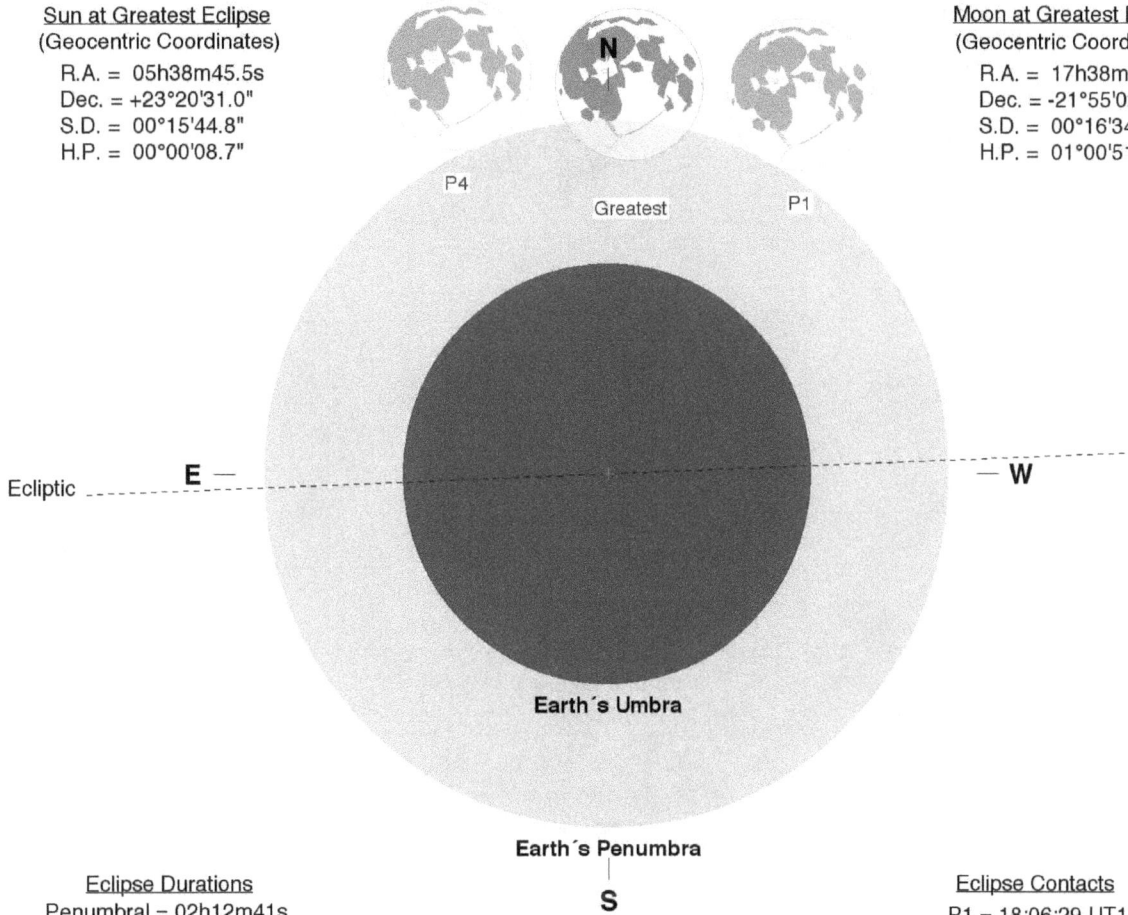

N

P4
Greatest
P1

Ecliptic

E —

— W

Earth's Umbra

Earth's Penumbra

S

Eclipse Durations
Penumbral = 02h12m41s

Eclipse Contacts
P1 = 18:06:29 UT1
P4 = 20:19:09 UT1

Eph. = JPL DE430
Rule = Herald-Sinnott
ΔT = 85 s

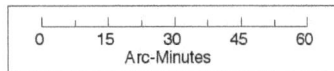

0	15	30	45	60

Arc-Minutes

©2020 F. Espenak, www.EclipseWise.com

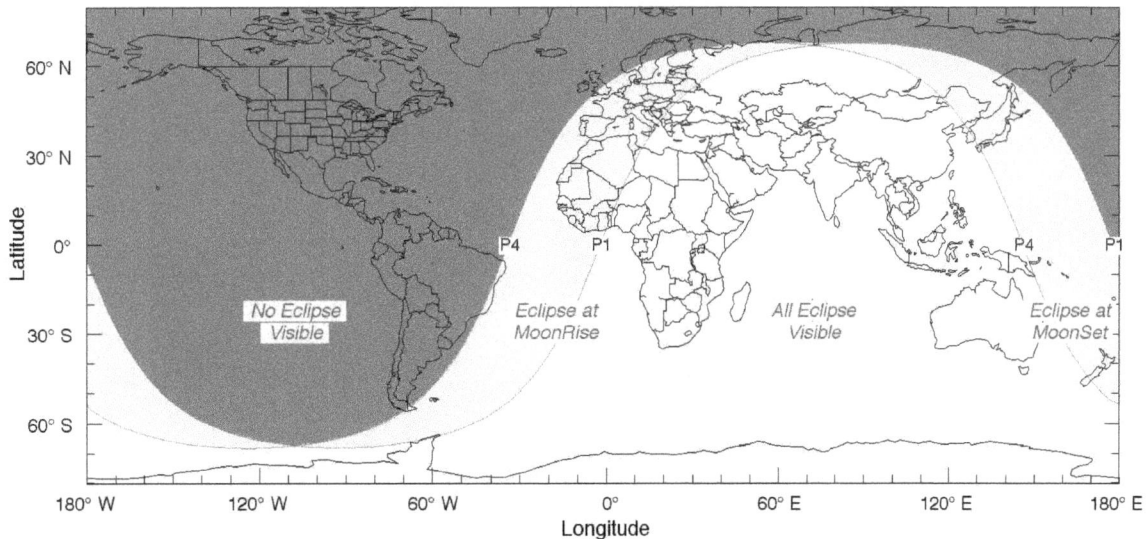

No Eclipse Visible

Eclipse at MoonRise

All Eclipse Visible

Eclipse at MoonSet

P4 P1   P4 P1

Latitude

Longitude

Penumbral Lunar Eclipse of 2049 Nov 09

Greatest Eclipse = 15:52:10.7 TD   (= 15:50:45.7 UT1)

Penumbral Magnitude = 0.6821	Gamma = 1.1965	Saros Series = 117
Umbral Magnitude = -0.3541	Axis = 1.1404°	Saros Member = 54 of 71

Sun at Greatest Eclipse
(Geocentric Coordinates)
R.A. = 15h00m53.5s
Dec. = -17°06'00.6"
S.D. = 00°16'08.8"
H.P. = 00°00'08.9"

Moon at Greatest Eclipse
(Geocentric Coordinates)
R.A. = 03h00m00.0s
Dec. = +18°13'14.6"
S.D. = 00°15'35.1"
H.P. = 00°57'11.8"

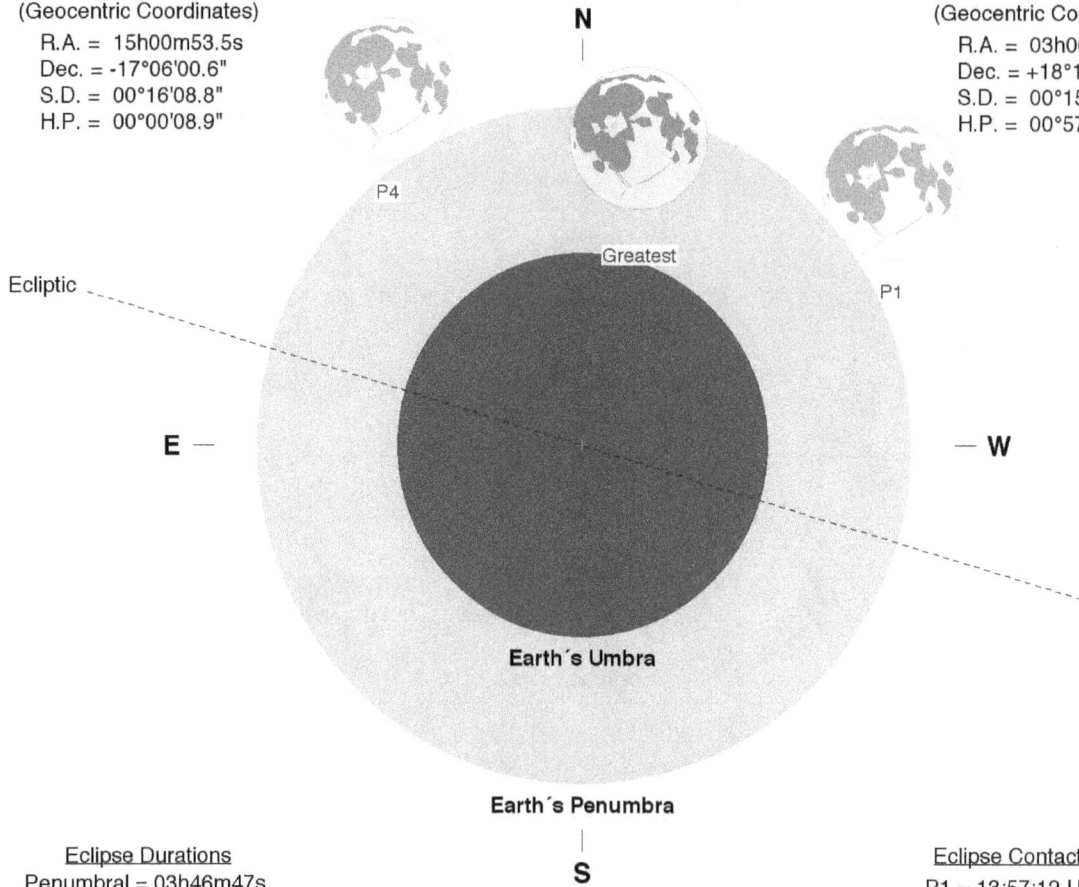

N

P4

Greatest

Ecliptic

P1

E

W

Earth's Umbra

Earth's Penumbra

S

Eclipse Durations
Penumbral = 03h46m47s

Eclipse Contacts
P1 = 13:57:12 UT1
P4 = 17:43:59 UT1

Eph. = JPL DE430
Rule = Herald-Sinnott
ΔT =   85 s

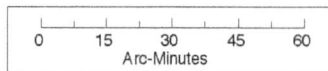

0	15	30	45	60

Arc-Minutes

©2020 F. Espenak, www.EclipseWise.com

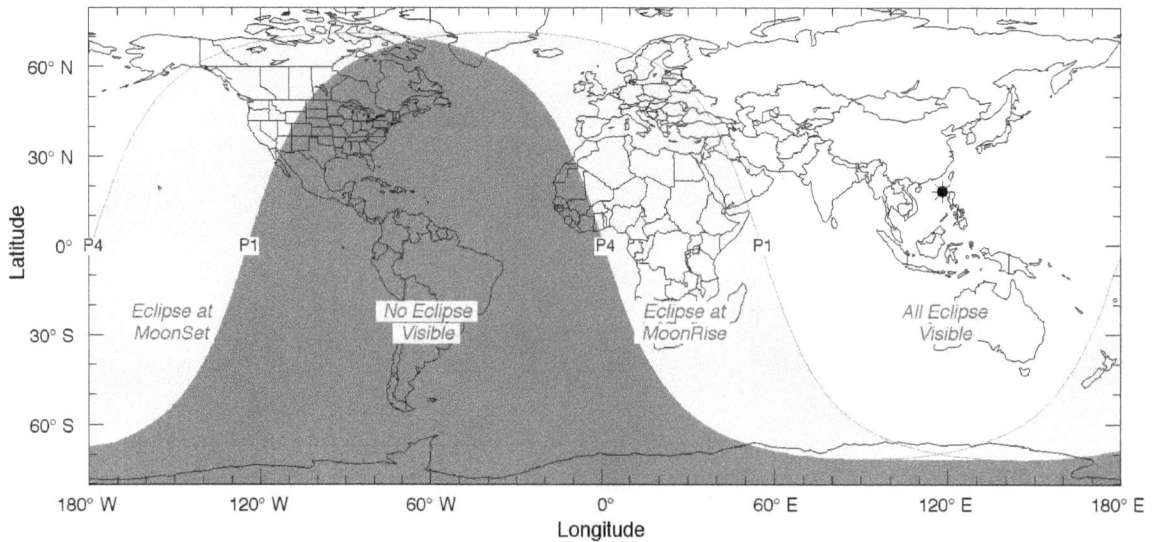

Eclipse at MoonSet

No Eclipse Visible

Eclipse at MoonRise

All Eclipse Visible

P4   P1

P4   P1

Latitude

Longitude

Total Lunar Eclipse of 2050 May 06

Greatest Eclipse = 22:32:01.6 TD (= 22:30:36.3 UT1)

Penumbral Magnitude = 2.1064	Gamma = -0.4181	Saros Series = 122
Umbral Magnitude = 1.0779	Axis = 0.3942°	Saros Member = 58 of 74

Sun at Greatest Eclipse
(Geocentric Coordinates)

R.A. = 02h56m30.8s
Dec. = +16°47'28.5"
S.D. = 00°15'51.3"
H.P. = 00°00'08.7"

N

Earth's Penumbra

Earth's Umbra

E

W

Ecliptic

P1

U1

Greatest   U2

U3

U4

P4

S

Moon at Greatest Eclipse
(Geocentric Coordinates)

R.A. = 14h56m12.1s
Dec. = -17°10'41.9"
S.D. = 00°15'24.9"
H.P. = 00°56'34.4"

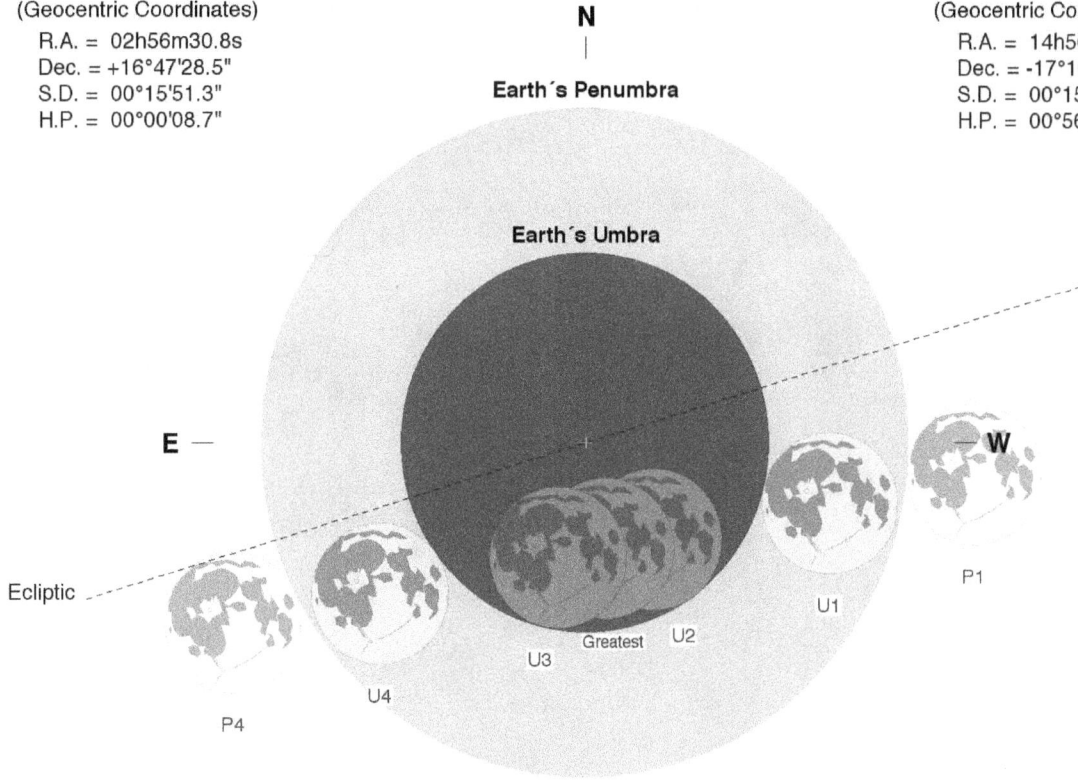

Eclipse Durations
Penumbral = 05h40m55s
Umbral = 03h26m46s
Total = 00h43m49s

Eph. = JPL DE430
Rule = Herald-Sinnott
ΔT = 85 s

0	15	30	45	60

Arc-Minutes

©2020 F. Espenak, www.EclipseWise.com

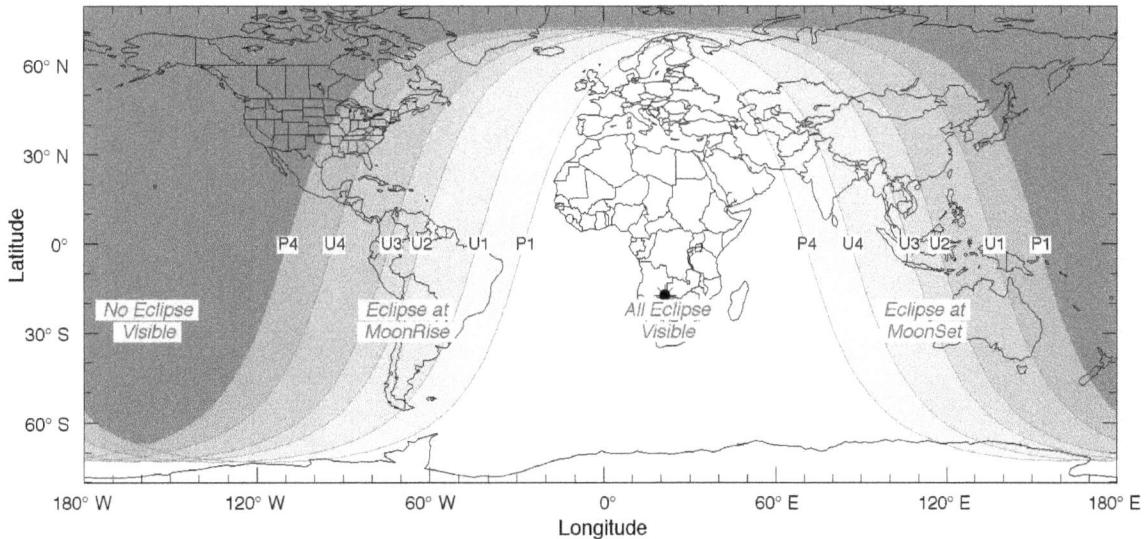

Eclipse Contacts
P1 = 19:40:04 UT1
U1 = 20:47:12 UT1
U2 = 22:08:36 UT1
U3 = 22:52:25 UT1
U4 = 00:13:58 UT1
P4 = 01:20:59 UT1

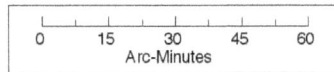

No Eclipse
Visible

Eclipse at
MoonRise

All Eclipse
Visible

Eclipse at
MoonSet

P4 U4 U3 U2 U1 P1   P4 U4 U3 U2 U1 P1

Total Lunar Eclipse of 2050 Oct 30

Greatest Eclipse = 03:21:46.8 TD   (= 03:20:21.2 UT1)

Penumbral Magnitude = 2.0356	Gamma = 0.4435	Saros Series = 127
Umbral Magnitude = 1.0549	Axis = 0.4454°	Saros Member = 44 of 72

Sun at Greatest Eclipse
(Geocentric Coordinates)
R.A. = 14h18m15.4s
Dec. = -13°48'46.9"
S.D. = 00°16'06.2"
H.P. = 00°00'08.9"

Moon at Greatest Eclipse
(Geocentric Coordinates)
R.A. = 02h17m49.7s
Dec. = +14°14'46.2"
S.D. = 00°16'25.2"
H.P. = 01°00'15.6"

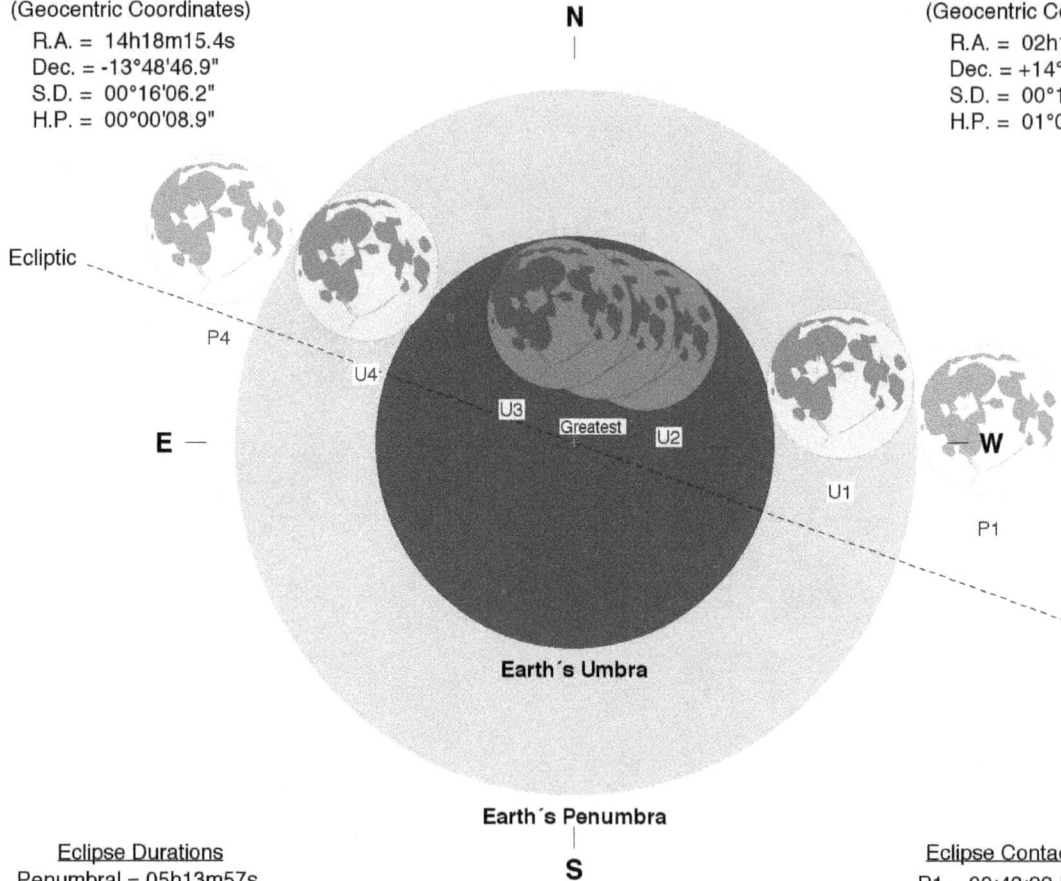

N

Ecliptic

P4

U4

U3

Greatest

U2

U1

P1

E

W

Earth's Umbra

Earth's Penumbra

S

Eclipse Durations
Penumbral = 05h13m57s
Umbral = 03h13m34s
Total = 00h35m05s

Eph. = JPL DE430
Rule = Herald-Sinnott
ΔT = 86 s

0   15   30   45   60
Arc-Minutes

©2020 F. Espenak, www.EclipseWise.com

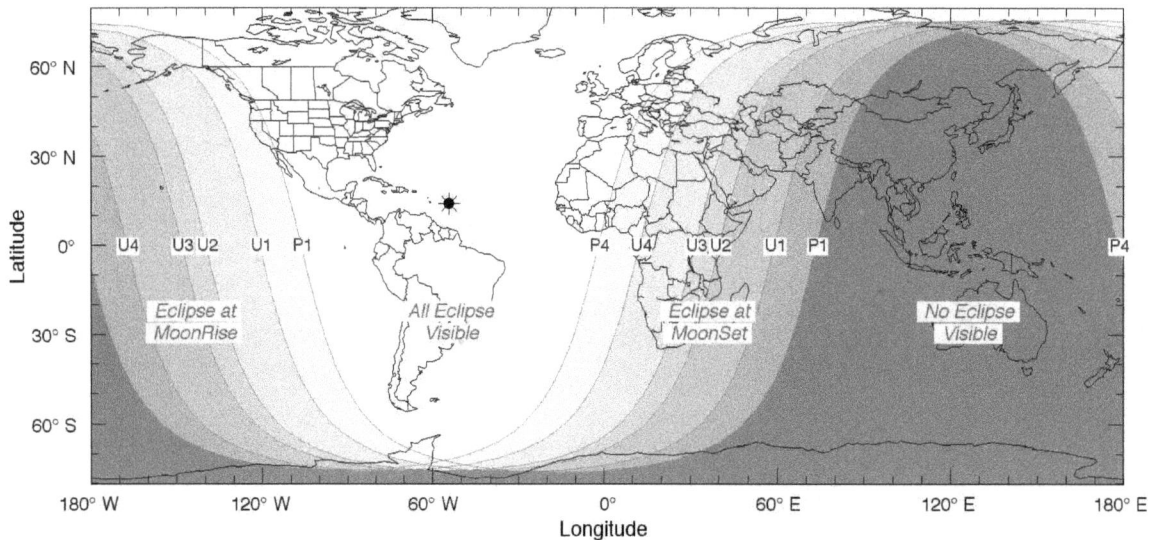

Eclipse Contacts
P1 = 00:43:22 UT1
U1 = 01:43:29 UT1
U2 = 03:02:39 UT1
U3 = 03:37:44 UT1
U4 = 04:57:03 UT1
P4 = 05:57:18 UT1

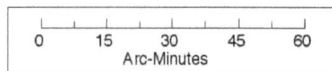

Eclipse at MoonRise

All Eclipse Visible

Eclipse at MoonSet

No Eclipse Visible

U4 U3 U2 U1 P1   P4 U4 U3 U2   U1 P1   P4

Total Lunar Eclipse of 2051 Apr 26

Greatest Eclipse = 02:16:27.8 TD (= 02:15:01.8 UT1)

Penumbral Magnitude = 2.2785	Gamma = 0.3371	Saros Series = 132
Umbral Magnitude = 1.2034	Axis = 0.3049°	Saros Member = 32 of 71

Sun at Greatest Eclipse
(Geocentric Coordinates)

R.A. = 02h14m06.4s
Dec. = +13°27'39.8"
S.D. = 00°15'53.9"
H.P. = 00°00'08.7"

Moon at Greatest Eclipse
(Geocentric Coordinates)

R.A. = 14h14m24.0s
Dec. = -13°09'52.9"
S.D. = 00°14'47.3"
H.P. = 00°54'16.4"

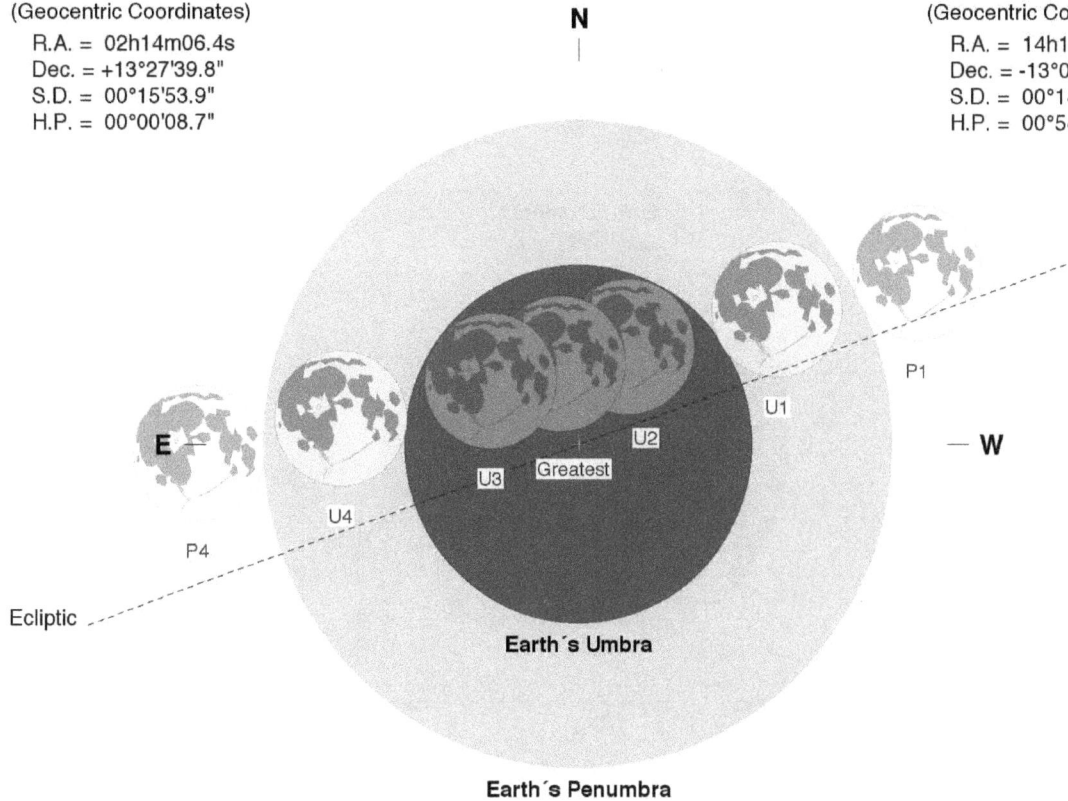

N

P1

U1

U2

Greatest

U3

E

W

U4

P4

Ecliptic

Earth's Umbra

Earth's Penumbra

S

Eclipse Durations
Penumbral = 06h05m43s
Umbral = 03h41m40s
Total = 01h10m15s

Eph. = JPL DE430
Rule = Herald-Sinnott
ΔT =   86 s

0	15	30	45	60

Arc-Minutes

©2020 F. Espenak, www.EclipseWise.com

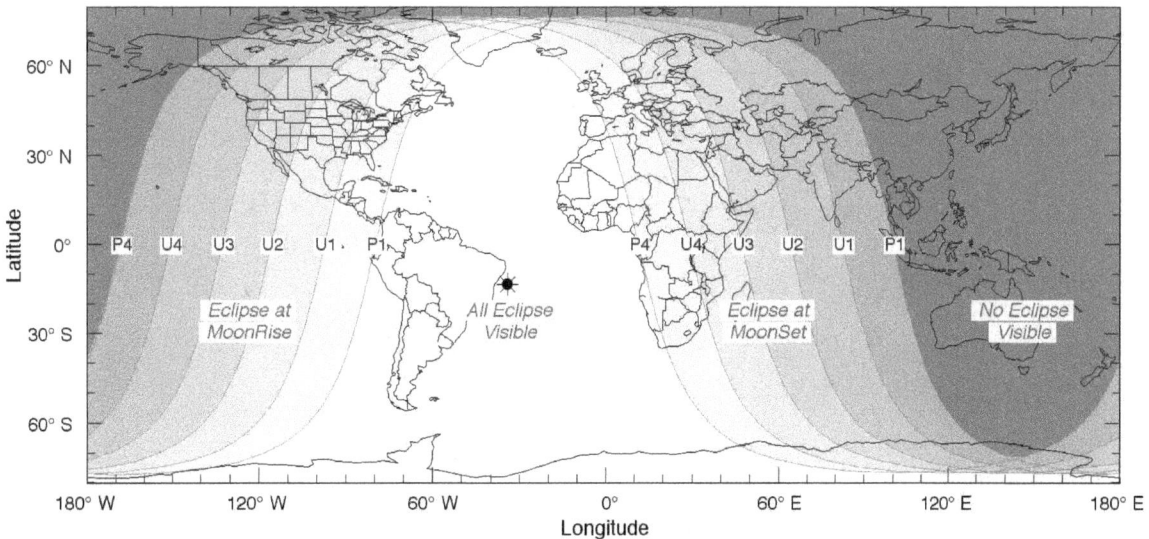

Eclipse Contacts
P1 = 23:12:11 UT1
U1 = 00:24:16 UT1
U2 = 01:40:03 UT1
U3 = 02:50:18 UT1
U4 = 04:05:56 UT1
P4 = 05:17:54 UT1

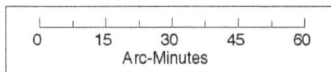

Eclipse at MoonRise • All Eclipse Visible • Eclipse at MoonSet • No Eclipse Visible

Total Lunar Eclipse of 2051 Oct 19

Greatest Eclipse = 19:11:49.6 TD   (= 19:10:23.3 UT1)

Penumbral Magnitude = 2.3719	Gamma = -0.2542	Saros Series = 137
Umbral Magnitude = 1.4130	Axis = 0.2603°	Saros Member = 28 of 78

Sun at Greatest Eclipse
(Geocentric Coordinates)

R.A. = 13h37m47.1s
Dec. = -10°10'03.3"
S.D. = 00°16'03.4"
H.P. = 00°00'08.8"

N

Earth's Penumbra

Ecliptic

Earth's Umbra

Moon at Greatest Eclipse
(Geocentric Coordinates)

R.A. = 01h38m04.1s
Dec. = +09°55'00.4"
S.D. = 00°16'44.6"
H.P. = 01°01'27.1"

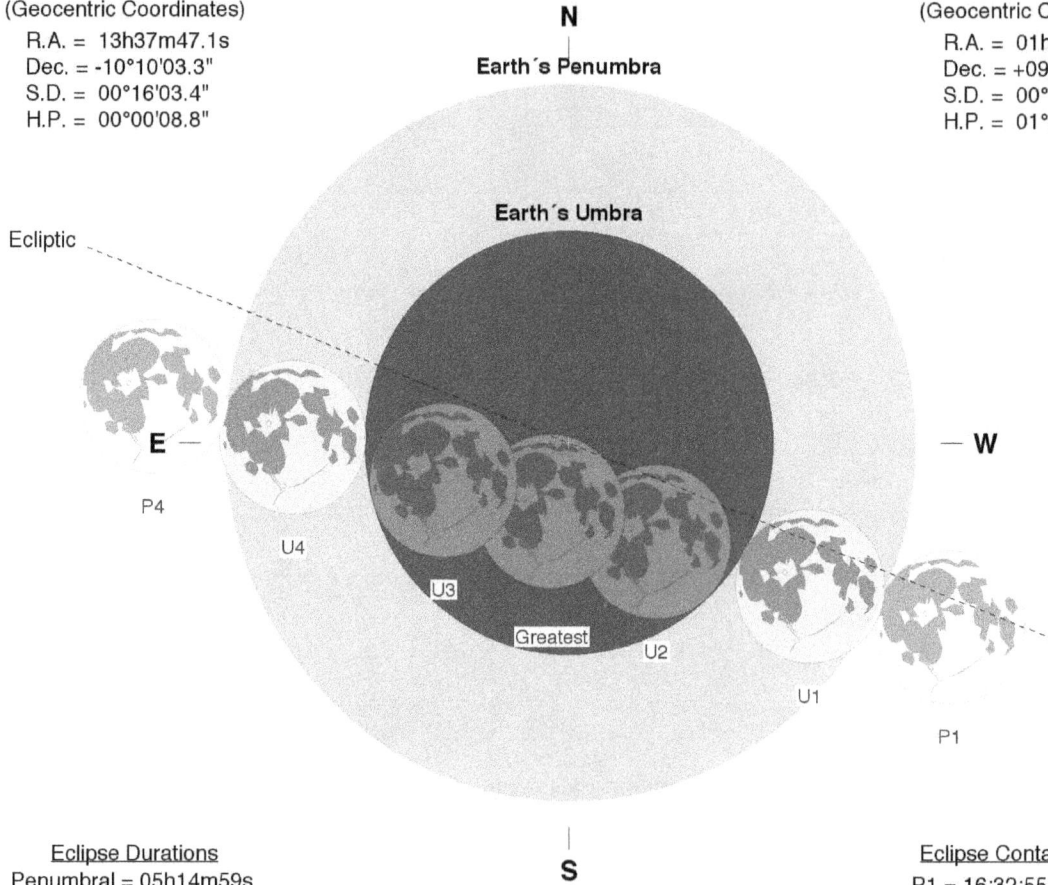

E

P4

U4

U3

Greatest

U2

W

U1

P1

S

Eclipse Durations
Penumbral = 05h14m59s
Umbral = 03h25m00s
Total = 01h24m12s

Eph. = JPL DE430
Rule = Herald-Sinnott
ΔT =   86 s

0   15   30   45   60
Arc-Minutes

©2020 F. Espenak, www.EclipseWise.com

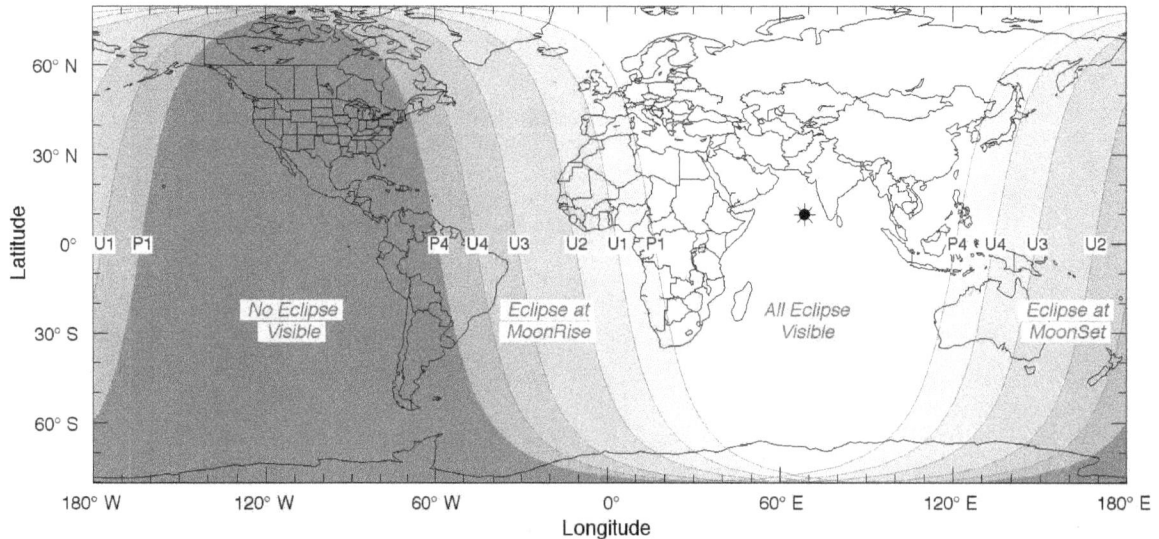

Eclipse Contacts
P1 = 16:32:55 UT1
U1 = 17:27:56 UT1
U2 = 18:28:23 UT1
U3 = 19:52:35 UT1
U4 = 20:52:56 UT1
P4 = 21:47:55 UT1

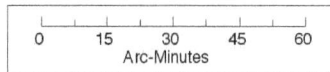

Penumbral Lunar Eclipse of 2052 Apr 14

Greatest Eclipse = 02:18:06.3 TD   (= 02:16:39.7 UT1)

Penumbral Magnitude = 0.9478	Gamma = 1.0629	Saros Series = 142
Umbral Magnitude = -0.1294	Axis = 0.9625°	Saros Member = 20 of 73

Sun at Greatest Eclipse
(Geocentric Coordinates)

R.A. = 01h32m05.6s
Dec. = +09°37'10.9"
S.D. = 00°15'56.9"
H.P. = 00°00'08.8"

Moon at Greatest Eclipse
(Geocentric Coordinates)

R.A. = 13h33m09.3s
Dec. = -08°41'36.6"
S.D. = 00°14'48.3"
H.P. = 00°54'20.2"

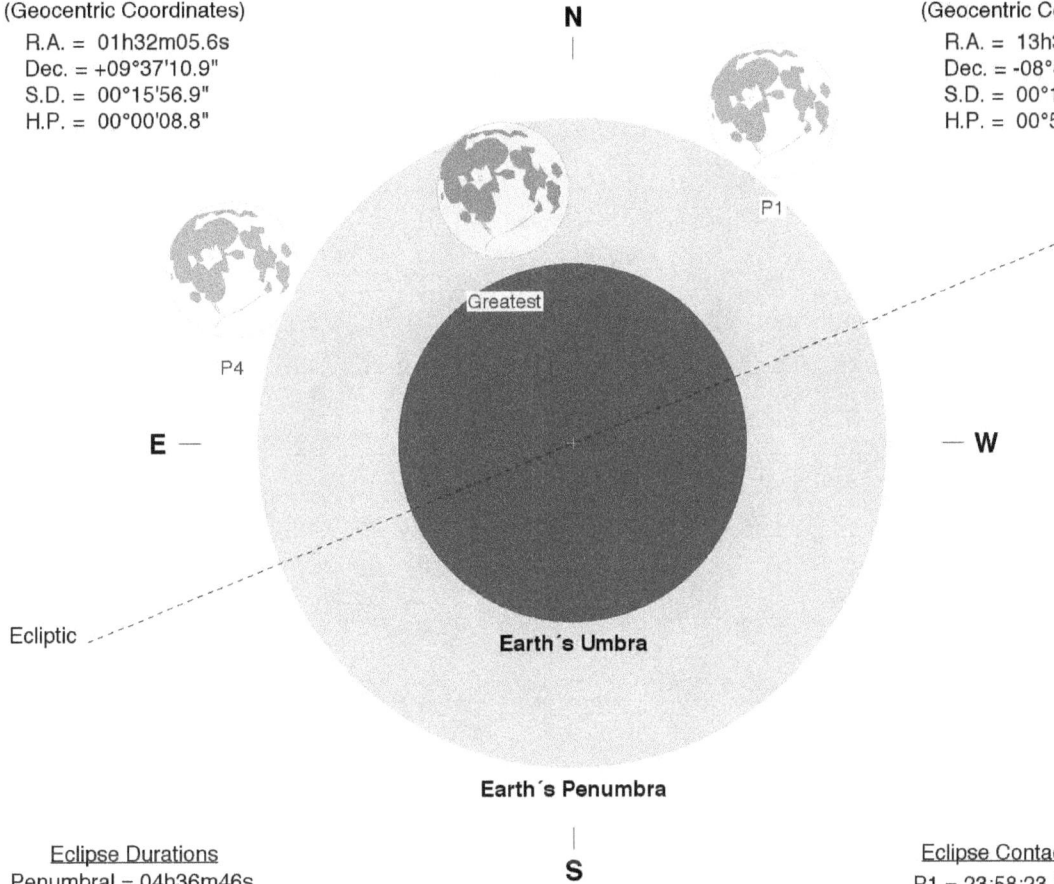

N

P1

Greatest

P4

E —

— W

Earth's Umbra

Ecliptic

Earth's Penumbra

S

Eclipse Durations
Penumbral = 04h36m46s

Eclipse Contacts
P1 = 23:58:23 UT1
P4 = 04:35:09 UT1

Eph. = JPL DE430
Rule = Herald-Sinnott
ΔT =   87 s

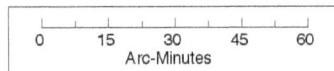

0	15	30	45	60

Arc-Minutes

©2020 F. Espenak, www.EclipseWise.com

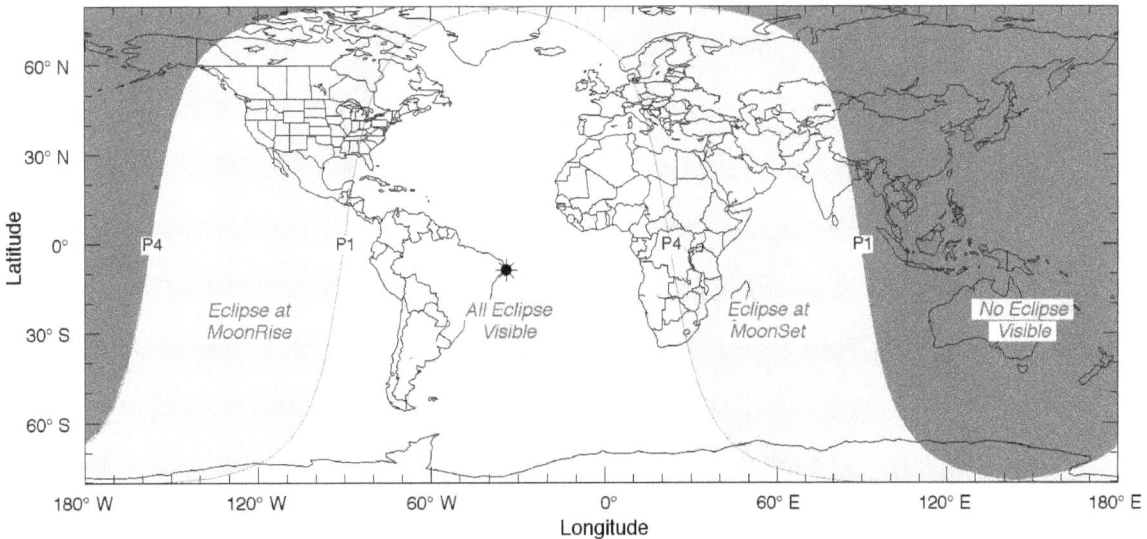

Eclipse at
MoonRise

All Eclipse
Visible

Eclipse at
MoonSet

No Eclipse
Visible

P4   P1   P4   P1

Partial Lunar Eclipse of 2052 Oct 08

Greatest Eclipse = 10:45:58.0 TD   (= 10:44:31.1 UT1)

Penumbral Magnitude = 1.0653	Gamma = -0.9727	Saros Series = 147
Umbral Magnitude = 0.0832	Axis = 0.9697°	Saros Member = 10 of 70

Sun at Greatest Eclipse
(Geocentric Coordinates)

R.A. = 12h58m28.0s
Dec. = -06°14'27.6"
S.D. = 00°16'00.5"
H.P. = 00°00'08.8"

Moon at Greatest Eclipse
(Geocentric Coordinates)

R.A. = 00h59m36.6s
Dec. = +05°18'49.9"
S.D. = 00°16'18.0"
H.P. = 00°59'49.3"

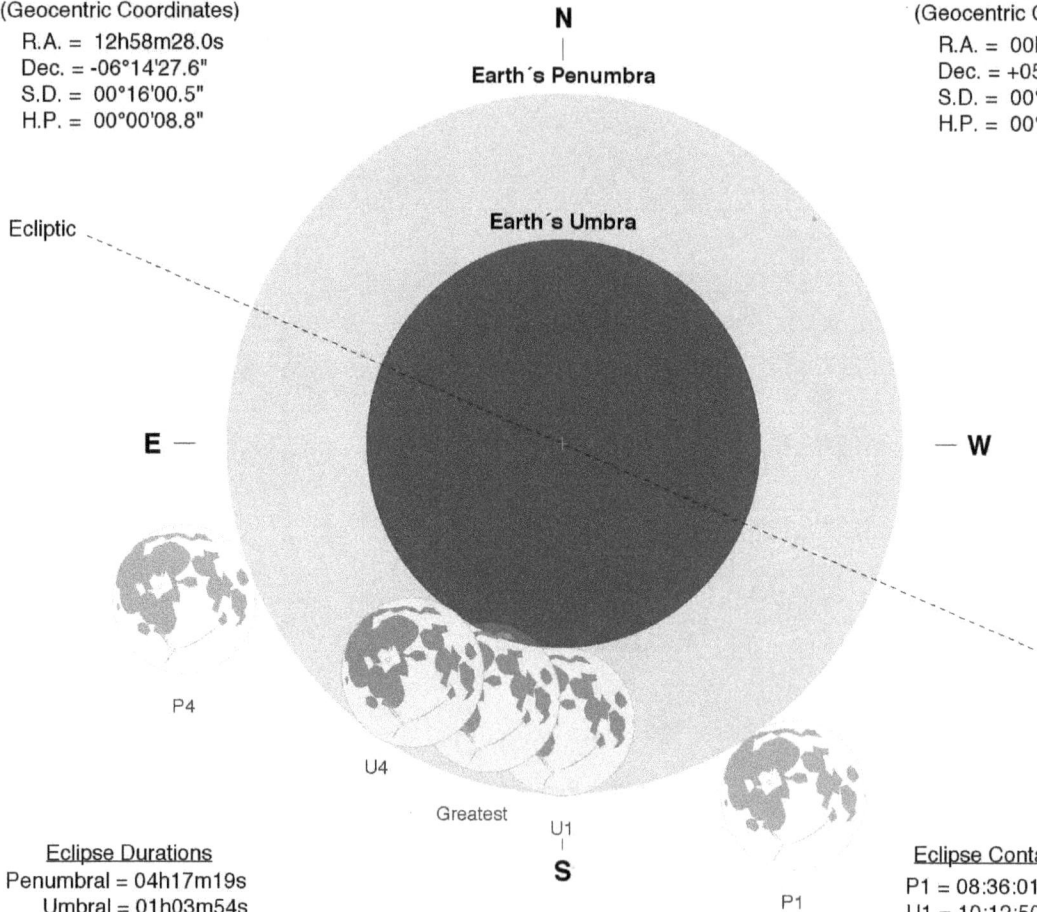

N

Earth's Penumbra

Earth's Umbra

Ecliptic

E

W

P4

U4

Greatest

U1

S

P1

Eclipse Durations
Penumbral = 04h17m19s
Umbral = 01h03m54s

Eph. = JPL DE430
Rule = Herald-Sinnott
ΔT = 87 s

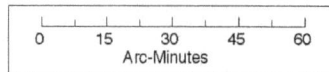

Eclipse Contacts
P1 = 08:36:01 UT1
U1 = 10:12:50 UT1
U4 = 11:16:44 UT1
P4 = 12:53:20 UT1

0   15   30   45   60
Arc-Minutes

©2020 F. Espenak, www.EclipseWise.com

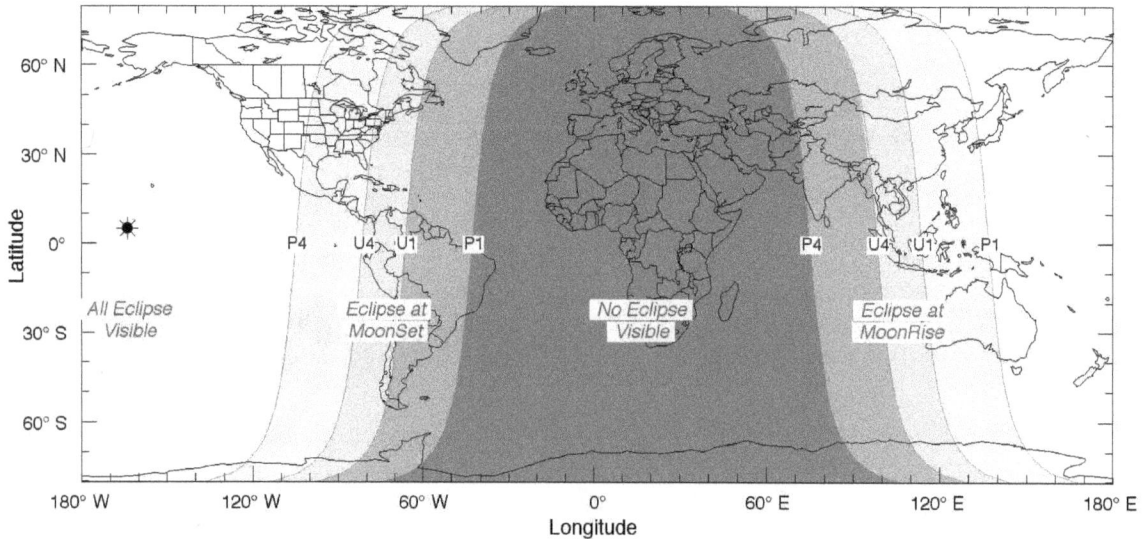

All Eclipse Visible

Eclipse at MoonSet

No Eclipse Visible

Eclipse at MoonRise

P4   U4   U1   P1   P4   U4   U1   P1

Penumbral Lunar Eclipse of 2053 Mar 04

Greatest Eclipse = 17:22:09.3 TD (= 17:20:42.1 UT1)

Penumbral Magnitude = 0.9334	Gamma = -1.0531	Saros Series = 114
Umbral Magnitude = -0.0796	Axis = 1.0255°	Saros Member = 61 of 71

Sun at Greatest Eclipse
(Geocentric Coordinates)

R.A. = 23h03m14.8s
Dec. = -06°03'47.9"
S.D. = 00°16'07.7"
H.P. = 00°00'08.9"

Moon at Greatest Eclipse
(Geocentric Coordinates)

R.A. = 11h02m02.1s
Dec. = +05°04'58.9"
S.D. = 00°15'55.3"
H.P. = 00°58'26.0"

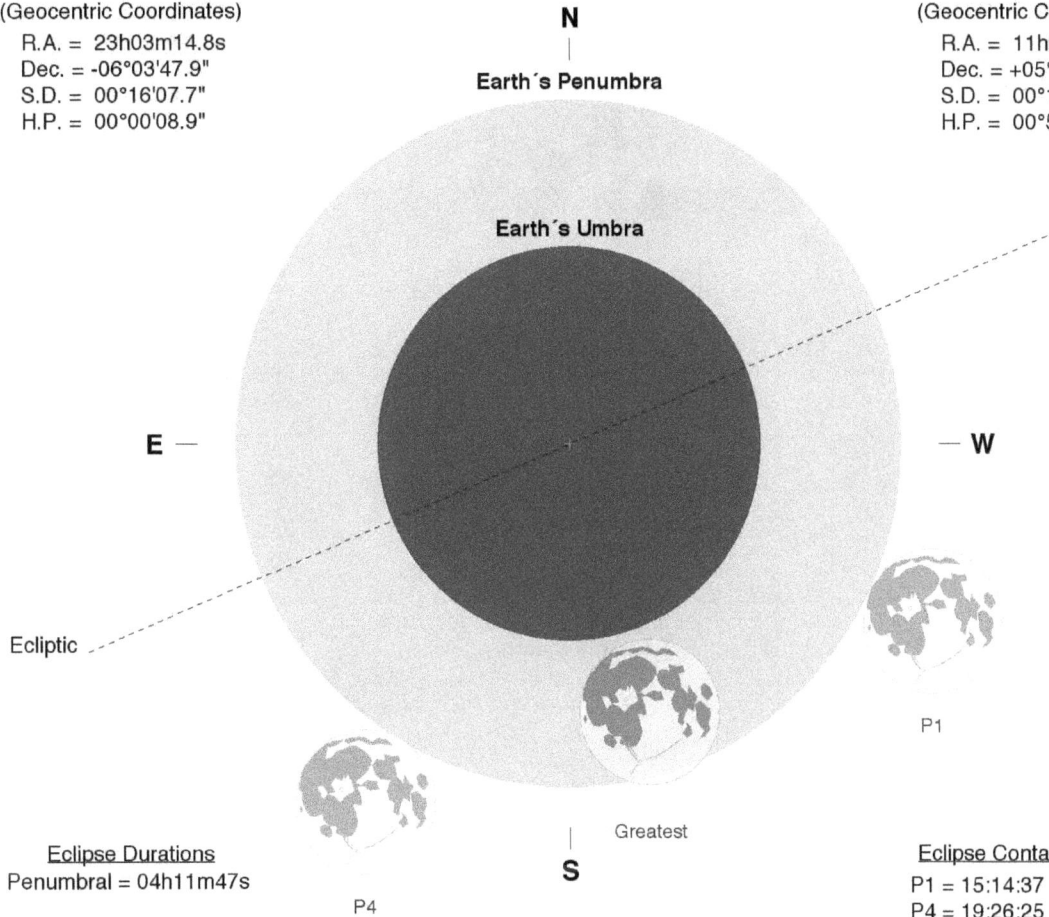

N

Earth´s Penumbra

Earth´s Umbra

E

W

Ecliptic

P1

Greatest

S

P4

Eclipse Durations
Penumbral = 04h11m47s

Eclipse Contacts
P1 = 15:14:37 UT1
P4 = 19:26:25 UT1

Eph. = JPL DE430
Rule = Herald-Sinnott
ΔT = 87 s

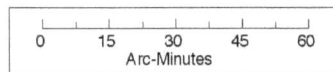

0   15   30   45   60
Arc-Minutes

©2020 F. Espenak, www.EclipseWise.com

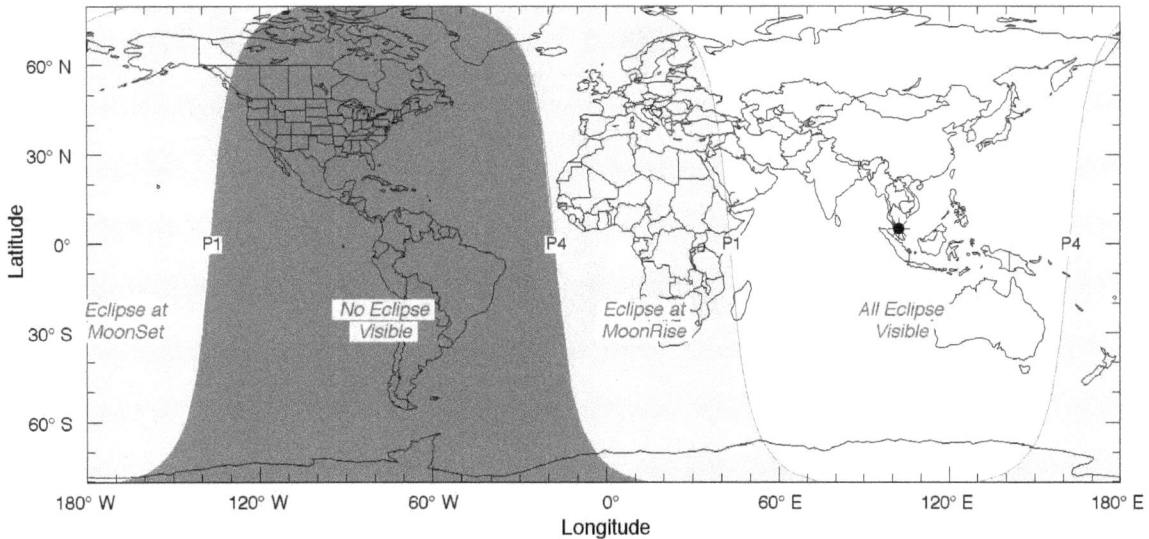

Eclipse at MoonSet

No Eclipse Visible

Eclipse at MoonRise

All Eclipse Visible

P1

P4

P4

Penumbral Lunar Eclipse of 2053 Aug 29

Greatest Eclipse = 08:05:49.8 TD (= 08:04:22.3 UT1)

Penumbral Magnitude = 1.0203	Gamma = 1.0165	Saros Series = 119
Umbral Magnitude = -0.0319	Axis = 0.9358°	Saros Member = 63 of 82

Sun at Greatest Eclipse
(Geocentric Coordinates)
R.A. = 10h32m52.4s
Dec. = +09°08'07.1"
S.D. = 00°15'50.2"
H.P. = 00°00'08.7"

Moon at Greatest Eclipse
(Geocentric Coordinates)
R.A. = 22h31m49.7s
Dec. = -08°14'09.0"
S.D. = 00°15'03.1"
H.P. = 00°55'14.5"

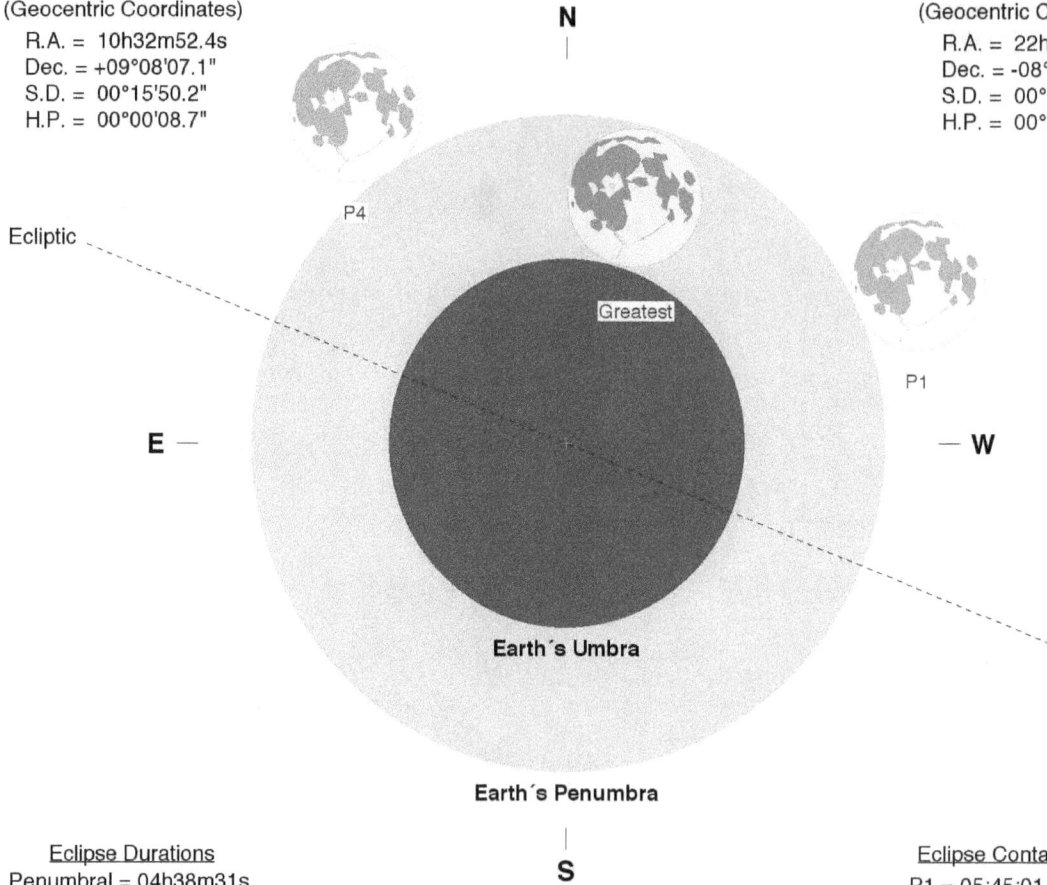

N

Ecliptic

P4

Greatest

P1

E

W

Earth's Umbra

Earth's Penumbra

S

Eclipse Durations
Penumbral = 04h38m31s

Eclipse Contacts
P1 = 05:45:01 UT1
P4 = 10:23:32 UT1

Eph. = JPL DE430
Rule = Herald-Sinnott
ΔT =   87 s

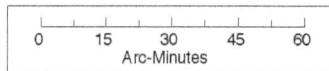

0	15	30	45	60

Arc-Minutes

©2020 F. Espenak, www.EclipseWise.com

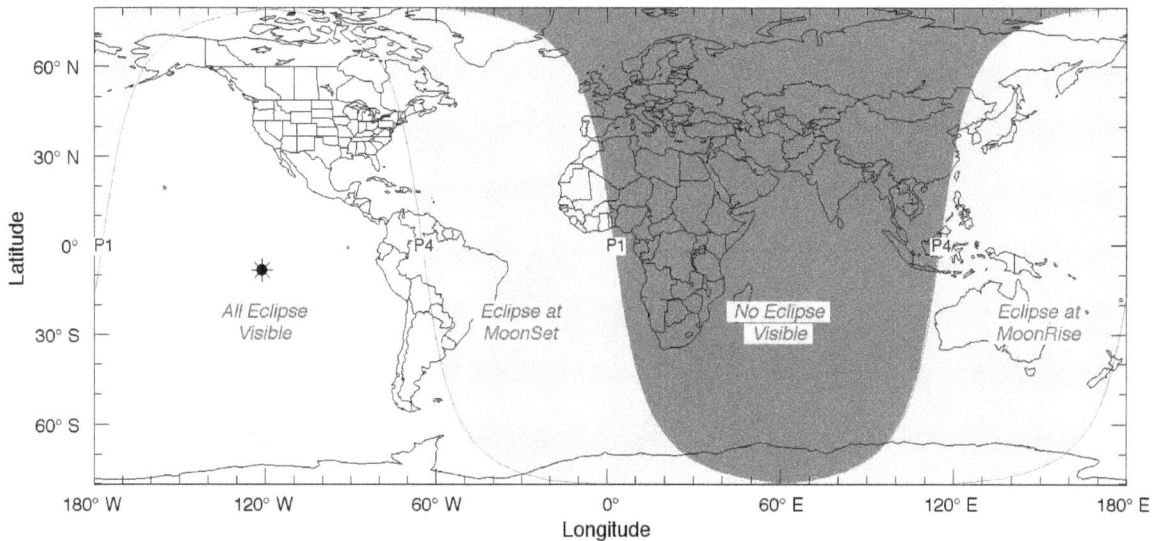

All Eclipse
Visible

Eclipse at
MoonSet

No Eclipse
Visible

Eclipse at
MoonRise

P1

P4

P1

P4

Total Lunar Eclipse of 2054 Feb 22

Greatest Eclipse = 06:51:27.0 TD   (= 06:49:59.1 UT1)

Penumbral Magnitude = 2.2502	Gamma = -0.3242	Saros Series = 124
Umbral Magnitude = 1.2780	Axis = 0.3298°	Saros Member = 51 of 73

Sun at Greatest Eclipse
(Geocentric Coordinates)
R.A. = 22h23m02.4s
Dec. = -10°05'18.4"
S.D. = 00°16'10.2"
H.P. = 00°00'08.9"

Moon at Greatest Eclipse
(Geocentric Coordinates)
R.A. = 10h22m40.9s
Dec. = +09°46'14.8"
S.D. = 00°16'38.0"
H.P. = 01°01'02.7"

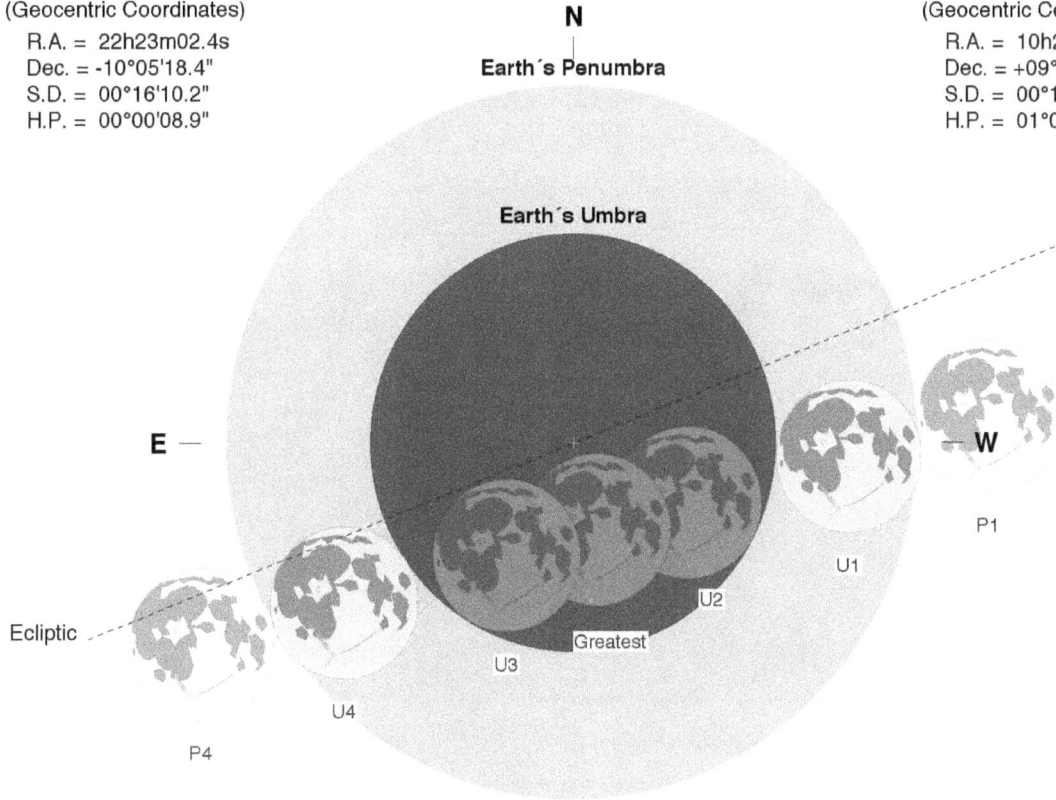

N

Earth´s Penumbra

Earth´s Umbra

E

W

Ecliptic

P1

U1

U2

Greatest

U3

U4

P4

S

Eclipse Durations
Penumbral = 05h15m33s
Umbral = 03h21m36s
Total = 01h12m44s

Eph. = JPL DE430
Rule = Herald-Sinnott
ΔT =   88 s

0	15	30	45	60

Arc-Minutes

©2020 F. Espenak, www.EclipseWise.com

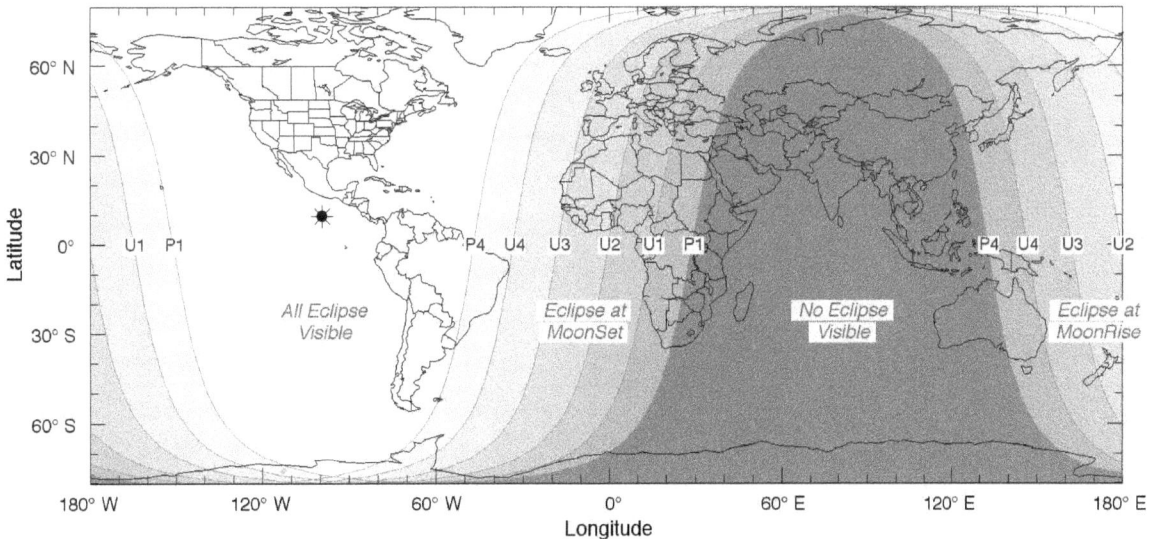

Eclipse Contacts
P1 = 04:12:11 UT1
U1 = 05:09:07 UT1
U2 = 06:13:29 UT1
U3 = 07:26:14 UT1
U4 = 08:30:43 UT1
P4 = 09:27:44 UT1

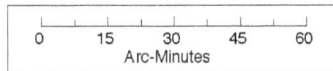

All Eclipse Visible

Eclipse at MoonSet

No Eclipse Visible

Eclipse at MoonRise

Total Lunar Eclipse of 2054 Aug 18

Greatest Eclipse = 09:26:30.1 TD (= 09:25:01.9 UT1)

Penumbral Magnitude = 2.3817	Gamma = 0.2806	Saros Series = 129
Umbral Magnitude = 1.3074	Axis = 0.2524°	Saros Member = 40 of 71

Sun at Greatest Eclipse
(Geocentric Coordinates)
R.A. = 09h51m47.0s
Dec. = +12°57'08.8"
S.D. = 00°15'48.0"
H.P. = 00°00'08.7"

Moon at Greatest Eclipse
(Geocentric Coordinates)
R.A. = 21h51m32.0s
Dec. = -12°42'26.7"
S.D. = 00°14'42.4"
H.P. = 00°53'58.4"

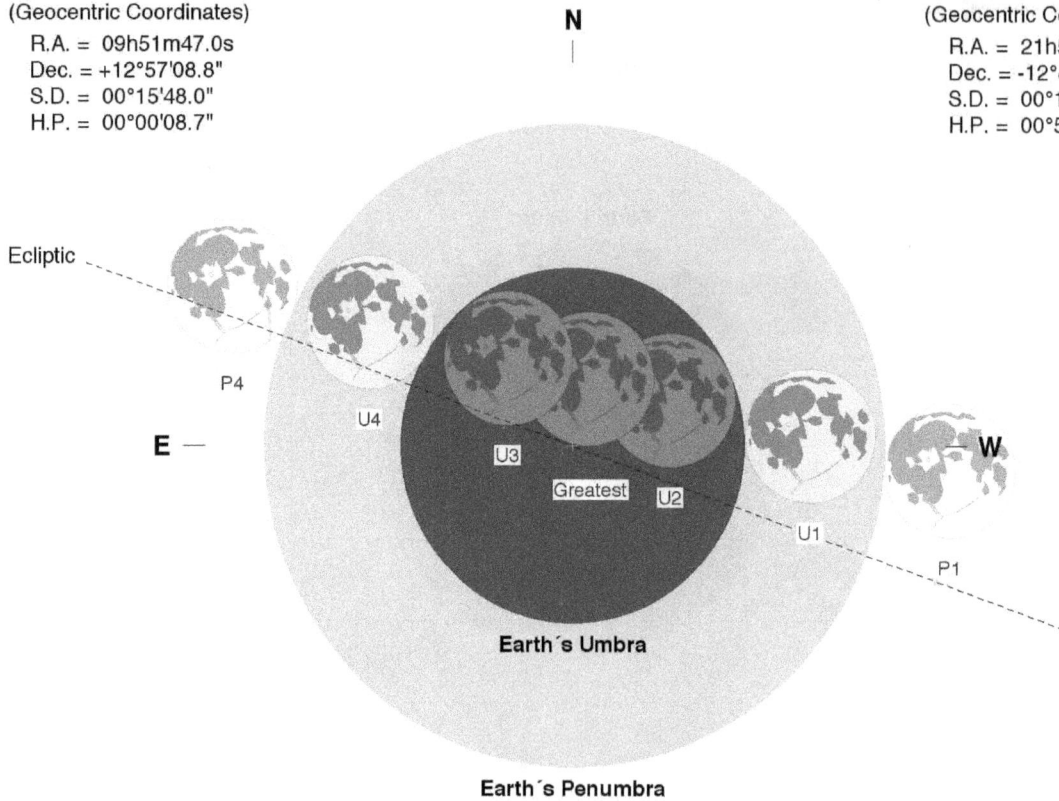

N

Ecliptic

P4
U4
E —
U3
Greatest
U2
U1
P1
— W

Earth's Umbra

Earth's Penumbra

S

Eclipse Durations
Penumbral = 06h10m22s
Umbral = 03h47m21s
Total = 01h23m40s

Eph. = JPL DE430
Rule = Herald-Sinnott
ΔT =   88 s

0	15	30	45	60
Arc-Minutes

©2020 F. Espenak, www.EclipseWise.com

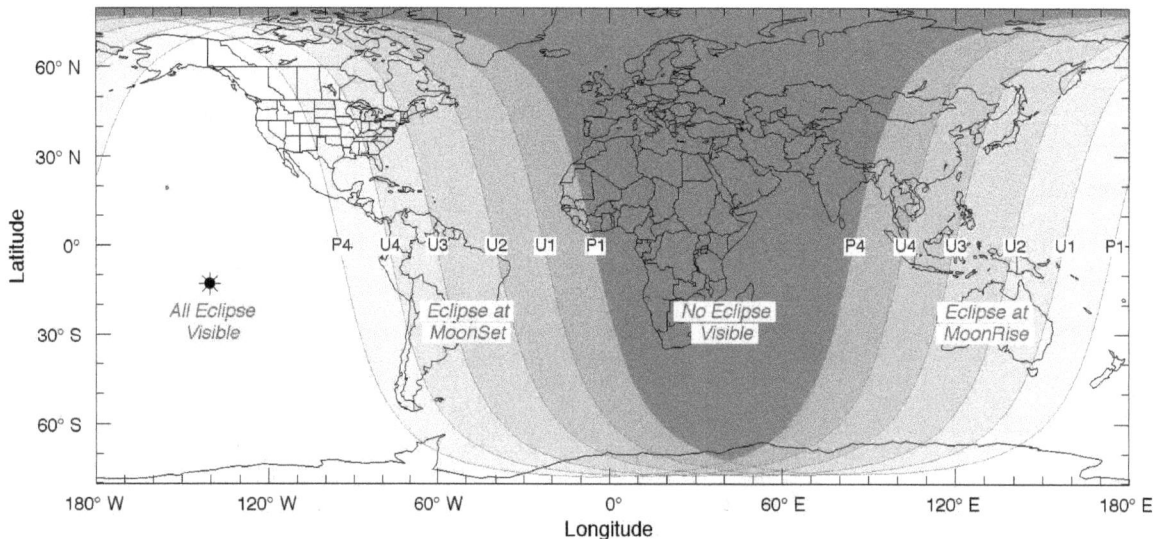

Eclipse Contacts
P1 = 06:19:49 UT1
U1 = 07:31:18 UT1
U2 = 08:43:05 UT1
U3 = 10:06:45 UT1
U4 = 11:18:39 UT1
P4 = 12:30:11 UT1

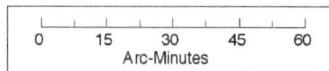

All Eclipse Visible
Eclipse at MoonSet
No Eclipse Visible
Eclipse at MoonRise

Total Lunar Eclipse of 2055 Feb 11

Greatest Eclipse = 22:46:17.5 TD   (= 22:44:49.0 UT1)

Penumbral Magnitude = 2.1982	Gamma = 0.3526	Saros Series = 134
Umbral Magnitude = 1.2258	Axis = 0.3594°	Saros Member = 29 of 72

<u>Sun at Greatest Eclipse</u>
(Geocentric Coordinates)

R.A. = 21h42m03.5s
Dec. = -13°47'10.8"
S.D. = 00°16'12.3"
H.P. = 00°00'08.9"

<u>Moon at Greatest Eclipse</u>
(Geocentric Coordinates)

R.A. = 09h42m24.2s
Dec. = +14°08'09.1"
S.D. = 00°16'39.9"
H.P. = 01°01'09.9"

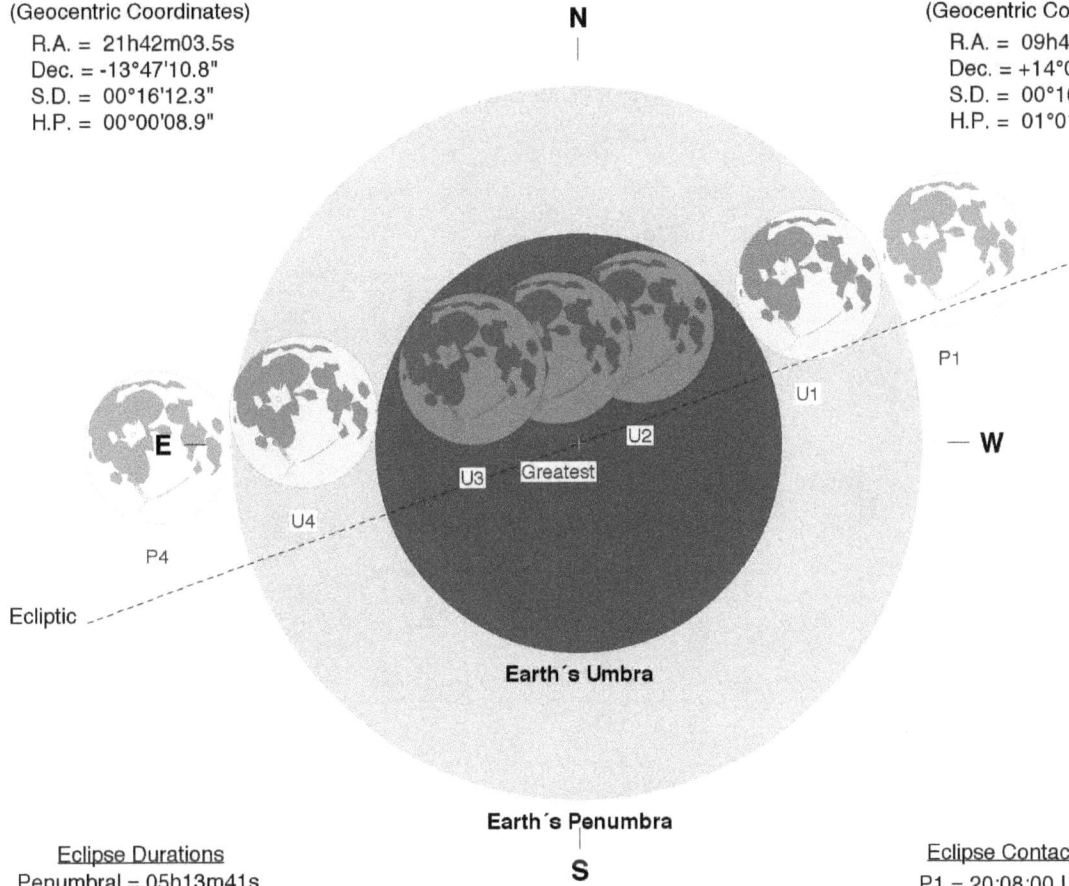

N

E — W

P1

U1

U2

U3 Greatest

U4

P4

Ecliptic

Earth's Umbra

Earth's Penumbra

S

<u>Eclipse Durations</u>
Penumbral = 05h13m41s
Umbral = 03h19m09s
Total = 01h06m35s

Eph. = JPL DE430
Rule = Herald-Sinnott
ΔT =   88 s

0   15   30   45   60
Arc-Minutes

©2020 F. Espenak, www.EclipseWise.com

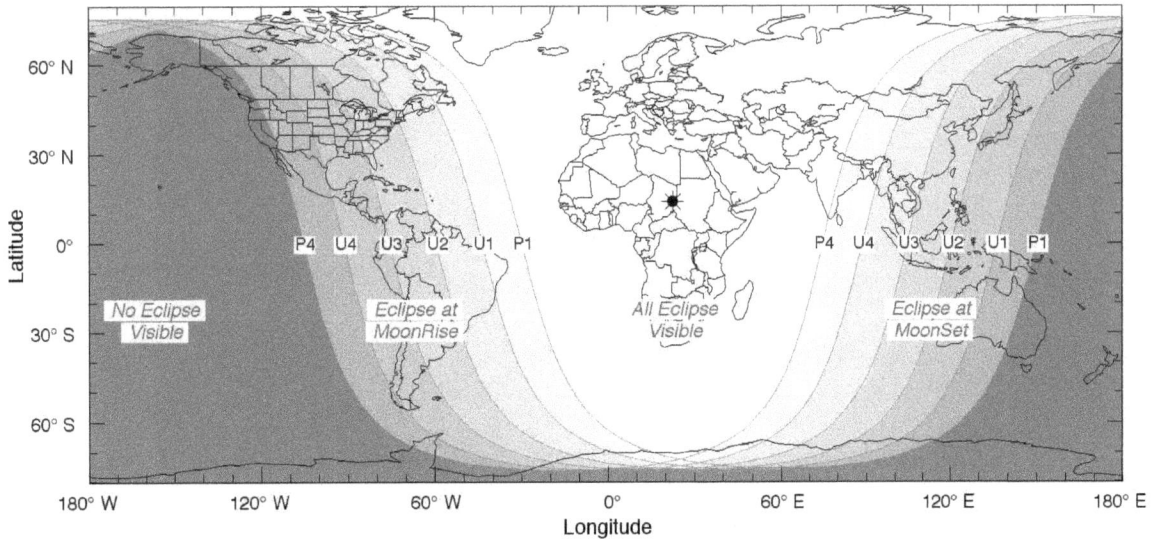

<u>Eclipse Contacts</u>
P1 = 20:08:00 UT1
U1 = 21:05:18 UT1
U2 = 22:11:39 UT1
U3 = 23:18:14 UT1
U4 = 00:24:27 UT1
P4 = 01:21:40 UT1

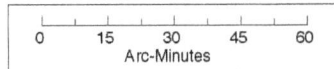

No Eclipse
Visible

Eclipse at
MoonRise

All Eclipse
Visible

Eclipse at
MoonSet

P4 U4 U3 U2 U1 P1   P4 U4 U3 U2 U1 P1

Latitude
60° N
30° N
0°
30° S
60° S

180° W   120° W   60° W   0°   60° E   120° E   180° E
Longitude

Partial Lunar Eclipse of 2055 Aug 07

Greatest Eclipse = 10:53:18.2 TD (= 10:51:49.4 UT1)

Penumbral Magnitude = 2.0081	Gamma = -0.4769	Saros Series = 139
Umbral Magnitude = 0.9606	Axis = 0.4391°	Saros Member = 23 of 79

Sun at Greatest Eclipse
(Geocentric Coordinates)
R.A. = 09h09m39.9s
Dec. = +16°20'36.9"
S.D. = 00°15'46.2"
H.P. = 00°00'08.7"

Moon at Greatest Eclipse
(Geocentric Coordinates)
R.A. = 21h10m01.5s
Dec. = -16°46'26.9"
S.D. = 00°15'03.3"
H.P. = 00°55'15.1"

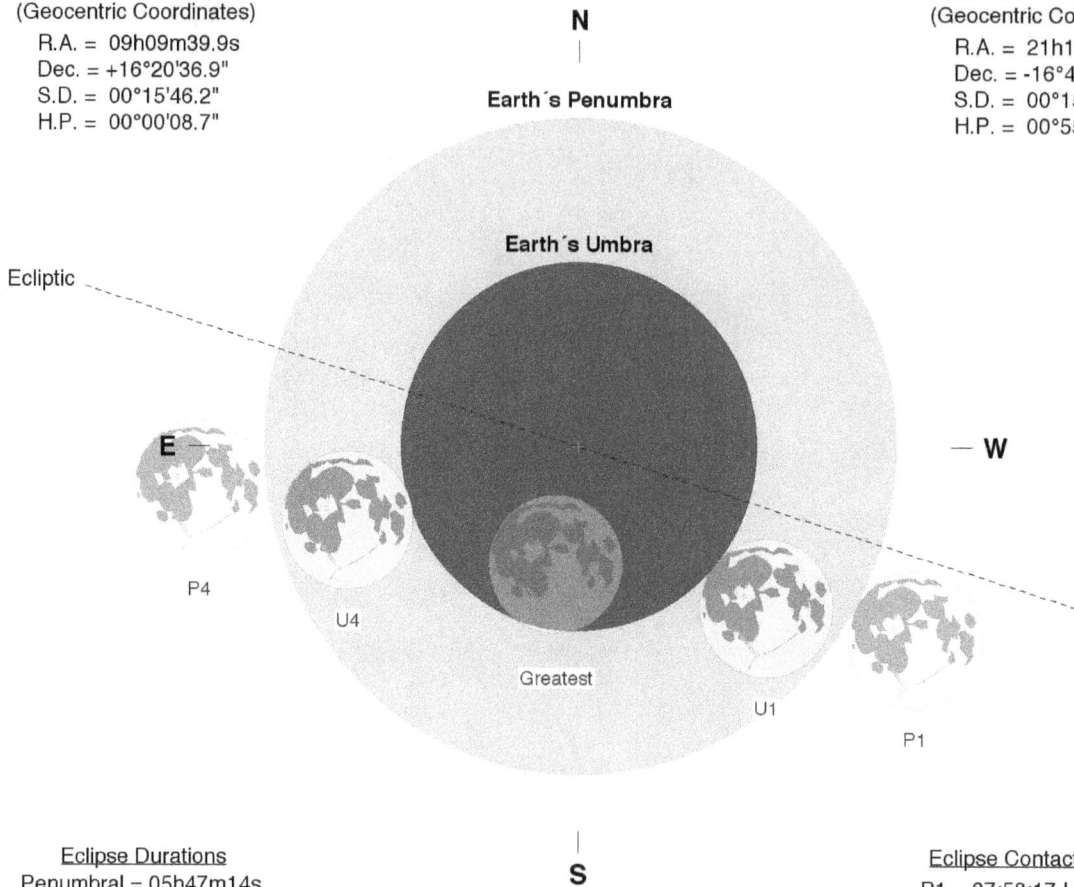

N

Earth's Penumbra

Earth's Umbra

Ecliptic

E

W

P4

U4

Greatest

U1

P1

S

Eclipse Durations
Penumbral = 05h47m14s
Umbral = 03h24m10s

Eph. = JPL DE430
Rule = Herald-Sinnott
ΔT =   89 s

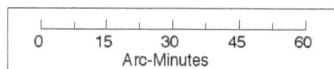

0	15	30	45	60

Arc-Minutes

©2020 F. Espenak, www.EclipseWise.com

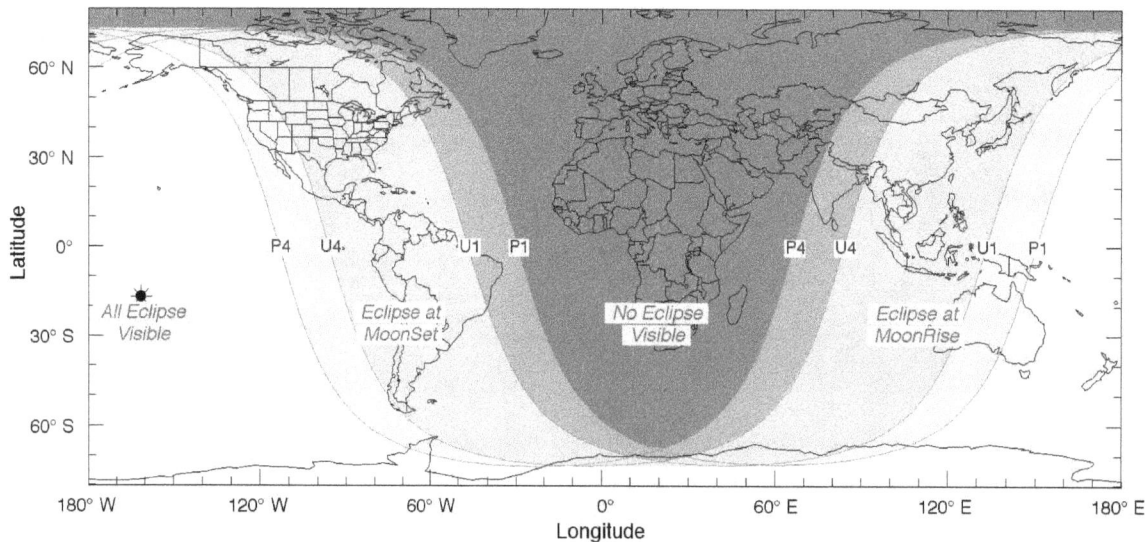

Eclipse Contacts
P1 = 07:58:17 UT1
U1 = 09:09:47 UT1
U4 = 12:33:56 UT1
P4 = 13:45:31 UT1

All Eclipse
Visible

Eclipse at
MoonSet

No Eclipse
Visible

Eclipse at
MoonRise

P4 U4   U1 P1   P4 U4   U1 P1

Penumbral Lunar Eclipse of 2056 Feb 01

Greatest Eclipse = 12:26:06.2 TD (= 12:24:37.1 UT1)

Penumbral Magnitude = 0.9069	Gamma = 1.0682	Saros Series = 144
Umbral Magnitude = -0.1084	Axis = 1.0447°	Saros Member = 18 of 71

Sun at Greatest Eclipse
(Geocentric Coordinates)
R.A. = 20h59m18.0s
Dec. = -17°05'10.5"
S.D. = 00°16'14.0"
H.P. = 00°00'08.9"

Moon at Greatest Eclipse
(Geocentric Coordinates)
R.A. = 09h00m07.0s
Dec. = +18°06'45.8"
S.D. = 00°15'59.4"
H.P. = 00°58'41.2"

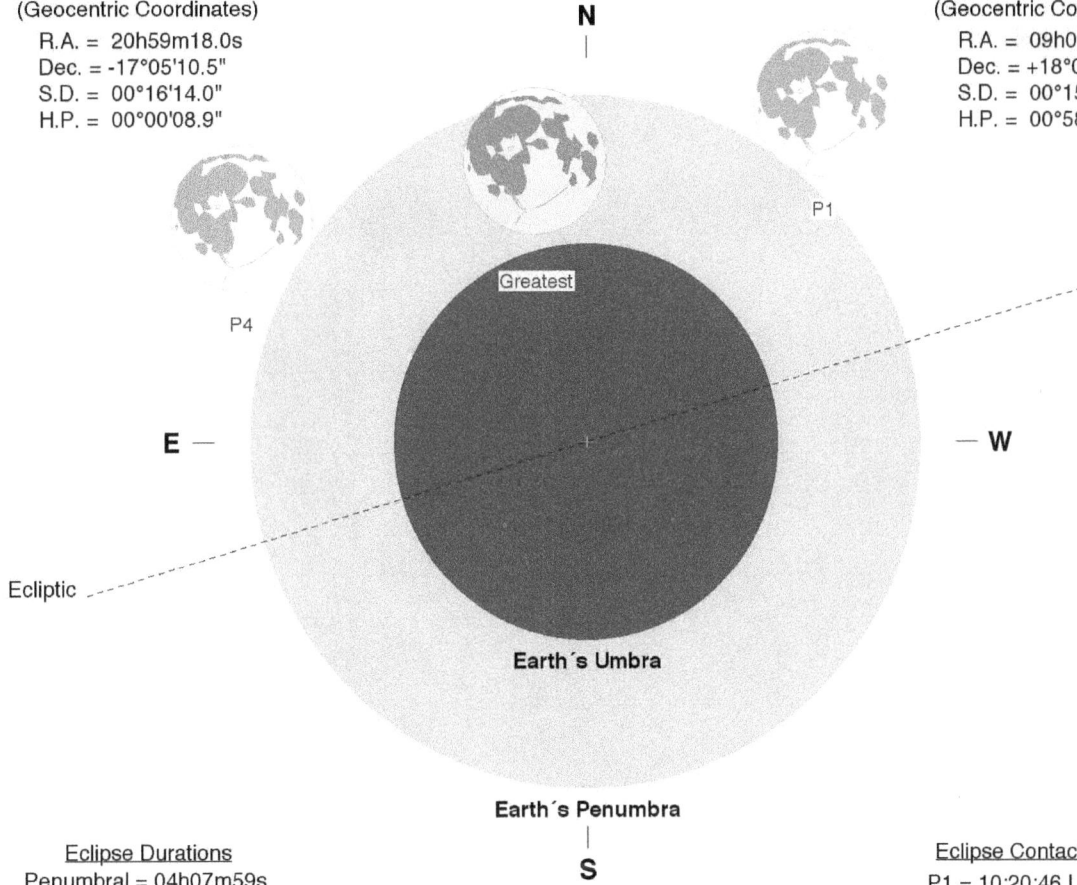

N

Greatest

P1

P4

E —

— W

Ecliptic

Earth's Umbra

Earth's Penumbra

S

Eclipse Durations
Penumbral = 04h07m59s

Eclipse Contacts
P1 = 10:20:46 UT1
P4 = 14:28:44 UT1

Eph. = JPL DE430
Rule = Herald-Sinnott
ΔT =   89 s

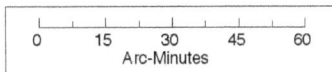

0   15   30   45   60
Arc-Minutes

©2020 F. Espenak, www.EclipseWise.com

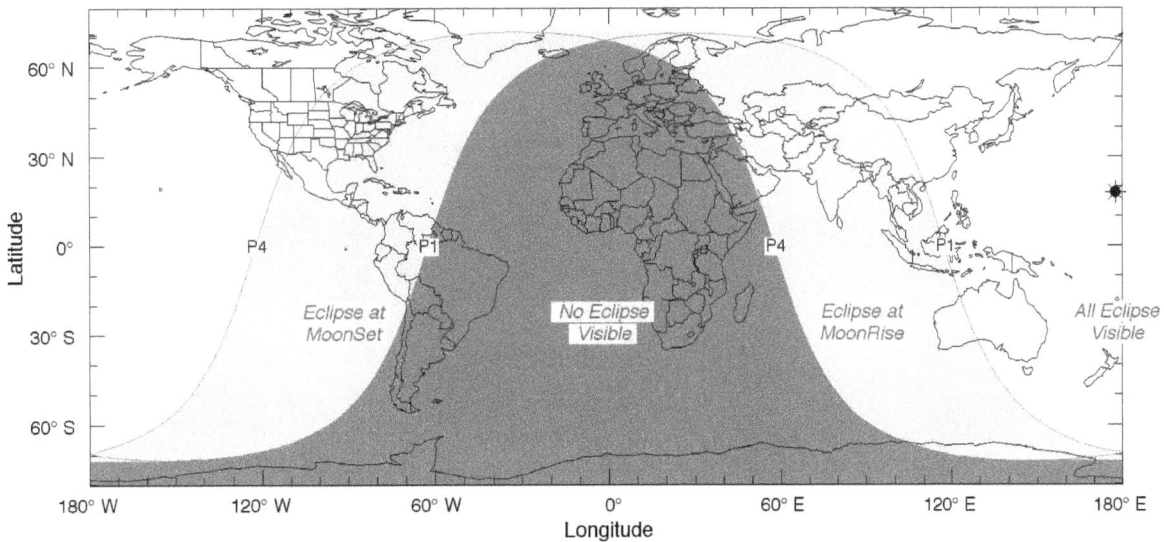

60° N

30° N

Latitude 0°

30° S

60° S

P4   P1   P4   P1

Eclipse at
MoonSet

No Eclipse
Visible

Eclipse at
MoonRise

All Eclipse
Visible

180° W   120° W   60° W   0°   60° E   120° E   180° E
Longitude

Penumbral Lunar Eclipse of 2056 Jun 27

Greatest Eclipse = 10:03:09.3 TD   (= 10:01:39.9 UT1)

Penumbral Magnitude = 0.3158	Gamma = 1.3770	Saros Series = 111
Umbral Magnitude = -0.6504	Axis = 1.3714°	Saros Member = 69 of 71

Sun at Greatest Eclipse
(Geocentric Coordinates)
R.A. = 06h28m18.7s
Dec. = +23°16'17.2"
S.D. = 00°15'44.0"
H.P. = 00°00'08.7"

N

Moon at Greatest Eclipse
(Geocentric Coordinates)
R.A. = 18h28m35.0s
Dec. = -21°54'04.7"
S.D. = 00°16'17.0"
H.P. = 00°59'45.8"

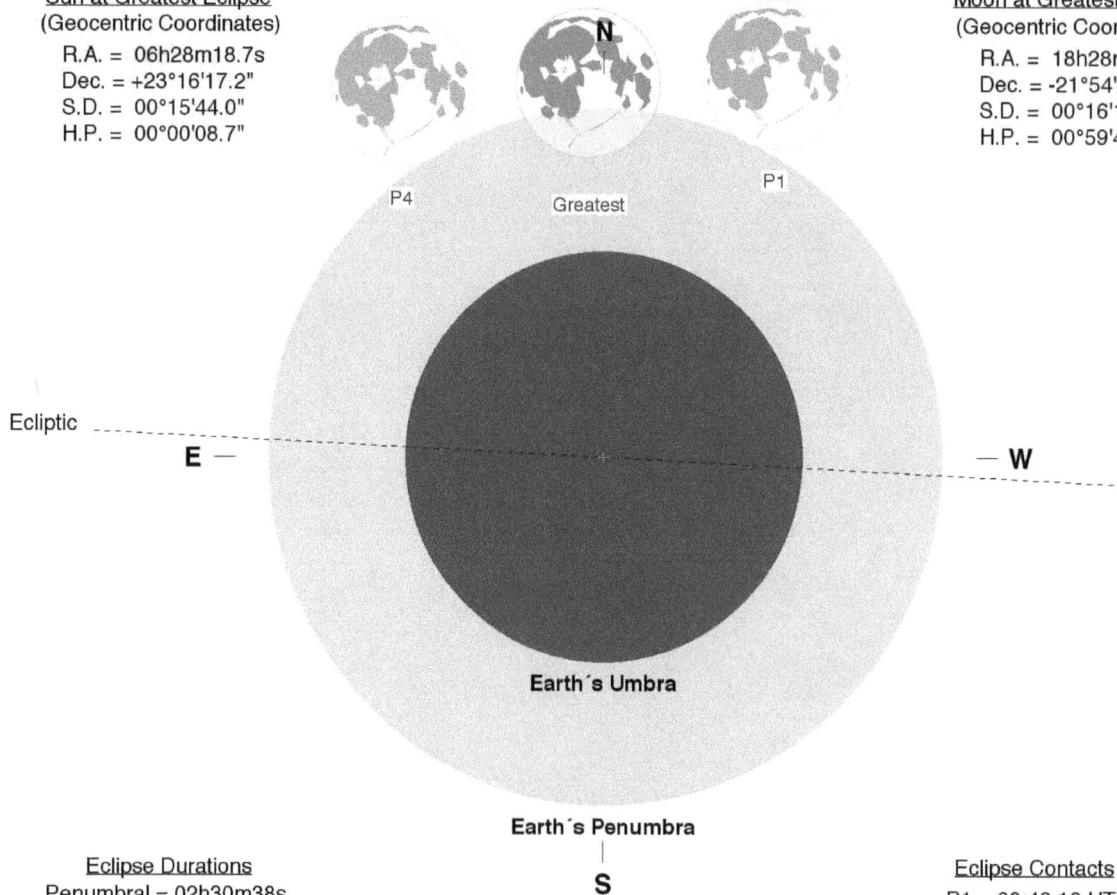

P4   Greatest   P1

Ecliptic

E —   — W

Earth's Umbra

Earth's Penumbra

S

Eclipse Durations
Penumbral = 02h30m38s

Eclipse Contacts
P1 = 08:46:16 UT1
P4 = 11:16:54 UT1

Eph. = JPL DE430
Rule = Herald-Sinnott
ΔT = 89 s

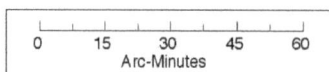

0	15	30	45	60

Arc-Minutes

©2020 F. Espenak, www.EclipseWise.com

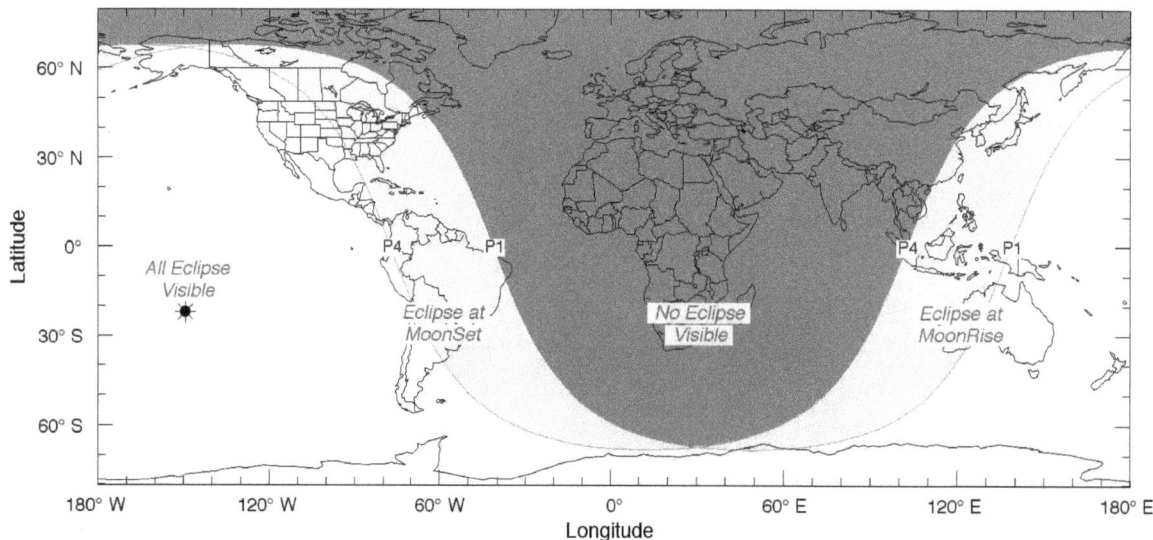

All Eclipse Visible

P4   P1   No Eclipse Visible   P4   P1

Eclipse at MoonSet   Eclipse at MoonRise

Longitude

Penumbral Lunar Eclipse of 2056 Jul 26

Greatest Eclipse = 18:43:24.9 TD (= 18:41:55.4 UT1)

Penumbral Magnitude = 0.6448	Gamma = -1.2048	Saros Series = 149
Umbral Magnitude = -0.3477	Axis = 1.1693°	Saros Member = 5 of 71

Sun at Greatest Eclipse
(Geocentric Coordinates)

R.A. = 08h27m22.5s
Dec. = +19°07'40.1"
S.D. = 00°15'44.9"
H.P. = 00°00'08.7"

Moon at Greatest Eclipse
(Geocentric Coordinates)

R.A. = 20h28m05.5s
Dec. = -20°17'05.8"
S.D. = 00°15'52.1"
H.P. = 00°58'14.4"

N

Earth´s Penumbra

Earth´s Umbra

Ecliptic

E

W

P4

Greatest S

P1

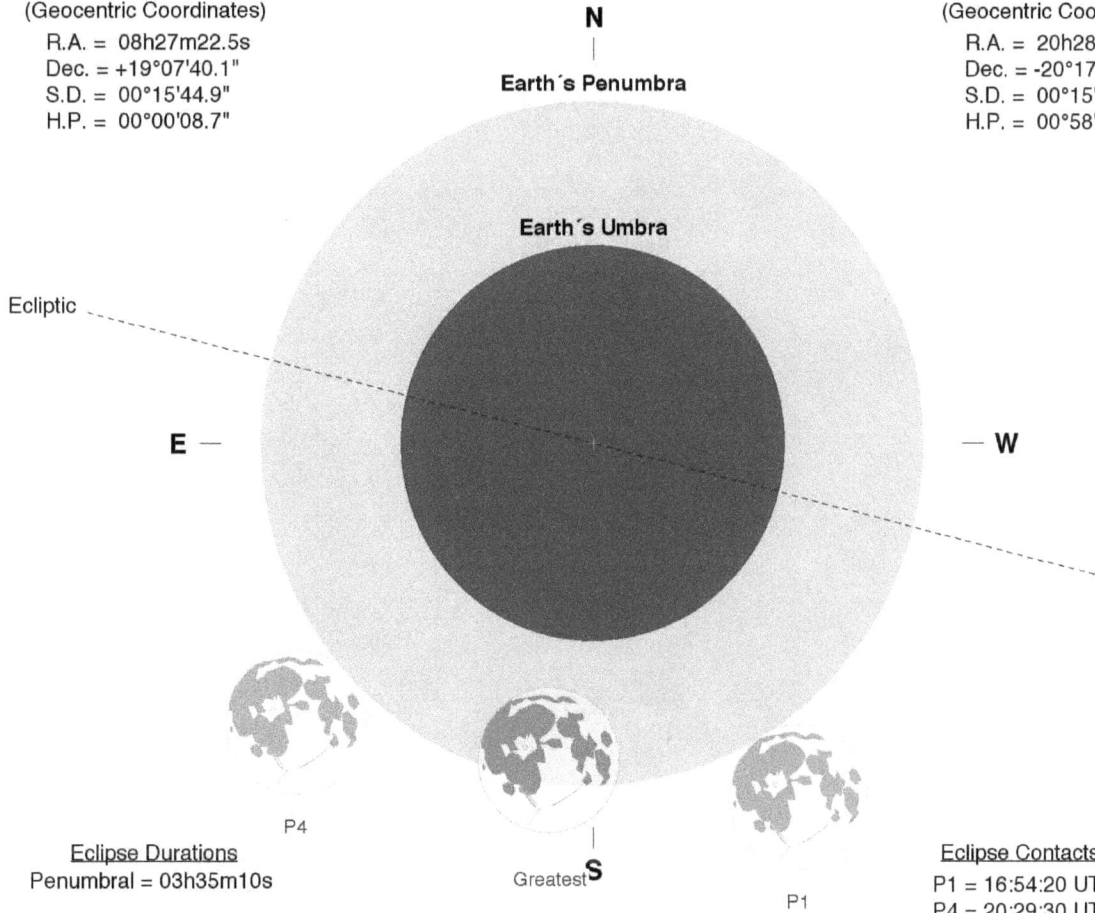

Eclipse Durations
Penumbral = 03h35m10s

Eclipse Contacts
P1 = 16:54:20 UT1
P4 = 20:29:30 UT1

Eph. = JPL DE430
Rule = Herald-Sinnott
ΔT = 89 s

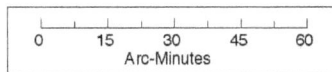

0   15   30   45   60
Arc-Minutes

©2020 F. Espenak, www.EclipseWise.com

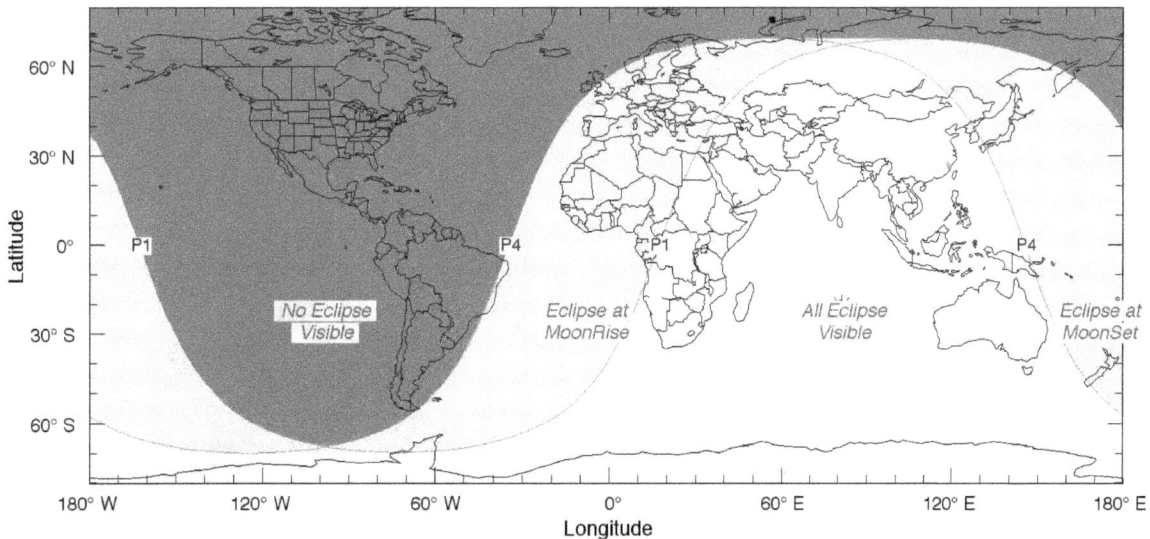

No Eclipse Visible

Eclipse at MoonRise

All Eclipse Visible

Eclipse at MoonSet

P1

P4

P1

P4

Penumbral Lunar Eclipse of 2056 Dec 22

Greatest Eclipse = 01:48:56.0 TD   (= 01:47:26.2 UT1)

Penumbral Magnitude = 0.7872	Gamma = -1.1560	Saros Series = 116
Umbral Magnitude = -0.3093	Axis = 1.0483°	Saros Member = 60 of 73

Sun at Greatest Eclipse
(Geocentric Coordinates)
R.A. = 18h04m03.3s
Dec. = -23°25'40.9"
S.D. = 00°16'15.5"
H.P. = 00°00'08.9"

N

Earth's Penumbra

Earth's Umbra

Moon at Greatest Eclipse
(Geocentric Coordinates)
R.A. = 06h04m28.2s
Dec. = +22°23'02.6"
S.D. = 00°14'49.6"
H.P. = 00°54'24.9"

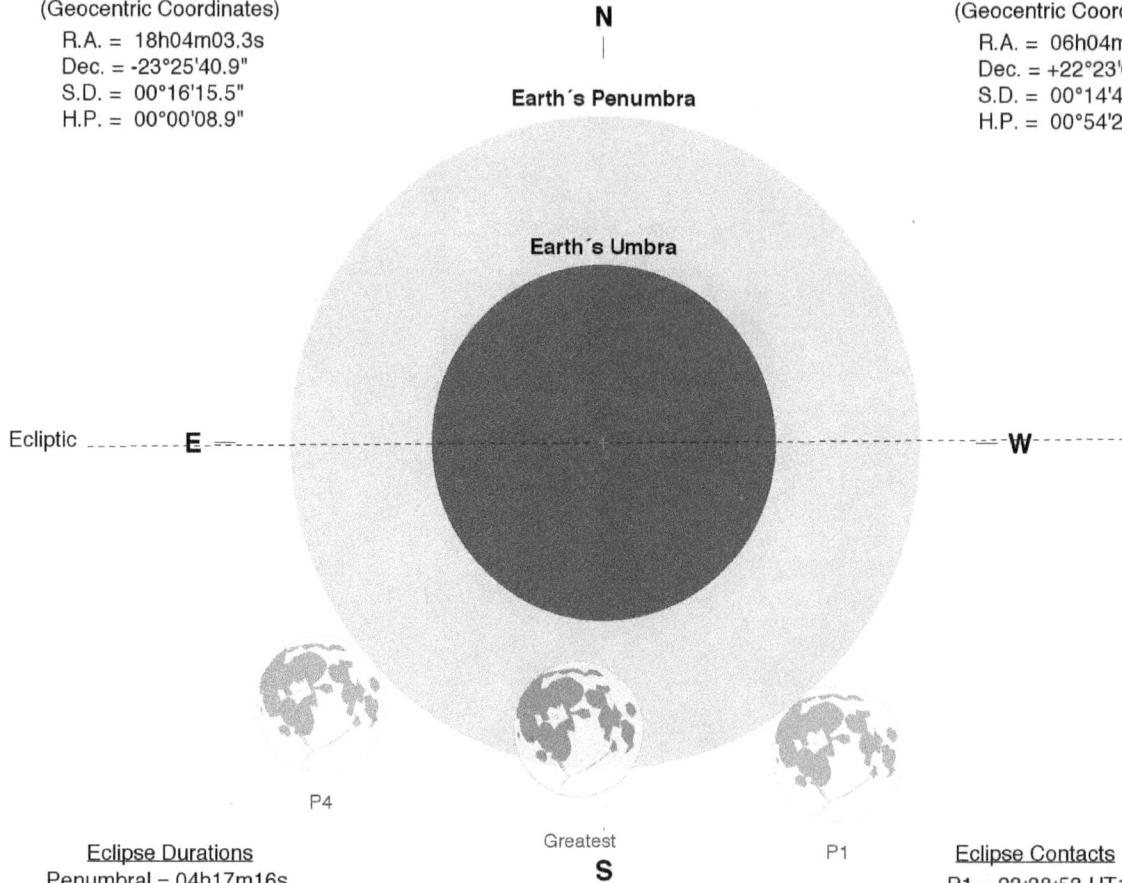

Ecliptic ---- **E** --------------------------------- **W** ----

P4

Greatest
S

P1

Eclipse Durations
Penumbral = 04h17m16s

Eclipse Contacts
P1 = 23:38:53 UT1
P4 = 03:56:09 UT1

Eph. = JPL DE430
Rule = Herald-Sinnott
ΔT =   90 s

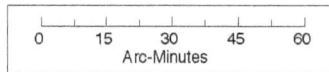

0	15	30	45	60

Arc-Minutes

©2020 F. Espenak, www.EclipseWise.com

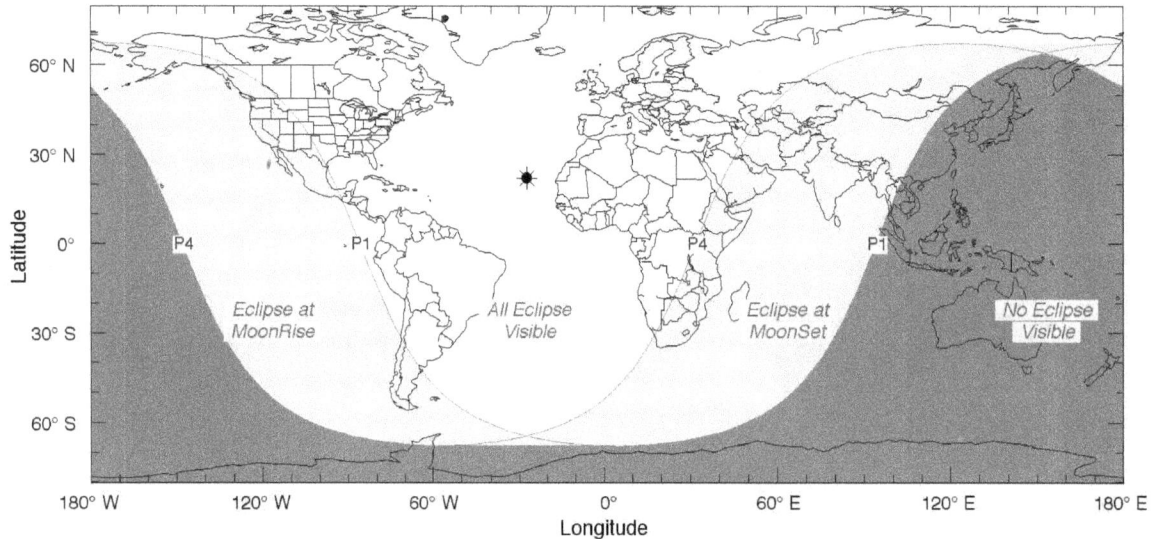

P4   P1   P4   P1

Eclipse at
MoonRise

All Eclipse
Visible

Eclipse at
MoonSet

No Eclipse
Visible

Latitude / Longitude

Partial Lunar Eclipse of 2057 Jun 17

Greatest Eclipse = 02:26:20.1 TD   (= 02:24:50.0 UT1)

Penumbral Magnitude = 1.6982	Gamma = 0.6168	Saros Series = 121
Umbral Magnitude = 0.7570	Axis = 0.6310°	Saros Member = 57 of 82

Sun at Greatest Eclipse
(Geocentric Coordinates)

R.A. = 05h44m25.1s
Dec. = +23°22'59.4"
S.D. = 00°15'44.7"
H.P. = 00°00'08.7"

Moon at Greatest Eclipse
(Geocentric Coordinates)

R.A. = 17h44m45.6s
Dec. = -22°45'25.3"
S.D. = 00°16'43.7"
H.P. = 01°01'23.6"

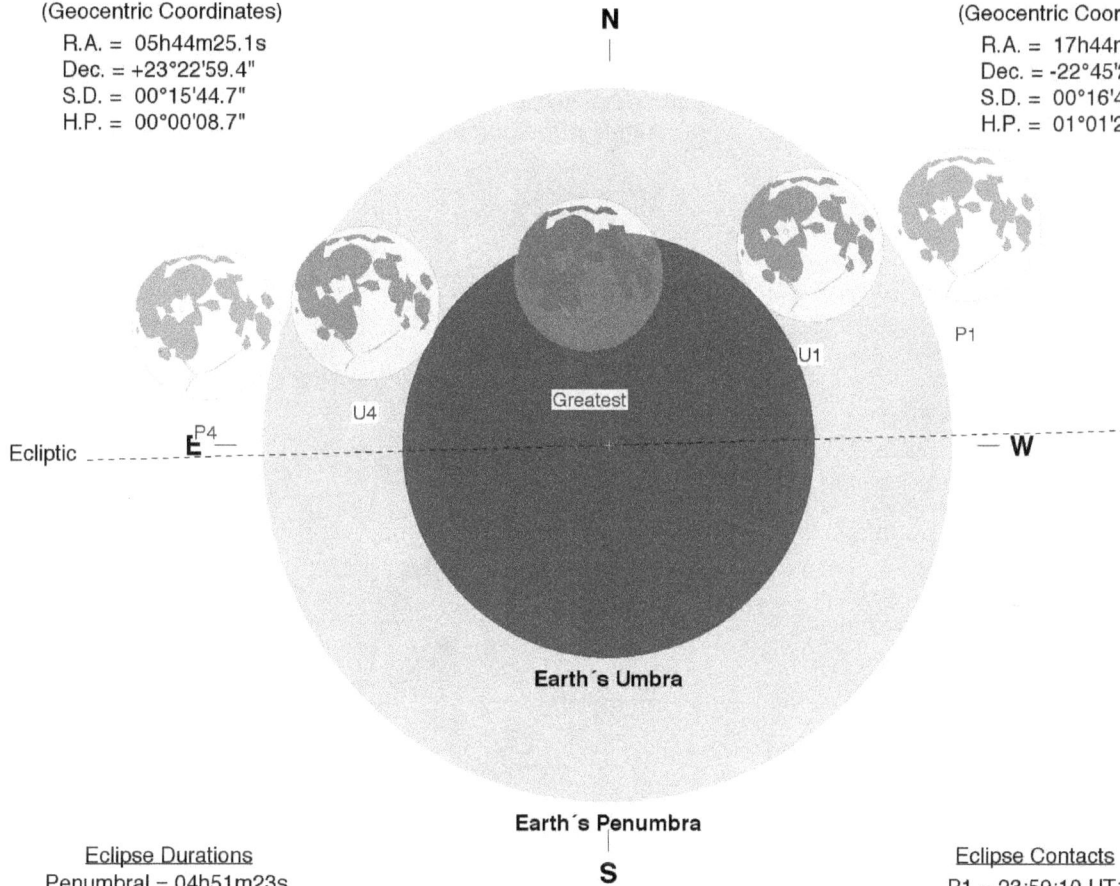

N

P1

U1

Greatest

U4

P4

Ecliptic

E

W

Earth's Umbra

Earth's Penumbra

S

Eclipse Durations
Penumbral = 04h51m23s
Umbral = 02h50m00s

Eph. = JPL DE430
Rule = Herald-Sinnott
ΔT =   90 s

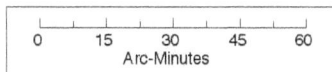

0	15	30	45	60

Arc-Minutes

©2020 F. Espenak, www.EclipseWise.com

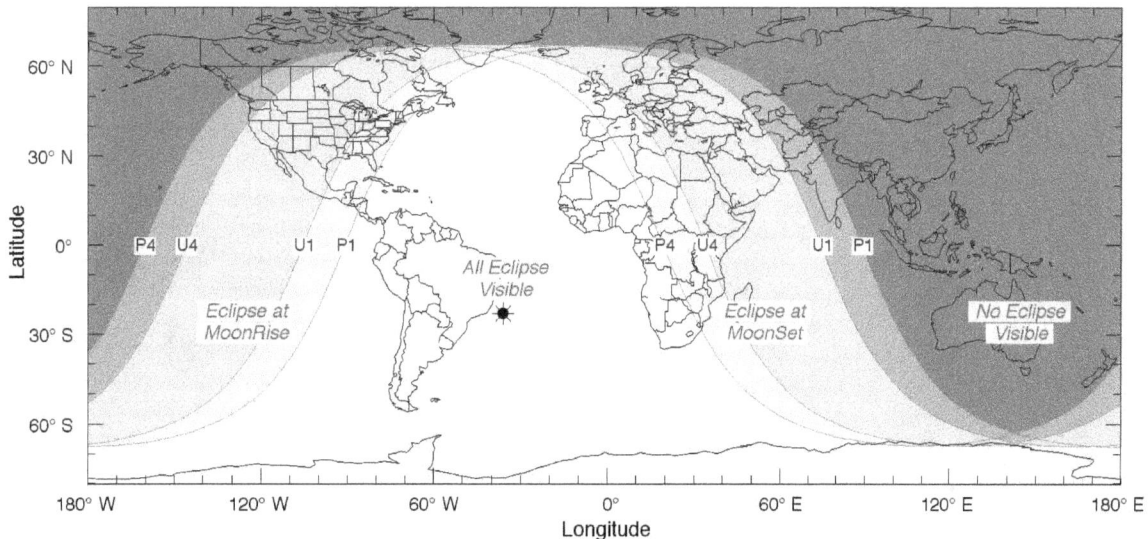

Eclipse Contacts
P1 = 23:59:10 UT1
U1 = 00:59:52 UT1
U4 = 03:49:53 UT1
P4 = 04:50:33 UT1

All Eclipse
Visible

Eclipse at
MoonRise

Eclipse at
MoonSet

No Eclipse
Visible

P4 U4   U1 P1   P4 U4   U1 P1

Partial Lunar Eclipse of 2057 Dec 11

Greatest Eclipse = 00:53:38.0 TD   (= 00:52:07.6 UT1)

Penumbral Magnitude = 2.0194	Gamma = -0.4853	Saros Series = 126
Umbral Magnitude = 0.9197	Axis = 0.4384°	Saros Member = 47 of 70

Sun at Greatest Eclipse
(Geocentric Coordinates)

R.A. = 17h14m07.3s
Dec. = -23°00'46.9"
S.D. = 00°16'14.5"
H.P. = 00°00'08.9"

N

Earth´s Penumbra

Earth´s Umbra

Ecliptic

E

W

P4

U4

Greatest

U1

P1

Moon at Greatest Eclipse
(Geocentric Coordinates)

R.A. = 05h14m27.7s
Dec. = +22°34'54.1"
S.D. = 00°14'46.1"
H.P. = 00°54'12.2"

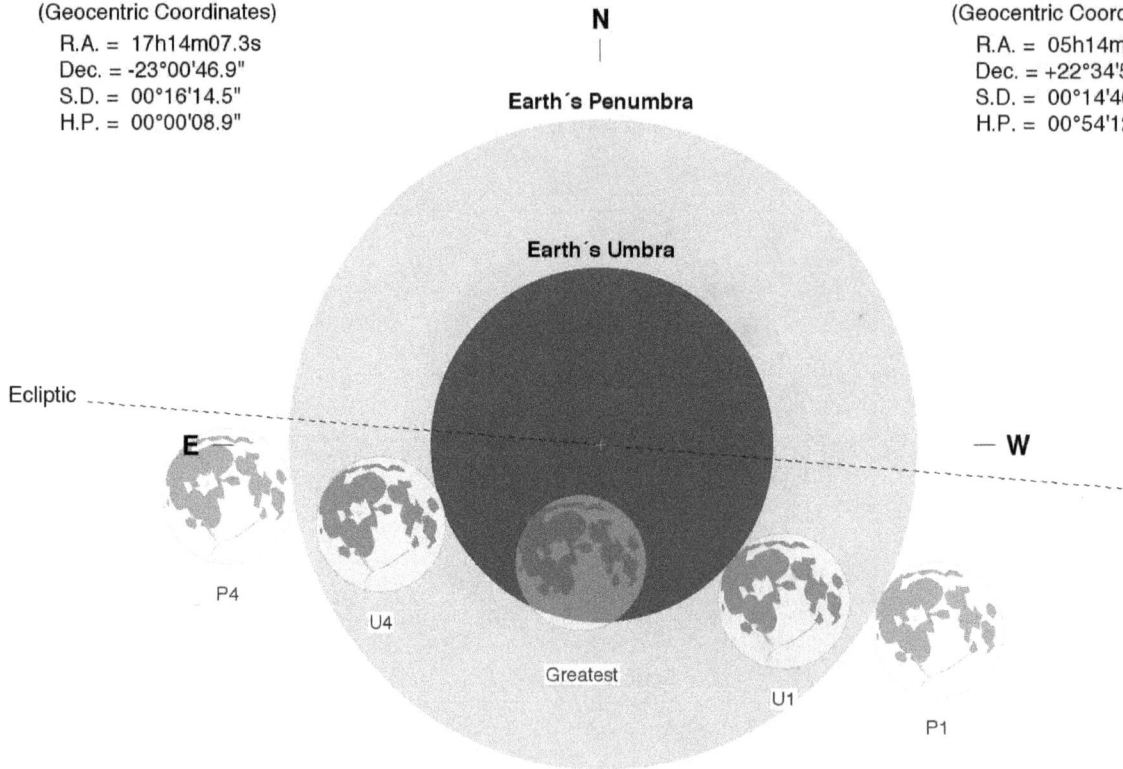

S

Eclipse Durations
Penumbral = 05h59m42s
Umbral = 03h24m50s

Eph. = JPL DE430
Rule = Herald-Sinnott
ΔT =   90 s

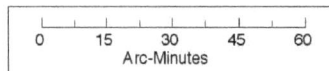

0	15	30	45	60

Arc-Minutes

©2020 F. Espenak, www.EclipseWise.com

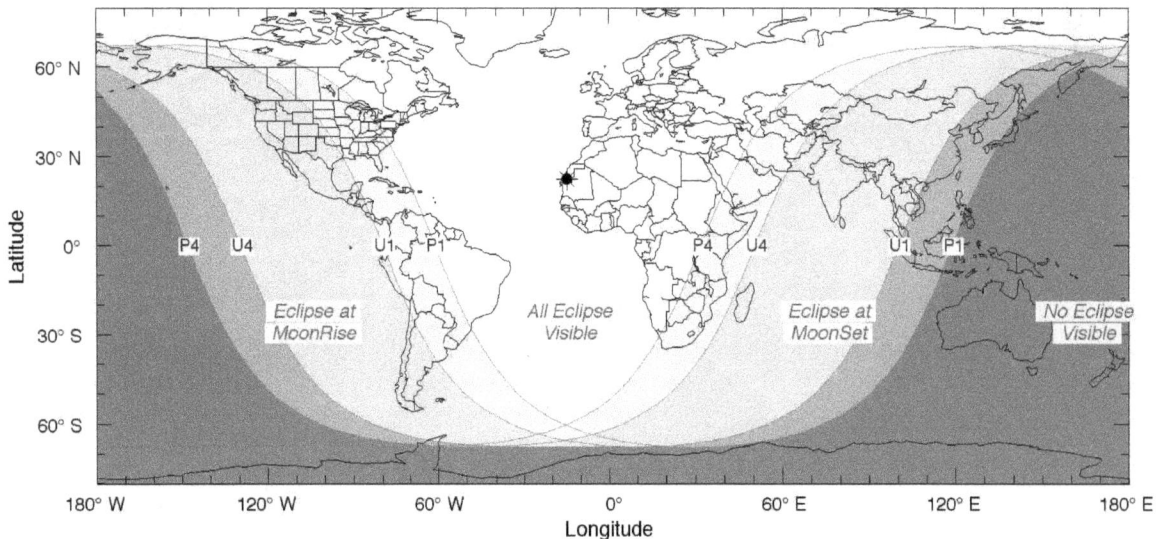

Eclipse Contacts
P1 = 21:52:20 UT1
U1 = 23:09:45 UT1
U4 = 02:34:35 UT1
P4 = 03:52:02 UT1

P4   U4   U1   P1   P4   U4   U1   P1

Eclipse at
MoonRise

All Eclipse
Visible

Eclipse at
MoonSet

No Eclipse
Visible

Total Lunar Eclipse of 2058 Jun 06

Greatest Eclipse = 19:15:48.0 TD   (= 19:14:17.3 UT1)

Penumbral Magnitude = 2.6226   Gamma = -0.1181   Saros Series = 131
Umbral Magnitude = 1.6628   Axis = 0.1186°   Saros Member = 36 of 72

<u>Sun at Greatest Eclipse</u>
(Geocentric Coordinates)
R.A. = 05h00m41.7s
Dec. = +22°43'57.0"
S.D. = 00°15'45.8"
H.P. = 00°00'08.7"

<u>Moon at Greatest Eclipse</u>
(Geocentric Coordinates)
R.A. = 17h00m35.5s
Dec. = -22°50'55.4"
S.D. = 00°16'25.3"
H.P. = 01°00'16.2"

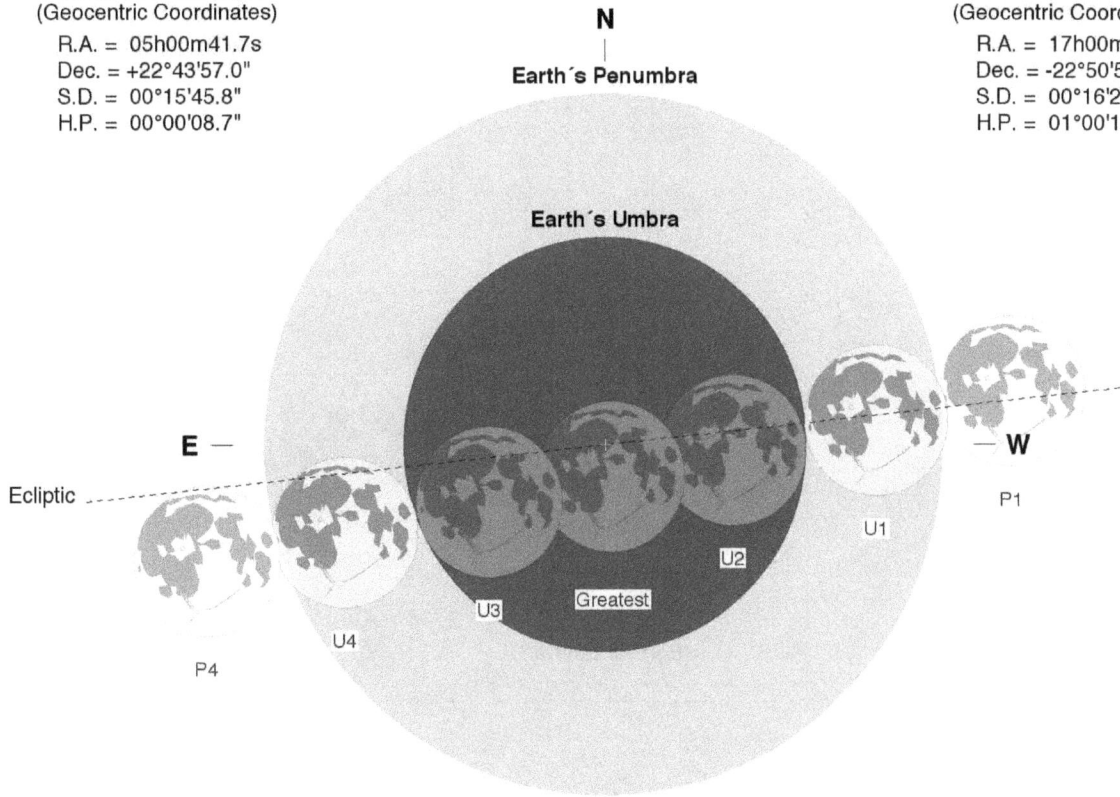

N

Earth's Penumbra

Earth's Umbra

E
W

Ecliptic

P1

U1

U2

Greatest

U3

U4

P4

S

<u>Eclipse Durations</u>
Penumbral = 05h24m28s
Umbral = 03h34m09s
Total = 01h38m03s

Eph. = JPL DE430
Rule = Herald-Sinnott
ΔT = 91 s

0   15   30   45   60
Arc-Minutes

©2020 F. Espenak, www.EclipseWise.com

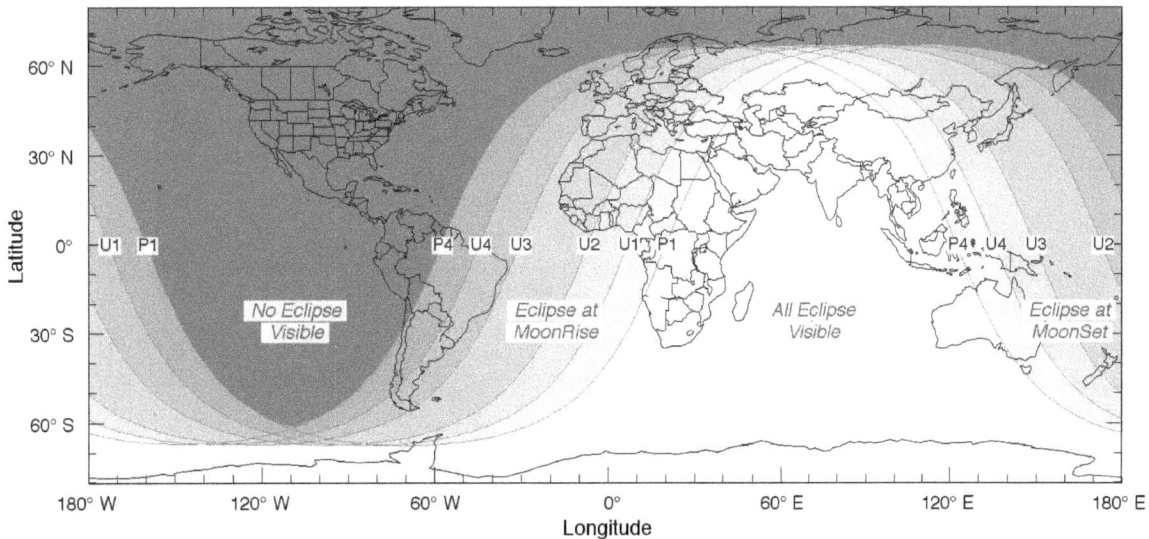

<u>Eclipse Contacts</u>
P1 = 16:32:01 UT1
U1 = 17:27:13 UT1
U2 = 18:25:14 UT1
U3 = 20:03:17 UT1
U4 = 21:01:22 UT1
P4 = 21:56:29 UT1

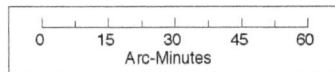

189

Total Lunar Eclipse of 2058 Nov 30

Greatest Eclipse = 03:16:17.9 TD (= 03:14:46.8 UT1)

Penumbral Magnitude = 2.4819	Gamma = 0.2208	Saros Series = 136
Umbral Magnitude = 1.4277	Axis = 0.2077°	Saros Member = 22 of 72

Sun at Greatest Eclipse
(Geocentric Coordinates)
R.A. = 16h25m34.1s
Dec. = -21°39'35.3"
S.D. = 00°16'12.9"
H.P. = 00°00'08.9"

Moon at Greatest Eclipse
(Geocentric Coordinates)
R.A. = 04h25m20.3s
Dec. = +21°51'37.8"
S.D. = 00°15'22.8"
H.P. = 00°56'26.8"

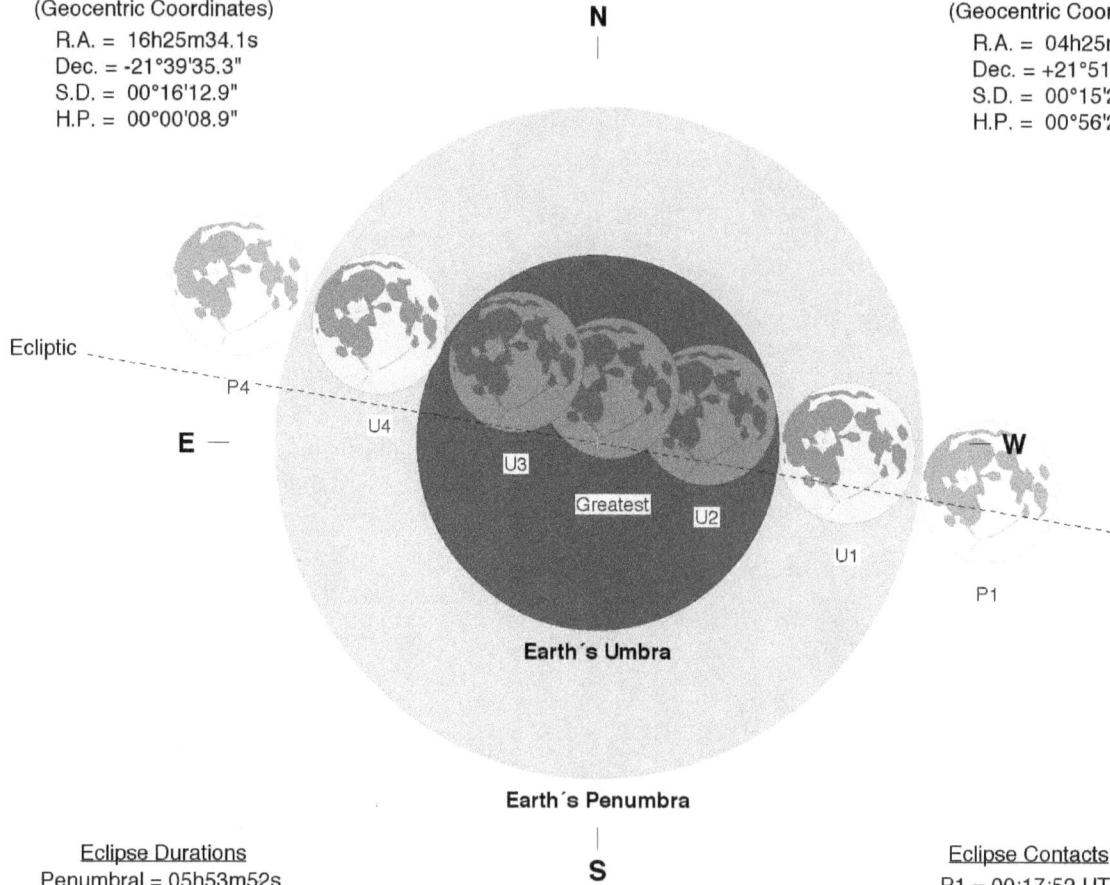

N

Ecliptic

P4

E

U4

U3

Greatest

U2

U1

P1

W

Earth's Umbra

Earth's Penumbra

S

Eclipse Durations
Penumbral = 05h53m52s
Umbral = 03h41m28s
Total = 01h30m25s

Eph. = JPL DE430
Rule = Herald-Sinnott
ΔT = 91 s

0   15   30   45   60
Arc-Minutes

©2020 F. Espenak, www.EclipseWise.com

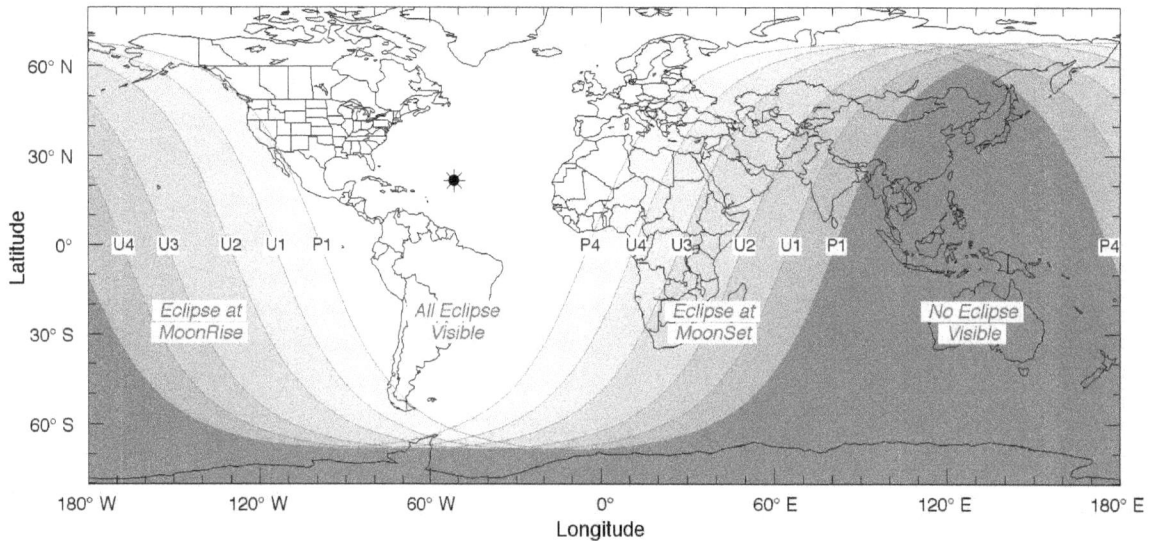

Eclipse Contacts
P1 = 00:17:52 UT1
U1 = 01:23:59 UT1
U2 = 02:29:28 UT1
U3 = 03:59:54 UT1
U4 = 05:05:27 UT1
P4 = 06:11:45 UT1

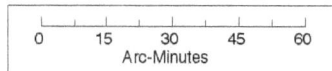

Eclipse at
MoonRise

All Eclipse
Visible

Eclipse at
MoonSet

No Eclipse
Visible

Partial Lunar Eclipse of 2059 May 27

Greatest Eclipse = 07:55:34.3 TD (= 07:54:02.8 UT1)

Penumbral Magnitude = 1.1963	Gamma = -0.9098	Saros Series = 141
Umbral Magnitude = 0.1846	Axis = 0.8684°	Saros Member = 26 of 72

Sun at Greatest Eclipse
(Geocentric Coordinates)
R.A. = 04h16m47.9s
Dec. = +21°18'58.7"
S.D. = 00°15'47.3"
H.P. = 00°00'08.7"

Moon at Greatest Eclipse
(Geocentric Coordinates)
R.A. = 16h15m47.5s
Dec. = -22°09'09.5"
S.D. = 00°15'36.4"
H.P. = 00°57'16.6"

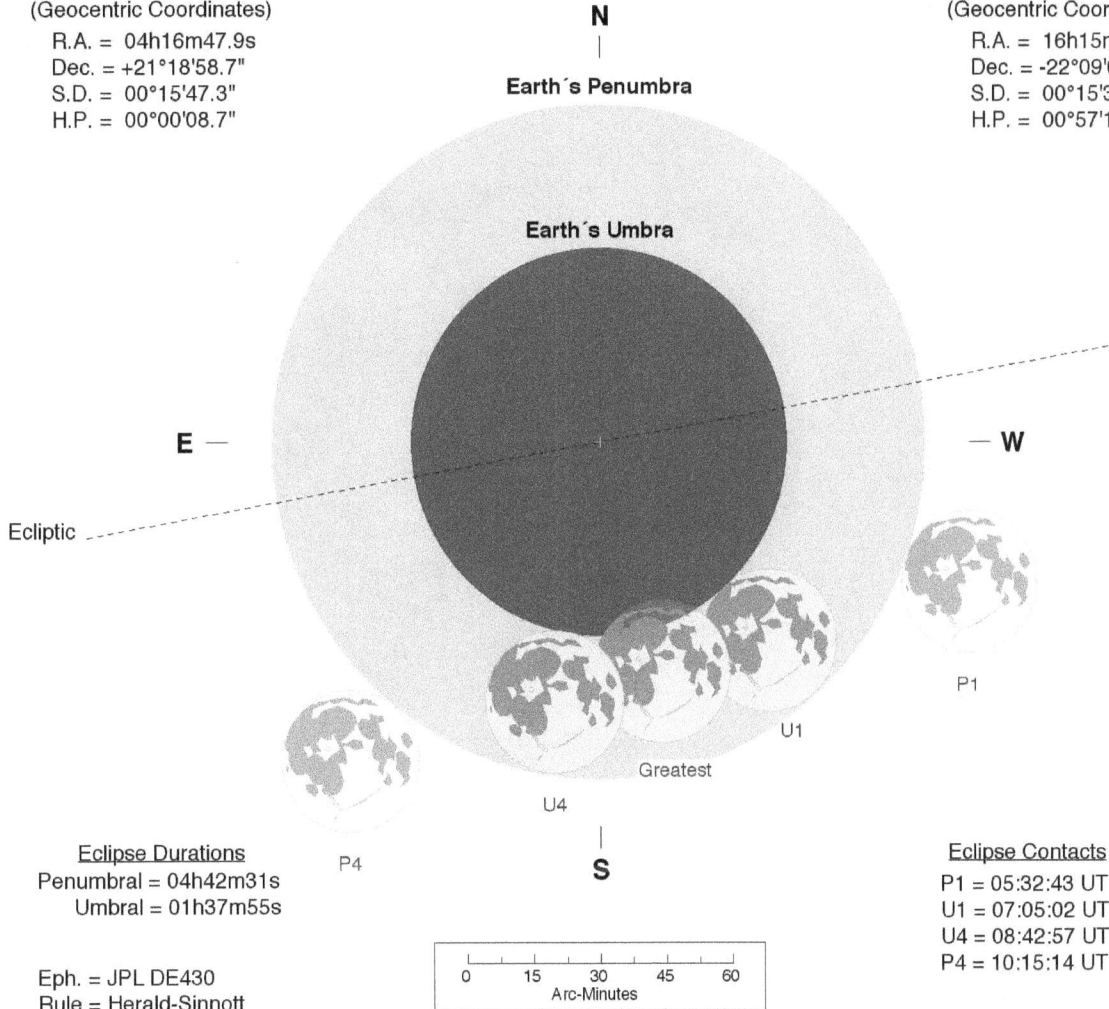

N

Earth's Penumbra

Earth's Umbra

E —

— W

Ecliptic

P1

U1

Greatest

U4

P4

S

Eclipse Durations
Penumbral = 04h42m31s
Umbral = 01h37m55s

Eph. = JPL DE430
Rule = Herald-Sinnott
ΔT = 91 s

0   15   30   45   60
Arc-Minutes

©2020 F. Espenak, www.EclipseWise.com

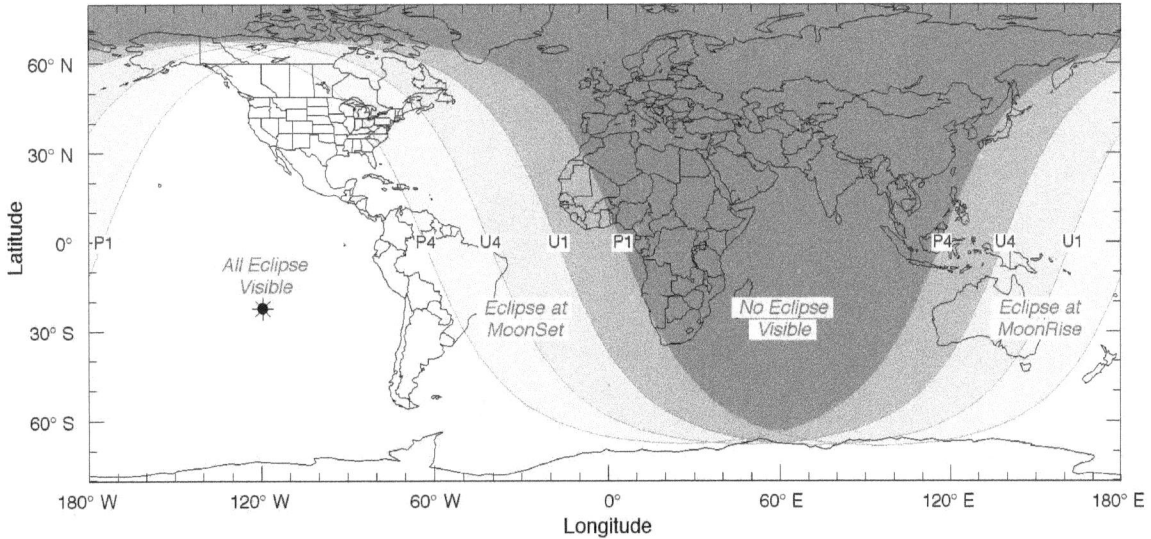

Eclipse Contacts
P1 = 05:32:43 UT1
U1 = 07:05:02 UT1
U4 = 08:42:57 UT1
P4 = 10:15:14 UT1

All Eclipse Visible

Eclipse at MoonSet

No Eclipse Visible

Eclipse at MoonRise

P1   P4   U4   U1   P1   P4   U4   U1

Partial Lunar Eclipse of 2059 Nov 19

Greatest Eclipse = 13:01:35.6 TD   (= 13:00:03.8 UT1)

Penumbral Magnitude = 1.2055	Gamma = 0.9004	Saros Series = 146
Umbral Magnitude = 0.2097	Axis = 0.8949°	Saros Member = 13 of 72

Sun at Greatest Eclipse
(Geocentric Coordinates)
R.A. = 15h39m38.2s
Dec. = -19°31'36.5"
S.D. = 00°16'10.9"
H.P. = 00°00'08.9"

Moon at Greatest Eclipse
(Geocentric Coordinates)
R.A. = 03h38m24.3s
Dec. = +20°22'25.1"
S.D. = 00°16'15.0"
H.P. = 00°59'38.3"

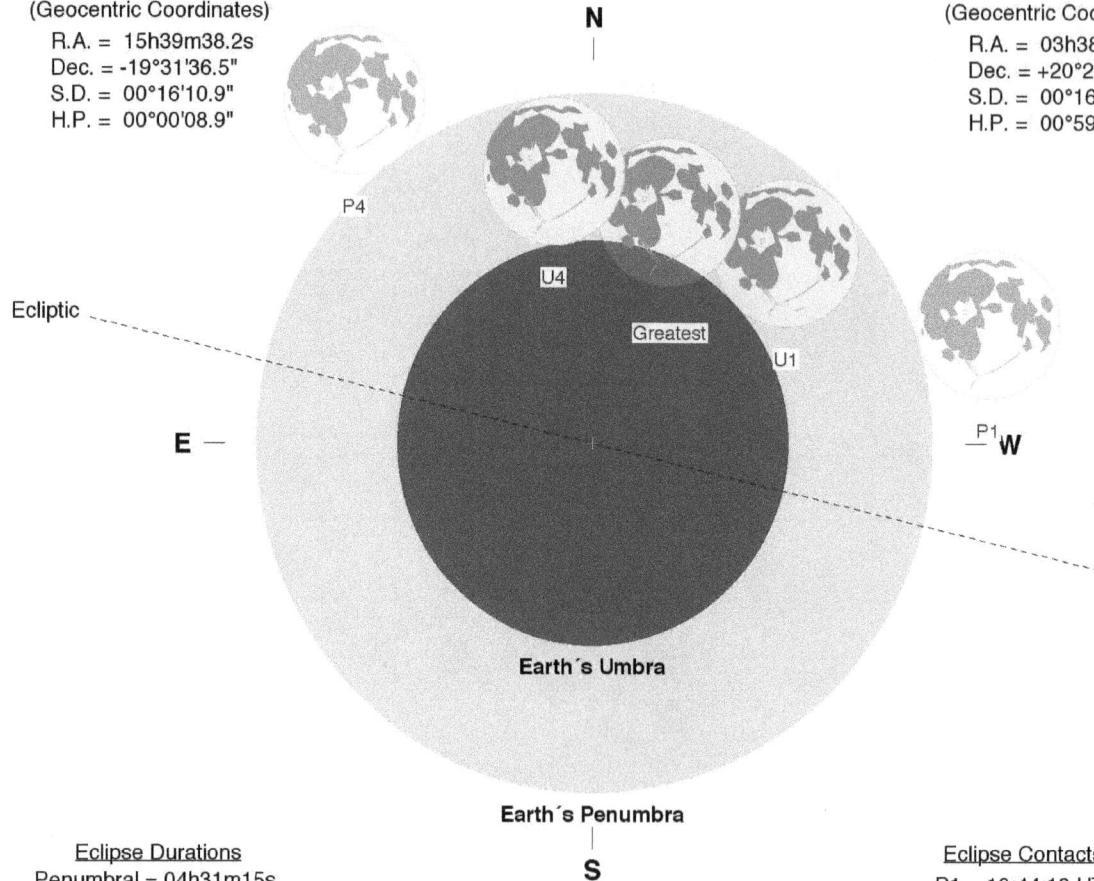

N

P4

U4

Greatest

U1

Ecliptic

E

P1 W

Earth's Umbra

Earth's Penumbra

S

Eclipse Durations
Penumbral = 04h31m15s
Umbral = 01h39m56s

Eph. = JPL DE430
Rule = Herald-Sinnott
ΔT =   92 s

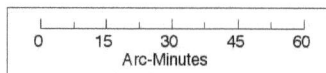

0	15	30	45	60
Arc-Minutes

©2020 F. Espenak, www.EclipseWise.com

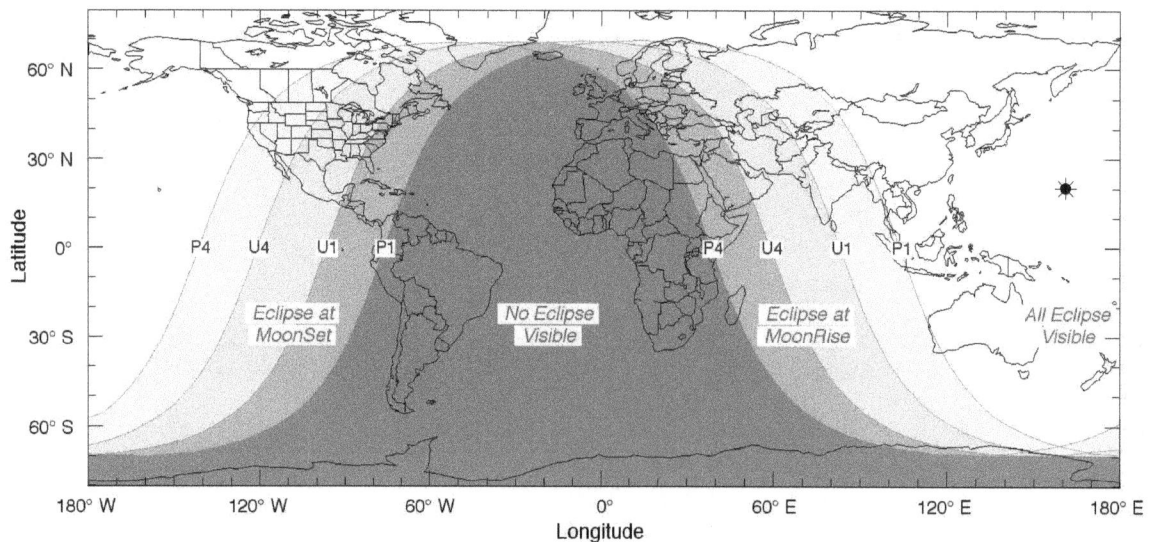

Eclipse Contacts
P1 = 10:44:18 UT1
U1 = 12:09:51 UT1
U4 = 13:49:47 UT1
P4 = 15:15:33 UT1

Eclipse at MoonSet

No Eclipse Visible

Eclipse at MoonRise

All Eclipse Visible

Penumbral Lunar Eclipse of 2060 Apr 15

Greatest Eclipse = 21:37:04.9 TD (= 21:35:32.8 UT1)

Penumbral Magnitude = 0.7694	Gamma = 1.1622	Saros Series = 113
Umbral Magnitude = -0.3136	Axis = 1.0462°	Saros Member = 66 of 71

Sun at Greatest Eclipse
(Geocentric Coordinates)

R.A. = 01h38m58.9s
Dec. = +10°17'00.0"
S.D. = 00°15'56.4"
H.P. = 00°00'08.8"

N

Moon at Greatest Eclipse
(Geocentric Coordinates)

R.A. = 13h40m53.7s
Dec. = -09°20'57.2"
S.D. = 00°14'43.1"
H.P. = 00°54'01.1"

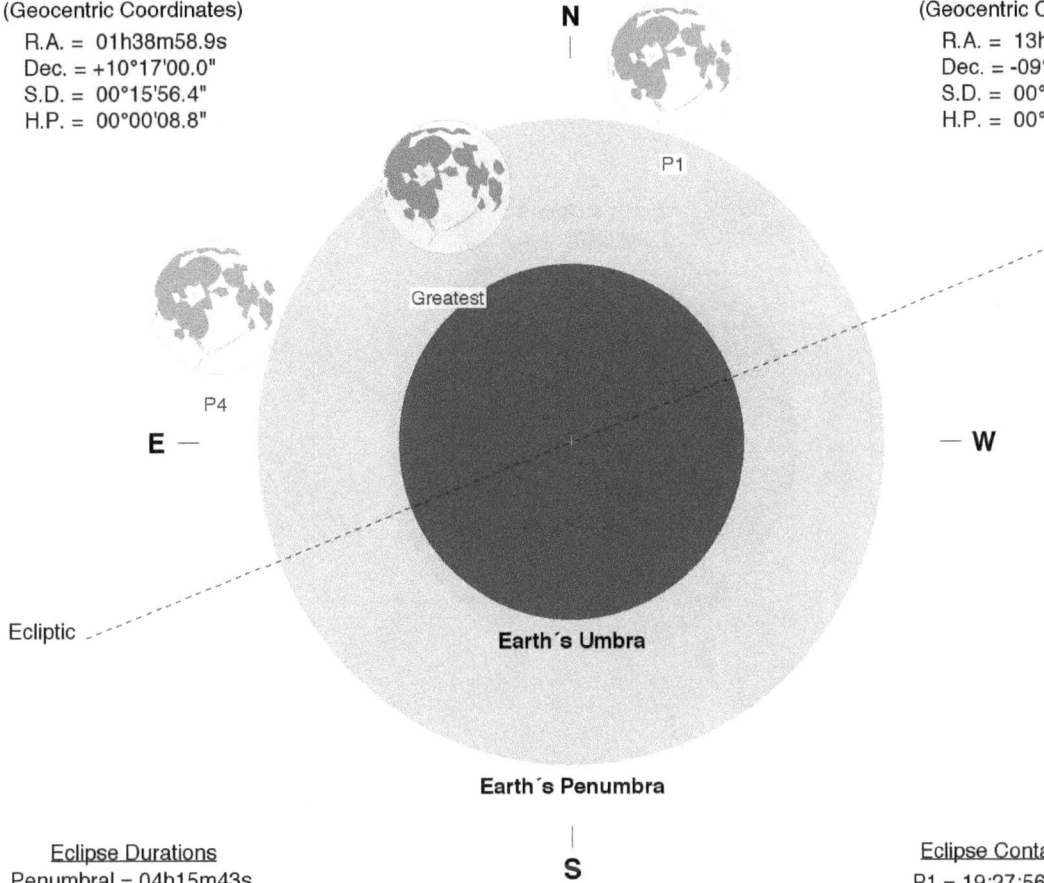

P1

Greatest

P4

Earth's Umbra

Ecliptic

E

W

Earth's Penumbra

S

Eclipse Durations
Penumbral = 04h15m43s

Eclipse Contacts
P1 = 19:27:56 UT1
P4 = 23:43:39 UT1

Eph. = JPL DE430
Rule = Herald-Sinnott
ΔT =   92 s

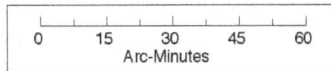

0	15	30	45	60

Arc-Minutes

©2020 F. Espenak, www.EclipseWise.com

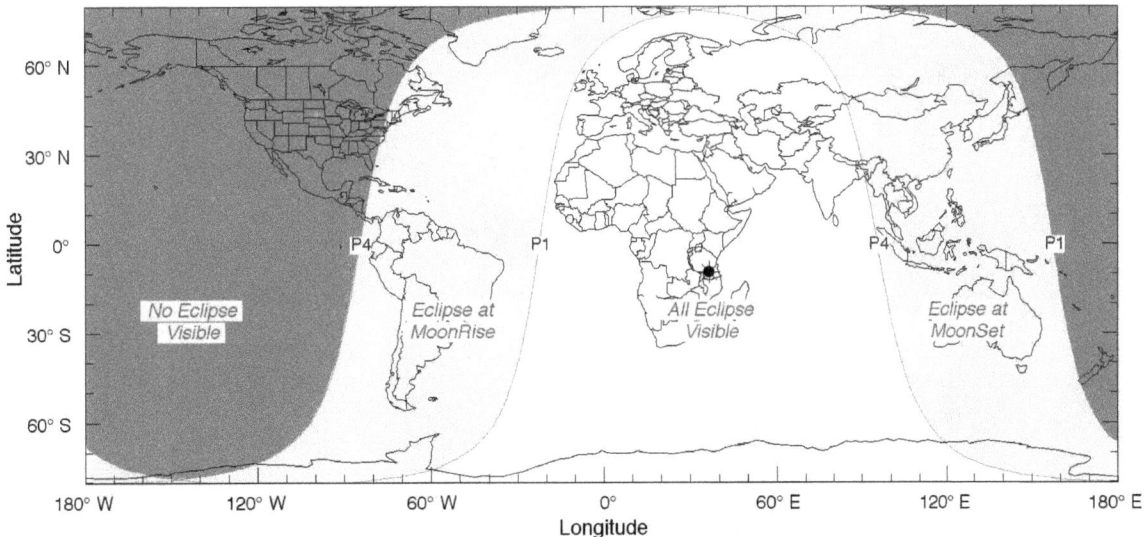

No Eclipse Visible

Eclipse at MoonRise

All Eclipse Visible

Eclipse at MoonSet

Penumbral Lunar Eclipse of 2060 Oct 09

Greatest Eclipse = 18:53:32.7 TD (= 18:52:00.2 UT1)

Penumbral Magnitude = 0.8816	Gamma = -1.0671	Saros Series = 118
Umbral Magnitude = -0.0779	Axis = 1.0892°	Saros Member = 54 of 73

Sun at Greatest Eclipse
(Geocentric Coordinates)
R.A. = 13h03m35.8s
Dec. = -06°46'17.7"
S.D. = 00°16'00.8"
H.P. = 00°00'08.8"

N

Earth's Penumbra

Moon at Greatest Eclipse
(Geocentric Coordinates)
R.A. = 01h05m39.2s
Dec. = +05°48'35.1"
S.D. = 00°16'41.3"
H.P. = 01°01'15.0"

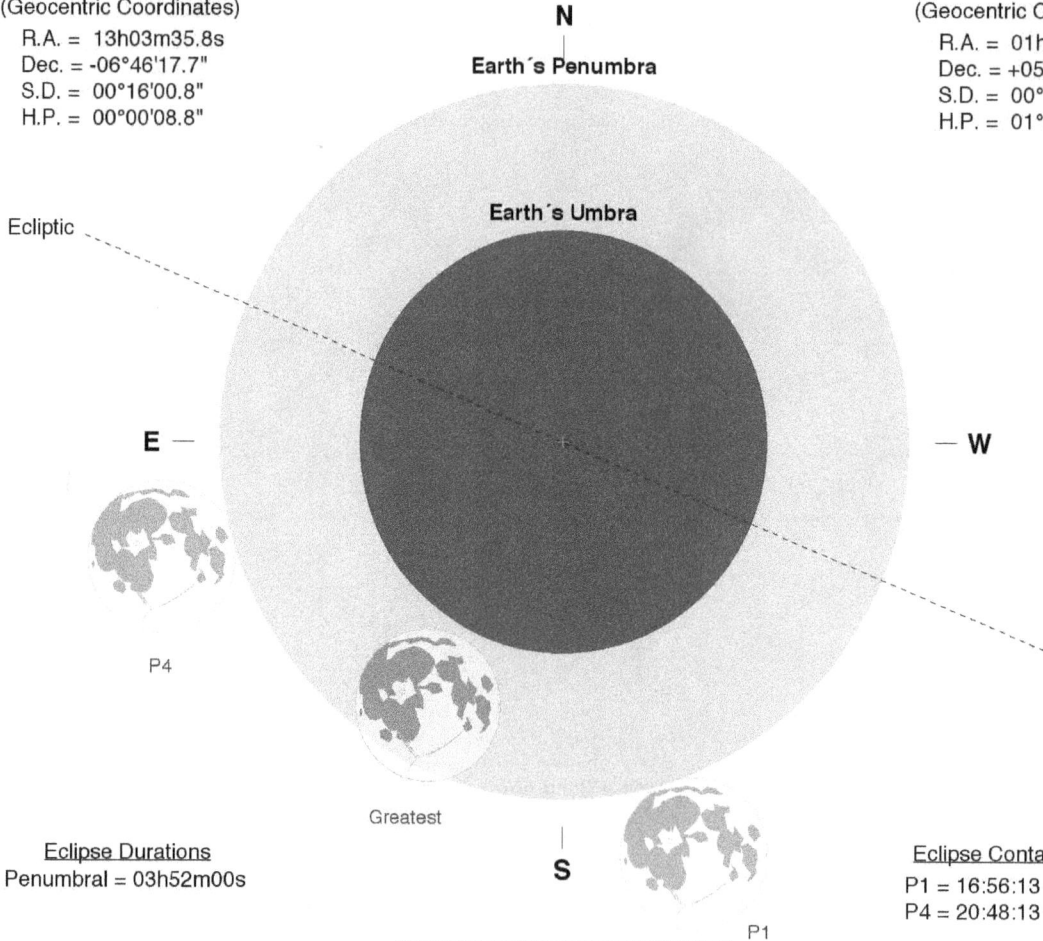

Ecliptic

Earth's Umbra

E

W

P4

Greatest

S

Eclipse Durations
Penumbral = 03h52m00s

P1

Eclipse Contacts
P1 = 16:56:13 UT1
P4 = 20:48:13 UT1

Eph. = JPL DE430
Rule = Herald-Sinnott
ΔT = 92 s

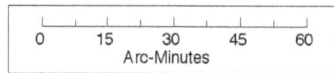

0	15	30	45	60

Arc-Minutes

©2020 F. Espenak, www.EclipseWise.com

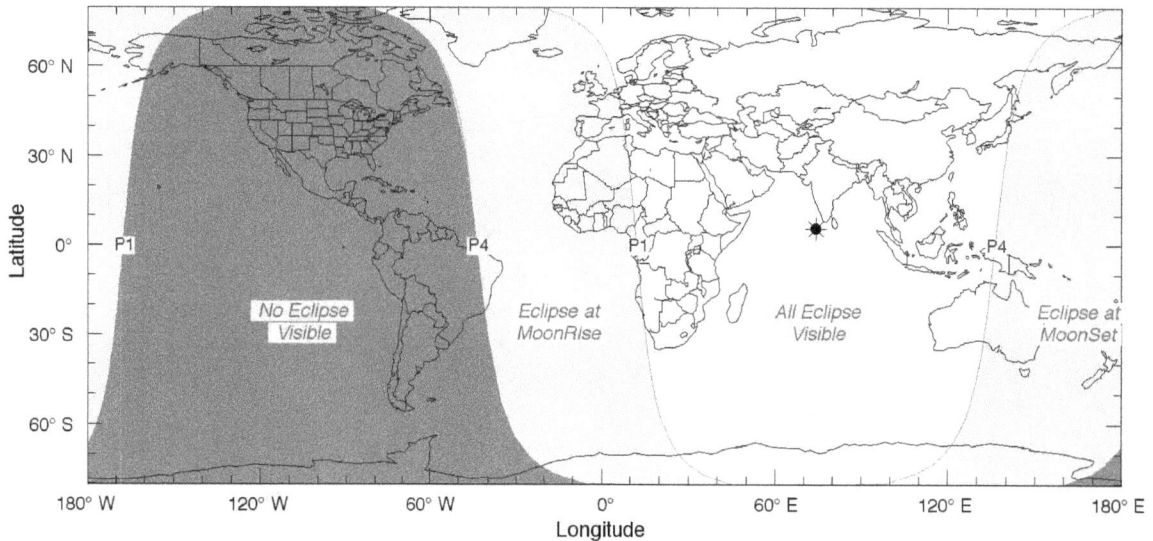

No Eclipse Visible

Eclipse at MoonRise

All Eclipse Visible

Eclipse at MoonSet

P1

P4

P1

P4

Penumbral Lunar Eclipse of 2060 Nov 08

Greatest Eclipse = 04:04:12.8 TD   (= 04:02:40.3 UT1)

Penumbral Magnitude = 0.0286	Gamma = 1.5332	Saros Series = 156
Umbral Magnitude = -0.9356	Axis = 1.5699°	Saros Member = 1 of 81

Sun at Greatest Eclipse
(Geocentric Coordinates)

R.A. = 14h56m11.8s
Dec. = -16°46'13.7"
S.D. = 00°16'08.5"
H.P. = 00°00'08.9"

Moon at Greatest Eclipse
(Geocentric Coordinates)

R.A. = 02h53m43.2s
Dec. = +18°13'31.2"
S.D. = 00°16'44.5"
H.P. = 01°01'26.6"

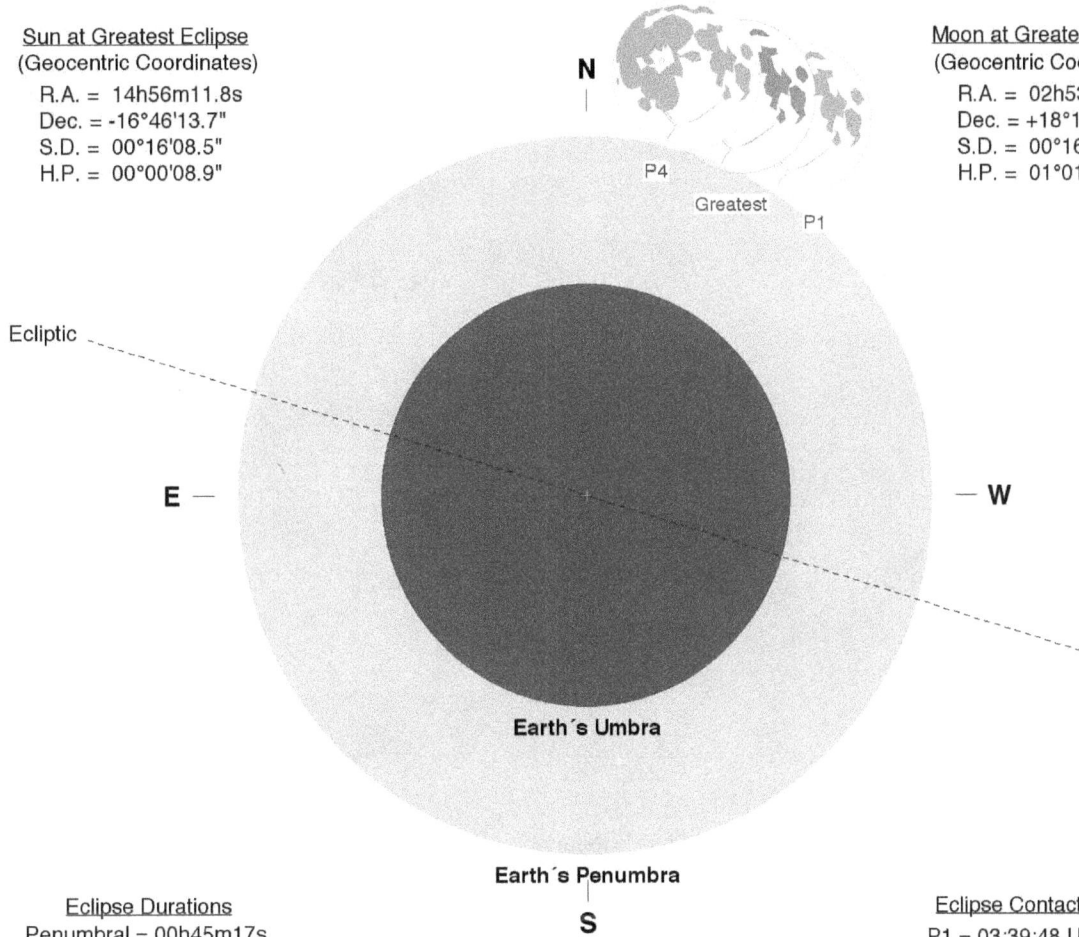

N

P4

Greatest

P1

Ecliptic

E

W

Earth´s Umbra

Earth´s Penumbra

S

Eclipse Durations

Penumbral = 00h45m17s

Eclipse Contacts

P1 = 03:39:48 UT1
P4 = 04:25:04 UT1

Eph. = JPL DE430
Rule = Herald-Sinnott
ΔT =   93 s

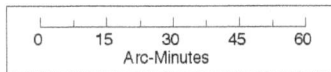

0	15	30	45	60

Arc-Minutes

©2020 F. Espenak, www.EclipseWise.com

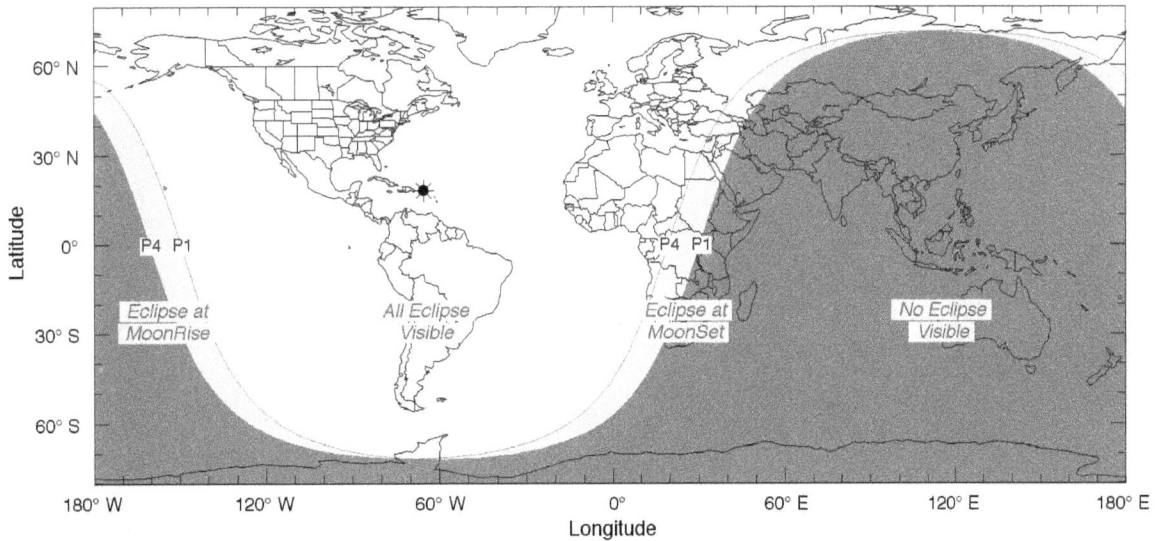

P4 P1

P4 P1

Eclipse at MoonRise

All Eclipse Visible

Eclipse at MoonSet

No Eclipse Visible

Total Lunar Eclipse of 2061 Apr 04

Greatest Eclipse = 21:54:04.8 TD   (= 21:52:32.0 UT1)

Penumbral Magnitude = 2.1064	Gamma = 0.4300	Saros Series = 123
Umbral Magnitude = 1.0360	Axis = 0.3929°	Saros Member = 55 of 72

Sun at Greatest Eclipse
(Geocentric Coordinates)

R.A. = 00h57m38.4s
Dec. = +06°09'23.4"
S.D. = 00°15'59.4"
H.P. = 00°00'08.8"

Moon at Greatest Eclipse
(Geocentric Coordinates)

R.A. = 12h58m23.4s
Dec. = -05°48'38.8"
S.D. = 00°14'56.4"
H.P. = 00°54'49.7"

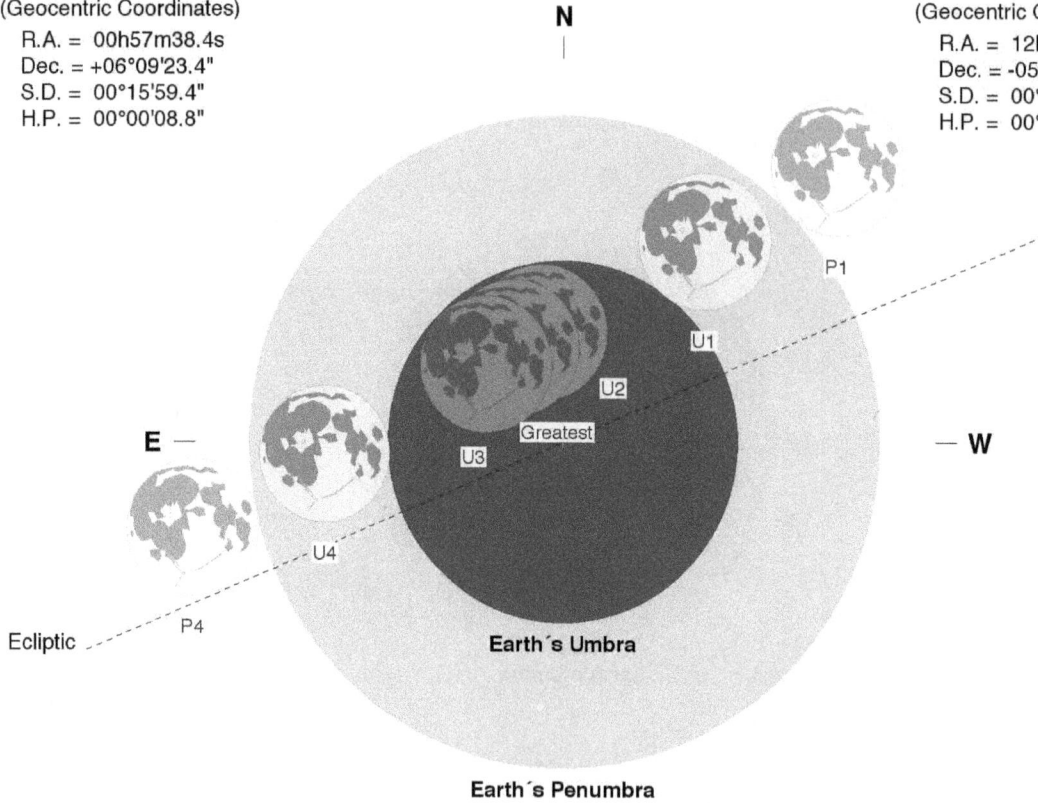

N

E

W

P1

U1

U2

Greatest

U3

U4

P4

Ecliptic

Earth's Umbra

Earth's Penumbra

S

Eclipse Durations

Penumbral = 05h55m48s
Umbral = 03h30m19s
Total = 00h30m53s

Eph. = JPL DE430
Rule = Herald-Sinnott
ΔT =   93 s

0   15   30   45   60
Arc-Minutes

©2020 F. Espenak, www.EclipseWise.com

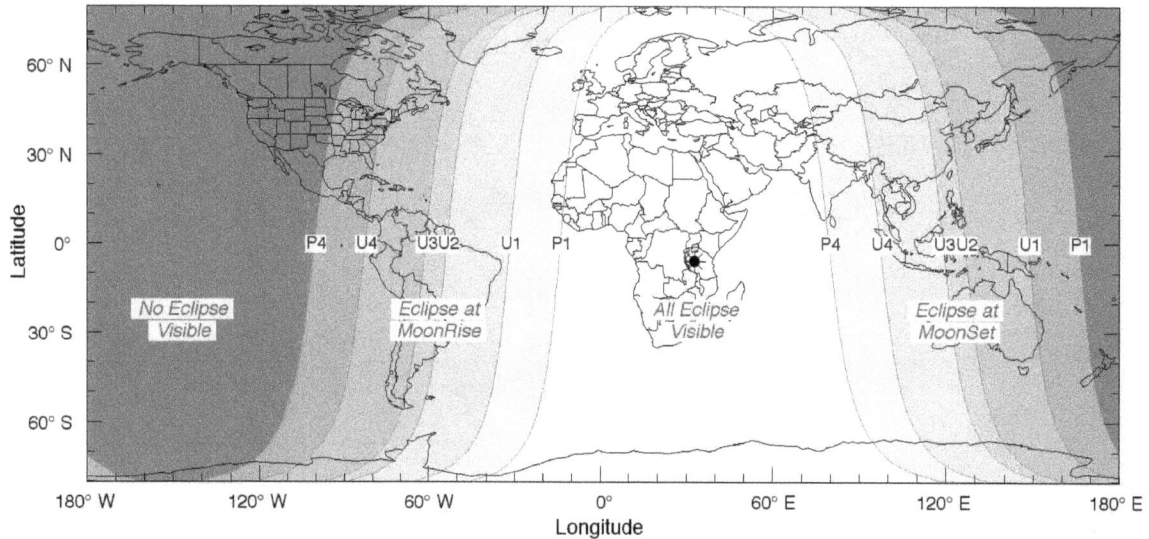

Eclipse Contacts

P1 = 18:54:45 UT1
U1 = 20:07:30 UT1
U2 = 21:37:23 UT1
U3 = 22:08:16 UT1
U4 = 23:37:48 UT1
P4 = 00:50:33 UT1

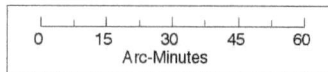

No Eclipse Visible

Eclipse at MoonRise

All Eclipse Visible

Eclipse at MoonSet

P4   U4   U3 U2   U1   P1   P4   U4   U3 U2   U1   P1

Total Lunar Eclipse of 2061 Sep 29

Greatest Eclipse = 09:38:13.1 TD (= 09:36:40.0 UT1)

Penumbral Magnitude = 2.1576	Gamma = -0.3810	Saros Series = 128
Umbral Magnitude = 1.1640	Axis = 0.3745°	Saros Member = 43 of 71

Sun at Greatest Eclipse
(Geocentric Coordinates)
R.A. = 12h24m54.1s
Dec. = -02°41'28.5"
S.D. = 00°15'57.9"
H.P. = 00°00'08.8"

Moon at Greatest Eclipse
(Geocentric Coordinates)
R.A. = 00h25m37.7s
Dec. = +02°21'48.8"
S.D. = 00°16'04.1"
H.P. = 00°58'58.5"

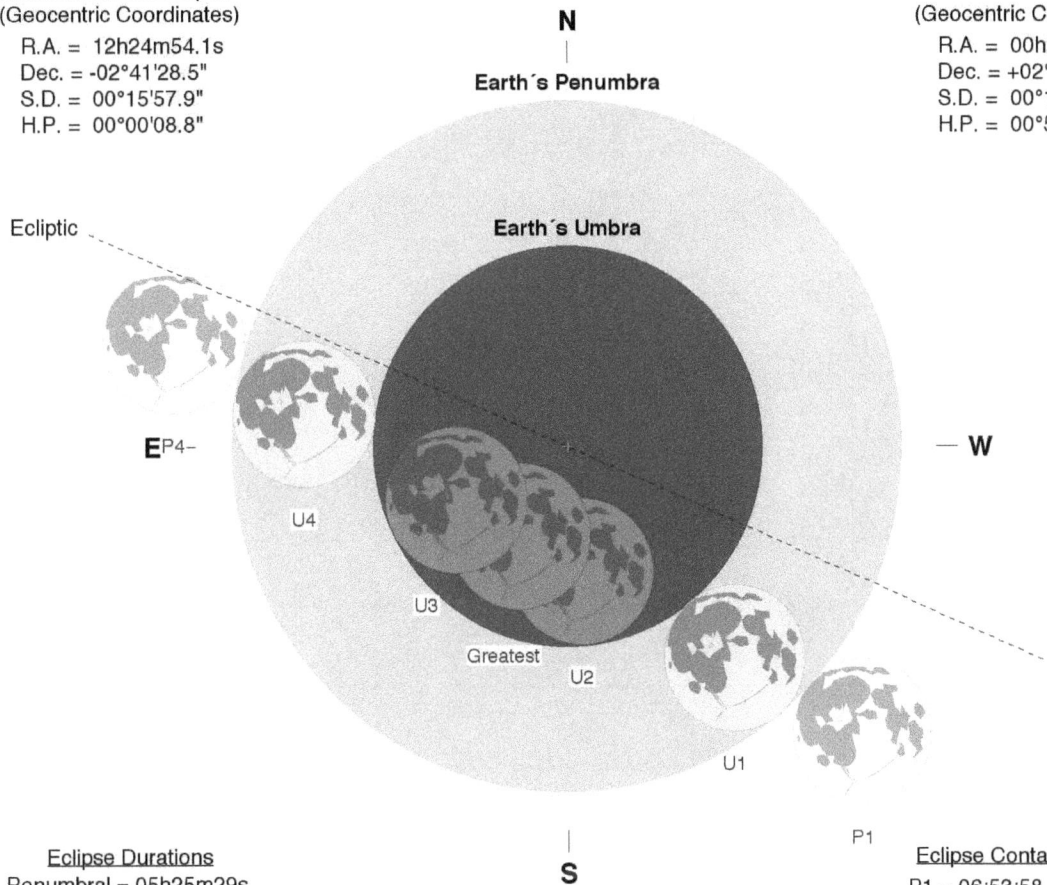

N

Earth´s Penumbra

Ecliptic

Earth´s Umbra

E P4–

W

U4

U3

Greatest

U2

U1

P1

S

Eclipse Durations
Penumbral = 05h25m29s
Umbral = 03h23m03s
Total = 00h59m38s

Eph. = JPL DE430
Rule = Herald-Sinnott
ΔT = 93 s

0 15 30 45 60
Arc-Minutes

©2020 F. Espenak, www.EclipseWise.com

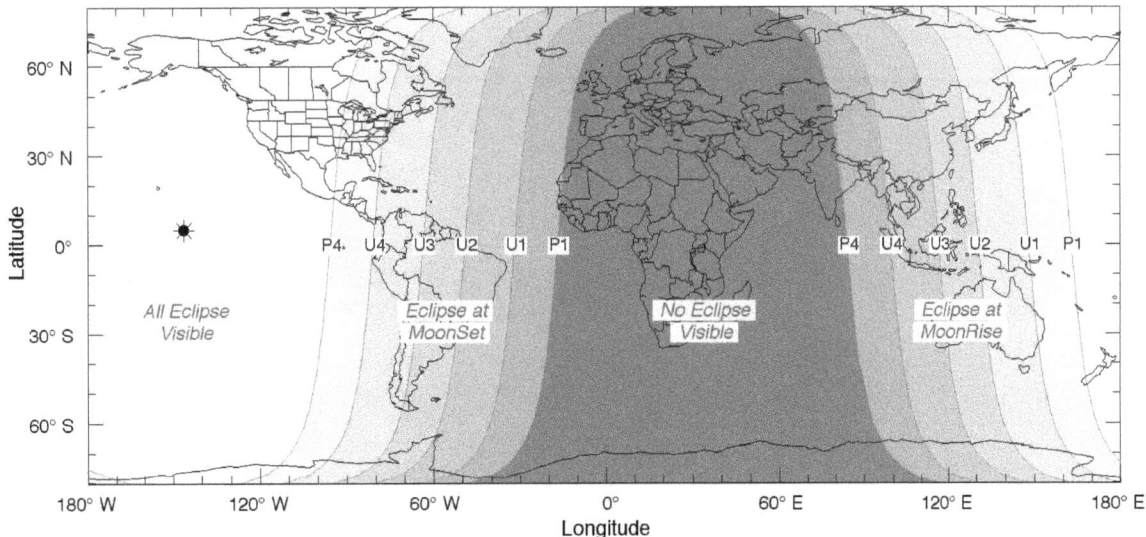

Eclipse Contacts
P1 = 06:53:58 UT1
U1 = 07:55:17 UT1
U2 = 09:07:07 UT1
U3 = 10:06:45 UT1
U4 = 11:18:20 UT1
P4 = 12:19:27 UT1

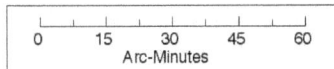

All Eclipse Visible

Eclipse at MoonSet

No Eclipse Visible

Eclipse at MoonRise

P4 U4 U3 U2 U1 P1   P4 U4 U3 U2 U1 P1

Total Lunar Eclipse of 2062 Mar 25

Greatest Eclipse = 03:33:50.4 TD   (= 03:32:16.9 UT1)

Penumbral Magnitude = 2.2925	Gamma = -0.3150	Saros Series = 133
Umbral Magnitude = 1.2715	Axis = 0.3027°	Saros Member = 29 of 71

Sun at Greatest Eclipse
(Geocentric Coordinates)
R.A. = 00h17m31.2s
Dec. = +01°53'45.5"
S.D. = 00°16'02.5"
H.P. = 00°00'08.8"

Moon at Greatest Eclipse
(Geocentric Coordinates)
R.A. = 12h16m55.9s
Dec. = -02°09'38.2"
S.D. = 00°15'42.7"
H.P. = 00°57'39.7"

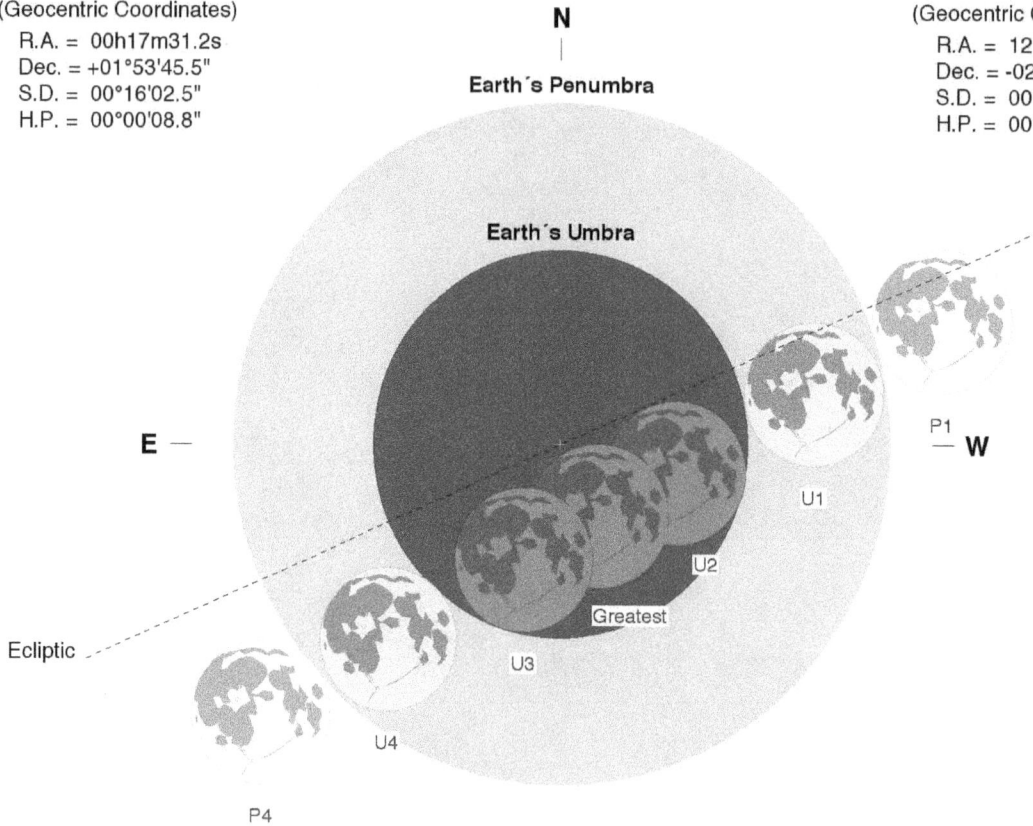

N

Earth's Penumbra

Earth's Umbra

E

W

P1

U1

U2

Greatest

U3

U4

P4

Ecliptic

S

Eclipse Durations
Penumbral = 05h39m03s
Umbral = 03h31m58s
Total = 01h15m16s

Eph. = JPL DE430
Rule = Herald-Sinnott
ΔT =   94 s

0	15	30	45	60

Arc-Minutes

©2020 F. Espenak, www.EclipseWise.com

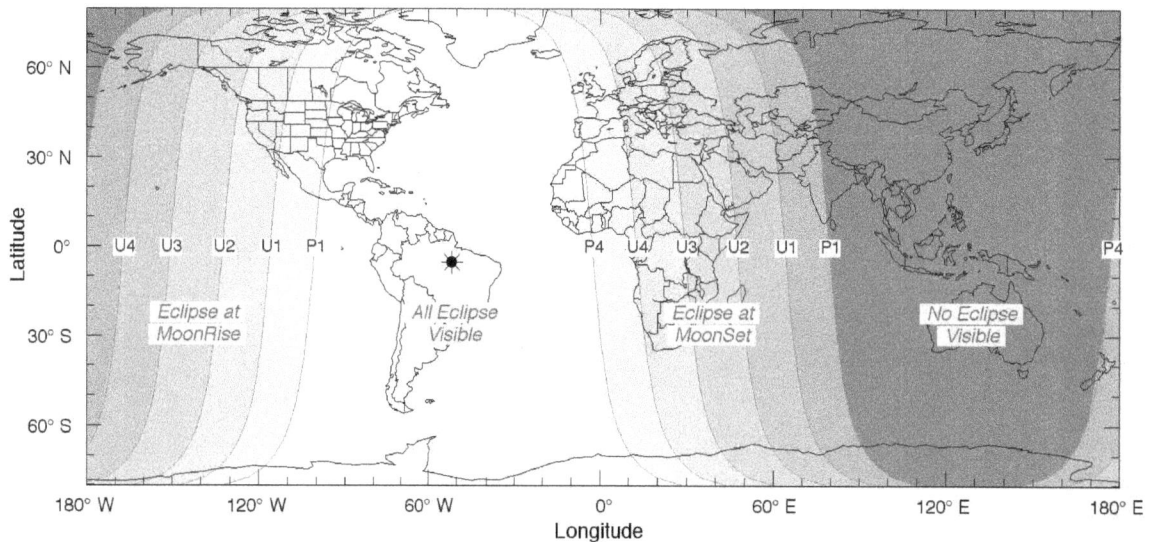

Eclipse Contacts
P1 = 00:42:44 UT1
U1 = 01:46:10 UT1
U2 = 02:54:25 UT1
U3 = 04:09:41 UT1
U4 = 05:18:08 UT1
P4 = 06:21:47 UT1

Total Lunar Eclipse of 2062 Sep 18

Greatest Eclipse = 18:34:02.1 TD (= 18:32:28.2 UT1)

Penumbral Magnitude = 2.1979	Gamma = 0.3736	Saros Series = 138
Umbral Magnitude = 1.1515	Axis = 0.3476°	Saros Member = 31 of 82

Sun at Greatest Eclipse
(Geocentric Coordinates)

R.A. = 11h45m50.3s
Dec. = +01°31'59.3"
S.D. = 00°15'55.0"
H.P. = 00°00'08.8"

Moon at Greatest Eclipse
(Geocentric Coordinates)

R.A. = 23h45m09.8s
Dec. = -01°13'45.9"
S.D. = 00°15'12.7"
H.P. = 00°55'49.7"

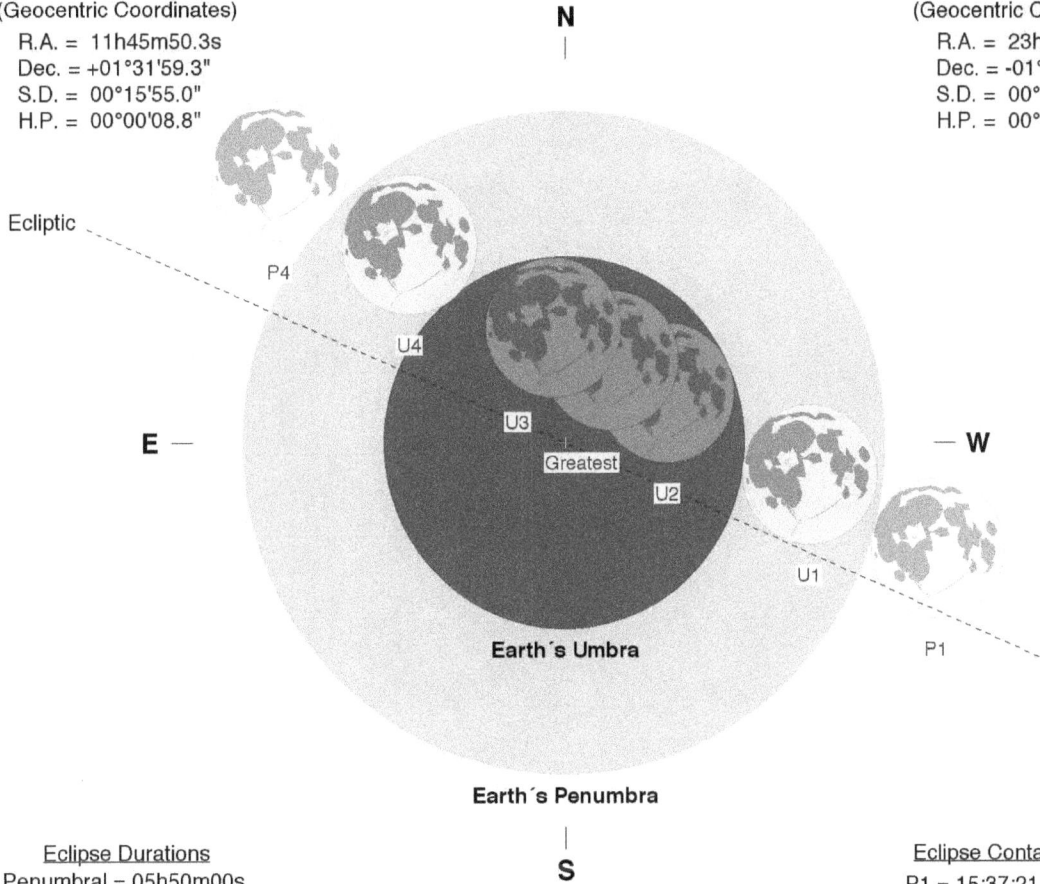

N

Ecliptic

P4
U4
U3
Greatest
U2
U1
P1

E

W

Earth's Umbra

Earth's Penumbra

S

Eclipse Durations
Penumbral = 05h50m00s
Umbral = 03h33m06s
Total = 01h00m11s

Eph. = JPL DE430
Rule = Herald-Sinnott
ΔT = 94 s

0   15   30   45   60
Arc-Minutes

©2020 F. Espenak, www.EclipseWise.com

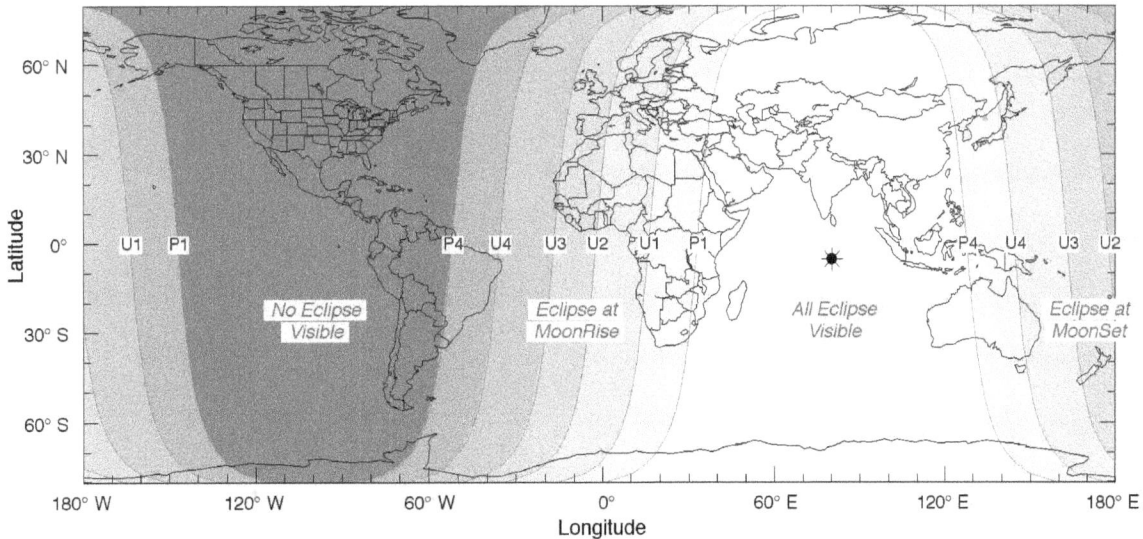

Eclipse Contacts
P1 = 15:37:21 UT1
U1 = 16:45:49 UT1
U2 = 18:02:08 UT1
U3 = 19:02:19 UT1
U4 = 20:18:56 UT1
P4 = 21:27:21 UT1

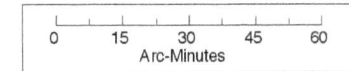

U1 P1   P4 U4   U3 U2   U1 P1   P4 U4   U3 U2

No Eclipse
Visible

Eclipse at
MoonRise

All Eclipse
Visible

Eclipse at
MoonSet

Partial Lunar Eclipse of 2063 Mar 14

Greatest Eclipse = 16:05:48.6 TD (= 16:04:14.4 UT1)

Penumbral Magnitude = 1.0108	Gamma = -1.0008	Saros Series = 143
Umbral Magnitude = 0.0363	Axis = 1.0105°	Saros Member = 20 of 72

Sun at Greatest Eclipse
(Geocentric Coordinates)
R.A. = 23h38m23.2s
Dec. = -02°20'14.4"
S.D. = 00°16'05.4"
H.P. = 00°00'08.8"

Moon at Greatest Eclipse
(Geocentric Coordinates)
R.A. = 11h36m26.0s
Dec. = +01°27'08.7"
S.D. = 00°16'30.6"
H.P. = 01°00'35.4"

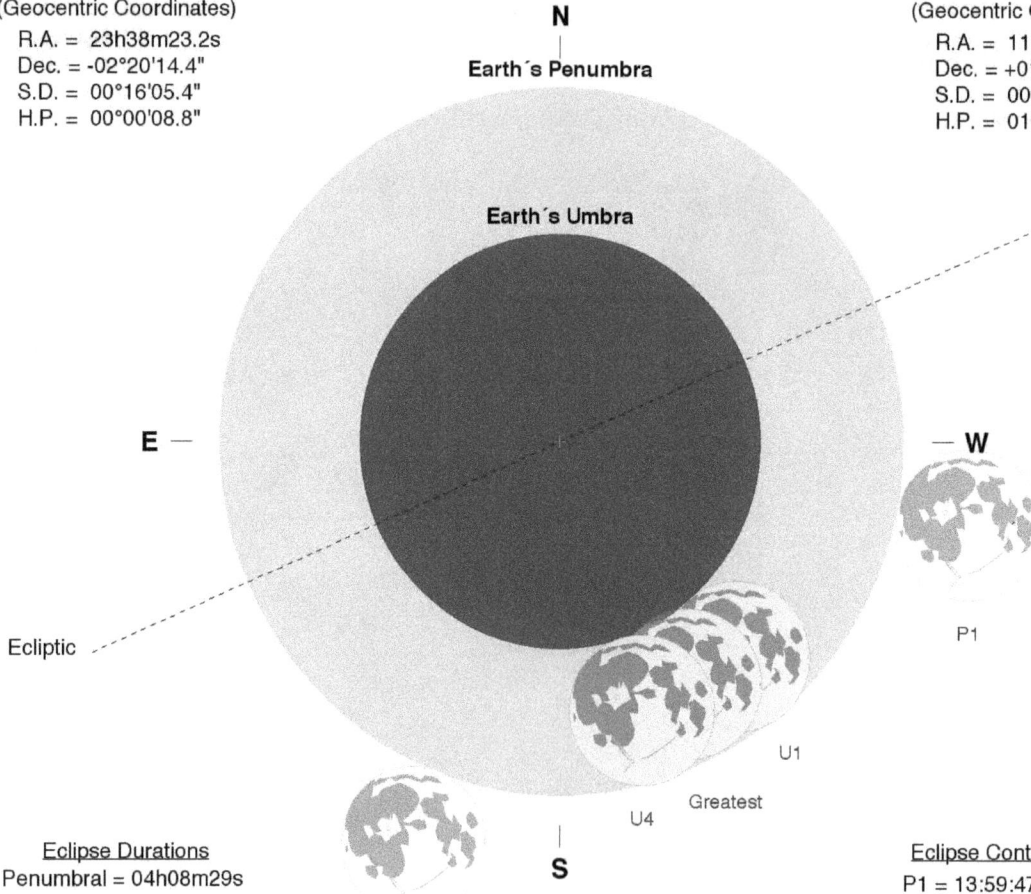

N

Earth's Penumbra

Earth's Umbra

E

W

Ecliptic

P1

U1

U4

Greatest

S

Eclipse Durations
Penumbral = 04h08m29s
Umbral = 00h41m54s

Eph. = JPL DE430
Rule = Herald-Sinnott
ΔT =   94 s

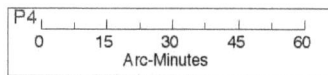

P4
0   15   30   45   60
Arc-Minutes

©2020 F. Espenak, www.EclipseWise.com

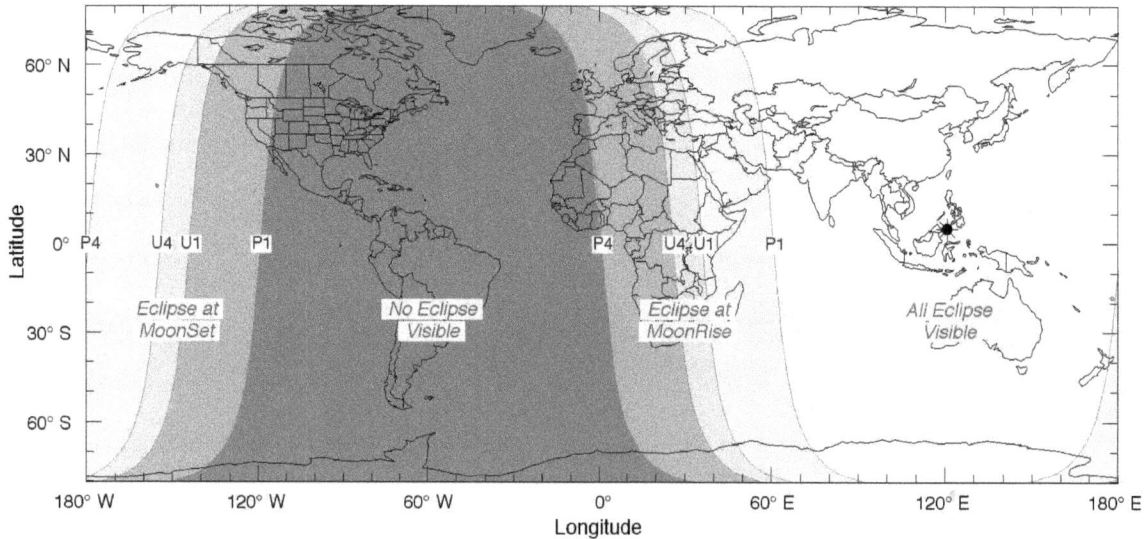

Eclipse Contacts
P1 = 13:59:47 UT1
U1 = 15:42:56 UT1
U4 = 16:24:50 UT1
P4 = 18:08:16 UT1

Eclipse at MoonSet

No Eclipse Visible

Eclipse at MoonRise

All Eclipse Visible

Penumbral Lunar Eclipse of 2063 Sep 07

Greatest Eclipse = 20:41:10.9 TD   (= 20:39:36.3 UT1)

Penumbral Magnitude = 0.8121	Gamma = 1.1375	Saros Series = 148
Umbral Magnitude = -0.2657	Axis = 1.0243°	Saros Member = 6 of 70

Sun at Greatest Eclipse
(Geocentric Coordinates)
R.A. = 11h05m48.9s
Dec. = +05°47'49.4"
S.D. = 00°15'52.2"
H.P. = 00°00'08.7"

Moon at Greatest Eclipse
(Geocentric Coordinates)
R.A. = 23h03m51.6s
Dec. = -04°53'44.4"
S.D. = 00°14'43.4"
H.P. = 00°54'02.2"

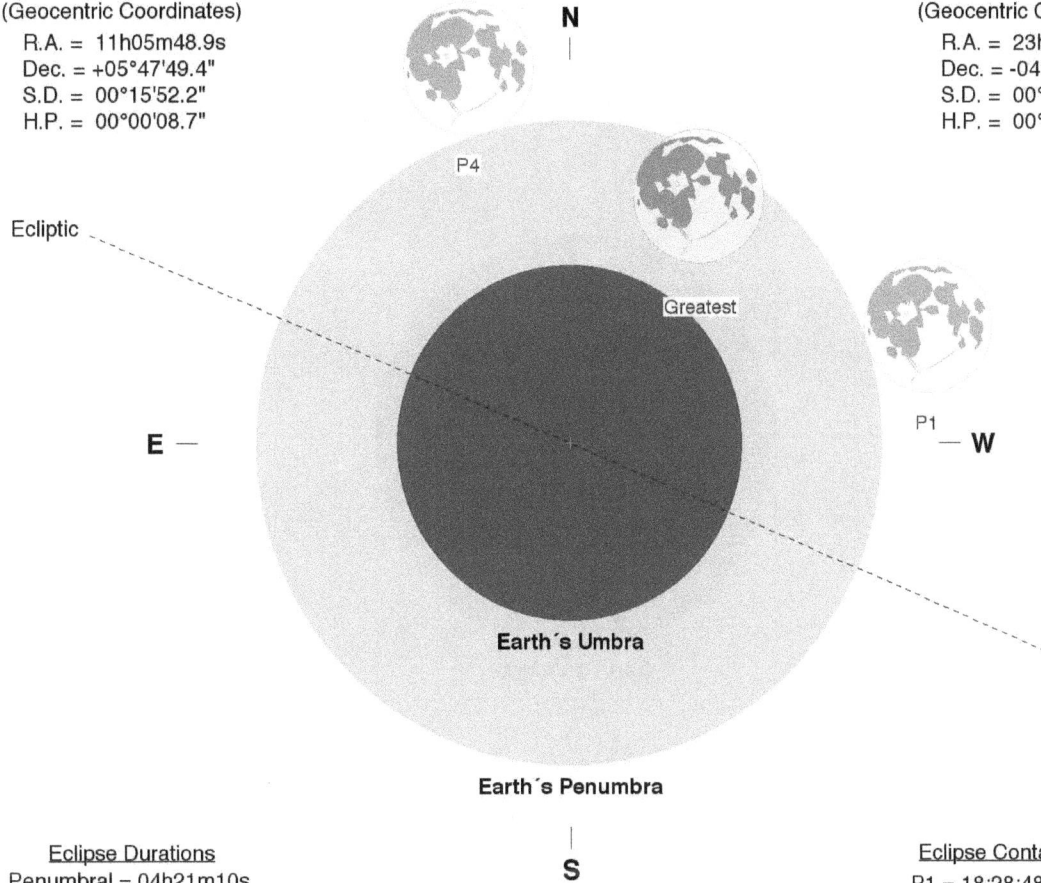

N

P4

Ecliptic

Greatest

E

P1

W

Earth´s Umbra

Earth´s Penumbra

S

Eclipse Durations
Penumbral = 04h21m10s

Eclipse Contacts
P1 = 18:28:48 UT1
P4 = 22:49:58 UT1

Eph. = JPL DE430
Rule = Herald-Sinnott
ΔT = 95 s

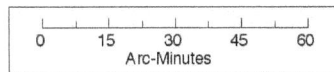

0   15   30   45   60
Arc-Minutes

©2020 F. Espenak, www.EclipseWise.com

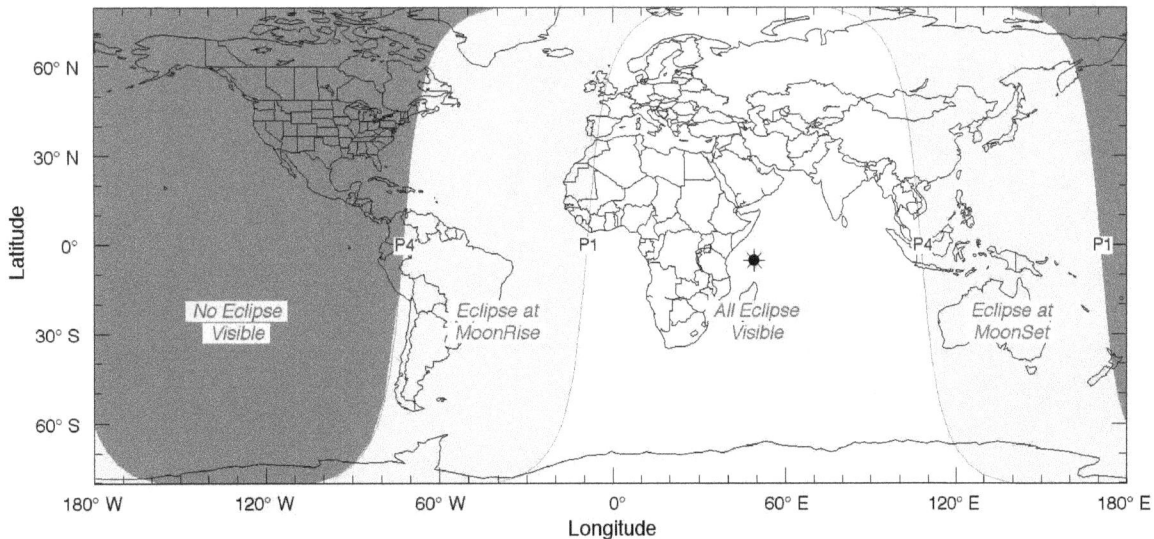

No Eclipse Visible

Eclipse at MoonRise

All Eclipse Visible

Eclipse at MoonSet

P4   P1   P4   P1

Partial Lunar Eclipse of 2064 Feb 02

Greatest Eclipse = 21:48:57.3 TD   (= 21:47:22.4 UT1)

Penumbral Magnitude = 1.0215	Gamma = 0.9969	Saros Series = 115
Umbral Magnitude = 0.0395	Axis = 1.0078°	Saros Member = 60 of 72

Sun at Greatest Eclipse
(Geocentric Coordinates)
R.A. = 21h05m13.1s
Dec. = -16°40'07.5"
S.D. = 00°16'13.8"
H.P. = 00°00'08.9"

Moon at Greatest Eclipse
(Geocentric Coordinates)
R.A. = 09h06m49.3s
Dec. = +17°36'03.4"
S.D. = 00°16'31.7"
H.P. = 01°00'39.5"

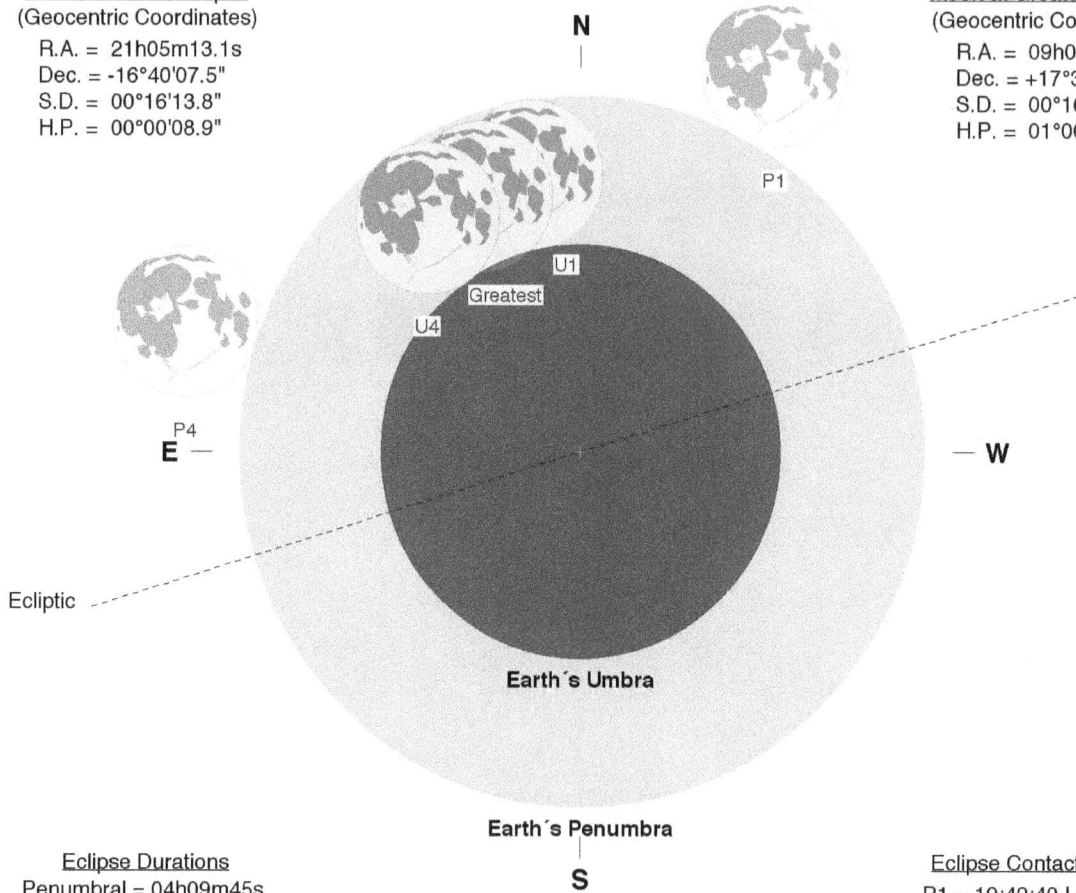

N

P1

U1

Greatest

U4

P4

E

W

Ecliptic

Earth's Umbra

Earth's Penumbra

S

Eclipse Durations
Penumbral = 04h09m45s
Umbral = 00h43m41s

Eph. = JPL DE430
Rule = Herald-Sinnott
ΔT =   95 s

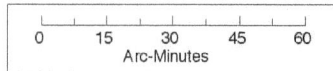

0	15	30	45	60
Arc-Minutes

Eclipse Contacts
P1 = 19:42:40 UT1
U1 = 21:25:49 UT1
U4 = 22:09:30 UT1
P4 = 23:52:25 UT1

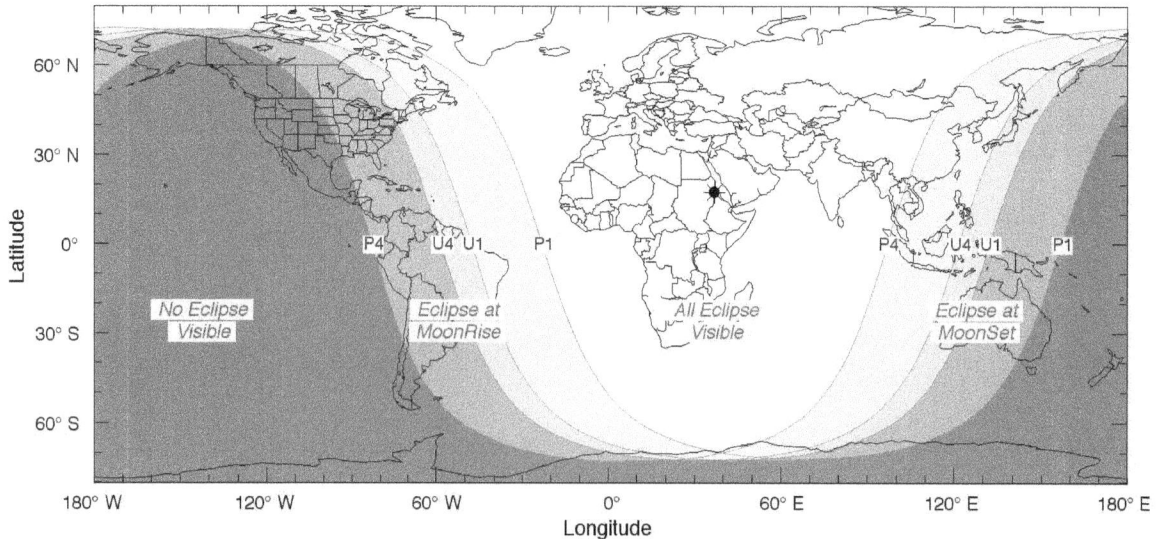

No Eclipse
Visible

Eclipse at
MoonRise

All Eclipse
Visible

Eclipse at
MoonSet

Partial Lunar Eclipse of 2064 Jul 28

Greatest Eclipse = 07:52:48.2 TD (= 07:51:12.9 UT1)

Penumbral Magnitude = 1.1378	Gamma = -0.9473	Saros Series = 120
Umbral Magnitude = 0.1055	Axis = 0.8840°	Saros Member = 60 of 83

Sun at Greatest Eclipse
(Geocentric Coordinates)

R.A. = 08h33m43.4s
Dec. = +18°45'12.2"
S.D. = 00°15'45.1"
H.P. = 00°00'08.7"

Moon at Greatest Eclipse
(Geocentric Coordinates)

R.A. = 20h35m00.2s
Dec. = -19°35'03.3"
S.D. = 00°15'15.5"
H.P. = 00°55'59.9"

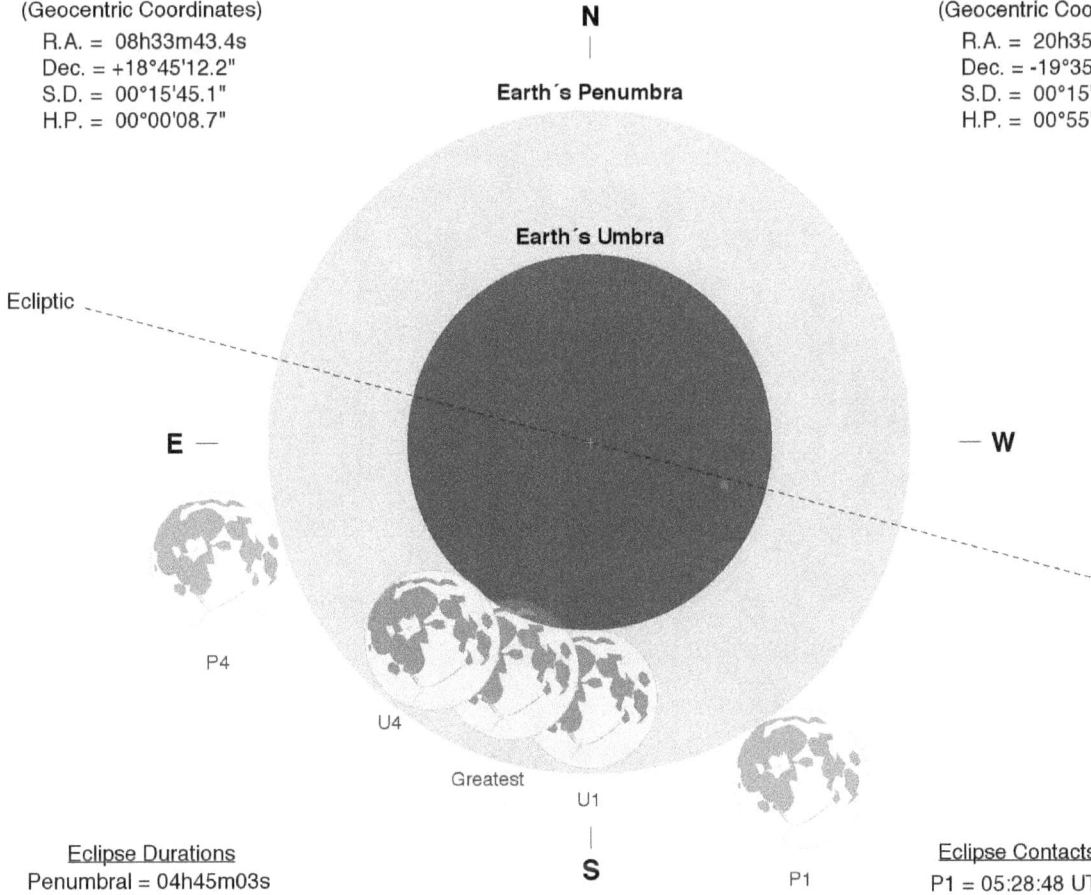

N

Earth's Penumbra

Earth's Umbra

Ecliptic

E —

— **W**

P4

U4

Greatest

U1

S

P1

Eclipse Durations
Penumbral = 04h45m03s
Umbral = 01h16m33s

Eph. = JPL DE430
Rule = Herald-Sinnott
ΔT = 95 s

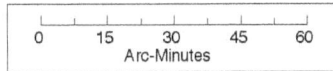

0	15	30	45	60
Arc-Minutes

©2020 F. Espenak, www.EclipseWise.com

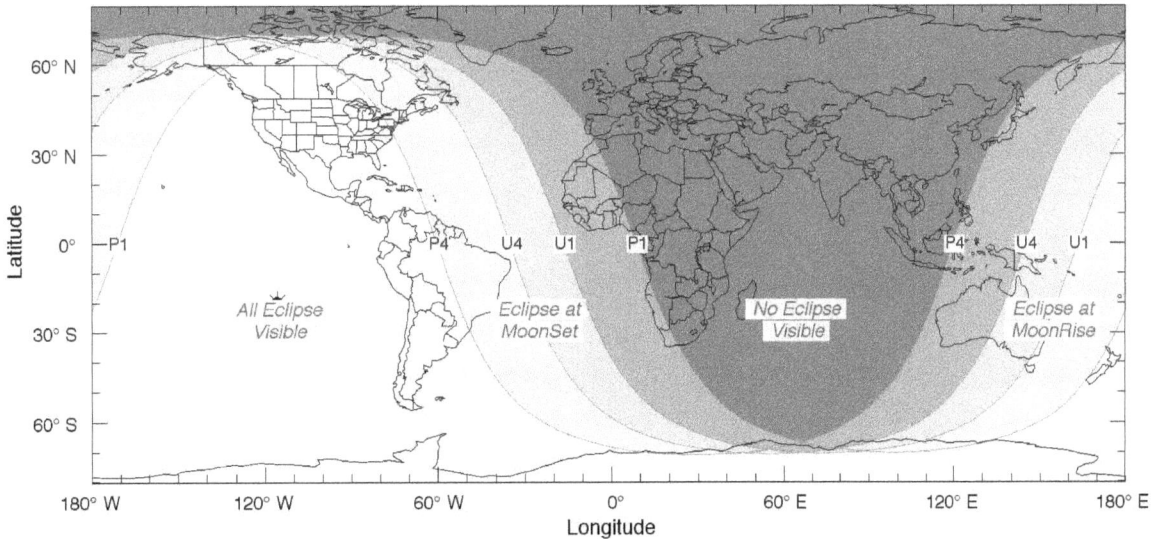

Eclipse Contacts
P1 = 05:28:48 UT1
U1 = 07:13:03 UT1
U4 = 08:29:36 UT1
P4 = 10:13:51 UT1

All Eclipse Visible

Eclipse at MoonSet

No Eclipse Visible

Eclipse at MoonRise

Total Lunar Eclipse of 2065 Jan 22

Greatest Eclipse = 09:58:58.5 TD   (= 09:57:22.9 UT1)

Penumbral Magnitude = 2.2579	Gamma = 0.3371	Saros Series = 125
Umbral Magnitude = 1.2248	Axis = 0.3243°	Saros Member = 51 of 72

Sun at Greatest Eclipse
(Geocentric Coordinates)
R.A. = 20h20m52.4s
Dec. = -19°29'52.6"
S.D. = 00°16'15.1"
H.P. = 00°00'08.9"

Moon at Greatest Eclipse
(Geocentric Coordinates)
R.A. = 08h21m19.3s
Dec. = +19°48'16.7"
S.D. = 00°15'43.9"
H.P. = 00°57'44.2"

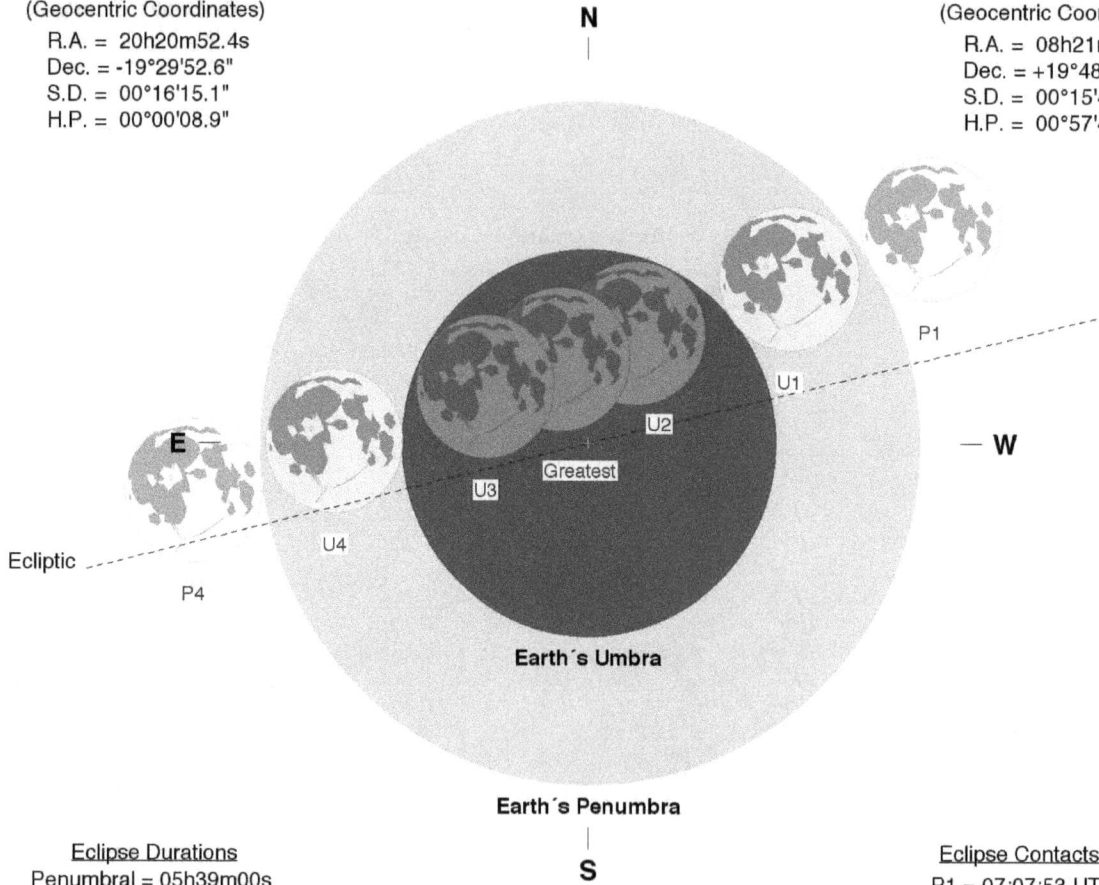

N

E — — W

Ecliptic

P1

U1

U2

Greatest

U3

U4

P4

Earth's Umbra

Earth's Penumbra

S

Eclipse Durations
Penumbral = 05h39m00s
Umbral = 03h29m43s
Total = 01h09m26s

Eph. = JPL DE430
Rule = Herald-Sinnott
ΔT =   96 s

0   15   30   45   60
Arc-Minutes

©2020 F. Espenak, www.EclipseWise.com

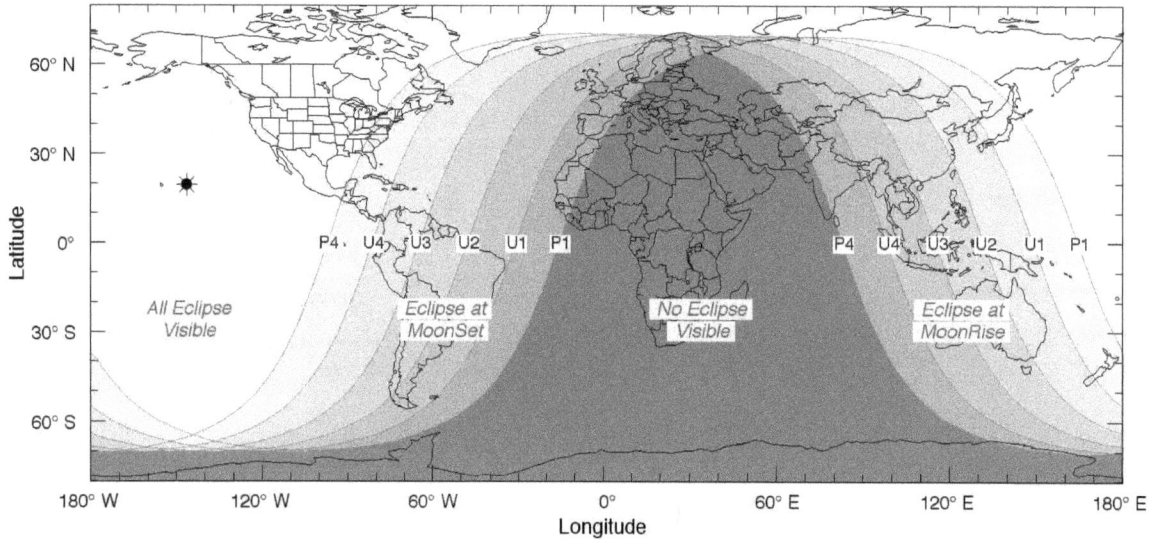

Eclipse Contacts
P1 = 07:07:53 UT1
U1 = 08:12:37 UT1
U2 = 09:22:50 UT1
U3 = 10:32:16 UT1
U4 = 11:42:20 UT1
P4 = 12:46:53 UT1

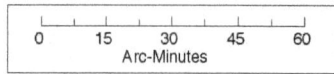

All Eclipse Visible

Eclipse at MoonSet

No Eclipse Visible

Eclipse at MoonRise

P4 U4 U3 U2 U1 P1

P4 U4 U3 U2 U1 P1

Latitude

Longitude

Total Lunar Eclipse of 2065 Jul 17

Greatest Eclipse = 17:48:40.5 TD (= 17:47:04.5 UT1)

Penumbral Magnitude = 2.5907	Gamma = -0.1402	Saros Series = 130
Umbral Magnitude = 1.6138	Axis = 0.1382°	Saros Member = 37 of 71

Sun at Greatest Eclipse
(Geocentric Coordinates)

R.A. = 07h50m48.2s
Dec. = +20°59'34.9"
S.D. = 00°15'44.3"
H.P. = 00°00'08.7"

Moon at Greatest Eclipse
(Geocentric Coordinates)

R.A. = 19h50m58.2s
Dec. = -21°07'32.3"
S.D. = 00°16'06.6"
H.P. = 00°59'07.5"

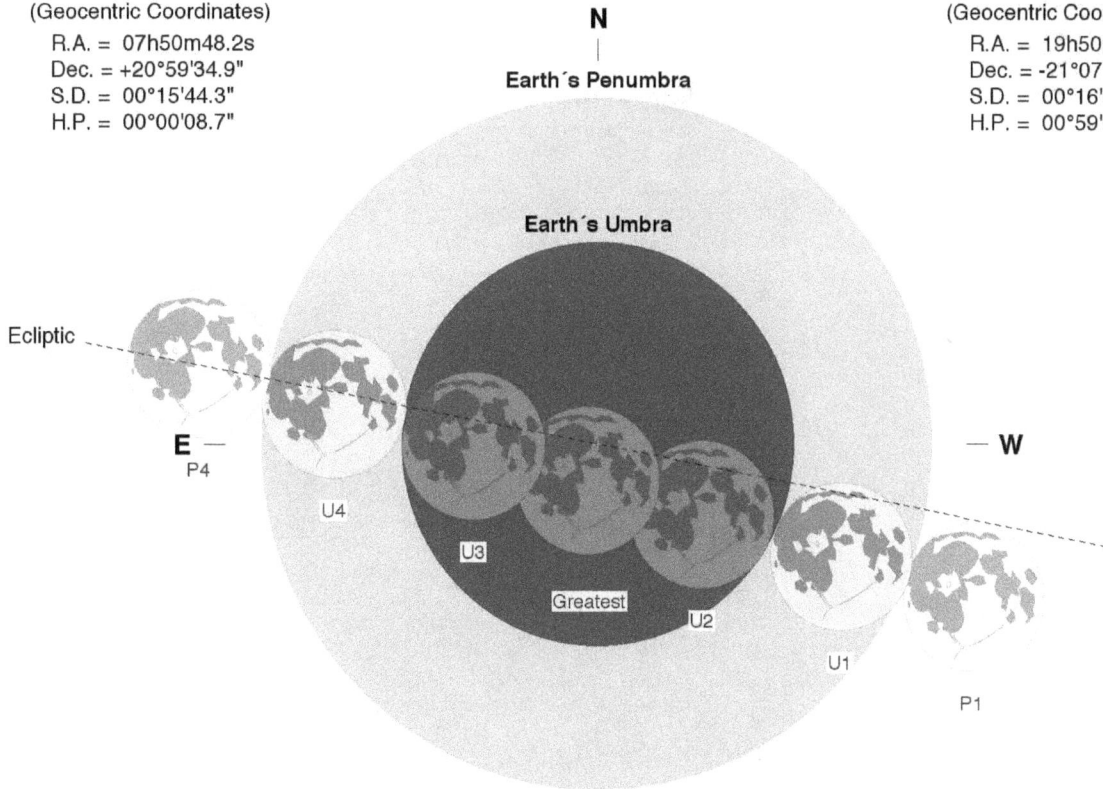

N

Earth´s Penumbra

Earth´s Umbra

Ecliptic

E

P4

U4

U3

Greatest

U2

U1

P1

W

S

Eclipse Durations
Penumbral = 05h31m53s
Umbral = 03h37m02s
Total = 01h37m43s

Eph. = JPL DE430
Rule = Herald-Sinnott
ΔT =   96 s

0   15   30   45   60
Arc-Minutes

©2020 F. Espenak, www.EclipseWise.com

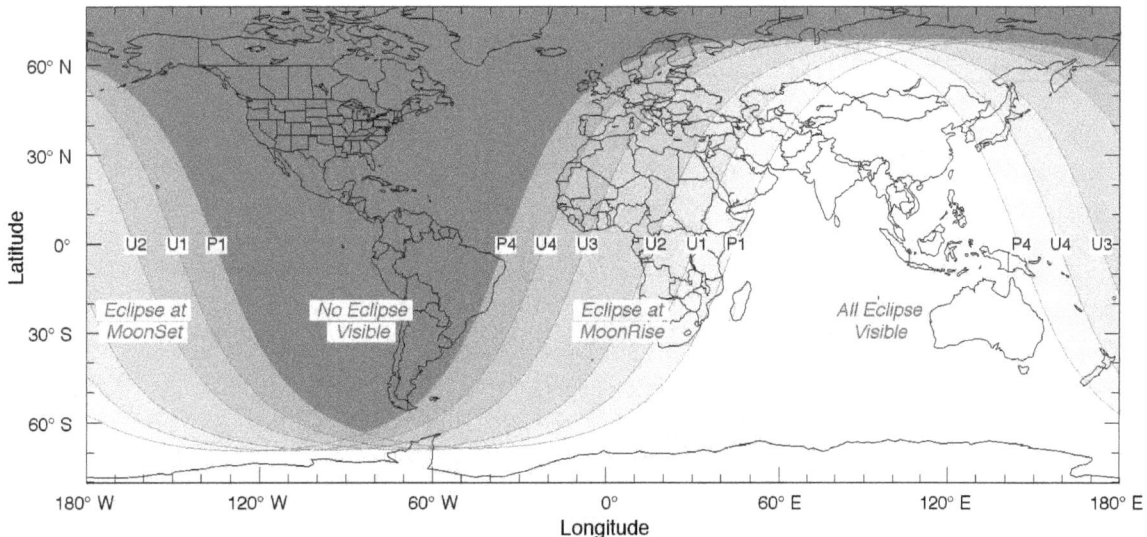

Eclipse Contacts
P1 = 15:01:12 UT1
U1 = 15:58:34 UT1
U2 = 16:58:15 UT1
U3 = 18:35:59 UT1
U4 = 19:35:36 UT1
P4 = 20:33:05 UT1

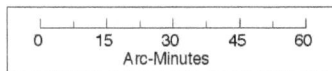

205

Total Lunar Eclipse of 2066 Jan 11

Greatest Eclipse = 15:04:47.3 TD   (= 15:03:10.9 UT1)

| Penumbral Magnitude = 2.2276 | Gamma = -0.3687 | Saros Series = 135 |
| Umbral Magnitude = 1.1395 | Axis = 0.3371° | Saros Member = 26 of 71 |

Sun at Greatest Eclipse
(Geocentric Coordinates)
R.A. = 19h33m47.1s
Dec. = -21°40'59.0"
S.D. = 00°16'15.8"
H.P. = 00°00'08.9"

N

Earth's Penumbra

Earth's Umbra

Moon at Greatest Eclipse
(Geocentric Coordinates)
R.A. = 07h33m24.8s
Dec. = +21°21'26.1"
S.D. = 00°14'56.8"
H.P. = 00°54'51.2"

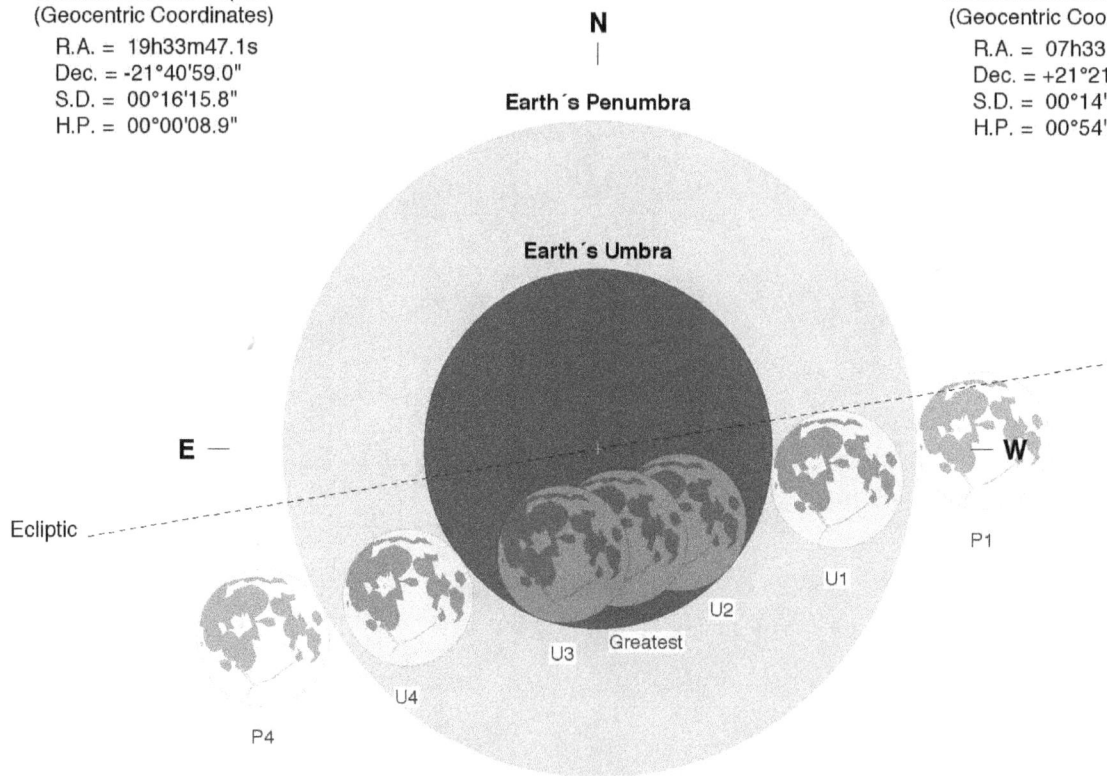

E

W

Ecliptic

P1

U1

U2

Greatest

U3

U4

P4

S

Eclipse Durations
Penumbral = 06h01m37s
Umbral = 03h35m59s
Total = 00h58m39s

Eph. = JPL DE430
Rule = Herald-Sinnott
ΔT = 96 s

0   15   30   45   60
Arc-Minutes

©2020 F. Espenak, www.EclipseWise.com

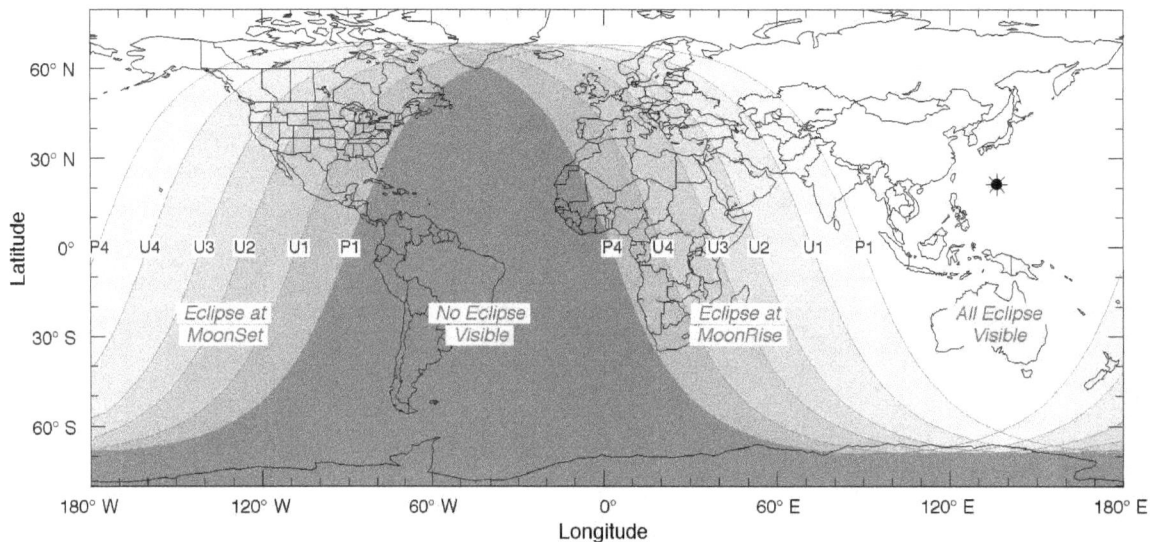

Eclipse Contacts
P1 = 12:02:18 UT1
U1 = 13:15:09 UT1
U2 = 14:33:44 UT1
U3 = 15:32:23 UT1
U4 = 16:51:08 UT1
P4 = 18:03:55 UT1

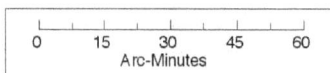

P4 U4 U3 U2 U1 P1   P4 U4 U3 U2 U1 P1

Eclipse at MoonSet

No Eclipse Visible

Eclipse at MoonRise

All Eclipse Visible

Latitude

Longitude

Partial Lunar Eclipse of 2066 Jul 07

Greatest Eclipse = 09:30:29.3 TD   (= 09:28:52.6 UT1)

Penumbral Magnitude = 1.7196	Gamma = 0.6056	Saros Series = 140
Umbral Magnitude = 0.7770	Axis = 0.6181°	Saros Member = 27 of 77

Sun at Greatest Eclipse
(Geocentric Coordinates)
R.A. = 07h07m48.8s
Dec. = +22°30'58.3"
S.D. = 00°15'43.9"
H.P. = 00°00'08.7"

Moon at Greatest Eclipse
(Geocentric Coordinates)
R.A. = 19h07m14.9s
Dec. = -21°54'43.6"
S.D. = 00°16'41.4"
H.P. = 01°01'15.2"

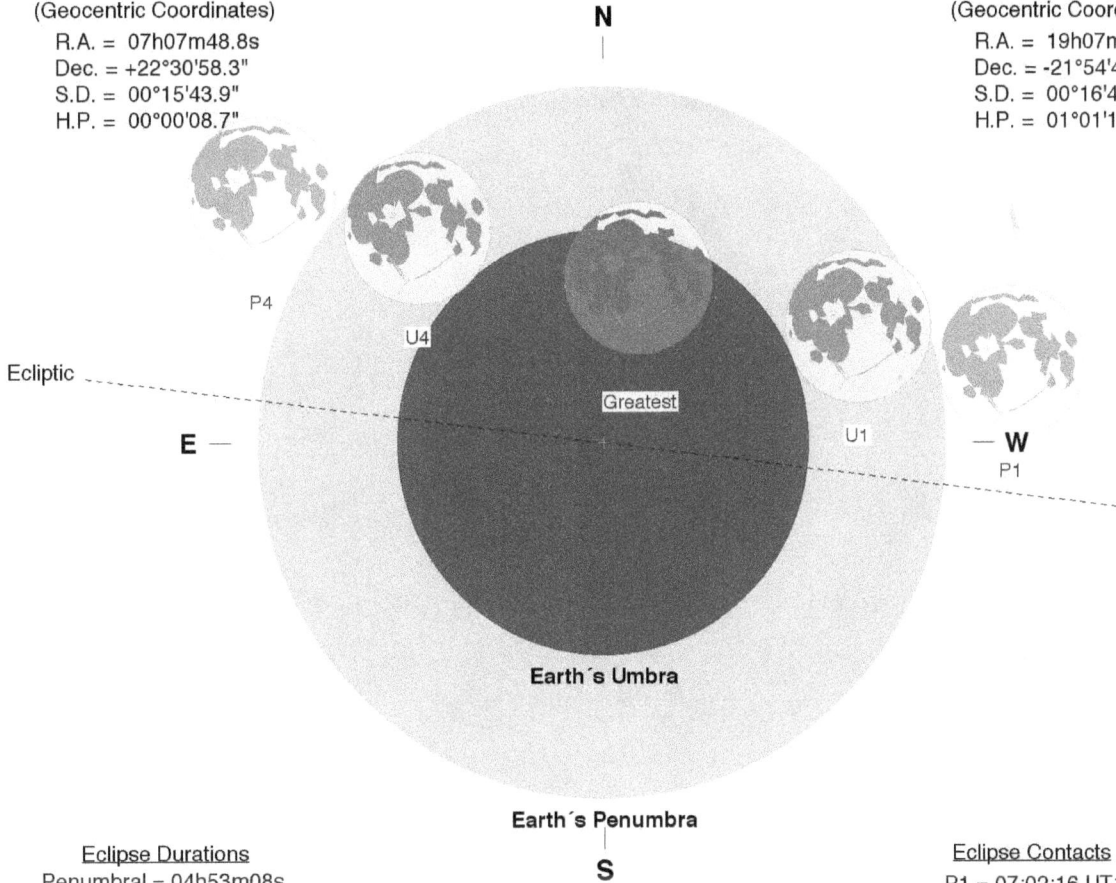

N

P4

U4

Ecliptic

E

Greatest

U1

W

P1

Earth's Umbra

Earth's Penumbra

S

Eclipse Durations
Penumbral = 04h53m08s
Umbral = 02h52m01s

Eph. = JPL DE430
Rule = Herald-Sinnott
ΔT =   97 s

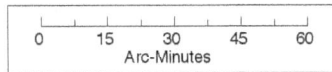

0	15	30	45	60

Arc-Minutes

©2020 F. Espenak, www.EclipseWise.com

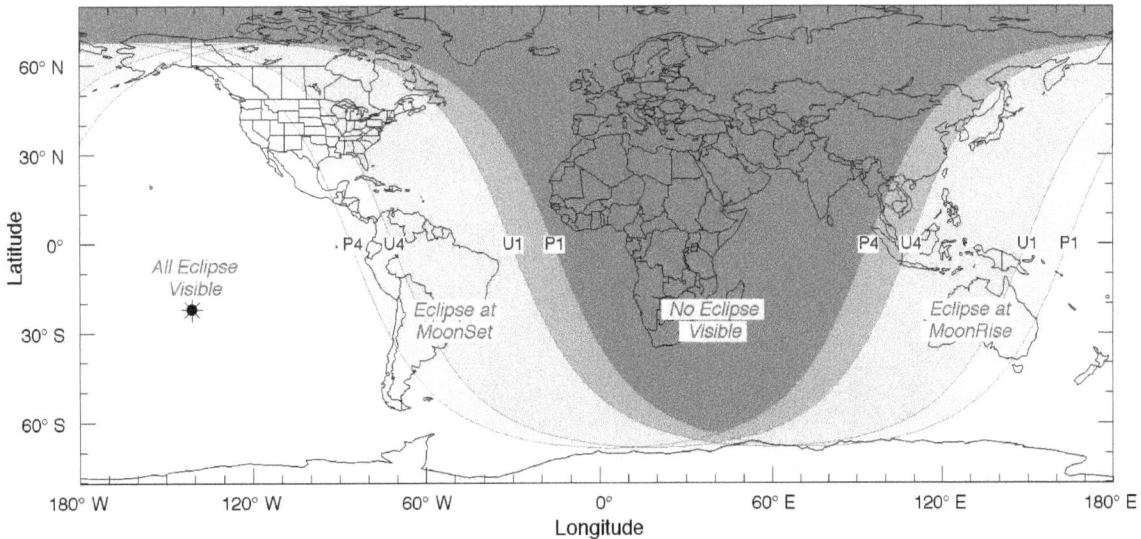

Eclipse Contacts
P1 = 07:02:16 UT1
U1 = 08:02:47 UT1
U4 = 10:54:48 UT1
P4 = 11:55:24 UT1

All Eclipse
Visible

Eclipse at
MoonSet

No Eclipse
Visible

Eclipse at
MoonRise

Latitude / Longitude

Penumbral Lunar Eclipse of 2066 Dec 31

Greatest Eclipse = 14:30:09.5 TD (= 14:28:32.4 UT1)

Penumbral Magnitude = 0.9789	Gamma = -1.0540	Saros Series = 145
Umbral Magnitude = -0.1264	Axis = 0.9485°	Saros Member = 14 of 71

Sun at Greatest Eclipse
(Geocentric Coordinates)
R.A. = 18h44m27.3s
Dec. = -23°02'13.7"
S.D. = 00°16'15.9"
H.P. = 00°00'08.9"

N

Earth's Penumbra

Earth's Umbra

Moon at Greatest Eclipse
(Geocentric Coordinates)
R.A. = 06h43m44.2s
Dec. = +22°06'11.2"
S.D. = 00°14'42.9"
H.P. = 00°54'00.3"

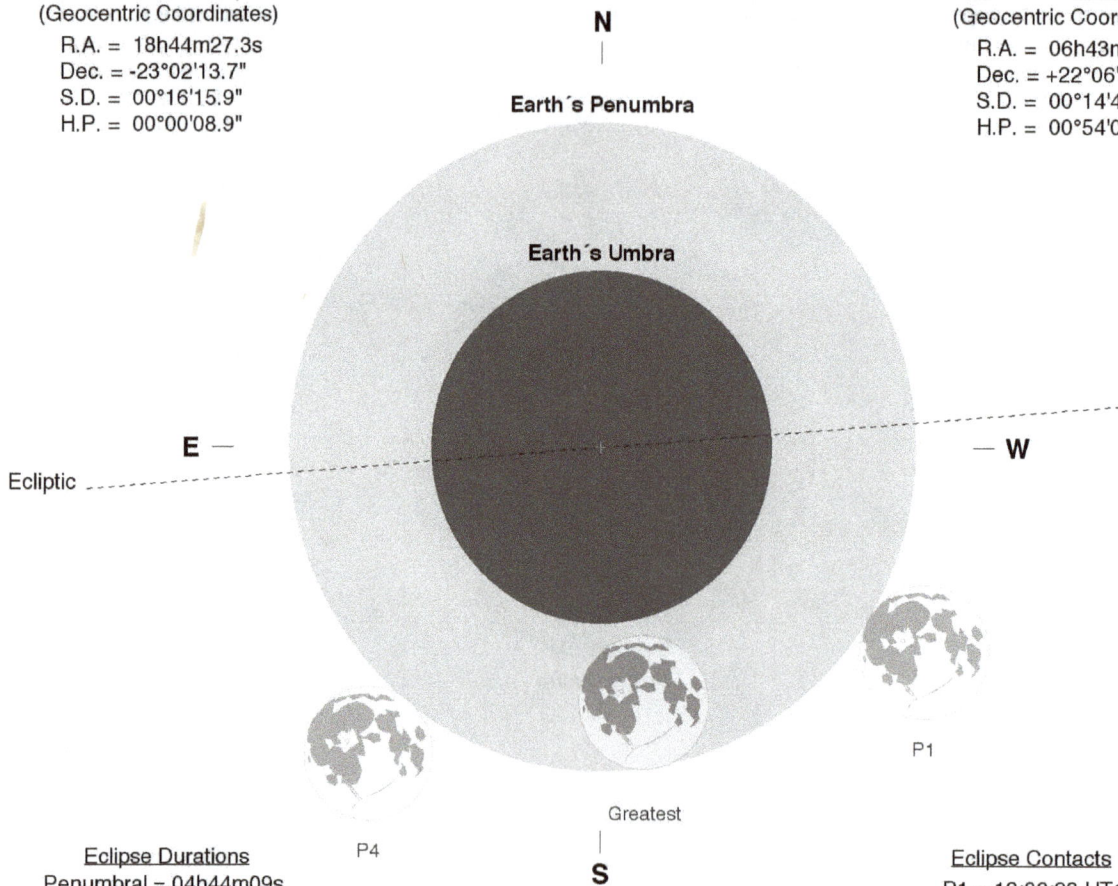

E

Ecliptic

W

P1

Greatest

S

P4

Eclipse Durations
Penumbral = 04h44m09s

Eclipse Contacts
P1 = 12:06:23 UT1
P4 = 16:50:32 UT1

Eph. = JPL DE430
Rule = Herald-Sinnott
ΔT = 97 s

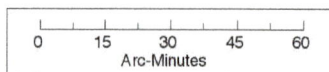

0 15 30 45 60
Arc-Minutes

©2020 F. Espenak, www.EclipseWise.com

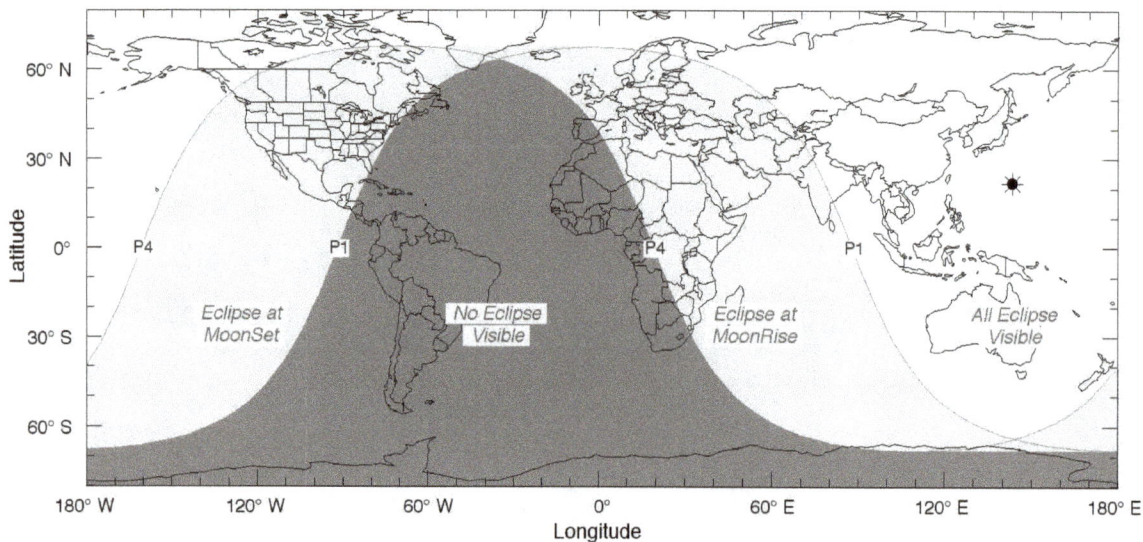

Eclipse at MoonSet

No Eclipse Visible

Eclipse at MoonRise

All Eclipse Visible

P4 P1 P4 P1

Penumbral Lunar Eclipse of 2067 May 28

Greatest Eclipse = 18:56:07.3 TD   (= 18:54:29.9 UT1)

Penumbral Magnitude = 0.6416	Gamma = -1.2013	Saros Series = 112
Umbral Magnitude = -0.3316	Axis = 1.1917°	Saros Member = 68 of 72

Sun at Greatest Eclipse
(Geocentric Coordinates)
R.A. = 04h23m02.8s
Dec. = +21°33'42.4"
S.D. = 00°15'47.0"
H.P. = 00°00'08.7"

N
Earth's Penumbra

Earth's Umbra

Moon at Greatest Eclipse
(Geocentric Coordinates)
R.A. = 16h22m41.8s
Dec. = -22°45'02.7"
S.D. = 00°16'13.2"
H.P. = 00°59'31.5"

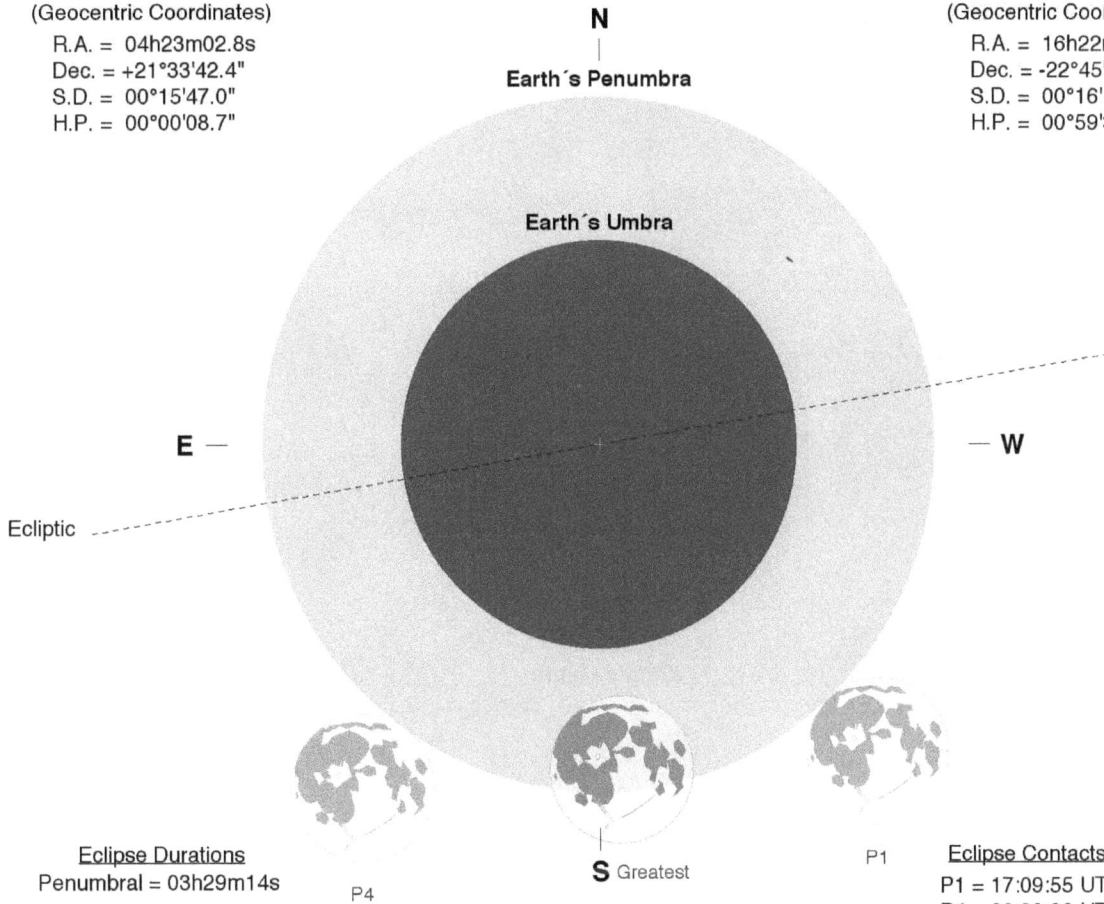

E

W

Ecliptic

S Greatest

P4

P1

Eclipse Durations
Penumbral = 03h29m14s

Eclipse Contacts
P1 = 17:09:55 UT1
P4 = 20:39:09 UT1

0   15   30   45   60
Arc-Minutes

Eph. = JPL DE430
Rule = Herald-Sinnott
ΔT =   97 s

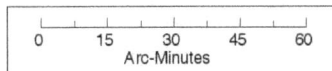

©2020 F. Espenak, www.EclipseWise.com

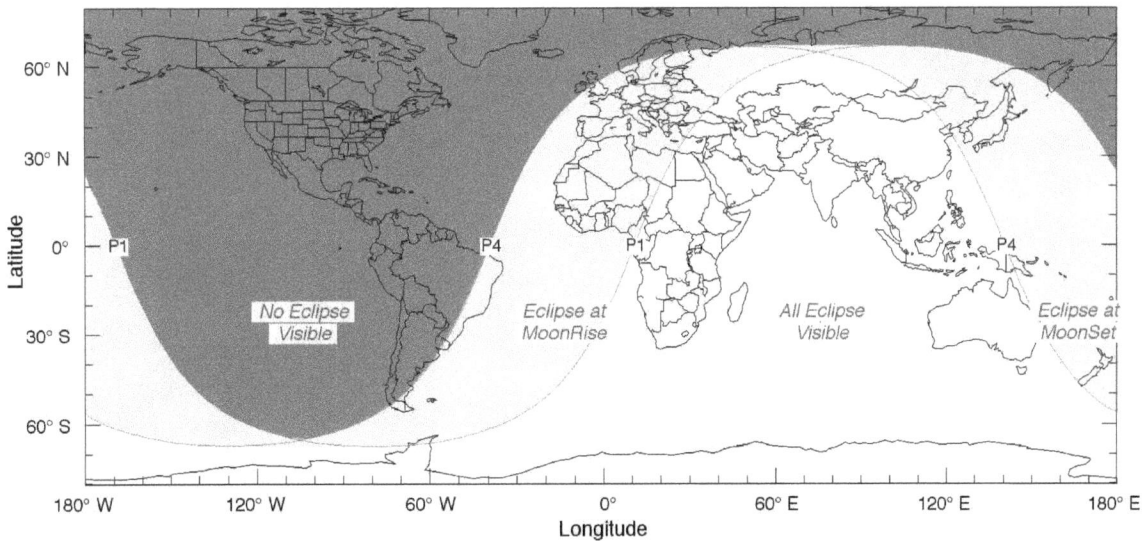

60° N

30° N

0°

Latitude

30° S

60° S

P1   P4   P1   P4

No Eclipse
Visible

Eclipse at
MoonRise

All Eclipse
Visible

Eclipse at
MoonSet

180° W   120° W   60° W   0°   60° E   120° E   180° E
Longitude

Penumbral Lunar Eclipse of 2067 Jun 27

Greatest Eclipse = 02:41:05.5 TD   (= 02:39:28.0 UT1)

Penumbral Magnitude = 0.3770	Gamma = 1.3394	Saros Series = 150
Umbral Magnitude = -0.5736	Axis = 1.3559°	Saros Member = 4 of 71

Sun at Greatest Eclipse
(Geocentric Coordinates)
R.A. = 06h24m21.9s
Dec. = +23°18'42.3"
S.D. = 00°15'44.1"
H.P. = 00°00'08.7"

Moon at Greatest Eclipse
(Geocentric Coordinates)
R.A. = 18h23m33.9s
Dec. = -21°58'06.1"
S.D. = 00°16'33.1"
H.P. = 01°00'44.6"

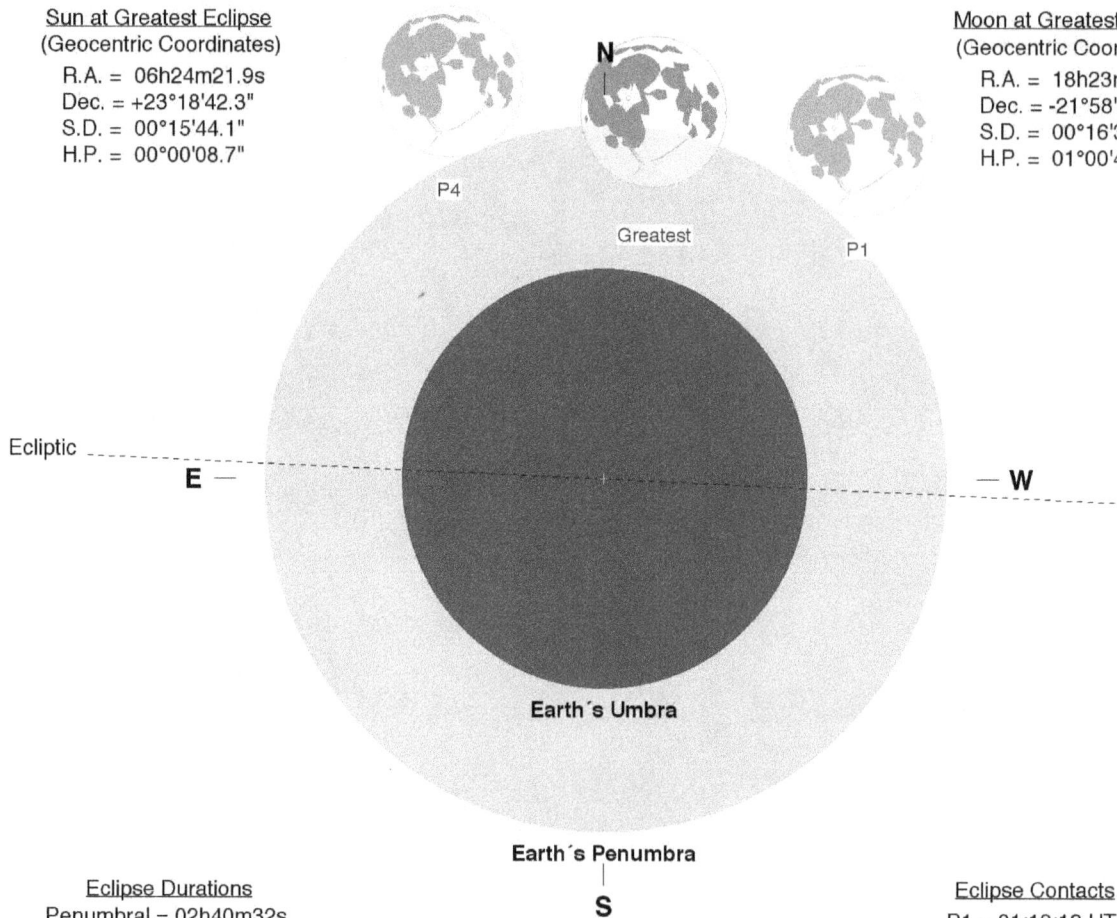

N

P4

Greatest

P1

Ecliptic

E —

— W

Earth´s Umbra

Earth´s Penumbra

S

Eclipse Durations
Penumbral = 02h40m32s

Eclipse Contacts
P1 = 01:19:12 UT1
P4 = 03:59:44 UT1

Eph. = JPL DE430
Rule = Herald-Sinnott
ΔT =   97 s

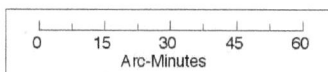

0	15	30	45	60

Arc-Minutes

©2020 F. Espenak, www.EclipseWise.com

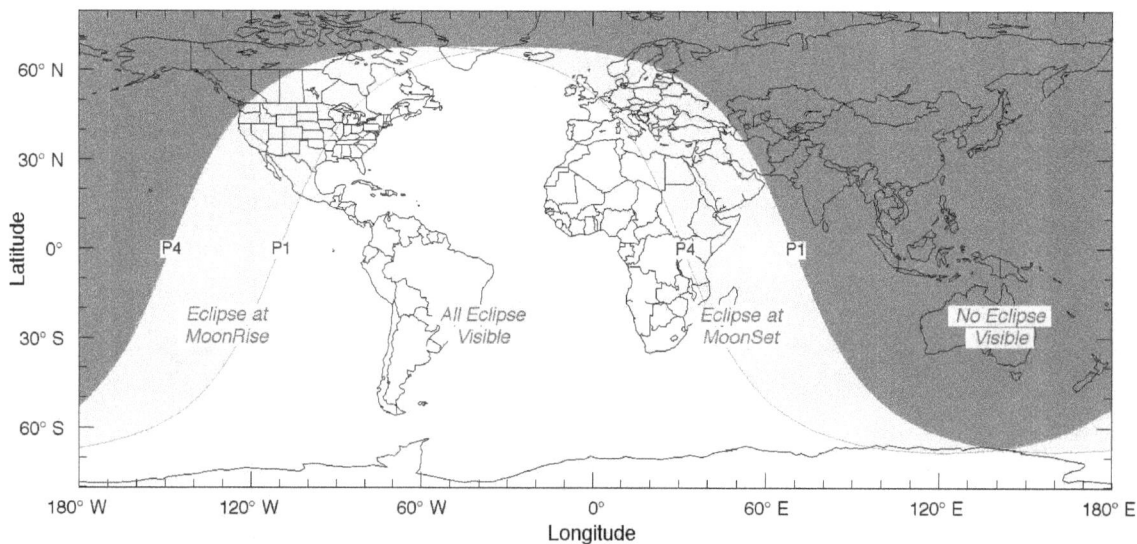

Eclipse at
MoonRise

All Eclipse
Visible

Eclipse at
MoonSet

No Eclipse
Visible

P4   P1

P4   P1

Penumbral Lunar Eclipse of 2067 Nov 21

Greatest Eclipse = 00:04:41.6 TD (= 00:03:03.8 UT1)

Penumbral Magnitude = 0.6557	Gamma = 1.2107	Saros Series = 117
Umbral Magnitude = -0.3798	Axis = 1.1575°	Saros Member = 55 of 71

Sun at Greatest Eclipse
(Geocentric Coordinates)

R.A. = 15h46m01.0s
Dec. = -19°52'08.2"
S.D. = 00°16'11.2"
H.P. = 00°00'08.9"

Moon at Greatest Eclipse
(Geocentric Coordinates)

R.A. = 03h45m24.1s
Dec. = +21°01'03.2"
S.D. = 00°15'37.9"
H.P. = 00°57'22.2"

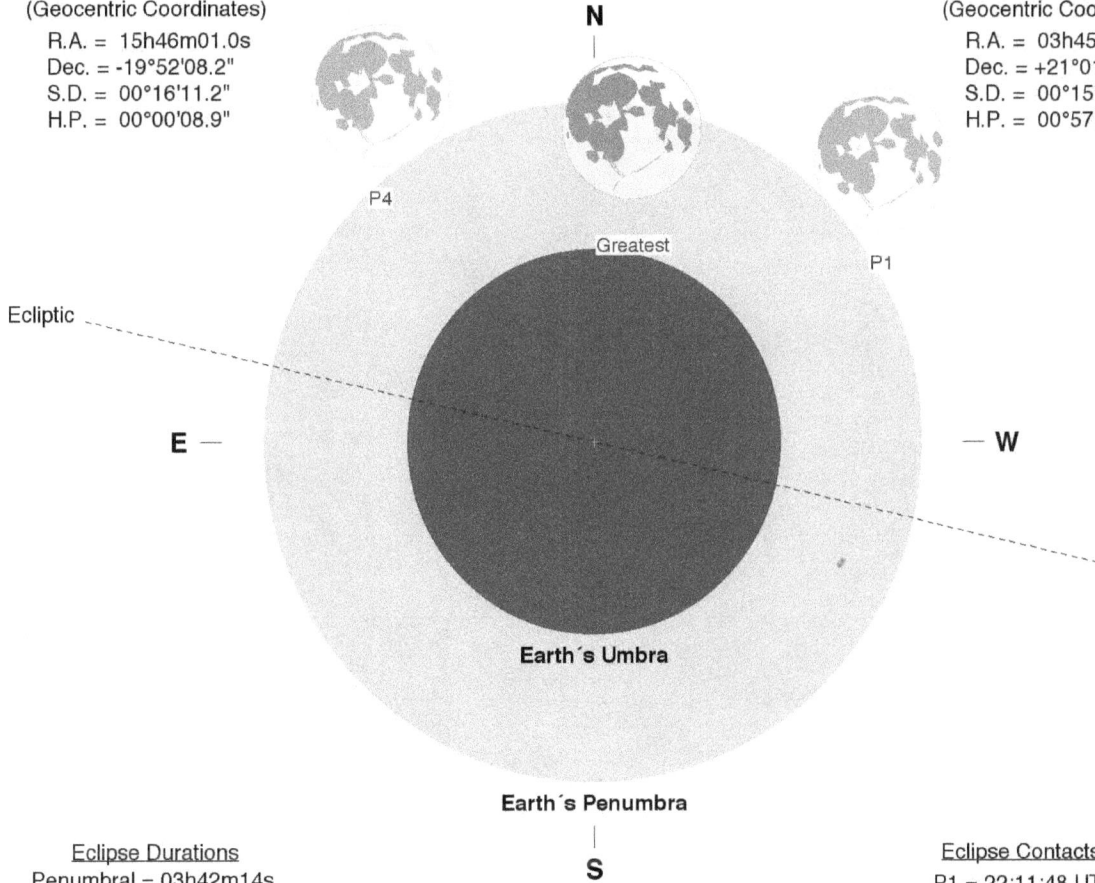

N

P4

Greatest

P1

Ecliptic

E —

— W

Earth's Umbra

Earth's Penumbra

S

Eclipse Durations
Penumbral = 03h42m14s

Eclipse Contacts
P1 = 22:11:48 UT1
P4 = 01:54:02 UT1

Eph. = JPL DE430
Rule = Herald-Sinnott
ΔT =   98 s

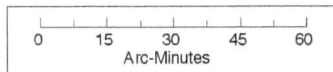

0	15	30	45	60

Arc-Minutes

©2020 F. Espenak, www.EclipseWise.com

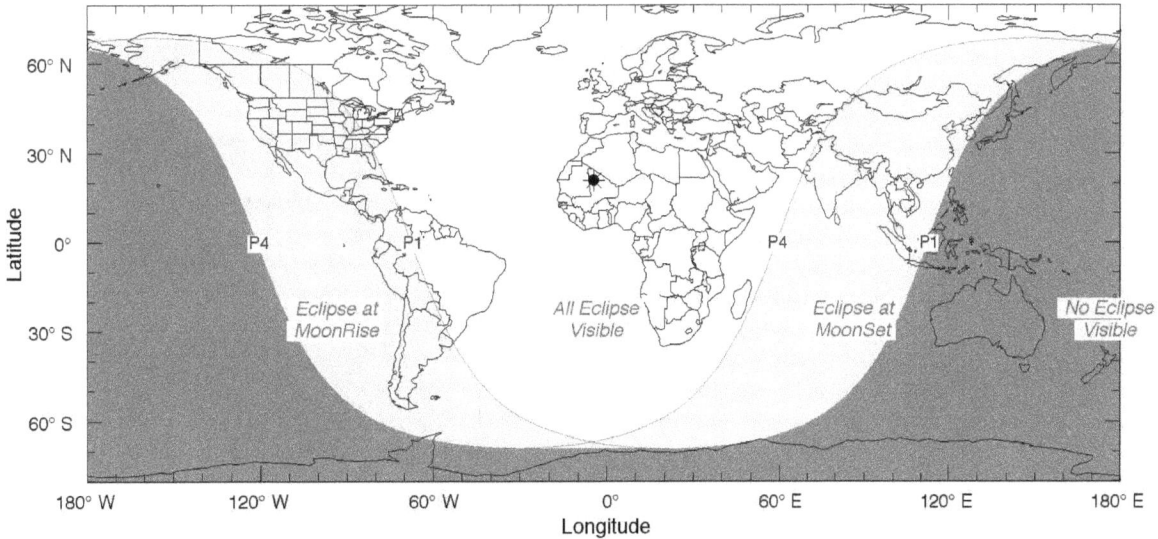

Eclipse at MoonRise

All Eclipse Visible

Eclipse at MoonSet

No Eclipse Visible

P4

P1

P4

P1

Partial Lunar Eclipse of 2068 May 17

Greatest Eclipse = 05:42:17.0 TD   (= 05:40:38.8 UT1)

Penumbral Magnitude = 1.9839	Gamma = -0.4852	Saros Series = 122
Umbral Magnitude = 0.9545	Axis = 0.4559°	Saros Member = 59 of 74

Sun at Greatest Eclipse
(Geocentric Coordinates)
R.A. = 03h39m33.2s
Dec. = +19°31'07.8"
S.D. = 00°15'49.0"
H.P. = 00°00'08.7"

N

Earth´s Penumbra

Earth´s Umbra

E

W

Ecliptic

U1

P1

U4

Greatest

P4

Moon at Greatest Eclipse
(Geocentric Coordinates)
R.A. = 15h39m17.9s
Dec. = -19°58'14.9"
S.D. = 00°15'21.9"
H.P. = 00°56'23.4"

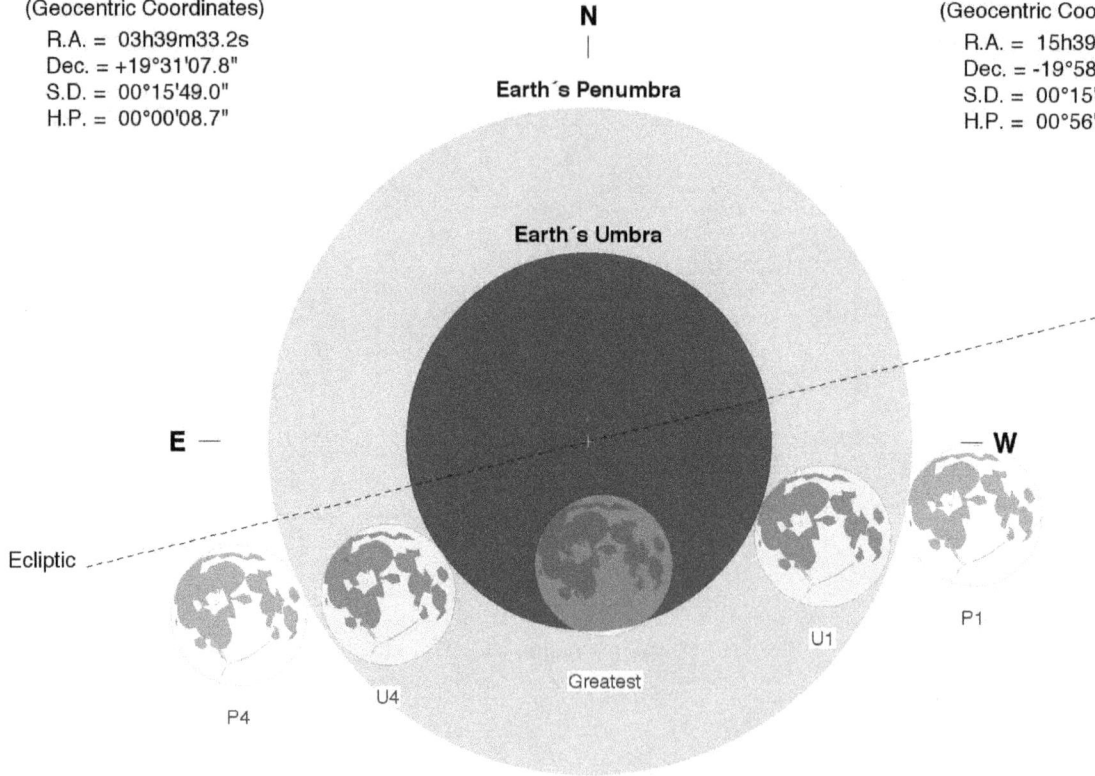

S

Eclipse Durations
Penumbral = 05h37m33s
Umbral = 03h19m46s

Eph. = JPL DE430
Rule = Herald-Sinnott
ΔT = 98 s

0	15	30	45	60

Arc-Minutes

©2020 F. Espenak, www.EclipseWise.com

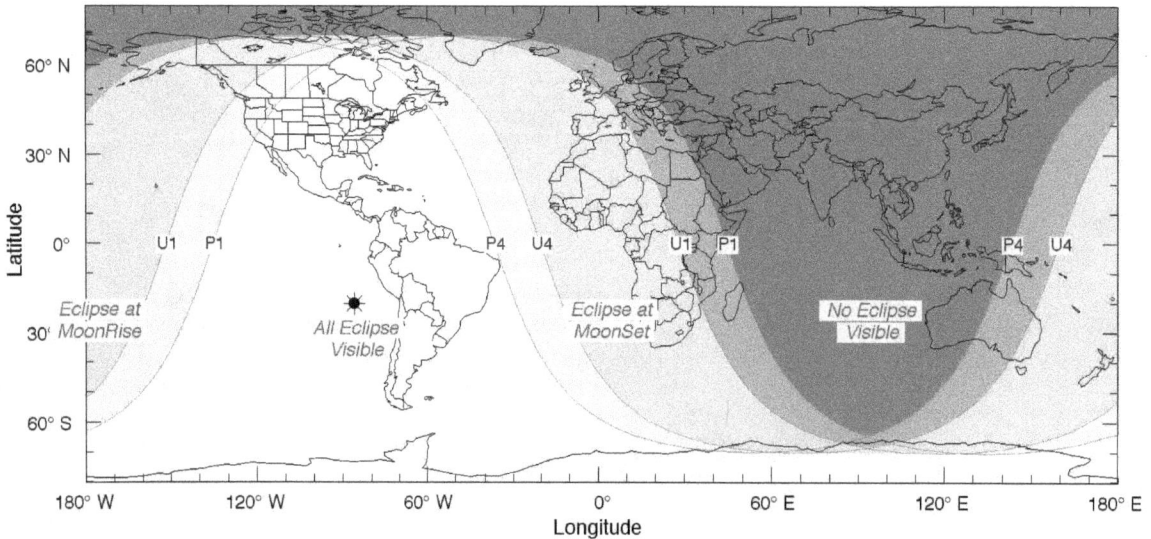

Eclipse Contacts
P1 = 02:51:49 UT1
U1 = 04:00:46 UT1
U4 = 07:20:31 UT1
P4 = 08:29:21 UT1

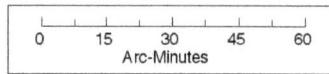

Total Lunar Eclipse of 2068 Nov 09

Greatest Eclipse = 11:46:59.7 TD (= 11:45:21.2 UT1)

Penumbral Magnitude = 1.9974	Gamma = 0.4645	Saros Series = 127
Umbral Magnitude = 1.0161	Axis = 0.4675°	Saros Member = 45 of 72

Sun at Greatest Eclipse
(Geocentric Coordinates)
R.A. = 15h01m47.2s
Dec. = -17°09'37.3"
S.D. = 00°16'08.8"
H.P. = 00°00'08.9"

Moon at Greatest Eclipse
(Geocentric Coordinates)
R.A. = 03h01m25.7s
Dec. = +17°37'12.0"
S.D. = 00°16'27.4"
H.P. = 01°00'23.7"

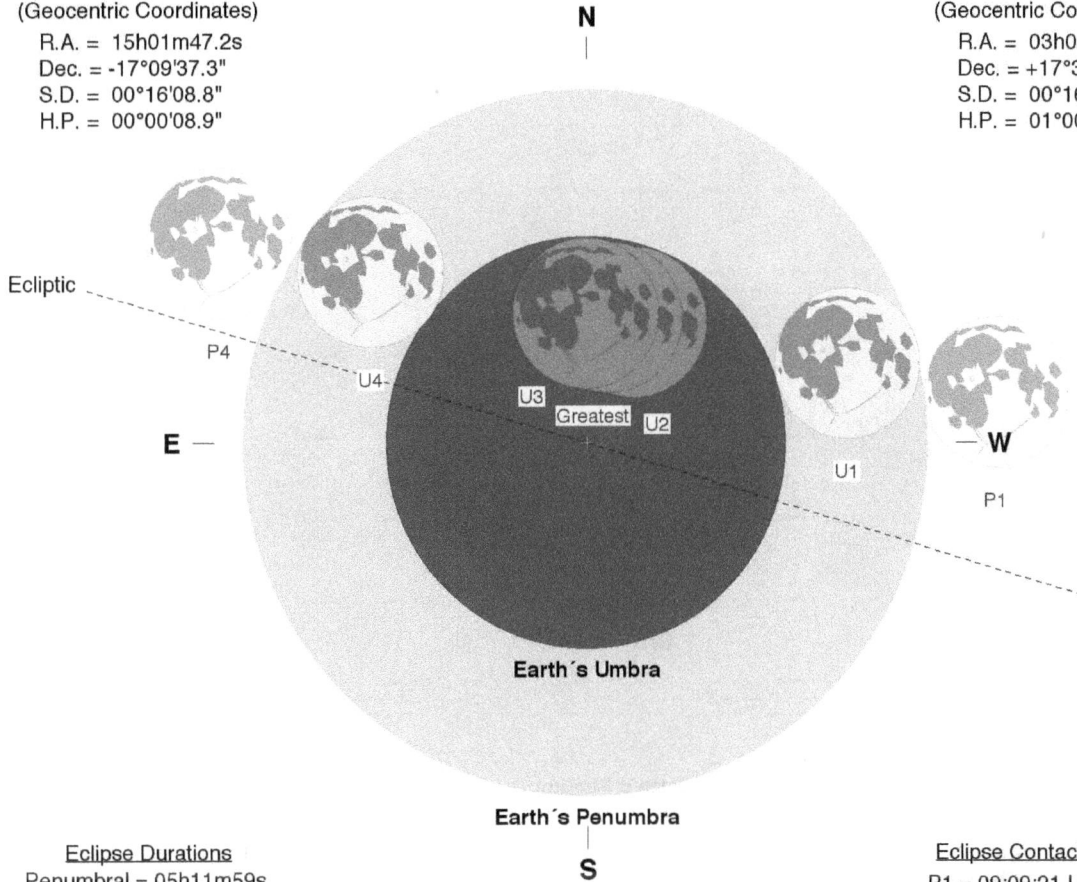

N

Ecliptic

P4

U4

U3

Greatest

U2

E

U1

W

P1

Earth's Umbra

Earth's Penumbra

S

Eclipse Durations
Penumbral = 05h11m59s
Umbral = 03h10m56s
Total = 00h19m12s

Eph. = JPL DE430
Rule = Herald-Sinnott
ΔT = 99 s

0 15 30 45 60
Arc-Minutes

©2020 F. Espenak, www.EclipseWise.com

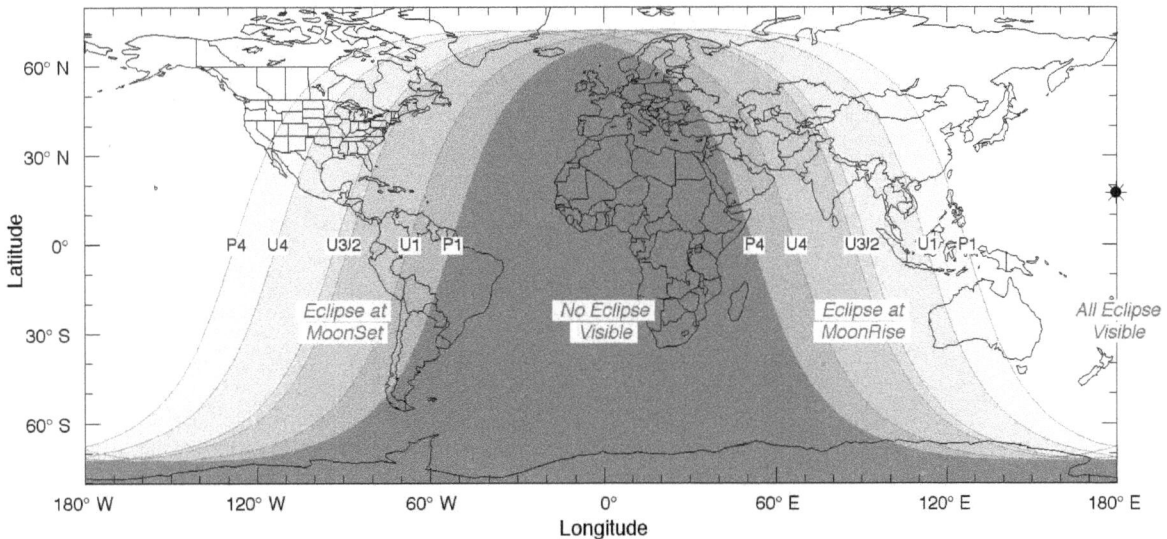

Eclipse Contacts
P1 = 09:09:21 UT1
U1 = 10:09:48 UT1
U2 = 11:35:37 UT1
U3 = 11:54:49 UT1
U4 = 13:20:45 UT1
P4 = 14:21:20 UT1

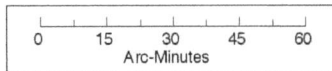

P4 U4 U3/2 U1 P1   P4 U4 U3/2 U1 P1

Eclipse at MoonSet   No Eclipse Visible   Eclipse at MoonRise   All Eclipse Visible

Total Lunar Eclipse of 2069 May 06

Greatest Eclipse = 09:09:56.6 TD (= 09:08:17.6 UT1)

| Penumbral Magnitude = 2.3977 | Gamma = 0.2717 | Saros Series = 132 |
| Umbral Magnitude = 1.3242 | Axis = 0.2454° | Saros Member = 33 of 71 |

Sun at Greatest Eclipse
(Geocentric Coordinates)

R.A. = 02h55m56.2s
Dec. = +16°44'53.2"
S.D. = 00°15'51.4"
H.P. = 00°00'08.7"

Moon at Greatest Eclipse
(Geocentric Coordinates)

R.A. = 14h56m07.8s
Dec. = -16°30'25.4"
S.D. = 00°14'46.2"
H.P. = 00°54'12.3"

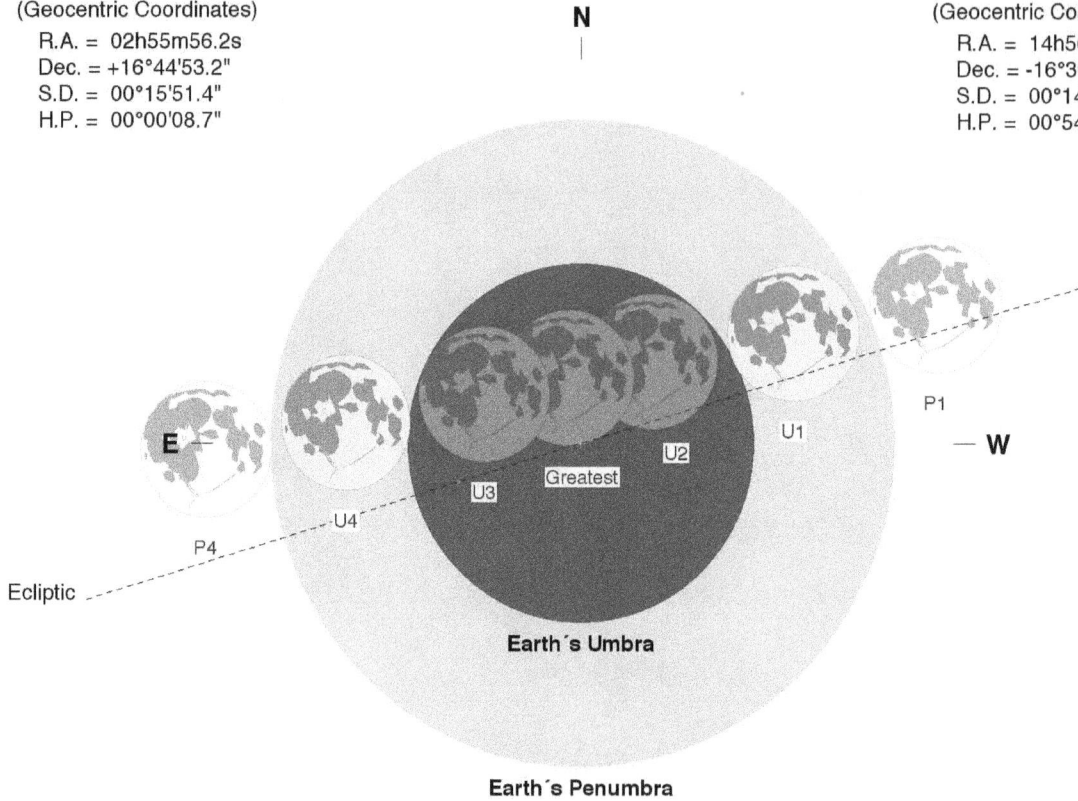

N

E — — W

Greatest
U3 U2 U1
U4
P4   P1
Ecliptic

Earth's Umbra

Earth's Penumbra

S

Eclipse Durations
Penumbral = 06h09m04s
Umbral = 03h47m01s
Total = 01h25m02s

Eph. = JPL DE430
Rule = Herald-Sinnott
ΔT =   99 s

0   15   30   45   60
Arc-Minutes

©2020 F. Espenak, www.EclipseWise.com

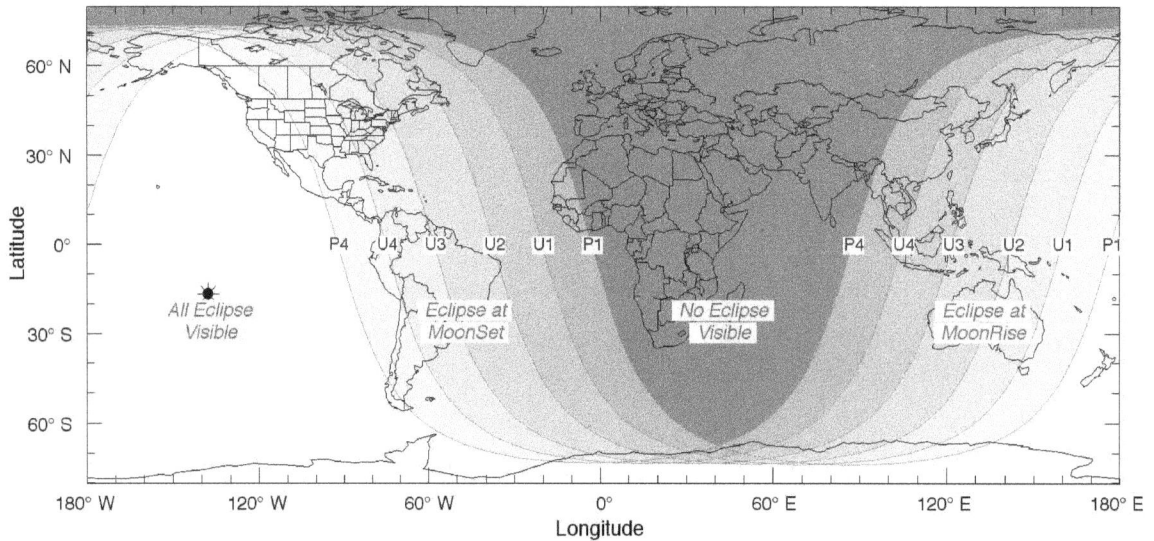

Eclipse Contacts
P1 = 06:03:46 UT1
U1 = 07:14:50 UT1
U2 = 08:25:52 UT1
U3 = 09:50:54 UT1
U4 = 11:01:51 UT1
P4 = 12:12:49 UT1

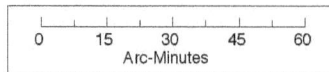

All Eclipse Visible

Eclipse at MoonSet

No Eclipse Visible

Eclipse at MoonRise

P4 U4 U3 U2 U1 P1   P4 U4 U3 U2 U1 P1

Total Lunar Eclipse of 2069 Oct 30

Greatest Eclipse = 03:35:05.7 TD   (= 03:33:26.4 UT1)

Penumbral Magnitude = 2.4247	Gamma = -0.2263	Saros Series = 137
Umbral Magnitude = 1.4628	Axis = 0.2317°	Saros Member = 29 of 78

Sun at Greatest Eclipse
(Geocentric Coordinates)

R.A. = 14h19m49.6s
Dec. = -13°56'35.4"
S.D. = 00°16'06.2"
H.P. = 00°00'08.9"

Moon at Greatest Eclipse
(Geocentric Coordinates)

R.A. = 02h20m02.8s
Dec. = +13°43'03.8"
S.D. = 00°16'44.5"
H.P. = 01°01'26.5"

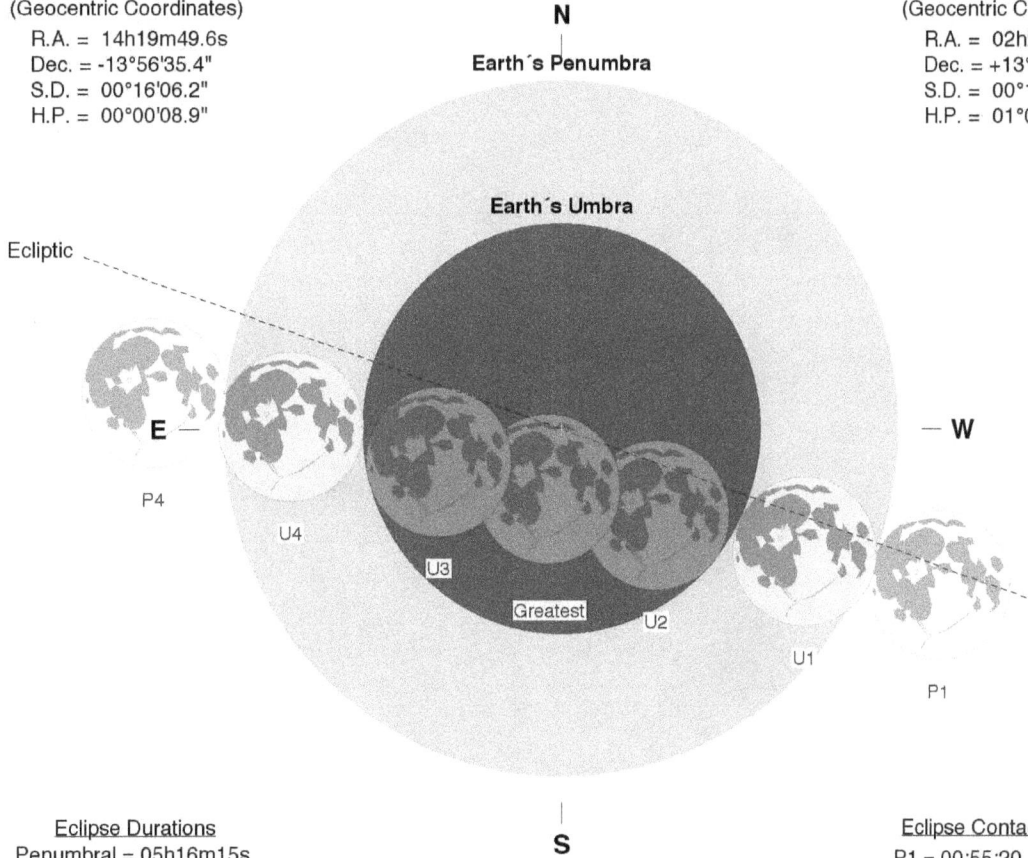

N

Earth's Penumbra

Earth's Umbra

Ecliptic

E

W

P4

U4

U3

Greatest

U2

U1

P1

S

Eclipse Durations
Penumbral = 05h16m15s
Umbral = 03h26m21s
Total = 01h27m26s

Eph. = JPL DE430
Rule = Herald-Sinnott
ΔT =   99 s

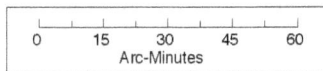

0   15   30   45   60
Arc-Minutes

©2020 F. Espenak, www.EclipseWise.com

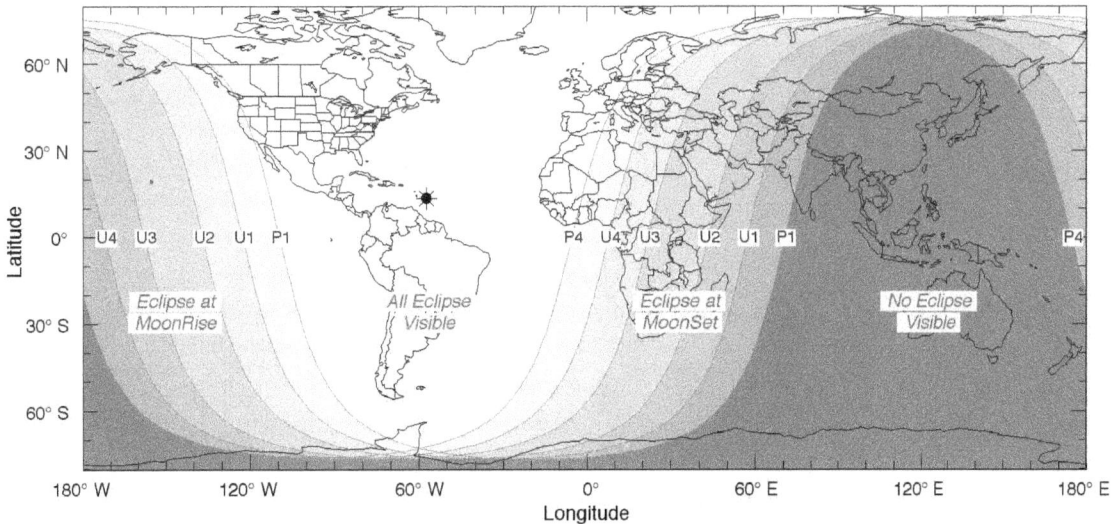

Eclipse Contacts
P1 = 00:55:20 UT1
U1 = 01:50:18 UT1
U2 = 02:49:48 UT1
U3 = 04:17:14 UT1
U4 = 05:16:39 UT1
P4 = 06:11:35 UT1

U4 U3 U2 U1 P1

Eclipse at
MoonRise

All Eclipse
Visible

P4 U4 U3 U2 U1 P1

Eclipse at
MoonSet

No Eclipse
Visible

P4

Penumbral Lunar Eclipse of 2070 Apr 25

Greatest Eclipse = 09:21:24.5 TD   (= 09:19:44.8 UT1)

Penumbral Magnitude = 1.0527	Gamma = 1.0044	Saros Series = 142
Umbral Magnitude = -0.0197	Axis = 0.9109°	Saros Member = 21 of 73

Sun at Greatest Eclipse
(Geocentric Coordinates)
R.A. = 02h12m57.7s
Dec. = +13°21'41.4"
S.D. = 00°15'54.1"
H.P. = 00°00'08.7"

Moon at Greatest Eclipse
(Geocentric Coordinates)
R.A. = 14h13m51.0s
Dec. = -12°28'35.7"
S.D. = 00°14'49.7"
H.P. = 00°54'25.3"

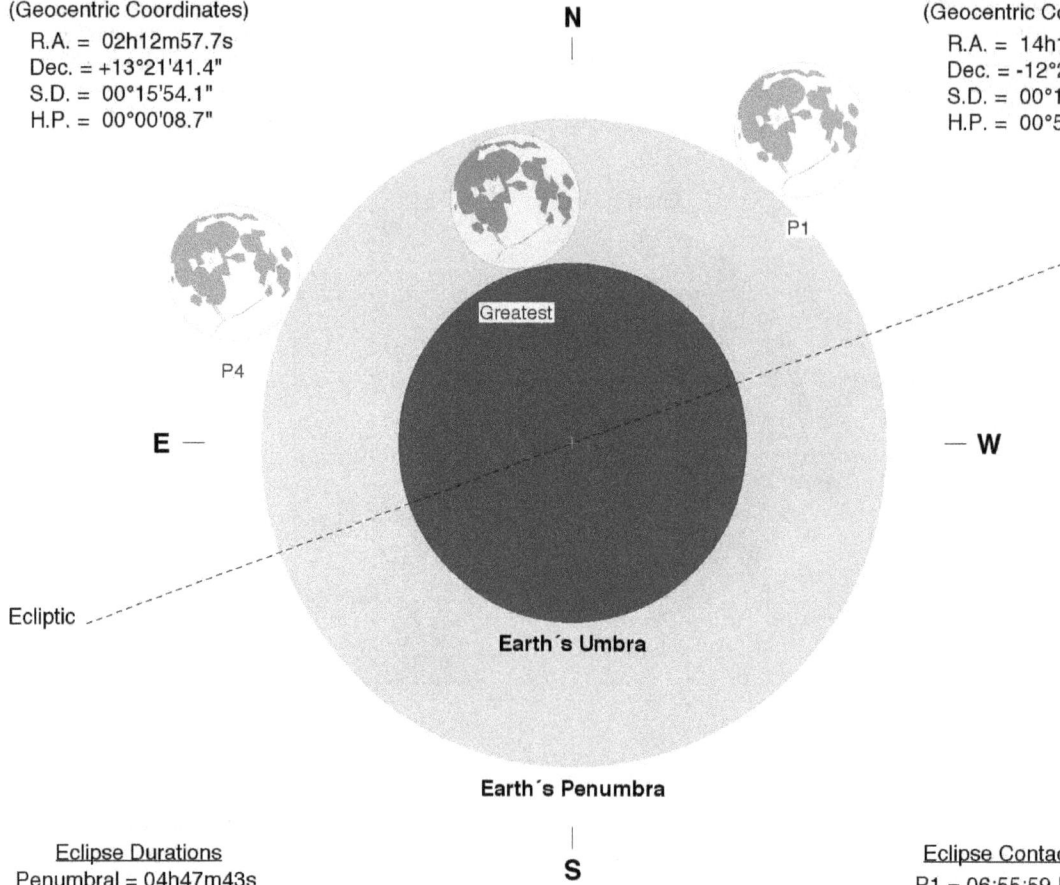

N

P1

Greatest

P4

E —

— W

Ecliptic

Earth's Umbra

Earth's Penumbra

S

Eclipse Durations
Penumbral = 04h47m43s

Eclipse Contacts
P1 = 06:55:59 UT1
P4 = 11:43:42 UT1

Eph. = JPL DE430
Rule = Herald-Sinnott
ΔT =   100 s

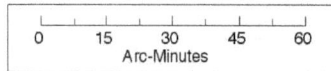

0	15	30	45	60

Arc-Minutes

©2020 F. Espenak, www.EclipseWise.com

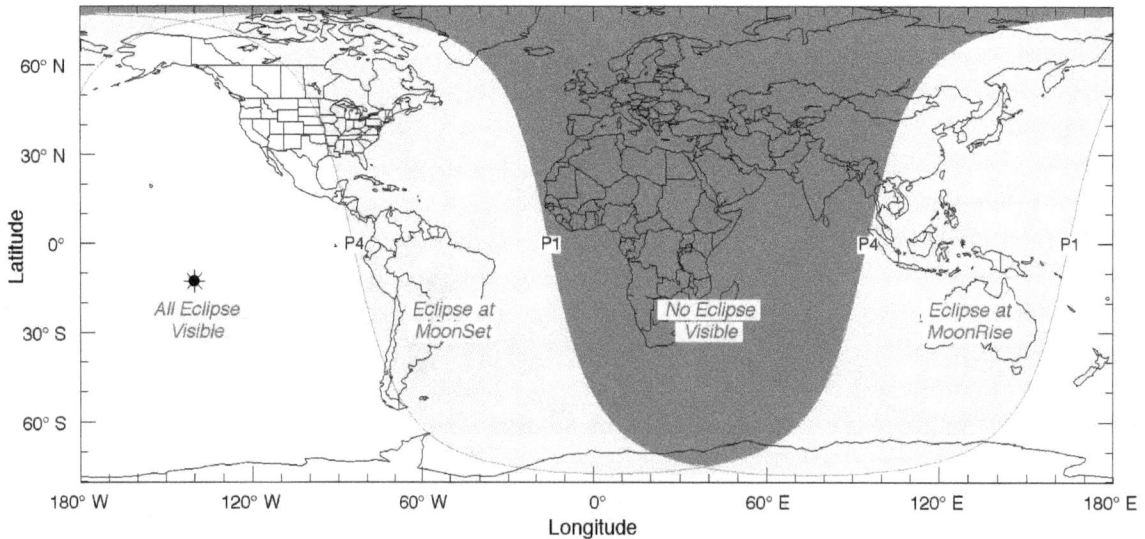

All Eclipse
Visible

Eclipse at
MoonSet

No Eclipse
Visible

Eclipse at
MoonRise

P4

P1

P4

P1

Partial Lunar Eclipse of 2070 Oct 19

Greatest Eclipse = 18:51:11.9 TD (= 18:49:31.8 UT1)

Penumbral Magnitude = 1.1270	Gamma = -0.9406	Saros Series = 147
Umbral Magnitude = 0.1395	Axis = 0.9354°	Saros Member = 11 of 70

Sun at Greatest Eclipse
(Geocentric Coordinates)

R.A. = 13h39m13.7s
Dec. = -10°18'14.5"
S.D. = 00°16'03.4"
H.P. = 00°00'08.8"

N

Earth's Penumbra

Ecliptic

Earth's Umbra

Moon at Greatest Eclipse
(Geocentric Coordinates)

R.A. = 01h40m14.8s
Dec. = +09°24'10.4"
S.D. = 00°16'15.6"
H.P. = 00°59'40.4"

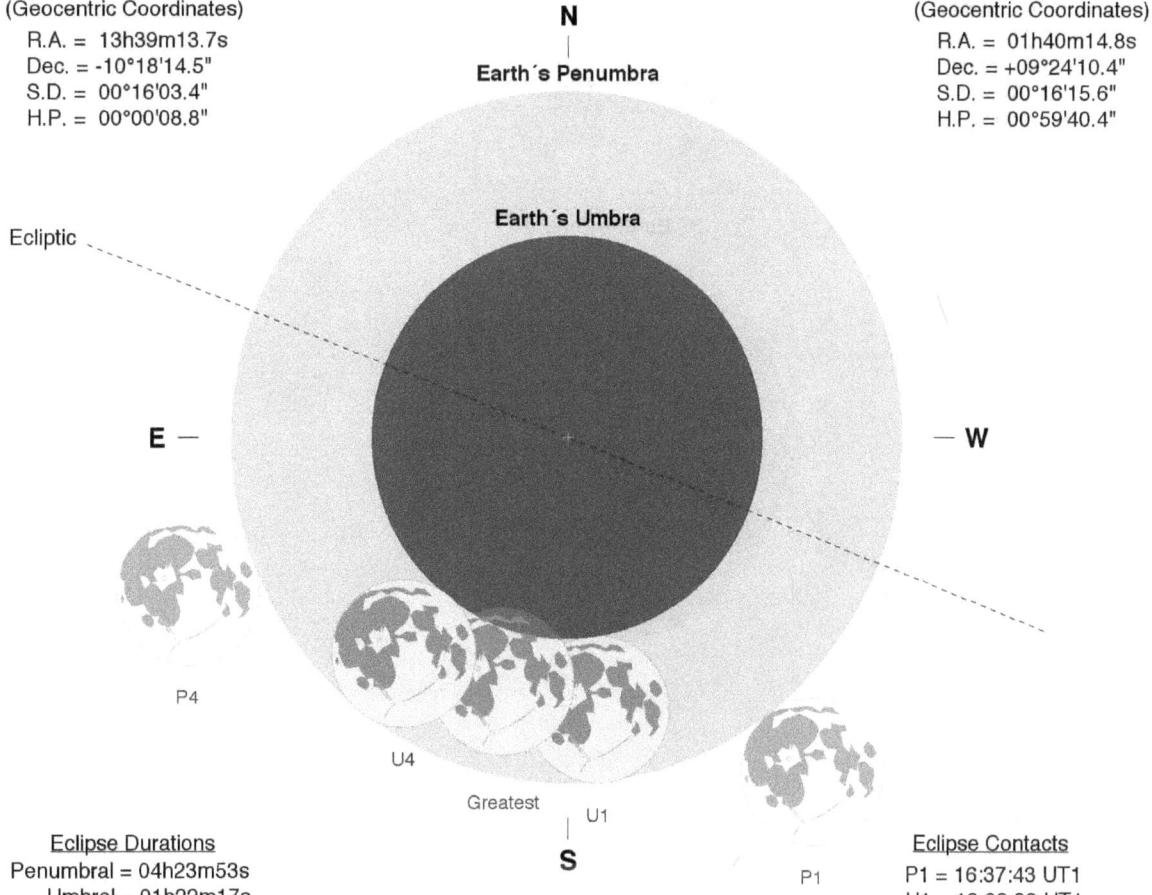

E —

— **W**

P4

U4

Greatest   U1

S

P1

Eclipse Durations
Penumbral = 04h23m53s
Umbral = 01h22m17s

Eph. = JPL DE430
Rule = Herald-Sinnott
ΔT = 100 s

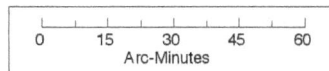

0	15	30	45	60

Arc-Minutes

©2020 F. Espenak, www.EclipseWise.com

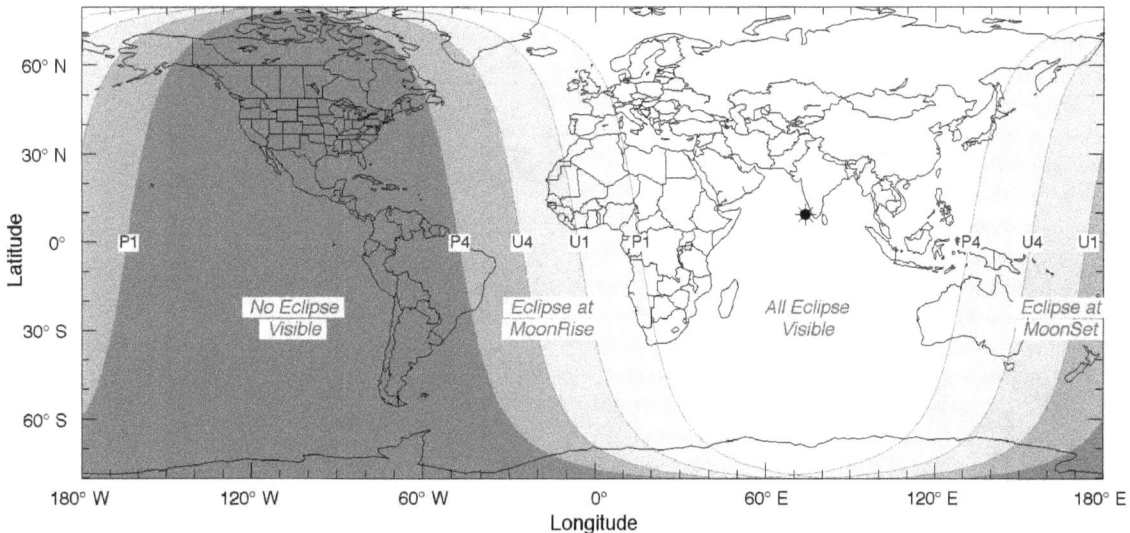

Eclipse Contacts
P1 = 16:37:43 UT1
U1 = 18:08:38 UT1
U4 = 19:30:55 UT1
P4 = 21:01:36 UT1

Penumbral Lunar Eclipse of 2071 Mar 16

Greatest Eclipse = 01:31:08.7 TD (= 01:29:28.3 UT1)

Penumbral Magnitude = 0.8890	Gamma = -1.0757	Saros Series = 114
Umbral Magnitude = -0.1183	Axis = 1.0506°	Saros Member = 62 of 71

Sun at Greatest Eclipse
(Geocentric Coordinates)
R.A. = 23h43m41.9s
Dec. = -01°45'49.9"
S.D. = 00°16'05.1"
H.P. = 00°00'08.8"

Moon at Greatest Eclipse
(Geocentric Coordinates)
R.A. = 11h42m24.8s
Dec. = +00°45'48.7"
S.D. = 00°15'58.1"
H.P. = 00°58'36.2"

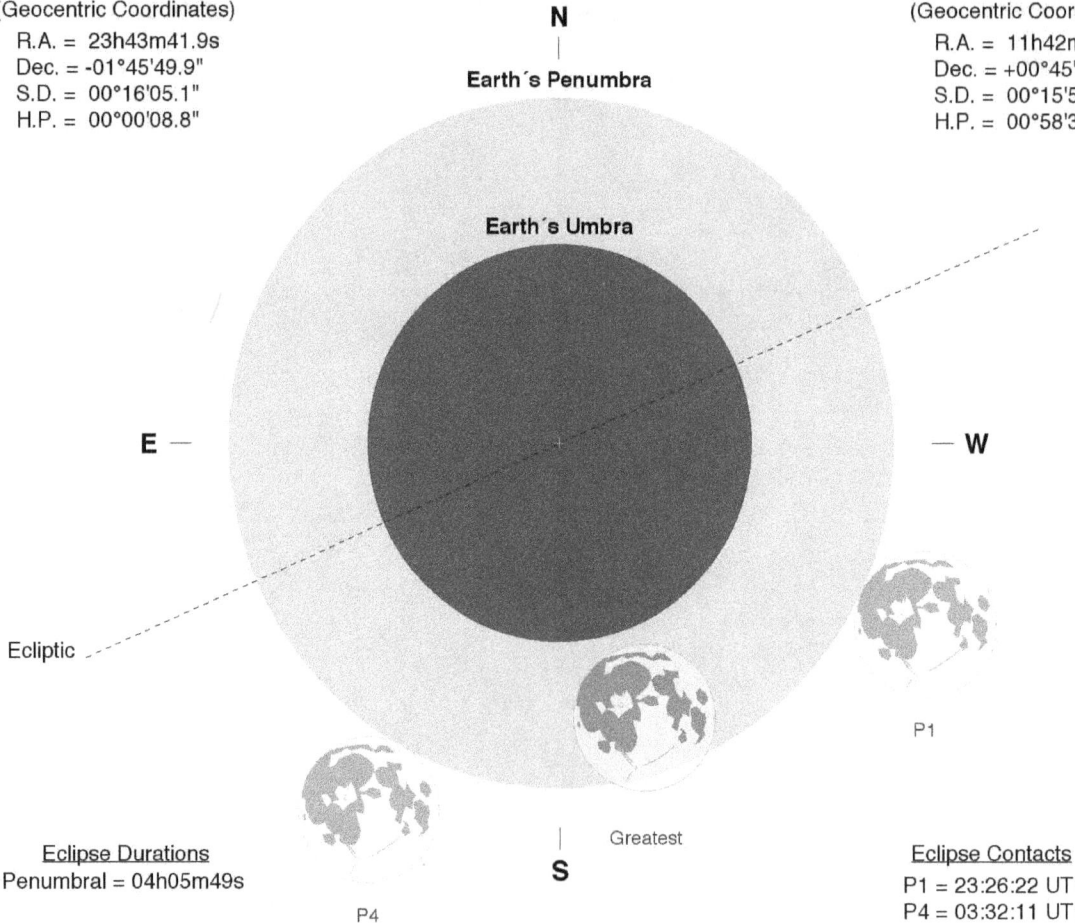

N

Earth's Penumbra

Earth's Umbra

E

W

Ecliptic

P1

Greatest

S

P4

Eclipse Durations
Penumbral = 04h05m49s

Eclipse Contacts
P1 = 23:26:22 UT1
P4 = 03:32:11 UT1

Eph. = JPL DE430
Rule = Herald-Sinnott
ΔT = 100 s

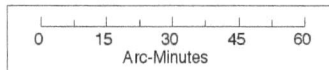

0	15	30	45	60

Arc-Minutes

©2020 F. Espenak, www.EclipseWise.com

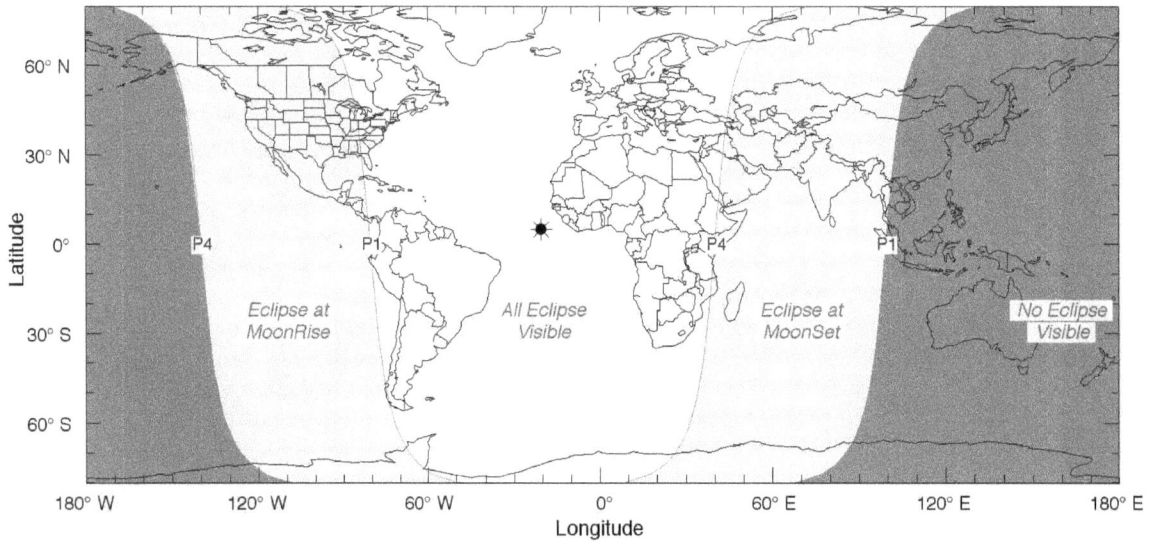

P4

P1

P4

P1

Eclipse at
MoonRise

All Eclipse
Visible

Eclipse at
MoonSet

No Eclipse
Visible

Penumbral Lunar Eclipse of 2071 Sep 09

Greatest Eclipse = 15:05:40.1 TD (= 15:03:59.3 UT1)

Penumbral Magnitude = 0.9000	Gamma = 1.0835	Saros Series = 119
Umbral Magnitude = -0.1575	Axis = 0.9949°	Saros Member = 64 of 82

Sun at Greatest Eclipse
(Geocentric Coordinates)
R.A. = 11h12m24.0s
Dec. = +05°06'21.1"
S.D. = 00°15'52.6"
H.P. = 00°00'08.7"

Moon at Greatest Eclipse
(Geocentric Coordinates)
R.A. = 23h11m12.9s
Dec. = -04°09'20.4"
S.D. = 00°15'00.8"
H.P. = 00°55'06.1"

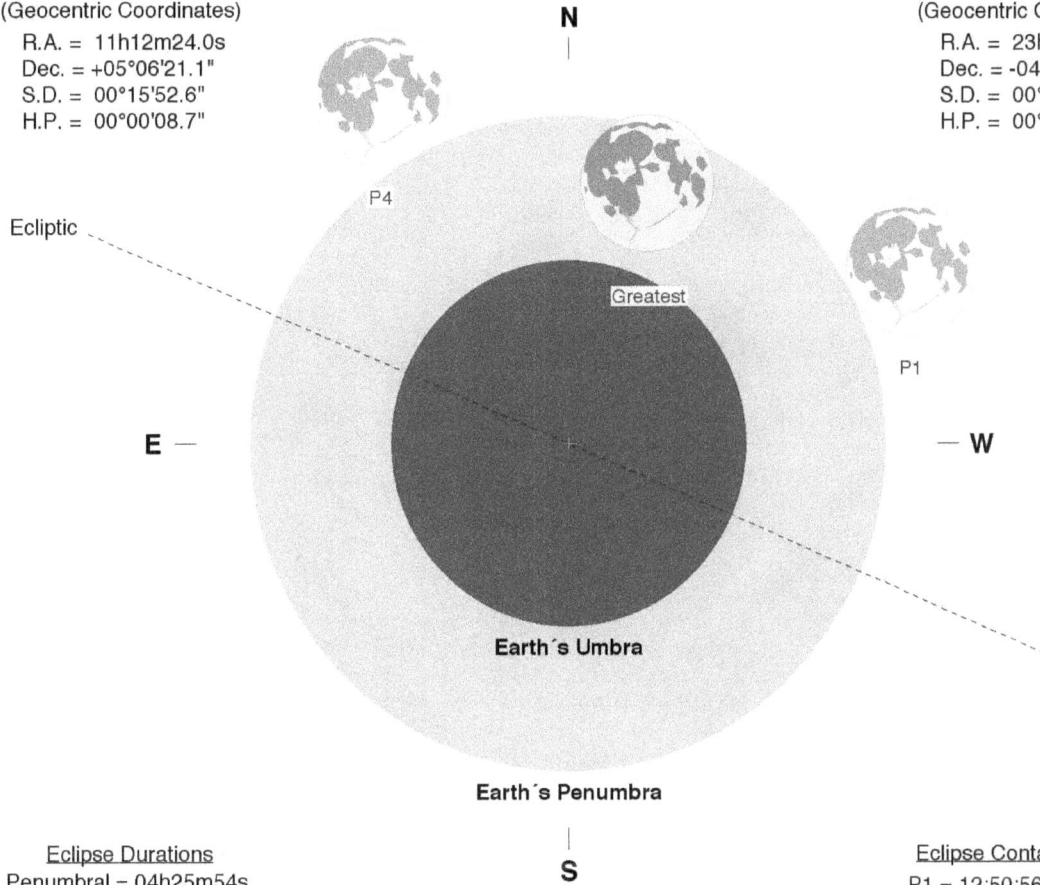

N

P4

Greatest

Ecliptic

P1

E —

— W

Earth's Umbra

Earth's Penumbra

S

Eclipse Durations
Penumbral = 04h25m54s

Eclipse Contacts
P1 = 12:50:56 UT1
P4 = 17:16:50 UT1

Eph. = JPL DE430
Rule = Herald-Sinnott
ΔT =   101 s

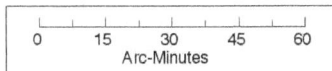

0   15   30   45   60
Arc-Minutes

©2020 F. Espenak, www.EclipseWise.com

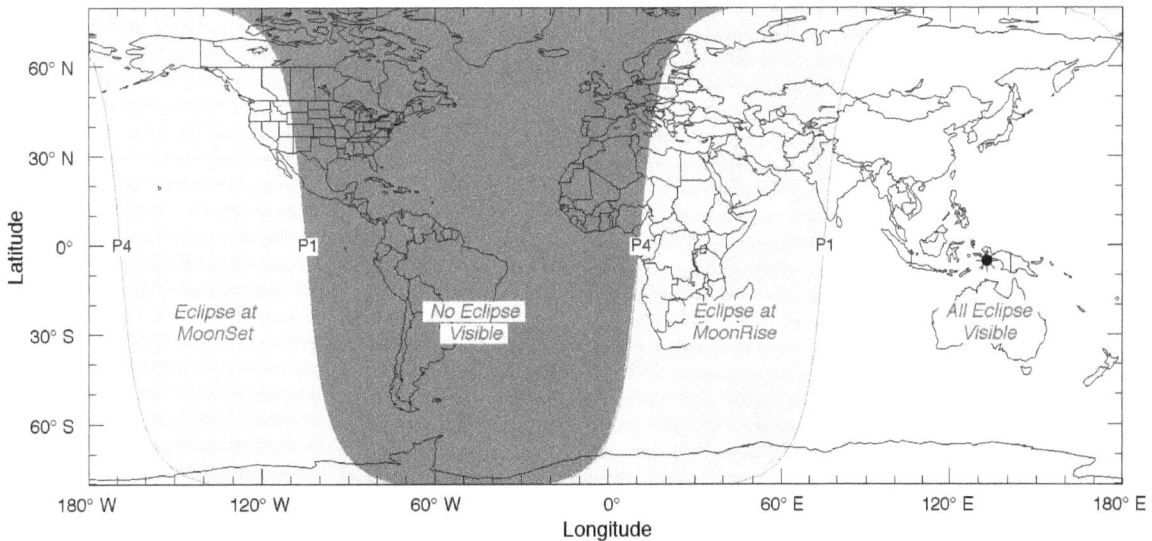

Eclipse at MoonSet

No Eclipse Visible

Eclipse at MoonRise

All Eclipse Visible

Total Lunar Eclipse of 2072 Mar 04

Greatest Eclipse = 15:23:07.0 TD (= 15:21:25.8 UT1)

Penumbral Magnitude = 2.2137	Gamma = -0.3431	Saros Series = 124
Umbral Magnitude = 1.2452	Axis = 0.3494°	Saros Member = 52 of 73

Sun at Greatest Eclipse
(Geocentric Coordinates)
R.A. = 23h04m23.3s
Dec. = -05°56'38.2"
S.D. = 00°16'07.7"
H.P. = 00°00'08.9"

Moon at Greatest Eclipse
(Geocentric Coordinates)
R.A. = 11h03m58.6s
Dec. = +05°36'35.5"
S.D. = 00°16'39.2"
H.P. = 01°01'07.1"

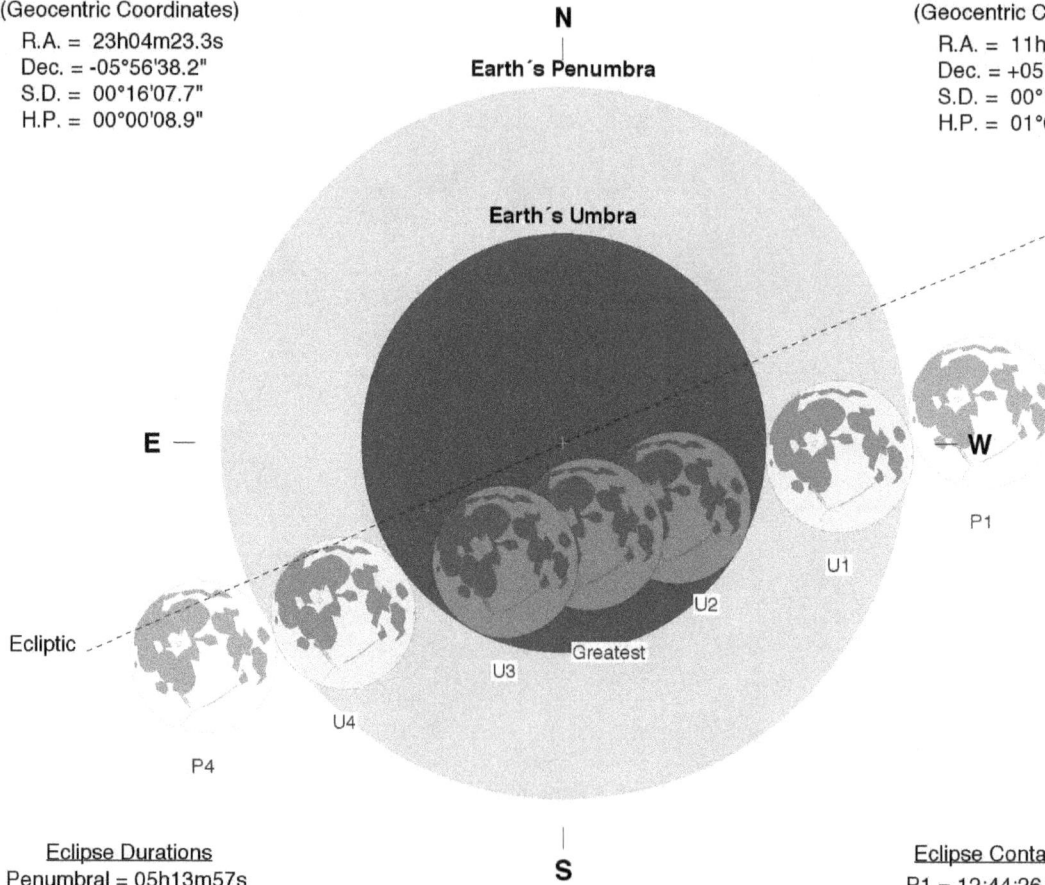

N

Earth's Penumbra

Earth's Umbra

E —

— W

P1

U1

U2

Greatest

U3

U4

P4

Ecliptic

S

Eclipse Durations
Penumbral = 05h13m57s
Umbral = 03h20m07s
Total = 01h09m02s

Eph. = JPL DE430
Rule = Herald-Sinnott
ΔT = 101 s

0   15   30   45   60
Arc-Minutes

©2020 F. Espenak, www.EclipseWise.com

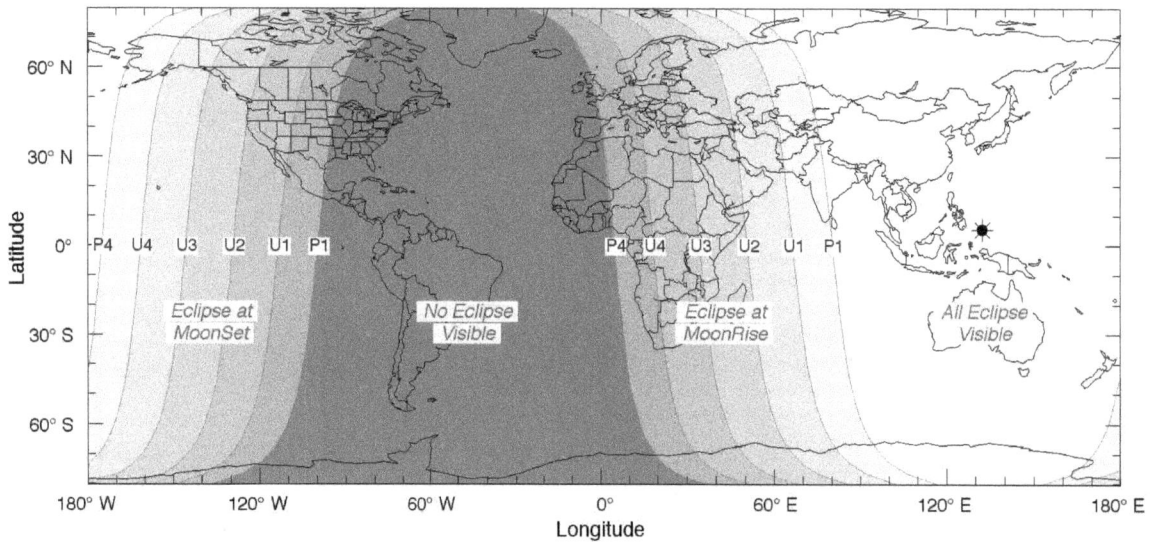

Eclipse Contacts
P1 = 12:44:26 UT1
U1 = 13:41:18 UT1
U2 = 14:46:46 UT1
U3 = 15:55:48 UT1
U4 = 17:01:25 UT1
P4 = 17:58:23 UT1

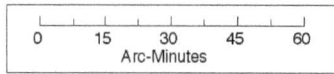

P4 U4 U3 U2 U1 P1   P4 U4 U3 U2 U1 P1

Eclipse at
MoonSet

No Eclipse
Visible

Eclipse at
MoonRise

All Eclipse
Visible

Longitude

Total Lunar Eclipse of 2072 Aug 28

Greatest Eclipse = 16:05:42.1 TD   (= 16:04:00.5 UT1)

Penumbral Magnitude = 2.2439	Gamma = 0.3563	Saros Series = 129
Umbral Magnitude = 1.1673	Axis = 0.3205°	Saros Member = 41 of 71

Sun at Greatest Eclipse
(Geocentric Coordinates)
R.A. = 10h31m55.6s
Dec. = +09°13'37.7"
S.D. = 00°15'50.1"
H.P. = 00°00'08.7"

Moon at Greatest Eclipse
(Geocentric Coordinates)
R.A. = 22h31m34.3s
Dec. = -08°55'07.7"
S.D. = 00°14'42.5"
H.P. = 00°53'58.7"

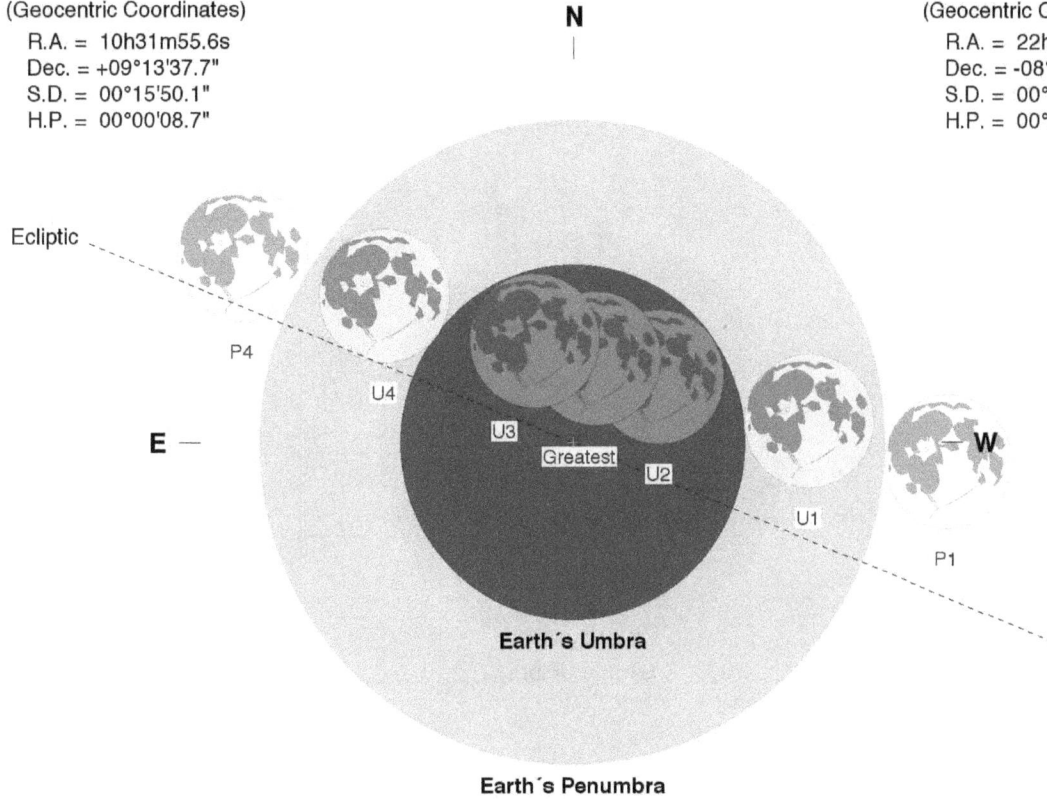

N

Ecliptic

P4

U4

U3

Greatest

U2

E

W

U1

P1

Earth's Umbra

Earth's Penumbra

S

Eclipse Durations
Penumbral = 06h06m53s
Umbral = 03h41m07s
Total = 01h04m50s

Eph. = JPL DE430
Rule = Herald-Sinnott
ΔT = 102 s

0   15   30   45   60
Arc-Minutes

©2020 F. Espenak, www.EclipseWise.com

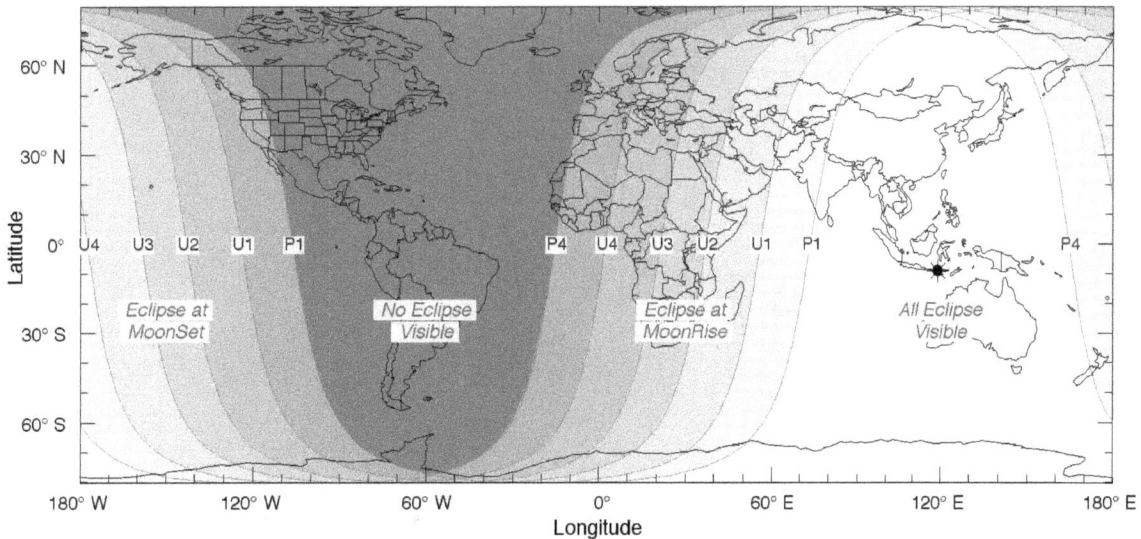

Eclipse Contacts
P1 = 13:00:31 UT1
U1 = 14:13:23 UT1
U2 = 15:31:26 UT1
U3 = 16:36:16 UT1
U4 = 17:54:29 UT1
P4 = 19:07:25 UT1

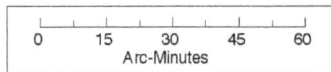

60° N

30° N

0°   U4   U3   U2   U1   P1   P4 U4 U3 U2   U1   P1   P4

Latitude

Eclipse at
MoonSet

No Eclipse
Visible

Eclipse at
MoonRise

All Eclipse
Visible

30° S

60° S

180° W   120° W   60° W   0°   60° E   120° E   180° E
Longitude

Total Lunar Eclipse of 2073 Feb 22

Greatest Eclipse = 07:24:52.8 TD (= 07:23:10.8 UT1)

Penumbral Magnitude = 2.2230	Gamma = 0.3389	Saros Series = 134
Umbral Magnitude = 1.2514	Axis = 0.3449°	Saros Member = 30 of 72

Sun at Greatest Eclipse
(Geocentric Coordinates)
R.A. = 22h24m36.1s
Dec. = -09°56'15.6"
S.D. = 00°16'10.2"
H.P. = 00°00'08.9"

Moon at Greatest Eclipse
(Geocentric Coordinates)
R.A. = 10h24m58.8s
Dec. = +10°16'11.2"
S.D. = 00°16'38.6"
H.P. = 01°01'05.0"

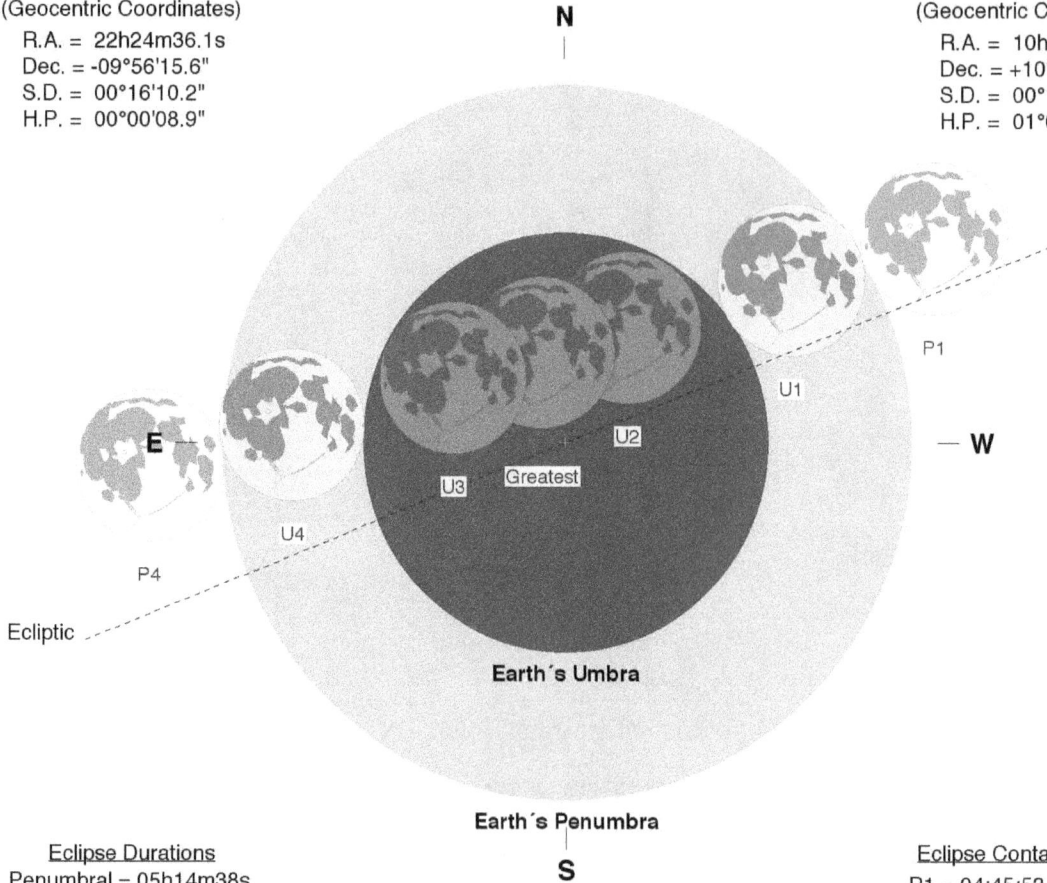

N

E — — W

Greatest

U1

U2

U3

U4

P1

P4

Ecliptic

Earth's Umbra

Earth's Penumbra

S

Eclipse Durations
Penumbral = 05h14m38s
Umbral = 03h20m27s
Total = 01h09m46s

Eph. = JPL DE430
Rule = Herald-Sinnott
ΔT =   102 s

0   15   30   45   60
Arc-Minutes

©2020 F. Espenak, www.EclipseWise.com

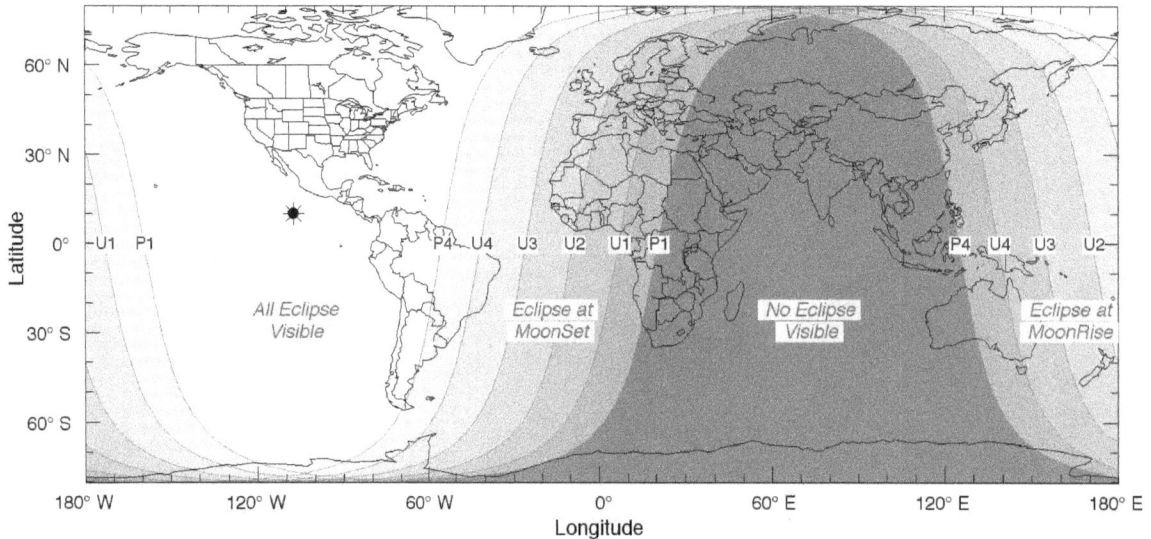

Eclipse Contacts
P1 = 04:45:53 UT1
U1 = 05:43:02 UT1
U2 = 06:48:26 UT1
U3 = 07:58:12 UT1
U4 = 09:03:28 UT1
P4 = 10:00:31 UT1

All Eclipse Visible

Eclipse at MoonSet

No Eclipse Visible

Eclipse at MoonRise

U1 P1   P4 U4 U3 U2 U1 P1   P4 U4 U3 U2

Total Lunar Eclipse of 2073 Aug 17

Greatest Eclipse = 17:42:41.2 TD   (= 17:40:58.8 UT1)

Penumbral Magnitude = 2.1490	Gamma = -0.3998	Saros Series = 139
Umbral Magnitude = 1.1024	Axis = 0.3691°	Saros Member = 24 of 79

Sun at Greatest Eclipse
(Geocentric Coordinates)

R.A. = 09h50m51.6s
Dec. = +13°01'53.3"
S.D. = 00°15'47.9"
H.P. = 00°00'08.7"

N

Earth's Penumbra

Ecliptic

Earth's Umbra

Moon at Greatest Eclipse
(Geocentric Coordinates)

R.A. = 21h51m13.5s
Dec. = -13°23'23.0"
S.D. = 00°15'05.7"
H.P. = 00°55'24.0"

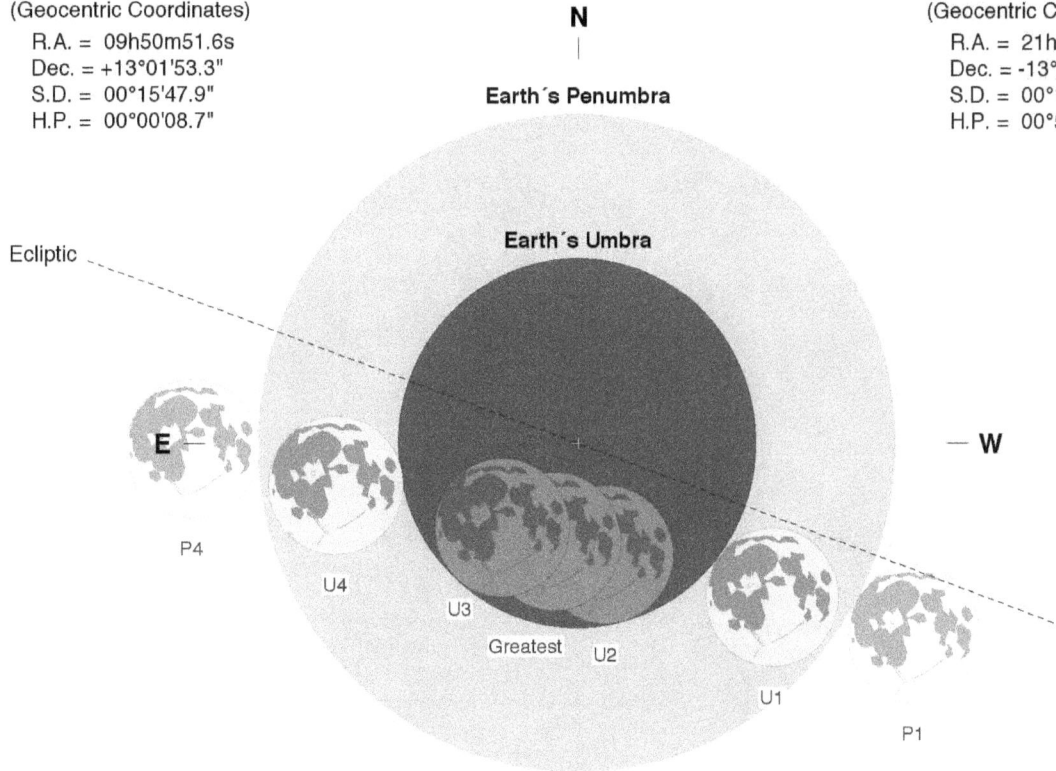

E ←

— **W**

P4

U4

U3

Greatest   U2

U1

P1

S

Eclipse Durations
Penumbral = 05h51m21s
Umbral = 03h32m20s
Total = 00h50m45s

Eph. = JPL DE430
Rule = Herald-Sinnott
ΔT =   102 s

Arc-Minutes
0   15   30   45   60

©2020 F. Espenak, www.EclipseWise.com

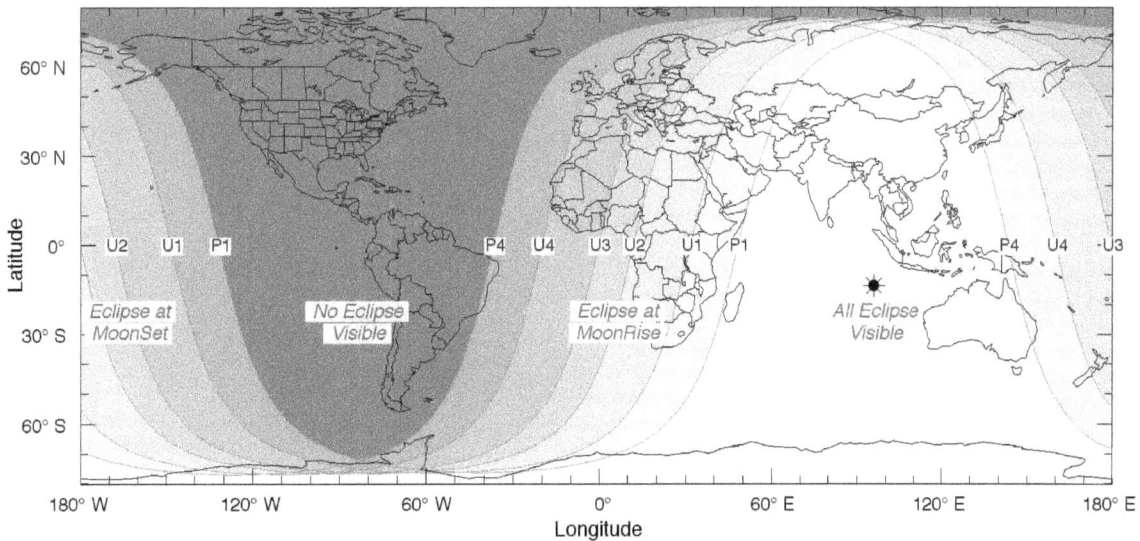

Eclipse Contacts
P1 = 14:45:23 UT1
U1 = 15:54:51 UT1
U2 = 17:15:44 UT1
U3 = 18:06:29 UT1
U4 = 19:27:11 UT1
P4 = 20:36:44 UT1

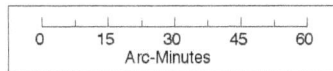

Eclipse at MoonSet

No Eclipse Visible

Eclipse at MoonRise

All Eclipse Visible

Penumbral Lunar Eclipse of 2074 Feb 11

Greatest Eclipse = 20:55:58.1 TD   (= 20:54:15.2 UT1)

Penumbral Magnitude = 0.9203	Gamma = 1.0612	Saros Series = 144
Umbral Magnitude = -0.0960	Axis = 1.0348°	Saros Member = 19 of 71

Sun at Greatest Eclipse
(Geocentric Coordinates)
R.A. = 21h43m16.1s
Dec. = -13°40'57.8"
S.D. = 00°16'12.3"
H.P. = 00°00'08.9"

Moon at Greatest Eclipse
(Geocentric Coordinates)
R.A. = 09h44m16.6s
Dec. = +14°41'18.0"
S.D. = 00°15'56.7"
H.P. = 00°58'31.1"

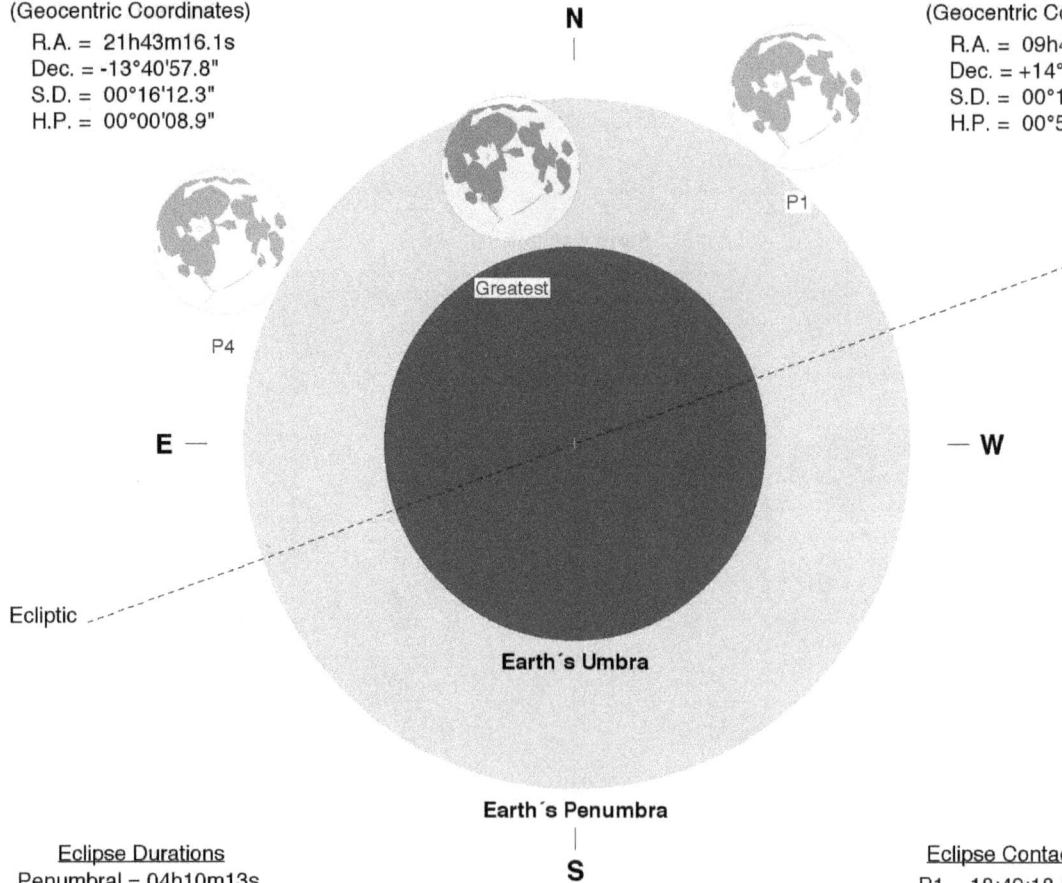

N

P1

Greatest

P4

E

W

Earth's Umbra

Earth's Penumbra

S

Ecliptic

Eclipse Durations
Penumbral = 04h10m13s

Eclipse Contacts
P1 = 18:49:18 UT1
P4 = 22:59:31 UT1

Eph. = JPL DE430
Rule = Herald-Sinnott
ΔT =   103 s

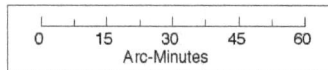

0	15	30	45	60
		Arc-Minutes		

©2020 F. Espenak, www.EclipseWise.com

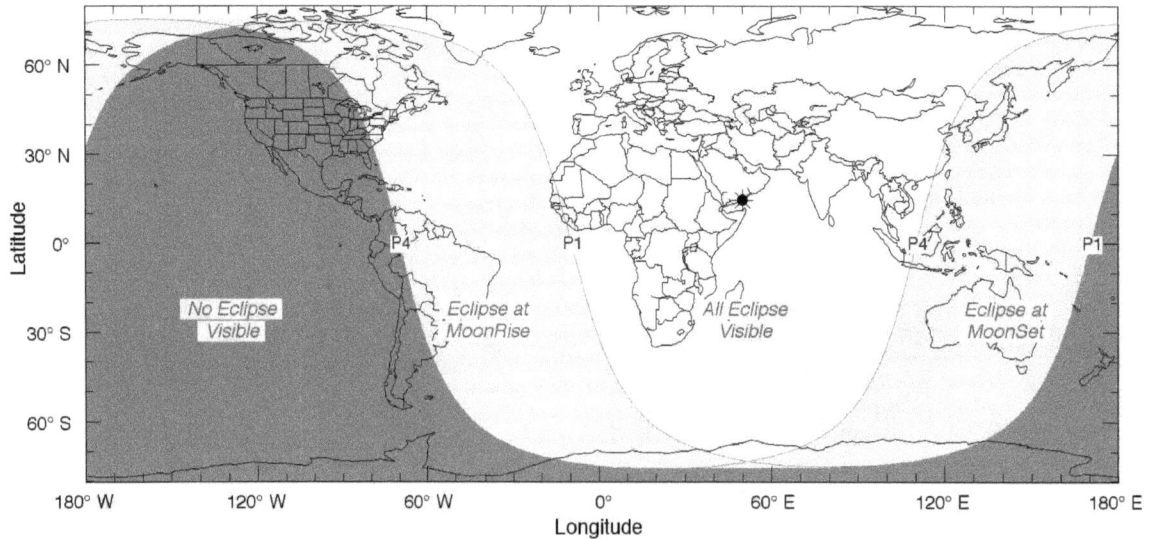

N

60° N

30° N

Latitude

0°

P4

P1

All Eclipse
Visible

P4

P1

No Eclipse
Visible

Eclipse at
MoonRise

30° S

Eclipse at
MoonSet

60° S

180° W   120° W   60° W   0°   60° E   120° E   180° E

Longitude

Penumbral Lunar Eclipse of 2074 Jul 08

Greatest Eclipse = 17:21:37.4 TD (= 17:19:54.3 UT1)

Penumbral Magnitude = 0.1884	Gamma = 1.4457	Saros Series = 111
Umbral Magnitude = -0.7751	Axis = 1.4438°	Saros Member = 70 of 71

Sun at Greatest Eclipse
(Geocentric Coordinates)

R.A. = 07h13m28.6s
Dec. = +22°21'16.2"
S.D. = 00°15'44.0"
H.P. = 00°00'08.7"

Moon at Greatest Eclipse
(Geocentric Coordinates)

R.A. = 19h13m17.1s
Dec. = -20°54'40.5"
S.D. = 00°16'19.8"
H.P. = 00°59'55.7"

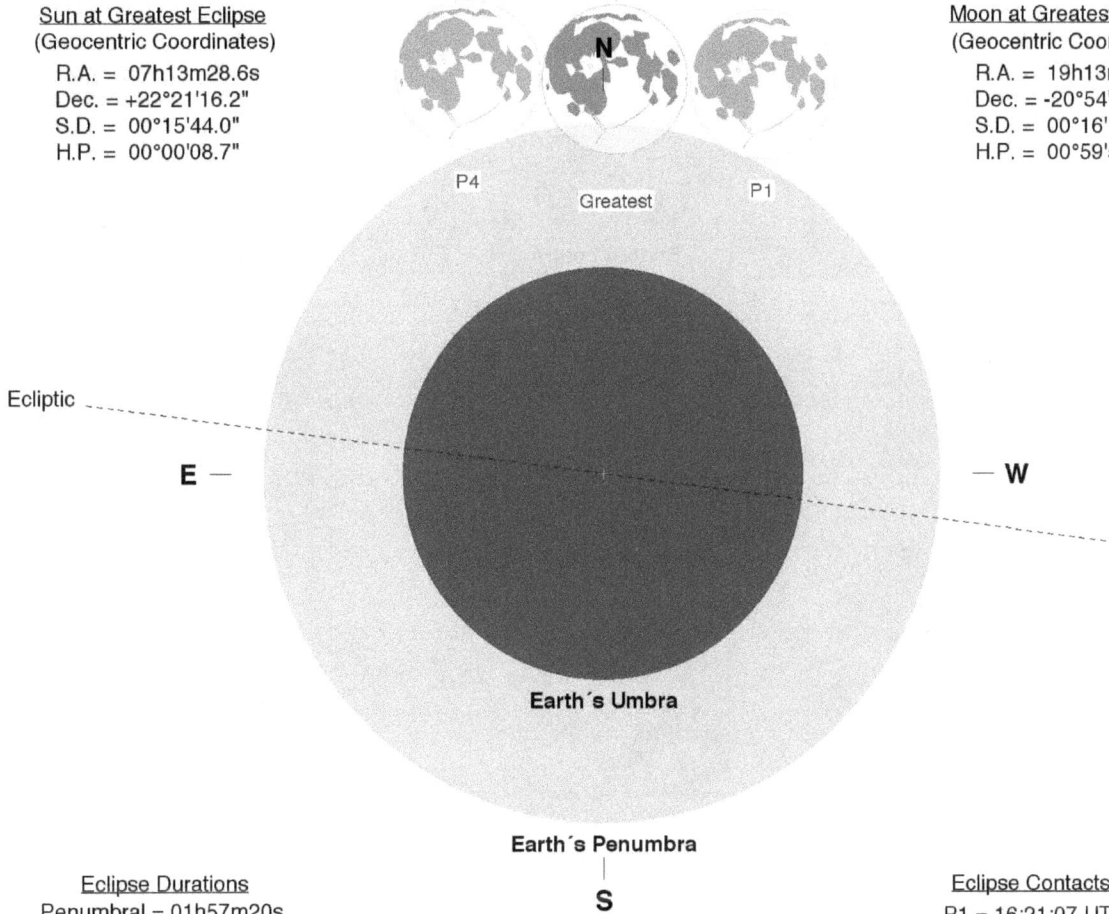

N

P4   Greatest   P1

Ecliptic

E

W

Earth's Umbra

Earth's Penumbra

S

Eclipse Durations
Penumbral = 01h57m20s

Eclipse Contacts
P1 = 16:21:07 UT1
P4 = 18:18:27 UT1

Eph. = JPL DE430
Rule = Herald-Sinnott
ΔT = 103 s

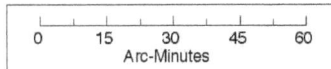

0	15	30	45	60

Arc-Minutes

©2020 F. Espenak, www.EclipseWise.com

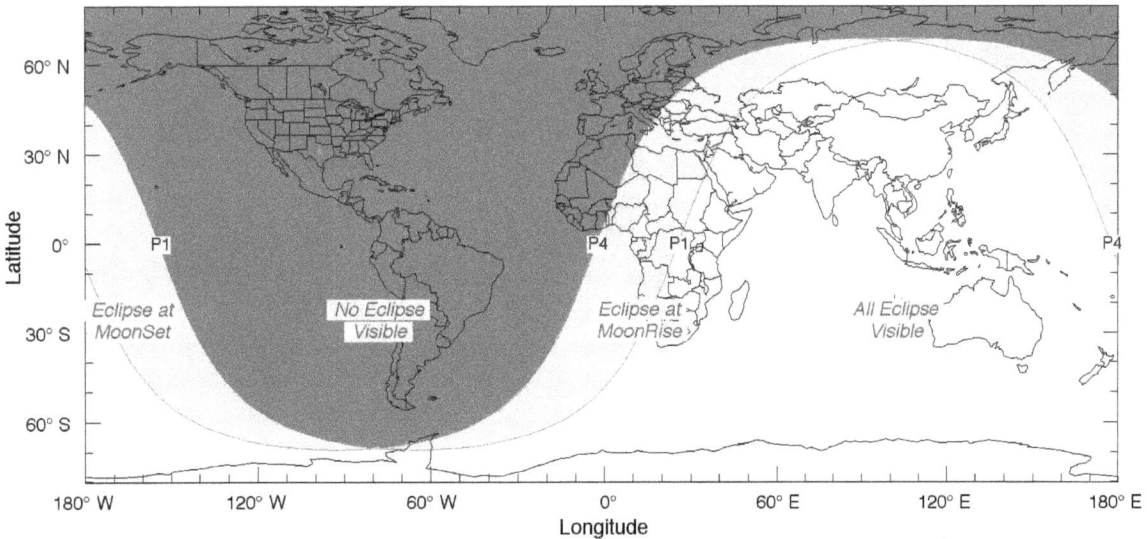

Eclipse at
MoonSet

No Eclipse
Visible

Eclipse at
MoonRise

All Eclipse
Visible

P1

P4   P1

P4

Penumbral Lunar Eclipse of 2074 Aug 07

Greatest Eclipse = 01:56:04.0 TD (= 01:54:20.8 UT1)

Penumbral Magnitude = 0.7826	Gamma = -1.1291	Saros Series = 149
Umbral Magnitude = -0.2079	Axis = 1.0996°	Saros Member = 6 of 71

Sun at Greatest Eclipse
(Geocentric Coordinates)
R.A. = 09h09m47.8s
Dec. = +16°19'56.5"
S.D. = 00°15'46.2"
H.P. = 00°00'08.7"

N

Earth´s Penumbra

Moon at Greatest Eclipse
(Geocentric Coordinates)
R.A. = 21h10m42.9s
Dec. = -17°24'35.5"
S.D. = 00°15'55.4"
H.P. = 00°58'26.2"

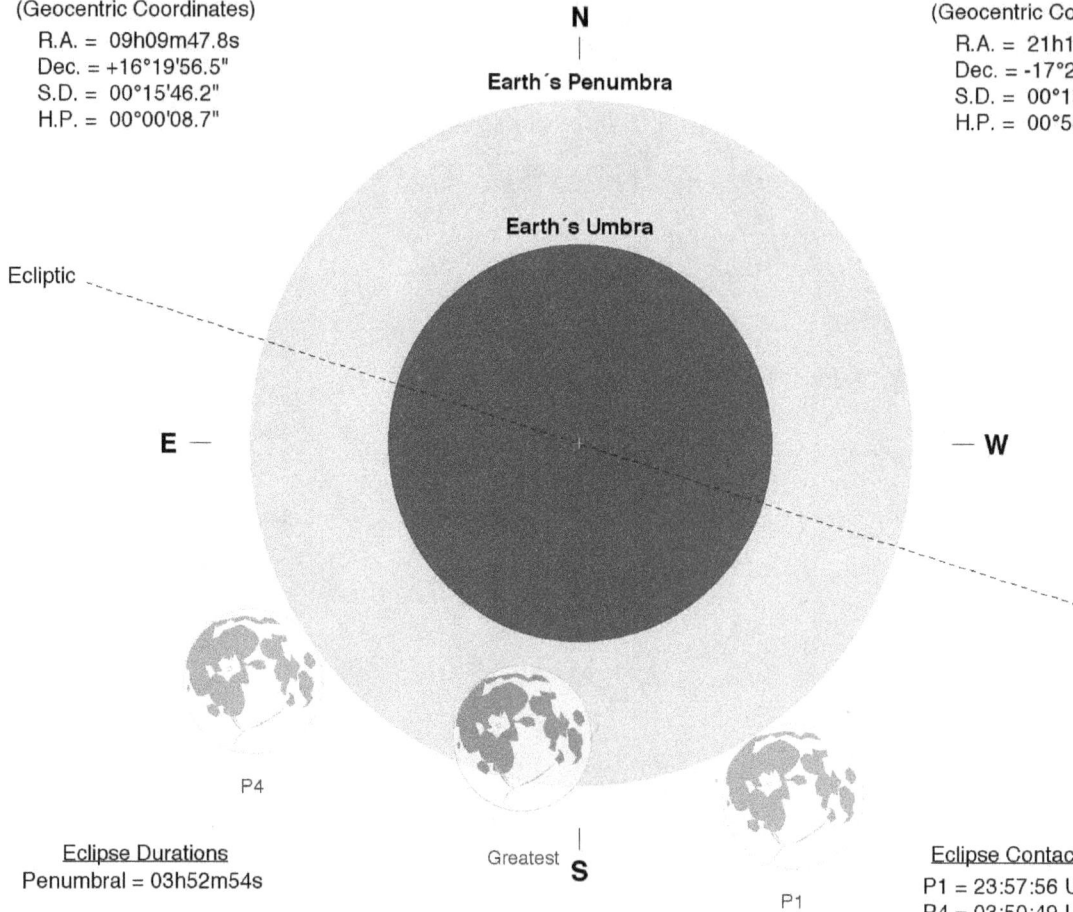

Earth´s Umbra

Ecliptic

E

W

P4

Greatest

S

P1

Eclipse Durations
Penumbral = 03h52m54s

Eclipse Contacts
P1 = 23:57:56 UT1
P4 = 03:50:49 UT1

Eph. = JPL DE430
Rule = Herald-Sinnott
ΔT =   103 s

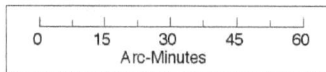

0   15   30   45   60
Arc-Minutes

©2020 F. Espenak, www.EclipseWise.com

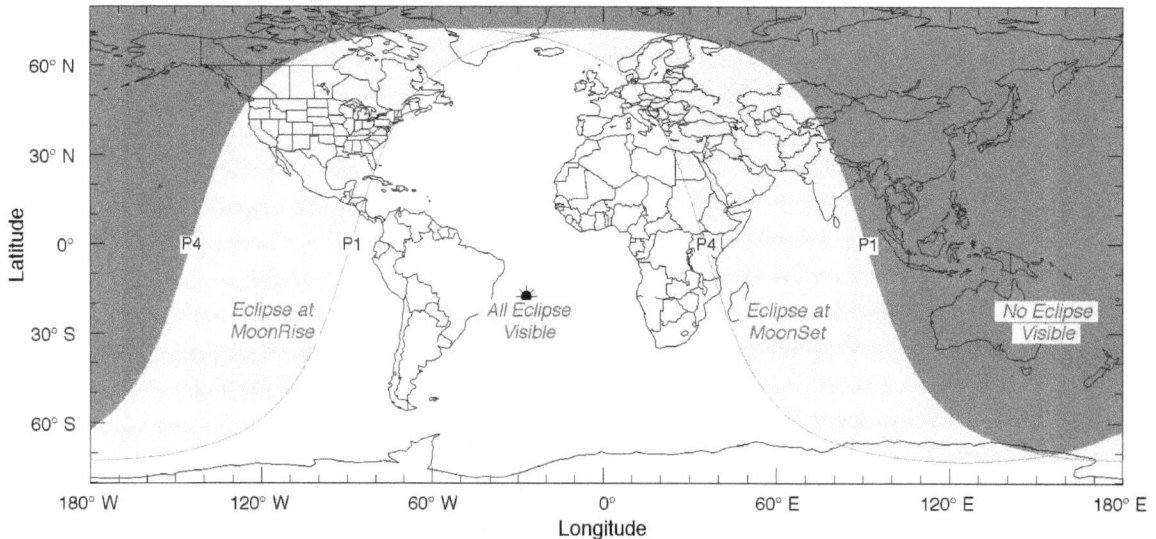

60° N

30° N

0°

30° S

60° S

P4   P1

All Eclipse
Visible

Eclipse at
MoonRise

P4   P1

Eclipse at
MoonSet

No Eclipse
Visible

Latitude

180° W   120° W   60° W   0°   60° E   120° E   180° E
Longitude

Penumbral Lunar Eclipse of 2075 Jan 02

Greatest Eclipse = 09:55:03.2 TD (= 09:53:19.7 UT1)

Penumbral Magnitude = 0.7729	Gamma = -1.1643	Saros Series = 116
Umbral Magnitude = -0.3256	Axis = 1.0543°	Saros Member = 61 of 73

Sun at Greatest Eclipse
(Geocentric Coordinates)

R.A. = 18h52m38.7s
Dec. = -22°52'37.0"
S.D. = 00°16'15.9"
H.P. = 00°00'08.9"

N

Earth's Penumbra

Moon at Greatest Eclipse
(Geocentric Coordinates)

R.A. = 06h52m40.9s
Dec. = +21°49'21.3"
S.D. = 00°14'48.4"
H.P. = 00°54'20.5"

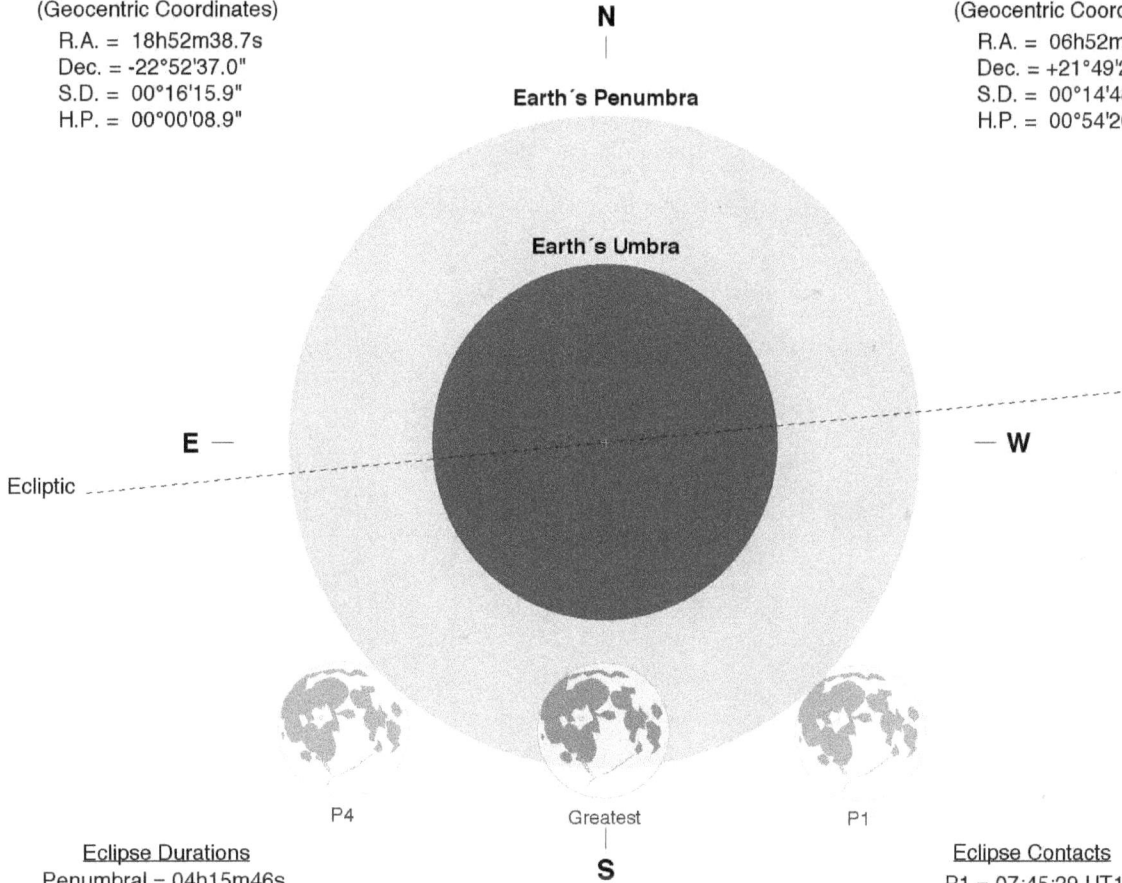

Earth's Umbra

E —

— W

Ecliptic

P4

Greatest

S

P1

Eclipse Durations
Penumbral = 04h15m46s

Eclipse Contacts
P1 = 07:45:29 UT1
P4 = 12:01:15 UT1

Eph. = JPL DE430
Rule = Herald-Sinnott
ΔT =   104 s

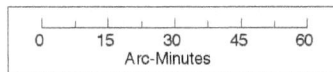

0	15	30	45	60

Arc-Minutes

©2020 F. Espenak, www.EclipseWise.com

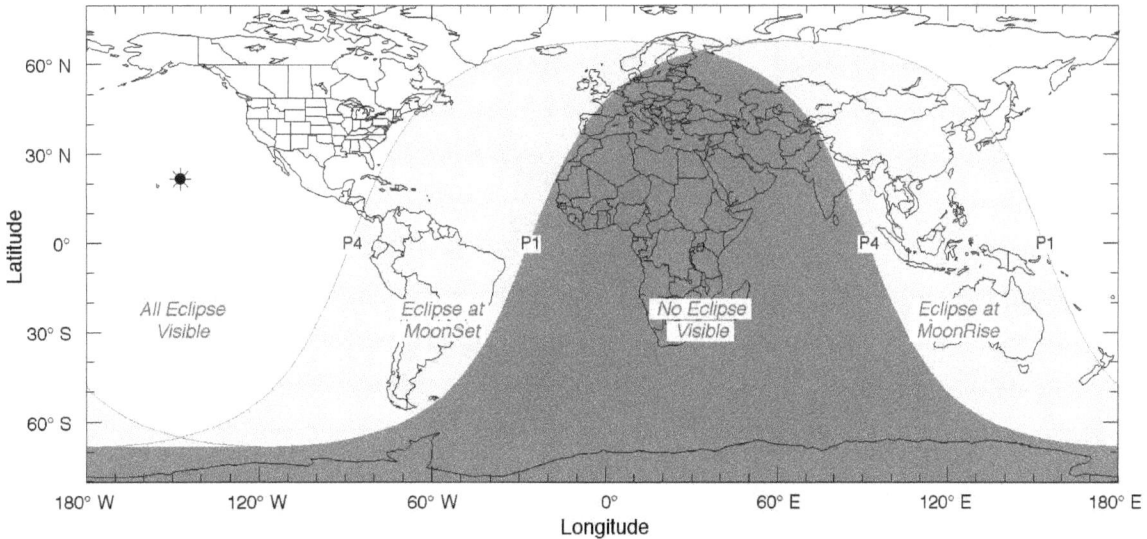

All Eclipse
Visible

Eclipse at
MoonSet

P4   P1

No Eclipse
Visible

P4   P1

Eclipse at
MoonRise

Partial Lunar Eclipse of 2075 Jun 28

Greatest Eclipse = 09:55:35.5 TD   (= 09:53:51.6 UT1)

Penumbral Magnitude = 1.5639	Gamma = 0.6897	Saros Series = 121
Umbral Magnitude = 0.6235	Axis = 0.7058°	Saros Member = 58 of 82

Sun at Greatest Eclipse
(Geocentric Coordinates)
R.A. = 06h29m58.9s
Dec. = +23°14'59.1"
S.D. = 00°15'44.0"
H.P. = 00°00'08.7"

Moon at Greatest Eclipse
(Geocentric Coordinates)
R.A. = 18h30m07.3s
Dec. = -22°32'40.9"
S.D. = 00°16'43.9"
H.P. = 01°01'24.3"

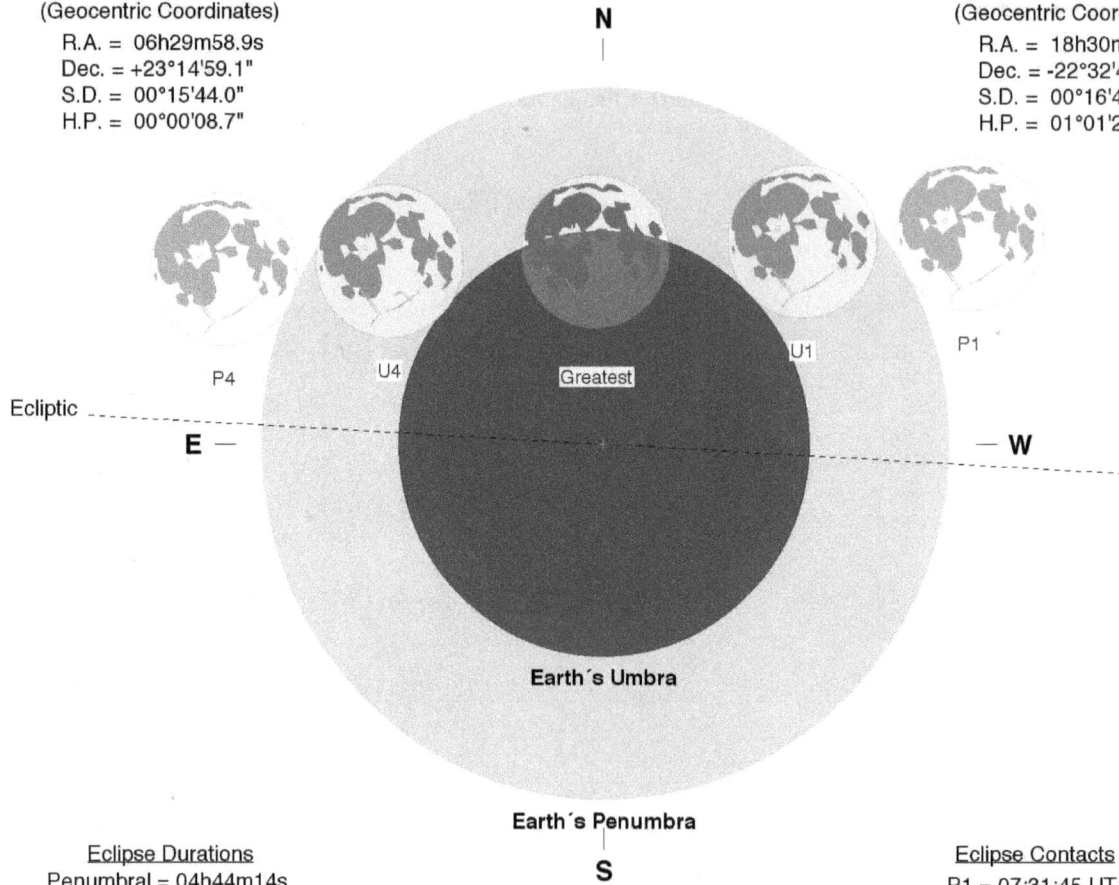

N

P4   U4   Greatest   U1   P1

Ecliptic

E —   — W

Earth's Umbra

Earth's Penumbra

S

Eclipse Durations
Penumbral = 04h44m14s
Umbral = 02h37m39s

Eph. = JPL DE430
Rule = Herald-Sinnott
ΔT =   104 s

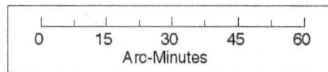

0   15   30   45   60
Arc-Minutes

©2020 F. Espenak, www.EclipseWise.com

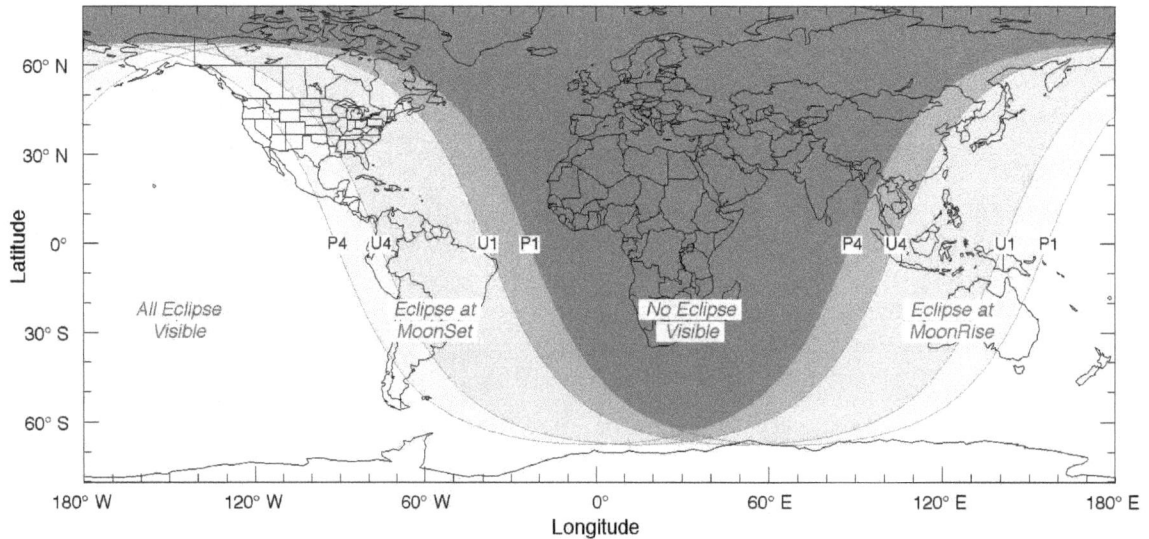

Eclipse Contacts
P1 = 07:31:45 UT1
U1 = 08:35:03 UT1
U4 = 11:12:42 UT1
P4 = 12:15:59 UT1

P4 U4   U1 P1   P4 U4   U1 P1

All Eclipse
Visible

Eclipse at
MoonSet

No Eclipse
Visible

Eclipse at
MoonRise

Partial Lunar Eclipse of 2075 Dec 22

Greatest Eclipse = 08:55:55.1 TD   (= 08:54:10.7 UT1)

Penumbral Magnitude = 2.0024	Gamma = -0.4945	Saros Series = 126
Umbral Magnitude = 0.9028	Axis = 0.4472°	Saros Member = 48 of 70

Sun at Greatest Eclipse
(Geocentric Coordinates)

R.A. = 18h02m40.3s
Dec. = -23°25'39.1"
S.D. = 00°16'15.5"
H.P. = 00°00'08.9"

N

Earth's Penumbra

Earth's Umbra

Moon at Greatest Eclipse
(Geocentric Coordinates)

R.A. = 06h02m51.4s
Dec. = +22°58'56.6"
S.D. = 00°14'47.1"
H.P. = 00°54'15.8"

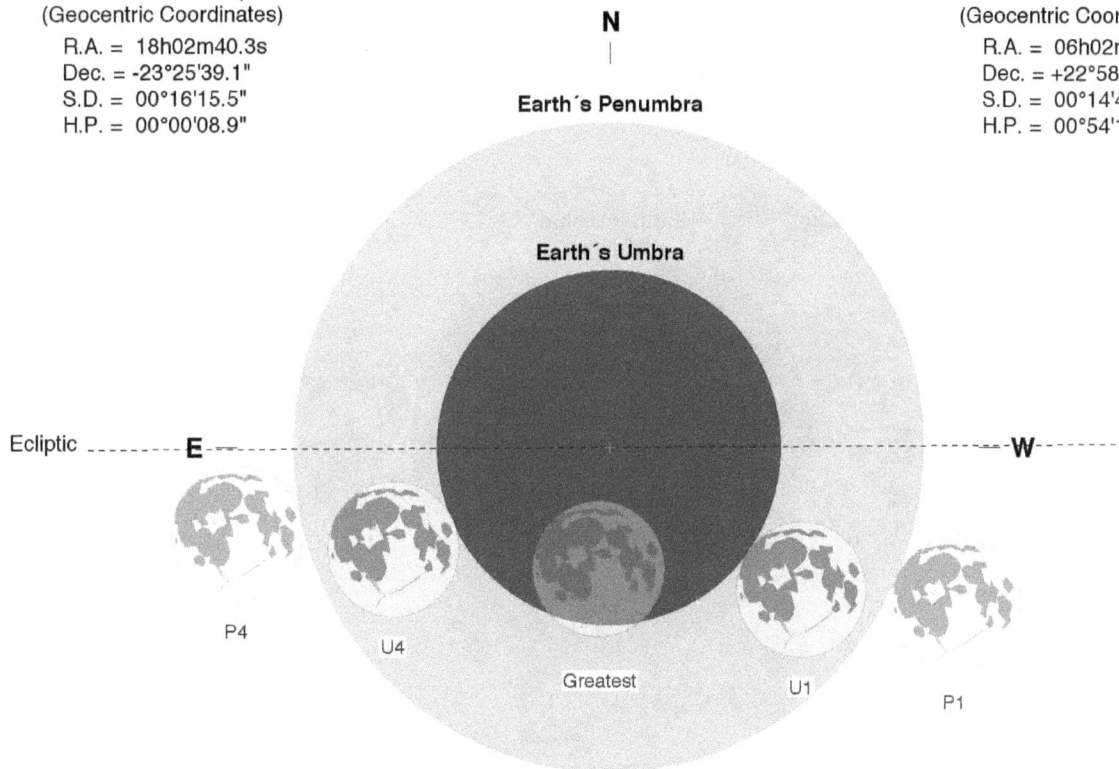

Ecliptic   **E**   **W**

P4   U4   Greatest   U1   P1

S

Eclipse Durations
Penumbral = 05h58m35s
Umbral = 03h23m21s

Eph. = JPL DE430
Rule = Herald-Sinnott
ΔT =   104 s

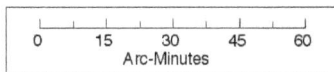

0   15   30   45   60
Arc-Minutes

©2020 F. Espenak, www.EclipseWise.com

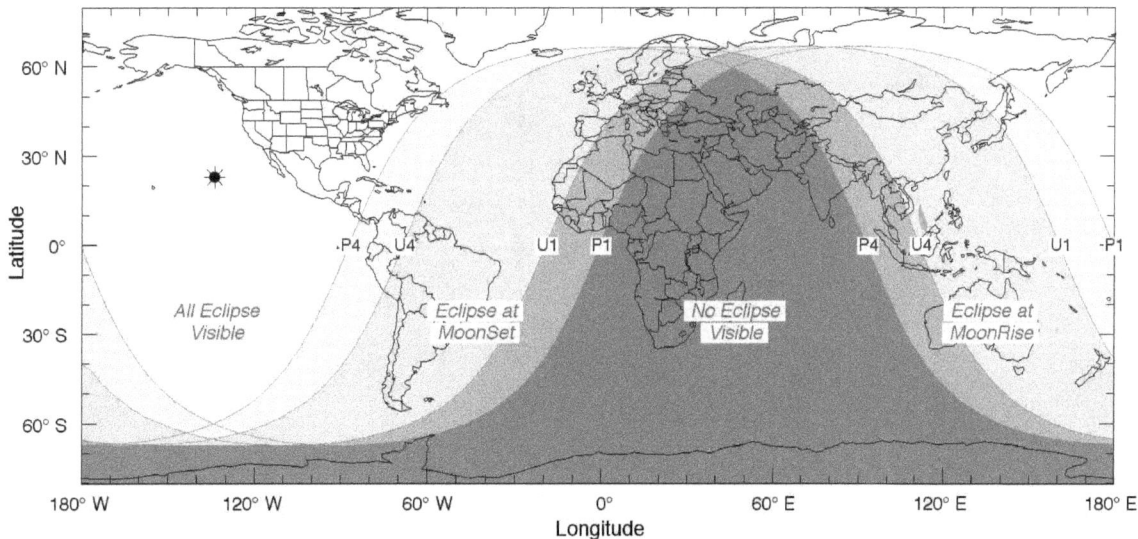

Eclipse Contacts
P1 = 05:54:55 UT1
U1 = 07:12:31 UT1
U4 = 10:35:52 UT1
P4 = 11:53:31 UT1

All Eclipse Visible

Eclipse at MoonSet

No Eclipse Visible

Eclipse at MoonRise

P4   U4   U1   P1   P4   U4   U1   -P1

Total Lunar Eclipse of 2076 Jun 17

Greatest Eclipse = 02:39:46.5 TD (= 02:38:01.7 UT1)

Penumbral Magnitude = 2.7570	Gamma = -0.0452	Saros Series = 131
Umbral Magnitude = 1.7959	Axis = 0.0453°	Saros Member = 37 of 72

Sun at Greatest Eclipse
(Geocentric Coordinates)
R.A. = 05h46m08.2s
Dec. = +23°23'27.6"
S.D. = 00°15'44.6"
H.P. = 00°00'08.7"

Moon at Greatest Eclipse
(Geocentric Coordinates)
R.A. = 17h46m06.8s
Dec. = -23°26'09.4"
S.D. = 00°16'22.8"
H.P. = 01°00'07.0"

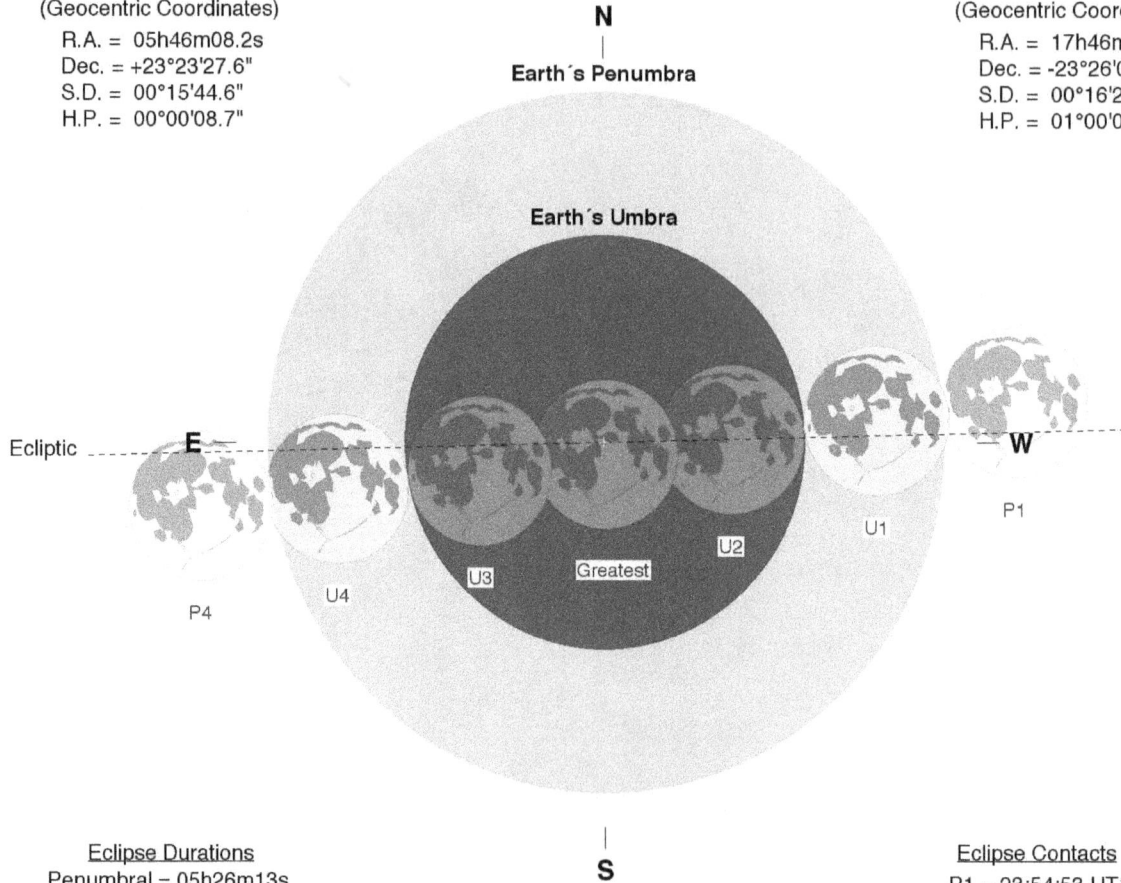

N

Earth's Penumbra

Earth's Umbra

Ecliptic

E

W

U3 Greatest U2 U1 P1

P4 U4

S

Eclipse Durations
Penumbral = 05h26m13s
Umbral = 03h35m52s
Total = 01h40m55s

Eph. = JPL DE430
Rule = Herald-Sinnott
ΔT = 105 s

0   15   30   45   60
Arc-Minutes

©2020 F. Espenak, www.EclipseWise.com

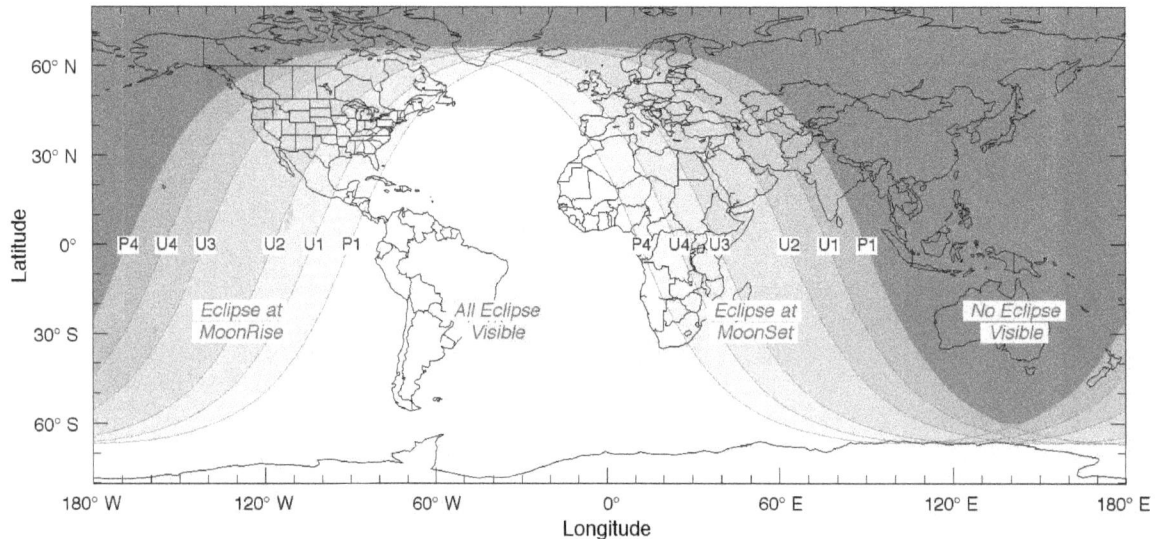

Eclipse Contacts
P1 = 23:54:53 UT1
U1 = 00:50:06 UT1
U2 = 01:47:34 UT1
U3 = 03:28:29 UT1
U4 = 04:25:58 UT1
P4 = 05:21:06 UT1

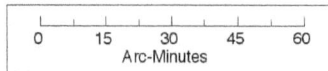

P4 U4 U3   U2 U1 P1   P4 U4 U3   U2 U1 P1

Eclipse at MoonRise   All Eclipse Visible   Eclipse at MoonSet   No Eclipse Visible

Total Lunar Eclipse of 2076 Dec 10

Greatest Eclipse = 11:34:51.5 TD   (= 11:33:06.3 UT1)

Penumbral Magnitude = 2.5006	Gamma = 0.2102	Saros Series = 136
Umbral Magnitude = 1.4476	Axis = 0.1983°	Saros Member = 23 of 72

Sun at Greatest Eclipse
(Geocentric Coordinates)

R.A. = 17h13m25.1s
Dec. = -22°59'52.7"
S.D. = 00°16'14.5"
H.P. = 00°00'08.9"

Moon at Greatest Eclipse
(Geocentric Coordinates)

R.A. = 05h13m15.8s
Dec. = +23°11'35.0"
S.D. = 00°15'25.4"
H.P. = 00°56'36.4"

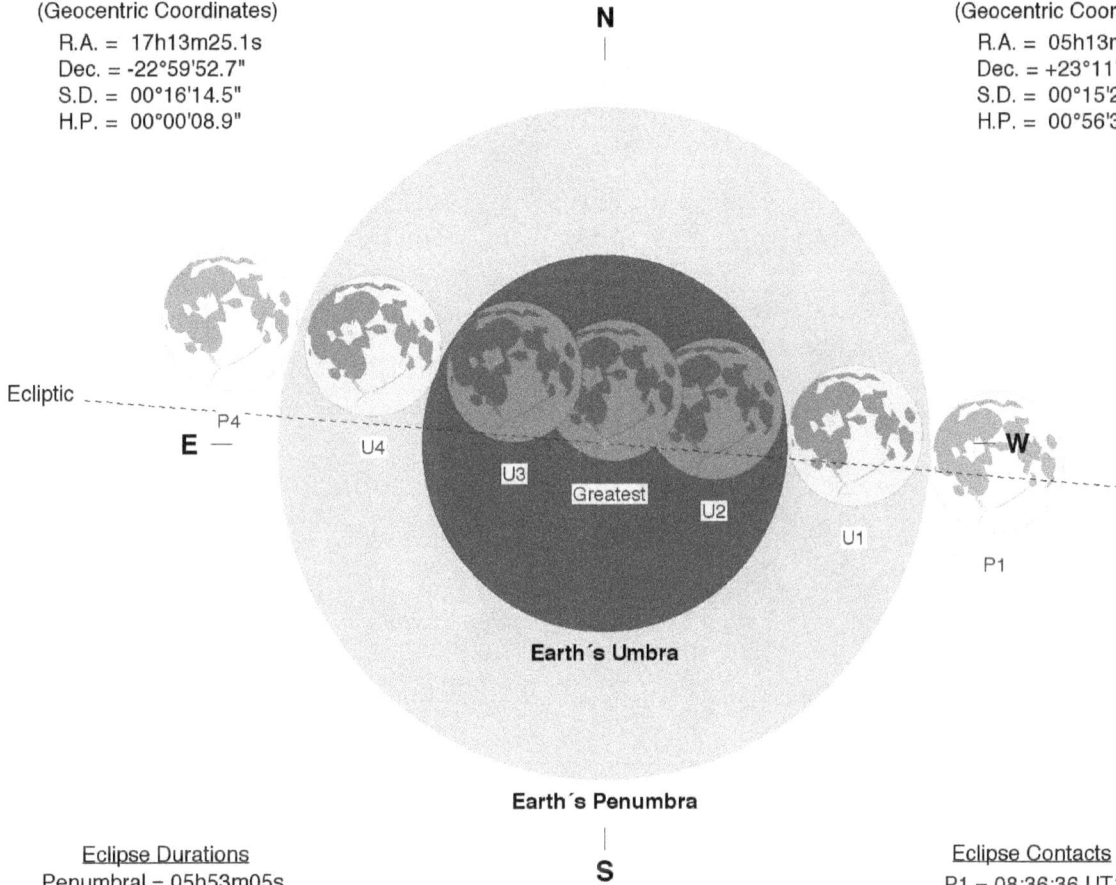

N

Ecliptic

E

P4

U4

U3

Greatest

U2

U1

P1

W

Earth's Umbra

Earth's Penumbra

S

Eclipse Durations
Penumbral = 05h53m05s
Umbral = 03h41m26s
Total = 01h31m33s

Eph. = JPL DE430
Rule = Herald-Sinnott
ΔT =   105 s

0	15	30	45	60

Arc-Minutes

©2020 F. Espenak, www.EclipseWise.com

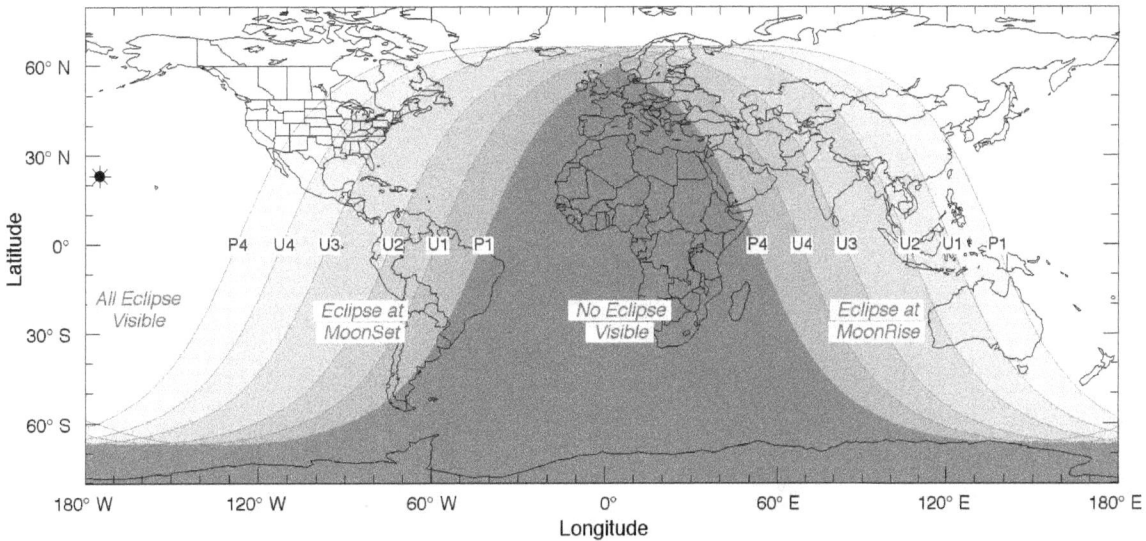

Eclipse Contacts
P1 = 08:36:36 UT1
U1 = 09:42:20 UT1
U2 = 10:47:16 UT1
U3 = 12:18:49 UT1
U4 = 13:23:47 UT1
P4 = 14:29:41 UT1

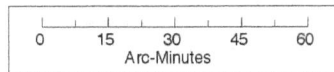

60° N

30° N

0°

Latitude

30° S

60° S

P4   U4   U3   U2   U1   P1

P4   U4   U3   U2   U1   P1

All Eclipse
Visible

Eclipse at
MoonSet

No Eclipse
Visible

Eclipse at
MoonRise

180° W   120° W   60° W   0°   60° E   120° E   180° E

Longitude

231

Partial Lunar Eclipse of 2077 Jun 06

Greatest Eclipse = 14:59:51.9 TD   (= 14:58:06.2 UT1)

Penumbral Magnitude = 1.3274	Gamma = -0.8388	Saros Series = 141
Umbral Magnitude = 0.3139	Axis = 0.7979°	Saros Member = 27 of 72

Sun at Greatest Eclipse
(Geocentric Coordinates)
R.A. = 05h01m37.4s
Dec. = +22°45'07.6"
S.D. = 00°15'45.7"
H.P. = 00°00'08.7"

N

Earth's Penumbra

Earth's Umbra

Moon at Greatest Eclipse
(Geocentric Coordinates)
R.A. = 17h00m56.3s
Dec. = -23°32'03.3"
S.D. = 00°15'33.2"
H.P. = 00°57'04.9"

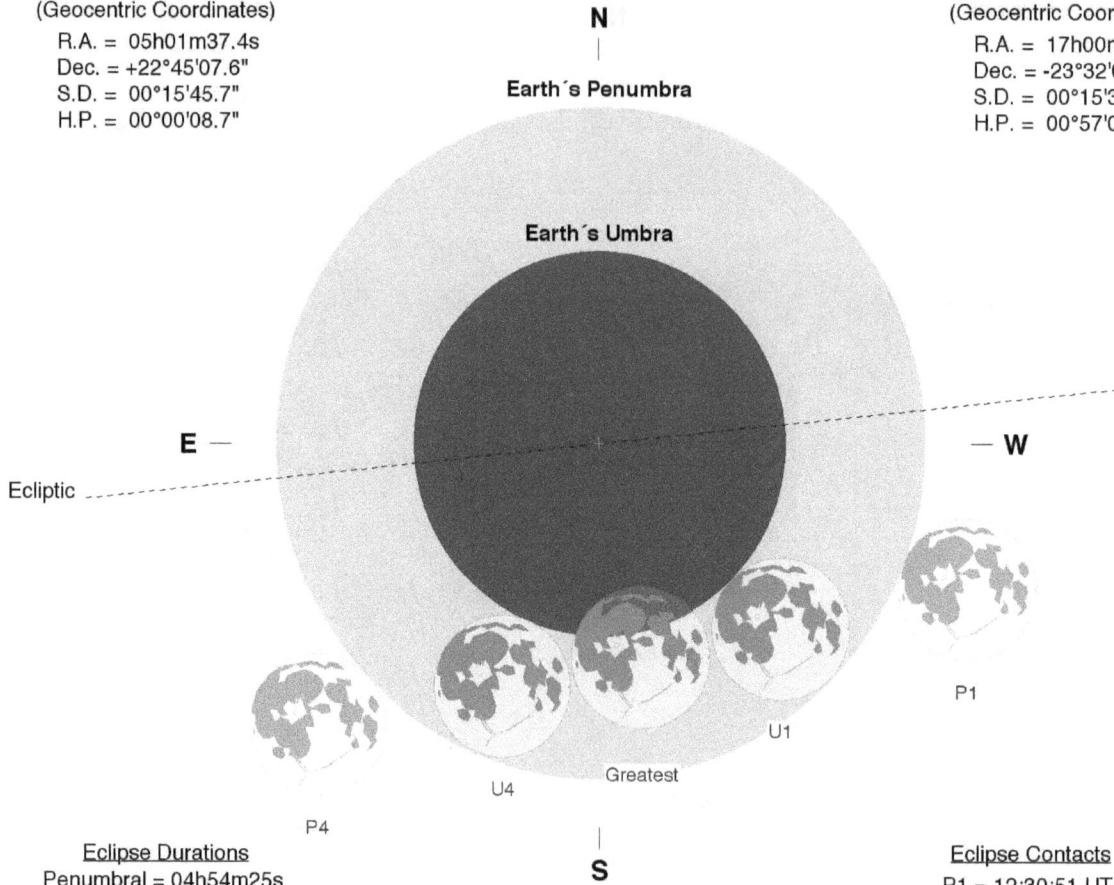

E —

Ecliptic

— **W**

P1

U1

Greatest

U4

P4

S

Eclipse Durations
Penumbral = 04h54m25s
Umbral = 02h05m45s

Eph. = JPL DE430
Rule = Herald-Sinnott
ΔT =   106 s

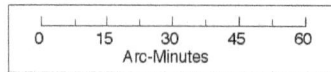

0	15	30	45	60

Arc-Minutes

©2020 F. Espenak, www.EclipseWise.com

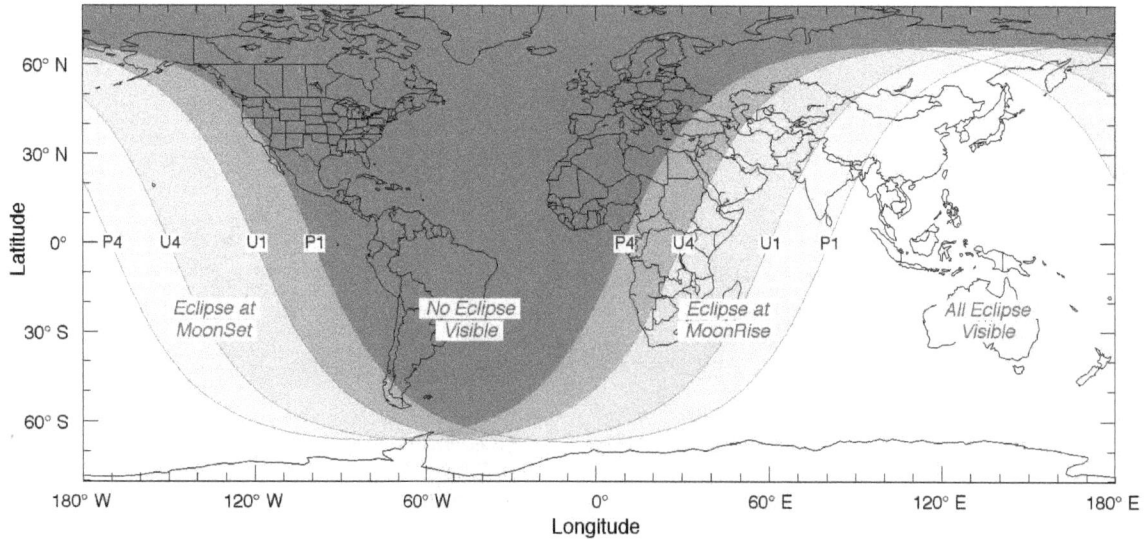

Eclipse Contacts
P1 = 12:30:51 UT1
U1 = 13:55:13 UT1
U4 = 16:00:57 UT1
P4 = 17:25:15 UT1

60° N

30° N

0° P4 U4 U1 P1   P4 U4 U1 P1

Latitude

30° S

60° S

Eclipse at MoonSet   No Eclipse Visible   *Eclipse at MoonRise*   *All Eclipse Visible*

180° W   120° W   60° W   0°   60° E   120° E   180° E

Longitude

Partial Lunar Eclipse of 2077 Nov 29

Greatest Eclipse = 21:35:52.8 TD (= 21:34:06.8 UT1)

Penumbral Magnitude = 1.2326	Gamma = 0.8855	Saros Series = 146
Umbral Magnitude = 0.2372	Axis = 0.8823°	Saros Member = 14 of 72

Sun at Greatest Eclipse
(Geocentric Coordinates)

R.A. = 16h26m14.3s
Dec. = -21°40'58.6"
S.D. = 00°16'12.9"
H.P. = 00°00'08.9"

Moon at Greatest Eclipse
(Geocentric Coordinates)

R.A. = 04h25m16.1s
Dec. = +22°32'10.3"
S.D. = 00°16'17.5"
H.P. = 00°59'47.4"

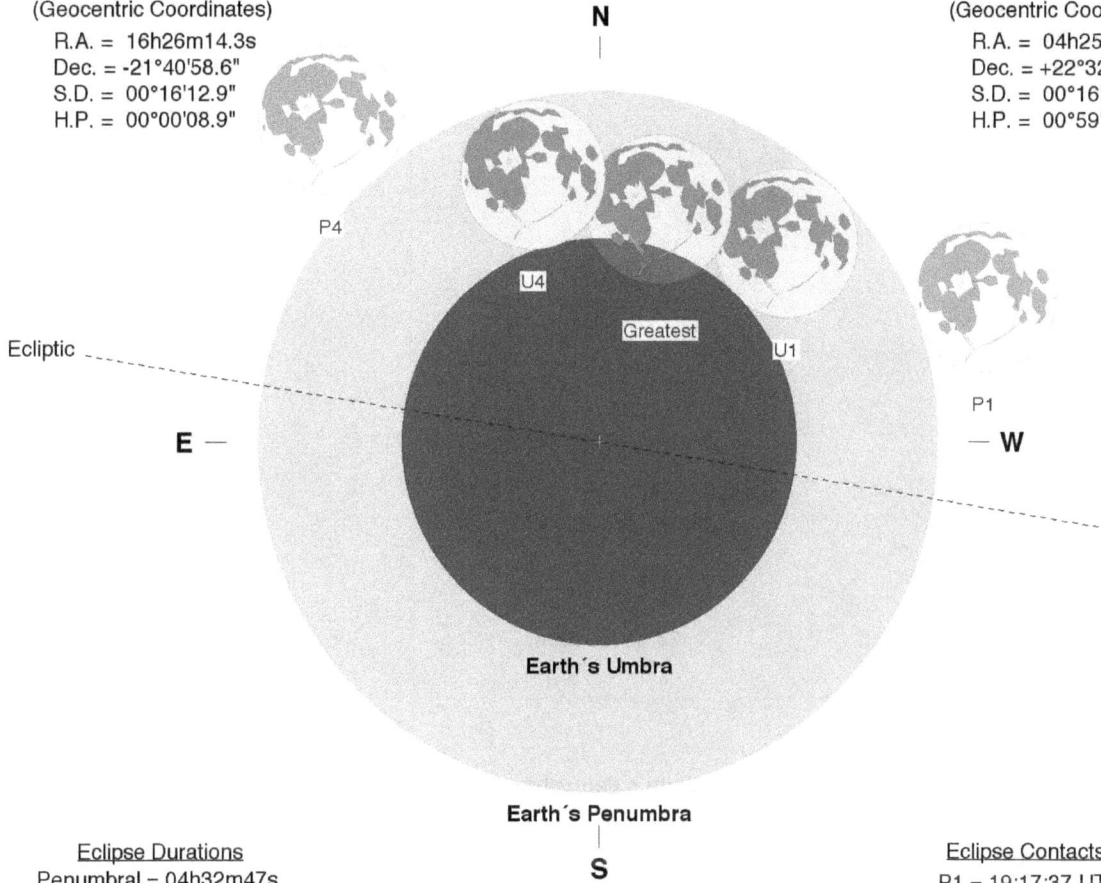

N

P4

U4

Greatest

U1

P1

Ecliptic

E

W

Earth's Umbra

Earth's Penumbra

S

Eclipse Durations
Penumbral = 04h32m47s
Umbral = 01h45m38s

Eph. = JPL DE430
Rule = Herald-Sinnott
ΔT = 106 s

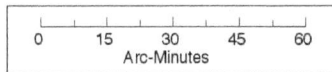

0 15 30 45 60
Arc-Minutes

©2020 F. Espenak, www.EclipseWise.com

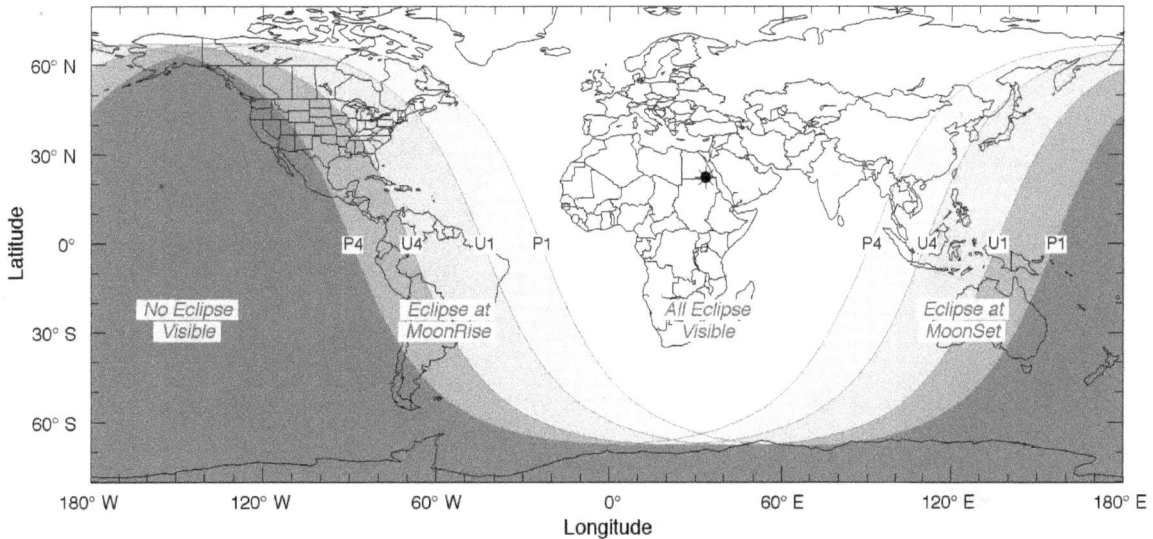

Eclipse Contacts
P1 = 19:17:37 UT1
U1 = 20:41:05 UT1
U4 = 22:26:44 UT1
P4 = 23:50:24 UT1

No Eclipse Visible

Eclipse at MoonRise

All Eclipse Visible

Eclipse at MoonSet

Penumbral Lunar Eclipse of 2078 Apr 27

Greatest Eclipse = 04:35:45.4 TD   (= 04:33:59.0 UT1)

Penumbral Magnitude = 0.6577	Gamma = 1.2223	Saros Series = 113
Umbral Magnitude = -0.4227	Axis = 1.0998°	Saros Member = 67 of 71

Sun at Greatest Eclipse
(Geocentric Coordinates)
R.A. = 02h19m59.8s
Dec. = +13°57'32.8"
S.D. = 00°15'53.7"
H.P. = 00°00'08.7"

Moon at Greatest Eclipse
(Geocentric Coordinates)
R.A. = 14h21m52.7s
Dec. = -12°57'31.9"
S.D. = 00°14'42.7"
H.P. = 00°53'59.6"

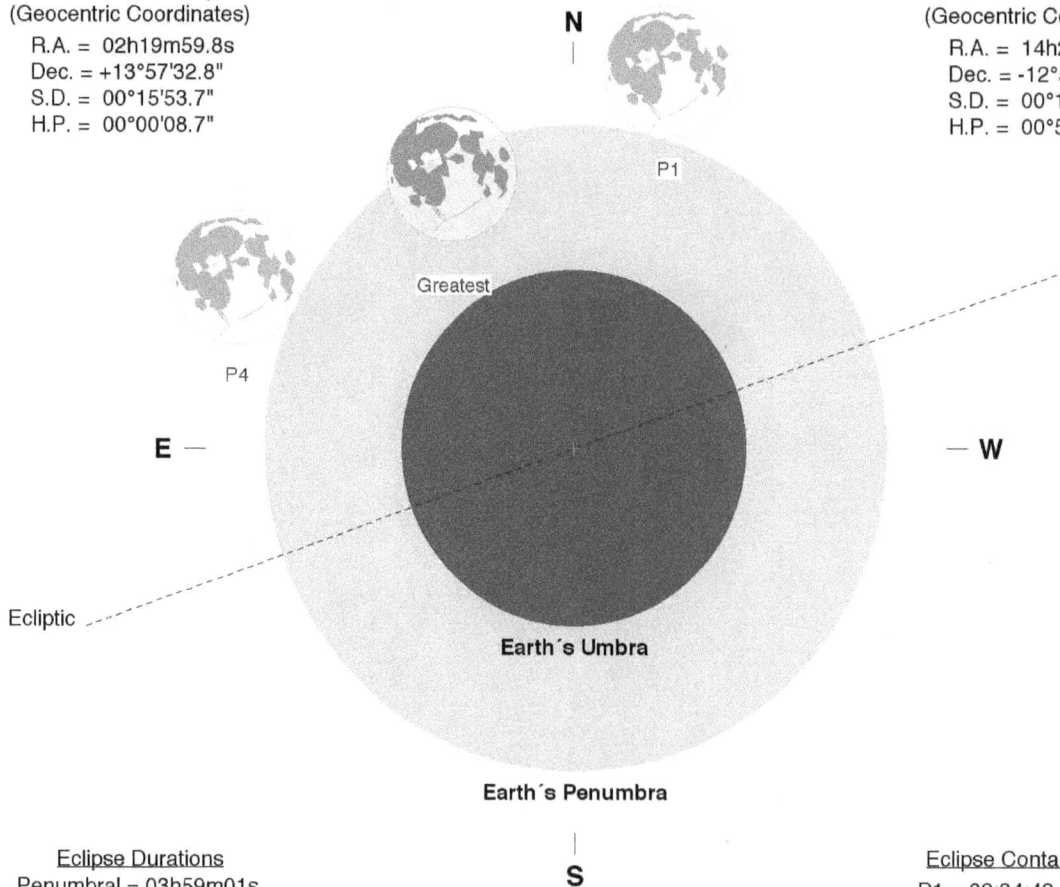

N

P1

Greatest

P4

E —

— W

Ecliptic

Earth's Umbra

Earth's Penumbra

S

Eclipse Durations
Penumbral = 03h59m01s

Eclipse Contacts
P1 = 02:34:42 UT1
P4 = 06:33:43 UT1

Eph. = JPL DE430
Rule = Herald-Sinnott
ΔT =   106 s

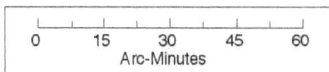

0	15	30	45	60

Arc-Minutes

©2020 F. Espenak, www.EclipseWise.com

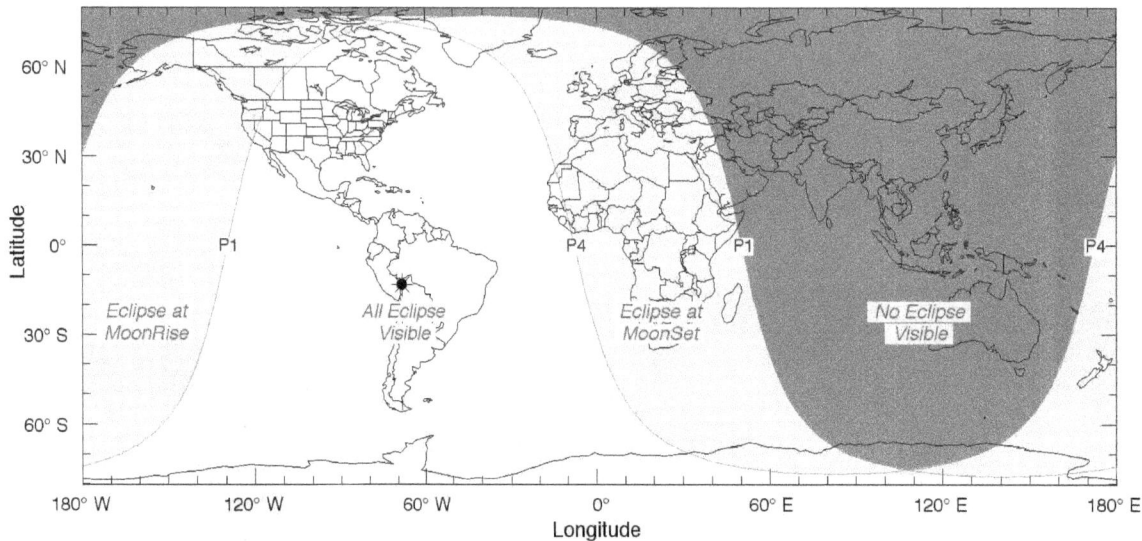

Eclipse at MoonRise

All Eclipse Visible

Eclipse at MoonSet

No Eclipse Visible

P1 · P4 · P1 · P4

Longitude

Penumbral Lunar Eclipse of 2078 Oct 21

Greatest Eclipse = 03:08:03.6 TD (= 03:06:16.8 UT1)

Penumbral Magnitude = 0.8191	Gamma = -1.1022	Saros Series = 118
Umbral Magnitude = -0.1442	Axis = 1.1239°	Saros Member = 55 of 73

Sun at Greatest Eclipse
(Geocentric Coordinates)

R.A. = 13h44m30.2s
Dec. = -10°48'15.9"
S.D. = 00°16'03.8"
H.P. = 00°00'08.8"

Moon at Greatest Eclipse
(Geocentric Coordinates)

R.A. = 01h46m32.1s
Dec. = +09°47'51.3"
S.D. = 00°16'40.4"
H.P. = 01°01'11.6"

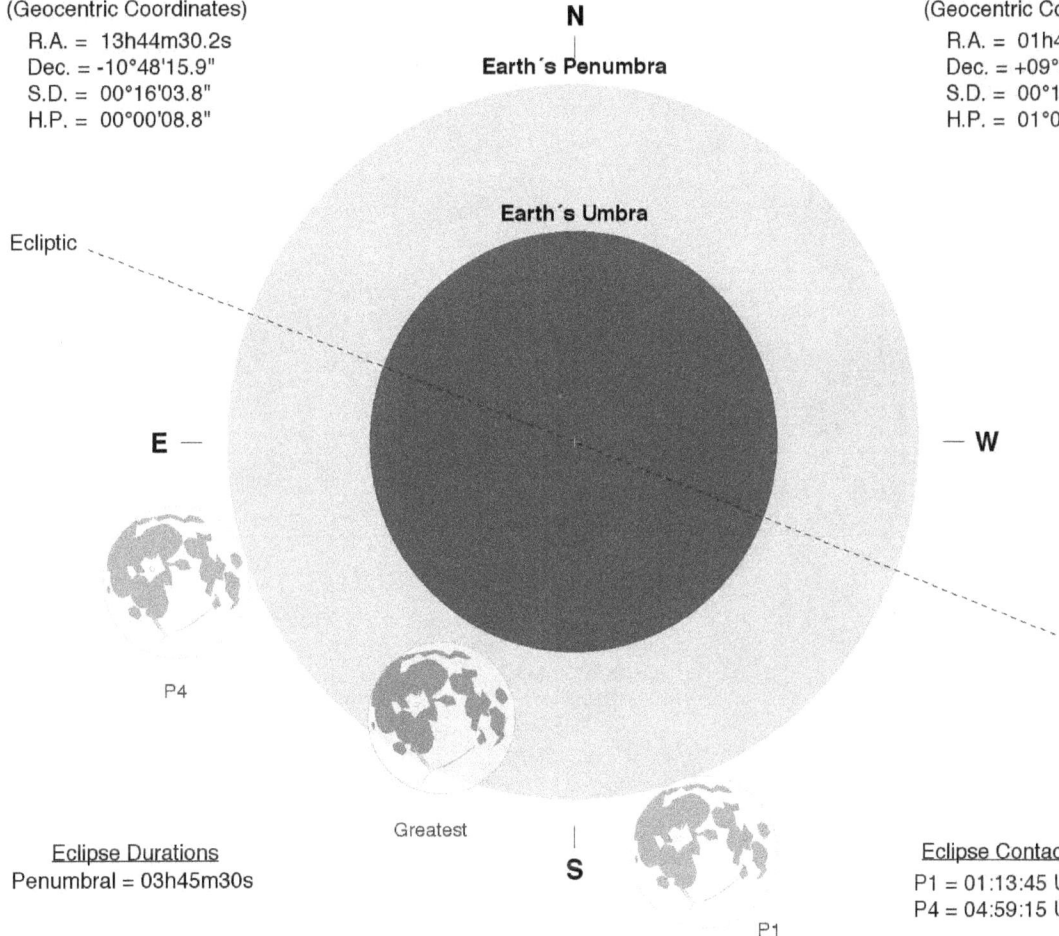

N

Earth's Penumbra

Earth's Umbra

Ecliptic

E —

— W

P4

Greatest

S

P1

Eclipse Durations
Penumbral = 03h45m30s

Eph. = JPL DE430
Rule = Herald-Sinnott
ΔT =   107 s

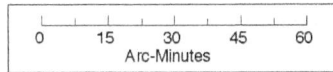

Eclipse Contacts
P1 = 01:13:45 UT1
P4 = 04:59:15 UT1

0	15	30	45	60

Arc-Minutes

©2020 F. Espenak, www.EclipseWise.com

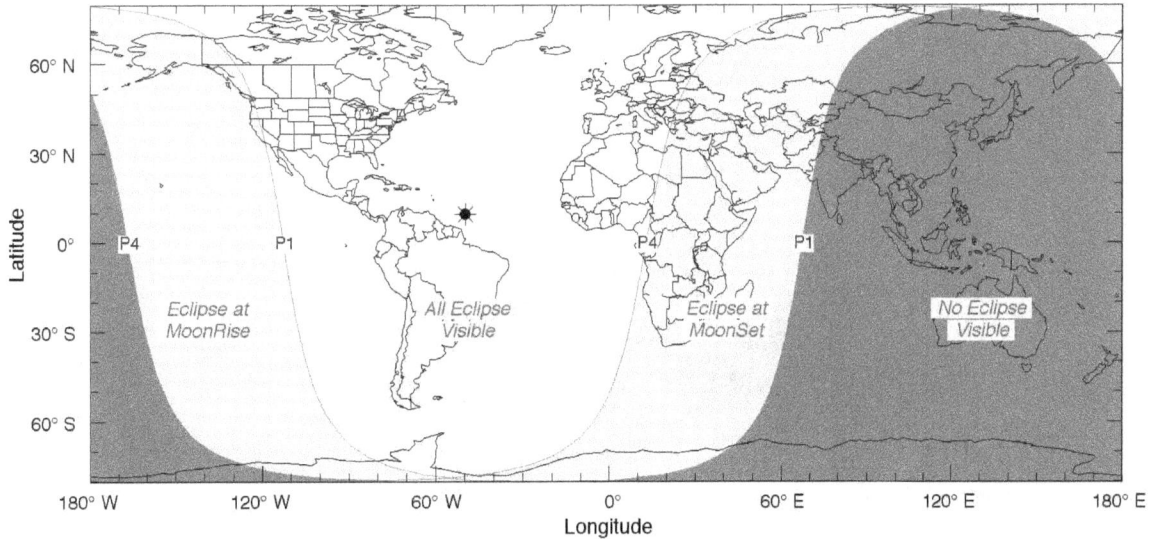

60° N

30° N

0°   P4   P1

Latitude

30° S

60° S

Eclipse at
MoonRise

P4   P1

All Eclipse
Visible

Eclipse at
MoonSet

No Eclipse
Visible

180° W   120° W   60° W   0°   60° E   120° E   180° E

Longitude

Penumbral Lunar Eclipse of 2078 Nov 19

Greatest Eclipse = 12:40:01.7 TD   (= 12:38:14.8 UT1)

Penumbral Magnitude = 0.0633	Gamma = 1.5148	Saros Series = 156
Umbral Magnitude = -0.9028	Axis = 1.5517°	Saros Member = 2 of 81

Sun at Greatest Eclipse
(Geocentric Coordinates)
R.A. = 15h41m12.4s
Dec. = -19°36'40.9"
S.D. = 00°16'10.9"
H.P. = 00°00'08.9"

N

P4

Greatest

P1

Moon at Greatest Eclipse
(Geocentric Coordinates)
R.A. = 03h39m05.9s
Dec. = +21°04'56.9"
S.D. = 00°16'45.0"
H.P. = 01°01'28.3"

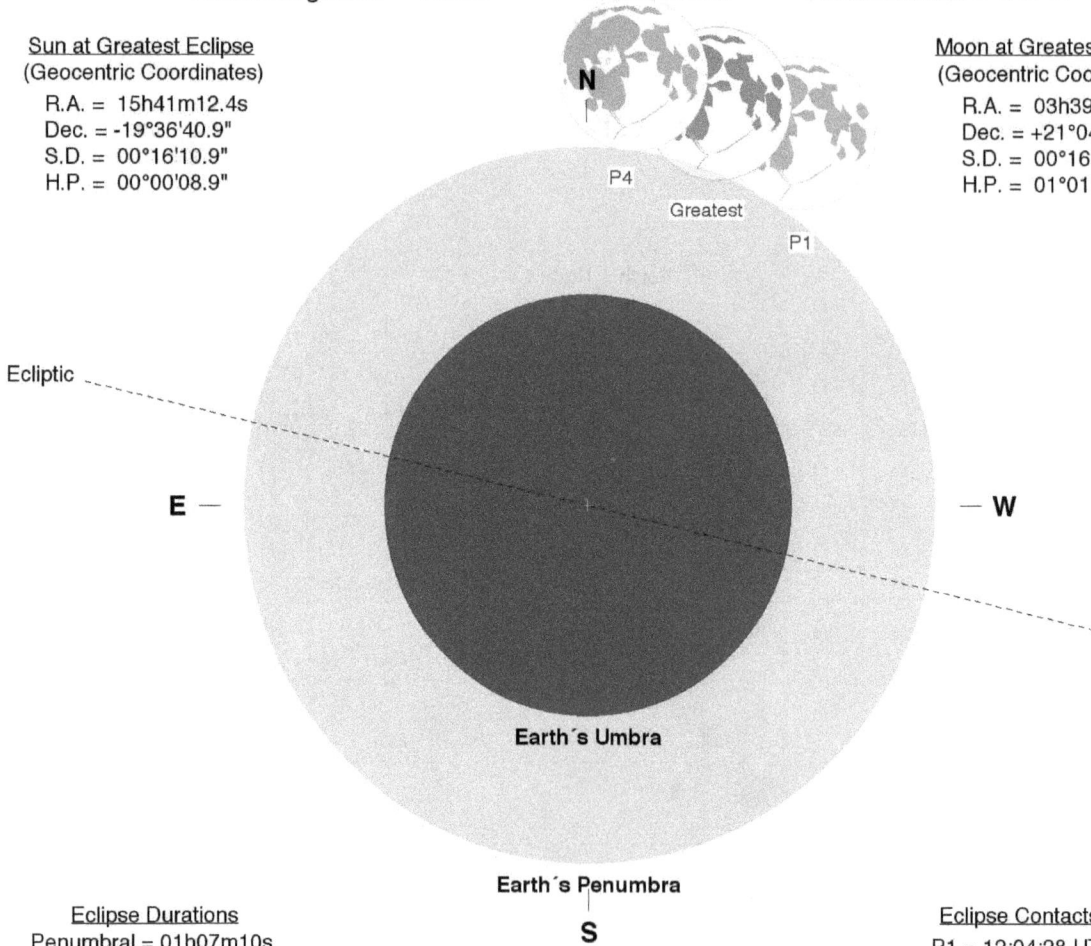

Ecliptic

E

W

Earth's Umbra

Earth's Penumbra

S

Eclipse Durations
Penumbral = 01h07m10s

Eclipse Contacts
P1 = 12:04:28 UT1
P4 = 13:11:38 UT1

Eph. = JPL DE430
Rule = Herald-Sinnott
ΔT = 107 s

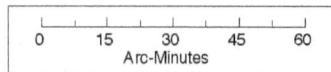

0	15	30	45	60

Arc-Minutes

©2020 F. Espenak, www.EclipseWise.com

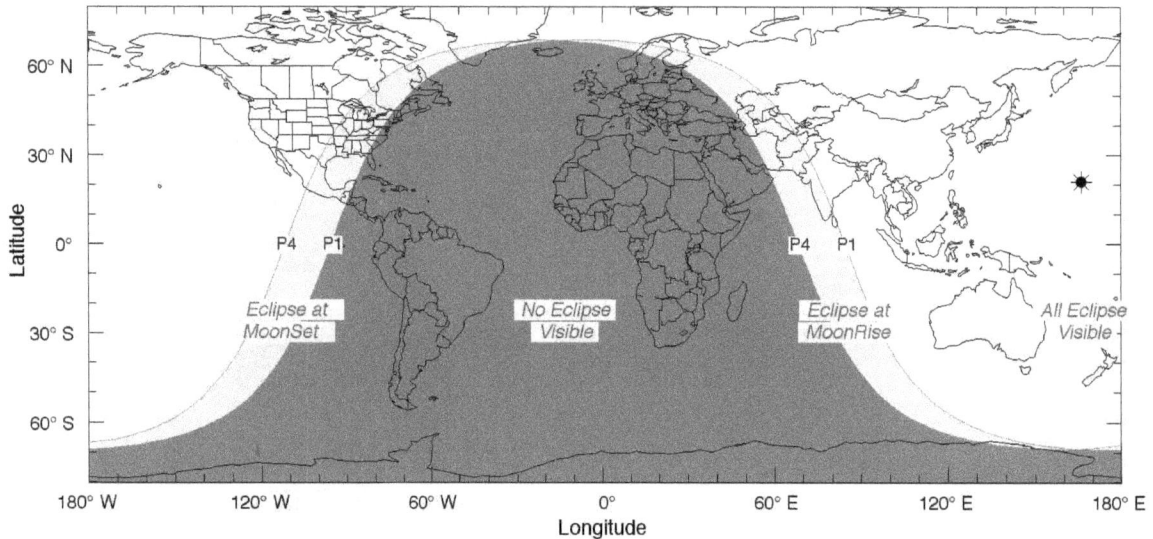

Eclipse at
MoonSet

No Eclipse
Visible

Eclipse at
MoonRise

All Eclipse
Visible

P4 P1

P4 P1

Latitude

Longitude

Partial Lunar Eclipse of 2079 Apr 16

Greatest Eclipse = 05:10:45.1 TD   (= 05:08:57.9 UT1)

Penumbral Magnitude = 2.0119	Gamma = 0.4800	Saros Series = 123
Umbral Magnitude = 0.9471	Axis = 0.4395°	Saros Member = 56 of 72

Sun at Greatest Eclipse
(Geocentric Coordinates)

R.A. = 01h37m57.0s
Dec. = +10°11'02.0"
S.D. = 00°15'56.6"
H.P. = 00°00'08.8"

Moon at Greatest Eclipse
(Geocentric Coordinates)

R.A. = 13h38m45.5s
Dec. = -09°47'31.2"
S.D. = 00°14'58.3"
H.P. = 00°54'56.9"

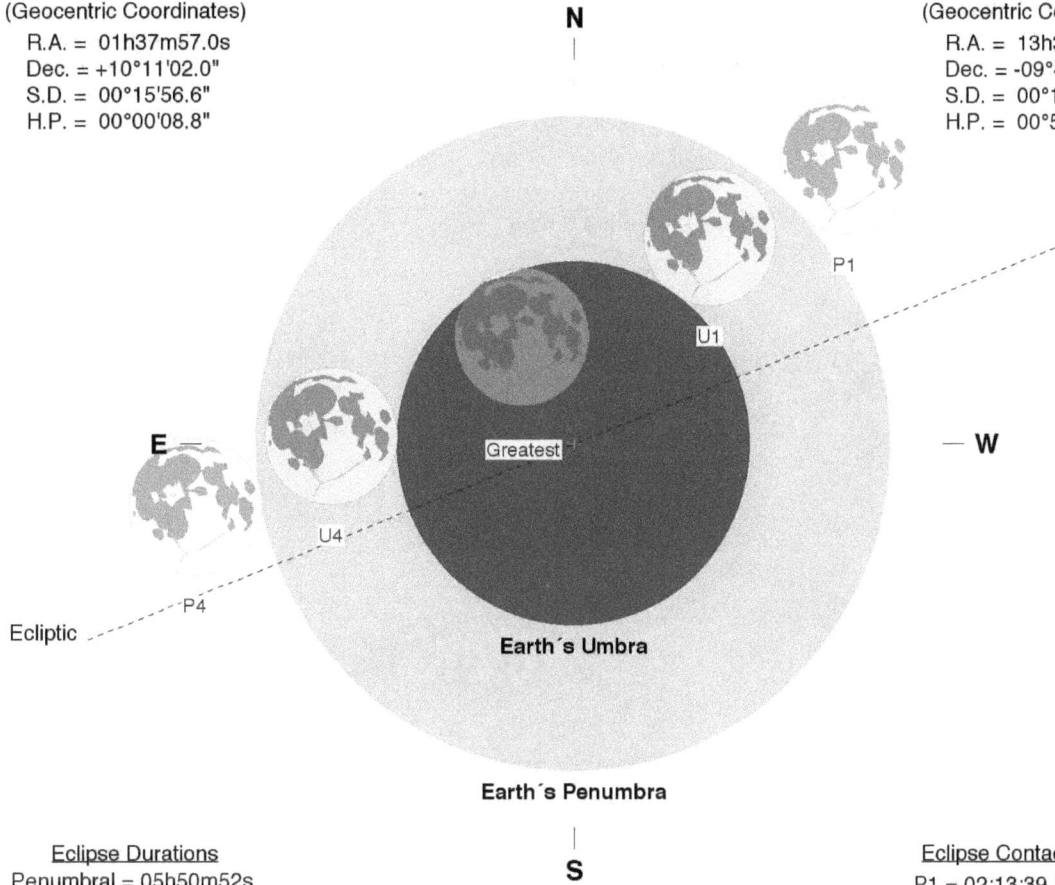

N

P1

U1

E

Greatest

W

U4

P4

Ecliptic

Earth's Umbra

Earth's Penumbra

S

Eclipse Durations
Penumbral = 05h50m52s
Umbral = 03h24m05s

Eph. = JPL DE430
Rule = Herald-Sinnott
ΔT =   107 s

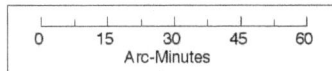

Arc-Minutes
0   15   30   45   60

©2020 F. Espenak, www.EclipseWise.com

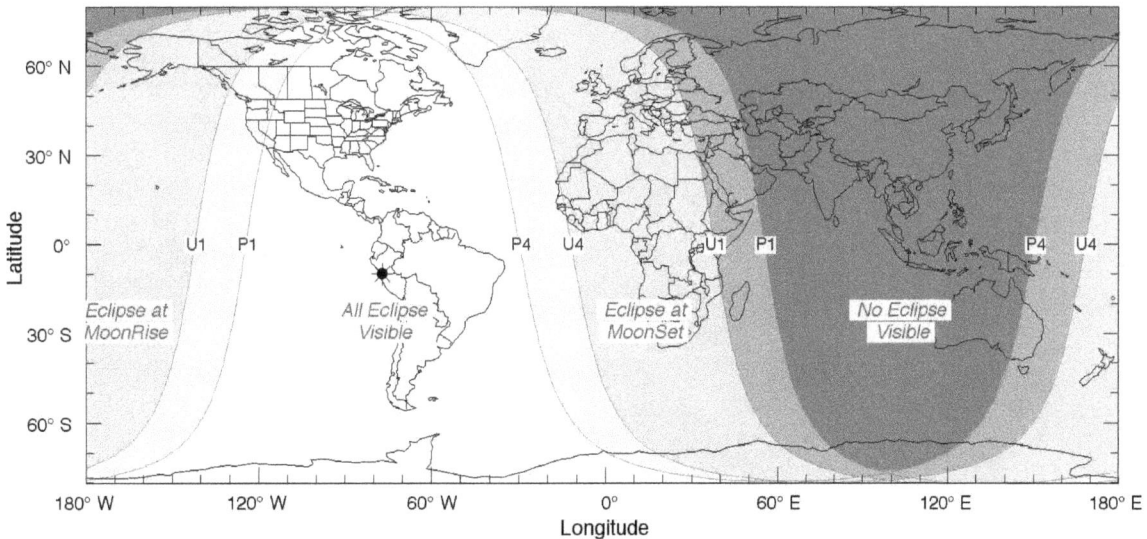

Eclipse Contacts
P1 = 02:13:39 UT1
U1 = 03:27:03 UT1
U4 = 06:51:07 UT1
P4 = 08:04:31 UT1

60° N

30° N

0°

30° S

60° S

Latitude

U1 P1   P4 U4   U1 P1   P4 U4

Eclipse at
MoonRise

All Eclipse
Visible

Eclipse at
MoonSet

No Eclipse
Visible

180° W   120° W   60° W   0°   60° E   120° E   180° E

Longitude

Total Lunar Eclipse of 2079 Oct 10

Greatest Eclipse = 17:30:30.0 TD   (= 17:28:42.3 UT1)

Penumbral Magnitude = 2.0806	Gamma = -0.4246	Saros Series = 128
Umbral Magnitude = 1.0811	Axis = 0.4161°	Saros Member = 44 of 71

<u>Sun at Greatest Eclipse</u>
(Geocentric Coordinates)
R.A. = 13h04m50.3s
Dec. = -06°53'53.3"
S.D. = 00°16'00.8"
H.P. = 00°00'08.8"

<u>Moon at Greatest Eclipse</u>
(Geocentric Coordinates)
R.A. = 01h05m37.6s
Dec. = +06°31'52.0"
S.D. = 00°16'01.3"
H.P. = 00°58'48.1"

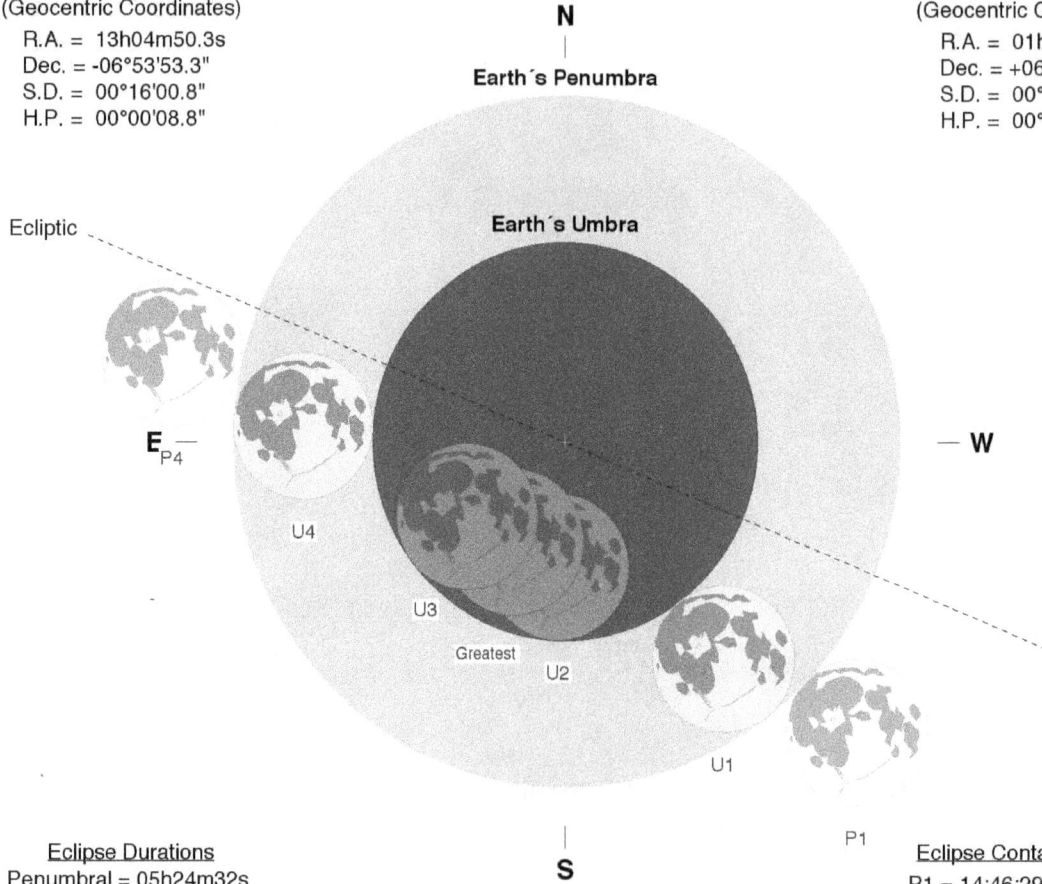

N

Earth´s Penumbra

Earth´s Umbra

Ecliptic

E
P4

W

U4

U3

Greatest

U2

U1

P1

S

<u>Eclipse Durations</u>
Penumbral = 05h24m32s
Umbral = 03h19m21s
Total = 00h43m07s

Eph. = JPL DE430
Rule = Herald-Sinnott
ΔT = 108 s

0   15   30   45   60
Arc-Minutes

©2020 F. Espenak, www.EclipseWise.com

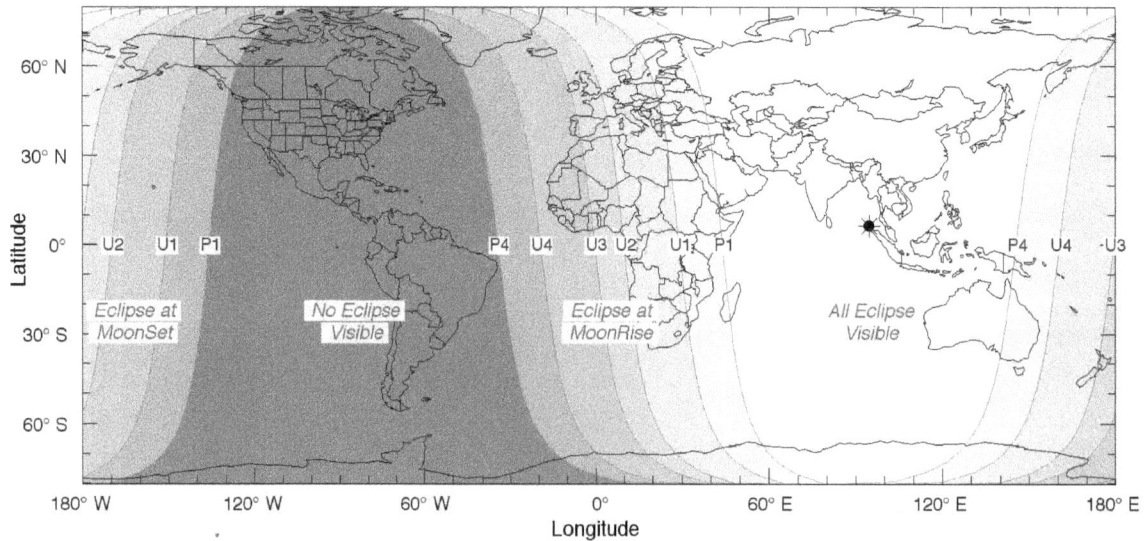

<u>Eclipse Contacts</u>
P1 = 14:46:29 UT1
U1 = 15:49:11 UT1
U2 = 17:07:27 UT1
U3 = 17:50:33 UT1
U4 = 19:08:32 UT1
P4 = 20:11:01 UT1

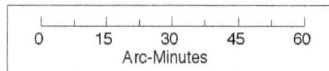

Eclipse at MoonSet

No Eclipse Visible

Eclipse at MoonRise

All Eclipse Visible

Total Lunar Eclipse of 2080 Apr 04

Greatest Eclipse = 11:23:38.4 TD   (= 11:21:50.3 UT1)

Penumbral Magnitude = 2.3626	Gamma = -0.2751	Saros Series = 133
Umbral Magnitude = 1.3479	Axis = 0.2651°	Saros Member = 30 of 71

Sun at Greatest Eclipse
(Geocentric Coordinates)
R.A. = 00h57m31.8s
Dec. = +06°08'40.0"
S.D. = 00°15'59.6"
H.P. = 00°00'08.8"

Moon at Greatest Eclipse
(Geocentric Coordinates)
R.A. = 12h57m01.5s
Dec. = -06°22'40.0"
S.D. = 00°15'45.6"
H.P. = 00°57'50.5"

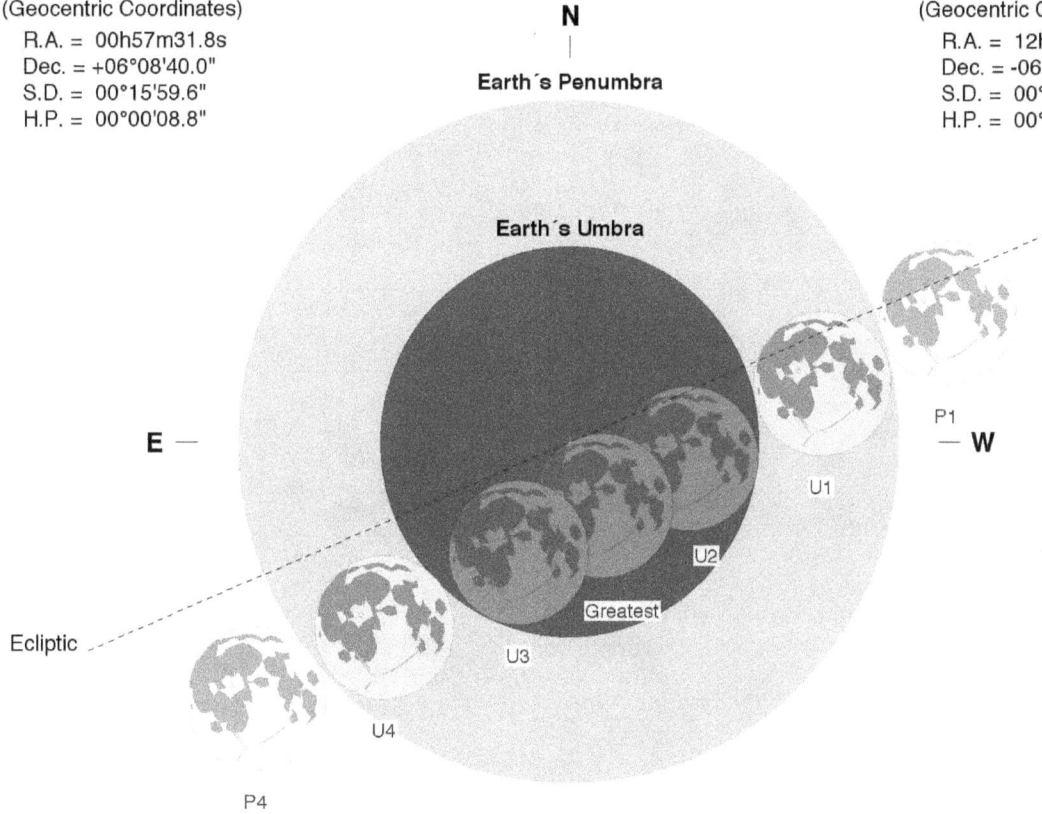

N

Earth´s Penumbra

Earth´s Umbra

E

W

Greatest

P1

U1

U2

U3

U4

P4

Ecliptic

S

Eclipse Durations
Penumbral = 05h39m02s
Umbral = 03h34m14s
Total = 01h22m46s

Eph. = JPL DE430
Rule = Herald-Sinnott
ΔT =   108 s

0   15   30   45   60
Arc-Minutes

©2020 F. Espenak, www.EclipseWise.com

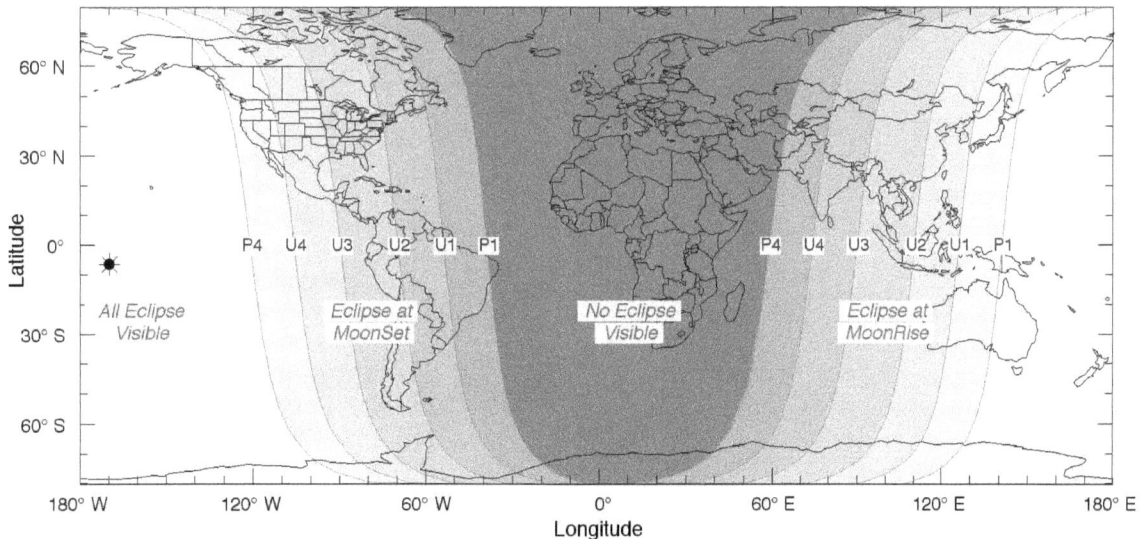

Eclipse Contacts
P1 = 08:32:19 UT1
U1 = 09:34:37 UT1
U2 = 10:40:16 UT1
U3 = 12:03:01 UT1
U4 = 13:08:51 UT1
P4 = 14:11:21 UT1

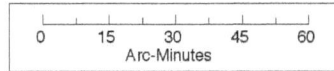

All Eclipse Visible

Eclipse at MoonSet

No Eclipse Visible

Eclipse at MoonRise

P4 U4 U3 U2 U1 P1   P4 U4 U3 U2 U1 P1

180° W   120° W   60° W   0°   60° E   120° E   180° E
Longitude

60° N
30° N
0°
30° S
60° S

Latitude

Total Lunar Eclipse of 2080 Sep 29

Greatest Eclipse = 01:52:41.8 TD (= 01:50:53.2 UT1)

Penumbral Magnitude = 2.2986	Gamma = 0.3203	Saros Series = 138
Umbral Magnitude = 1.2462	Axis = 0.2972°	Saros Member = 32 of 82

Sun at Greatest Eclipse
(Geocentric Coordinates)

R.A. = 12h25m10.2s
Dec. = -02°43'10.3"
S.D. = 00°15'57.8"
H.P. = 00°00'08.8"

Moon at Greatest Eclipse
(Geocentric Coordinates)

R.A. = 00h24m35.6s
Dec. = +02°58'45.9"
S.D. = 00°15'10.2"
H.P. = 00°55'40.3"

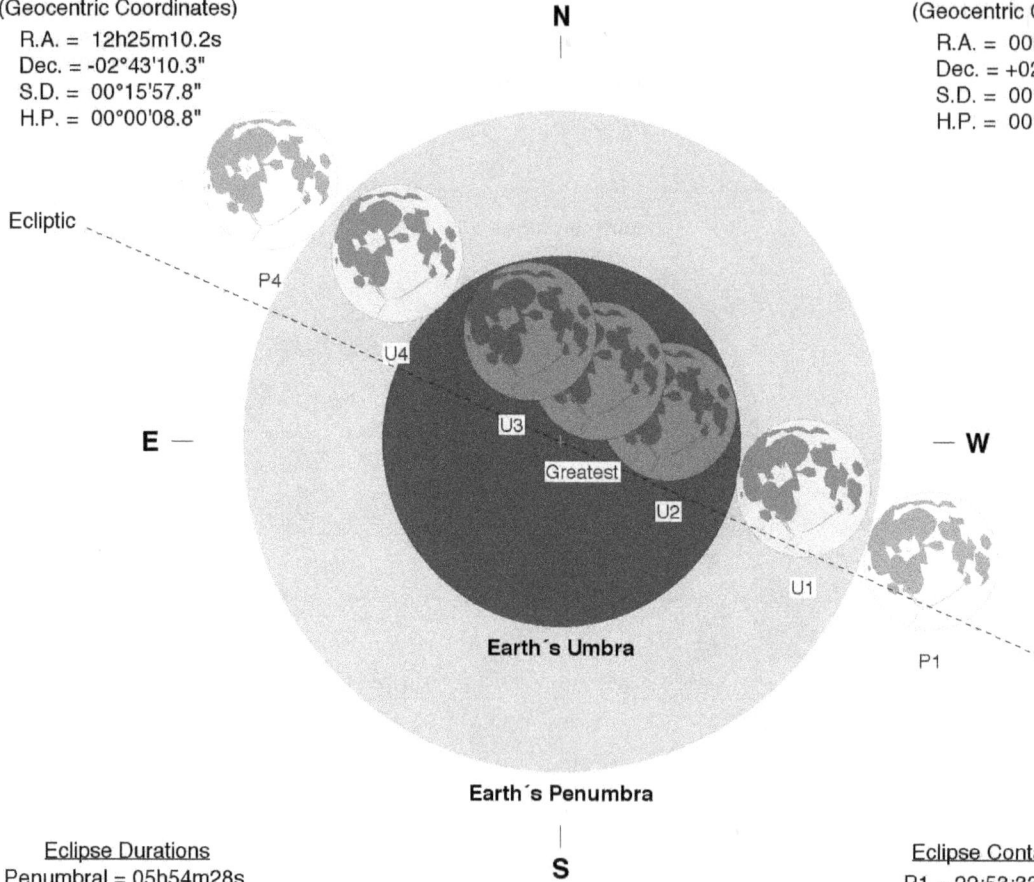

N

Ecliptic

P4

U4

U3

Greatest

E

W

U2

U1

Earth's Umbra

P1

Earth's Penumbra

S

Eclipse Durations

Penumbral = 05h54m28s
Umbral = 03h38m02s
Total = 01h14m25s

Eph. = JPL DE430
Rule = Herald-Sinnott
ΔT =   109 s

0   15   30   45   60
Arc-Minutes

©2020 F. Espenak, www.EclipseWise.com

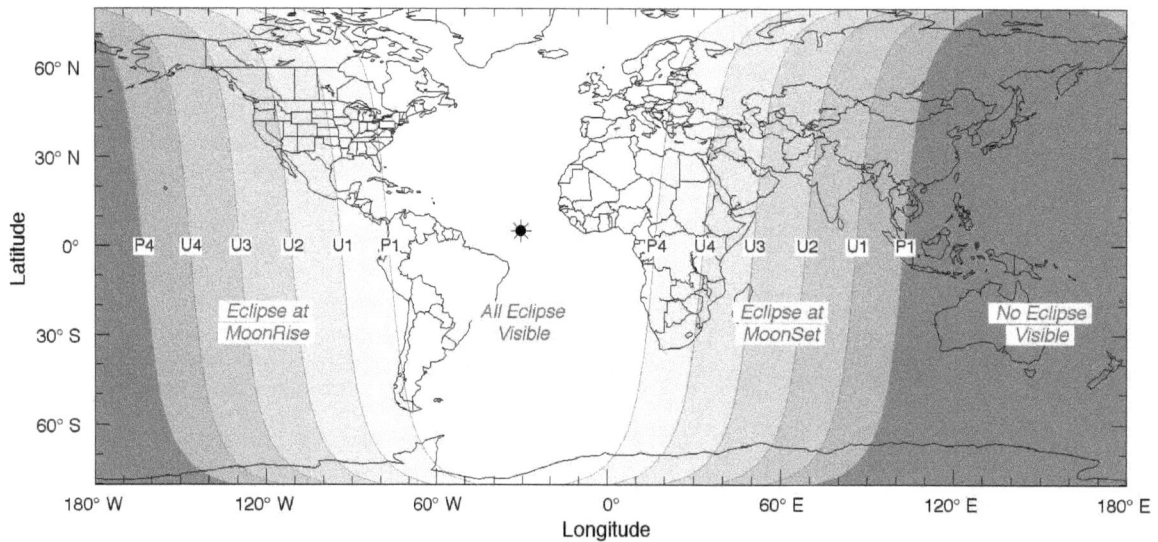

Eclipse Contacts

P1 = 22:53:33 UT1
U1 = 00:01:47 UT1
U2 = 01:13:28 UT1
U3 = 02:27:53 UT1
U4 = 03:39:49 UT1
P4 = 04:48:01 UT1

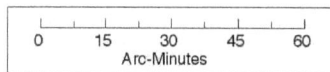

Eclipse at MoonRise

All Eclipse Visible

Eclipse at MoonSet

No Eclipse Visible

Partial Lunar Eclipse of 2081 Mar 25

Greatest Eclipse = 00:22:00.6 TD (= 00:20:11.6 UT1)

Penumbral Magnitude = 1.0673	Gamma = -0.9688	Saros Series = 143
Umbral Magnitude = 0.0973	Axis = 0.9799°	Saros Member = 21 of 72

Sun at Greatest Eclipse
(Geocentric Coordinates)

R.A. = 00h18m29.3s
Dec. = +01°59'59.9"
S.D. = 00°16'02.5"
H.P. = 00°00'08.8"

N

Earth's Penumbra

Earth's Umbra

E —

— **W**

P1

U1

Greatest

Ecliptic

U4

S

P4

Moon at Greatest Eclipse
(Geocentric Coordinates)

R.A. = 12h16m35.4s
Dec. = -02°51'27.7"
S.D. = 00°16'32.3"
H.P. = 01°00'41.9"

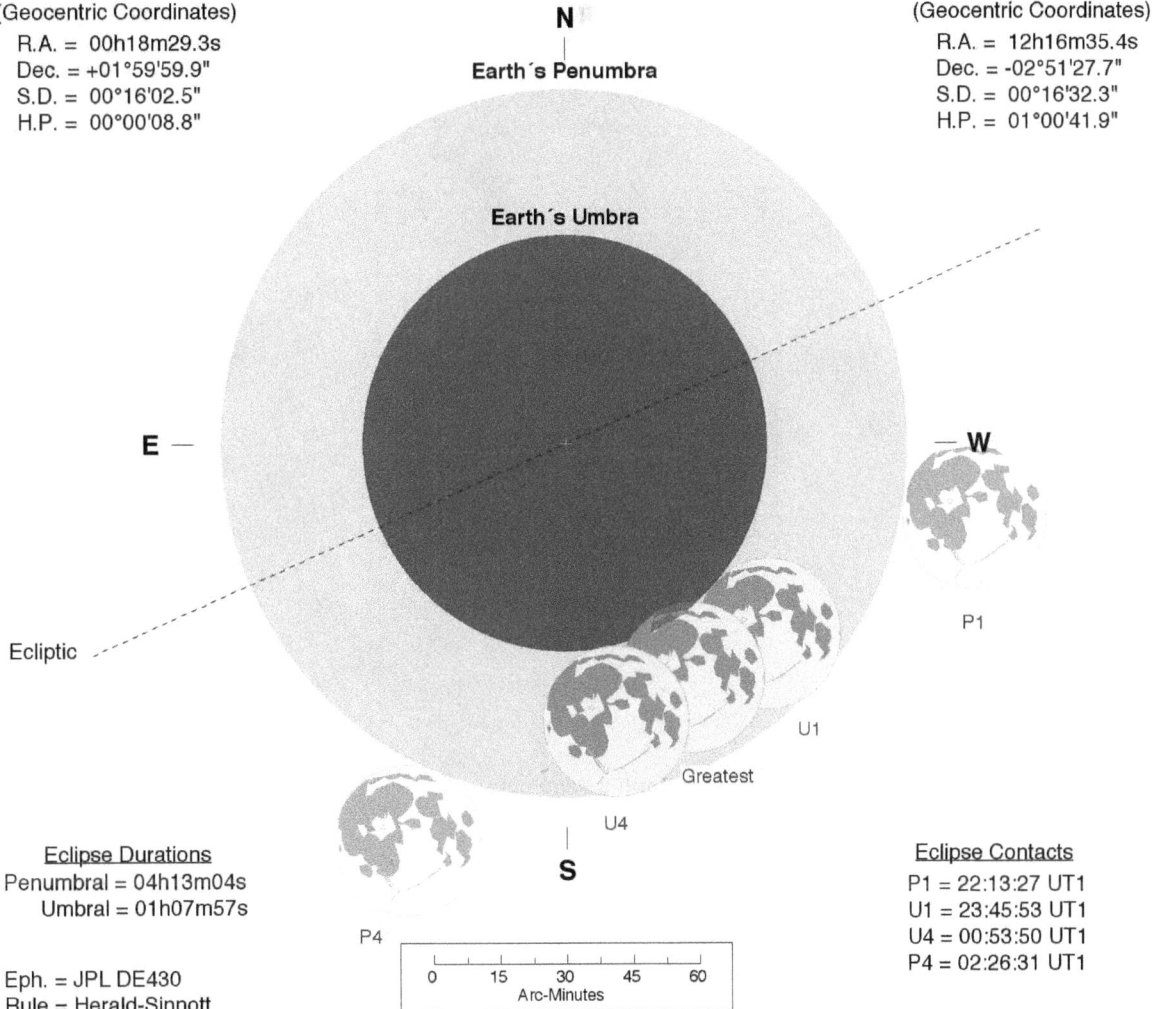

Eclipse Durations
Penumbral = 04h13m04s
Umbral = 01h07m57s

Eph. = JPL DE430
Rule = Herald-Sinnott
ΔT = 109 s

0	15	30	45	60

Arc-Minutes

©2020 F. Espenak, www.EclipseWise.com

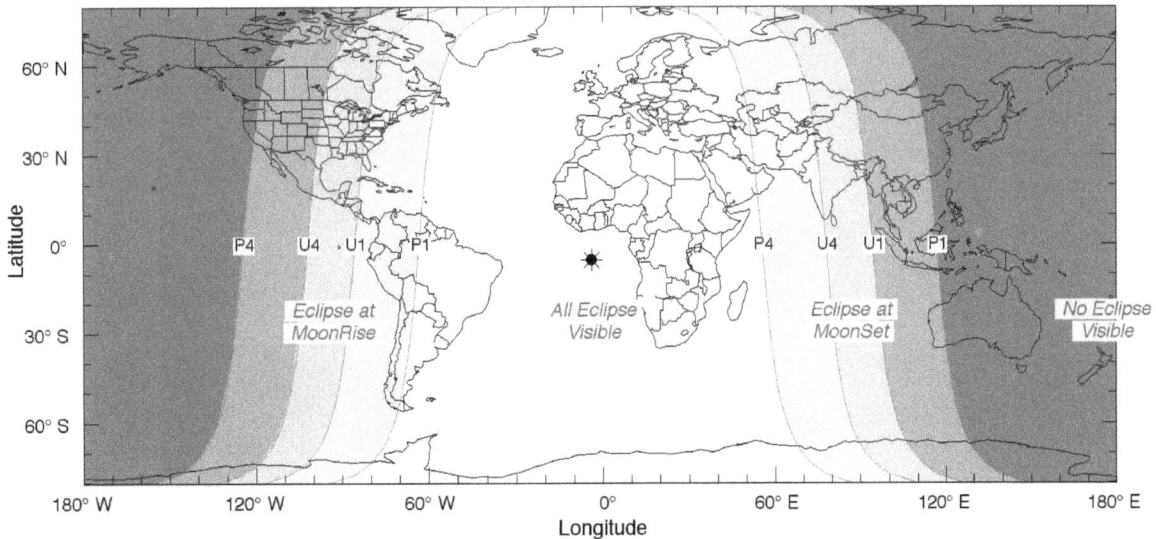

Eclipse Contacts
P1 = 22:13:27 UT1
U1 = 23:45:53 UT1
U4 = 00:53:50 UT1
P4 = 02:26:31 UT1

Eclipse at MoonRise

All Eclipse Visible

Eclipse at MoonSet

No Eclipse Visible

Penumbral Lunar Eclipse of 2081 Sep 18

Greatest Eclipse = 03:35:24.9 TD   (= 03:33:35.5 UT1)

Penumbral Magnitude = 0.9291	Gamma = 1.0747	Saros Series = 148
Umbral Magnitude = -0.1524	Axis = 0.9673°	Saros Member = 7 of 70

Sun at Greatest Eclipse
(Geocentric Coordinates)
R.A. = 11h45m03.1s
Dec. = +01°37'05.2"
S.D. = 00°15'54.9"
H.P. = 00°00'08.8"

Moon at Greatest Eclipse
(Geocentric Coordinates)
R.A. = 23h43m10.4s
Dec. = -00°46'20.5"
S.D. = 00°14'42.9"
H.P. = 00°54'00.3"

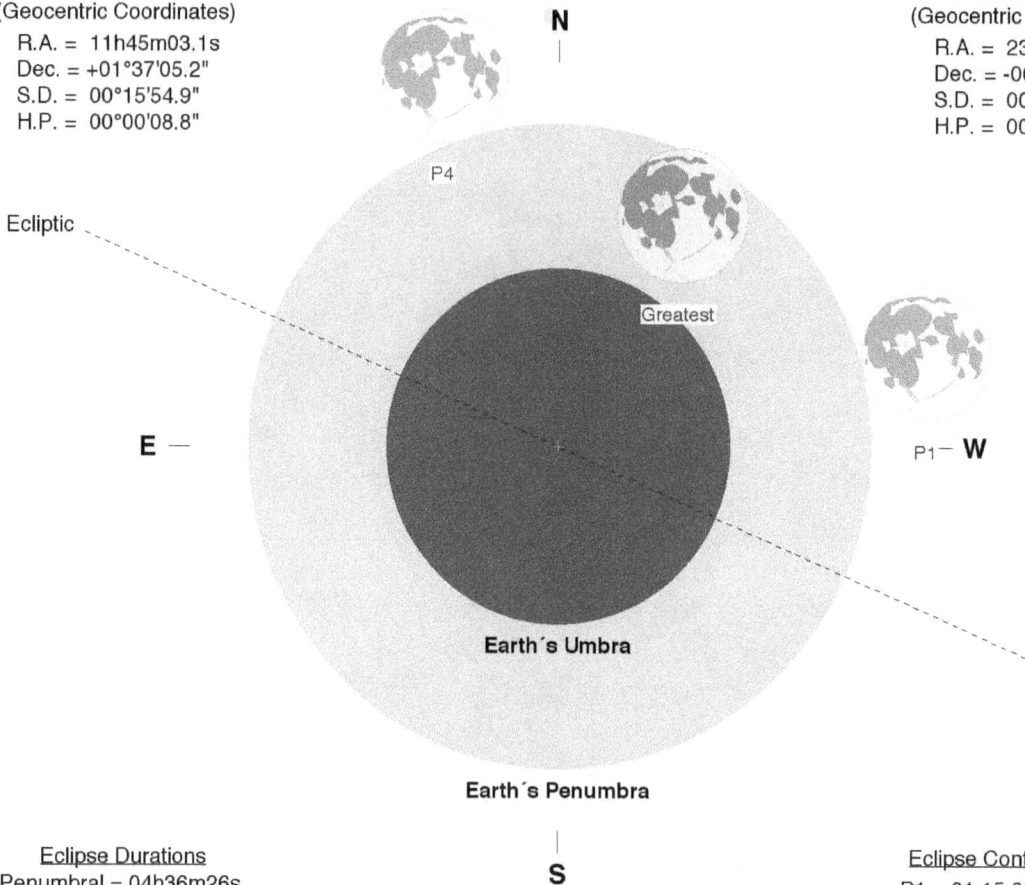

N

P4

Ecliptic

Greatest

E —

P1 — W

Earth's Umbra

Earth's Penumbra

S

Eclipse Durations
Penumbral = 04h36m26s

Eclipse Contacts
P1 = 01:15:09 UT1
P4 = 05:51:36 UT1

Eph. = JPL DE430
Rule = Herald-Sinnott
ΔT =   109 s

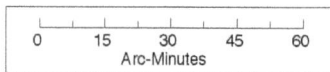

0	15	30	45	60

Arc-Minutes

©2020 F. Espenak, www.EclipseWise.com

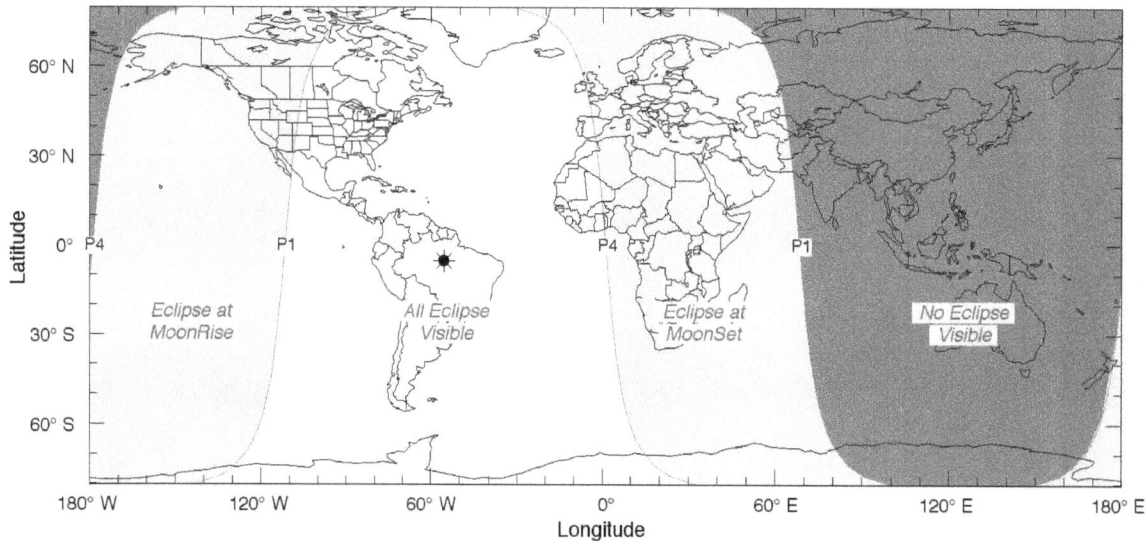

Eclipse at MoonRise

All Eclipse Visible

Eclipse at MoonSet

No Eclipse Visible

Partial Lunar Eclipse of 2082 Feb 13

Greatest Eclipse = 06:29:19.6 TD (= 06:27:29.9 UT1)

Penumbral Magnitude = 0.9974	Gamma = 1.0101	Saros Series = 115
Umbral Magnitude = 0.0153	Axis = 1.0191°	Saros Member = 61 of 72

Sun at Greatest Eclipse
(Geocentric Coordinates)
R.A. = 21h49m00.2s
Dec. = -13°11'39.6"
S.D. = 00°16'12.1"
H.P. = 00°00'08.9"

Moon at Greatest Eclipse
(Geocentric Coordinates)
R.A. = 09h50m47.2s
Dec. = +14°07'00.7"
S.D. = 00°16'29.8"
H.P. = 01°00'32.5"

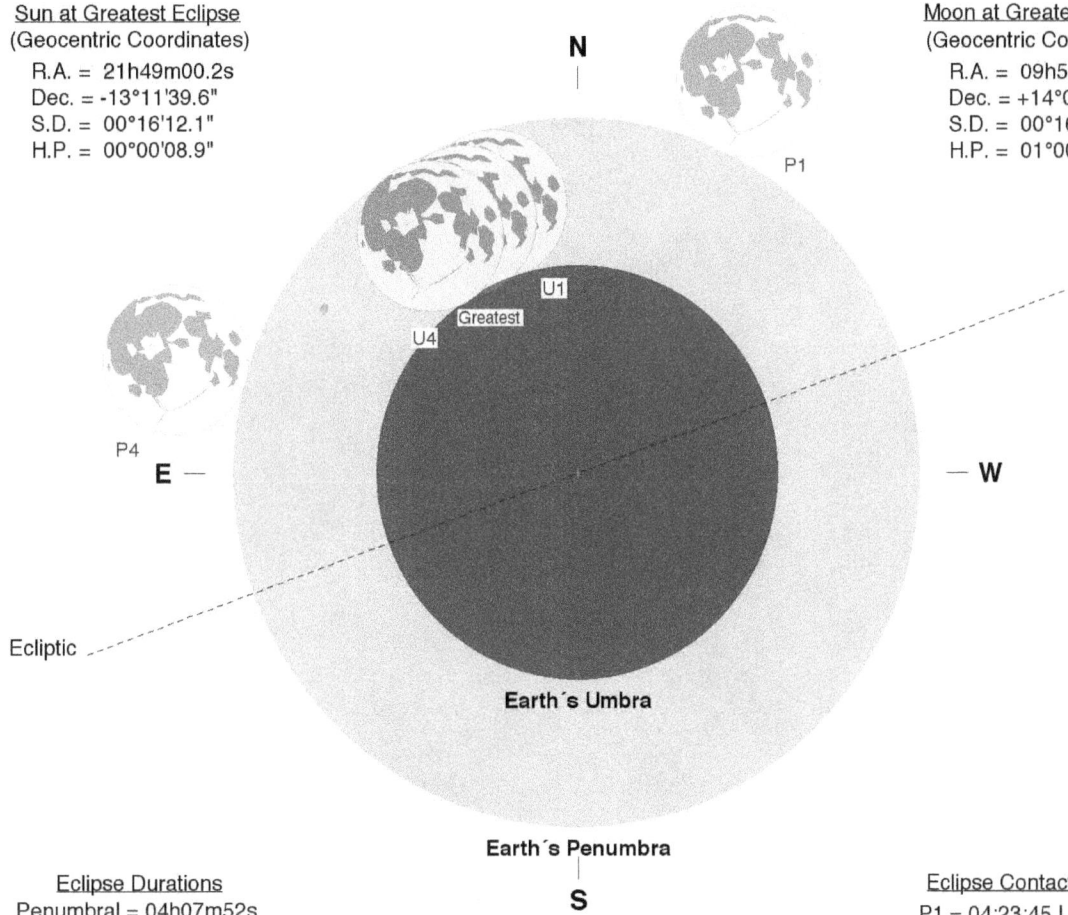

N

P1

U1

Greatest

U4

E —

— W

P4

Ecliptic

Earth´s Umbra

Earth´s Penumbra

S

Eclipse Durations
Penumbral = 04h07m52s
Umbral = 00h27m20s

Eph. = JPL DE430
Rule = Herald-Sinnott
ΔT =   110 s

0	15	30	45	60

Arc-Minutes

©2020 F. Espenak, www.EclipseWise.com

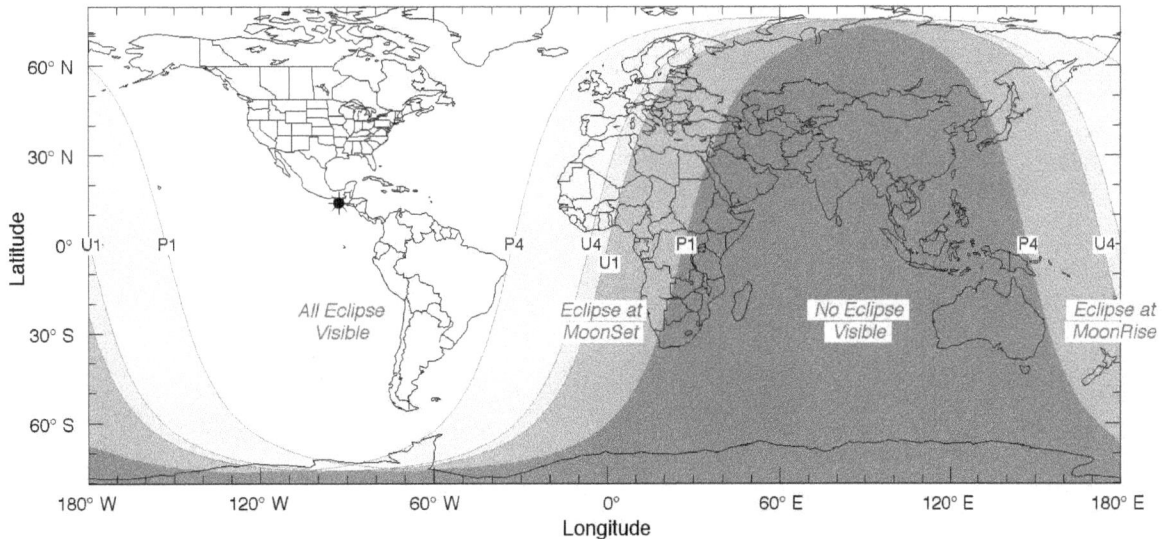

Eclipse Contacts
P1 = 04:23:45 UT1
U1 = 06:14:09 UT1
U4 = 06:41:29 UT1
P4 = 08:31:38 UT1

All Eclipse Visible

Eclipse at MoonSet

No Eclipse Visible

Eclipse at MoonRise

Penumbral Lunar Eclipse of 2082 Aug 08

Greatest Eclipse = 14:46:42.3 TD   (= 14:44:52.1 UT1)

Penumbral Magnitude = 1.0030	Gamma = -1.0204	Saros Series = 120
Umbral Magnitude = -0.0275	Axis = 0.9552°	Saros Member = 61 of 83

Sun at Greatest Eclipse
(Geocentric Coordinates)

R.A. = 09h15m56.4s
Dec. = +15°52'33.9"
S.D. = 00°15'46.4"
H.P. = 00°00'08.7"

Moon at Greatest Eclipse
(Geocentric Coordinates)

R.A. = 21h17m30.2s
Dec. = -16°45'16.7"
S.D. = 00°15'18.4"
H.P. = 00°56'10.5"

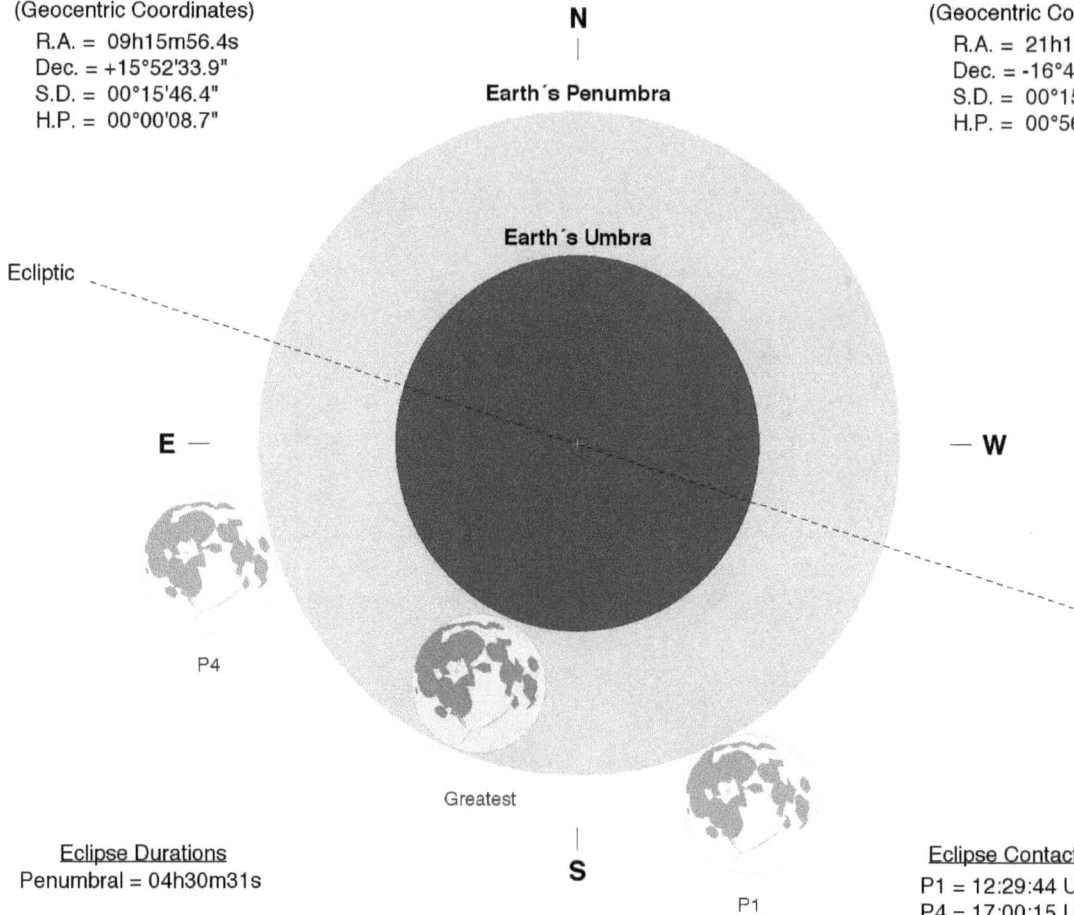

N

Earth's Penumbra

Earth's Umbra

Ecliptic

E

W

P4

Greatest

S

P1

Eclipse Durations

Penumbral = 04h30m31s

Eclipse Contacts

P1 = 12:29:44 UT1
P4 = 17:00:15 UT1

Eph. = JPL DE430
Rule = Herald-Sinnott
ΔT =   110 s

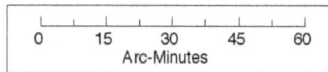

0	15	30	45	60

Arc-Minutes

©2020 F. Espenak, www.EclipseWise.com

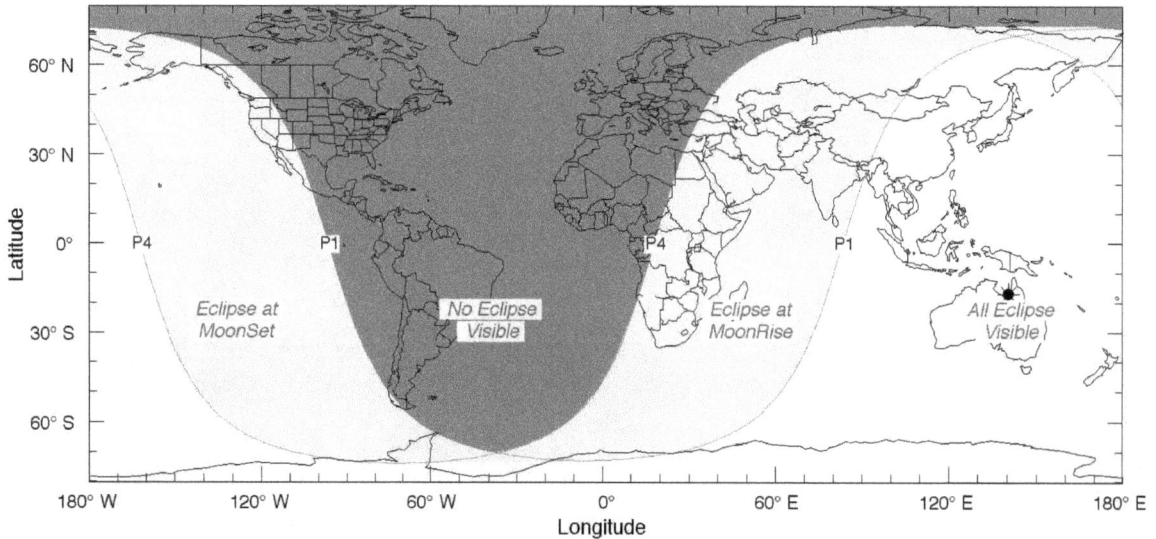

Total Lunar Eclipse of 2083 Feb 02

Greatest Eclipse = 18:26:45.7 TD (= 18:24:55.1 UT1)

Penumbral Magnitude = 2.2418	Gamma = 0.3464	Saros Series = 125
Umbral Magnitude = 1.2070	Axis = 0.3323°	Saros Member = 52 of 72

<u>Sun at Greatest Eclipse</u>
(Geocentric Coordinates)

R.A. = 21h06m12.7s
Dec. = -16°35'42.2"
S.D. = 00°16'13.8"
H.P. = 00°00'08.9"

<u>Moon at Greatest Eclipse</u>
(Geocentric Coordinates)

R.A. = 09h06m44.6s
Dec. = +16°54'07.4"
S.D. = 00°15'41.1"
H.P. = 00°57'34.0"

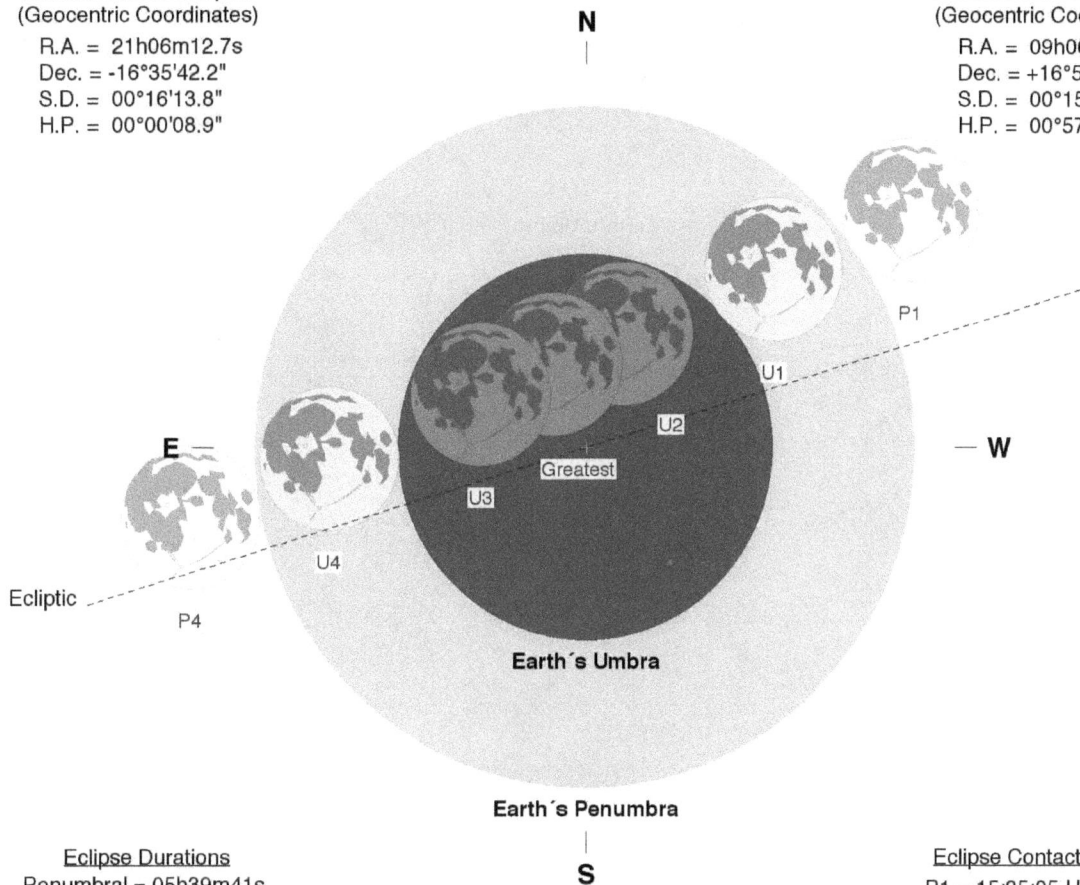

N

E — — W

Greatest

U1
U2
U3
U4
P1
P4

Ecliptic

Earth's Umbra

Earth's Penumbra

S

<u>Eclipse Durations</u>
Penumbral = 05h39m41s
Umbral = 03h29m30s
Total = 01h07m11s

Eph. = JPL DE430
Rule = Herald-Sinnott
ΔT =   111 s

0   15   30   45   60
Arc-Minutes

©2020 F. Espenak, www.EclipseWise.com

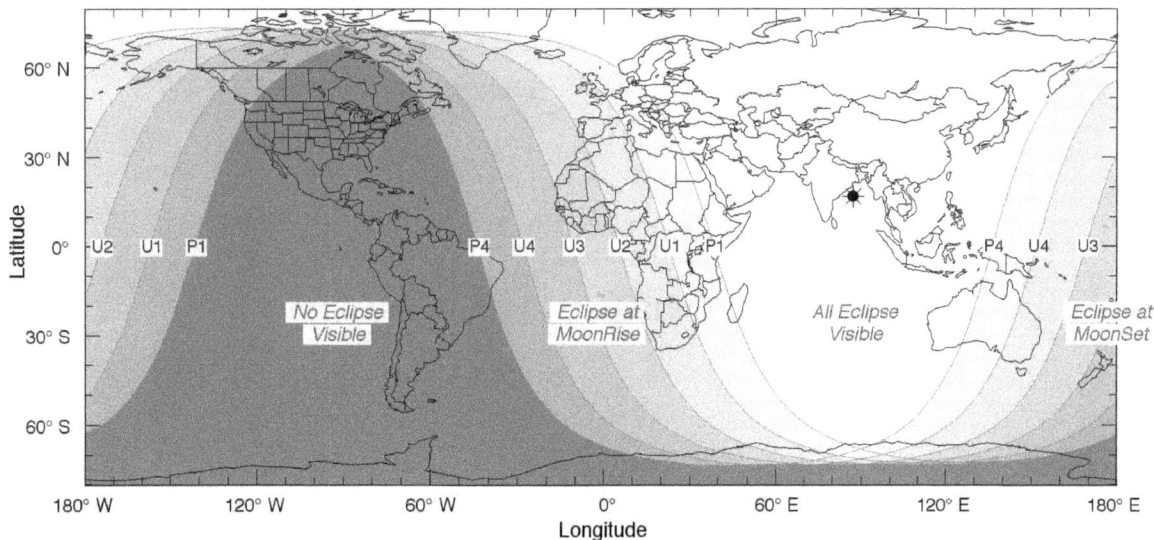

<u>Eclipse Contacts</u>
P1 = 15:35:05 UT1
U1 = 16:40:17 UT1
U2 = 17:51:32 UT1
U3 = 18:58:43 UT1
U4 = 20:09:47 UT1
P4 = 21:14:46 UT1

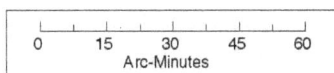

No Eclipse Visible

Eclipse at MoonRise

All Eclipse Visible

Eclipse at MoonSet

Total Lunar Eclipse of 2083 Jul 29

Greatest Eclipse = 01:05:34.2 TD (= 01:03:43.2 UT1)

| Penumbral Magnitude = 2.4537 | Gamma = -0.2143 | Saros Series = 130 |
| Umbral Magnitude = 1.4791 | Axis = 0.2118° | Saros Member = 38 of 71 |

Sun at Greatest Eclipse
(Geocentric Coordinates)
R.A. = 08h34m15.1s
Dec. = +18°43'08.7"
S.D. = 00°15'45.0"
H.P. = 00°00'08.7"

Moon at Greatest Eclipse
(Geocentric Coordinates)
R.A. = 20h34m33.5s
Dec. = -18°55'05.1"
S.D. = 00°16'09.6"
H.P. = 00°59'18.6"

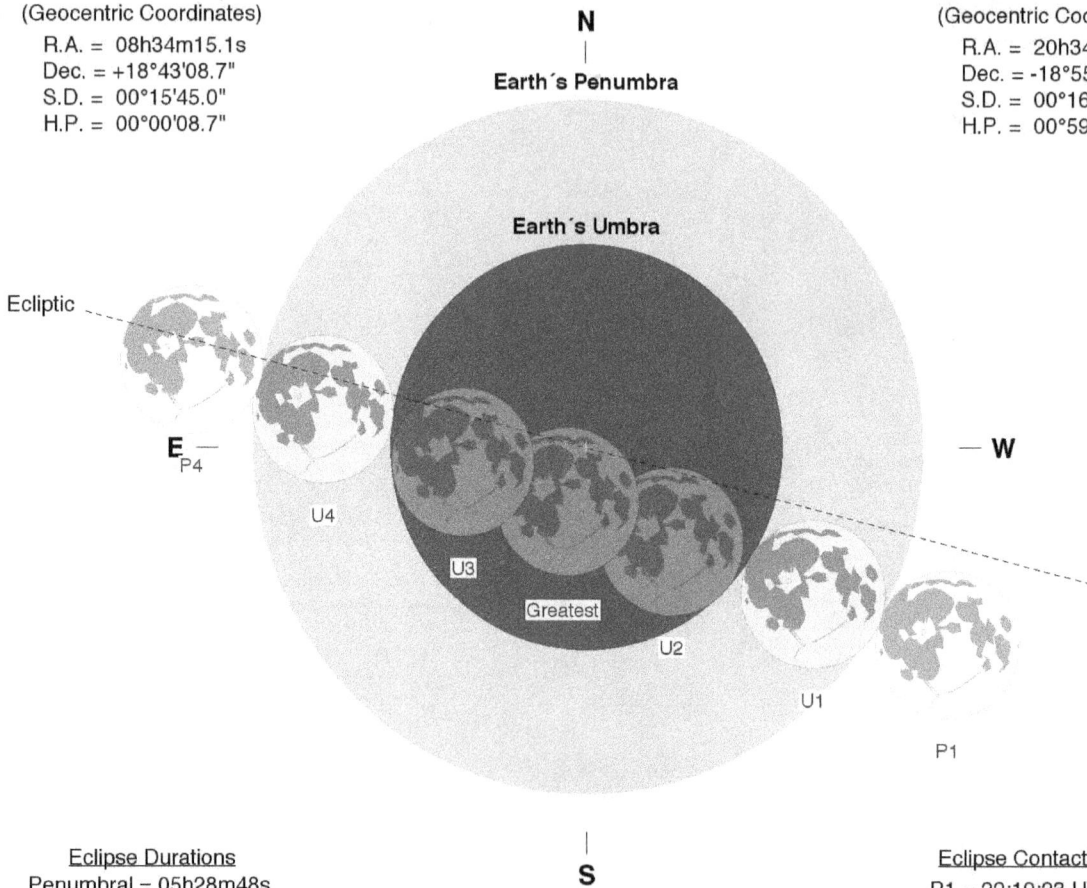

N

Earth's Penumbra

Earth's Umbra

Ecliptic

E
P4

W

U4

U3

Greatest

U2

U1

P1

S

Eclipse Durations
Penumbral = 05h28m48s
Umbral = 03h33m37s
Total = 01h31m06s

Eph. = JPL DE430
Rule = Herald-Sinnott
ΔT = 111 s

0   15   30   45   60
Arc-Minutes

©2020 F. Espenak, www.EclipseWise.com

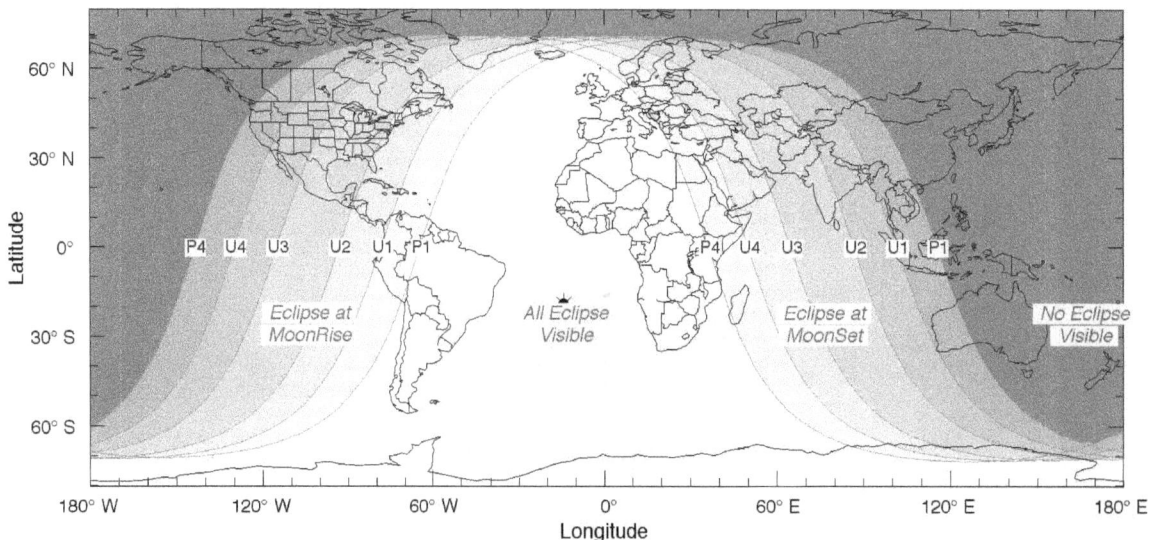

Eclipse Contacts
P1 = 22:19:23 UT1
U1 = 23:16:56 UT1
U2 = 00:18:15 UT1
U3 = 01:49:21 UT1
U4 = 02:50:33 UT1
P4 = 03:48:11 UT1

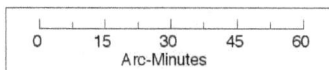

P4 U4 U3 U2 U1 P1

P4 U4   U3   U2 U1 P1

Eclipse at
MoonRise

All Eclipse
Visible

Eclipse at
MoonSet

No Eclipse
Visible

Total Lunar Eclipse of 2084 Jan 22

Greatest Eclipse = 23:13:00.0 TD   (= 23:11:08.5 UT1)

Penumbral Magnitude = 2.2425	Gamma = -0.3610	Saros Series = 135
Umbral Magnitude = 1.1531	Axis = 0.3294°	Saros Member = 27 of 71

Sun at Greatest Eclipse
(Geocentric Coordinates)

R.A. = 20h20m36.4s
Dec. = -19°30'37.0"
S.D. = 00°16'15.1"
H.P. = 00°00'08.9"

Moon at Greatest Eclipse
(Geocentric Coordinates)

R.A. = 08h20m09.1s
Dec. = +19°11'55.7"
S.D. = 00°14'55.1"
H.P. = 00°54'45.0"

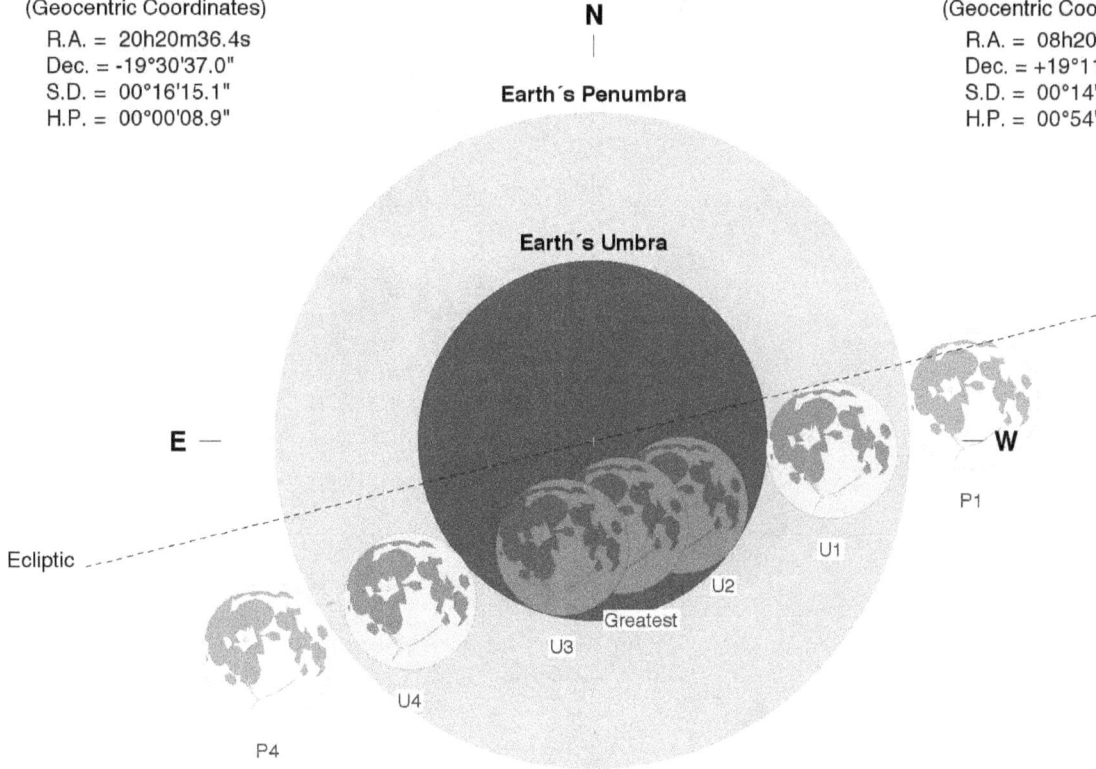

N

Earth´s Penumbra

Earth´s Umbra

E

W

P1

U1

U2

Greatest

U3

Ecliptic

U4

P4

S

Eclipse Durations

Penumbral = 06h02m52s
Umbral = 03h37m03s
Total = 01h01m15s

Eph. = JPL DE430
Rule = Herald-Sinnott
ΔT =   111 s

0	15	30	45	60

Arc-Minutes

©2020 F. Espenak, www.EclipseWise.com

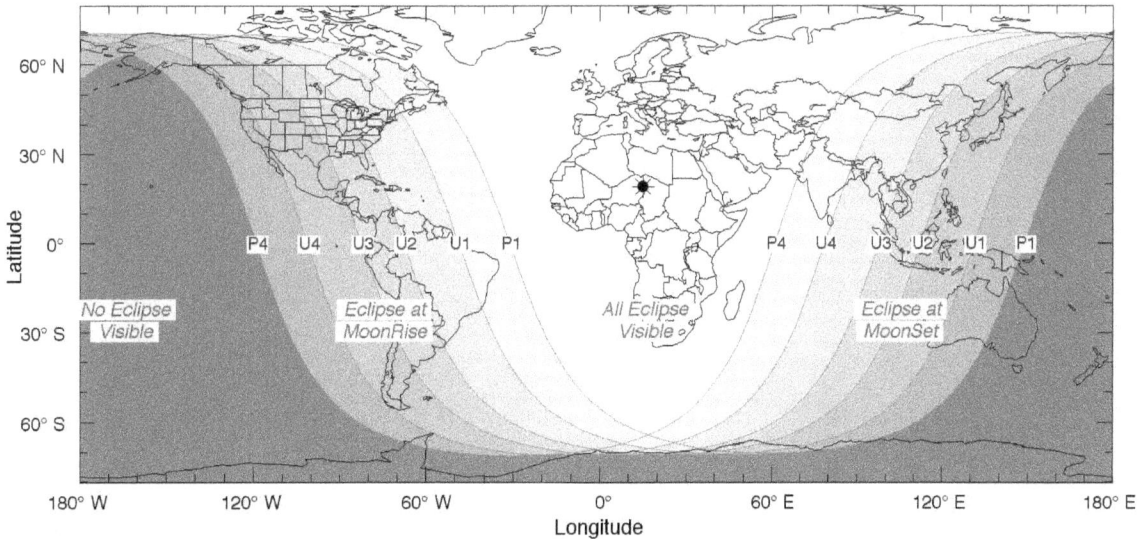

Eclipse Contacts

P1 = 20:09:37 UT1
U1 = 21:22:34 UT1
U2 = 22:40:21 UT1
U3 = 23:41:37 UT1
U4 = 00:59:36 UT1
P4 = 02:12:30 UT1

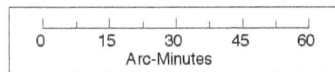

No Eclipse Visible

Eclipse at MoonRise

All Eclipse Visible

Eclipse at MoonSet

P4 U4 U3 U2 U1 P1   P4 U4 U3 U2 U1 P1

Partial Lunar Eclipse of 2084 Jul 17

Greatest Eclipse = 16:58:50.9 TD   (= 16:56:59.0 UT1)

| Penumbral Magnitude = 1.8557 | Gamma = 0.5313 | Saros Series = 140 |
| Umbral Magnitude = 0.9136 | Axis = 0.5428° | Saros Member = 28 of 77 |

Sun at Greatest Eclipse
(Geocentric Coordinates)
R.A. = 07h52m19.0s
Dec. = +20°55'24.9"
S.D. = 00°15'44.3"
H.P. = 00°00'08.7"

Moon at Greatest Eclipse
(Geocentric Coordinates)
R.A. = 19h51m39.7s
Dec. = -20°24'10.2"
S.D. = 00°16'42.3"
H.P. = 01°01'18.4"

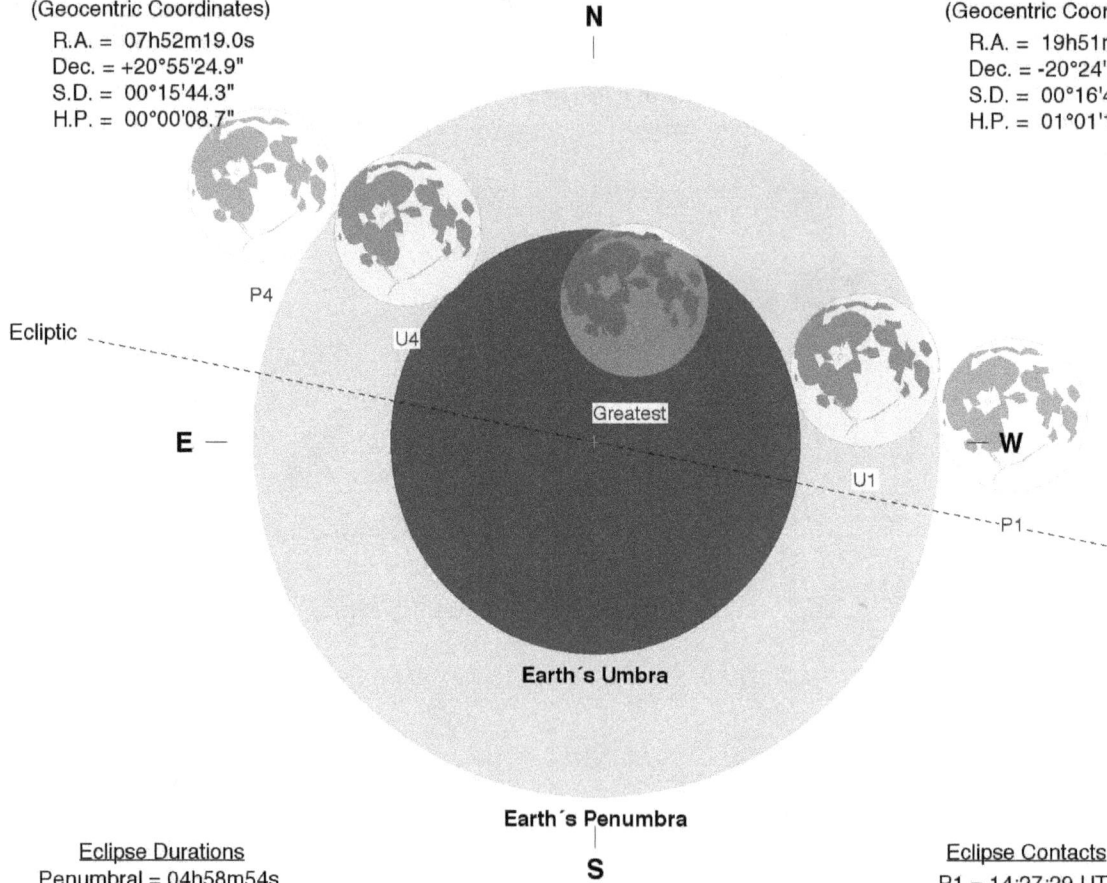

N

Ecliptic

P4
U4

Greatest

E

W

U1

P1

Earth's Umbra

Earth's Penumbra

S

Eclipse Durations
Penumbral = 04h58m54s
Umbral = 03h02m03s

Eph. = JPL DE430
Rule = Herald-Sinnott
ΔT =   112 s

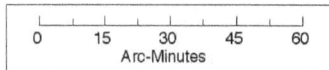

0   15   30   45   60
Arc-Minutes

©2020 F. Espenak, www.EclipseWise.com

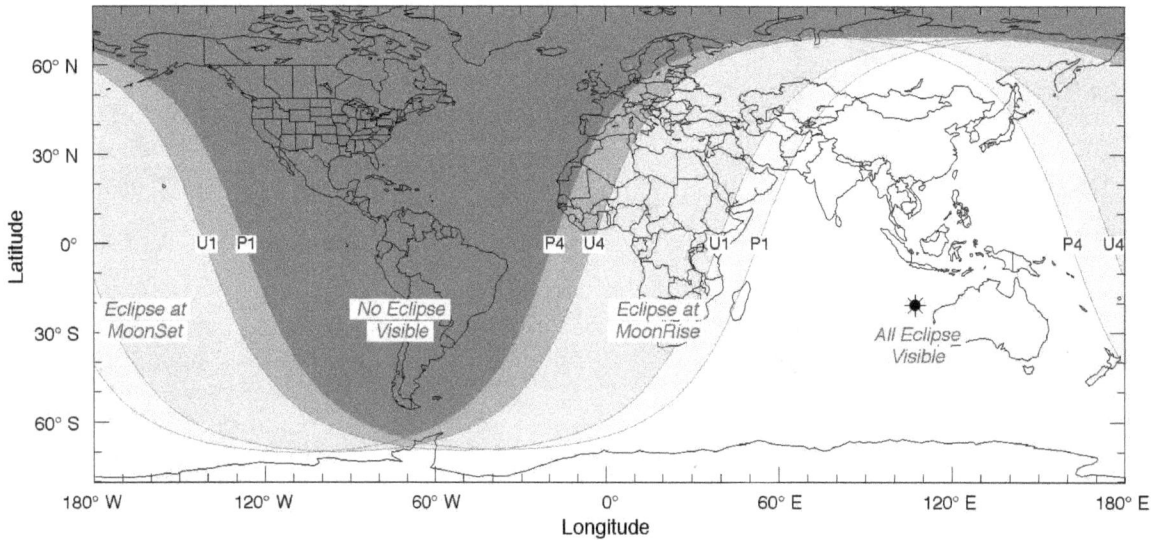

Eclipse Contacts
P1 = 14:27:29 UT1
U1 = 15:25:52 UT1
U4 = 18:27:55 UT1
P4 = 19:26:23 UT1

Eclipse at MoonSet
No Eclipse Visible
Eclipse at MoonRise
All Eclipse Visible

Penumbral Lunar Eclipse of 2085 Jan 10

Greatest Eclipse = 22:32:29.0 TD (= 22:30:36.7 UT1)

Penumbral Magnitude = 0.9944	Gamma = -1.0453	Saros Series = 145
Umbral Magnitude = -0.1102	Axis = 0.9413°	Saros Member = 15 of 71

Sun at Greatest Eclipse
(Geocentric Coordinates)
R.A. = 19h32m28.5s
Dec. = -21°43'43.7"
S.D. = 00°16'15.8"
H.P. = 00°00'08.9"

N

Earth´s Penumbra

Earth´s Umbra

Moon at Greatest Eclipse
(Geocentric Coordinates)
R.A. = 07h31m27.3s
Dec. = +20°49'04.6"
S.D. = 00°14'43.4"
H.P. = 00°54'02.0"

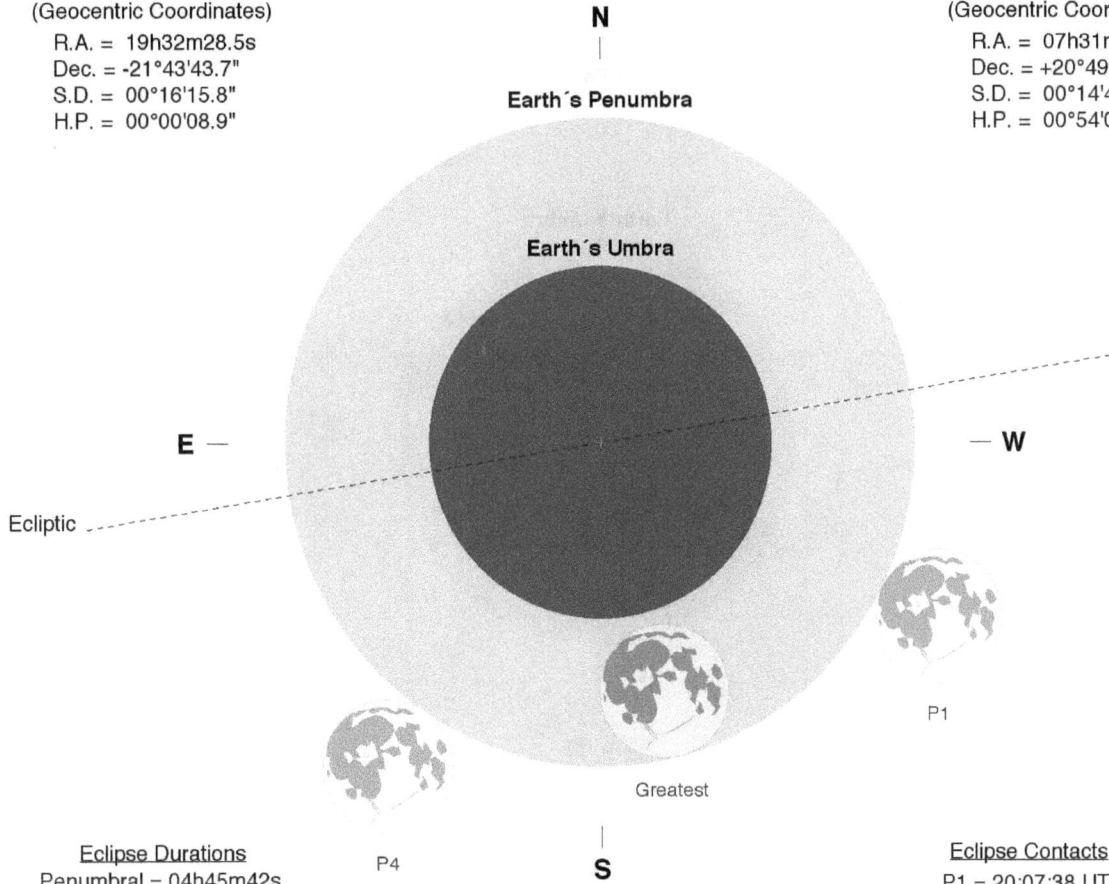

E —

Ecliptic

— **W**

P1

Greatest

P4

S

Eclipse Durations
Penumbral = 04h45m42s

Eclipse Contacts
P1 = 20:07:38 UT1
P4 = 00:53:21 UT1

Eph. = JPL DE430
Rule = Herald-Sinnott
ΔT =   112 s

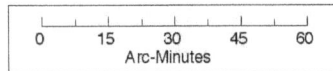

0	15	30	45	60

Arc-Minutes

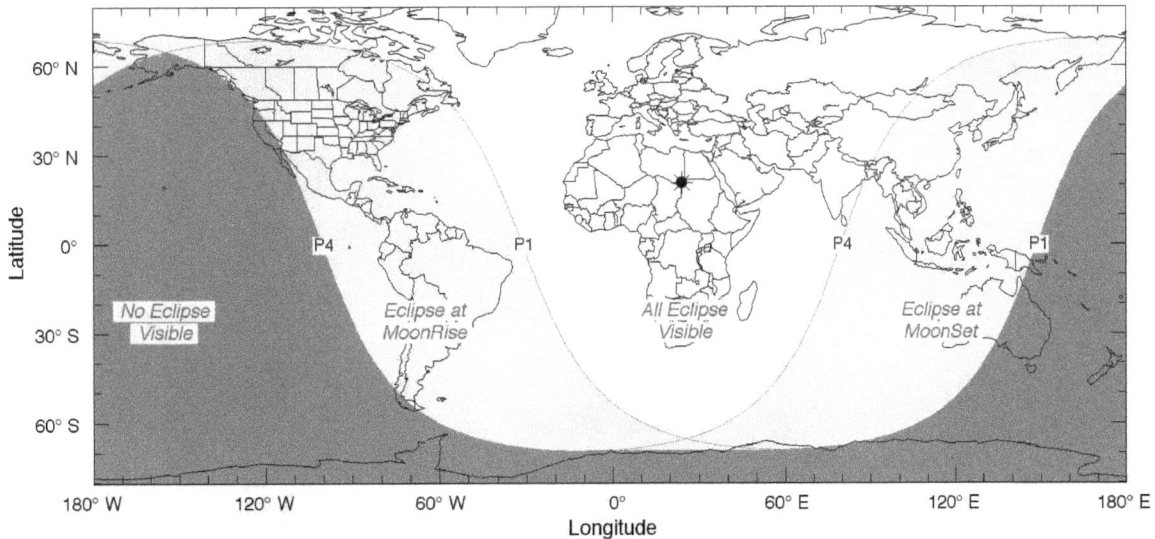

No Eclipse Visible

Eclipse at MoonRise

All Eclipse Visible

Eclipse at MoonSet

P4   P1   P4   P1

Penumbral Lunar Eclipse of 2085 Jun 08

Greatest Eclipse = 02:17:36.2 TD   (= 02:15:43.5 UT1)

Penumbral Magnitude = 0.5079	Gamma = -1.2746	Saros Series = 112
Umbral Magnitude = -0.4668	Axis = 1.2606°	Saros Member = 69 of 72

Sun at Greatest Eclipse
(Geocentric Coordinates)

R.A. = 05h08m01.1s
Dec. = +22°53'26.1"
S.D. = 00°15'45.6"
H.P. = 00°00'08.7"

Moon at Greatest Eclipse
(Geocentric Coordinates)

R.A. = 17h08m03.2s
Dec. = -24°09'04.5"
S.D. = 00°16'10.2"
H.P. = 00°59'20.7"

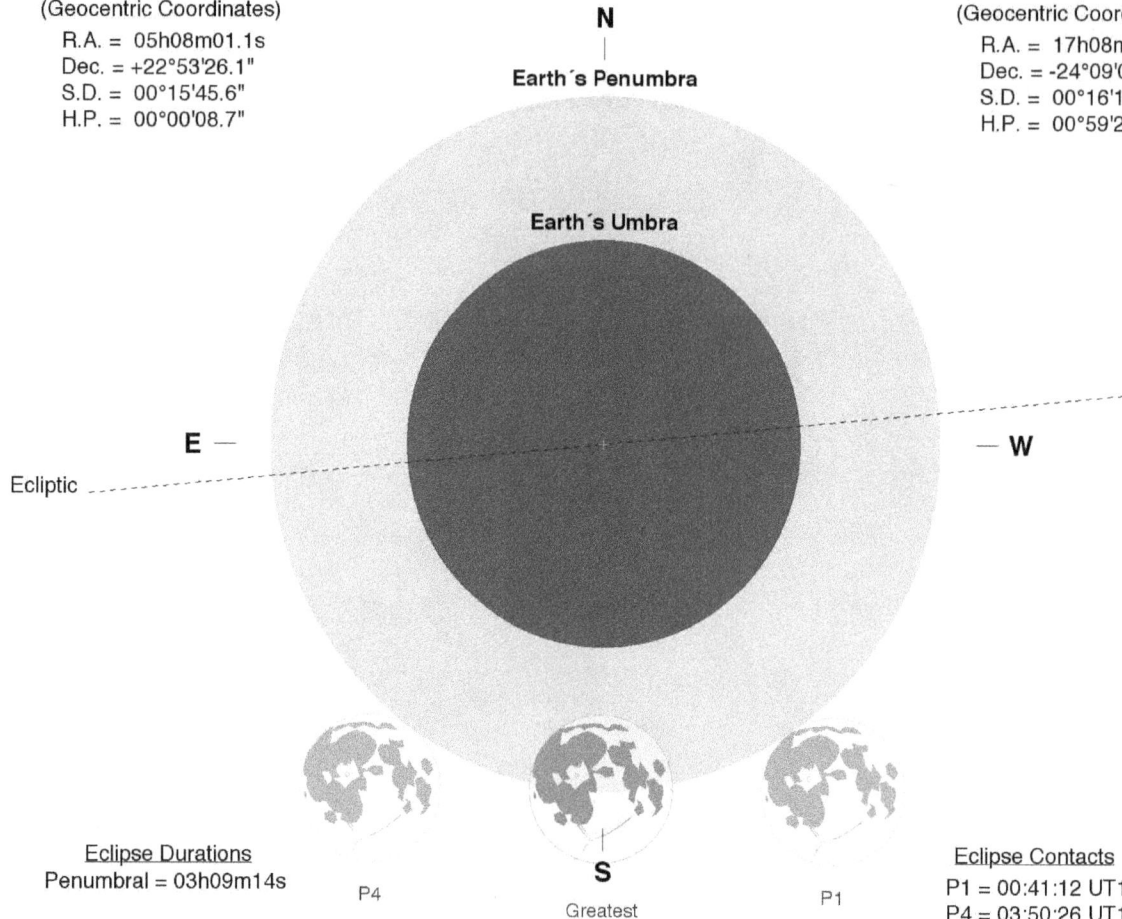

N

Earth's Penumbra

Earth's Umbra

E

Ecliptic

W

S

P4   Greatest   P1

Eclipse Durations
Penumbral = 03h09m14s

Eclipse Contacts
P1 = 00:41:12 UT1
P4 = 03:50:26 UT1

Eph. = JPL DE430
Rule = Herald-Sinnott
ΔT = 113 s

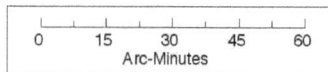

0	15	30	45	60

Arc-Minutes

©2020 F. Espenak, www.EclipseWise.com

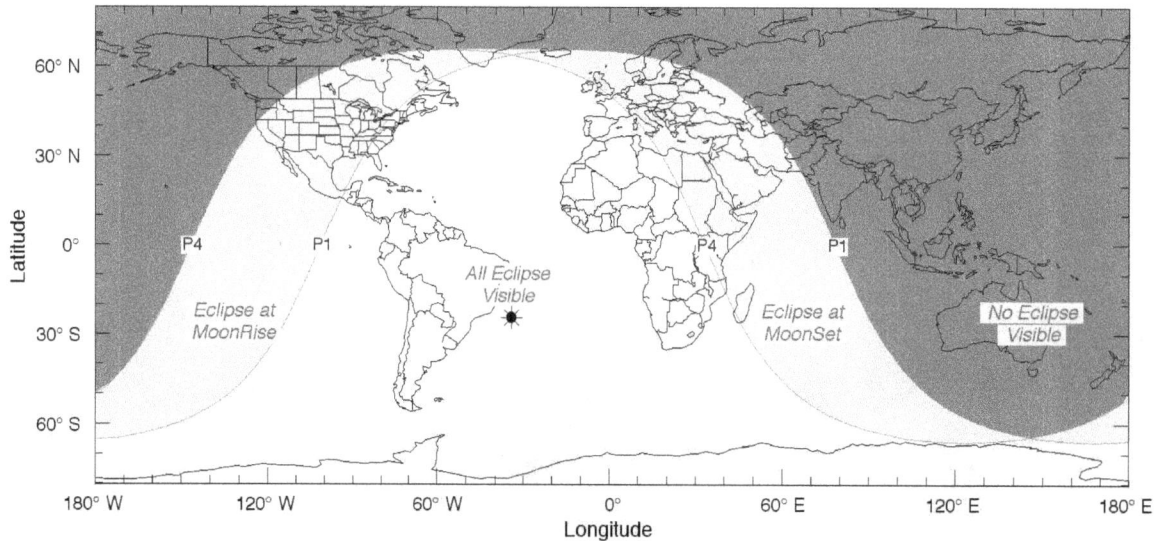

60° N

30° N

0°

30° S

60° S

Latitude

Eclipse at
MoonRise

P4   P1

All Eclipse
Visible

Eclipse at
MoonSet

P4   P1

No Eclipse
Visible

180° W   120° W   60° W   0°   60° E   120° E   180° E

Longitude

Penumbral Lunar Eclipse of 2085 Jul 07

Greatest Eclipse = 10:04:39.1 TD   (= 10:02:46.3 UT1)

Penumbral Magnitude = 0.5064	Gamma = 1.2695	Saros Series = 150
Umbral Magnitude = -0.4461	Axis = 1.2824°	Saros Member = 5 of 71

Sun at Greatest Eclipse
(Geocentric Coordinates)
R.A. = 07h09m34.5s
Dec. = +22°27'55.5"
S.D. = 00°15'44.0"
H.P. = 00°00'08.7"

Moon at Greatest Eclipse
(Geocentric Coordinates)
R.A. = 19h08m24.2s
Dec. = -21°12'43.2"
S.D. = 00°16'31.1"
H.P. = 01°00'37.2"

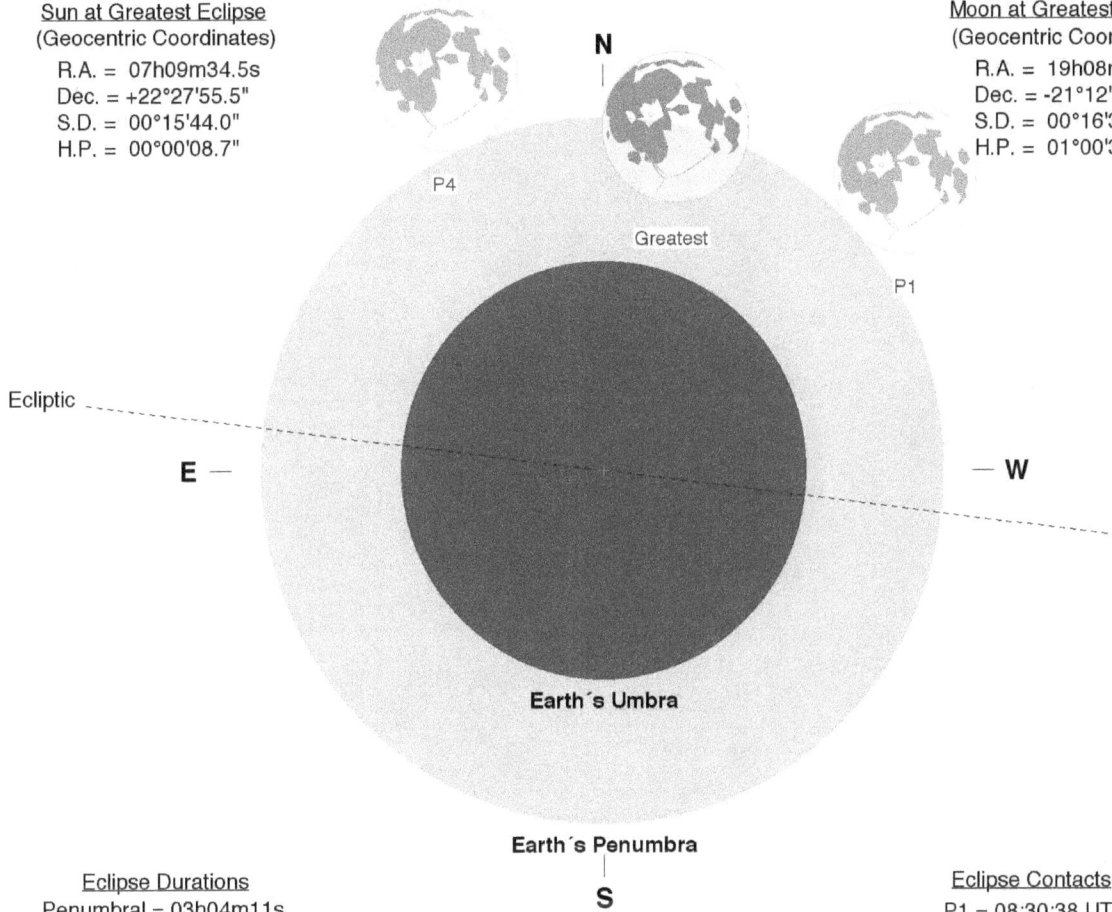

N

P4

Greatest

P1

Ecliptic

E

W

Earth's Umbra

Earth's Penumbra

S

Eclipse Durations
Penumbral = 03h04m11s

Eclipse Contacts
P1 = 08:30:38 UT1
P4 = 11:34:49 UT1

Eph. = JPL DE430
Rule = Herald-Sinnott
ΔT =   113 s

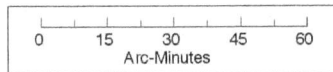

0   15   30   45   60
Arc-Minutes

©2020 F. Espenak, www.EclipseWise.com

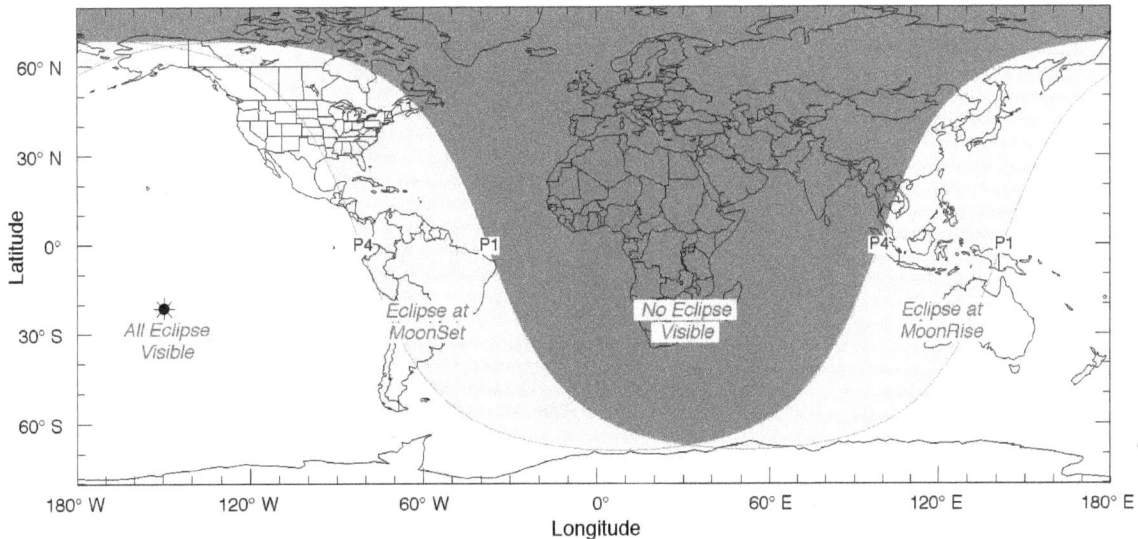

P4   P1   P4   P1

Eclipse at
MoonSet

No Eclipse
Visible

Eclipse at
MoonRise

All Eclipse
Visible

Penumbral Lunar Eclipse of 2085 Dec 01

Greatest Eclipse = 08:25:35.2 TD (= 08:23:42.0 UT1)

Penumbral Magnitude = 0.6400	Gamma = 1.2190	Saros Series = 117
Umbral Magnitude = -0.3944	Axis = 1.1689°	Saros Member = 56 of 71

Sun at Greatest Eclipse
(Geocentric Coordinates)

R.A. = 16h32m47.2s
Dec. = -21°54'56.5"
S.D. = 00°16'13.1"
H.P. = 00°00'08.9"

Moon at Greatest Eclipse
(Geocentric Coordinates)

R.A. = 04h32m31.6s
Dec. = +23°04'59.3"
S.D. = 00°15'40.7"
H.P. = 00°57'32.5"

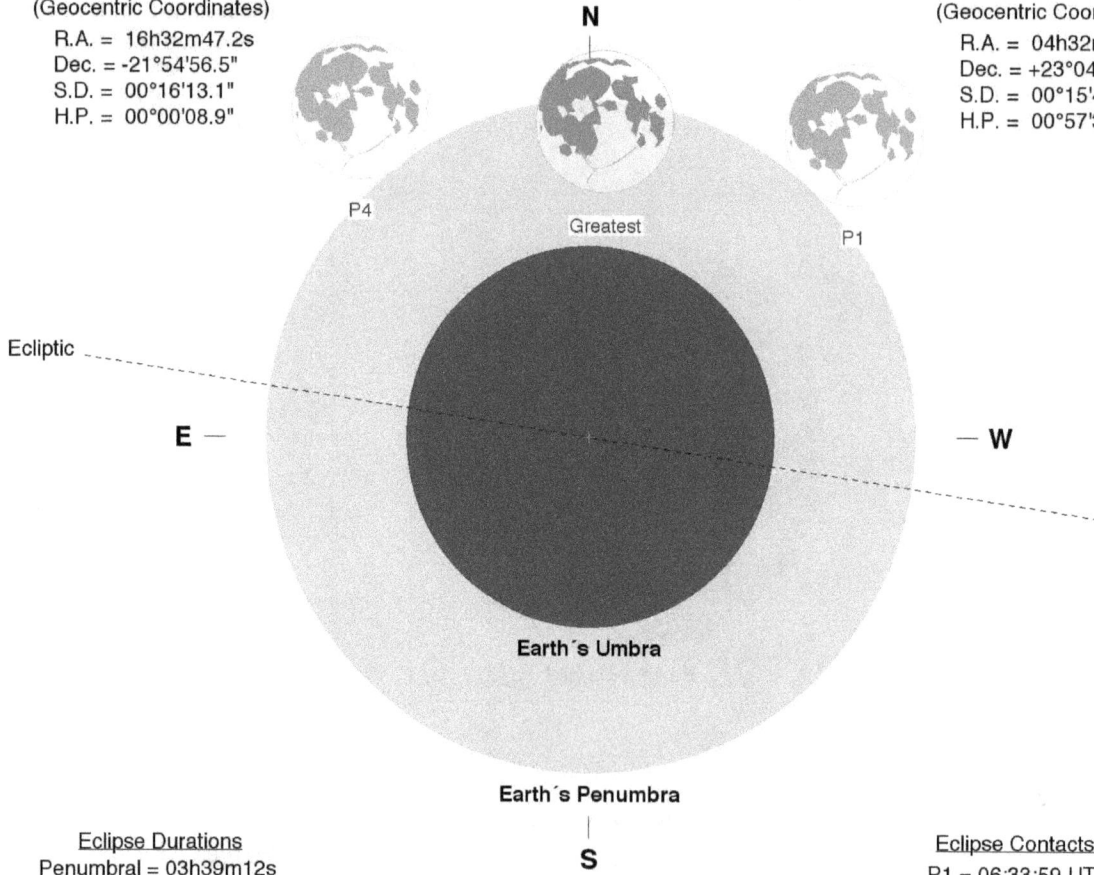

N

P4

Greatest

P1

Ecliptic

E

W

Earth's Umbra

Earth's Penumbra

S

Eclipse Durations
Penumbral = 03h39m12s

Eclipse Contacts
P1 = 06:33:59 UT1
P4 = 10:13:11 UT1

0	15	30	45	60

Arc-Minutes

Eph. = JPL DE430
Rule = Herald-Sinnott
ΔT = 113 s

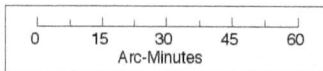

©2020 F. Espenak, www.EclipseWise.com

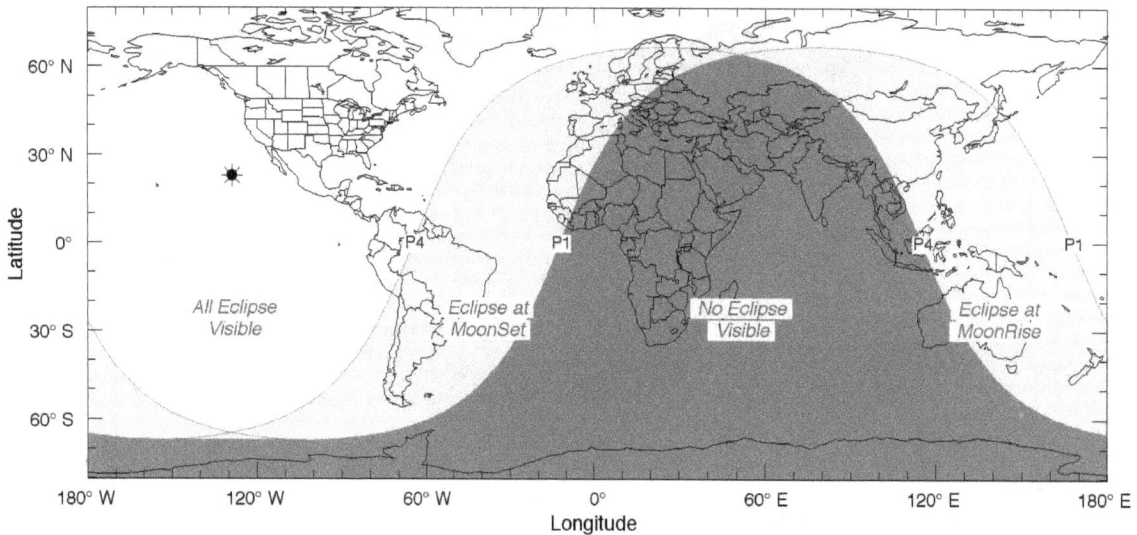

All Eclipse Visible

Eclipse at MoonSet

No Eclipse Visible

Eclipse at MoonRise

P4   P1   P4   P1

Partial Lunar Eclipse of 2086 May 28

Greatest Eclipse = 12:43:46.6 TD   (= 12:41:53.0 UT1)

Penumbral Magnitude = 1.8499	Gamma = -0.5585	Saros Series = 122
Umbral Magnitude = 0.8193	Axis = 0.5232°	Saros Member = 60 of 74

Sun at Greatest Eclipse
(Geocentric Coordinates)

R.A. = 04h23m38.1s
Dec. = +21°34'54.6"
S.D. = 00°15'47.1"
H.P. = 00°00'08.7"

Moon at Greatest Eclipse
(Geocentric Coordinates)

R.A. = 16h23m29.5s
Dec. = -22°06'14.3"
S.D. = 00°15'19.0"
H.P. = 00°56'12.6"

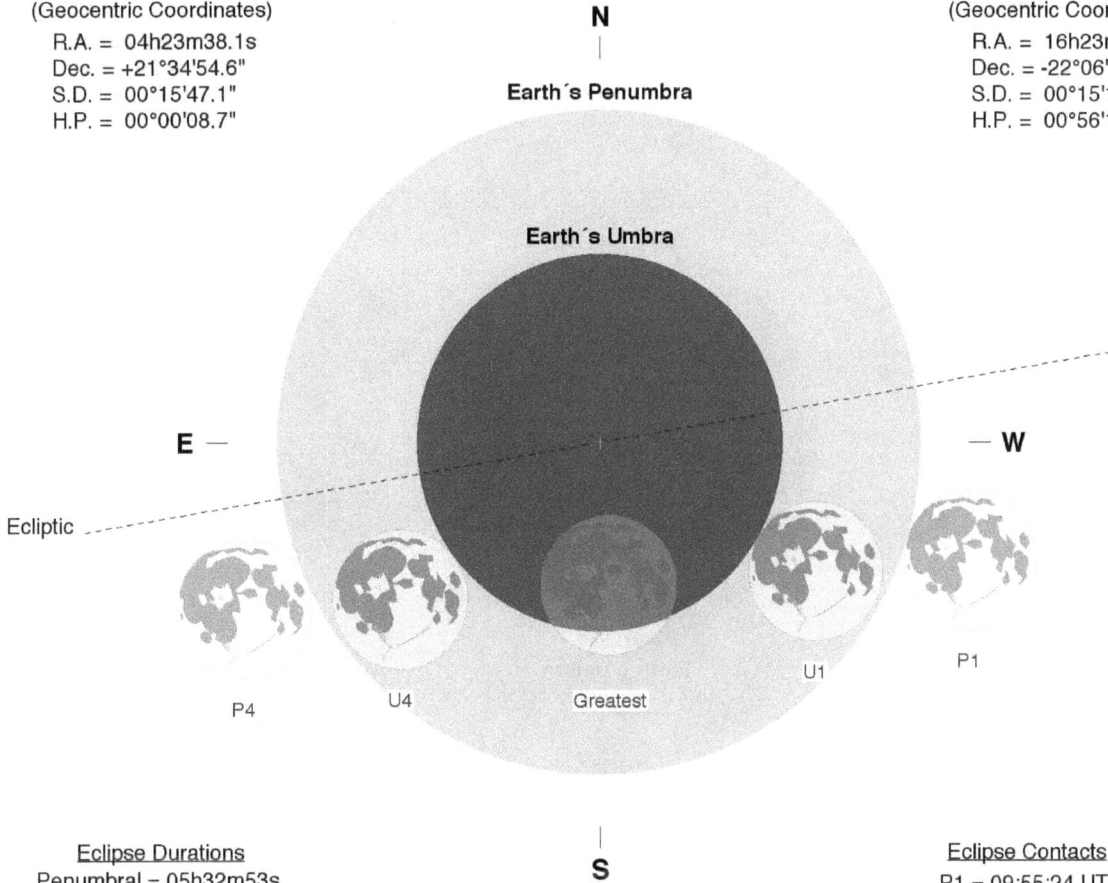

N

Earth's Penumbra

Earth's Umbra

E —

— **W**

Ecliptic

P4   U4   Greatest   U1   P1

S

Eclipse Durations
Penumbral = 05h32m53s
Umbral = 03h10m12s

Eph. = JPL DE430
Rule = Herald-Sinnott
ΔT =   114 s

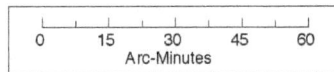

0   15   30   45   60
Arc-Minutes

©2020 F. Espenak, www.EclipseWise.com

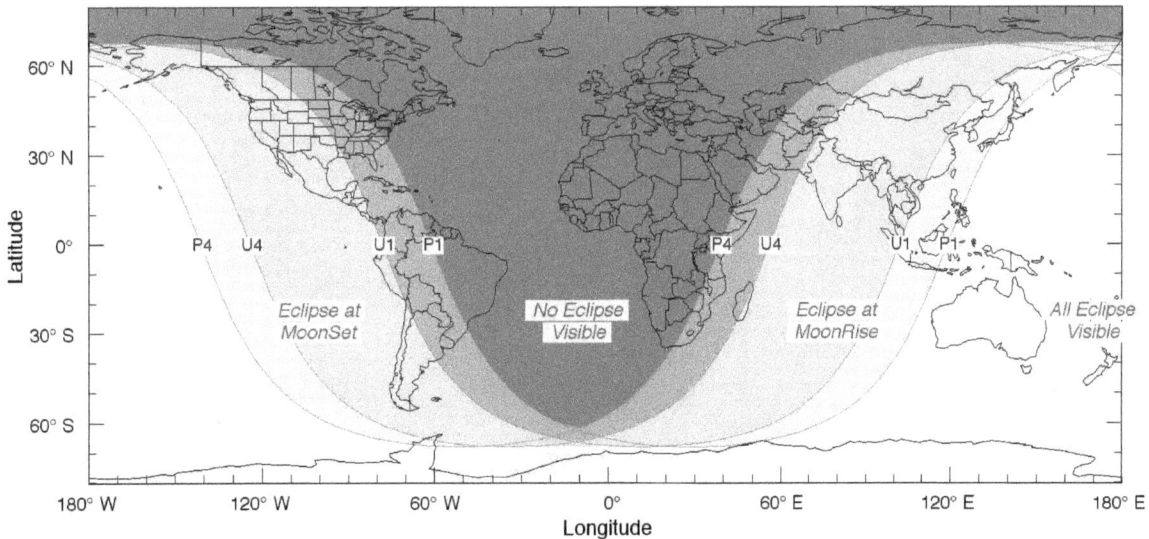

Eclipse Contacts
P1 = 09:55:24 UT1
U1 = 11:06:48 UT1
U4 = 14:17:00 UT1
P4 = 15:28:17 UT1

Eclipse at
MoonSet

No Eclipse
Visible

Eclipse at
MoonRise

All Eclipse
Visible

Partial Lunar Eclipse of 2086 Nov 20

Greatest Eclipse = 20:19:42.2 TD (= 20:17:48.2 UT1)

Penumbral Magnitude = 1.9692	Gamma = 0.4800	Saros Series = 127
Umbral Magnitude = 0.9877	Axis = 0.4841°	Saros Member = 46 of 72

Sun at Greatest Eclipse
(Geocentric Coordinates)

R.A. = 15h47m01.2s
Dec. = -19°55'10.1"
S.D. = 00°16'11.1"
H.P. = 00°00'08.9"

Moon at Greatest Eclipse
(Geocentric Coordinates)

R.A. = 03h46m46.2s
Dec. = +20°24'00.1"
S.D. = 00°16'29.4"
H.P. = 01°00'31.3"

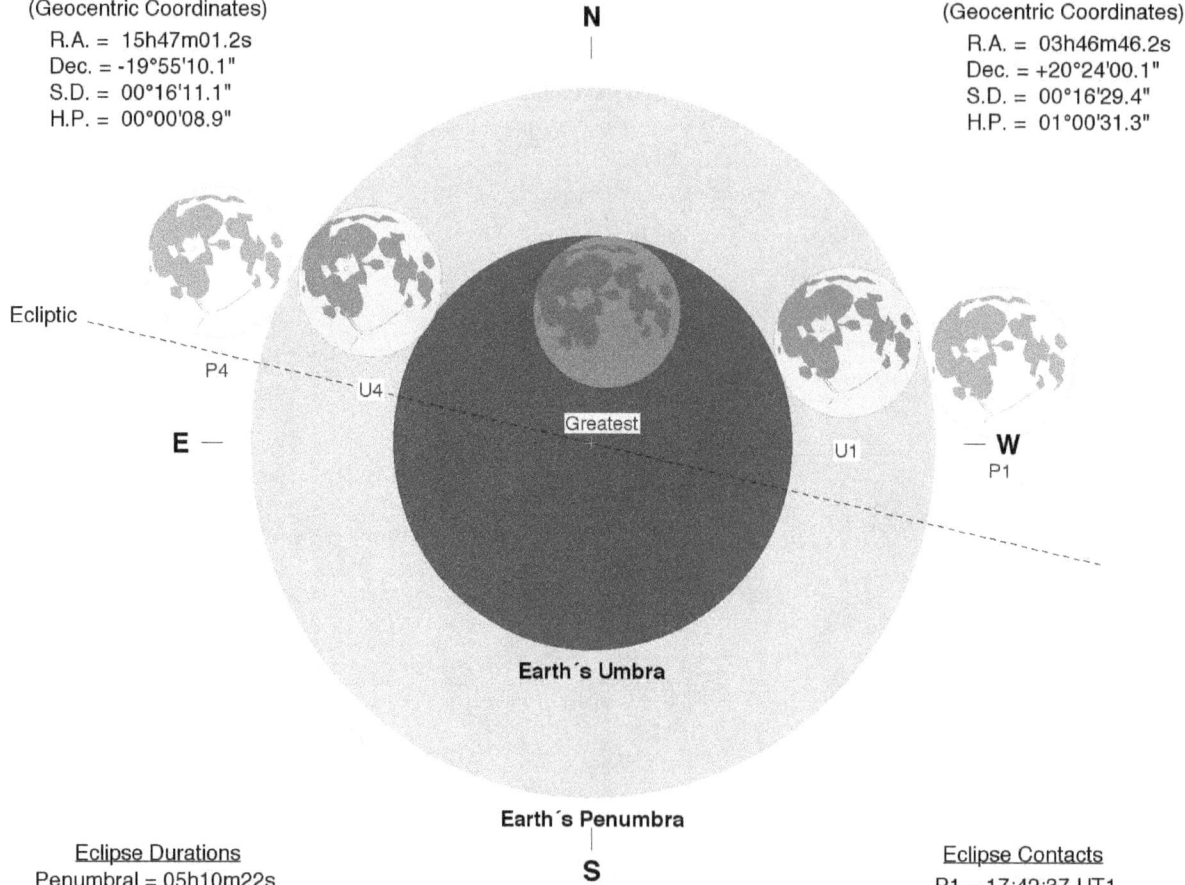

N

Ecliptic

P4

U4

E

Greatest

U1

W

P1

Earth's Umbra

Earth's Penumbra

S

Eclipse Durations
Penumbral = 05h10m22s
Umbral = 03h08m52s

Eph. = JPL DE430
Rule = Herald-Sinnott
ΔT = 114 s

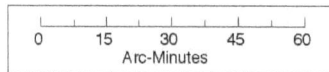

0	15	30	45	60

Arc-Minutes

©2020 F. Espenak, www.EclipseWise.com

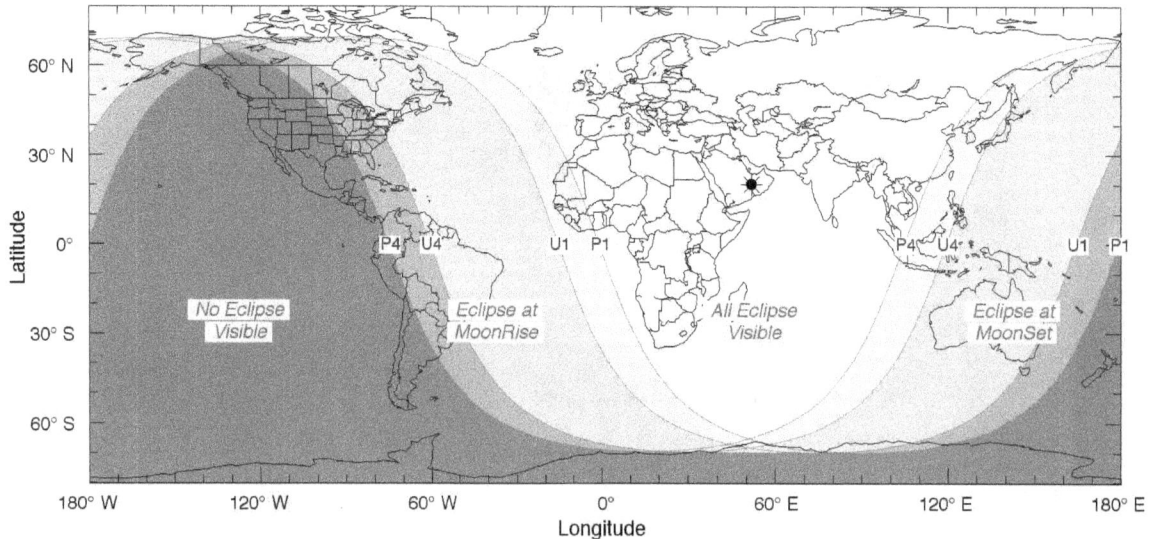

Eclipse Contacts
P1 = 17:42:37 UT1
U1 = 18:43:19 UT1
U4 = 21:52:10 UT1
P4 = 22:52:59 UT1

No Eclipse Visible

Eclipse at MoonRise

All Eclipse Visible

Eclipse at MoonSet

Total Lunar Eclipse of 2087 May 17

Greatest Eclipse = 15:55:20.2 TD   (= 15:53:25.7 UT1)

Penumbral Magnitude = 2.5289	Gamma = 0.1999	Saros Series = 132
Umbral Magnitude = 1.4568	Axis = 0.1803°	Saros Member = 34 of 71

Sun at Greatest Eclipse
(Geocentric Coordinates)

R.A. = 03h38m52.3s
Dec. = +19°28'43.2"
S.D. = 00°15'49.1"
H.P. = 00°00'08.7"

Moon at Greatest Eclipse
(Geocentric Coordinates)

R.A. = 15h38m58.3s
Dec. = -19°17'59.5"
S.D. = 00°14'45.2"
H.P. = 00°54'08.7"

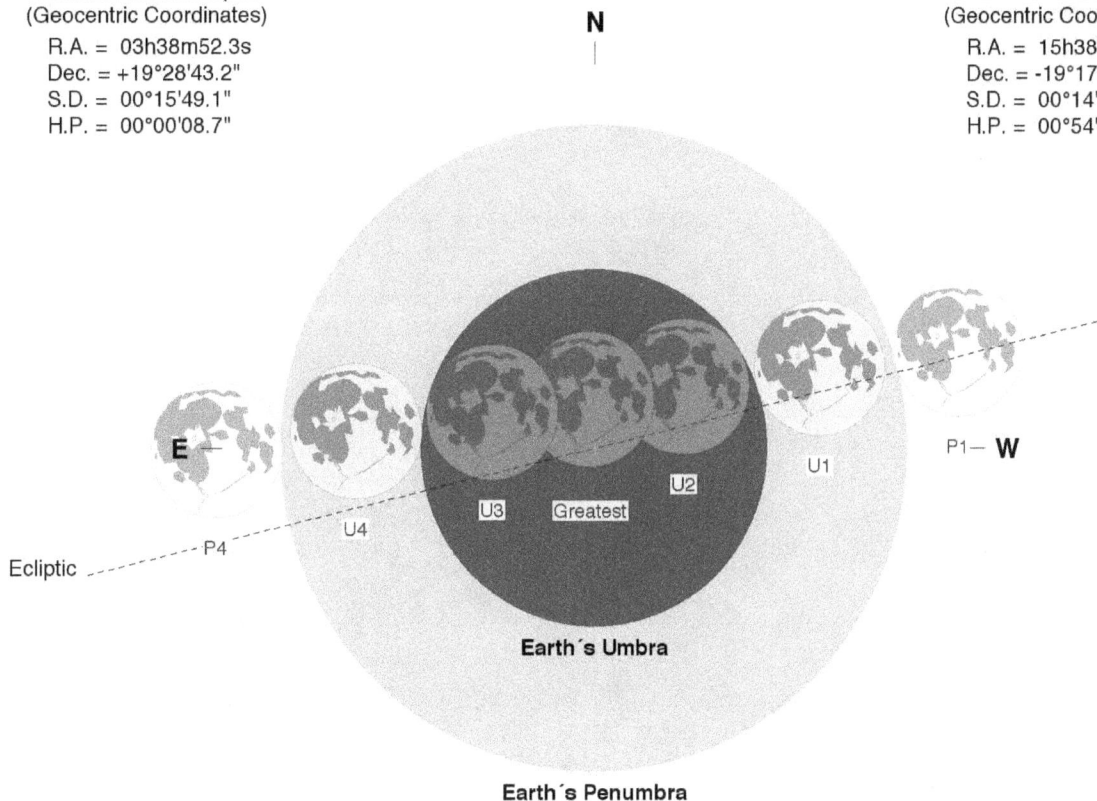

N

E

W

P1

U1

U2

Greatest

U3

U4

P4

Ecliptic

Earth´s Umbra

Earth´s Penumbra

S

Eclipse Durations

Penumbral = 06h11m55s
Umbral = 03h51m31s
Total = 01h35m54s

Eph. = JPL DE430
Rule = Herald-Sinnott
ΔT =   115 s

0	15	30	45	60

Arc-Minutes

©2020 F. Espenak, www.EclipseWise.com

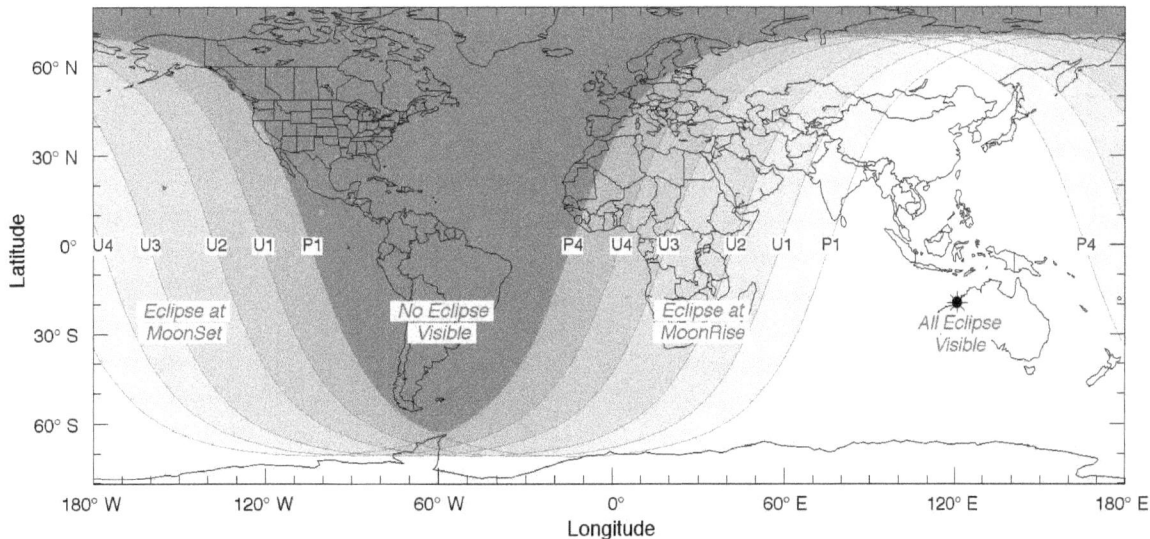

Eclipse Contacts

P1 = 12:47:28 UT1
U1 = 13:57:42 UT1
U2 = 15:05:32 UT1
U3 = 16:41:25 UT1
U4 = 17:49:13 UT1
P4 = 18:59:23 UT1

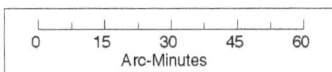

255

Total Lunar Eclipse of 2087 Nov 10

Greatest Eclipse = 12:05:33.4 TD   (= 12:03:38.4 UT1)

Penumbral Magnitude = 2.4667	Gamma = -0.2043	Saros Series = 137
Umbral Magnitude = 1.5018	Axis = 0.2091°	Saros Member = 30 of 78

Sun at Greatest Eclipse
(Geocentric Coordinates)

R.A. = 15h03m26.2s
Dec. = -17°16'19.9"
S.D. = 00°16'08.9"
H.P. = 00°00'08.9"

Moon at Greatest Eclipse
(Geocentric Coordinates)

R.A. = 03h03m35.7s
Dec. = +17°03'59.6"
S.D. = 00°16'44.2"
H.P. = 01°01'25.3"

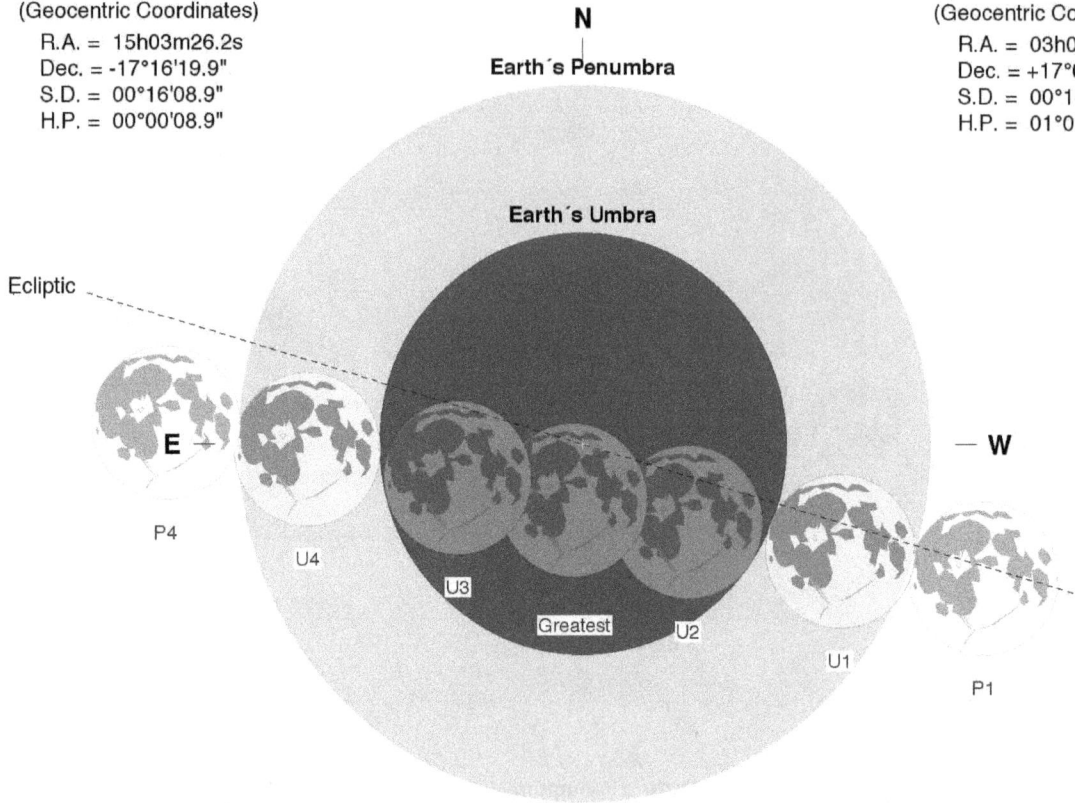

N

Earth´s Penumbra

Earth´s Umbra

Ecliptic

E

W

P4

U4

U3

Greatest

U2

U1

P1

S

Eclipse Durations

Penumbral = 05h17m17s
Umbral = 03h27m19s
Total = 01h29m38s

Eph. = JPL DE430
Rule = Herald-Sinnott
ΔT =   115 s

0   15   30   45   60
Arc-Minutes

©2020 F. Espenak, www.EclipseWise.com

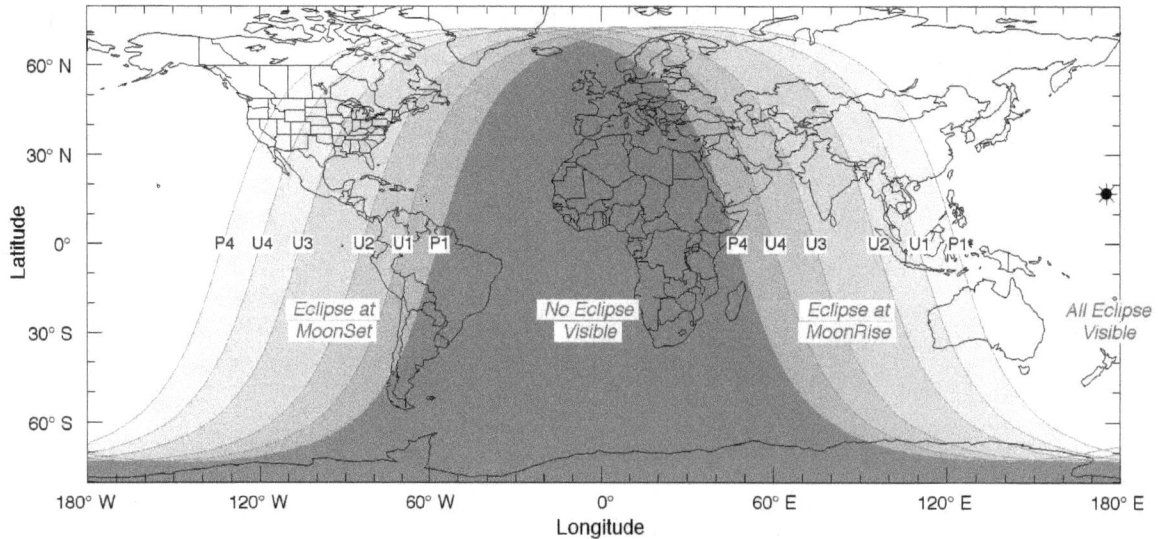

Eclipse Contacts

P1 = 09:25:00 UT1
U1 = 10:20:00 UT1
U2 = 11:18:52 UT1
U3 = 12:48:30 UT1
U4 = 13:47:19 UT1
P4 = 14:42:18 UT1

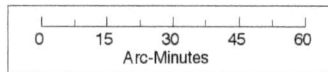

Partial Lunar Eclipse of 2088 May 05

Greatest Eclipse = 16:16:50.2 TD (= 16:14:54.7 UT1)

Penumbral Magnitude = 1.1708	Gamma = 0.9388	Saros Series = 142
Umbral Magnitude = 0.1032	Axis = 0.8529°	Saros Member = 22 of 73

Sun at Greatest Eclipse
(Geocentric Coordinates)

R.A. = 02h54m46.1s
Dec. = +16°39'43.9"
S.D. = 00°15'51.5"
H.P. = 00°00'08.7"

Moon at Greatest Eclipse
(Geocentric Coordinates)

R.A. = 14h55m26.8s
Dec. = -15°49'30.0"
S.D. = 00°14'51.3"
H.P. = 00°54'30.9"

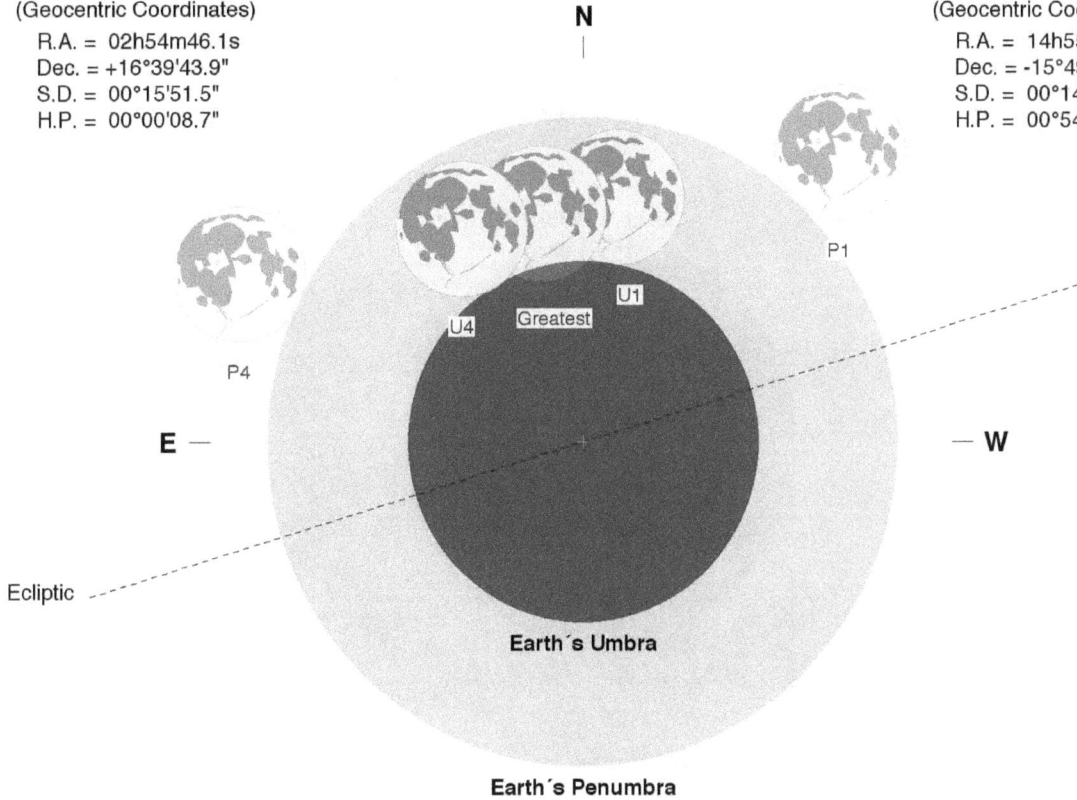

N

P1

U4 Greatest U1

E —

— W

Ecliptic

Earth's Umbra

Earth's Penumbra

S

Eclipse Durations
Penumbral = 04h58m44s
Umbral = 01h17m50s

Eph. = JPL DE430
Rule = Herald-Sinnott
ΔT =   115 s

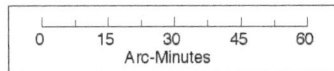

0   15   30   45   60
Arc-Minutes

©2020 F. Espenak, www.EclipseWise.com

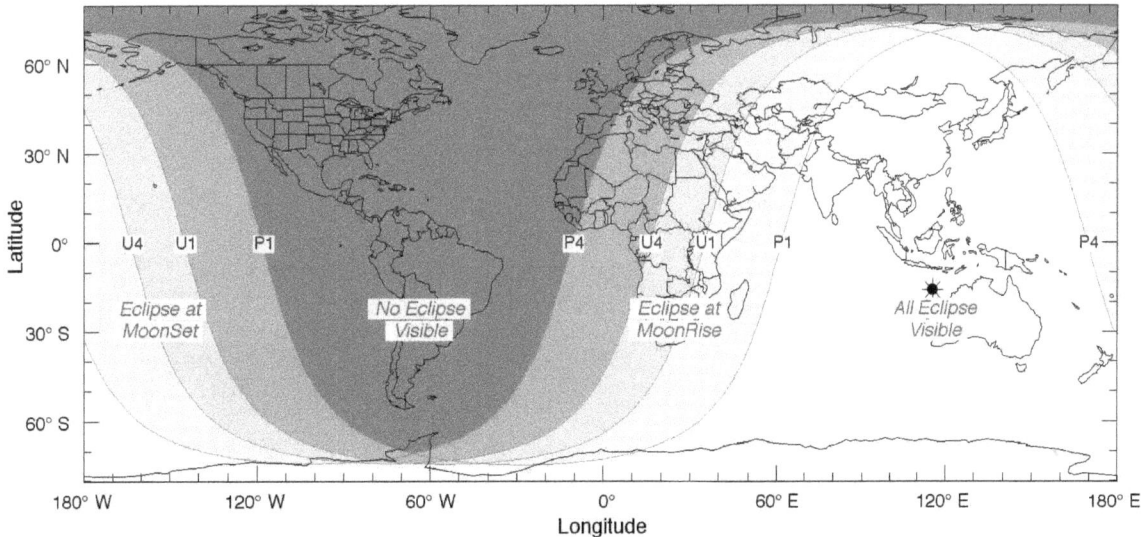

Eclipse Contacts
P1 = 13:45:37 UT1
U1 = 15:36:04 UT1
U4 = 16:53:54 UT1
P4 = 18:44:21 UT1

Eclipse at MoonSet

No Eclipse Visible

Eclipse at MoonRise

All Eclipse Visible

Partial Lunar Eclipse of 2088 Oct 30

Greatest Eclipse = 03:03:20.4 TD   (= 03:01:24.5 UT1)

Penumbral Magnitude = 1.1773	Gamma = -0.9147	Saros Series = 147
Umbral Magnitude = 0.1843	Axis = 0.9073°	Saros Member = 12 of 70

Sun at Greatest Eclipse
(Geocentric Coordinates)

R.A. = 14h21m16.5s
Dec. = -14°03'44.3"
S.D. = 00°16'06.3"
H.P. = 00°00'08.9"

Moon at Greatest Eclipse
(Geocentric Coordinates)

R.A. = 02h22m08.1s
Dec. = +13°10'45.7"
S.D. = 00°16'13.1"
H.P. = 00°59'31.3"

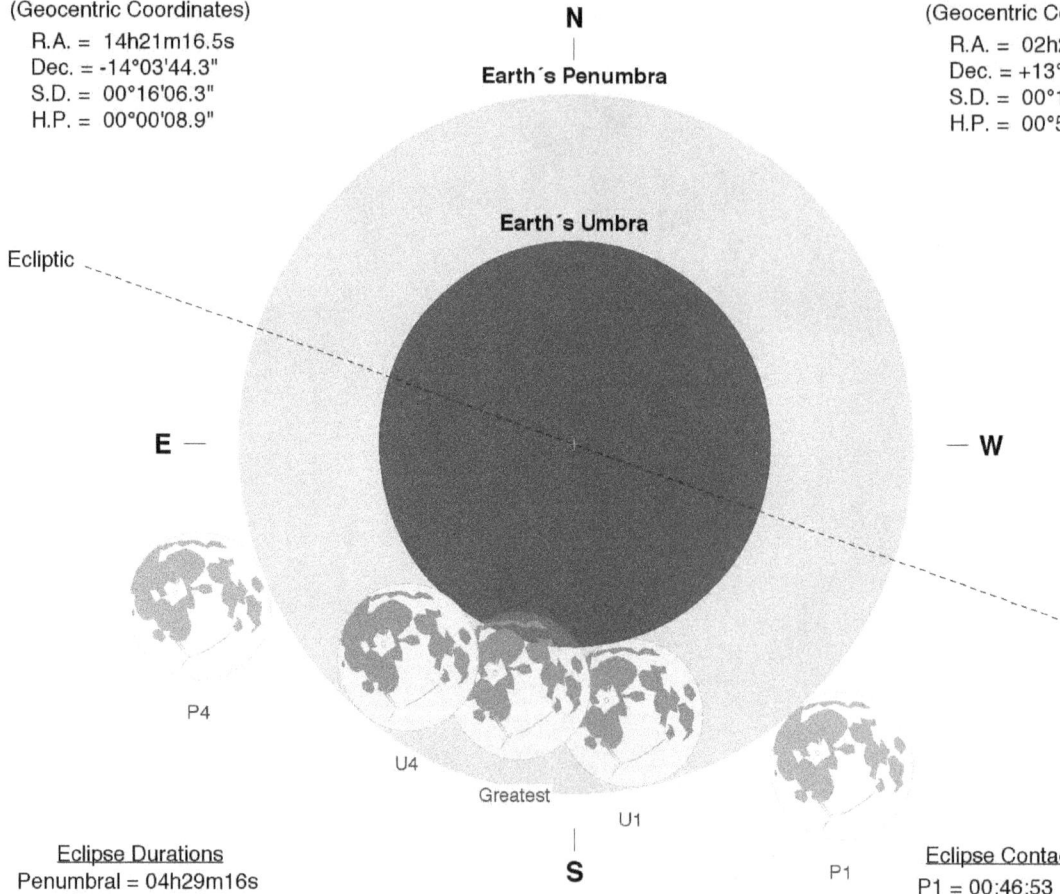

N

Earth's Penumbra

Earth's Umbra

Ecliptic

E

W

P4

U4

Greatest

U1

S

P1

Eclipse Durations

Penumbral = 04h29m16s
Umbral = 01h34m14s

Eph. = JPL DE430
Rule = Herald-Sinnott
ΔT =   116 s

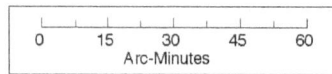

Eclipse Contacts

P1 = 00:46:53 UT1
U1 = 02:14:30 UT1
U4 = 03:48:45 UT1
P4 = 05:16:09 UT1

0	15	30	45	60

Arc-Minutes

©2020 F. Espenak, www.EclipseWise.com

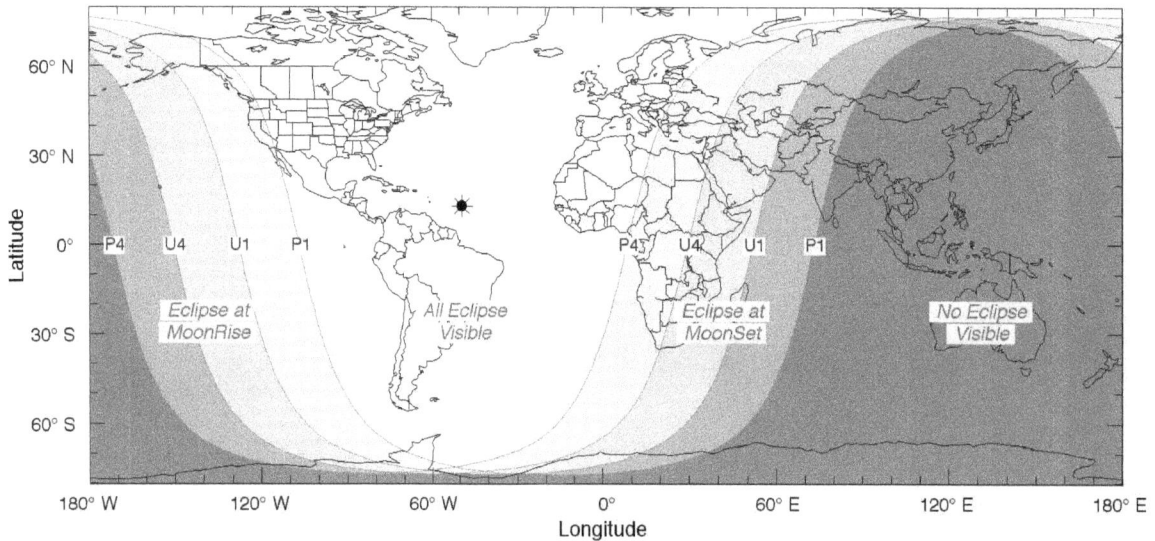

Penumbral Lunar Eclipse of 2089 Mar 26

Greatest Eclipse = 09:34:12.8 TD   (= 09:32:16.6 UT1)

Penumbral Magnitude = 0.8343	Gamma = -1.1039	Saros Series = 114
Umbral Magnitude = -0.1670	Axis = 1.0812°	Saros Member = 63 of 71

Sun at Greatest Eclipse
(Geocentric Coordinates)

R.A. = 00h23m44.9s
Dec. = +02°33'57.2"
S.D. = 00°16'02.1"
H.P. = 00°00'08.8"

Moon at Greatest Eclipse
(Geocentric Coordinates)

R.A. = 12h22m25.7s
Dec. = -03°35'44.6"
S.D. = 00°16'00.9"
H.P. = 00°58'46.5"

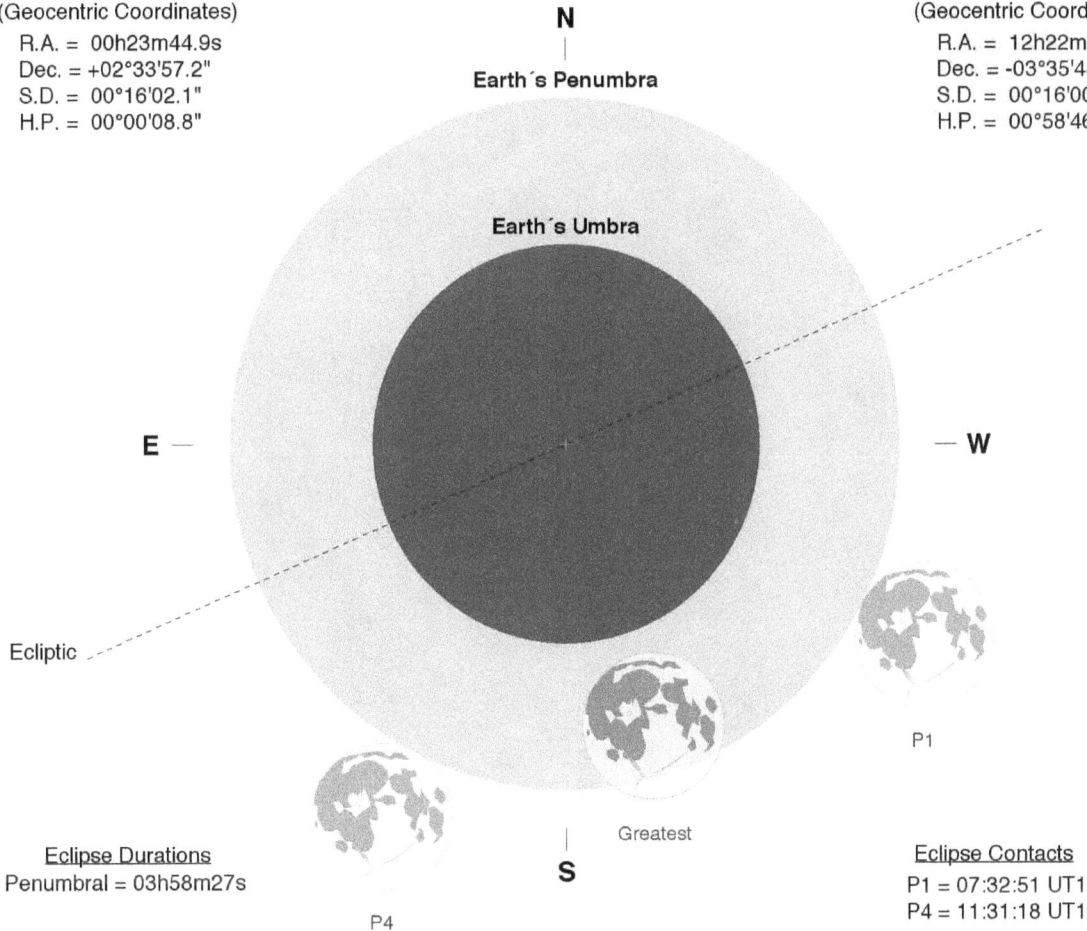

N

Earth´s Penumbra

Earth´s Umbra

E —

— **W**

Ecliptic

P1

Greatest

S

P4

Eclipse Durations
Penumbral = 03h58m27s

Eclipse Contacts
P1 = 07:32:51 UT1
P4 = 11:31:18 UT1

Eph. = JPL DE430
Rule = Herald-Sinnott
ΔT = 116 s

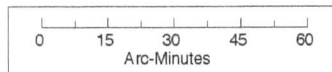

0   15   30   45   60
Arc-Minutes

©2020 F. Espenak, www.EclipseWise.com

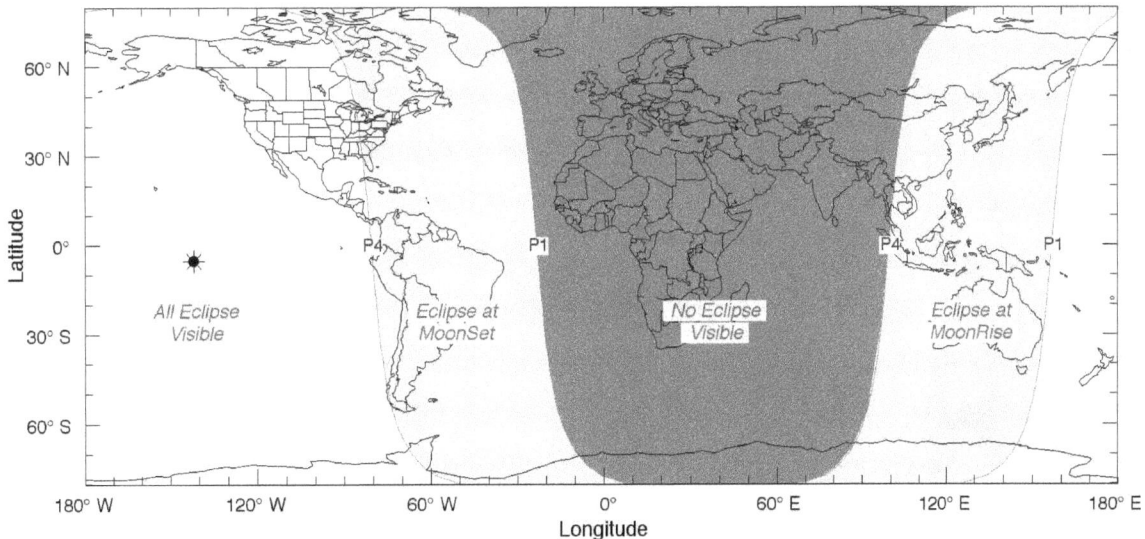

60° N

30° N

Latitude

0°

30° S

60° S

All Eclipse
Visible

Eclipse at
MoonSet

P4   P1

No Eclipse
Visible

P4   P1

Eclipse at
MoonRise

180° W   120° W   60° W   0°   60° E   120° E   180° E
Longitude

Penumbral Lunar Eclipse of 2089 Sep 19

Greatest Eclipse = 22:11:16.1 TD   (= 22:09:19.4 UT1)

Penumbral Magnitude = 0.7904	Gamma = 1.1448	Saros Series = 119
Umbral Magnitude = -0.2726	Axis = 1.0487°	Saros Member = 65 of 82

Sun at Greatest Eclipse
(Geocentric Coordinates)

R.A. = 11h51m38.2s
Dec. = +00°54'20.2"
S.D. = 00°15'55.3"
H.P. = 00°00'08.8"

Moon at Greatest Eclipse
(Geocentric Coordinates)

R.A. = 23h50m21.3s
Dec. = +00°05'34.8"
S.D. = 00°14'58.7"
H.P. = 00°54'58.2"

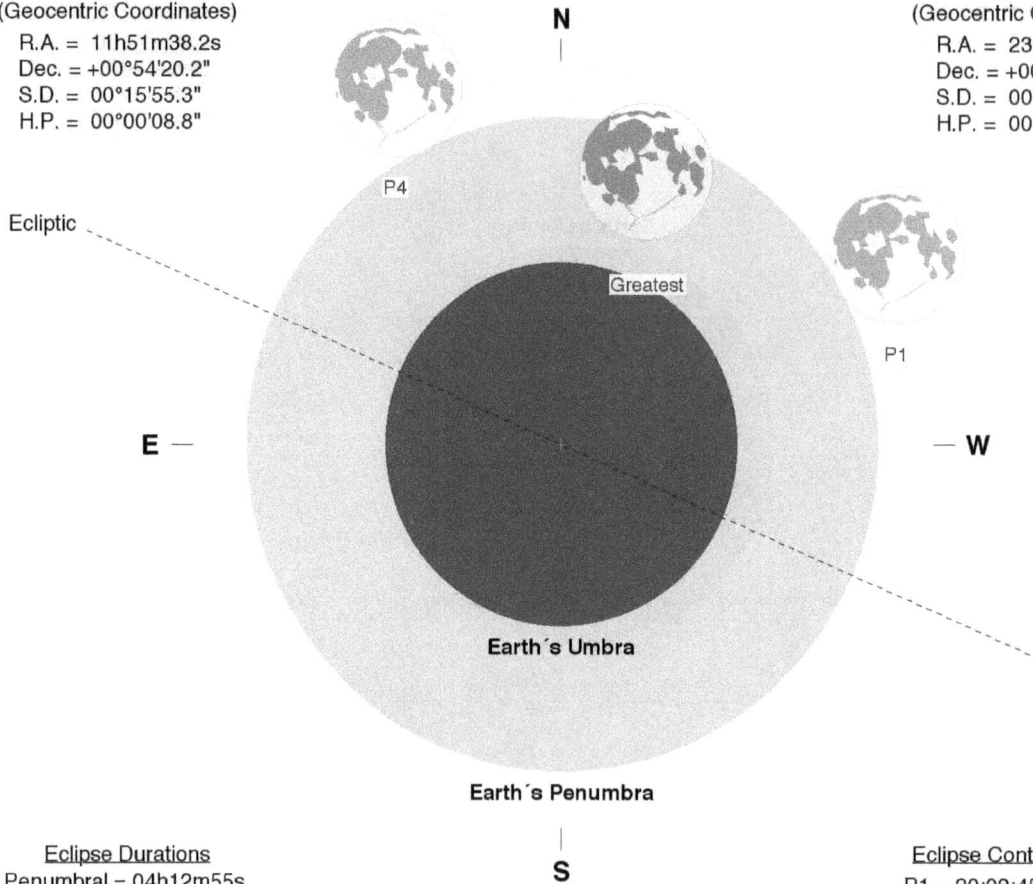

N

P4

Ecliptic

Greatest

P1

E

W

Earth's Umbra

Earth's Penumbra

S

Eclipse Durations
Penumbral = 04h12m55s

Eclipse Contacts
P1 = 20:02:45 UT1
P4 = 00:15:40 UT1

Eph. = JPL DE430
Rule = Herald-Sinnott
ΔT =   117 s

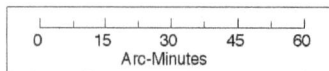

0	15	30	45	60
		Arc-Minutes		

©2020 F. Espenak, www.EclipseWise.com

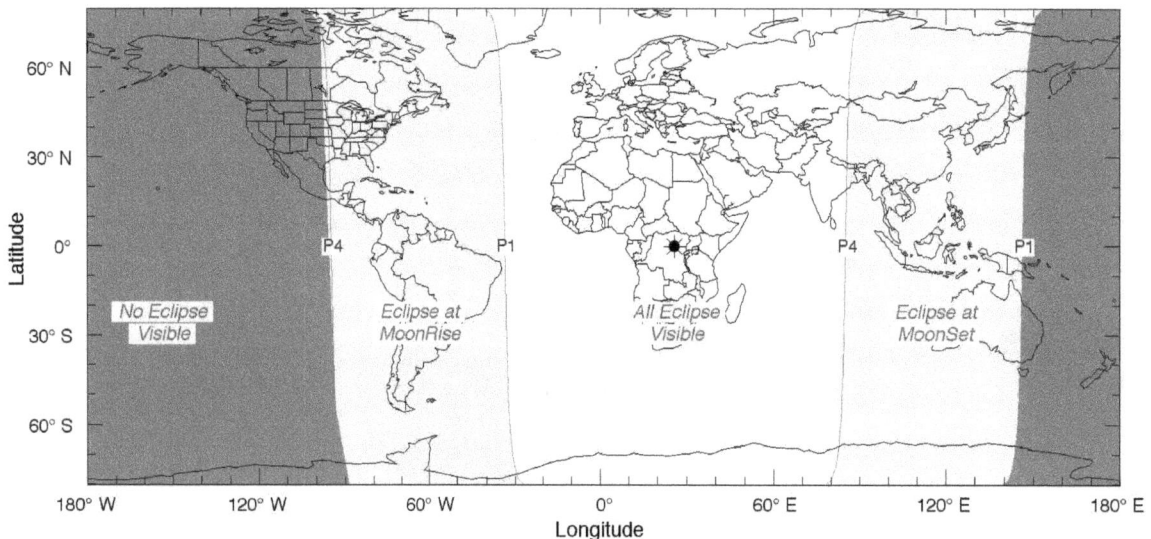

No Eclipse
Visible

Eclipse at
MoonRise

All Eclipse
Visible

Eclipse at
MoonSet

Total Lunar Eclipse of 2090 Mar 15

Greatest Eclipse = 23:48:30.6 TD (= 23:46:33.4 UT1)

Penumbral Magnitude = 2.1670	Gamma = -0.3675	Saros Series = 124
Umbral Magnitude = 1.0023	Axis = 0.3747°	Saros Member = 53 of 73

Sun at Greatest Eclipse
(Geocentric Coordinates)

R.A. = 23h44m53.4s
Dec. = -01°38'06.1"
S.D. = 00°16'05.0"
H.P. = 00°00'08.8"

Moon at Greatest Eclipse
(Geocentric Coordinates)

R.A. = 11h44m26.0s
Dec. = +01°16'41.2"
S.D. = 00°16'40.3"
H.P. = 01°01'11.0"

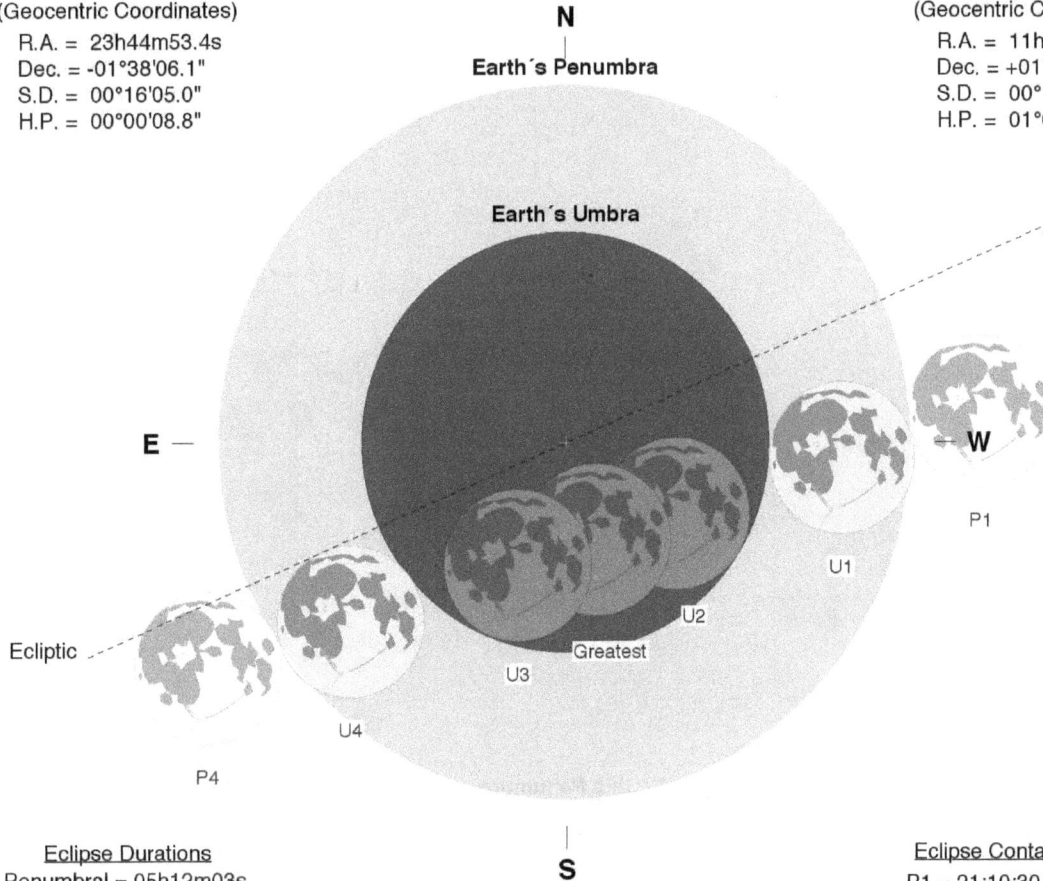

N

Earth's Penumbra

Earth's Umbra

E

W

Ecliptic

P1

U1

U2

Greatest

U3

U4

P4

S

Eclipse Durations

Penumbral = 05h12m03s
Umbral = 03h18m09s
Total = 01h03m34s

Eph. = JPL DE430
Rule = Herald-Sinnott
ΔT =   117 s

0	15	30	45	60

Arc-Minutes

©2020 F. Espenak, www.EclipseWise.com

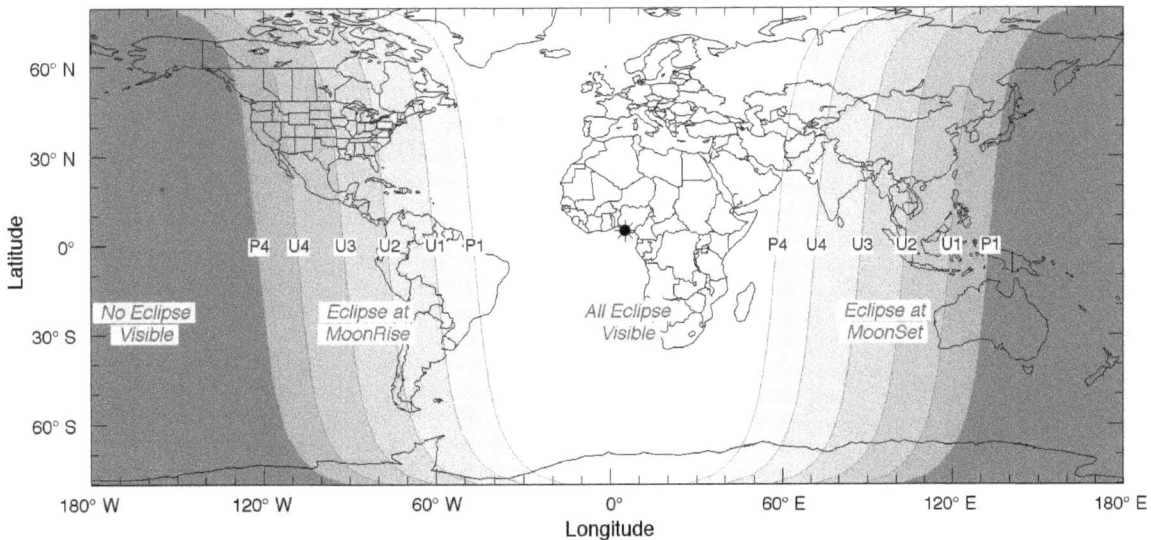

Eclipse Contacts

P1 = 21:10:30 UT1
U1 = 22:07:24 UT1
U2 = 23:14:37 UT1
U3 = 00:18:10 UT1
U4 = 01:25:33 UT1
P4 = 02:22:33 UT1

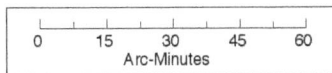

No Eclipse Visible

Eclipse at MoonRise

All Eclipse Visible

Eclipse at MoonSet

P4 U4 U3 U2 U1 P1   P4 U4 U3 U2 U1 P1

Total Lunar Eclipse of 2090 Sep 08

Greatest Eclipse = 22:52:29.3 TD   (= 22:50:31.6 UT1)

Penumbral Magnitude = 2.1178	Gamma = 0.4257	Saros Series = 129
Umbral Magnitude = 1.0387	Axis = 0.3830°	Saros Member = 42 of 71

Sun at Greatest Eclipse
(Geocentric Coordinates)
R.A. = 11h11m25.3s
Dec. = +05°12'29.2"
S.D. = 00°15'52.5"
H.P. = 00°00'08.7"

Moon at Greatest Eclipse
(Geocentric Coordinates)
R.A. = 23h10m58.1s
Dec. = -04°50'31.3"
S.D. = 00°14'42.7"
H.P. = 00°53'59.5"

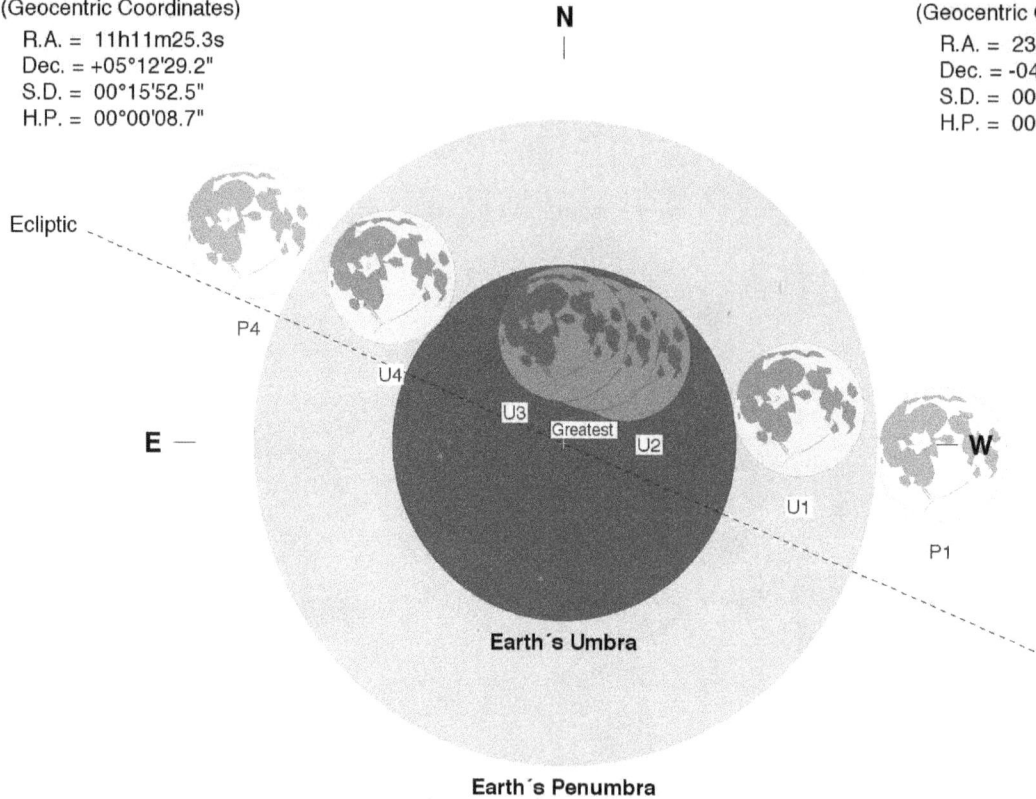

N

Ecliptic

P4

U4

U3

Greatest

U2

E

U1

W

P1

Earth's Umbra

Earth's Penumbra

S

Eclipse Durations
Penumbral = 06h02m52s
Umbral = 03h33m51s
Total = 00h32m31s

Eph. = JPL DE430
Rule = Herald-Sinnott
ΔT =   118 s

0	15	30	45	60

Arc-Minutes

Eclipse Contacts
P1 = 19:49:03 UT1
U1 = 21:03:30 UT1
U2 = 22:34:03 UT1
U3 = 23:06:34 UT1
U4 = 00:37:21 UT1
P4 = 01:51:54 UT1

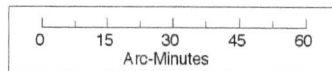

©2020 F. Espenak, www.EclipseWise.com

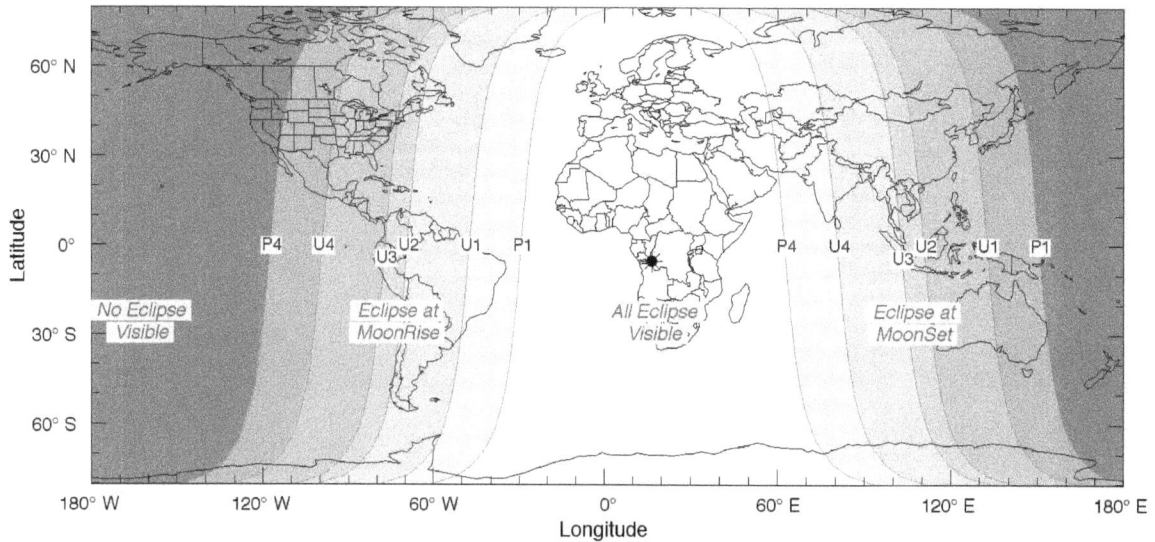

No Eclipse Visible

Eclipse at MoonRise

All Eclipse Visible

Eclipse at MoonSet

Total Lunar Eclipse of 2091 Mar 05

Greatest Eclipse = 15:58:22.4 TD (= 15:56:24.3 UT1)

Penumbral Magnitude = 2.2548	Gamma = 0.3212	Saros Series = 134
Umbral Magnitude = 1.2843	Axis = 0.3265°	Saros Member = 31 of 72

Sun at Greatest Eclipse
(Geocentric Coordinates)

R.A. = 23h05m57.5s
Dec. = -05°46'46.7"
S.D. = 00°16'07.7"
H.P. = 00°00'08.9"

Moon at Greatest Eclipse
(Geocentric Coordinates)

R.A. = 11h06m20.7s
Dec. = +06°05'30.3"
S.D. = 00°16'37.1"
H.P. = 01°00'59.5"

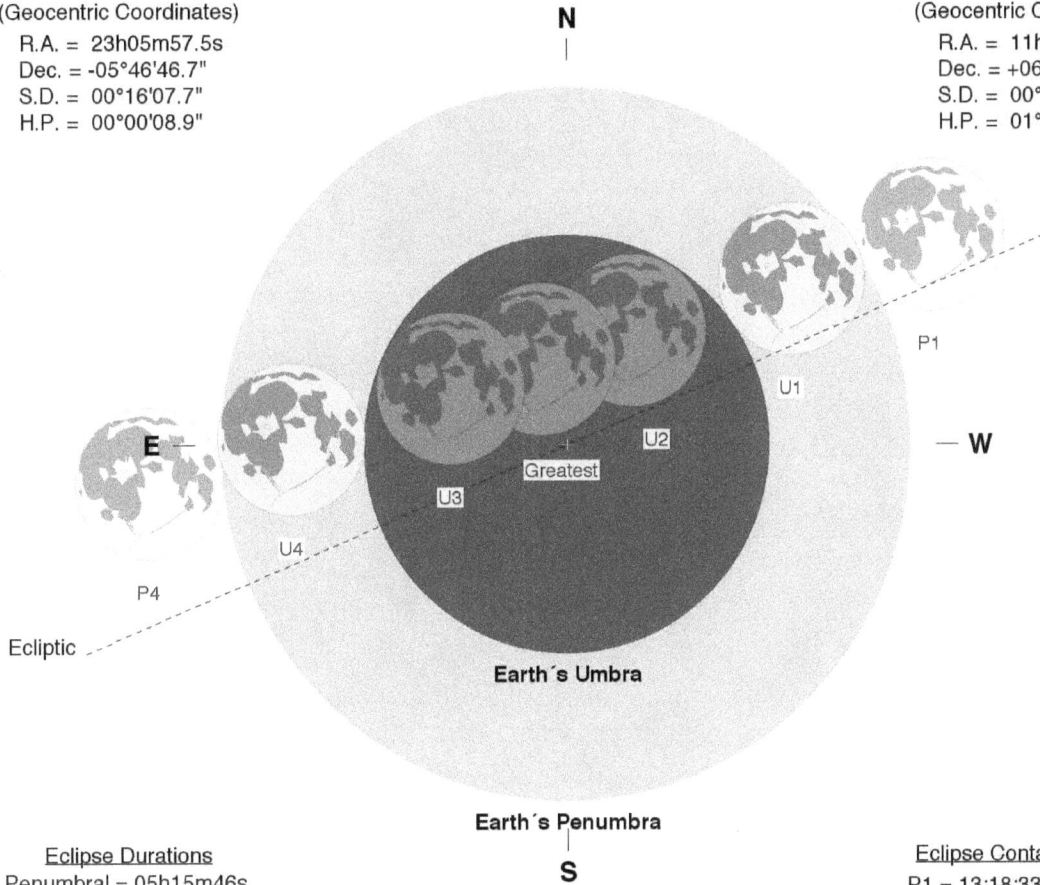

N

P1

U1

E

U2

W

Greatest

U3

U4

P4

Ecliptic

Earth's Umbra

Earth's Penumbra

S

Eclipse Durations

Penumbral = 05h15m46s
Umbral = 03h22m00s
Total = 01h13m28s

Eph. = JPL DE430
Rule = Herald-Sinnott
ΔT = 118 s

0	15	30	45	60

Arc-Minutes

©2020 F. Espenak, www.EclipseWise.com

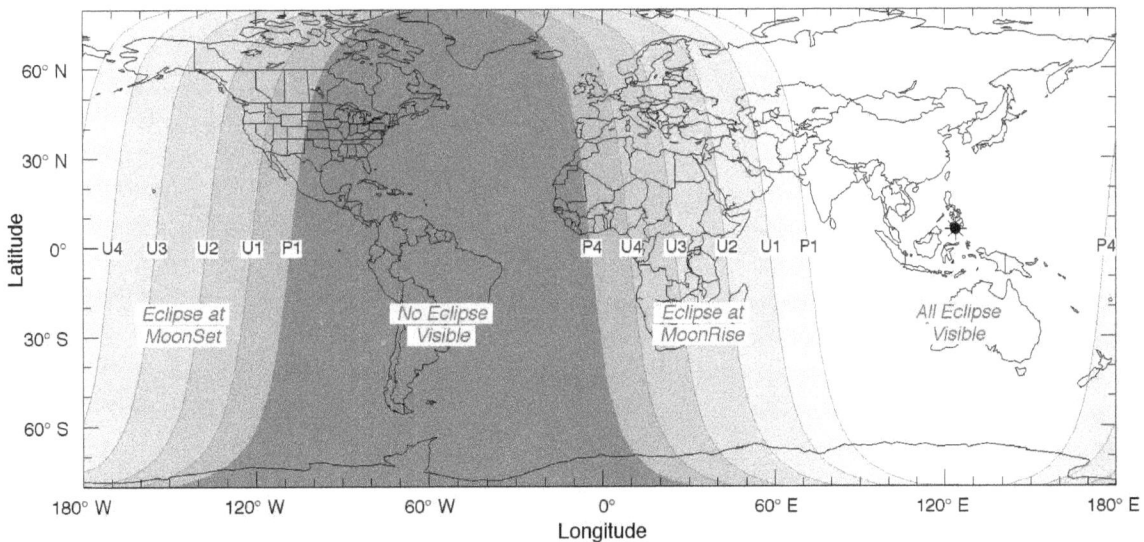

Eclipse Contacts

P1 = 13:18:33 UT1
U1 = 14:15:29 UT1
U2 = 15:19:49 UT1
U3 = 16:33:17 UT1
U4 = 17:37:29 UT1
P4 = 18:34:18 UT1

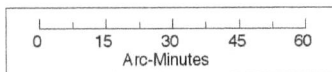

Total Lunar Eclipse of 2091 Aug 29

Greatest Eclipse = 00:38:24.8 TD (= 00:36:26.2 UT1)

Penumbral Magnitude = 2.2821	Gamma = -0.3270	Saros Series = 139
Umbral Magnitude = 1.2362	Axis = 0.3028°	Saros Member = 25 of 79

Sun at Greatest Eclipse
(Geocentric Coordinates)
R.A. = 10h31m03.4s
Dec. = +09°18'39.7"
S.D. = 00°15'49.9"
H.P. = 00°00'08.7"

Moon at Greatest Eclipse
(Geocentric Coordinates)
R.A. = 22h31m23.5s
Dec. = -09°36'08.1"
S.D. = 00°15'08.2"
H.P. = 00°55'33.3"

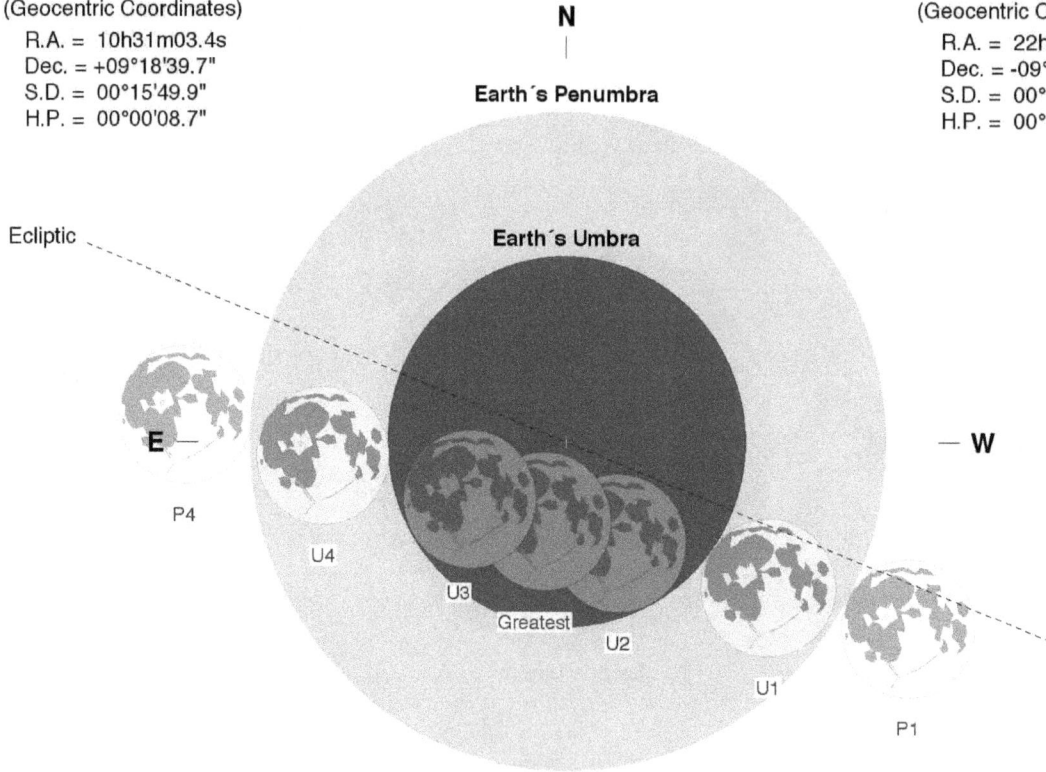

N

Earth's Penumbra

Ecliptic

Earth's Umbra

E

W

P4

U4

U3

Greatest

U2

U1

P1

S

Eclipse Durations
Penumbral = 05h54m16s
Umbral = 03h38m18s
Total = 01h13m31s

Eph. = JPL DE430
Rule = Herald-Sinnott
ΔT = 119 s

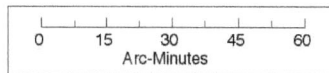

0   15   30   45   60
Arc-Minutes

©2020 F. Espenak, www.EclipseWise.com

Eclipse Contacts
P1 = 21:39:24 UT1
U1 = 22:47:20 UT1
U2 = 23:59:48 UT1
U3 = 01:13:19 UT1
U4 = 02:25:38 UT1
P4 = 03:33:39 UT1

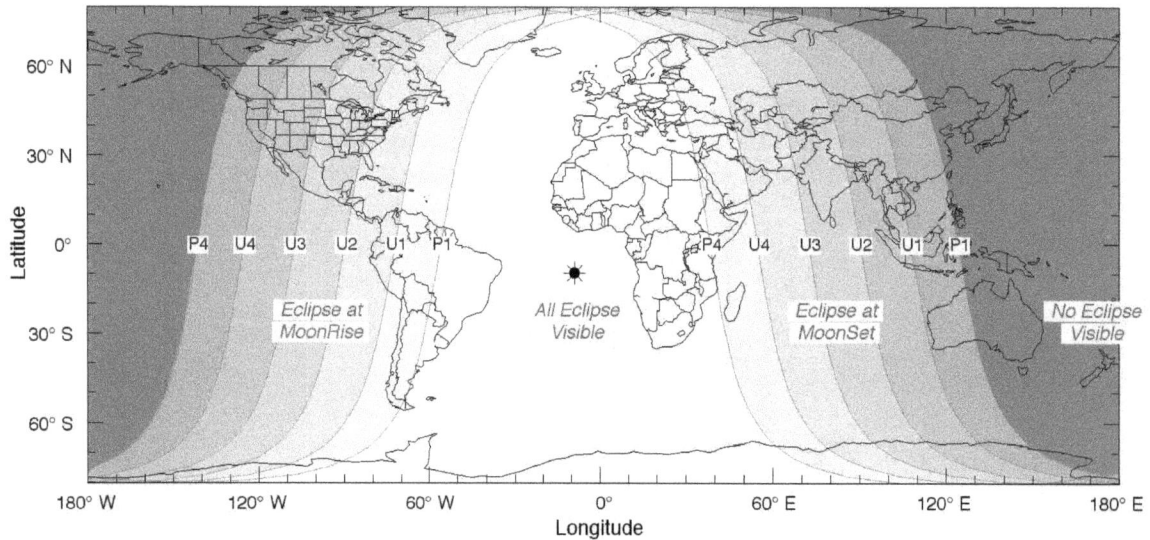

P4 U4 U3 U2 U1 P1   P4 U4 U3 U2 U1 P1

Eclipse at
MoonRise

All Eclipse
Visible

Eclipse at
MoonSet

No Eclipse
Visible

Penumbral Lunar Eclipse of 2092 Feb 23

Greatest Eclipse = 05:20:59.5 TD (= 05:19:00.4 UT1)

Penumbral Magnitude = 0.9395	Gamma = 1.0509	Saros Series = 144
Umbral Magnitude = -0.0777	Axis = 1.0218°	Saros Member = 20 of 71

Sun at Greatest Eclipse
(Geocentric Coordinates)

R.A. = 22h25m46.7s
Dec. = -09°49'23.6"
S.D. = 00°16'10.2"
H.P. = 00°00'08.9"

Moon at Greatest Eclipse
(Geocentric Coordinates)

R.A. = 10h26m54.6s
Dec. = +10°48'23.1"
S.D. = 00°15'53.9"
H.P. = 00°58'20.7"

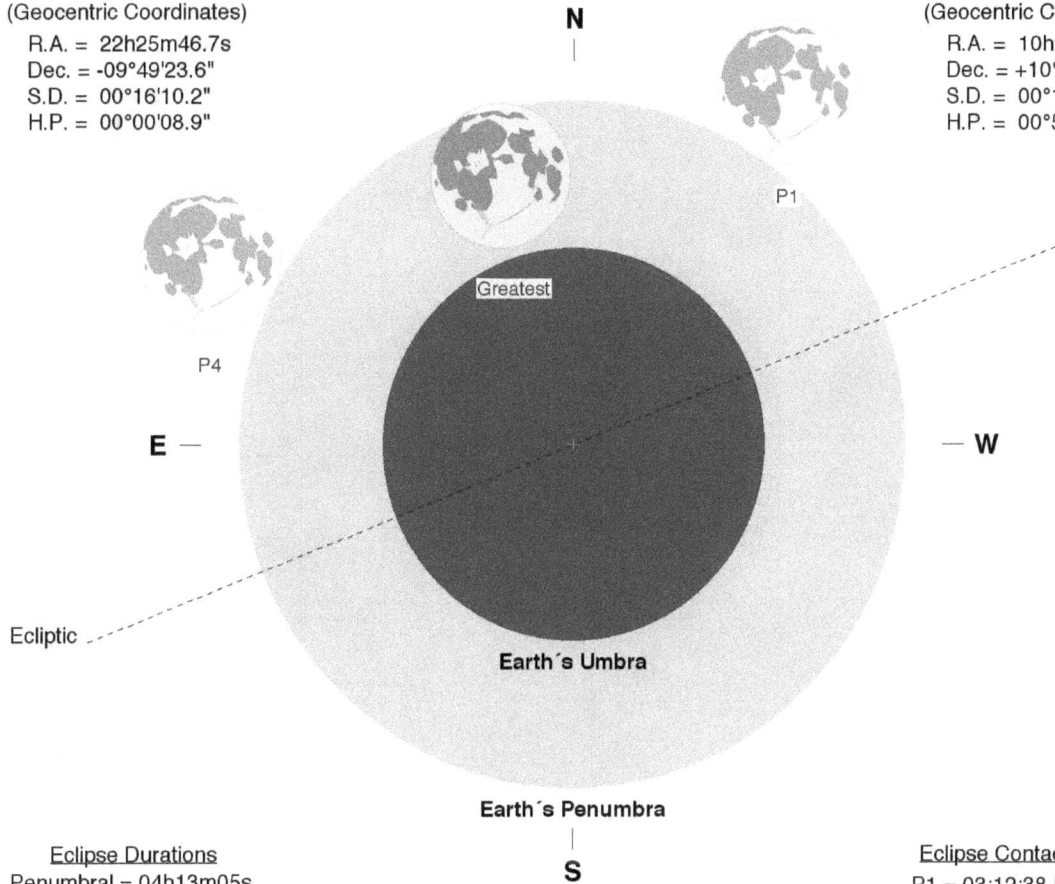

N

P1

Greatest

P4

E — — W

Ecliptic

Earth's Umbra

Earth's Penumbra

S

Eclipse Durations
Penumbral = 04h13m05s

Eclipse Contacts
P1 = 03:12:38 UT1
P4 = 07:25:43 UT1

Eph. = JPL DE430
Rule = Herald-Sinnott
ΔT = 119 s

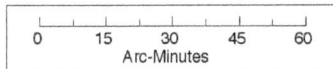

0	15	30	45	60

Arc-Minutes

©2020 F. Espenak, www.EclipseWise.com

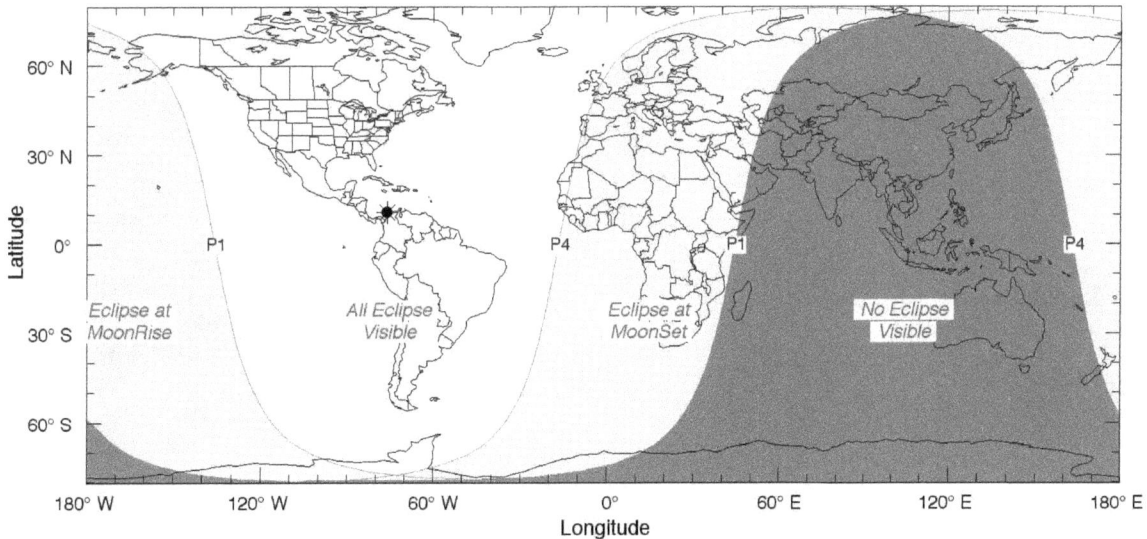

Eclipse at
MoonRise

All Eclipse
Visible

Eclipse at
MoonSet

No Eclipse
Visible

P1   P4   P1   P4

Penumbral Lunar Eclipse of 2092 Jul 19

Greatest Eclipse = 00:41:57.2 TD (= 00:39:57.7 UT1)

Penumbral Magnitude = 0.0634	Gamma = 1.5132	Saros Series = 111
Umbral Magnitude = -0.8979	Axis = 1.5152°	Saros Member = 71 of 71

Sun at Greatest Eclipse
(Geocentric Coordinates)
R.A. = 07h57m50.3s
Dec. = +20°40'07.4"
S.D. = 00°15'44.3"
H.P. = 00°00'08.7"

Moon at Greatest Eclipse
(Geocentric Coordinates)
R.A. = 19h57m10.8s
Dec. = -19°09'40.4"
S.D. = 00°16'22.4"
H.P. = 01°00'05.3"

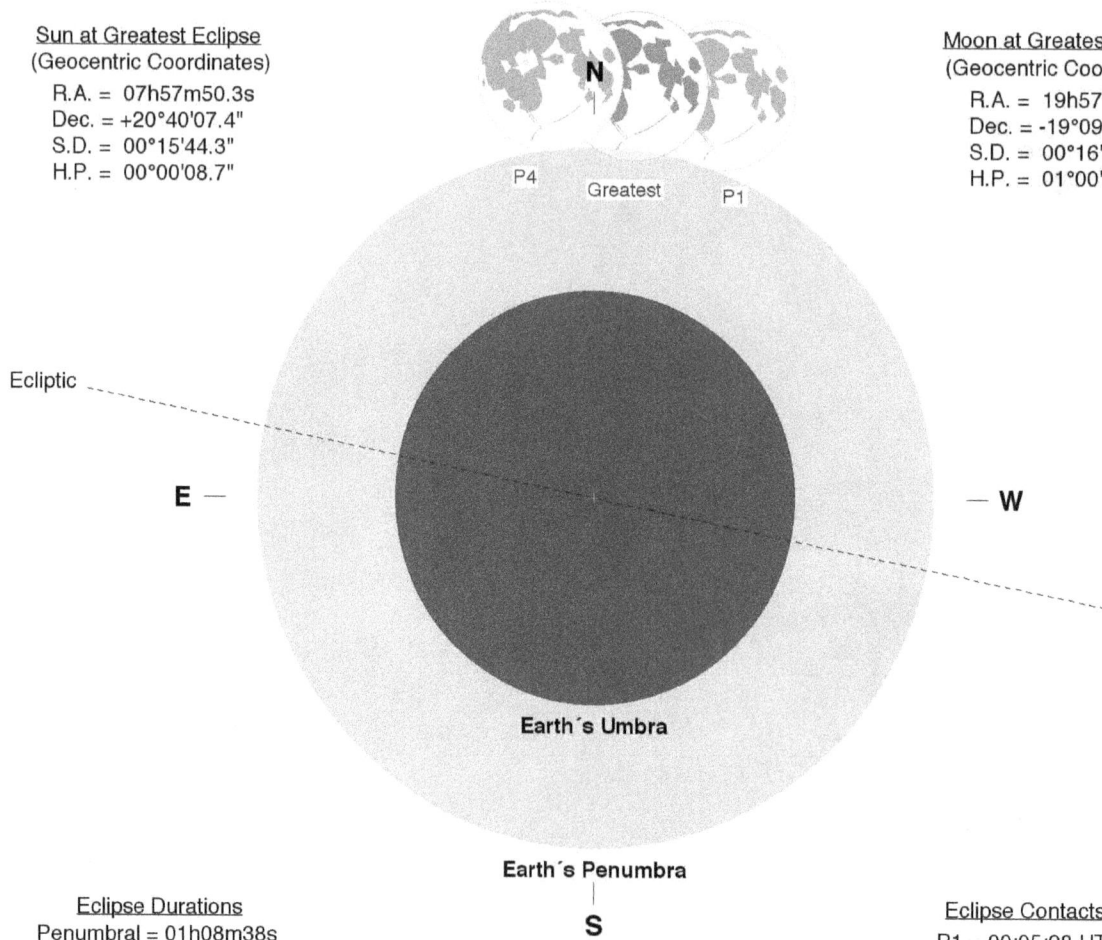

N

P4   Greatest   P1

Ecliptic

E —   — W

Earth's Umbra

Earth's Penumbra

S

Eclipse Durations
Penumbral = 01h08m38s

Eclipse Contacts
P1 = 00:05:28 UT1
P4 = 01:14:06 UT1

Eph. = JPL DE430
Rule = Herald-Sinnott
ΔT =   119 s

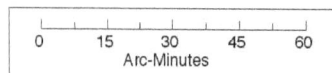

0	15	30	45	60
Arc-Minutes

©2020 F. Espenak, www.EclipseWise.com

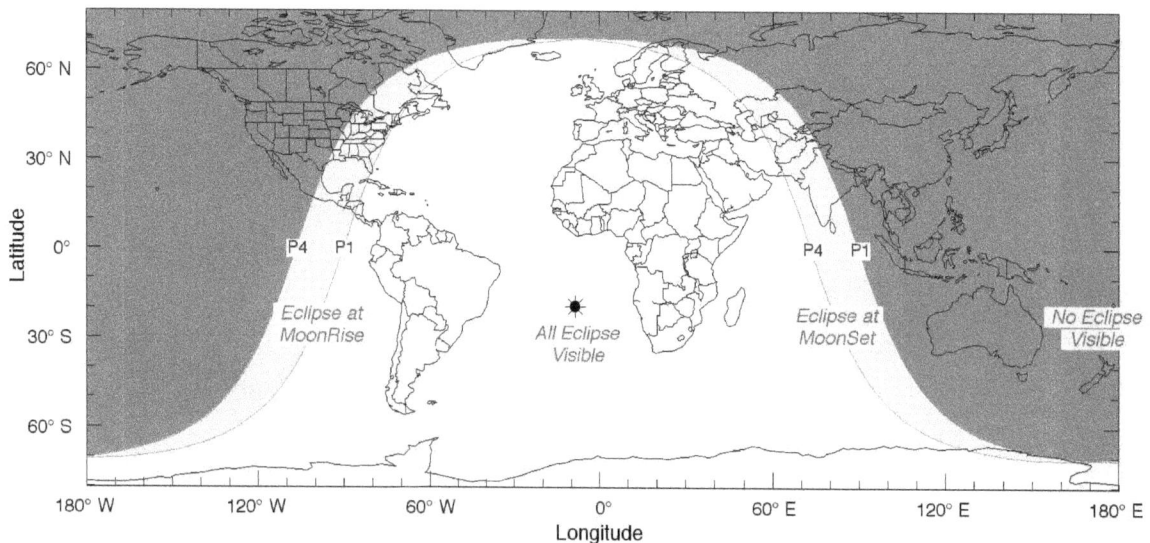

Eclipse at MoonRise   *All Eclipse Visible*   *Eclipse at MoonSet*   *No Eclipse Visible*

P4 P1   P4 P1

Penumbral Lunar Eclipse of 2092 Aug 17

Greatest Eclipse = 09:13:59.6 TD (= 09:12:00.0 UT1)

Penumbral Magnitude = 0.9143	Gamma = -1.0569	Saros Series = 149
Umbral Magnitude = -0.0746	Axis = 1.0327°	Saros Member = 7 of 71

Sun at Greatest Eclipse
(Geocentric Coordinates)

R.A. = 09h51m02.7s
Dec. = +13°00'50.2"
S.D. = 00°15'47.9"
H.P. = 00°00'08.7"

Moon at Greatest Eclipse
(Geocentric Coordinates)

R.A. = 21h52m04.9s
Dec. = -14°00'55.6"
S.D. = 00°15'58.6"
H.P. = 00°58'37.9"

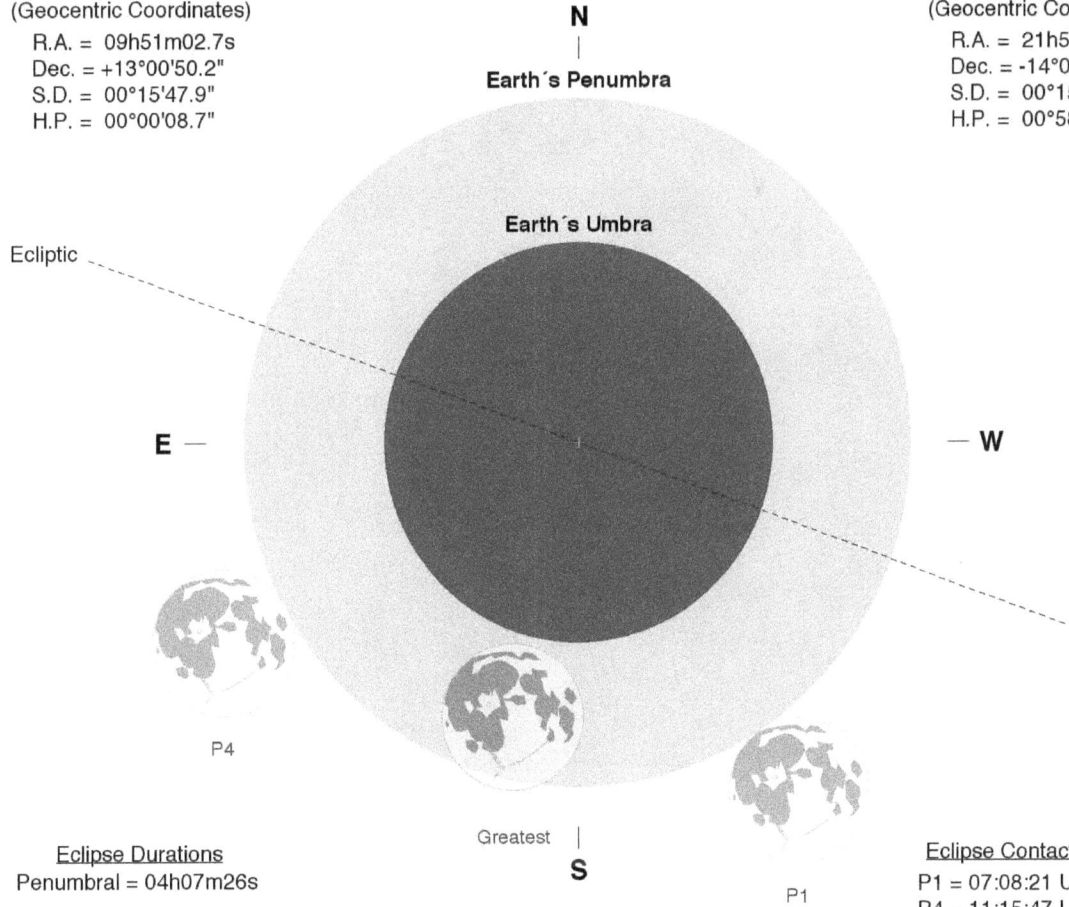

N

Earth's Penumbra

Earth's Umbra

Ecliptic

E

W

P4

Greatest

S

P1

Eclipse Durations

Penumbral = 04h07m26s

Eclipse Contacts

P1 = 07:08:21 UT1
P4 = 11:15:47 UT1

Eph. = JPL DE430
Rule = Herald-Sinnott
ΔT =   120 s

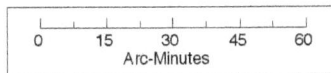

0	15	30	45	60

Arc-Minutes

©2020 F. Espenak, www.EclipseWise.com

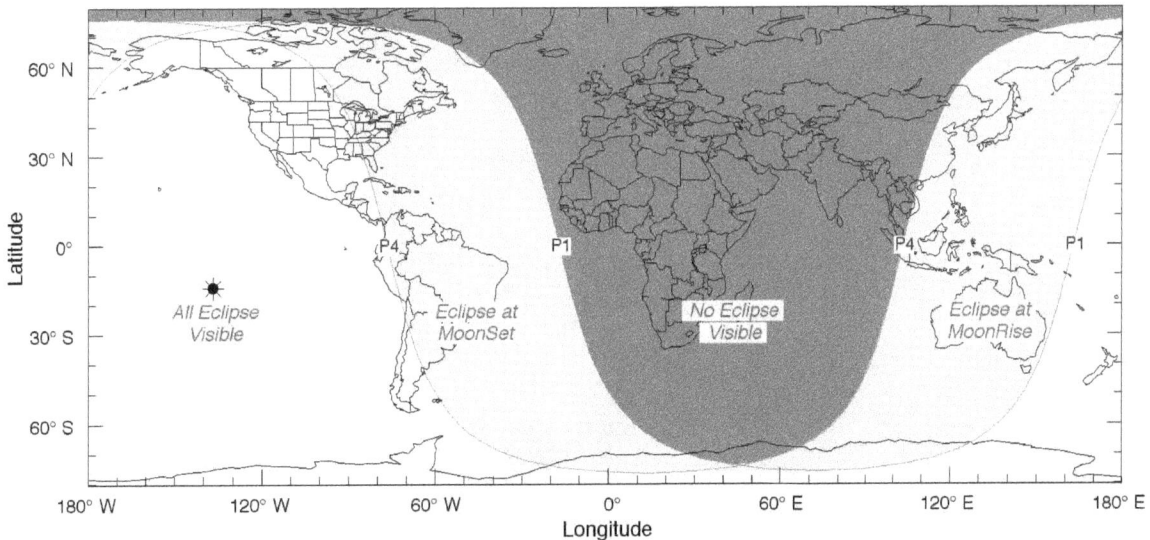

All Eclipse
Visible

Eclipse at
MoonSet

No Eclipse
Visible

Eclipse at
MoonRise

P4

P1

P4

P1

Penumbral Lunar Eclipse of 2093 Jan 12

Greatest Eclipse = 18:00:02.4 TD   (= 17:58:02.4 UT1)

Penumbral Magnitude = 0.7567	Gamma = -1.1734	Saros Series = 116
Umbral Magnitude = -0.3430	Axis = 1.0613°	Saros Member = 62 of 73

Sun at Greatest Eclipse
(Geocentric Coordinates)
R.A. = 19h40m30.6s
Dec. = -21°25'06.3"
S.D. = 00°16'15.7"
H.P. = 00°00'08.9"

Moon at Greatest Eclipse
(Geocentric Coordinates)
R.A. = 07h40m11.4s
Dec. = +20°21'35.1"
S.D. = 00°14'47.3"
H.P. = 00°54'16.4"

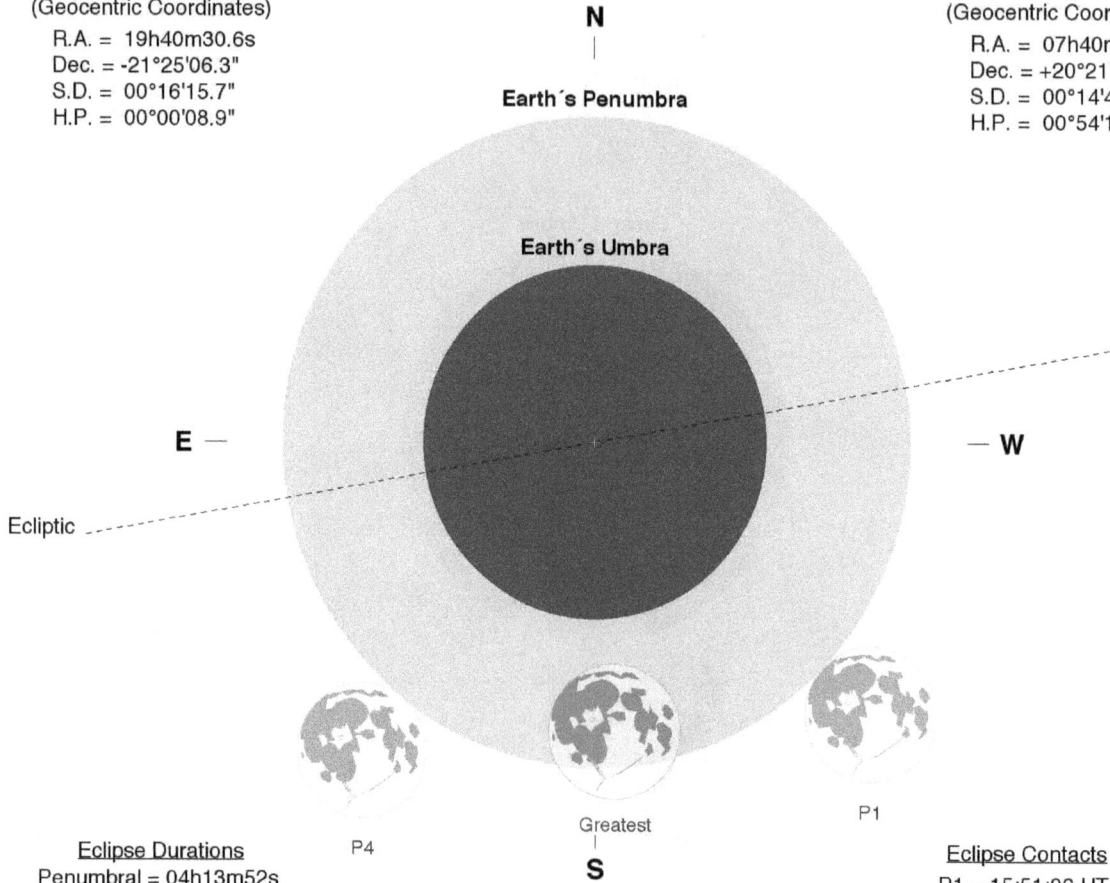

N

Earth's Penumbra

Earth's Umbra

E

W

Ecliptic

Greatest

P4

P1

S

Eclipse Durations
Penumbral = 04h13m52s

Eclipse Contacts
P1 = 15:51:06 UT1
P4 = 20:04:58 UT1

Eph. = JPL DE430
Rule = Herald-Sinnott
ΔT = 120 s

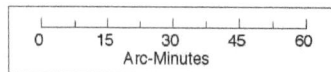

0   15   30   45   60
Arc-Minutes

©2020 F. Espenak, www.EclipseWise.com

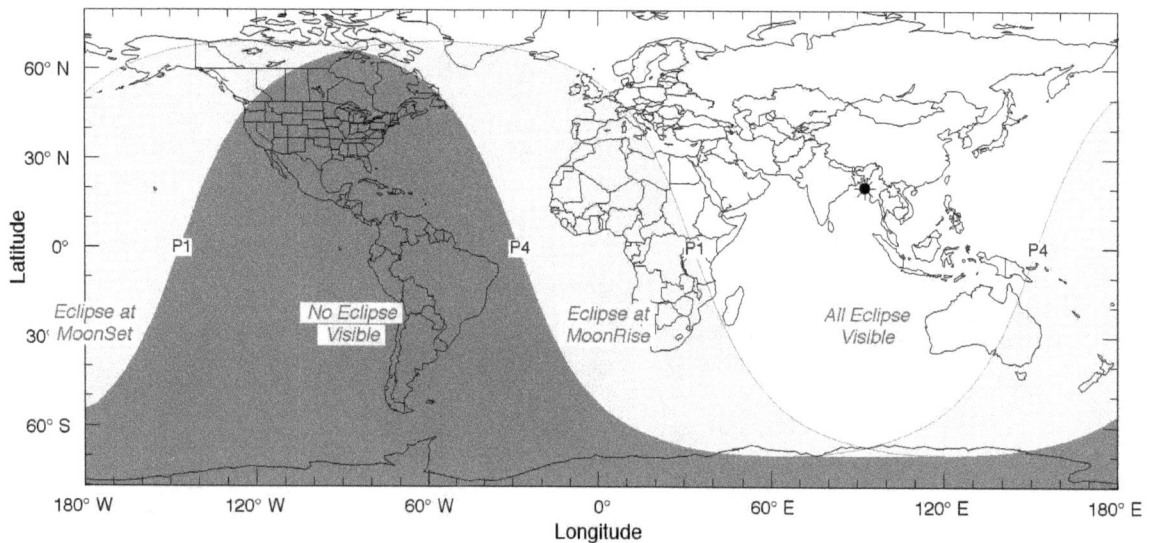

60° N

30° N

Latitude

0°

30° S

60° S

P1

P4

P1

P4

Eclipse at
MoonSet

No Eclipse
Visible

Eclipse at
MoonRise

All Eclipse
Visible

180° W   120° W   60° W   0°   60° E   120° E   180° E

Longitude

Partial Lunar Eclipse of 2093 Jul 08

Greatest Eclipse = 17:24:17.5 TD (= 17:22:17.1 UT1)

Penumbral Magnitude = 1.4289	Gamma = 0.7632	Saros Series = 121
Umbral Magnitude = 0.4886	Axis = 0.7810°	Saros Member = 59 of 82

Sun at Greatest Eclipse
(Geocentric Coordinates)
R.A. = 07h15m09.4s
Dec. = +22°18'09.9"
S.D. = 00°15'43.9"
H.P. = 00°00'08.7"

N

Moon at Greatest Eclipse
(Geocentric Coordinates)
R.A. = 19h15m03.2s
Dec. = -21°31'19.6"
S.D. = 00°16'43.9"
H.P. = 01°01'24.3"

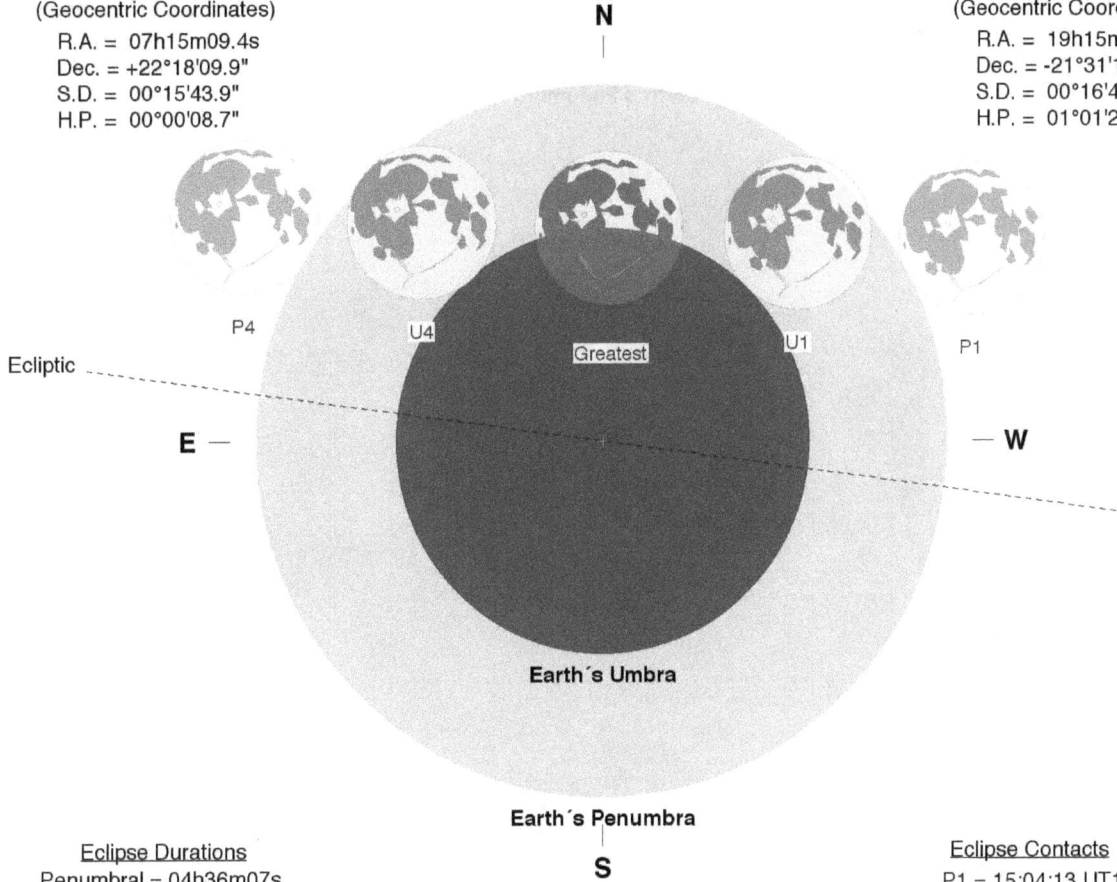

P4   U4   Greatest   U1   P1

Ecliptic

E —   — W

Earth's Umbra

Earth's Penumbra

S

Eclipse Durations
Penumbral = 04h36m07s
Umbral = 02h22m30s

Eph. = JPL DE430
Rule = Herald-Sinnott
ΔT = 120 s

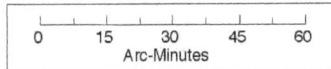

0   15   30   45   60
Arc-Minutes

©2020 F. Espenak, www.EclipseWise.com

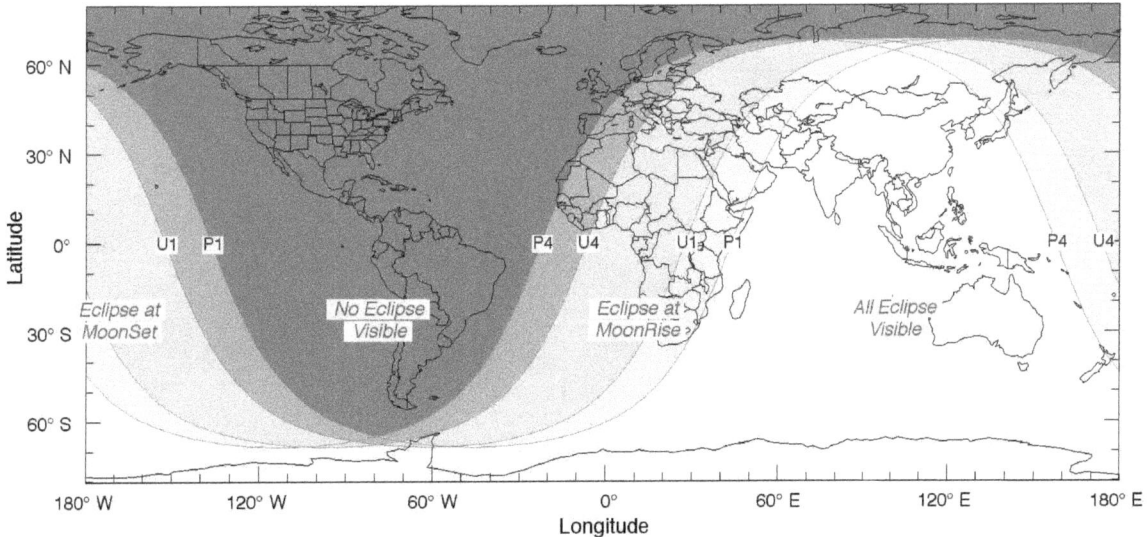

Eclipse Contacts
P1 = 15:04:13 UT1
U1 = 16:11:01 UT1
U4 = 18:33:32 UT1
P4 = 19:40:20 UT1

Eclipse at MoonSet

No Eclipse Visible

Eclipse at MoonRise

All Eclipse Visible

Partial Lunar Eclipse of 2094 Jan 01

Greatest Eclipse = 17:00:06.0 TD (= 16:58:05.1 UT1)

| Penumbral Magnitude = 1.9872 | Gamma = -0.5025 | Saros Series = 126 |
| Umbral Magnitude = 0.8886 | Axis = 0.4550° | Saros Member = 49 of 70 |

Sun at Greatest Eclipse
(Geocentric Coordinates)
R.A. = 18h51m13.7s
Dec. = -22°54'14.9"
S.D. = 00°16'15.9"
H.P. = 00°00'08.9"

Moon at Greatest Eclipse
(Geocentric Coordinates)
R.A. = 06h51m15.2s
Dec. = +22°26'57.1"
S.D. = 00°14'48.2"
H.P. = 00°54'19.9"

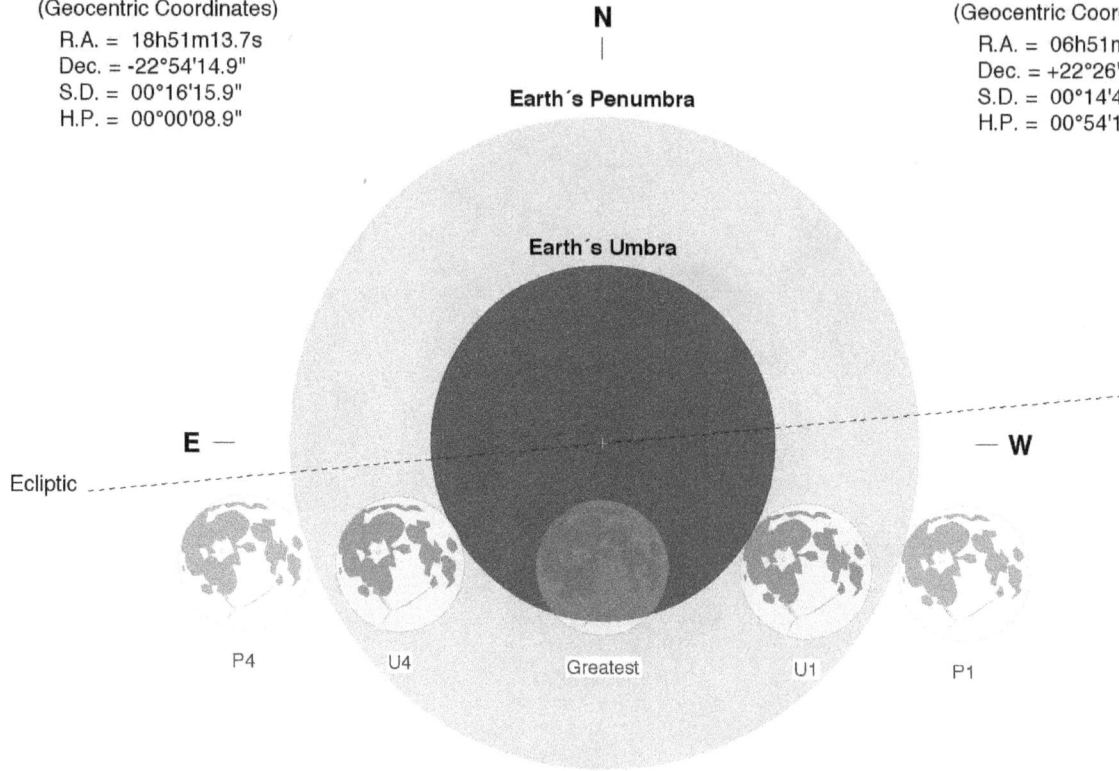

N

Earth's Penumbra

Earth's Umbra

E

W

Ecliptic

P4   U4   Greatest   U1   P1

S

Eclipse Durations
Penumbral = 05h57m24s
Umbral = 03h22m02s

Eph. = JPL DE430
Rule = Herald-Sinnott
ΔT = 121 s

0   15   30   45   60
Arc-Minutes

©2020 F. Espenak, www.EclipseWise.com

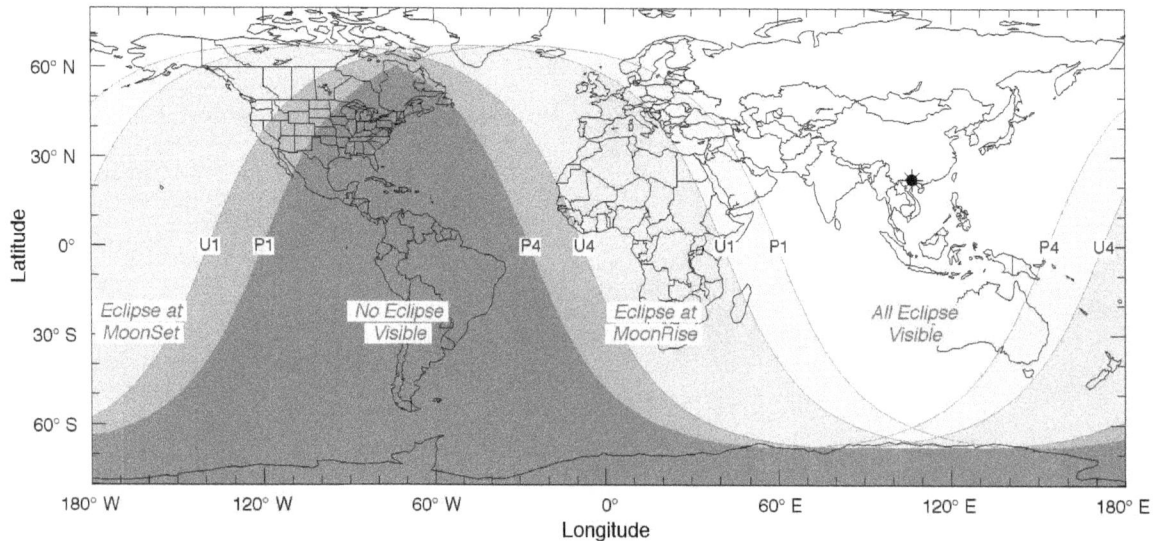

Eclipse Contacts
P1 = 13:59:24 UT1
U1 = 15:17:03 UT1
U4 = 18:39:05 UT1
P4 = 19:56:48 UT1

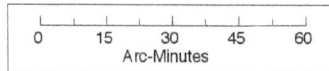

U1 P1   P4 U4   U1 P1   P4 U4

Eclipse at
MoonSet

No Eclipse
Visible

Eclipse at
MoonRise

All Eclipse
Visible

Latitude

Longitude

Total Lunar Eclipse of 2094 Jun 28

Greatest Eclipse = 10:01:56.7 TD   (= 09:59:55.3 UT1)

Penumbral Magnitude = 2.7879	Gamma = 0.0288	Saros Series = 131
Umbral Magnitude = 1.8249	Axis = 0.0288°	Saros Member = 38 of 72

Sun at Greatest Eclipse
(Geocentric Coordinates)

R.A. = 06h31m43.3s
Dec. = +23°13'34.7"
S.D. = 00°15'44.1"
H.P. = 00°00'08.7"

N

Moon at Greatest Eclipse
(Geocentric Coordinates)

R.A. = 18h31m43.6s
Dec. = -23°11'51.1"
S.D. = 00°16'20.2"
H.P. = 00°59'57.5"

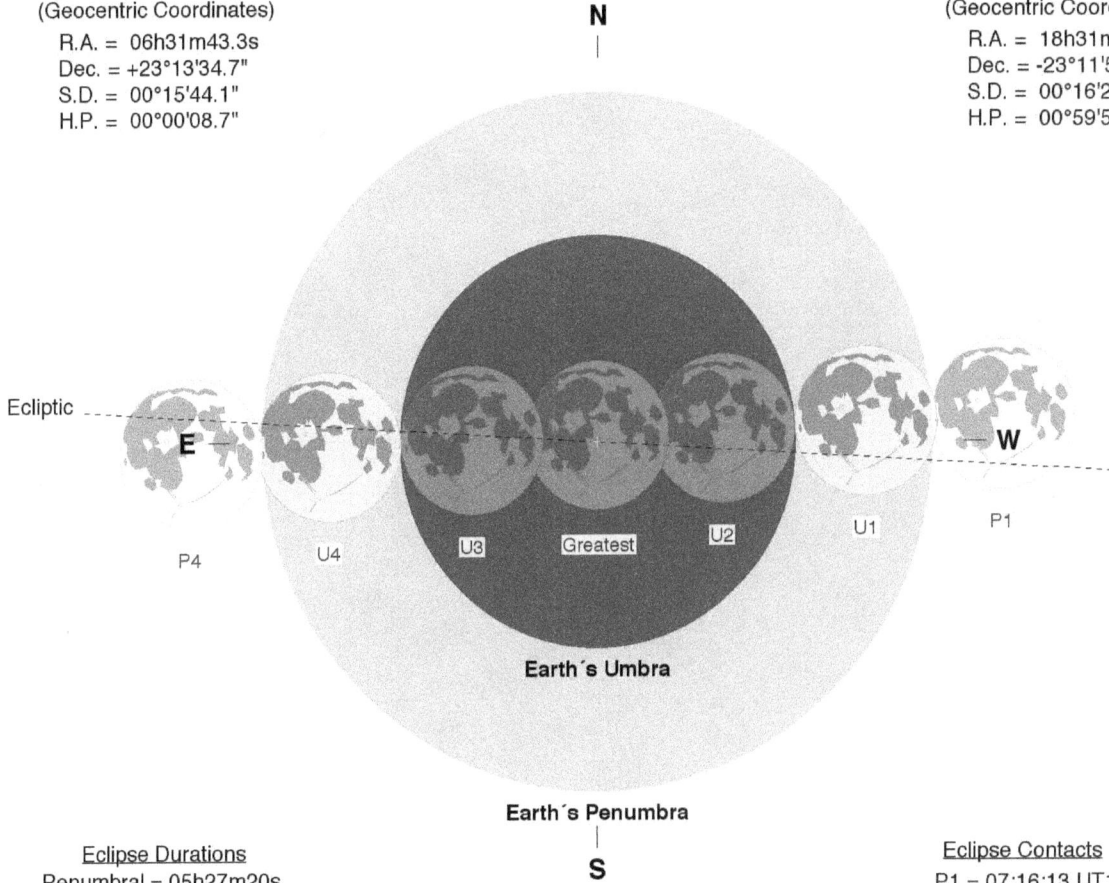

Ecliptic

E

W

P4   U4   U3   Greatest   U2   U1   P1

Earth's Umbra

Earth's Penumbra

S

Eclipse Durations
Penumbral = 05h27m20s
Umbral = 03h36m30s
Total = 01h41m22s

Eph. = JPL DE430
Rule = Herald-Sinnott
ΔT =   121 s

0   15   30   45   60
Arc-Minutes

©2020 F. Espenak, www.EclipseWise.com

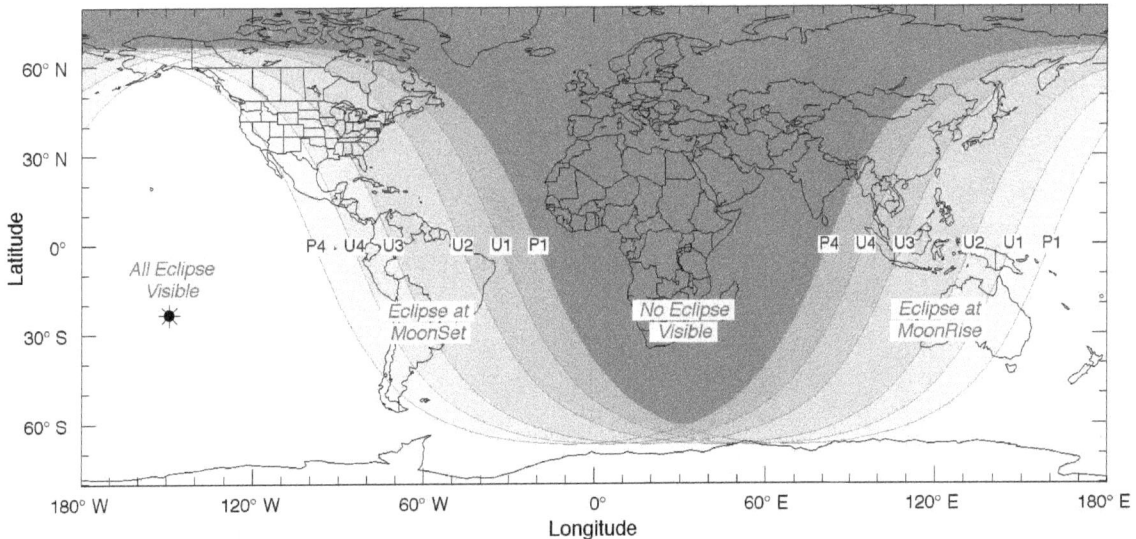

Eclipse Contacts
P1 = 07:16:13 UT1
U1 = 08:11:41 UT1
U2 = 09:09:15 UT1
U3 = 10:50:37 UT1
U4 = 11:48:11 UT1
P4 = 12:43:33 UT1

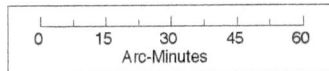

All Eclipse
Visible

Eclipse at
MoonSet

No Eclipse
Visible

Eclipse at
MoonRise

P4 U4 U3   U2 U1 P1   P4 U4 U3   U2 U1 P1

Total Lunar Eclipse of 2094 Dec 21

Greatest Eclipse = 19:56:32.1 TD   (= 19:54:30.3 UT1)

Penumbral Magnitude = 2.5153	Gamma = 0.2016	Saros Series = 136
Umbral Magnitude = 1.4642	Axis = 0.1907°	Saros Member = 24 of 72

Sun at Greatest Eclipse
(Geocentric Coordinates)
R.A. = 18h01m58.4s
Dec. = -23°25'33.8"
S.D. = 00°16'15.4"
H.P. = 00°00'08.9"

Moon at Greatest Eclipse
(Geocentric Coordinates)
R.A. = 06h01m53.5s
Dec. = +23°36'57.2"
S.D. = 00°15'28.1"
H.P. = 00°56'46.1"

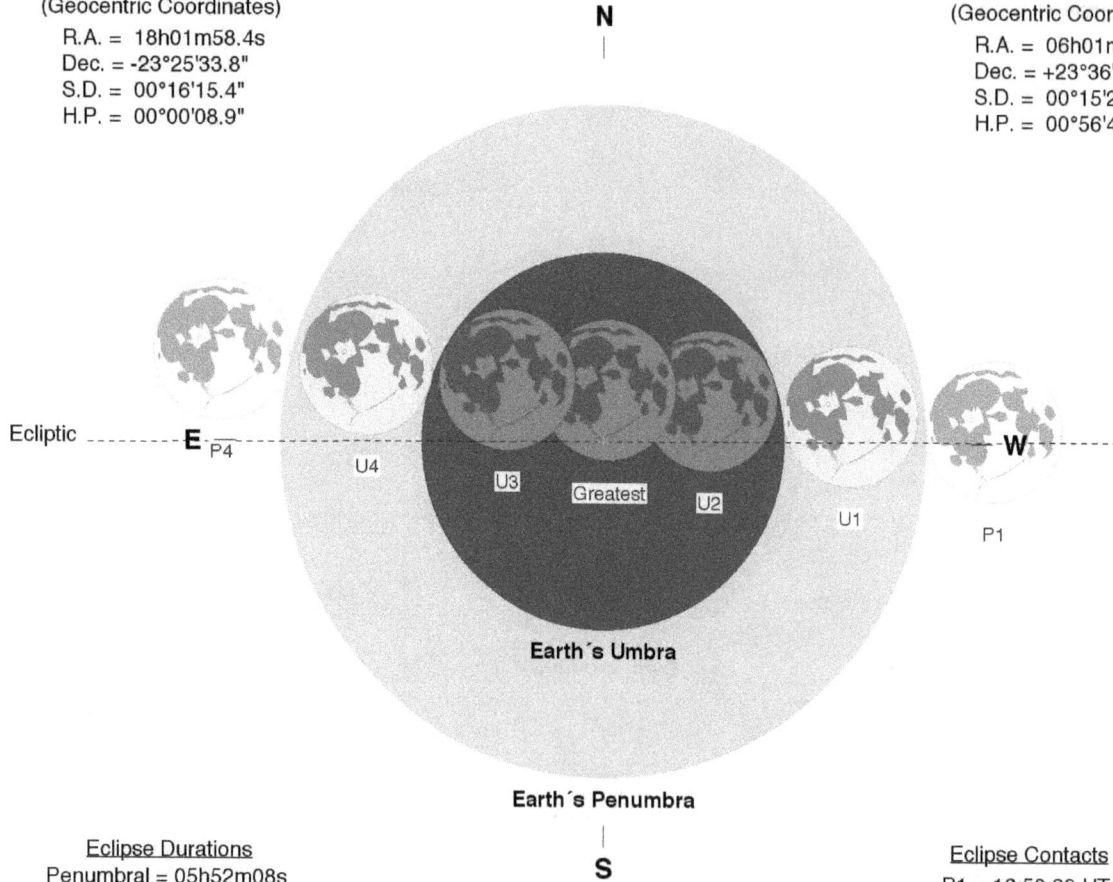

N

Ecliptic

E P4

U4

U3   Greatest   U2

U1

W

P1

Earth's Umbra

Earth's Penumbra

S

Eclipse Durations
Penumbral = 05h52m08s
Umbral = 03h41m18s
Total = 01h32m24s

Eph. = JPL DE430
Rule = Herald-Sinnott
ΔT =   122 s

0   15   30   45   60
Arc-Minutes

©2020 F. Espenak, www.EclipseWise.com

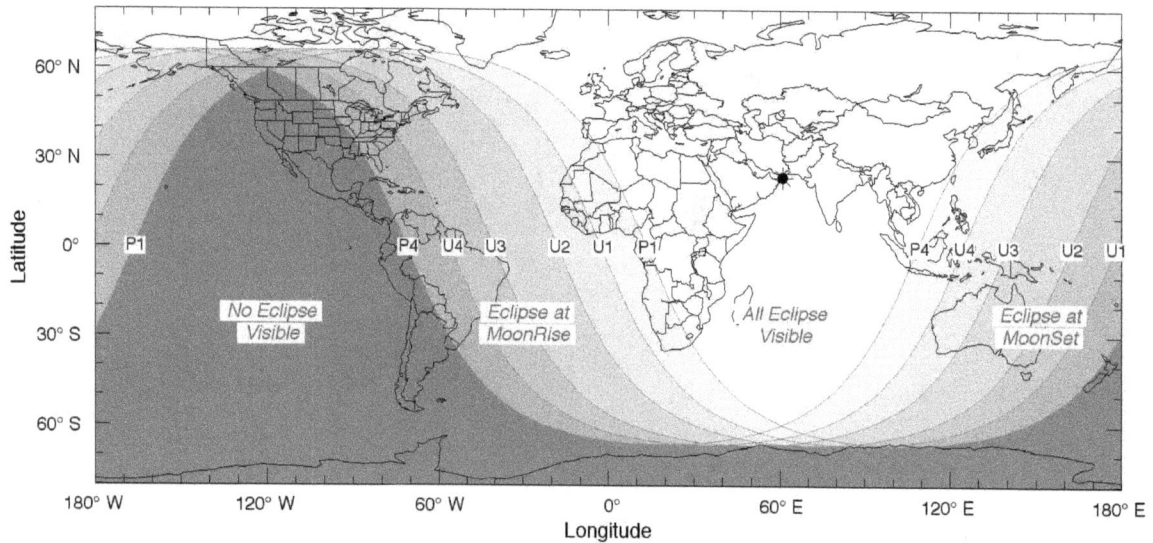

Eclipse Contacts
P1 = 16:58:29 UT1
U1 = 18:03:49 UT1
U2 = 19:08:16 UT1
U3 = 20:40:40 UT1
U4 = 21:45:07 UT1
P4 = 22:50:37 UT1

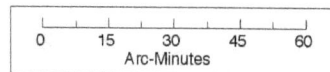

Partial Lunar Eclipse of 2095 Jun 17

Greatest Eclipse = 22:00:10.7 TD (= 21:58:08.3 UT1)

Penumbral Magnitude = 1.4632	Gamma = -0.7653	Saros Series = 141
Umbral Magnitude = 0.4474	Axis = 0.7255°	Saros Member = 28 of 72

Sun at Greatest Eclipse
(Geocentric Coordinates)

R.A. = 05h47m03.2s
Dec. = +23°23'37.9"
S.D. = 00°15'44.7"
H.P. = 00°00'08.7"

Moon at Greatest Eclipse
(Geocentric Coordinates)

R.A. = 17h46m40.2s
Dec. = -24°06'50.7"
S.D. = 00°15'30.0"
H.P. = 00°56'53.2"

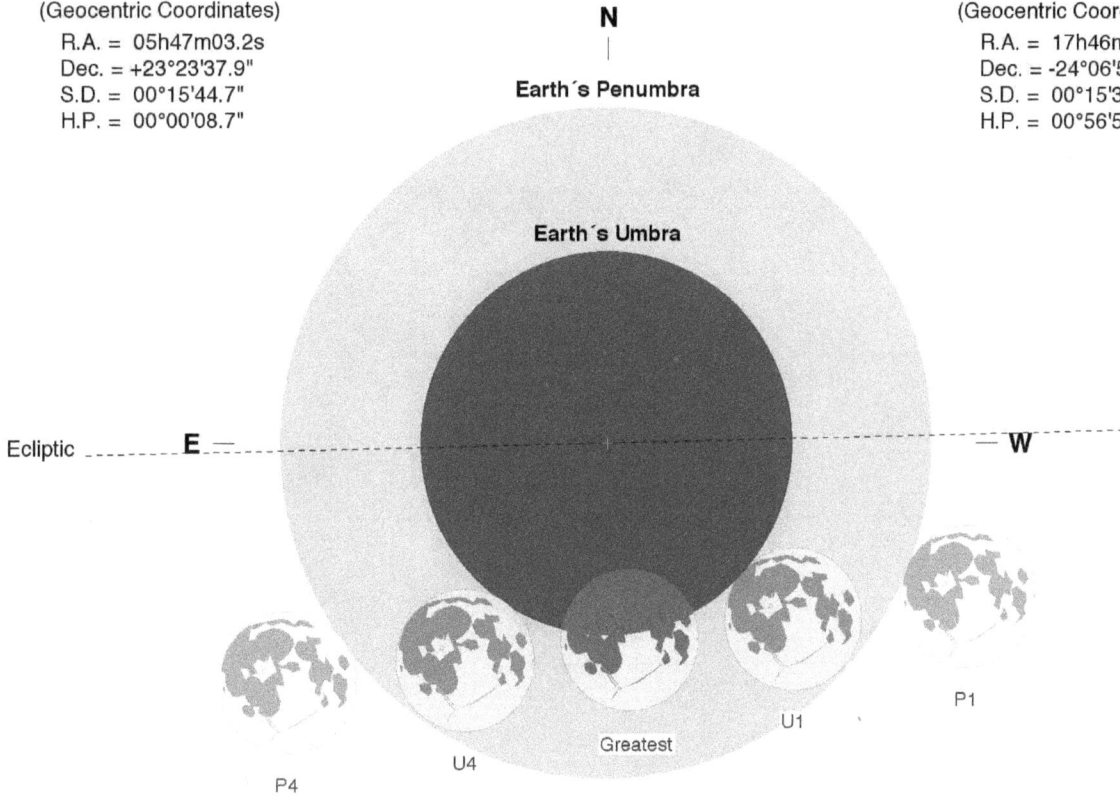

N

Earth's Penumbra

Earth's Umbra

Ecliptic

E

W

P4

U4

Greatest

U1

P1

S

Eclipse Durations

Penumbral = 05h05m32s
Umbral = 02h27m39s

Eph. = JPL DE430
Rule = Herald-Sinnott
ΔT = 122 s

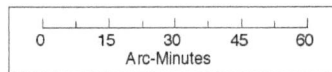

0 15 30 45 60
Arc-Minutes

©2020 F. Espenak, www.EclipseWise.com

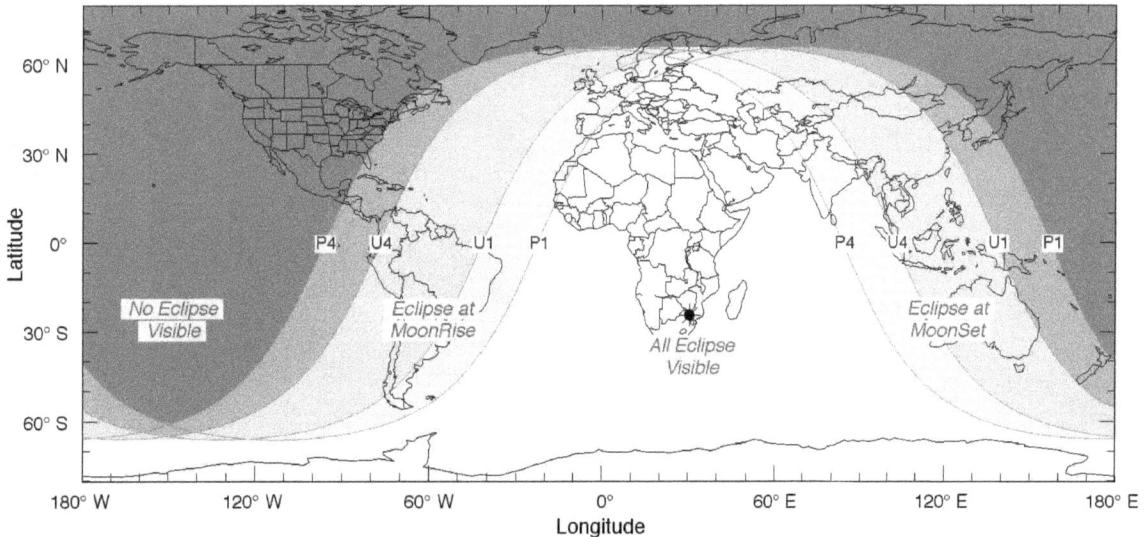

Eclipse Contacts

P1 = 19:25:20 UT1
U1 = 20:44:20 UT1
U4 = 23:11:59 UT1
P4 = 00:30:52 UT1

P4 U4 U1 P1   P4 U4 U1 P1

No Eclipse
Visible

Eclipse at
MoonRise

All Eclipse
Visible

Eclipse at
MoonSet

273

Partial Lunar Eclipse of 2095 Dec 11

Greatest Eclipse = 06:15:01.7 TD   (= 06:12:58.8 UT1)

Penumbral Magnitude = 1.2526	Gamma = 0.8743	Saros Series = 146
Umbral Magnitude = 0.2581	Axis = 0.8732°	Saros Member = 15 of 72

Sun at Greatest Eclipse
(Geocentric Coordinates)
  R.A. = 17h14m07.8s
  Dec. = -23°00'32.2"
  S.D. = 00°16'14.4"
  H.P. = 00°00'08.9"

Moon at Greatest Eclipse
(Geocentric Coordinates)
  R.A. = 05h13m27.5s
  Dec. = +23°52'06.6"
  S.D. = 00°16'19.8"
  H.P. = 00°59'56.1"

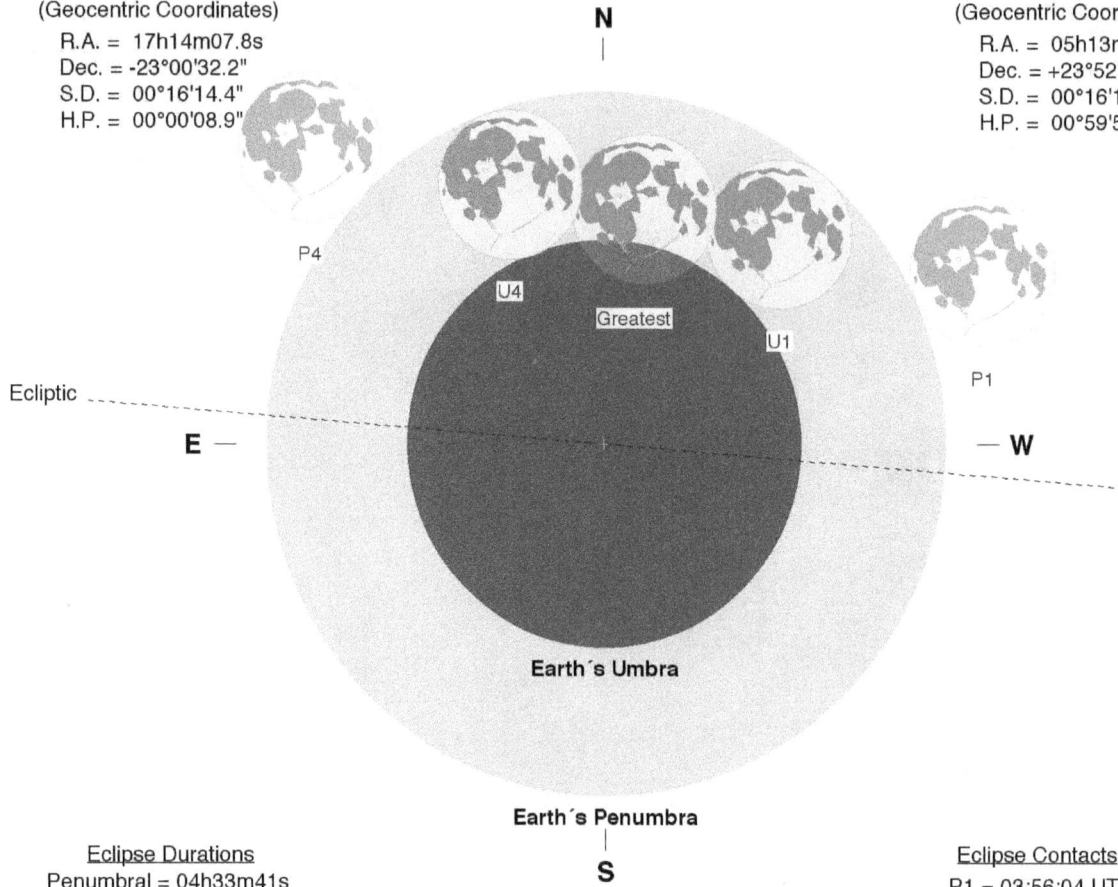

N

P4

U4

Greatest

U1

P1

Ecliptic

E

W

Earth's Umbra

Earth's Penumbra

S

Eclipse Durations
  Penumbral = 04h33m41s
  Umbral = 01h49m37s

Eph. = JPL DE430
Rule = Herald-Sinnott
ΔT =   123 s

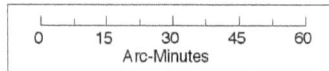

Eclipse Contacts
P1 = 03:56:04 UT1
U1 = 05:18:01 UT1
U4 = 07:07:38 UT1
P4 = 08:29:45 UT1

0	15	30	45	60
Arc-Minutes

©2020 F. Espenak, www.EclipseWise.com

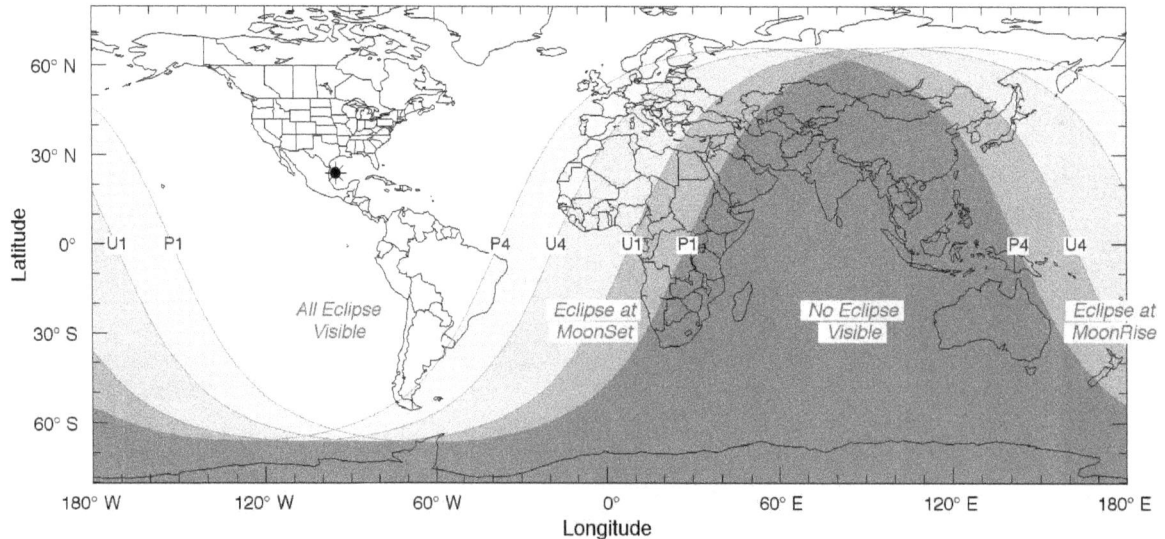

All Eclipse Visible

Eclipse at MoonSet

No Eclipse Visible

Eclipse at MoonRise

U1   P1

P4   U4   U1   P1

P4   U4

Penumbral Lunar Eclipse of 2096 May 07

Greatest Eclipse = 11:24:43.7 TD   (= 11:22:40.5 UT1)

Penumbral Magnitude = 0.5327	Gamma = 1.2897	Saros Series = 113
Umbral Magnitude = -0.5451	Axis = 1.1601°	Saros Member = 68 of 71

Sun at Greatest Eclipse
(Geocentric Coordinates)

R.A. = 03h01m57.6s
Dec. = +17°10'18.7"
S.D. = 00°15'51.1"
H.P. = 00°00'08.7"

Moon at Greatest Eclipse
(Geocentric Coordinates)

R.A. = 15h03m45.2s
Dec. = -16°05'38.7"
S.D. = 00°14'42.5"
H.P. = 00°53'58.7"

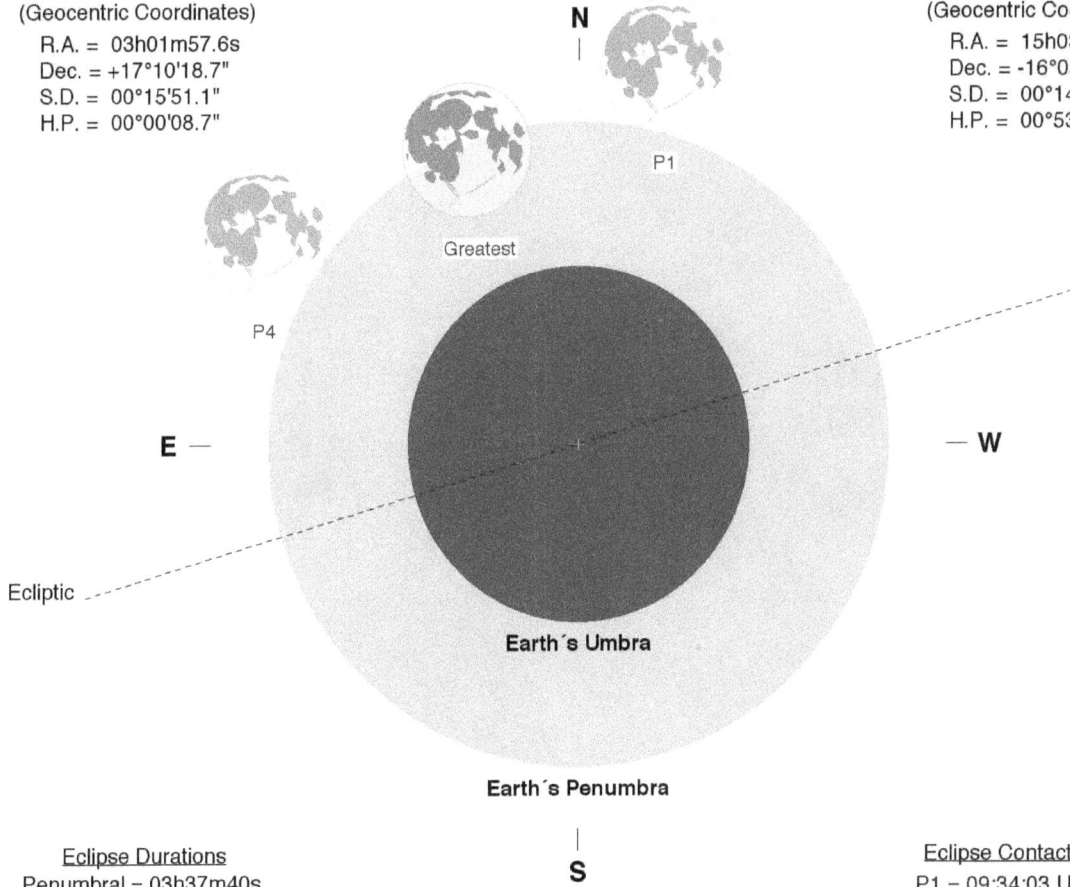

N

P1

Greatest

P4

E —

— W

Ecliptic

Earth's Umbra

Earth's Penumbra

S

Eclipse Durations
Penumbral = 03h37m40s

Eclipse Contacts
P1 = 09:34:03 UT1
P4 = 13:11:43 UT1

Eph. = JPL DE430
Rule = Herald-Sinnott
ΔT =   123 s

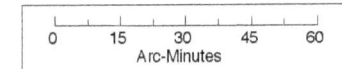

0   15   30   45   60
Arc-Minutes

©2020 F. Espenak, www.EclipseWise.com

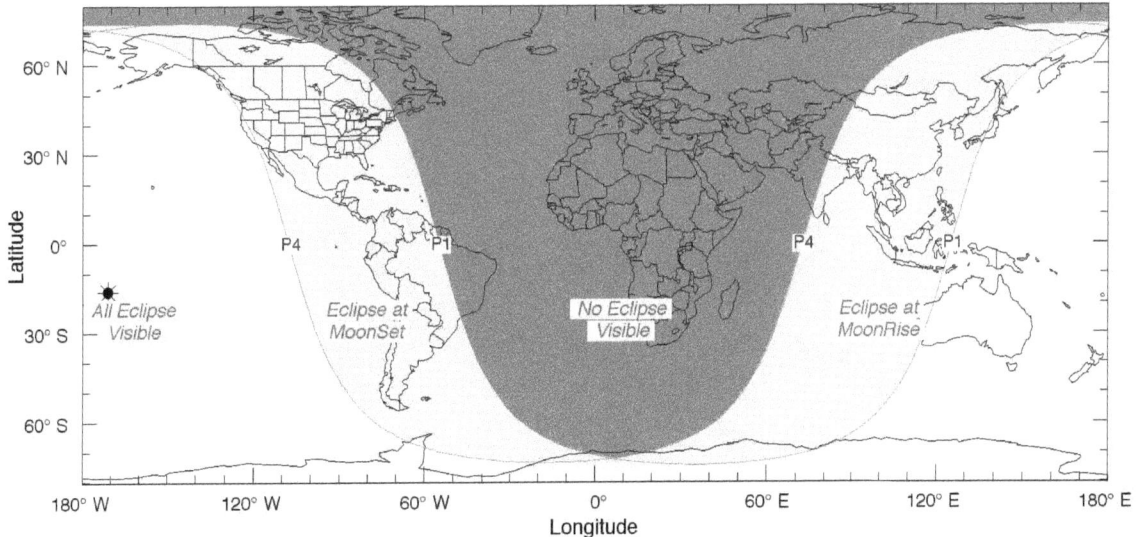

60° N

30° N

0°

30° S

60° S

Latitude

All Eclipse
Visible

P4   P1   P4   P1

Eclipse at
MoonSet

No Eclipse
Visible

Eclipse at
MoonRise

180° W   120° W   60° W   0°   60° E   120° E   180° E

Longitude

Penumbral Lunar Eclipse of 2096 Jun 06

Greatest Eclipse = 02:43:40.0 TD   (= 02:41:36.7 UT1)

Penumbral Magnitude = 0.0064	Gamma = -1.5724	Saros Series = 151
Umbral Magnitude = -1.0567	Axis = 1.4260°	Saros Member = 1 of 71

Sun at Greatest Eclipse
(Geocentric Coordinates)
R.A. = 05h01m12.3s
Dec. = +22°44'24.9"
S.D. = 00°15'45.8"
H.P. = 00°00'08.7"

N

Earth's Penumbra

Earth's Umbra

E —

Ecliptic

— **W**

Moon at Greatest Eclipse
(Geocentric Coordinates)
R.A. = 16h59m58.8s
Dec. = -24°08'18.4"
S.D. = 00°14'49.7"
H.P. = 00°54'25.1"

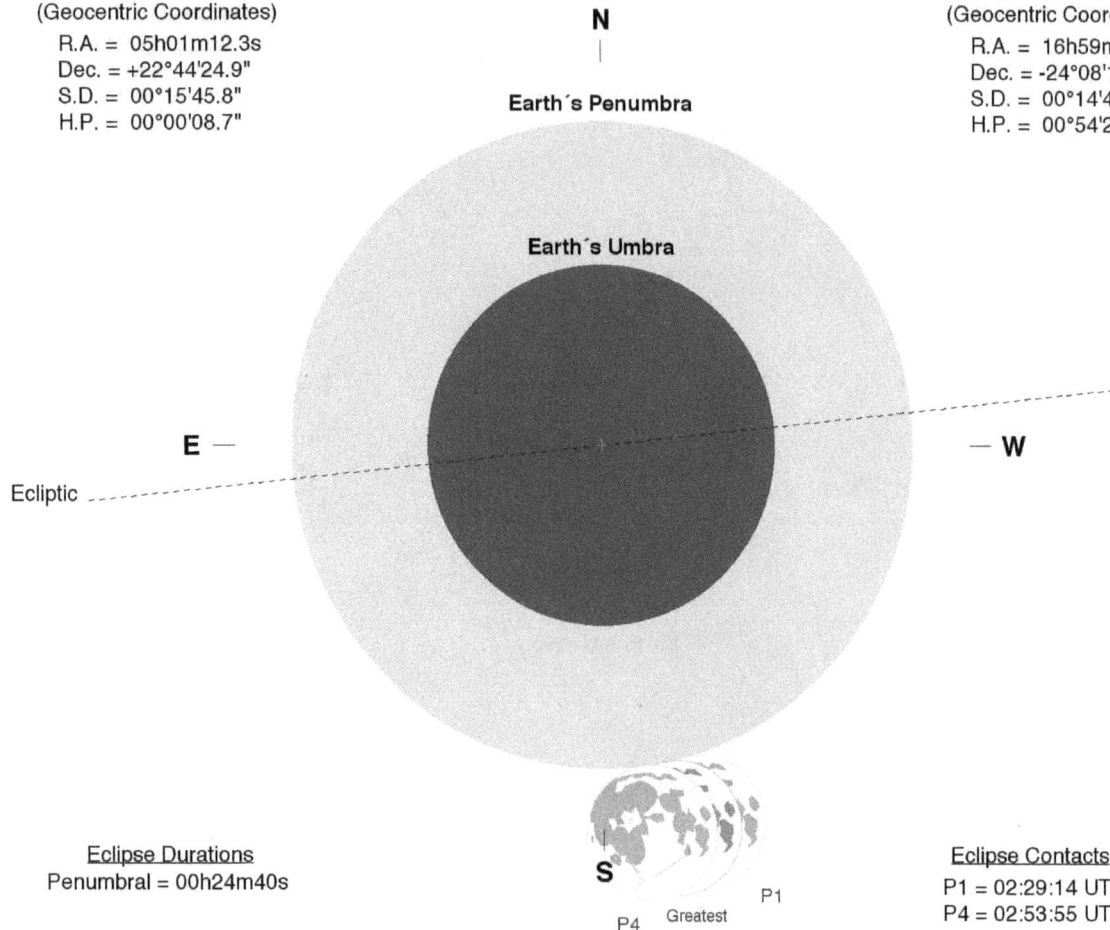

S

P1

P4   Greatest

0	15	30	45	60

Arc-Minutes

Eclipse Durations
Penumbral = 00h24m40s

Eph. = JPL DE430
Rule = Herald-Sinnott
ΔT =   123 s

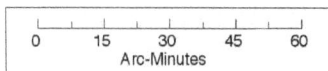

Eclipse Contacts
P1 = 02:29:14 UT1
P4 = 02:53:55 UT1

©2020 F. Espenak, www.EclipseWise.com

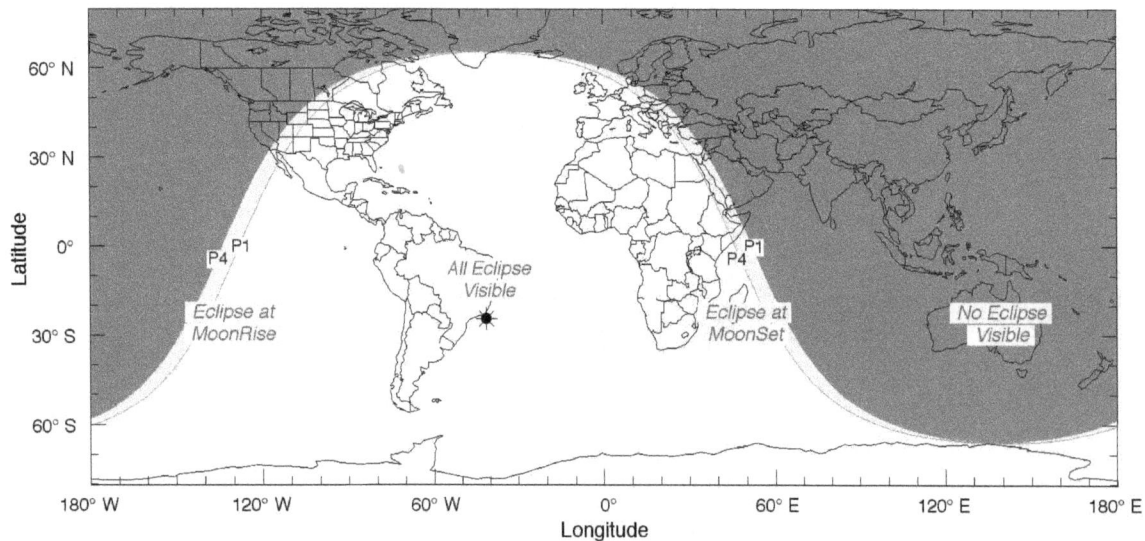

Penumbral Lunar Eclipse of 2096 Oct 31

Greatest Eclipse = 11:30:23.9 TD (= 11:28:20.1 UT1)

Penumbral Magnitude = 0.7685	Gamma = -1.1308	Saros Series = 118
Umbral Magnitude = -0.1987	Axis = 1.1519°	Saros Member = 56 of 73

Sun at Greatest Eclipse
(Geocentric Coordinates)

R.A. = 14h26m45.6s
Dec. = -14°30'56.1"
S.D. = 00°16'06.6"
H.P. = 00°00'08.9"

N

Earth's Penumbra

Earth's Umbra

Ecliptic

E —

— W

Moon at Greatest Eclipse
(Geocentric Coordinates)

R.A. = 02h28m41.7s
Dec. = +13°27'49.1"
S.D. = 00°16'39.3"
H.P. = 01°01'07.7"

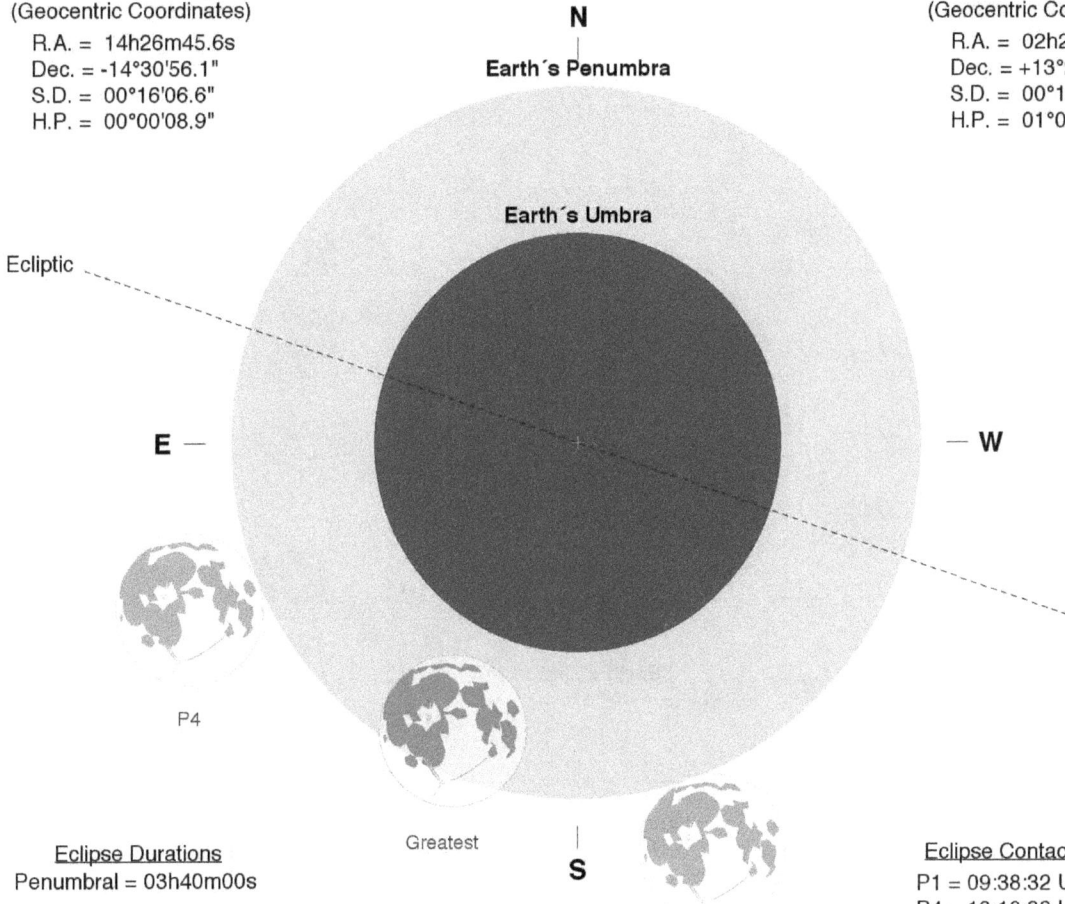

P4

Greatest

S

P1

Eclipse Durations
Penumbral = 03h40m00s

Eclipse Contacts
P1 = 09:38:32 UT1
P4 = 13:18:32 UT1

Eph. = JPL DE430
Rule = Herald-Sinnott
ΔT = 124 s

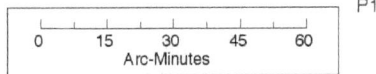

0	15	30	45	60

Arc-Minutes

©2020 F. Espenak, www.EclipseWise.com

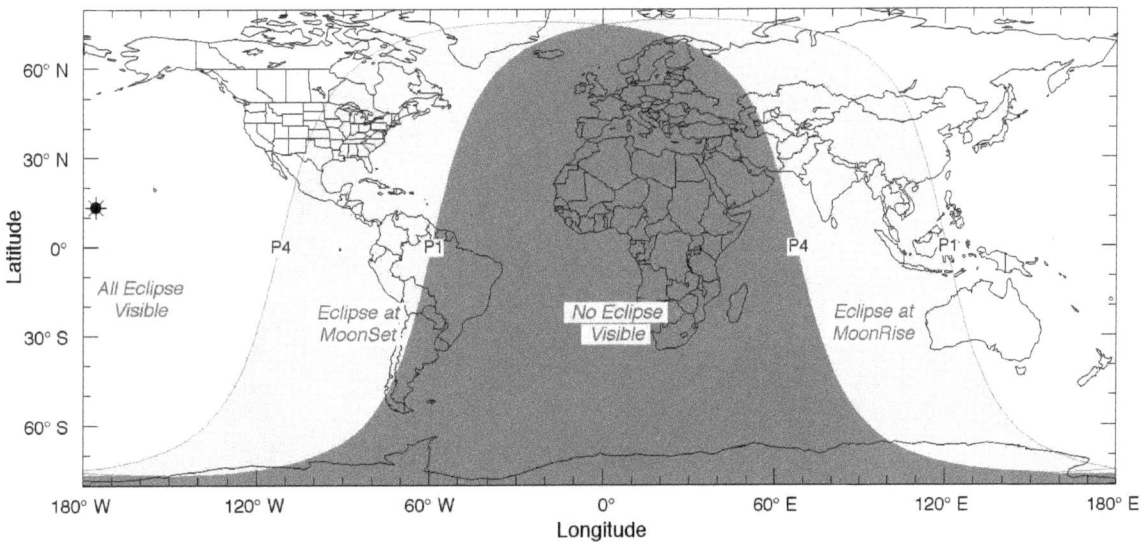

All Eclipse Visible

Eclipse at MoonSet

No Eclipse Visible

Eclipse at MoonRise

P4   P1   P4   P1

Penumbral Lunar Eclipse of 2096 Nov 29

Greatest Eclipse = 21:22:20.3 TD (= 21:20:16.5 UT1)

Penumbral Magnitude = 0.0879	Gamma = 1.5018	Saros Series = 156
Umbral Magnitude = -0.8799	Axis = 1.5388°	Saros Member = 3 of 81

Sun at Greatest Eclipse
(Geocentric Coordinates)
R.A. = 16h27m51.9s
Dec. = -21°44'27.4"
S.D. = 00°16'12.9"
H.P. = 00°00'08.9"

N

Moon at Greatest Eclipse
(Geocentric Coordinates)
R.A. = 04h26m12.3s
Dec. = +23°13'53.1"
S.D. = 00°16'45.2"
H.P. = 01°01'29.3"

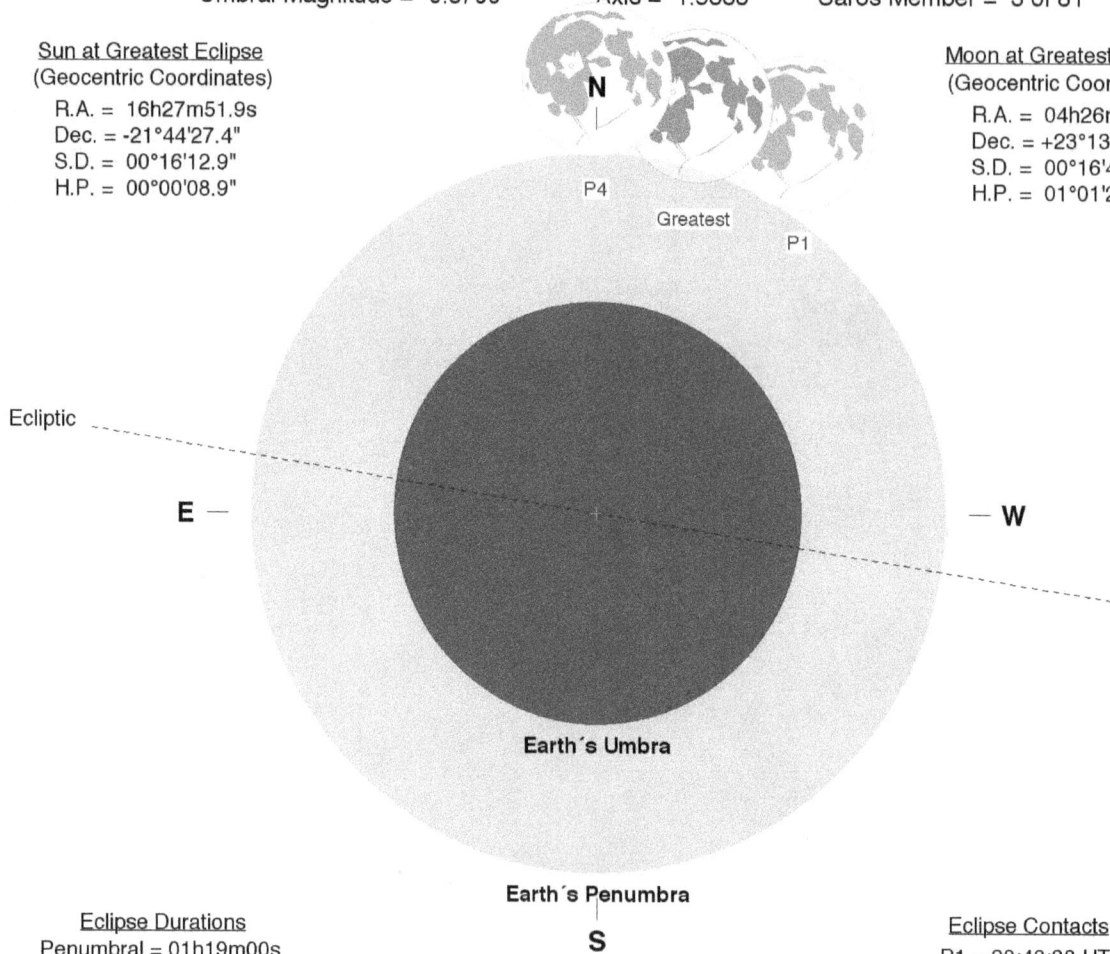

P4
Greatest
P1

Ecliptic

E —   — W

Earth´s Umbra

Earth´s Penumbra

S

Eclipse Durations
Penumbral = 01h19m00s

Eclipse Contacts
P1 = 20:40:38 UT1
P4 = 21:59:38 UT1

Eph. = JPL DE430
Rule = Herald-Sinnott
ΔT =   124 s

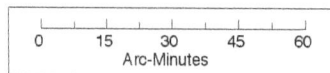

0	15	30	45	60
Arc-Minutes

©2020 F. Espenak, www.EclipseWise.com

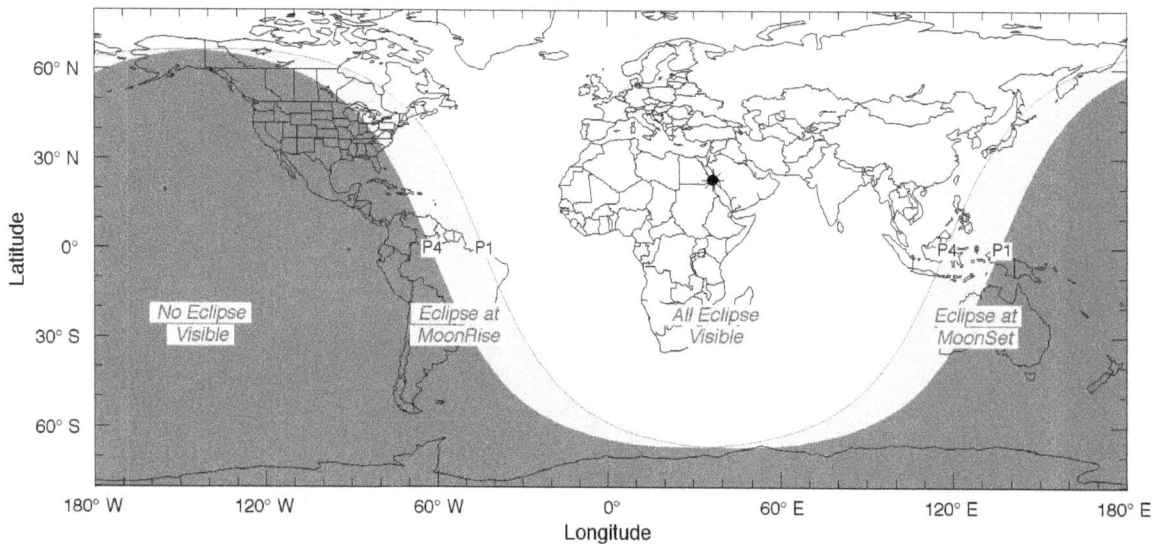

No Eclipse Visible

Eclipse at MoonRise

All Eclipse Visible

Eclipse at MoonSet

Partial Lunar Eclipse of 2097 Apr 26

Greatest Eclipse = 12:18:17.1 TD (= 12:16:12.9 UT1)

Penumbral Magnitude = 1.9032	Gamma = 0.5377	Saros Series = 123
Umbral Magnitude = 0.8439	Axis = 0.4935°	Saros Member = 57 of 72

Sun at Greatest Eclipse
(Geocentric Coordinates)
R.A. = 02h18m55.6s
Dec. = +13°52'05.0"
S.D. = 00°15'53.8"
H.P. = 00°00'08.7"

Moon at Greatest Eclipse
(Geocentric Coordinates)
R.A. = 14h19m46.7s
Dec. = -13°25'11.6"
S.D. = 00°15'00.4"
H.P. = 00°55'04.6"

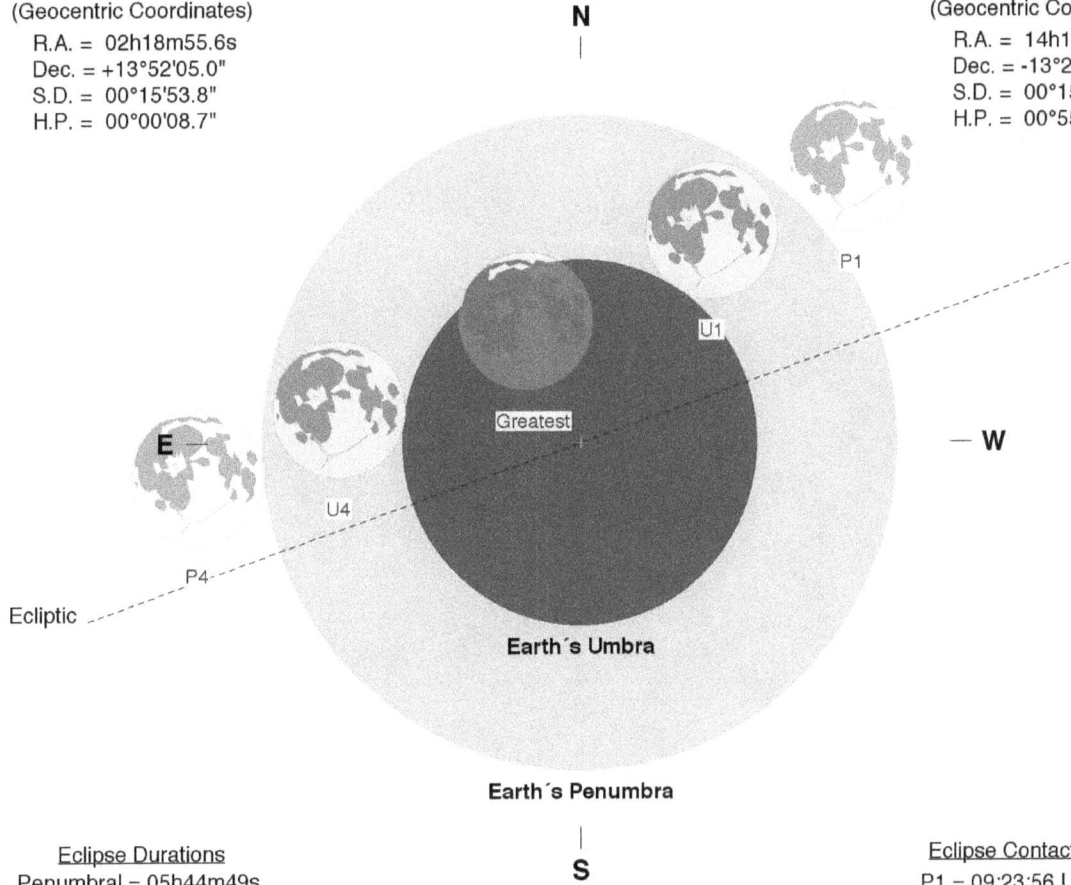

N

P1

U1

Greatest

E

W

U4

P4

Ecliptic

Earth's Umbra

Earth's Penumbra

S

Eclipse Durations
Penumbral = 05h44m49s
Umbral = 03h15m52s

Eph. = JPL DE430
Rule = Herald-Sinnott
ΔT =   124 s

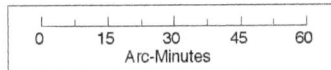

Eclipse Contacts
P1 = 09:23:56 UT1
U1 = 10:38:24 UT1
U4 = 13:54:16 UT1
P4 = 15:08:44 UT1

0   15   30   45   60
Arc-Minutes

©2020 F. Espenak, www.EclipseWise.com

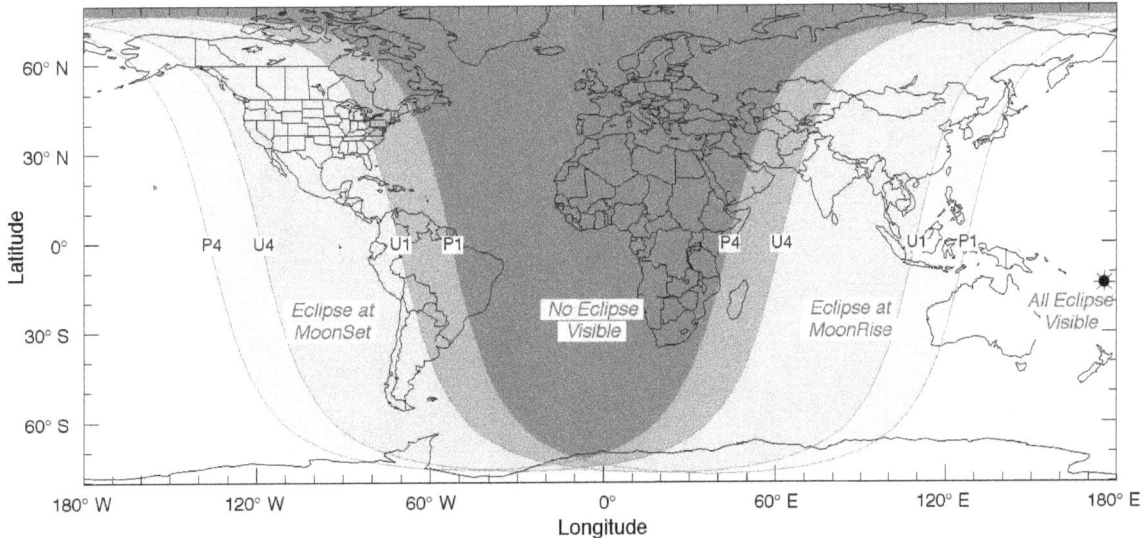

Total Lunar Eclipse of 2097 Oct 21

Greatest Eclipse = 01:30:55.2 TD (= 01:28:50.5 UT1)

Penumbral Magnitude = 2.0171	Gamma = -0.4608	Saros Series = 128
Umbral Magnitude = 1.0116	Axis = 0.4503°	Saros Member = 45 of 71

Sun at Greatest Eclipse
(Geocentric Coordinates)
R.A. = 13h45m46.4s
Dec. = -10°55'19.8"
S.D. = 00°16'03.7"
H.P. = 00°00'08.8"

Moon at Greatest Eclipse
(Geocentric Coordinates)
R.A. = 01h46m35.5s
Dec. = +10°31'09.1"
S.D. = 00°15'58.5"
H.P. = 00°58'37.8"

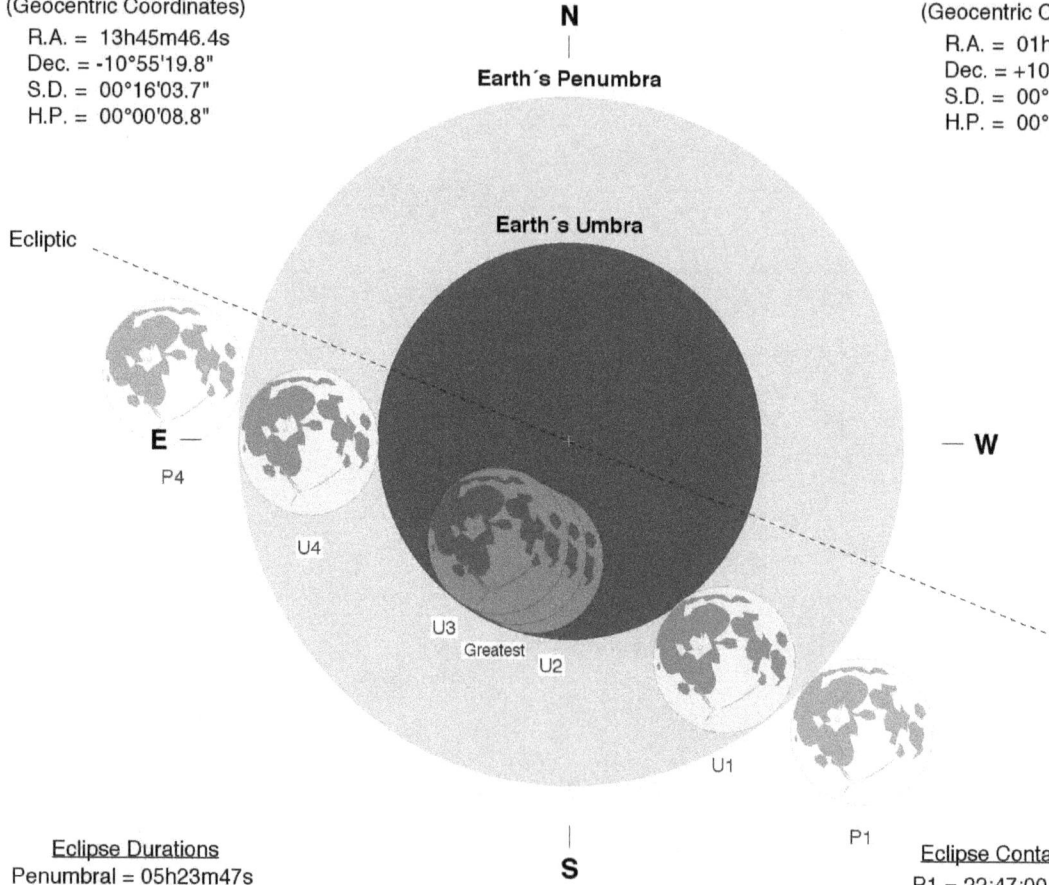

N

Earth's Penumbra

Earth's Umbra

Ecliptic

E

P4

U4

U3

Greatest

U2

U1

W

S

P1

Eclipse Durations
Penumbral = 05h23m47s
Umbral = 03h15m53s
Total = 00h16m43s

Eph. = JPL DE430
Rule = Herald-Sinnott
ΔT = 125 s

0   15   30   45   60
Arc-Minutes

©2020 F. Espenak, www.EclipseWise.com

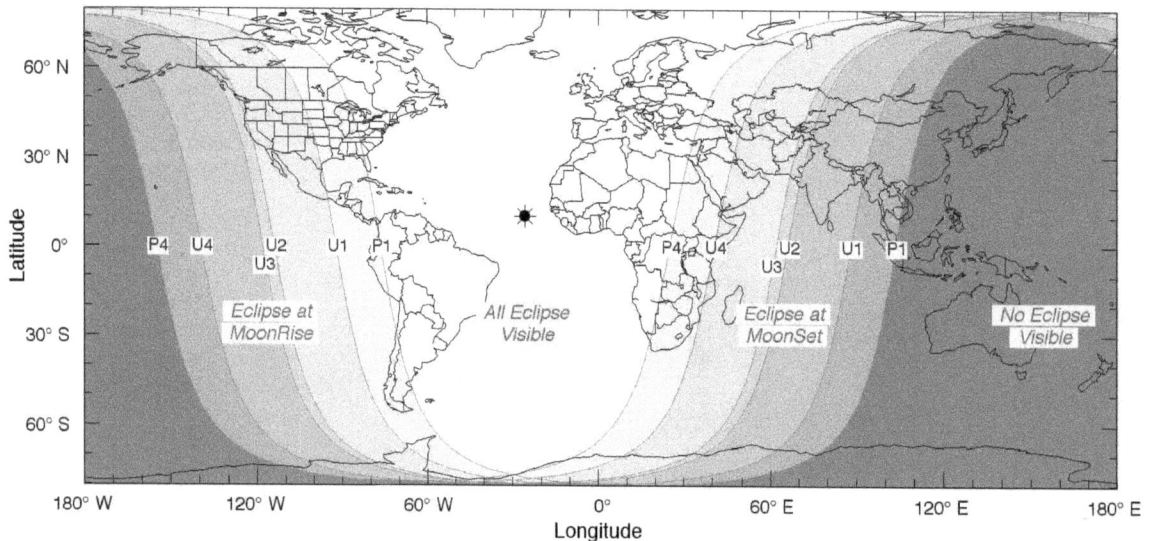

Eclipse Contacts
P1 = 22:47:00 UT1
U1 = 23:51:04 UT1
U2 = 01:20:47 UT1
U3 = 01:37:30 UT1
U4 = 03:06:57 UT1
P4 = 04:10:47 UT1

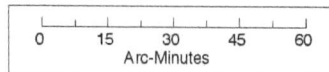

Total Lunar Eclipse of 2098 Apr 15

Greatest Eclipse = 19:04:47.9 TD (= 19:02:42.7 UT1)

Penumbral Magnitude = 2.4474	Gamma = -0.2272	Saros Series = 133
Umbral Magnitude = 1.4389	Axis = 0.2197°	Saros Member = 31 of 71

Sun at Greatest Eclipse
(Geocentric Coordinates)

R.A. = 01h37m51.1s
Dec. = +10°10'23.5"
S.D. = 00°15'56.7"
H.P. = 00°00'08.8"

N

Earth´s Penumbra

Earth´s Umbra

E

W

P1

U1

U2

Greatest

U3

U4

P4

Ecliptic

S

Moon at Greatest Eclipse
(Geocentric Coordinates)

R.A. = 13h37m26.8s
Dec. = -10°22'08.6"
S.D. = 00°15'48.6"
H.P. = 00°58'01.5"

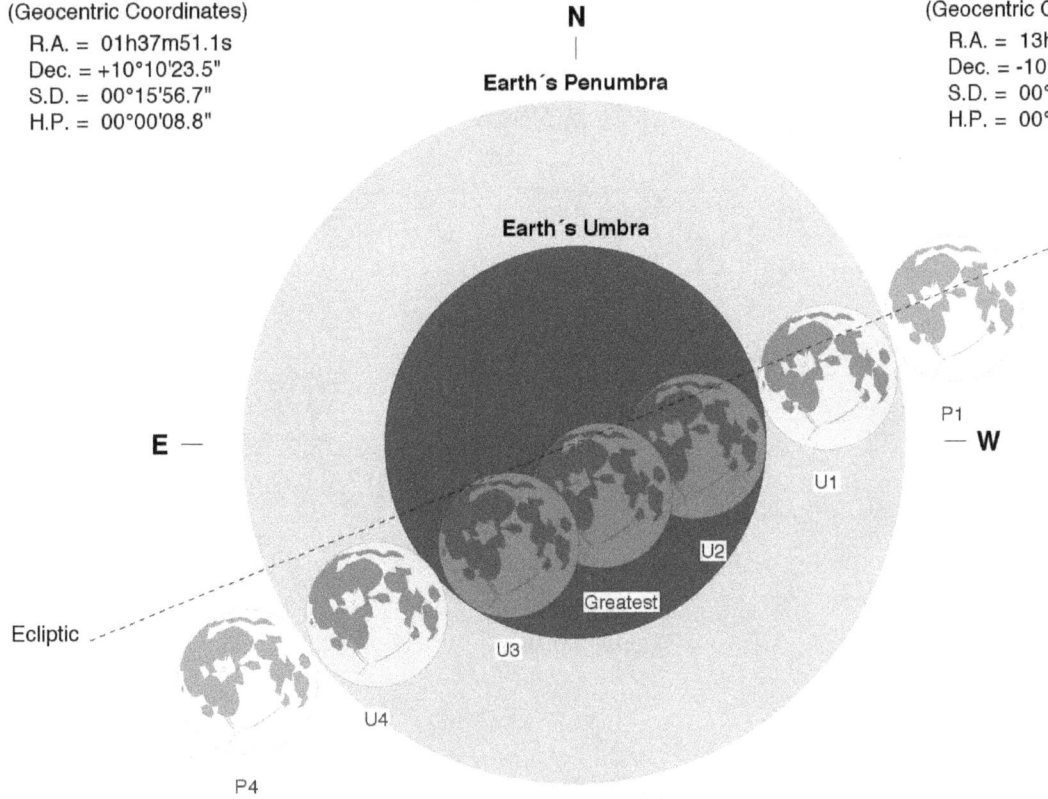

Eclipse Durations
Penumbral = 05h39m02s
Umbral = 03h36m29s
Total = 01h29m40s

Eph. = JPL DE430
Rule = Herald-Sinnott
ΔT =   125 s

0	15	30	45	60

Arc-Minutes

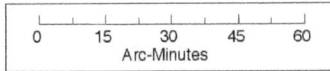

©2020 F. Espenak, www.EclipseWise.com

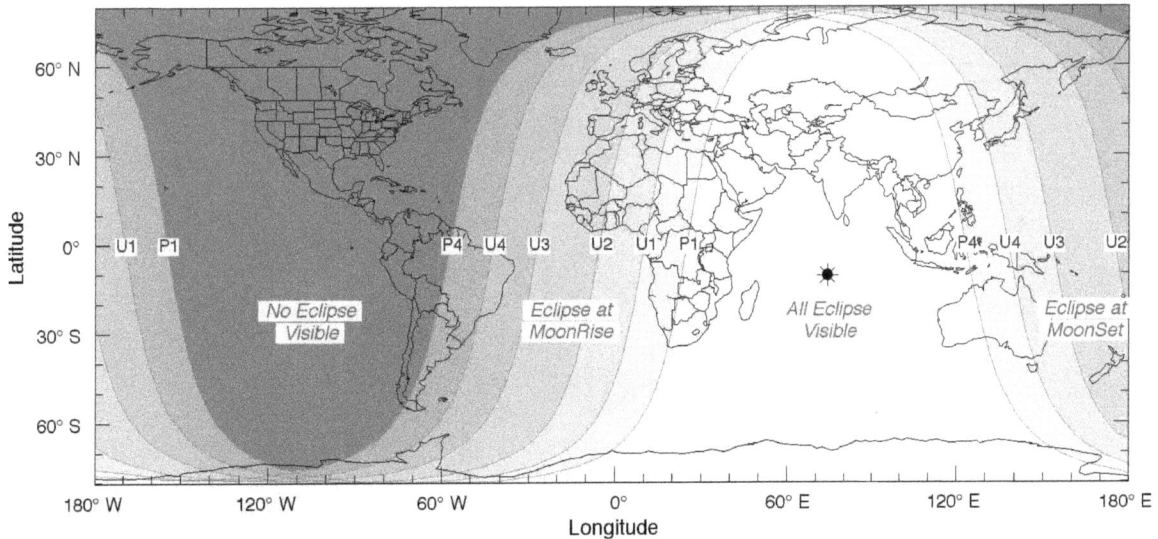

Eclipse Contacts
P1 = 16:13:12 UT1
U1 = 17:14:23 UT1
U2 = 18:17:44 UT1
U3 = 19:47:23 UT1
U4 = 20:50:52 UT1
P4 = 21:52:14 UT1

No Eclipse
Visible

Eclipse at
MoonRise

All Eclipse
Visible

Eclipse at
MoonSet

Total Lunar Eclipse of 2098 Oct 10

Greatest Eclipse = 09:19:58.3 TD   (= 09:17:52.6 UT1)

Penumbral Magnitude = 2.3850	Gamma = 0.2749	Saros Series = 138
Umbral Magnitude = 1.3266	Axis = 0.2543°	Saros Member = 33 of 82

Sun at Greatest Eclipse
(Geocentric Coordinates)
R.A. = 13h05m04.7s
Dec. = -06°55'18.7"
S.D. = 00°16'00.8"
H.P. = 00°00'08.8"

Moon at Greatest Eclipse
(Geocentric Coordinates)
R.A. = 01h04m35.6s
Dec. = +07°08'45.8"
S.D. = 00°15'07.7"
H.P. = 00°55'31.3"

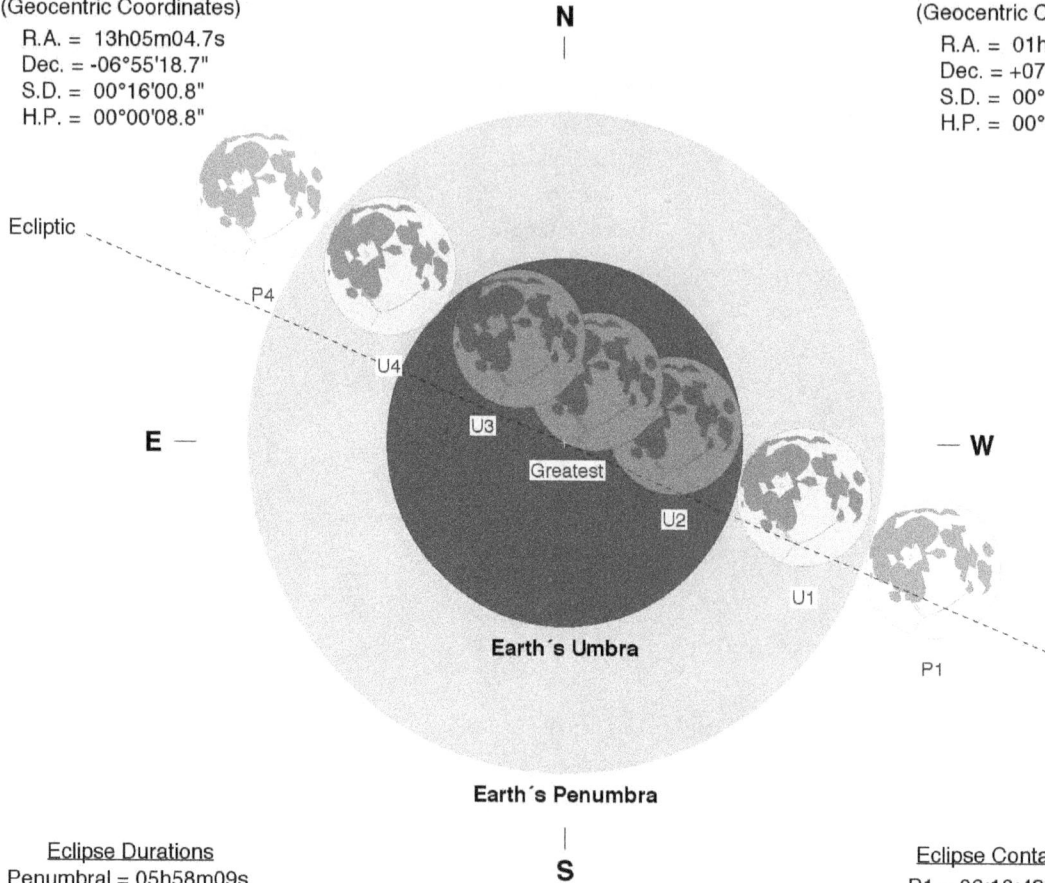

N

Ecliptic

P4
U4
U3
Greatest
U2
U1
P1

E

W

Earth's Umbra

Earth's Penumbra

S

Eclipse Durations
Penumbral = 05h58m09s
Umbral = 03h41m42s
Total = 01h23m22s

Eph. = JPL DE430
Rule = Herald-Sinnott
ΔT =   126 s

0	15	30	45	60

Arc-Minutes

©2020 F. Espenak, www.EclipseWise.com

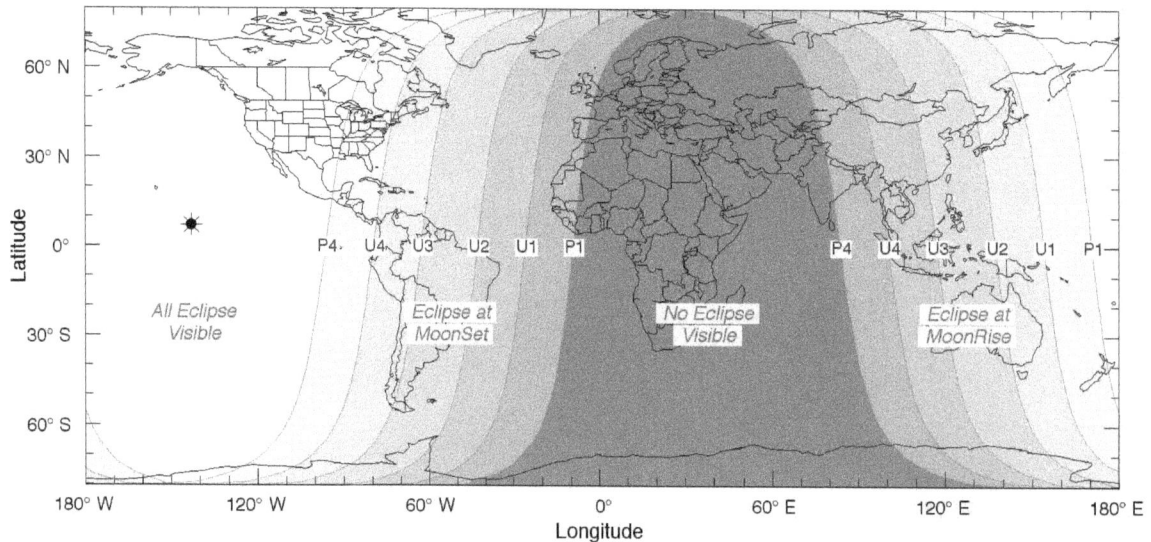

Eclipse Contacts
P1 = 06:18:42 UT1
U1 = 07:26:57 UT1
U2 = 08:36:01 UT1
U3 = 09:59:23 UT1
U4 = 11:08:40 UT1
P4 = 12:16:51 UT1

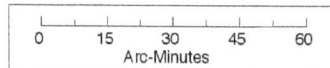

P4 U4 U3 U2 U1 P1   P4 U4 U3 U2 U1 P1

All Eclipse
Visible

Eclipse at
MoonSet

No Eclipse
Visible

Eclipse at
MoonRise

Longitude

Partial Lunar Eclipse of 2099 Apr 05

Greatest Eclipse = 08:30:55.7 TD (= 08:28:49.5 UT1)

Penumbral Magnitude = 1.1353	Gamma = -0.9304	Saros Series = 143
Umbral Magnitude = 0.1700	Axis = 0.9427°	Saros Member = 22 of 72

Sun at Greatest Eclipse
(Geocentric Coordinates)

R.A. = 00h58m32.6s
Dec. = +06°14'54.6"
S.D. = 00°15'59.6"
H.P. = 00°00'08.8"

Moon at Greatest Eclipse
(Geocentric Coordinates)

R.A. = 12h56m44.9s
Dec. = -07°04'45.1"
S.D. = 00°16'34.0"
H.P. = 01°00'48.0"

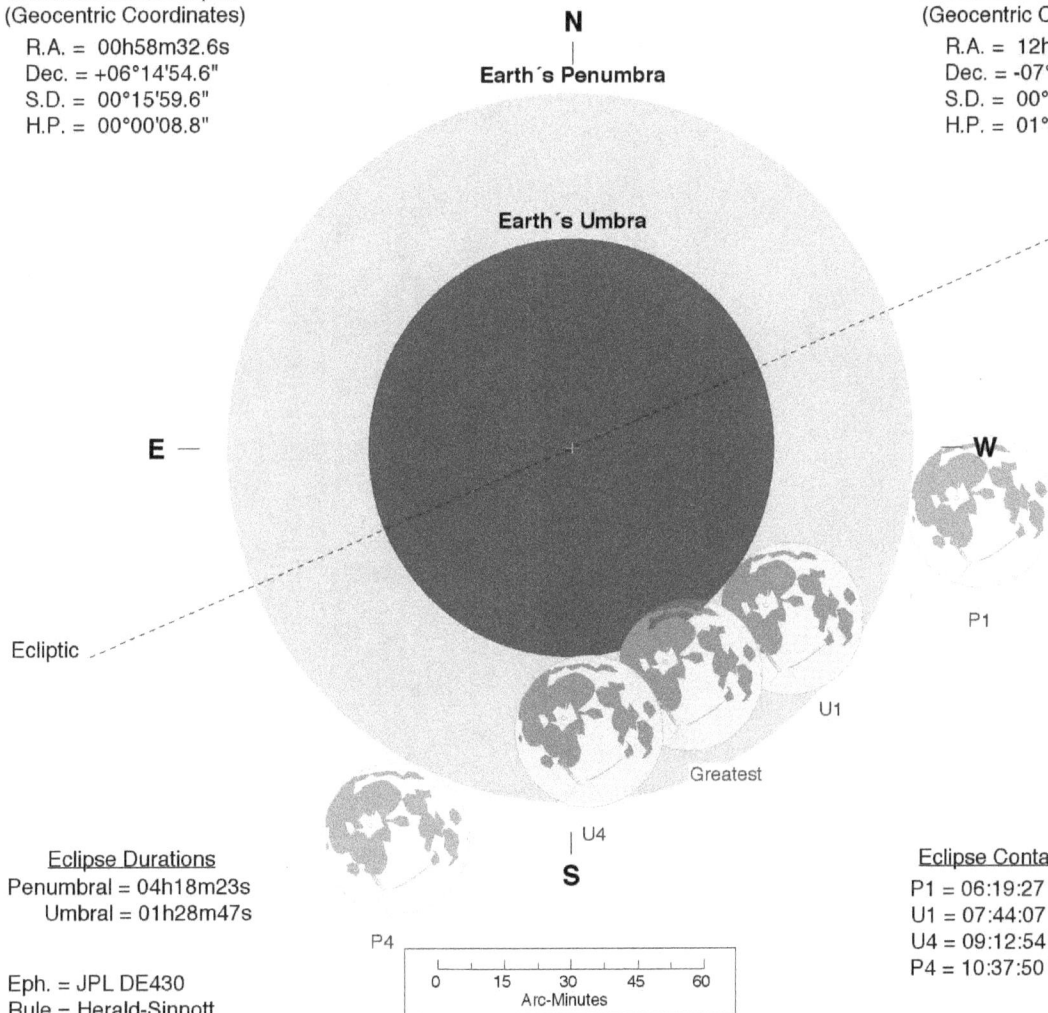

N

Earth's Penumbra

Earth's Umbra

E

W

P1

Ecliptic

U1

Greatest

U4

S

P4

Eclipse Durations
Penumbral = 04h18m23s
Umbral = 01h28m47s

Eph. = JPL DE430
Rule = Herald-Sinnott
ΔT =   126 s

0   15   30   45   60
Arc-Minutes

©2020 F. Espenak, www.EclipseWise.com

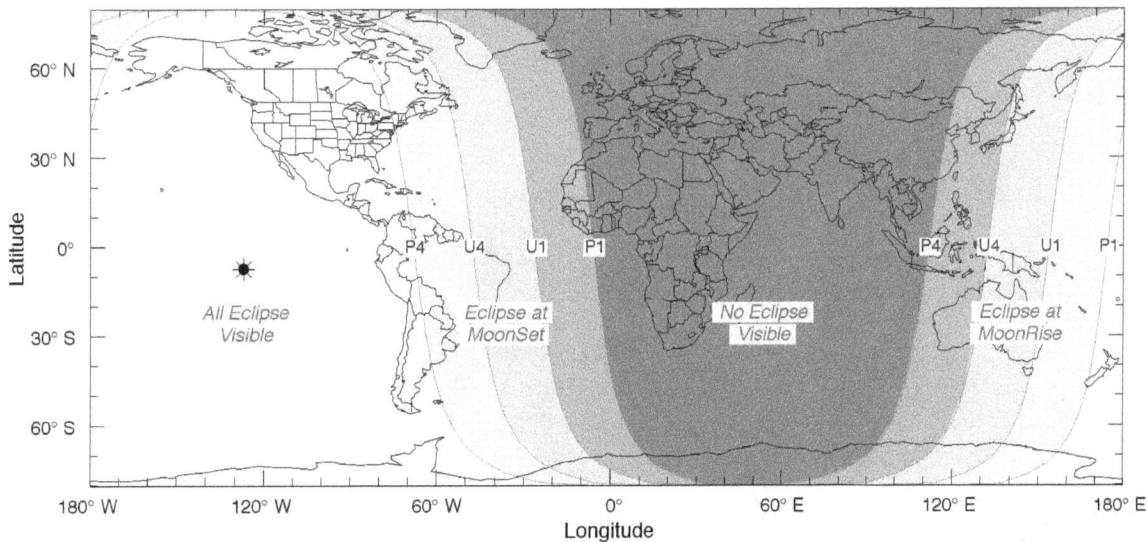

Eclipse Contacts
P1 = 06:19:27 UT1
U1 = 07:44:07 UT1
U4 = 09:12:54 UT1
P4 = 10:37:50 UT1

All Eclipse
Visible

Eclipse at
MoonSet

No Eclipse
Visible

Eclipse at
MoonRise

P4   U4   U1   P1

P4   U4   U1   P1

Penumbral Lunar Eclipse of 2099 Sep 29

Greatest Eclipse = 10:36:37.5 TD (= 10:34:30.8 UT1)

Penumbral Magnitude = 1.0360	Gamma = 1.0175	Saros Series = 148
Umbral Magnitude = -0.0491	Axis = 0.9153°	Saros Member = 8 of 70

Sun at Greatest Eclipse
(Geocentric Coordinates)
R.A. = 12h24m20.2s
Dec. = -02°37'46.3"
S.D. = 00°15'57.7"
H.P. = 00°00'08.8"

Moon at Greatest Eclipse
(Geocentric Coordinates)
R.A. = 00h22m33.6s
Dec. = +03°25'49.1"
S.D. = 00°14'42.6"
H.P. = 00°53'59.0"

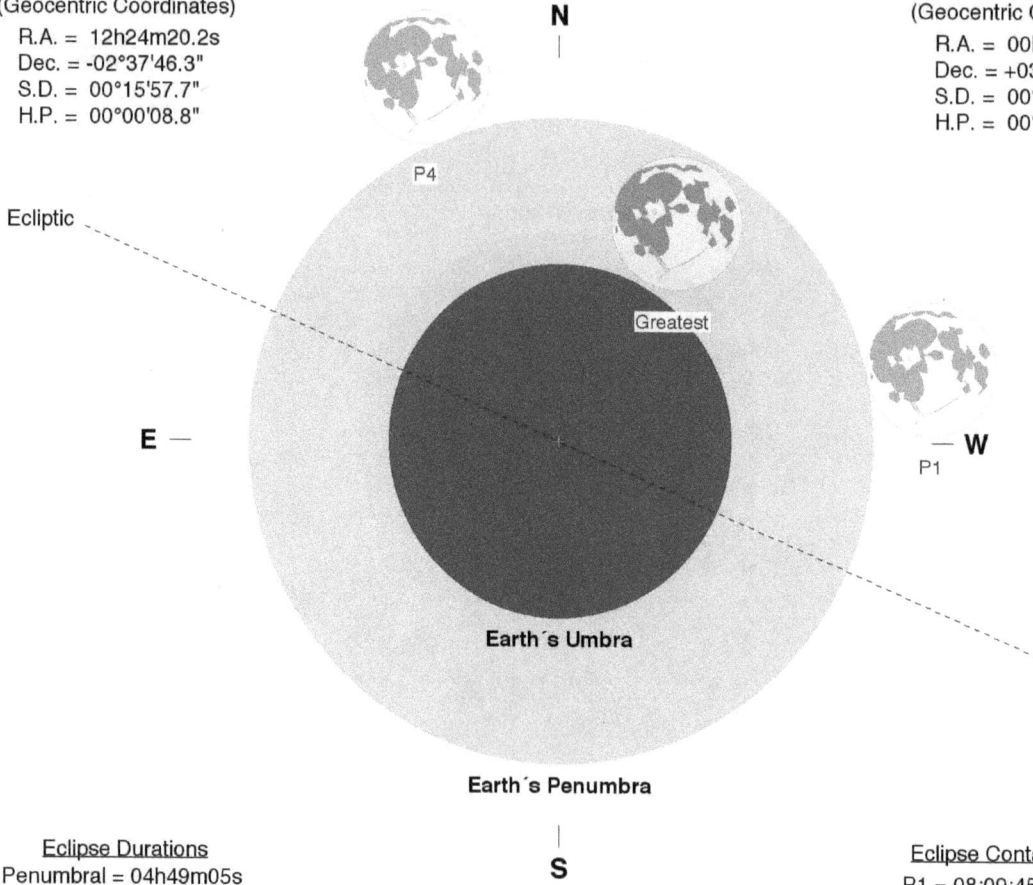

N

P4

Ecliptic

Greatest

E

W

P1

Earth's Umbra

Earth's Penumbra

S

Eclipse Durations
Penumbral = 04h49m05s

Eclipse Contacts
P1 = 08:09:45 UT1
P4 = 12:58:51 UT1

Eph. = JPL DE430
Rule = Herald-Sinnott
ΔT = 127 s

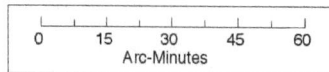

0	15	30	45	60

Arc-Minutes

©2020 F. Espenak, www.EclipseWise.com

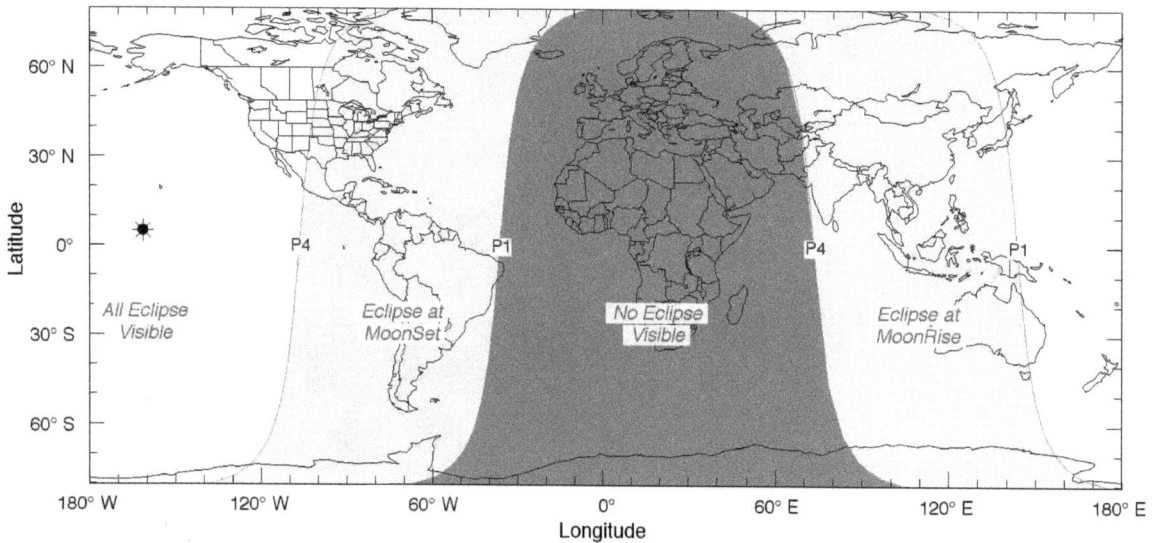

P4   P1   P4   P1

All Eclipse
Visible

Eclipse at
MoonSet

No Eclipse
Visible

Eclipse at
MoonRise

Latitude

Longitude

Penumbral Lunar Eclipse of 2100 Feb 24

Greatest Eclipse = 15:05:12.2 TD   (= 15:03:05.1 UT1)

Penumbral Magnitude = 0.9669	Gamma = 1.0267	Saros Series = 115
Umbral Magnitude = -0.0151	Axis = 1.0337°	Saros Member = 62 of 72

Sun at Greatest Eclipse
(Geocentric Coordinates)

R.A. = 22h31m21.3s
Dec. = -09°17'00.6"
S.D. = 00°16'09.9"
H.P. = 00°00'08.9"

Moon at Greatest Eclipse
(Geocentric Coordinates)

R.A. = 10h33m16.4s
Dec. = +10°12'10.6"
S.D. = 00°16'27.7"
H.P. = 01°00'25.0"

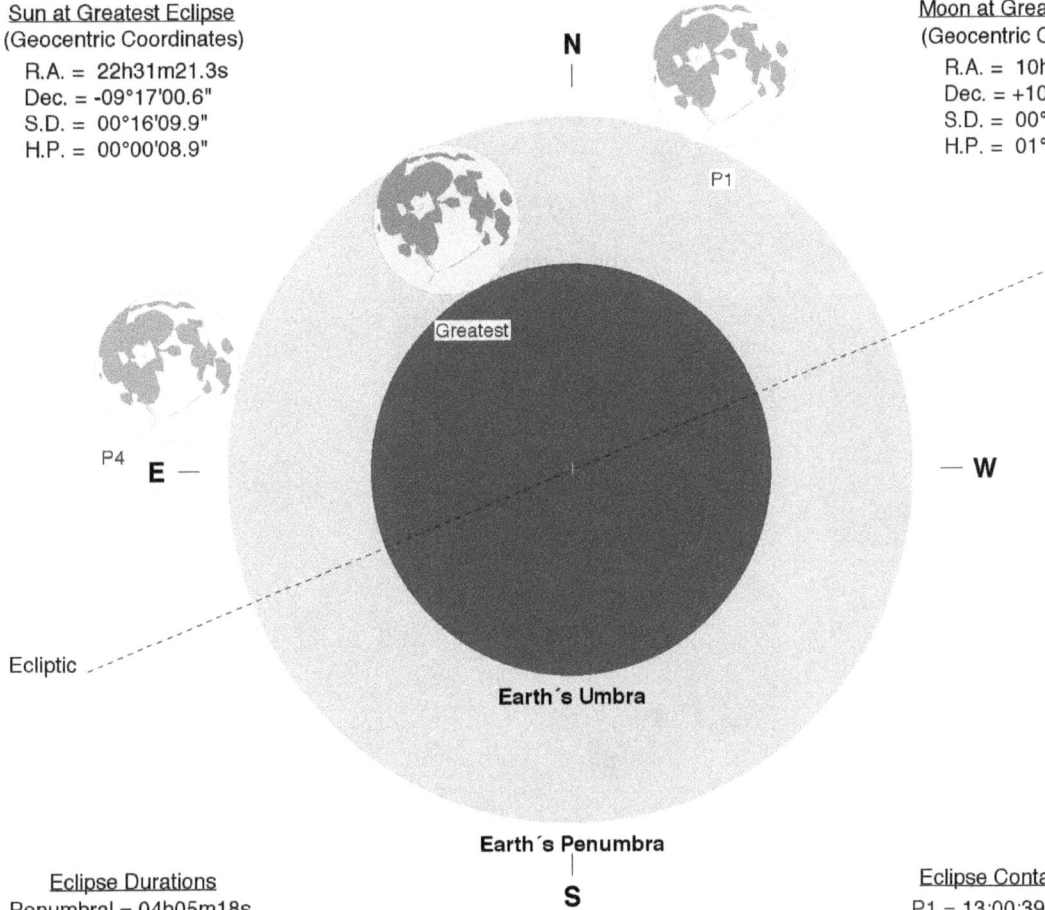

N

P1

Greatest

E

W

Ecliptic

Earth's Umbra

Earth's Penumbra

S

Eclipse Durations

Penumbral = 04h05m18s

Eclipse Contacts

P1 = 13:00:39 UT1
P4 = 17:05:57 UT1

Eph. = JPL DE430
Rule = Herald-Sinnott
ΔT = 127 s

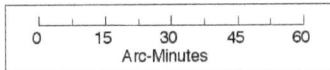

0   15   30   45   60
Arc-Minutes

©2020 F. Espenak, www.EclipseWise.com

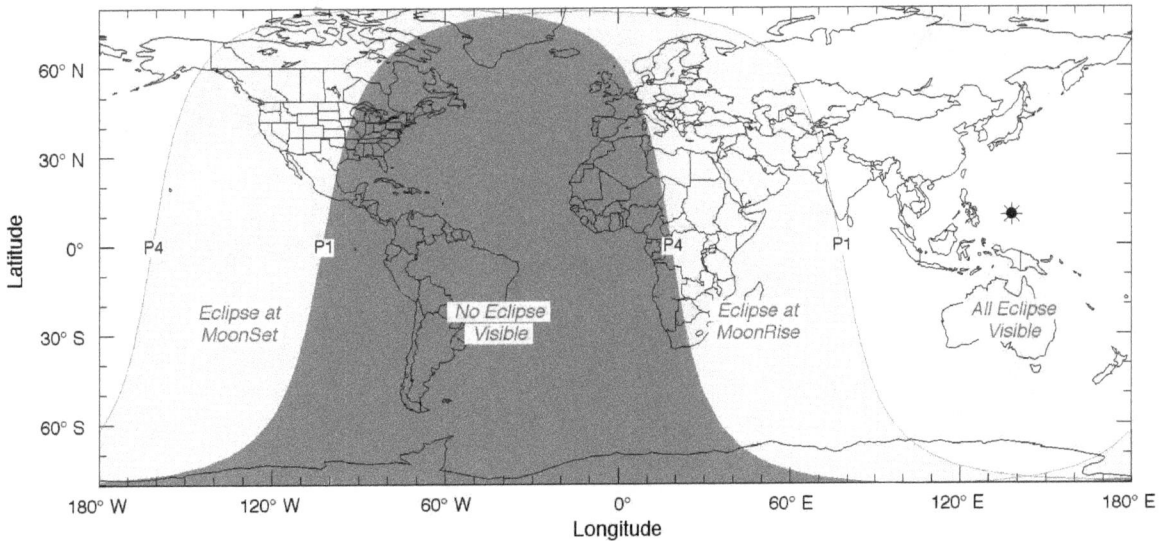

P4

P1

No Eclipse
Visible

P4

P1

Eclipse at
MoonSet

Eclipse at
MoonRise

All Eclipse
Visible

Penumbral Lunar Eclipse of 2100 Aug 19

Greatest Eclipse = 21:44:59.4 TD   (= 21:42:51.8 UT1)

Penumbral Magnitude = 0.8735	Gamma = -1.0906	Saros Series = 120
Umbral Magnitude = -0.1556	Axis = 1.0242°	Saros Member = 62 of 83

Sun at Greatest Eclipse
(Geocentric Coordinates)
R.A. = 09h56m57.4s
Dec. = +12°29'47.3"
S.D. = 00°15'48.1"
H.P. = 00°00'08.7"

N

Earth´s Penumbra

Earth´s Umbra

Ecliptic

E —

— **W**

P4

Greatest

S

P1

Moon at Greatest Eclipse
(Geocentric Coordinates)
R.A. = 21h58m46.4s
Dec. = -13°25'12.3"
S.D. = 00°15'21.3"
H.P. = 00°56'21.3"

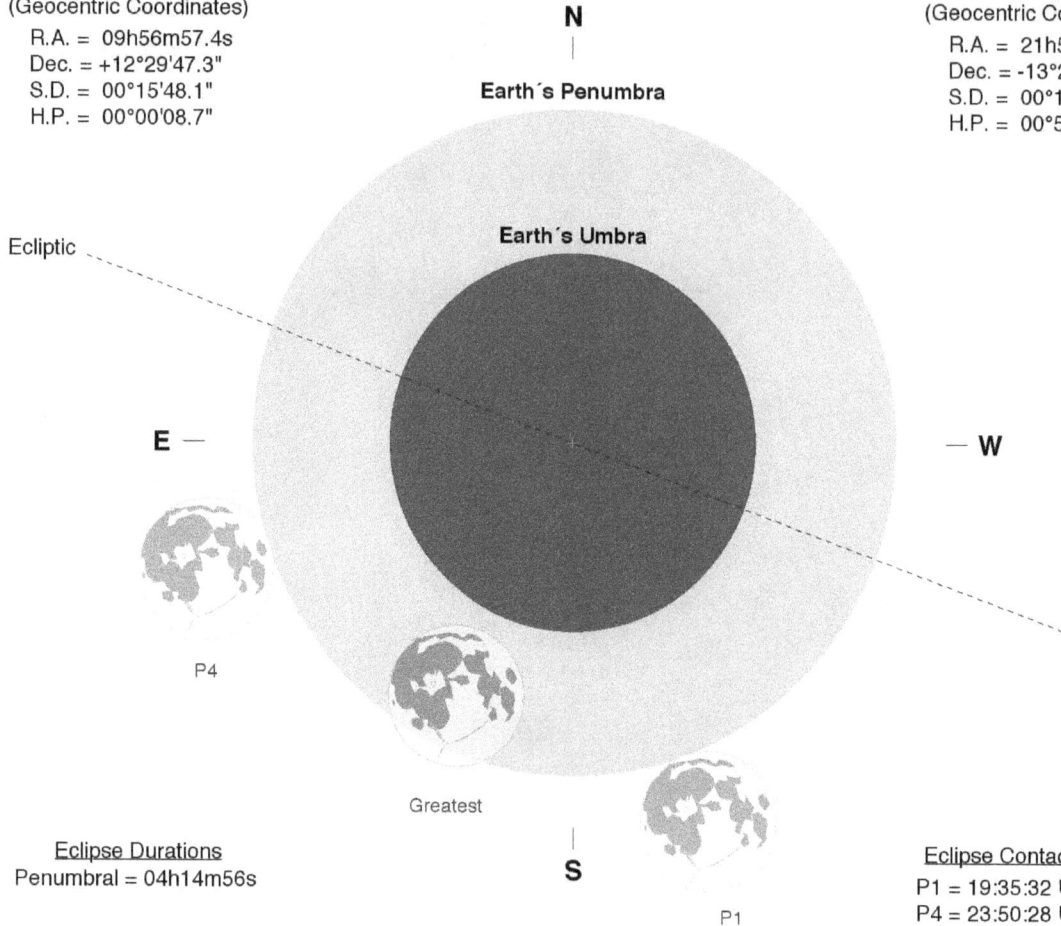

Eclipse Durations
Penumbral = 04h14m56s

Eph. = JPL DE430
Rule = Herald-Sinnott
ΔT =   128 s

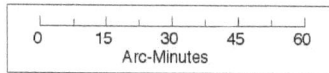

Eclipse Contacts
P1 = 19:35:32 UT1
P4 = 23:50:28 UT1

0	15	30	45	60

Arc-Minutes

©2020 F. Espenak, www.EclipseWise.com

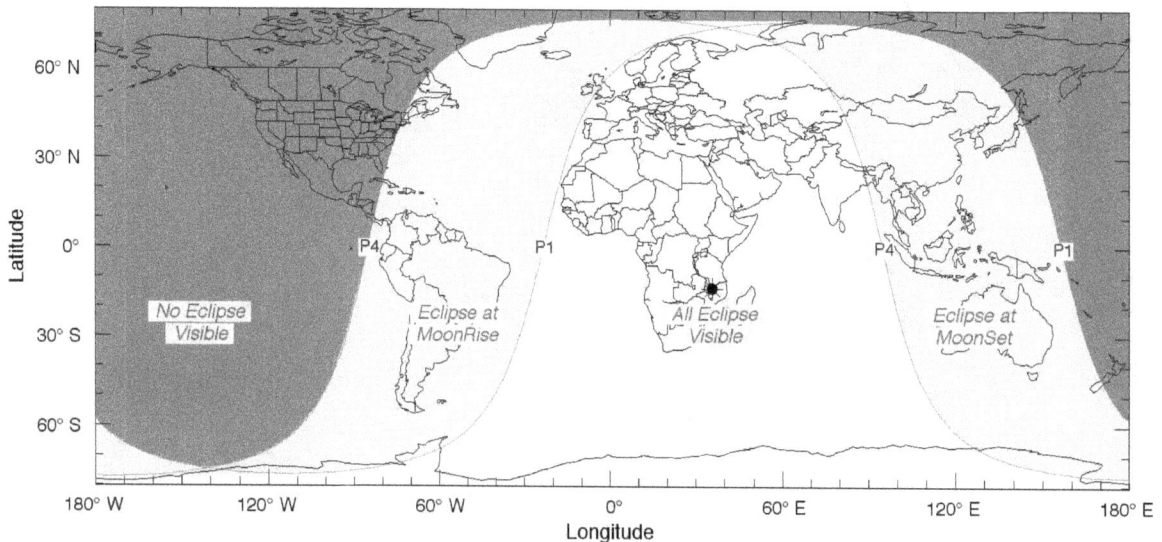

60° N

30° N

0°

30° S

60° S

No Eclipse
Visible

Eclipse at
MoonRise

P4   P1

All Eclipse
Visible

Eclipse at
MoonSet

P4   P1

Latitude

180° W   120° W   60° W   0°   60° E   120° E   180° E

Longitude

Astropixels Publications

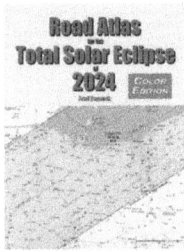

Road Atlas for the Total Solar Eclipse of 2024 (Fred Espenak) contains a comprehensive series of 26 high-resolution maps of the path of totality across the USA, Mexico and Canada. The large scale (1 inch = 22 miles) shows both major and minor roads, towns and cities, rivers, lakes, parks, national forests, wilderness areas and mountain ranges. The duration of totality is plotted in 30-second steps, making it easy to estimate the length of the total eclipse from any location in the eclipse path.

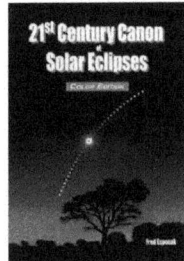

21st Century Canon of Solar Eclipses (Fred Espenak) contains maps and data for all 224 solar eclipses occurring during the 100-year period from 2001 through 2100. Appendix A is comprehensive catalog listing the essential characteristics of each eclipse. Appendix B contains maps depicting the geographic regions of visibility of each eclipse with 12 maps per page. Appendix C has detailed full-page maps of every eclipse from 2017 through 2066. Appendix D plots the track of every central eclipse (total, annular and hybrid) on large-scale maps with countries borders and major cities.

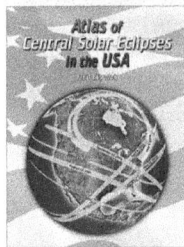

Atlas of Central Solar Eclipses in the USA (Fred Espenak) contains of a series of 499 global maps showing the track of every total and annular solar eclipse across the USA from 1001 through 3000. It is accompanied by a catalog that lists the major characteristics of each eclipse. A set of 20 detailed maps, each covering a 50-year period and centered on the lower 48 states, shows the path of every total and annular eclipse. The maps include state boundaries and major cities. These maps also cover southern Canada and northern Mexico.

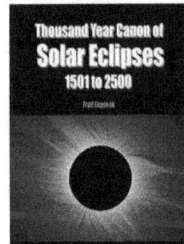

Thousand Year Canon of Solar Eclipses 1501 to 2500 (Fred Espenak) contains maps and data for each of the 2,389 solar eclipses occurring over the ten-century period centered on the present era. A comprehensive catalog lists the essential characteristics of each eclipse while a series of global maps show the exact geographic extent of each eclipse.

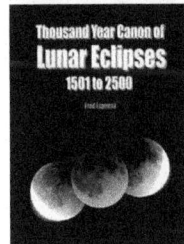

Thousand Year Canon of Lunar Eclipses 1501 to 2500 (Fred Espenak) contains diagrams, maps and data for each of the 2,424 lunar eclipses occurring over the ten-century period centered on the present era. A comprehensive catalog lists the essential characteristics of each eclipse while a series of diagrams and maps illustrate the Moon-shadow geometry and geographic regions of visibility of each eclipse.

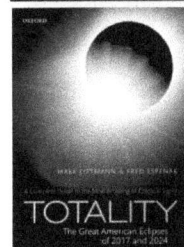

Totality: The Great American Eclipses of 2017 and 2024 (Mark Littmann & Fred Espenak) is the ultimate guide to the most stunning of celestial sights, total eclipses of the Sun The book provides information, photographs, and illustrations to help the public understand and safely enjoy all aspects of these eclipses including how to observe a total eclipse of the Sun, how to photograph and video record an eclipse, why solar eclipses happen, and more. Several chapters focus exclusively on the total eclipses of 2017 and 2024 though the USA.

For complete details on the above *Astropixels Publications*, visit:

astropixels.com/pubs/

www.ingramcontent.com/pod-product-compliance
Lightning Source LLC
Chambersburg PA
CBHW080233270326
41926CB00020B/4213